Understanding CONSTRUCTION DRAWINGS for Housing and Small Buildings

Third Edition

Tom Stephenson
George Brown College

NELSON

NELSON

Understanding Construction Drawings, Third Edition
by Tom Stephenson

Vice President, Editorial Higher Education:
Anne Williams

Publisher:
Jackie Wood

Marketing Manager:
Cara Cortese

Developmental Editor:
Katherine Goodes

Photo Researcher:
Rachael Eagen

Permissions Coordinator:
Rachael Eagen

Production Project Manager:
Jaime Smith

Production Service:
Integra Software Services Pvt. Ltd.

Copy Editor:
Susan James

Proofreader:
Susan James

Indexer:
Constance A. Angelo

Design Director:
Ken Phipps

Managing Designer:
Franca Amore

Interior Design:
Dave Murphy

Cover Design:
Dave Murphy

Cover Image:
Tom Stephenson

Compositor:
Integra Software Services Pvt. Ltd.

Printed and bound in the United States

6 7 8 9 21 20 19 18

For more information contact Nelson Education Ltd., 1120 Birchmount Road, Toronto, Ontario, M1K 5G4. Or you can visit our Internet site at nelson.com

Library and Archives Canada Cataloguing in Publication
Stephenson, Tom, 1962-, author

Understanding construction drawings for housing and small buildings / Tom Stephenson. — Third edition.

Includes bibliographical references and index.
ISBN-13: 978-0-17-653155-3 (pbk.)

1. Structural drawing—Textbooks. 2. Dwellings—Drawings—Textbooks. 3. Small buildings—Drawings—Textbooks. I. Title.

TH431.S74 2014 692′.1

C2014-902411-8

ISBN-13: 978-0-17-653155-3
ISBN-10: 0-17-653155-6

Contents

Foreword

The housing industry in Canada has weathered both strong economic markets and difficult conditions, but even Canada's Economic Action Plan recognizes housing as a strong engine for the economic well-being of the country. In addition, renovation activity continues to surpass new home construction, with annual expenditures over $25 billion and growing.

What does this mean? Quite simply, it means that the residential construction industry is strong and growing, and is providing opportunities for careers of all sorts. The construction industry, because it is comprised of many small companies and a huge network of support industries and services, is not well understood as a huge contributor to the economic well-being of the country. In fact, the Canadian housing industry is the barometer of the Canadian economy. It is essential to Canada's social and economic progress and to the vitality and competitiveness of our urban communities.

Over the last few years, in my leadership roles with the Canadian Home Builders' Association, I have been in a position to carry many important messages to governments at all levels, and I have also had the opportunity to speak to many groups, particularly students, about the many career opportunities offered within the industry.

If you think for a moment about a house and the many components and products required to create a functioning building, then consider all of things that go into that house to make it a home, then multiply those elements by the number of people and functions necessary to bring those pieces together, you will appreciate the long list of exciting roles within the construction industry. However, before any building can be created, there must be a plan, a blueprint, a construction drawing. The more complex the building, the more elements will appear on the construction drawings; many drawings have numerous pages to capture all of the necessary components. The ability to accurately read and understand construction drawings is a skill that is needed by not only the carpenters on the site but also the estimator who orders the lumber, the mason who determines the amount of brick, the roofing supplier estimating the shingle requirements, the window manufacturer supplying the windows, etc. Consider the understanding needed by the electrician, the plumber, and the heating contractor. If you are supplying the kitchen cabinets or floor coverings and tile, you will have to understand the construction drawings, in order to make your product fit. As you learn to read and understand construction drawings, you soon realize that you have learned to think in three dimensions.

Construction drawings are so much more than just a floor plan! For a custom builder they represent part of the contract; for a production builder they help guarantee quality and cost control. Municipal inspectors check the construction drawings to ensure code compliance before construction begins. Mortgage lenders and insurance adjusters often use the drawings in order to evaluate buildings not yet constructed.

In a market where many homes are purchased before they are constructed (presale) it is critical for even the salespeople to understand the drawings in order to explain the spaces and specifications to potential purchasers. The details of the construction drawings provide for every condition expected: sloping lots, soil conditions, setbacks, drainage service connections, etc. The construction drawings are referred to repeatedly throughout the construction process, when each element is added, and as the municipal building inspector checks the progress against the drawing to ensure compliance.

In recent years there has been increased attention given to consideration of the "envelope" of the building. This detail deals with insulation and moisture barriers as well as air flow. All of these concerns improve the performance of the building. Again, this detail is clearly part of the construction drawings.

As you develop the skill for reading construction drawings, remember not to read them in isolation. The lines and details on the page must create an image of the finished product in your mind. A construction drawing is a picture of a box drawn from all sides, inside and out, top and bottom, and all of the elements to be contained in the box. Once mastered, the skill for reading and understanding construction drawings lasts a lifetime, opening the door for numerous career choices not only on construction sites but also in other industries related to housing.

MARY LAWSON

V.P. General Manager, Dalerose Country Homes

2004 President—Canadian Home Builders' Association

1998 President—Greater Toronto Home Builders' Association

1988 President—Calgary Home Builders Association

Currently on the Foundation Board of George Brown College

Previously served on Board of Governors of George Brown College

Preface

I wrote this revised and updated text to respond to the need to educate students on print reading for the Canadian housing and renovation markets. Although the United States and Canada share many common practices, the number of differing practices makes it important for students to learn Canadian construction methods and applicable codes.

By teaching this course to the construction trades, architectural, construction management, and renovation programs, I have found it imperative that students understand and visualize the building process and the components used to build. Also, while portions of this content may be taught in more detail in other courses, this course lays the foundation for everything that follows.

This text is updated and references the National Building Code (2010 NBC) of Canada as its reference for code-related issues. There are numerous references to NBC parts provided throughout the text. Your province may have its own provincial code adapted from the NBC, and many of the reference parts may be the same. Additional content in this edition includes

- more details and information on sustainable building practices,
- information and photos about energy efficiency,
- an overview of building information modelling (BIM) and three-dimensional (3D) images of the model house used throughout the text,
- information on engineered flooring systems and ICF details,
- details about roof framing,
- detailed information on concrete strengths and conversion factors,
- NBC requirements regarding the notching and drilling of dimensional lumber,
- a new set of imperial drawings called the Brook, and
- over 160 additional imperial-based assignment and quiz questions.

Although print reading information can transcend numerous construction sectors, the primary focus of this text is housing and small buildings (Part 9 of the NBC).

Because this is a Canadian text it covers two systems of measurement. The first and official system is the metric system; the other is the imperial system. Five primary sets of drawings, as well as numerous section details, are used throughout this text. The Five complete drawings are:

1. **Doncaster Drawings**
   These are the main drawings used throughout the text and are of a production home. The actual house construction has been photographed from start to finish. Comparisons and references between the drawings and the construction methods used to build the house are made throughout the text. Added to this edition are a number of 3D Revit drawing images that display the structural components in such a way that the student can more readily visualize the components used to construct the building. This helps students better understand what the drawings mean and how this information is turned into reality, as well as showing the finished project. These drawings are drawn in metric and can be found in the back of the text and for instructors on the text's website in pdf format. (see **http://www.nelson.com/stephenson3e**).

2. **Whittington Drawings**
   This set of drawings provides students with another perspective and layout. Both the Whittington drawings and the Doncaster drawings are used in a number of the assignments and quizzes. These drawings are drawn in metric and can be found in the back of the text and for instructors on the text's website in pdf format. (see **http://www.nelson.com/stephenson3e**).

3. **The Brook Drawings**
   New to this edition, the Brook drawings provide students with a full set of drawings done in imperial measurements, allowing students to compare and contrast them with the Doncaster and Whittington drawings, which are drawn in metric measurement (the Brook drawings come attached to the textbook and have the name "Brook" in the title block). A number of assignments and have been added that utilize the Brook drawings in Chapters 6 through 13. Unique to the Brook drawings is that they are based on production home building, providing the student with the opportunity to view a design that includes two options for the front elevation of the house, Elevation A and Elevation B, which enables builders to conform to local architectural control requirements and to diversify the look of the neighbourhood they are creating. The Brook drawings also provide a number of options and details not indicated in the Doncaster or Whittington drawings, such as the optional side door for both Elevations A and B, and incorporates the use of manufactured joist systems for the floors. There is also a separate set of electrical plans provided for these drawings, enabling the student to better comprehend how building systems can be separated into their own drawings as their complexity increases. These drawings can also be found for instructors on the text's website in pdf format. (see **http://www.nelson.com/stephenson3e**).

4. **Khalmur Crescent Drawings**
   These drawings, found in Chapter 15, are of a simple addition and were drawn by the Toronto Area Chief Building Officials Committee (TACBOC) to help explain drawing requirements needed for a building permit. The Khalmur drawings help students compare some of the differences involved in reading plans and building a new home as opposed to building an addition, where the process involves matching materials that support and tie into the existing building. These drawings are drawn in imperial and for instructors can also be found on the text's website in pdf format. (see **http://www.nelson.com/stephenson3e**).

5. **Mini Mart Drawings**
   These drawings, discussed in Chapter 16, are for a small retail store. The building is structural steel and the drawings that were produced for educational purposes introduce you to architectural and engineering drawings using a structural grid. These drawings are in metric and can be found in the back of the text and for instructors on the text's website in pdf format. (see **http://www.nelson.com/stephenson3e**).

TEXT LAYOUT

Chapters 1 to 4 of this text cover the fundamentals of reading drawings, including visualization; understanding scales and measurement systems; types of drawings; construction notes; and understanding abbreviations, hatchings, and symbols.

Chapters 5 to 13 follow the typical sequence of constructing a house from initial zoning and permits to completion.

Chapter 14 provides valuable information on the purpose and use of specifications and contracts.

Chapter 15 is broken into two sections the first is on renovation practices and interpreting drawings for renovations. The second is on multiple-unit buildings (e.g., townhouses). This latter section focuses on understanding townhouse drawings, including issues of fire separation and sound transmission.

Chapter 16 references the Mini Mart drawings included with the text. Chapter 16 is meant to introduce you to other methods of construction, such as structural steel, and to how the buildings differ from housing and small buildings. Chapter 16 also introduces how the structural grid is used for orientation purposes on large residential and commercial drawings.

Chapters 14, 15, and 16 were placed at the end of the text recognizing that some instructors may not have time to cover all of the chapters and may therefore use them in another course. The chapters are designed such that an instructor can move some out of sequence to better suit teaching and program needs (e.g., Chapter 14, Specifications and Contracts, could be moved to follow Chapter 4, Abbreviations, Hatchings, Symbols, Construction Notes, and Detail Drawings).

FEATURES FOR STUDENTS

- Chapter Outcomes listed at the start of every chapter
- Comparison of production home building to custom home building
- Completely updated references to pertinent National Building Code of Canada (2010) sections
- Construction methods and materials addressed and related to construction drawings
- Green and sustainable building practices referenced throughout the textbook
- Details of construction process of actual house with matching drawings
- Five complete sets of drawings
- One to two Student Assignments per chapter with answers in appendix at back of text
- One Quiz Review per chapter
- Key terms in bold with list at end of chapter and definitions provided in glossary
- List of reference websites provided in appendix.

INSTRUCTOR RESOURCES

The **Nelson Education Teaching Advantage (NETA)** program delivers research-based instructor resources that promote student engagement and higher-order thinking to enable the success of Canadian students and educators. To ensure the high quality of these materials, all Nelson ancillaries have been professionally copy-edited. The ancillaries for this text are authored by the text author, Tom Stephenson, and are provided on the **Instructor Companion Site** at www.nelson.com/stephenson3e.

- Instructor's Manual offers chapter-specific Key Teaching Points, Teaching Tips, and Lecture Ideas as well as a list of Learning Objectives, Chapter Topics, and Key Terms.
- Quiz Review answers are available separately in Word and PDF format.
- The Test Bank is available in a new, cloud-based platform. **Testing Powered by Cognero®** is a secure online testing system that allows you to author, edit, and manage test bank content from any place you have Internet access. There are approximately 320 multiple-choice, 80 True/False, and 160 short-answer questions available in Cognero.
- An additional short answer, multiple choice, and essay type questions with answers are provided for each chapter in printable Word or rtf format.
- NETA PowerPoint lecture slides for each chapter are in Microsoft® PowerPoint®. There is an average of 55 slides per chapter, many featuring key figures, tables, and photographs from *Understanding Construction Drawings*, third edition. NETA principles of clear design

and engaging content have been incorporated throughout, making it simple for instructors to customize them for their courses.

- Additional drawings and photos from the text are also available in the Image Library
- The Doncaster, Whittington, Brook, Khalmur Crescent, and Mini Mart drawings from the back of the text are also provided in pdf format with the Instructor Resources for the text at **http://www.nelson.com/stephenson3e** for the instructor to use in the PowerPoint slides.

Acknowledgements

I would like to thank Wayne Cassidy of Cassidy & Co. for providing the core drawings used in this text and for his ongoing support of educating students in the construction sector, and I also want to thank Tribute Homes for providing access to its home building site to photograph the construction of the homes used in the text drawings and for providing background documents and information. Particular thanks to Steve Brockstein, construction project manager; John Mazzuca, senior site supervisor; and David Speigel, vice president of operations.

I also want to thank my many colleagues and students from George Brown College who have taught me so much over the years, particularly John Wills who originally helped me brainstorm the initial concept and provided some of the key contacts that helped make this text become a reality, and Rainier Blundel for his expertise in drawing the drain drawings used in Chapter 11. I also want to thank Clint Kissoon and the many other contributors to the Mini Mart drawings, with special thanks to Frank Osso for his help editing the CAD files. Also I want to thank Clint Kissoon for providing permission to use the BIM illustrations of the Doncaster house that is currently being researched and used in the College's BIM certificate and construction programs. Finally I want to thank Jonathan Gibb for his work for George Brown College on developing the BIM for the Doncaster house; Jonathan has made me see how quickly the teacher can become the student in a world that is constantly changing.

Special thanks to Mary Lawson for taking time to write a foreword, and to Michael Nauth for his technical check of the drawings.

Thank you to the reviewers who provided their feedback on the preliminary reviews: Luc Lapierre, Algonquin College; Martin Askes, Fanshawe College; Scott Fleming, Fleming College; and Ann Pepper, Georgian College.

I would like to thank the editorial team at Nelson Education Ltd., in particular, Anne Williams, Jackie Wood, Alexis Hood, Susan Calvert, Susan James, Claire Horsnell, Jaime Smith, and Rachel Eagan (who sought all the permissions). Thanks also to Katherine Goodes for guiding me through the development of this edition, keeping the project on track and on time.

Finally, I want to thank my lovely wife Sophie who is simply the best wife in the world, for not only being patient with me during this long process, but also encouraging, motivating, and inspiring me to write and complete the text. I could not have done this without you! I also want to thank my daughter Madeline the journalist for all her editorial and writing advice and my daughter Frances the lawyer for her help on the legal areas pertaining to construction contracts. This work is truly a family project.

About the Author

Tom is a full-time professor at George Brown College. He has a diverse background in teaching, having taught in the disciplines of Building Science, Construction Management and Construction Business Management, Architecture, Carpentry and Brick and Stone Masonry. Tom developed, the Building Renovation Technology program and teaches in the Construction Engineering Technology program and the Architectural Technology and Civil diploma programs, as well as in the Bachelor of Applied Technology, Construction Science and Management Degree program at George Brown.

Tom has worked with leaders in industry, such as Mattamy Homes in developing their in-house employee training program, and he currently consults with EllisDon Construction, assisting them in the development and delivery of their in-house project management. Prior to teaching, Tom owned, managed, and operated a mid-sized general contracting.

Tom holds two degrees, a Bachelor of Administrative Studies and a Master of Business Administration. He has been an active member at the local, provincial and national levels of the Canadian Home Building Association. He also teaches Project Management for the Toronto Construction, the Ontario General Contractors Association, and the Labourers' International Union of North America. Tom is the 2014 recipient of the Society for Teaching and Learning in Higher Education (STLHE) College Sector Educator Award.

Courtesy of Tom Stephenson

1 Construction Drawings: An Overview

LEARNING OBJECTIVES

After completing this chapter, you will be able to:

- Identify, compare, and contrast the different construction industry sectors.
- Identify different types of drawings and their best use as applied to construction.
- Identify the basic processes used to produce a set of drawings.
- Analyze, visualize, and sketch an orthographic drawing based on a three-dimensional drawing of an object.

INTRODUCTION

This chapter provides an overview of the construction industry, its many sectors, and its impact on the Canadian economy. The types of drawings used for construction projects are explained, as are key concepts that will help you develop the skill of analyzing and visualizing what you are trying to construct from a set of drawings. This chapter will also provide a brief overview of Building Information Modelling, referred to as BIM. The construction industry is evolving with exciting changes in technology that are taking place. It is expected within the next few years that there will be widespread adoption of BIM, which will advance the use of three-dimensional (3D) models, along with the integration of other disciplines such as estimating, project management, and facilities management in larger projects.

FIGURE 1.1
School being constructed for children from new housing being built at this location. New residential subdivisions require institutional buildings such as schools.

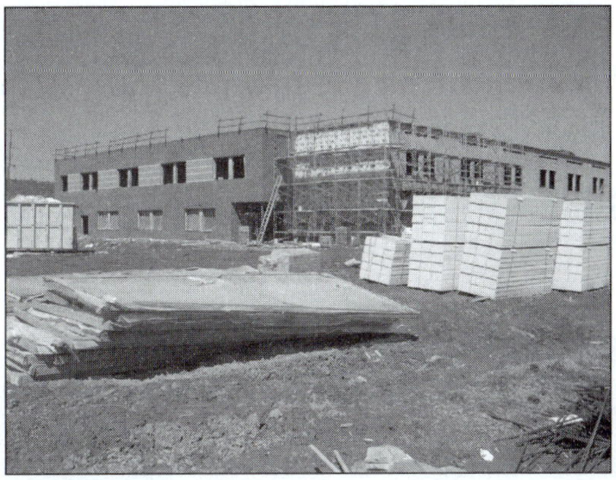

Courtesy of Tom Stephenson

FIGURE 1.2
Office building under construction.

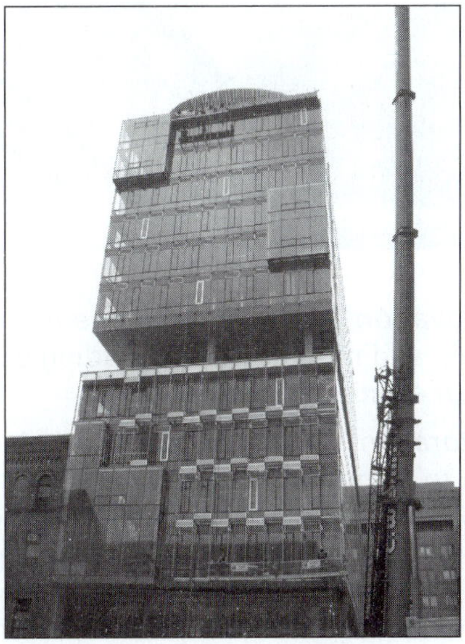

Courtesy of Tom Stephenson

THE CONSTRUCTION INDUSTRY

While this text will focus on "housing" and "small buildings" as defined in the **National Building Code of Canada (NBC),** it is important to appreciate how much we rely on construction in our daily lives. The construction industry plays a large role in the economic growth and prosperity of Canada. The industry comprises a number of sectors, including:

- *Institutional:* hospitals, universities, colleges, schools, libraries
- *Commercial:* office buildings, retail stores, restaurants

- *Industrial:* manufacturing and industrial-type buildings.

FIGURE 1.3
Industrial construction.

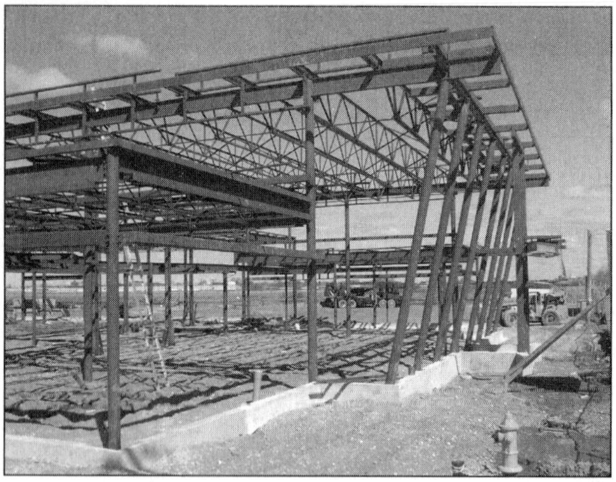

The acronym **ICI** is commonly used to identify the institutional, commercial, and industrial construction sectors.

- *Residential:* single, multiple low-rise, and high-rise residential buildings

FIGURE 1.4
High-rise condominium under construction with exterior curtain walls completed.

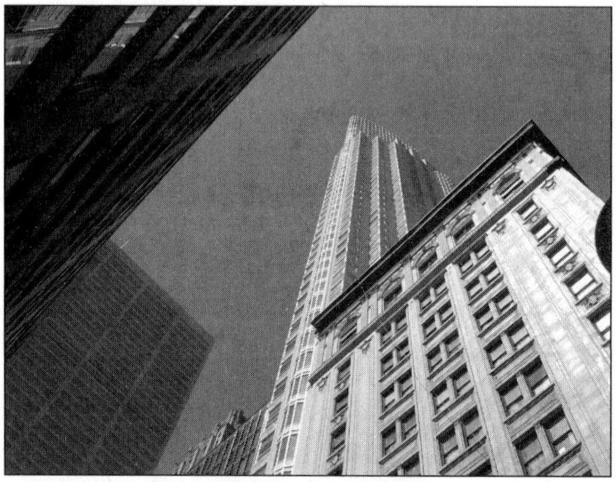

Courtesy of Tom Stephenson

- *Renovation and redevelopment:* renewal, repair, and modification, including change of use of existing structures
- *Restoration:* the restoration of existing buildings to bring the structure back to or above its original condition, including both exterior and interior work. Includes historical projects and projects where materials have worn to the extent that they require replacement or repair (for example, underground parking garages and balcony slabs on condominium buildings)

FIGURE 1.5
Residential sub-division under construction.

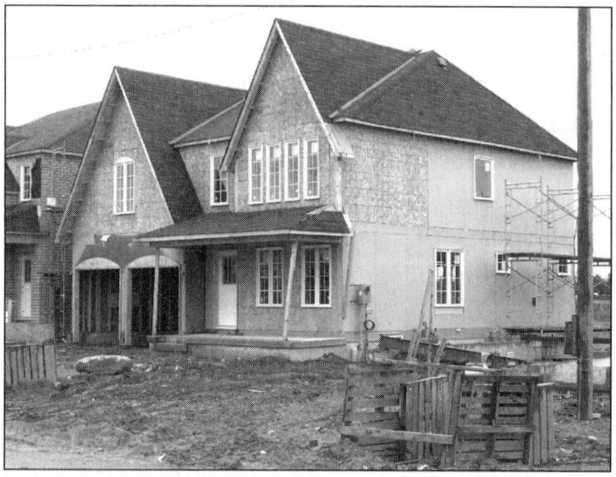

Courtesy of Tom Stephenson

FIGURE 1.6
Extensive renovations to an existing residential building.

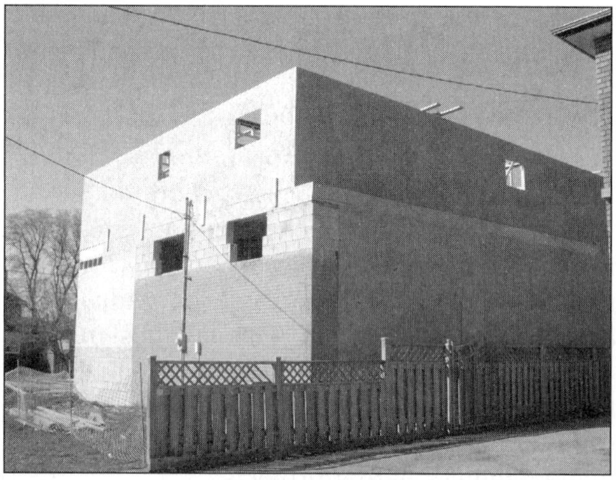

Courtesy of Tom Stephenson

FIGURE 1.7
Restoration of a historical building.

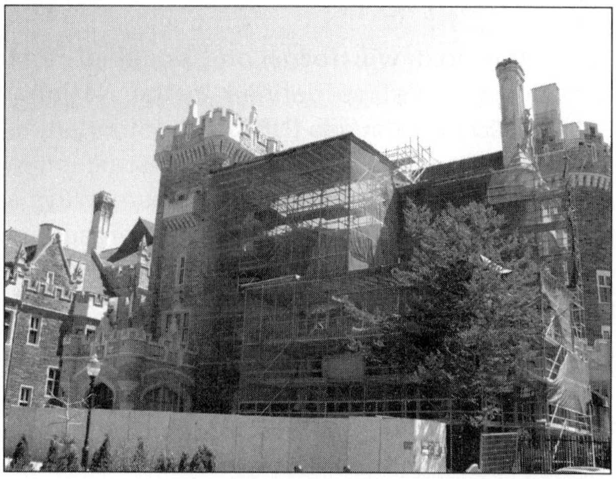

Courtesy of Tom Stephenson

- *Infrastructure:* civil projects such as roads, bridges, subways, water and sewage treatment facilities

These sectors do not necessarily remain separate and distinct. For example, a residential high-rise may have commercial office and retail space on the lower floors; an industrial manufacturing plant may include commercial office space; and, increasingly, hospitals, universities, and colleges may incorporate restaurants and retail space.

The renovation and restoration sectors involve all the other sectors, and all homes and buildings will eventually require modifications, upgrades, or retrofitting to maintain them and bring them up to their **highest and best use.** "Highest and best use" is a real estate development term that refers to the development of a property that will offer the best return because it is being put to the best and most desired use.

Another area that spans all the construction sectors listed above is the movement of society towards building more energy efficient and environmentally sustainable buildings and infrastructure. The World Commission on Environment and Development developed a definition of **sustainable development** that was included in its 1987 report:

> Humanity has the ability to make development sustainable—*to ensure that it meets the needs of the present without compromising the ability of future generations to meet their own needs.* The concept of sustainable development does imply limits—not absolute limits but limitations imposed by the present state of technology and social organization on environmental resources and by the ability of the biosphere to absorb the effects of human activity.[1]

The Green movement is affecting the way we design and construct our buildings. As a result, building design, management, methodology, and technology are changing and evolving to better meet societal requirements and demands. In order to be successful, Green Building requires that we do a better job of integrating our design and construction processes to ensure that these benefits are maximized and that the building functions as an integrated system. Although there are numerous definitions for what **Green Building** is, this text will define Green Building as building to a higher, more energy efficient and environmentally sustainable level than is currently standard practice (e.g. National Building Code regulations), with the intention of eventually becoming truly sustainable.

[1]From "Our Common Future," Report of the World Commission on Environment and Development, World Commission on Environment and Development, 1987. Published as Annex to General Assembly document A/42/427, Chapter 2.

THE PURPOSE OF DRAWINGS

In order for any construction project to be successful it must have a clearly thought out set of drawings and specifications. The drawings and specifications, also known as plans, blueprints, or prints, provide general contractors and subcontractors with the information required to accurately **tender** and bid for projects. The drawings and specifications provide the bidders with accurate, detailed information concerning the project. The general contractor is then able to accurately estimate the resources (materials, labour, and equipment) required to construct the project. The client will be able to select the successful bidder with the confidence that all the bidders submitted their bids based on the same drawings and specifications, which accurately reflect the project. The client can make a selection based on an apples-to-apples bid, as opposed to an apples-to-oranges bid, in which the client and contractor have different expectations for the project. Failure in this area is the source of many misunderstandings, cost overruns, poor construction, and lawsuits.

Construction drawings are also a requirement for applying for a building permit. They ensure that the building meets all building code requirements, including minimum standards with regard to safety and durability. The National Building Code (NBC) is the model code used in Canada; many provinces have developed their own building codes based on the NBC.

Architectural Offices

Drawings are developed by skilled architects, architectural technologists, and engineers (for the purposes of this text, unless noted for specialization purposes, we will refer to architects, architectural technologists, and engineers as designers); some smaller contractors may also perform this function, particularly for housing and small buildings (Part 9 of the National Building Code). Licensing requirements vary between provinces, and the intended use or occupancy of the building may require professional stamps of approval from different professional designations such as architects and engineers. Designers may work in small one-person practices, or in large multiperson offices that work on hundreds of projects a year. Traditionally, drawings were done by hand and run through a process that resulted in a blue background with white lines; these types of drawings, known as blueprints, are usually now found in city archive offices and historical records storage. From these roots, drawings received the name "blueprints" and to this day are often still referred to that way.

Today almost all construction drawings are drawn using computer-aided design **(CAD)** systems, which allow the designer to quickly design, revise, and send drawings electronically. However, some designers still use the traditional method of drawing by hand.

CAD drawings are printed on a **plotter**. The process is referred to as "plotting the drawings."

FIGURE 1.8
CAD designers at work at Cassidy & Co., an architectural technologists' practice specializing in production of home designs.

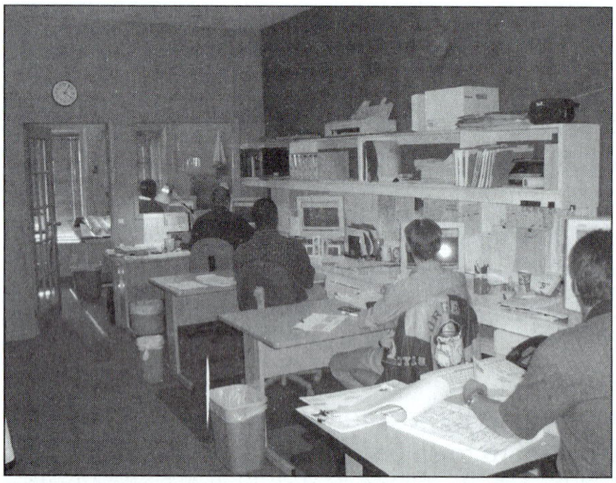

FIGURE 1.9
A CAD drawing being plotted. PDF files are also often sent via email for review and plotting purposes.

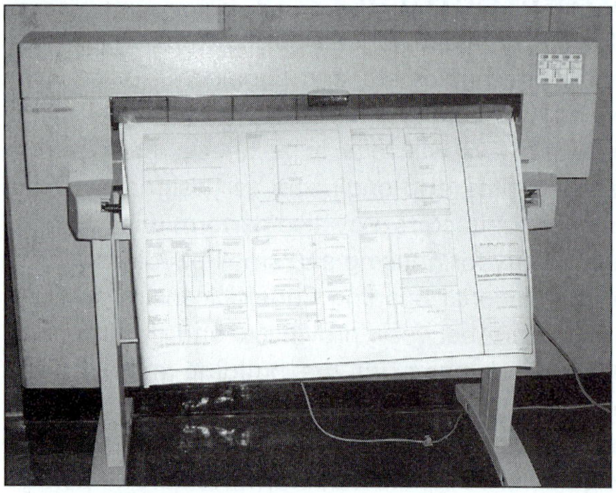

Once plotted, multiple copies can then be produced by either further plotting or photocopying. While still in its infancy, the use of tablets with touch screens to view drawings is gaining traction very quickly, offering the advantage of being able to zoom into view and annotate drawings, forming an effective means of sending requests for information (RFI) quickly and efficiently.

Imperial and Metric Standard Sheet Sizes

Drawings are plotted or drawn on many different sizes of sheets (see Tables 1.1 and 1.2 below). Many residential drawings for production home building are drawn on 11″ × 17″ or B-size sheets. The 11″ × 17″ size provides a low-cost alternative for copying. The size of sheet is determined by the size of the project, the scale, and the readability of the drawings at that particular scale (refer to Chapter 2 for more detailed information on scales).

TABLE 1.1
Drawing Sheet Sizes: Metric

Sheet Designation	Overall Size (mm)
A0	841 × 1189
A1	594 × 841
A2	420 × 594
A3	297 × 420
A4	210 × 297

Source: Reproduced with the permission of Canadian Standards Association **from *CSA Withdrawn Standard CAN3 B78.3-M77 - Building Drawings - Basic Engineering*** which is copyrighted by CSA, 5060 Spectrum Way, Mississauga ON, L4W 5N6 Canada.

TABLE 1.2
Drawing Sheet Sizes: Imperial

Sheet Designation	Overall Size (inches)
A	8½″ × 11″
B	11″ × 17″
C	17″ × 22″
D	22″ × 34″
E	34″ × 44″

Source: With the permission of Canadian Standards Association (operating as CSA Group), material is reproduced from CSA Group Standard, CAN3 B78.3-M77 entitled "Building Drawings - Basic Engineering" which is copyrighted by CSA Group, 5060 Spectrum Way, Suite 100, Mississauga ON, L4W 5N6. This material is not the complete and official position of CSA Group on the referenced subject, which is represented solely by the standard in its entirety. While use of the material has been authorized, CSA is not responsible for the manner in which the data is presented, nor for any interpretations thereof. For more information or to purchase standards from CSA Group, please visit http://shop.csa.ca/ or call 1-800-463-6727.

Types of Drawings

Construction drawings are laid out to most efficiently provide the information required to construct the building. To the novice, a set of construction drawings might seem confusing, and you may conclude that a three-dimensional drawing will provide the best means of illustrating the building while providing the information to build from. There are definite advantages to having a three-dimensional model for visualization and integration purposes (BIM). It is, however, impractical to build from three-dimensional drawings. For example, the detailed information required for the building process cannot be properly displayed, nor can we see the required information through the walls or floors. Therefore, a number of two-dimensional or orthographic drawings are required for a typical construction project. However, this does not mean that three-dimensional drawings are not important. They are used in a number of different ways, from on-site sketching to **architectural renderings**.

The kinds of three-dimensional drawings used most frequently in the construction industry are:

- Perspective drawings
- Isometric drawings
- Oblique drawings.

The type of two-dimensional drawings used most frequently in the construction industry is:

- Orthographic drawings.

FIGURE 1.10

Railway track illustration showing vanishing point as rails merge to one.

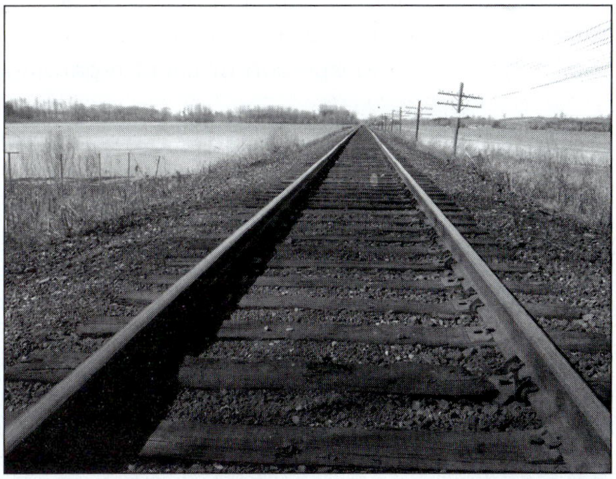

Courtesy of Tom Stephenson

FIGURE 1.11

Rectangular block drawn in perspective.

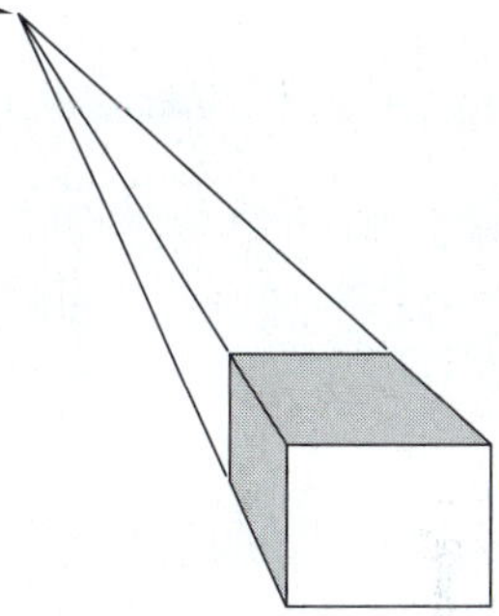

Courtesy of Tom Stephenson

Perspective Drawings

Perspective drawings provide the most realistic view of the object. A picture taken with a camera is a perspective view. These drawings are drawn to what is known as a vanishing point. (There may be several vanishing points on a single drawing.) This can be most easily understood if you imagine standing on a railroad track and looking off into the distance at the rails. You will see two rails joining to one in the distance, in essence travelling to a vanishing point.

Because this type of drawing has tapering views, it is not drawn to **scale** and is therefore not typically used for placing any kind of dimensions on the drawing. However, it is used by builders, developers, and designers as a means of showcasing the project. Typically, perspective drawings are used in sales and marketing materials such as brochures, newspapers, and posters. The builders use this architectural rendering format because they want their potential clients to see the project in an attractive and realistic format.

Isometric Drawings

Isometric drawings provide a three-dimensional look that allows the viewer to readily see the shape of the object. An advantage of isometric drawings over perspective drawings is that they can be drawn to scale, thus providing a means for adding dimensions. However, if the drawings are complex it will become difficult, if not impossible, to provide all the necessary information without excessively crowding the drawing.

Isometric drawings are drawn with all horizontal lines drawn at a 30-degree angle to the horizontal, all vertical lines drawn vertical, and with circles represented by drawing an ellipse to provide a three-dimensional look to the drawing.

Isometric drawings are used to provide a three-dimensional scale view of an object. Usually the objects do not require a great deal of explanation. In many cases, site superintendents will quickly sketch an isometric drawing to convey information to a tradesperson or client regarding a particular construction detail (for example, a framing detail or a cabinet detail). Inexpensive or free software such as Google SketchUp can also be used to quickly draw sketches on a tablet or laptop. You will also find isometric drawings used in many textbooks and magazines for the same purpose.

FIGURE 1.12
Perspective drawing used as a promotional rendering for sales and marketing of new homes, Tribute Homes, drawn by Cassidy & Co.

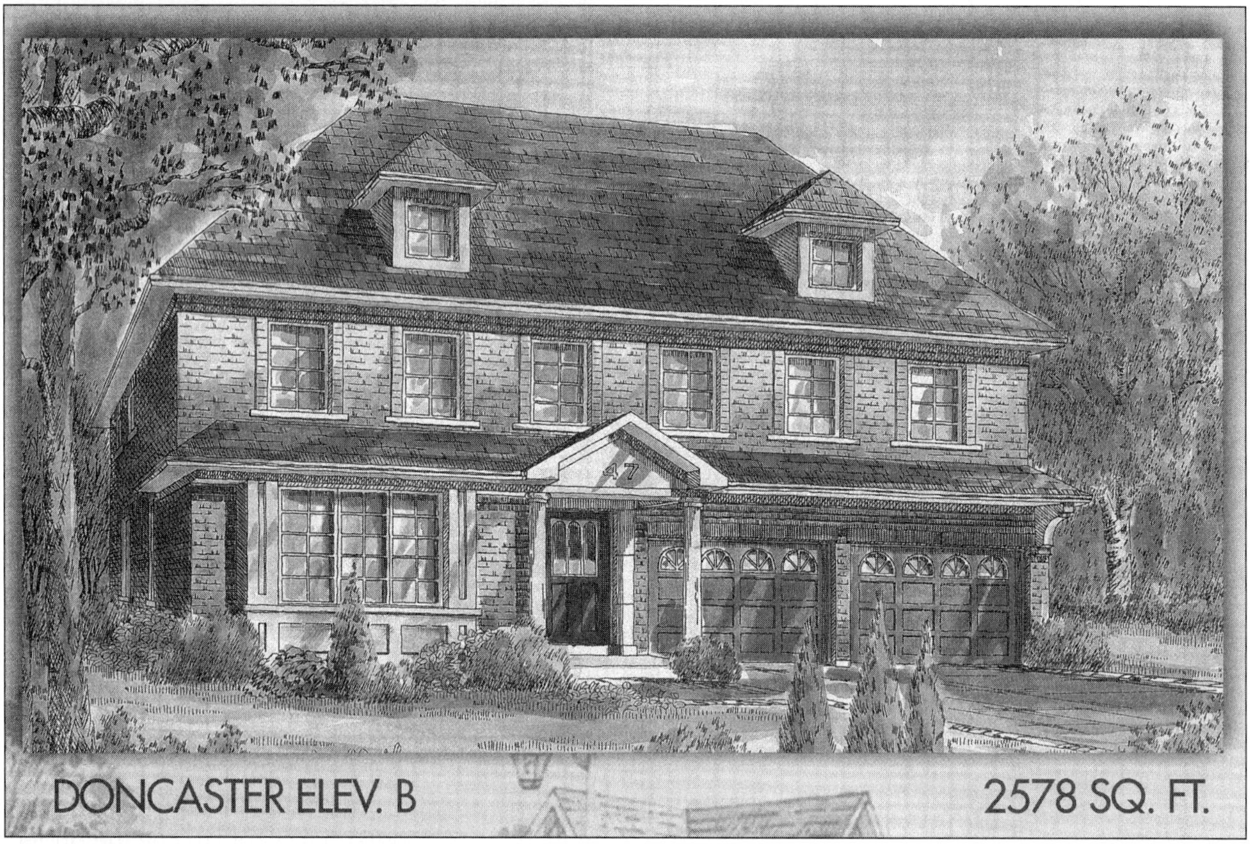

Source: Cassidy & Co. Architectural Technologists

FIGURE 1.13
Isometric sketch displaying horizontal lines at 30-degree angle.

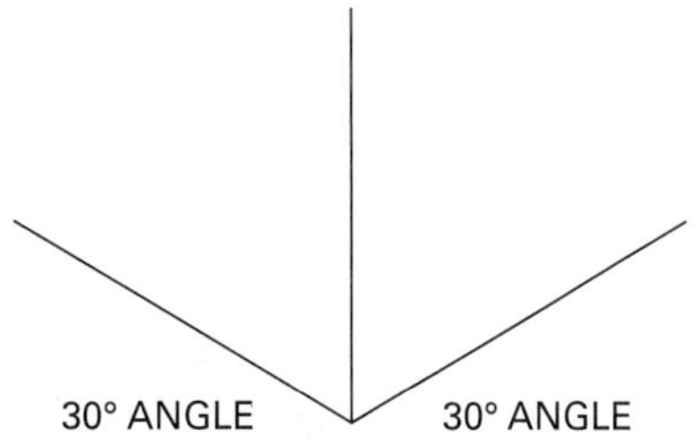

FIGURE 1.14
Isometric sketch of horizontal lines at 30 degrees for a rectangle.

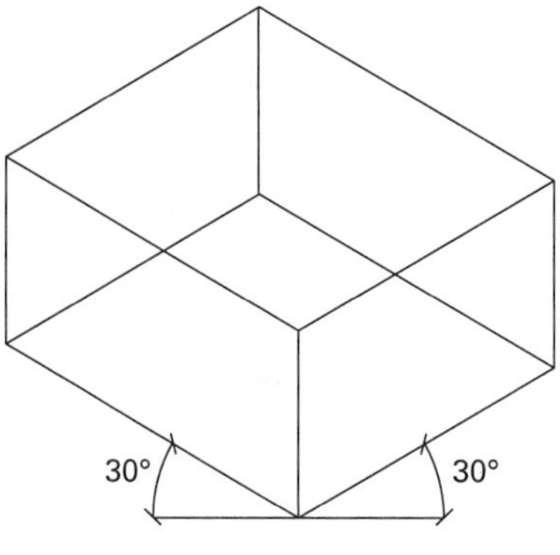

FIGURE 1.15
Isometric sketch used to illustrate a framing detail for a two-stud corner.

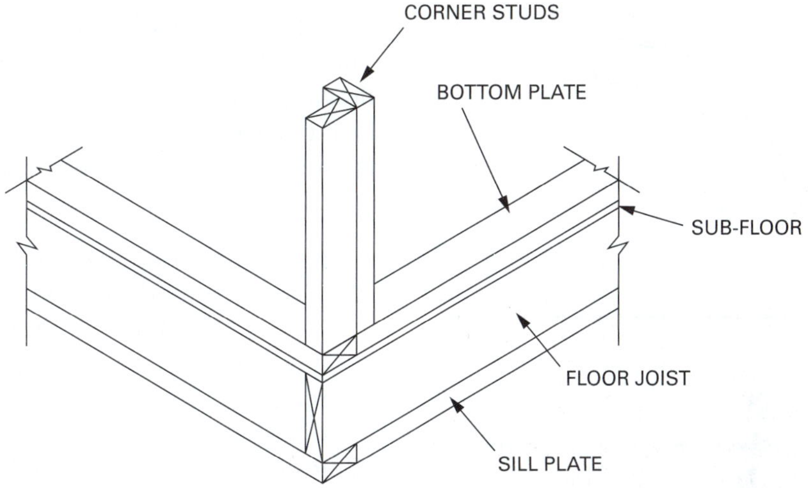

FIGURE 1.16
Isometric sketch used to illustrate a cabinet detail.

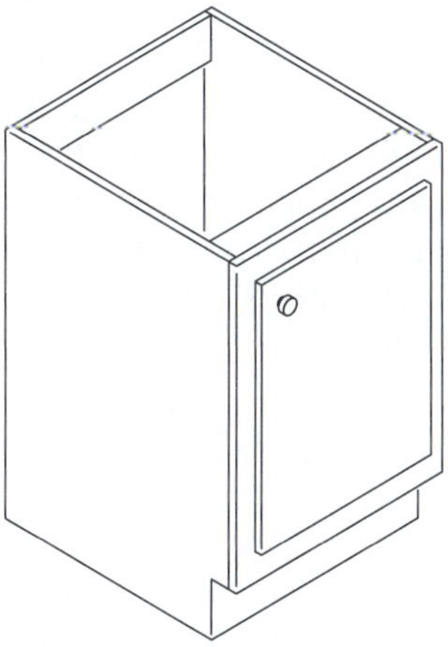

FIGURE 1.17
Oblique drawing of a rectangle or box.

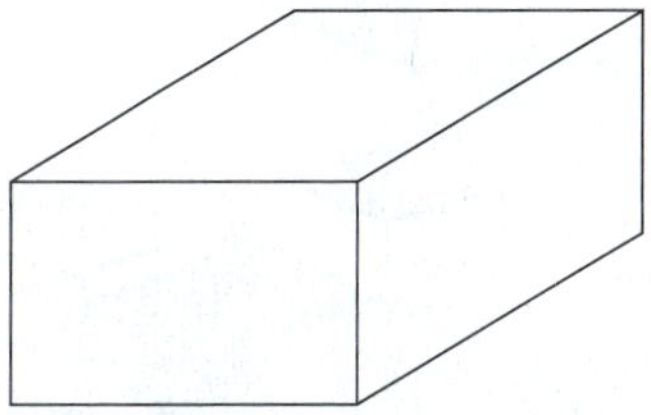

FIGURE 1.18
Oblique drawing of a simple house.

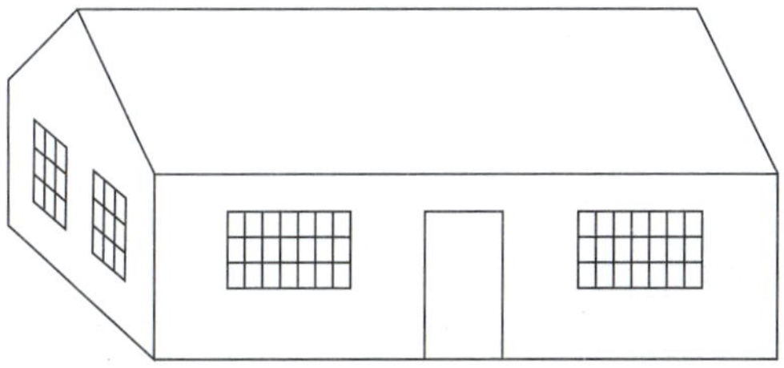

FIGURE 1.19
Single orthographic view of a crown moulding.

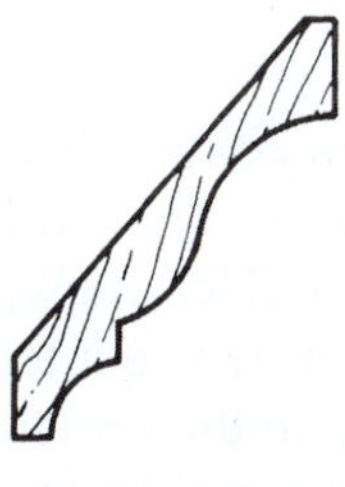

Oblique Drawings

Similar to isometric drawings, **oblique drawings** provide a three-dimensional view that provides readers with three-dimensional (visual) information regarding the object while showing the object as a whole. Oblique drawings are relatively easy to draw. They are drawn with the face of the object in an orthographic view (straight on) while the side of the object is drawn out at an angle, generally 45 degrees.

Oblique drawings are used effectively to display mouldings or trim details. A client may look at a section view of a moulding but may not be able to visualize it because it is not being viewed as three-dimensional. However, adding a few lines going out at 45 degrees from the sectional drawing can suddenly clarify the image, providing a three-dimensional representation of the object.

FIGURE 1.20

Oblique view of same moulding as shown in Figure 1.19; notice how much easier it is to visualize the crown moulding.

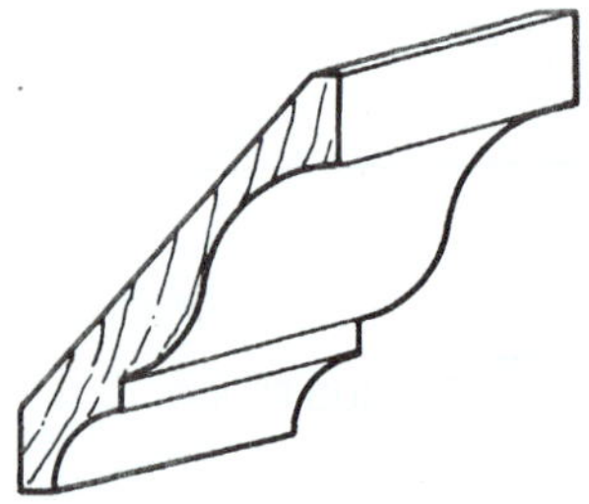

FIGURE 1.21

Simple house enclosed in a glass box; method of orthographic projection.

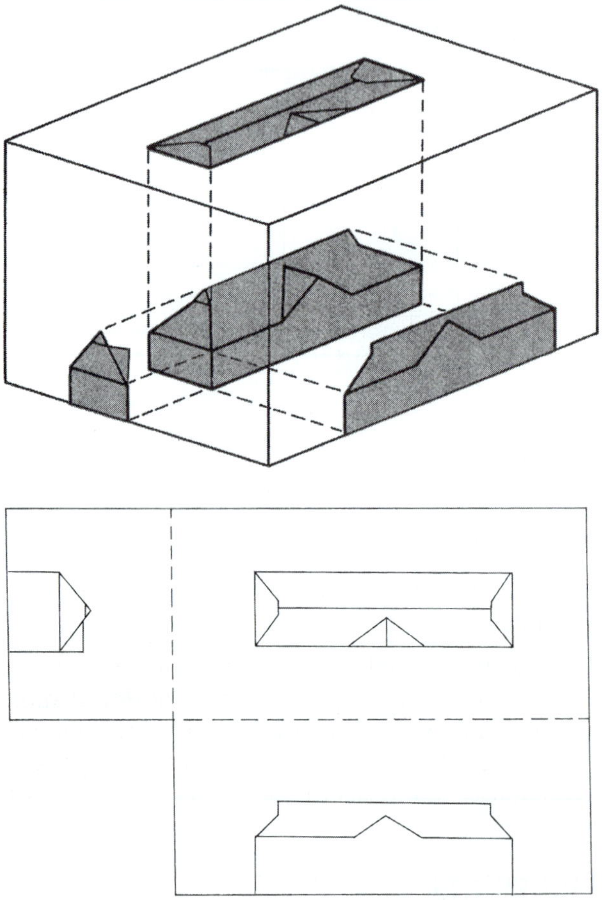

Manufacturers may use this technique to help sell their products, and designers, site superintendents, project managers, contractors, and tradespeople may use it to help their clients or others working on the project visualize an element such as crown moulding.

Orthographic Drawings

Orthographic drawings are the drawings most commonly used in construction. These drawings provide a single straight-on view of the object. The best way to visualize how orthographic drawings are drawn is to imagine that the object is in a glass box and that each face of the object is projected directly outward to the glass surface.

Advantages of using orthographic drawings:

- Provides straight-on view of the surface being studied. Readers therefore know exactly what they are looking at in that view.
- Can be drawn accurately to scale, which makes it easy to add dimensions to the drawing while keeping everything in the proper proportion.
- There is room to provide more detailed information about the object (for example, material used, joist sizes, beam sizes, door types, location of walls). Orthographic drawings are effective for all types of views, including elevations, plans, sections, and details.
- Helps readers focus on the different surfaces being looked at, one surface at a time. Readers also become accustomed to referring to specific drawings for the required information. With practice this becomes intuitive over time.
- They are the best method of using drawings for each view; they communicate all of the graphical information required for building. All the information required to lay out and construct the building can be inserted into the appropriate orthographic drawing.

Disadvantages of using orthographic drawings:

- By looking at one surface straight on, readers cannot distinguish depth (for example, is the garage connected to the house, is it set back from the house, or is it jutting forward from the house?). Readers cannot tell for sure by looking at just one orthographic drawing.
- In order for readers to fully comprehend the object, they must flip through the various types of orthographic drawings and form a three-dimensional view of the object in their minds. For the purposes of this text, we will refer to this as **visualization**.
- Orthographic drawings make it difficult to detect clashes between building systems components such as ductwork, piping, and beams. Three-dimensional modelling with BIM makes this much less problematic in more complex structures.

FIGURE 1.22

Orthographic drawing of front elevation of a house compared with isometric drawing of same house.

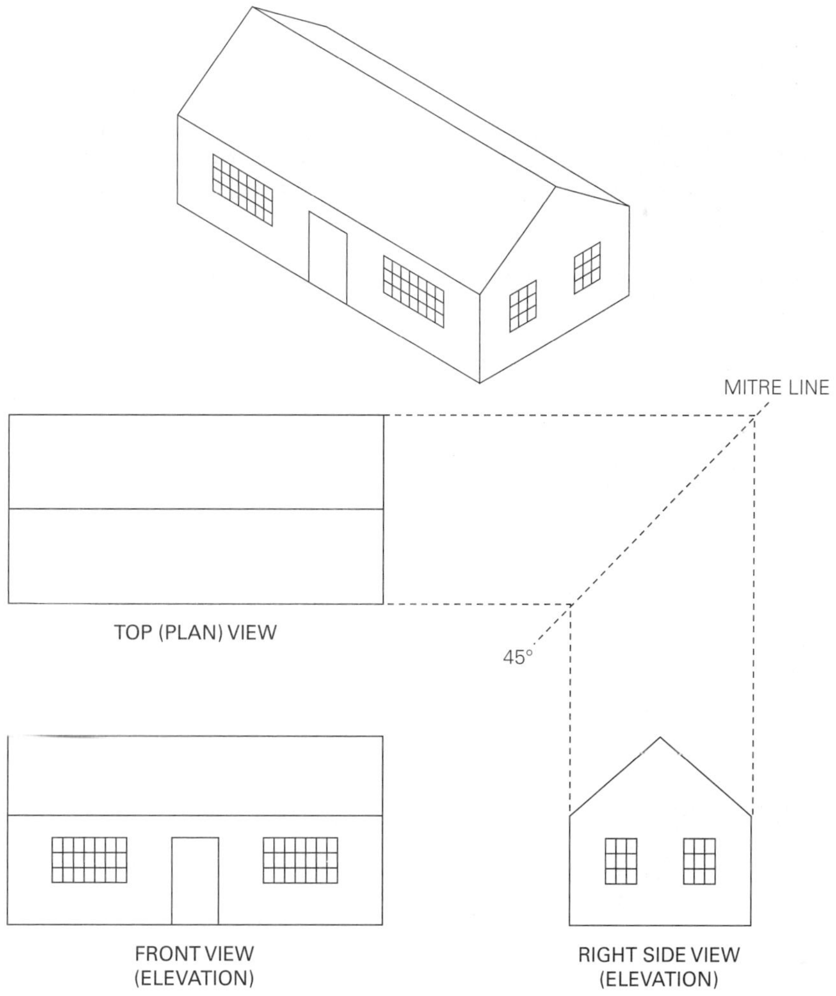

The ability of readers to visualize the building or object in their minds is a key principle in successfully interpreting construction drawings. And, whether you are a tradesperson, site superintendent, or project manager—or working in any facet of the construction industry—being able to visualize is an essential skill that will facilitate your success in this business. Developing the ability to visualize three-dimensional objects from single-plane objects does take time; nevertheless, with some practice you will develop this important skill.

Types of Orthographic Drawings A set of residential construction drawings will usually comprise the following orthographic drawings:

- *Site plan or plot plan:* A bird's-eye view of the property with the building located on it. The site plan will show building setbacks and elevations; it may show existing and future grades, as well as the location of utilities and services (refer to Chapter 5 for more detail).
- *Foundation plan:* A cross-sectional view displaying the information required to lay out and construct the foundation. The foundation drawing indicates locations of beams, columns, foundation walls, windows, stairs, footings, and directions of joists (refer to Chapters 3 & 6 for more detail).
- *First-floor plan:* A cross-sectional view displaying the required information to lay out and construct the first floor of the building. Similar to the foundation plan, the first-floor plan displays exterior walls, windows, stairs, kitchen layout, window and door locations, bathroom layout (if applicable), information about the ceiling joists or basic roof truss information, depending on whether or not there is a second floor (refer to Chapters 3, 7, and 8 for more detail).

FIGURE 1.23

These images are based on a 3D model of the Doncaster house drawings used in the textbook. Models of this type can be part of the BIM integration used for clash detection, for estimating, and for project management purposes: BIM 3D, 4D, and 5D.

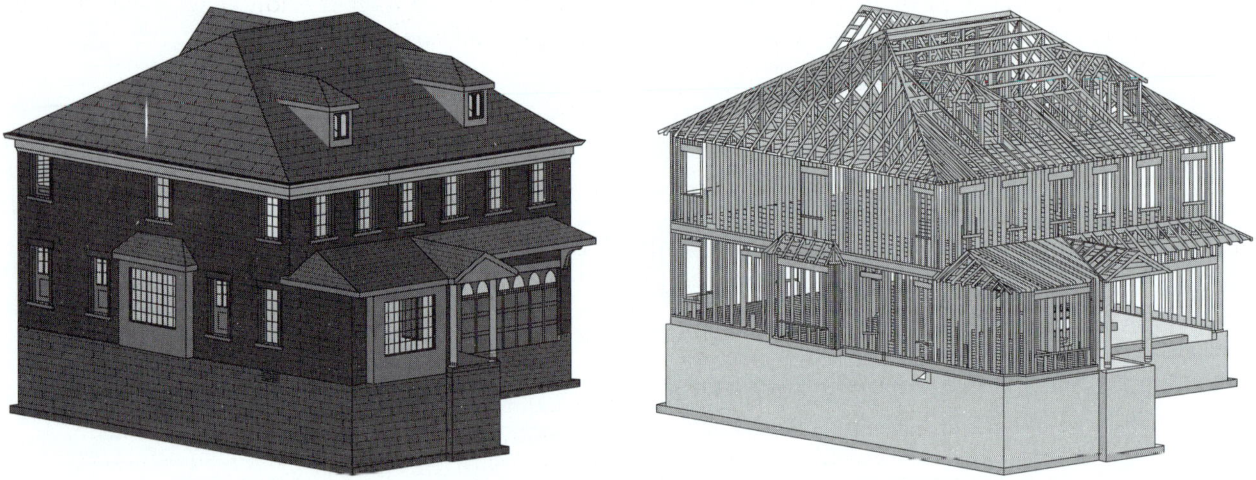

Source: Courtesy of Clint Kissoon, Chair, George Brown College, School of Architectural Studies.

- *Second-floor plan (if applicable):* A cross-sectional view displaying the required information to lay out and construct the second floor of the building. The second-floor plan is similar to the plans above, and displays information about the ceiling joists or basic roof truss information, depending on whether or not there is a third floor, in which case it would provide the joist information for the third floor (refer to Chapters 3 and 9 for more detail).
- *Roof plan:* This drawing details the major intersections of the roof, indicating the locations of the hips and valleys, and direction of slope. If the roof is complicated, this plan may also detail the location of all rafters. Separate, more detailed roof plans are provided by roof truss manufacturers (refer to Chapters 3 and 9 for more detail).
- *Four elevation drawings (front, left, rear, and right sides):* These drawings provide information regarding the exterior finishes and the elevations of walls, floors, doors, and windows (refer to Chapter 3 for more detail).
- *Detail drawings:* These drawings provide more detailed information on any components within the structure that cannot be fully explained or illustrated on the main drawings. For example, the specific requirements of a beam, column, and joist intersection may not be fully understandable from a floor plan, so a detail drawing will be drawn at a larger scale to provide more information so the contractor can construct it as it was designed (refer to Chapter 4 for more detail).
- *Mechanical, plumbing, and electrical drawings:* These drawings may be done separately or may be included in the floor plans above. These drawings provide the information required by the electrical, plumbing, and heating ventilation and air conditioning contractors (HVAC) (refer to Chapter 12 for more detail).

Introduction to Building Information Modelling (BIM)

As mentioned earlier in the chapter, **BIM** is an integration of new and improved technologies that facilitates better communication and documentation for the management and operation of construction projects. The construction industry as a whole has not been known for its quick adoption of new technologies, due often to the risks involved and in some cases the lack of technical expertise; however, in recent years the pace of change has been accelerating and adoption of BIM has quickly become more mainstream, starting with large, complex construction projects, particularly in the institutional, commercial, and industrial sectors of construction. In addition, a number of production builders from Newfoundland to Vancouver have also begun to adopt a number of BIM components into their operations. To what level a firm may adopt BIM (levels 3D–6D) depends on the type of work they do and the financial benefits that can be gained through adoption. One thing that is certain is that as the technology becomes easier to implement, more businesses will see the benefits of BIM and will be willing to adopt it into higher levels of integration as the processes become more defined, lower in cost, and more user friendly. The levels of BIM are generally defined as indicated in Table 1.3.

TABLE 1.3

BIM Dimensional Overview 2D–6D

	Building Information Modelling (BIM) Dimensional Overview
2D	A drawing with two dimensions is typically called an orthographic drawing (as explained earlier in the chapter) and can be plotted from a BIM model into working drawings. A 2D drawing is not a BIM type project but rather can be easily produced from a BIM model for site and construction purposes.
3D	A model of the building or structure in three dimensions, drawn using computer software that enables the viewer to rotate, zoom, and pass through walls to gain a full understanding of the object they are viewing. A 3D model enables the designers and project teams to detect clashes between components such as mechanical, structural, and electrical systems. This can greatly reduce onsite time and rework as errors can be worked out in the three-dimensional model before materials are shipped to site where, traditionally, adjustment and rework occurs due to problems with visualization that causes errors and omissions.
4D	Uses the 3D model for the purpose of phasing and sequencing the work. Animations and assemblies of buildings can be used to fully visualize the project while aiding in the sequencing and scheduling process. The model can be linked to the schedule. This can be used as a sales tool so that clients can better understand and visualize their projects. 4D BIM can also be used to assist the contractor in visualizing how the components will go together and assist in determining the best sequencing. This is particularly useful in complex and detailed assemblies.
5D	Uses the 3D model for the purpose of estimating. All objects on the model are defined, enabling the estimator to calculate the quantities of materials required for the project. The 5D representation can also be used to incorporate the 4D time element to display the effects of cost over time.
6D	Uses the as-built 3D model for the use of facilities management required for the life cycle of the building or structure. 6D facilitates the integration of the building systems into a database that is linked with the as-built 3D model to assist in the maintenance of the building systems, and can be used for future alterations and additions to the building or structure. The accurate, updated 6D model can also provide the as-built data to facilities for future referencing. While this advanced level may not be of great use in building a single residential house, it will provide great opportunities for long-term management of larger buildings and structures.

Source: Reproduced with the permission of Tom Stephenson.

ENDNOTE

1. *Our Common Future, Report of World Commission on Environment and Development*, World Commission on Environment and Development, 1987. Published as Annex to General Assembly document A/42/427, Chapter 2.

CHAPTER 1 ASSIGNMENT: ORTHOGRAPHIC VIEWS

This assignment will help you understand how a designer takes a three-dimensional object and draws it in orthographic views. You will quickly see that one orthographic view provides limited information about the object but multiple views provide all the necessary information.

From the isometric drawings shown, draw a front view, top view, and right side view. Each cube represents one square on the grid paper; ensure that you orient your views similar to the example provided in Figure 1.24. The front elevation/view is located in the bottom left corner, the top elevation/view is located in line and directly above the front view, while the right side elevation/view is located in line to the right of the front view.

The dashed lines represent the grid squares only, and do not need to be darkened or reproduced, with the exception of the heavy

FIGURE 1.24
Example of orthographic projection based on the isometric drawing at top right. Notice the orientation of the front, top, and right side views, similar to unfolding a box.

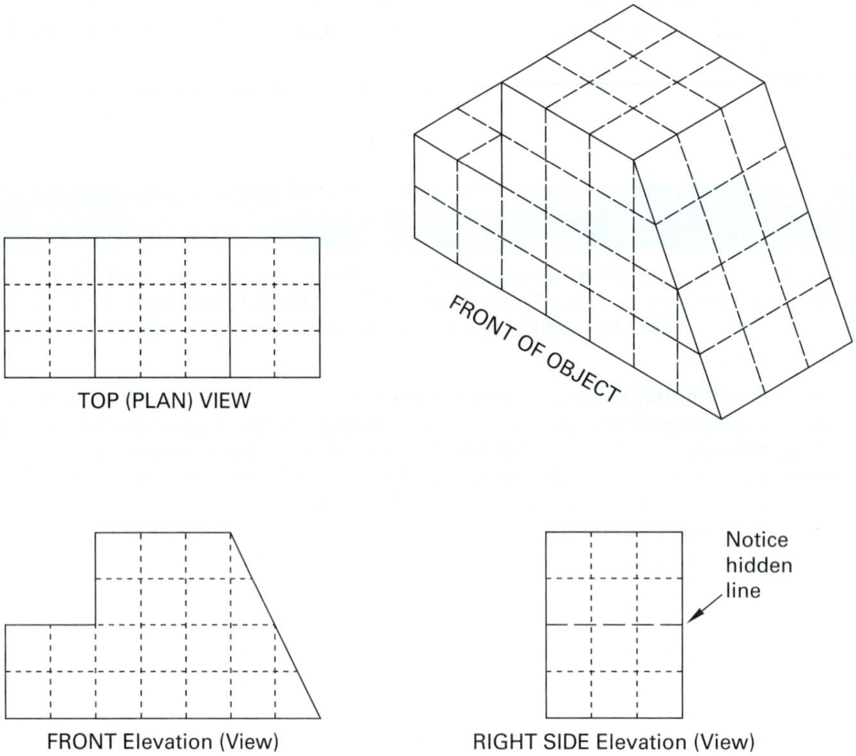

dashed line shown in the right side view. This line is called a hidden line and represents an object line that is hidden below or behind the surface; in this case it represents the indentation that runs from front to back at the rear of the object. You will find hidden lines used extensively in construction drawings.

KEY TERMS

Definitions for key terms appear in the glossary at the end of the text.

Architectural rendering (p. 5)
CAD (p. 4)
Green Building (p. 3)
Highest and best use (p. 3)
ICI (p. 2)
Isometric drawings (p. 5)
National Building Code of Canada (NBC) (p. 1)
Oblique drawings (p. 7)
Orthographic drawings (p. 8)
Perspective drawings (p. 5)
Plotter (p. 4)
Scale (p. 5)
Sustainable development (p. 3)
Tender (p. 3)
Visualization (p. 8)
Building Information Modelling (BIM) (p.10)

1.

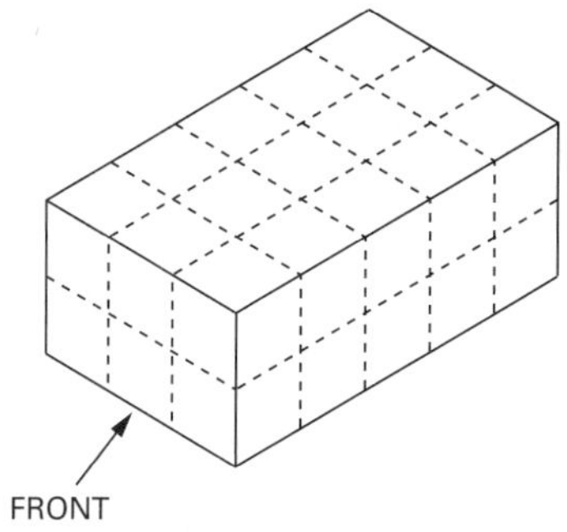

2.

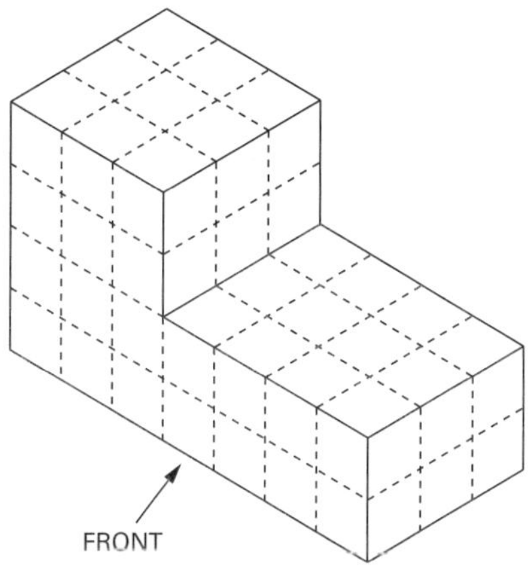

3.

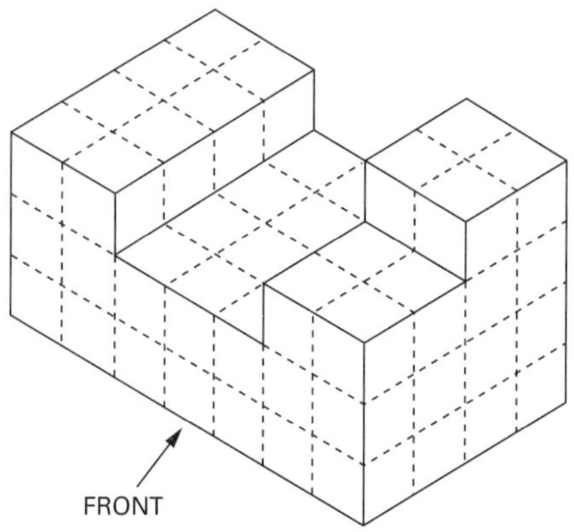

4.

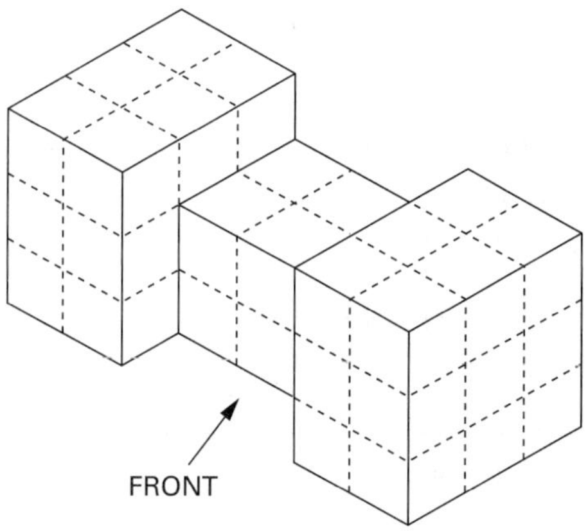

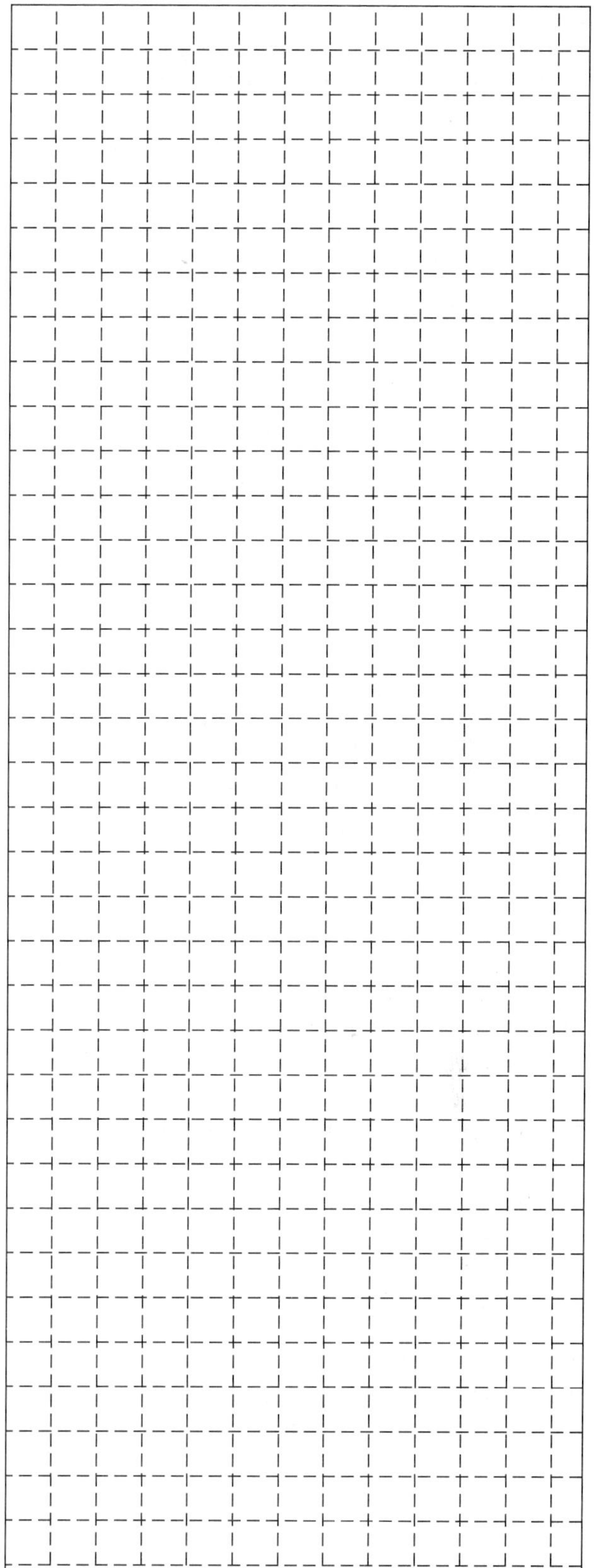

5.

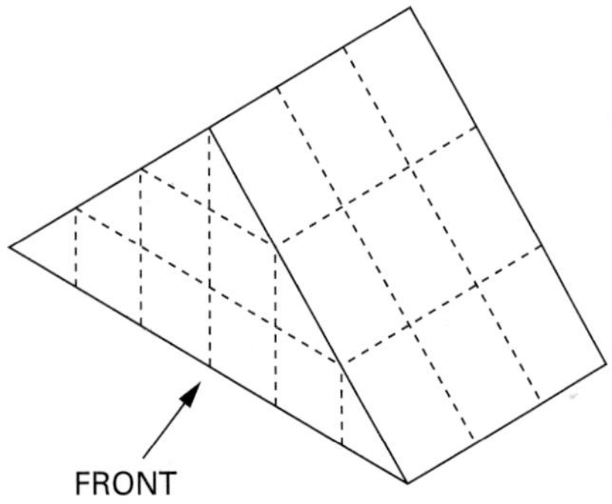

6.

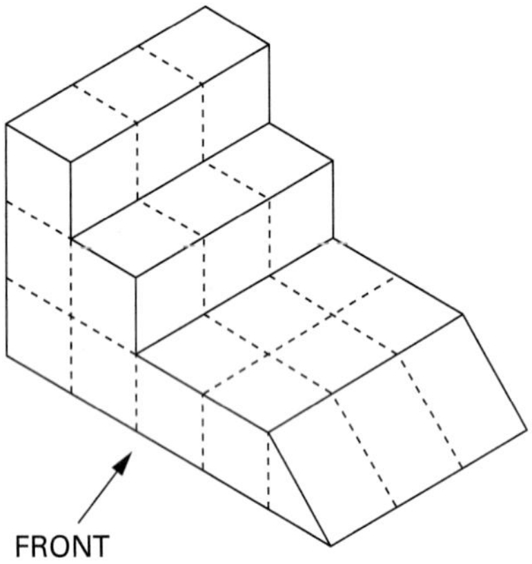

7.

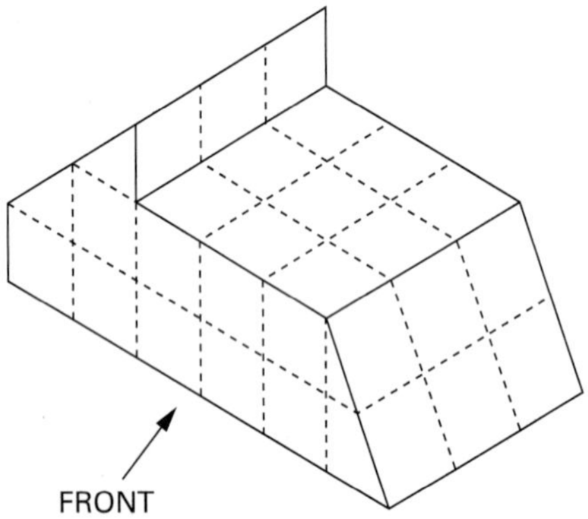

8.

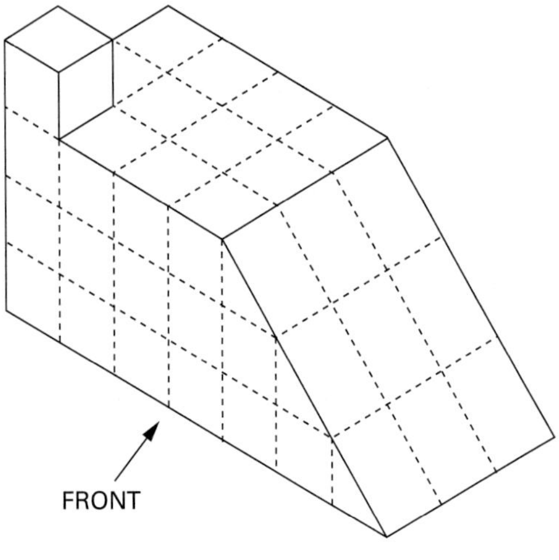

9.

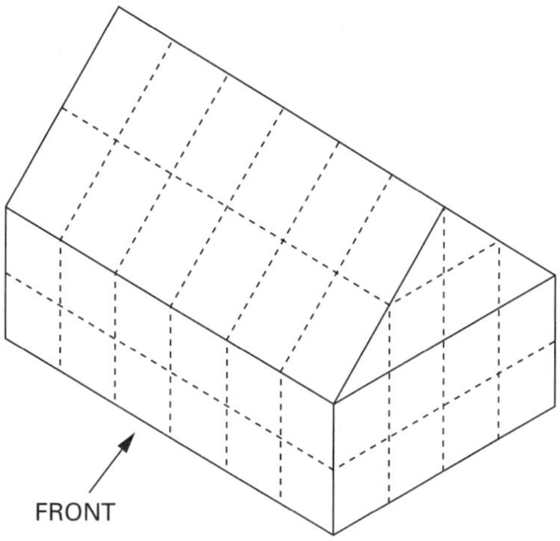

10.

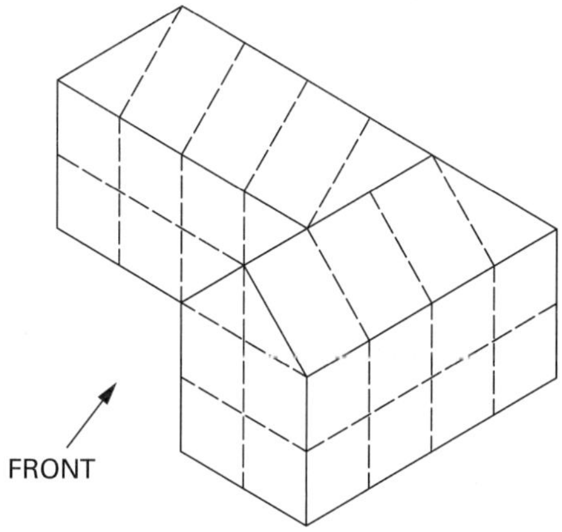

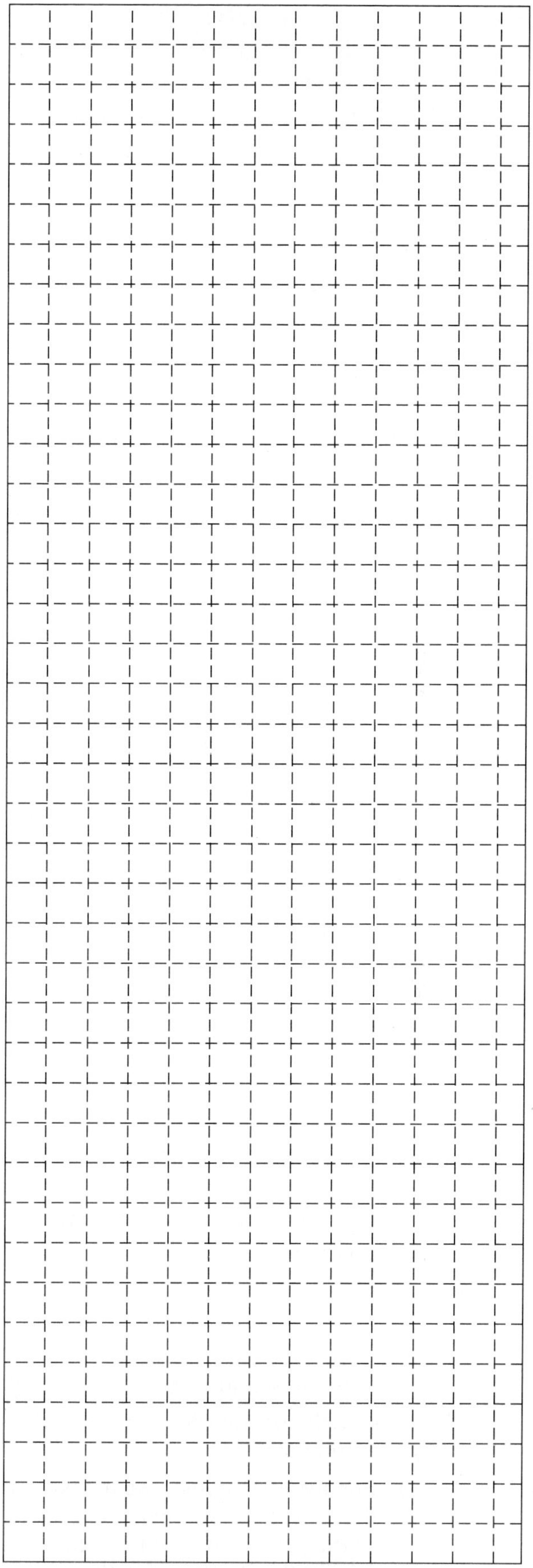

Chapter 1 Quiz Review: Construction industry and drawing types

Identify the choice that best answers the question.

_____ 1. What type of construction drawing is a roof plan?

a) isometric
b) perspective
c) orthographic
d) oblique

_____ 2. What does the phrase "highest and best use" describe?

a) rafters being placed on a roof that provides the highest and best slope
b) the selection of material that will perform the best
c) the development of a property that will offer the best return
d) the selection of a subtrade that will perform the best

_____ 3. How does the textbook define "Green Building"?

a) the same as sustainable development
b) building to a higher, more energy efficient and environmentally sustainable level than is currently standard practice
c) meeting building code requirements
d) providing the highest profitability while improving energy efficiency

_____ 4. What do construction drawings that are not complete or accurate result in?

a) happier clients
b) faster construction, as the work can be implemented quicker and at a lower cost
c) fewer litigation cases
d) rework, cost overruns, delayed schedules, and increased disputes

_____ 5. What is the definition of a sustainable development?

a) Sustainable development ensures that development meets the needs of the present without compromising the ability of future generations to meet their own needs.
b) Sustainable development is building to slightly better standards than are currently used today.
c) Sustainable development refers to large construction projects that take a long time to complete.
d) Sustainable development refers to small construction projects that don't impact the environment.

_____ 6. What type of drawing is often used to best visually represent different types of moulding, such as crown moulding?

a) orthographic
b) oblique
c) elevation
d) plan

_____ 7. What type of drawings have their lines drawn to a vanishing point?

a) isometric
b) perspective
c) oblique
d) orthographic

_____ 8. What type of drawings are drawn with the object's horizontal lines at 30 degrees, and the object's vertical lines vertical?

a) perspective
b) oblique
c) isometric
d) orthographic

_____ 9. Which of the following most closely describes an orthographic drawing?

a) All horizontal lines at 30 degrees.
b) A number of lines that go to a vanishing point.
c) The front of the object is horizontal, and the sides of the object move to the rear at a 45-degree angle.
d) The views are two-dimensional and may include a front, top, left, right, and rear view.

_____ 10. What does the acronym ICI stand for?

a) improved concrete insulation
b) industrial, commercial, institutional construction sectors
c) independent construction institute
d) integrated concrete I-beam system

_____ 11. Which of the following most closely describes the term "blueprint"?

a) the original method that was used for reproducing construction drawings and how construction drawings are still referred to by some people today

b) advanced drawings that are printed in the colour blue

c) artistic perspective drawings that are printed in the colour blue

d) the current method that is used for reproducing construction drawings and the term most people use when describing construction drawings

_____ 12. What construction sector could include work in all the other construction sectors?

a) institutional

b) renovation

c) industrial

d) commercial

_____ 13. Currently, how are most residential construction drawings drawn and plotted?

a) manually, by hand

b) orthographically, using computer software

c) three-dimensionally, using computer software

d) isometrically, by hand

_____ 14. What is a very popular sheet size for residential drawings used by production builders?

a) A

b) B

c) E

d) F

_____ 15. What level/dimension is a BIM for a project that includes clash detection?

a) 2D

b) 3D

c) 4D

d) 5D

_____ 16. What is the term used for producing original CAD drawings?

a) printing

b) copying

c) plotting

d) scanning

_____ 17. What level/dimension is a BIM for a project that includes estimating?

a) 3D

b) 4D

c) 5D

d) 6D

_____ 18. What type of information does a site plan provide?

a) detailed information about the roof of the building

b) information about the exterior finishes and window locations

c) information about the setbacks for the house, grading, and elevation heights

d) information about the interior wall locations

_____ 19. What type of drawing is an elevation construction drawing?

a) perspective

b) orthographic

c) oblique

d) isometric

_____ 20. What type of drawing provides information regarding the exterior finishes, and information about the heights between floors and ceilings, types of doors, and windows?

a) site plan

b) roof plan

c) floor plan

d) elevation drawing

2 Measurement Systems, Scales, and Line Types

LEARNING OBJECTIVES

After completing this chapter, you will be able to:

- Understand the concepts of drawing to scale.
- Use both metric and imperial scales for the purpose of interpreting information from drawings.
- Convert common units of measurement from one system of measurement to another.
- Interpret the measurements provided on the drawings and apply them to the construction project.
- Identify and interpret the various line types used to illustrate and dimension construction drawings.

INTRODUCTION

This chapter explains the systems of measurement typically used on construction drawings. The chapter also covers how drawings are drawn to scale, how to read a scale, and the types of scales used for different drawing types. The chapter also discusses the line types used to convey different types of information on the drawings.

DRAWING TO SCALE

It's feasible to draw an object full or actual size when constructing a small object, such as a jewellery box. However, it is not practical to draw construction projects full size. Instead we **scale** the drawings to suit both the level of detail required and the size of paper onto which we plan to plot the drawings. The term "scale" refers to the proportional shrinking or enlarging of an object. For example, a photograph is proportionally smaller or larger than the actual object being photographed, and a map is drawn to a scale that matches the area it represents; sketches are also drawn to an approximate scale.

The scale used for producing drawings is determined by a number of factors:

- *The size of the object being drawn:* Generally speaking, the larger the object, the greater the reduction scale required. For example, a floor-plan layout for a house may be drawn at $\frac{1}{50}$ or $\frac{1}{100}$ its original size or, as is the case with the floor plans for the Doncaster drawings that came with the textbook, at a scale of $\frac{1}{75}$.
- *The level of detail required by the drawing:* Complicated portions of drawings or components of drawings requiring a high level of detail will not be reduced as much as drawings that require less detail. For example, a beam detail will typically be reduced less than a floor plan.
- *The size of paper on which the drawing is being plotted:* As indicated in Chapter 1, drawings can be plotted onto a variety of paper sizes. Generally speaking, a larger sheet can accommodate a smaller scale. For example, a foundation plan drawn at $\frac{1}{50}$ may not fit on an 11" × 17" sheet but a $\frac{1}{75}$ scale of the same drawing may fit perfectly.
- *The system of measurement in which the drawing is being created:* Both systems of measurement (imperial and metric) have standard scales commonly used for construction drawings. For example, the Brook drawings that come with the textbook have a scale of $\frac{3}{16}$" per foot. In other words, in this case $\frac{3}{16}$" drawn on paper represents 12" or 1'-0" of building.

SYSTEMS OF MEASUREMENT

Since the 1970s, Canada has officially used the **metric system of measurement.** As a result, official documents such as building codes, surveys, and government reports use the metric system. However, many Canadian businesses are strongly influenced by our largest trading partner, the

United States; consequently, many of the goods manufactured and produced in Canada are still created using the **imperial system of measurement.**

Currently, the construction industry uses both methods of measurement, with some materials fully switched to metric and others still using the imperial system.

For example, an 8" masonry block is actually not 8" and has not been 8" since the mid-1970s. An 8" imperial size block used to be 7⅝" high × 15⅝" long × 7⅝" wide; it was ⅜" smaller than 8" to accommodate a standard mortar joint. The block and joint would add up to 8" × 8" × 16". An 8" block manufactured today is not 8"; it is 190 mm × 390 mm × 190 mm, with a 10 mm allowance for the mortar joint. The metric block when laid is 200mm × 400mm. 200mm when converted to imperial is 7⅞" not 8". The difference is small but when you consider a difference of ⅛" on every course it does not take long to add up. For example 8 courses of block laid in a wall will have a 1" difference in overall height between metric and imperial blocks. Tradespeople involved in renovation and restoration work are familiar with the difference because old blocks do not align properly with new blocks. Even now, however, blocks are still generally referred to as 8" blocks instead of 200 mm blocks.

Note, construction drawings will display imperial dimensions as follows:

8 inches is drawn as 8"

2 feet is drawn as 2' or 2'–0"

15 feet 6 inches is drawn as 15'–6"

Another example of imperial versus metric in construction is the on-centre spacing for **studs** and floor joists (the structural members that make up the walls and floors). The most commonly used spacing is 16" on centre; the building code has used a **hard conversion,** converting this number to 400 mm (approximately 15¾") on centre (OC). The conversion sounds straightforward enough. However, we must consider several factors in construction; for example, for a set of drawings drawn in metric, the framer will look at the floor plan and lay out the walls for their location on the floor using the metric layout provided. When the framer sees that the drawings call for the studs to be laid out at 400 mm OC, he/she will then proceed to lay out the wall plates for the studs using the hard-conversion equivalent, 16" OC. The framers are required to do this because the plywood and drywall for the floors and walls are manufactured in the standard imperial sizes of 4' × 8'. If the framer laid out the **floor joists** at 400 mm OC, the **subfloor** would not match at the joints, resulting in excessive cutting and fitting. Building code enforcement accepts this slight variation in practice due to the constraints of the materials.

Examples can be seen in Table 2.1. A **soft conversion** column is shown to display the exact conversion sizes.

FIGURE 2.1

Imperial block and metric block isometric drawing with dimensions.

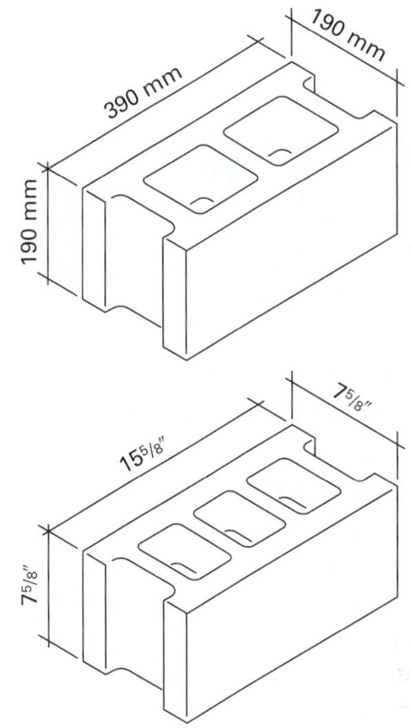

TABLE 2.1

Common Layout Sizes

Standard imperial spacing (sizes accepted as equivalent to NBC)	Soft conversion to metric from imperial measurement	Hard conversion to metric (sizes used by NBC)
12"	304.8 mm	300 mm
16"	406.4 mm	400 mm
24"	609.6 mm	600 mm

FIGURE 2.2
Imperial/metric tape measure; note the difference between 400 mm (indicated as 40 cm) and 16". Also note the diamond shape that indicates 19.2" on the tape; this spacing divides equally into 8', the standard size for sheet goods, and is being used more frequently now than in the past due to lumber savings and increased energy efficiency. Refer to Chapters 7, 8, and 9 for more detail on framing on centre spacing.

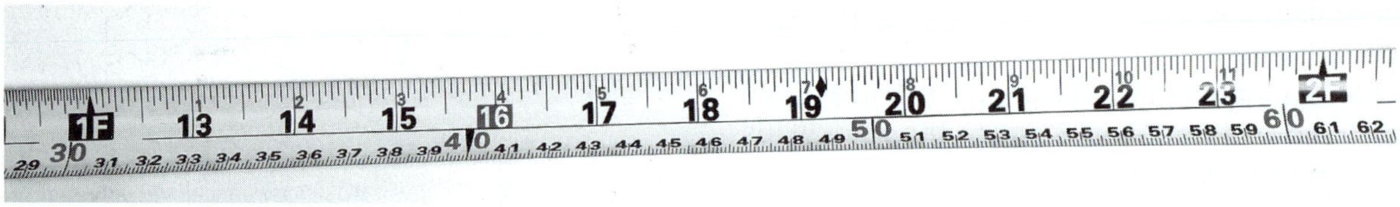

Courtesy of Tom Stephenson

There are numerous similar examples in construction; remember that success in construction in Canada requires you to be fluent in both systems of measurement, and you need to be able to convert one system to another as required. This text will refer to both systems of measurement with the understanding that metric is the official system of measurement. Many construction drawings use the imperial system, particularly in low rise residential construction and residential renovations, while institutional, commercial, high rise residential, and industrial typically use the metric system of measurement. Even when drawings are in metric, the process of constructing the project will also require the use of the imperial system of measurement and vice versa. An imperial/metric tape measure can be used to easily flip from one unit of measurement to the other (refer to Fig. 2.2).

SIMPLE MATHEMATICAL CONVERSION

Most residential drawings using the metric system of measurement use millimetres as the standard unit of measurement (e.g., 1250 mm) while residential drawings using imperial units of measurement use feet and inches (e.g., 4'–6"). To convert from one measurement to the other, use the conversion rate of 25.4 mm = 1". Therefore, to convert 10" to mm, multiply 10 × 25.4 = 254 mm. Conversely, to convert 254 mm to inches, divide by 25.4.

To convert 3'–4" to mm, first convert feet to inches: 3' = 3 × 12" then add 4" to arrive at 40". Next multiply 40" by 25.4 = 1016 mm.

For metric and imperial conversion factors, refer to Appendix B. You can also purchase or Google online metric conversion calculators or download a smart phone conversion calculator or construction calculator, if required.

SCALES

Metric Scales

A wide variety of scales are used in construction drawings. When drawings are drawn manually, the designer uses a triangular-shaped rule with six scales on it, two per side, for a total of six scales on a single metric scale rule. These scales are used by designers, estimators, project managers, and site supervisors. The use of CAD systems has enabled an unlimited number of scales to be used; however, most construction drawings will use a limited number of more commonly used scales.

Table 2.2 is a list of typical metric scales used for construction projects, along with some examples of where they may be used (note: scale size may vary depending on project). Try to visualize the proportions provided in Table 2.2 under the reduction size column.

Reading the Metric Scale

A metric scale is read from left to right, with the measurement designation provided on the first full number. For example, on the 1:50 scale 1 m represents 1 metre. There are no calculations required; the scale does the conversion.

There are 20 divisions on the 1:50 scale from 0 to 1 m; this means that each division equals 1/20th of a metre (1 metre = 1000 millimetres); therefore, 1000/20 = 50 mm or .05 m. Each scale will have a different value for each division. This can be easily calculated as demonstrated above, or you can break the divisions down visually. For example, using the 1:20 scale illustrated in Figure 2.5, you can see that the point from 0 to 200 mm represents 200 mm, the halfway point, with the longer division line therefore representing 100 mm and each space between 0 and the 100 mm

TABLE 2.2
Typical Metric Scales

Metric scale type	Typical drawing type	Reduction size
1:250	Site plans	1/250 of actual size; 1 mm drawing = 250 mm actual size
1:150	Site plans	1/150 of actual size; 1 mm drawing = 150 mm actual size
1:100	Floor plans, elevations, and site plans	1/100 of actual size; 1 mm drawing = 100 mm actual size
1:75	Floor plans and elevations	1/75 of actual size; 1 mm drawing = 75 mm actual size
1:50	Floor plans and elevations	1/50 of actual size; 1 mm drawing = 50 mm actual size
1:20	Interior elevations and cabinet details (e.g., kitchen, study, bathroom)	1/20 of actual size; 1 mm drawing = 20 mm actual size
1:10	Construction details	1/10 of actual size; 1 mm drawing = 10 mm actual size
1:5	Construction details	1/5 of actual size; 1 mm drawing = 5 mm actual size
1:2	Small objects that do not require much reduction	1/2 of actual size; 1 mm drawing = 2 mm actual size
1:1	Objects that can be drawn full size, usually detail drawings of small objects	full size

FIGURE 2.3
Metric scale.

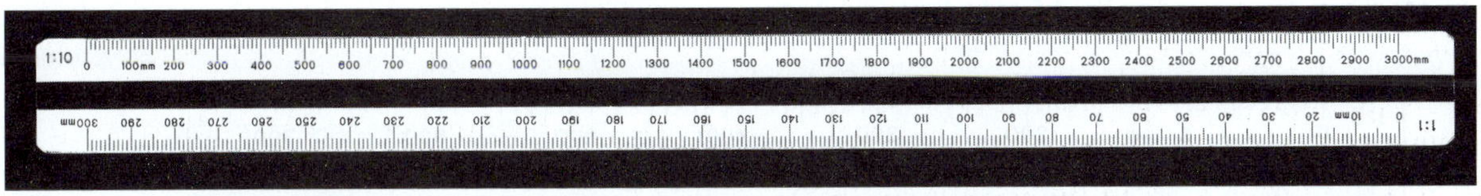

FIGURE 2.4
Example of 1:50 scale.

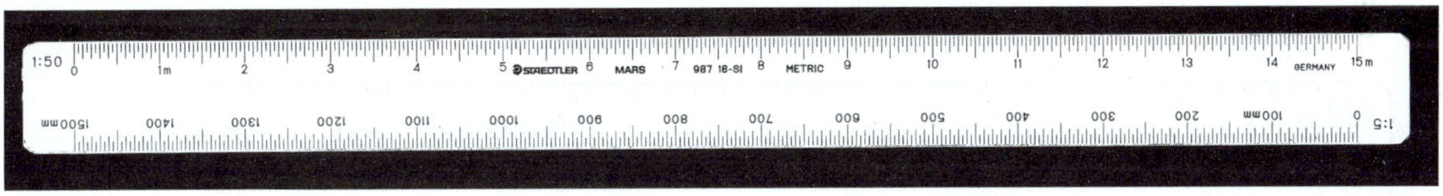

FIGURE 2.5
Example of 1:20 scale.

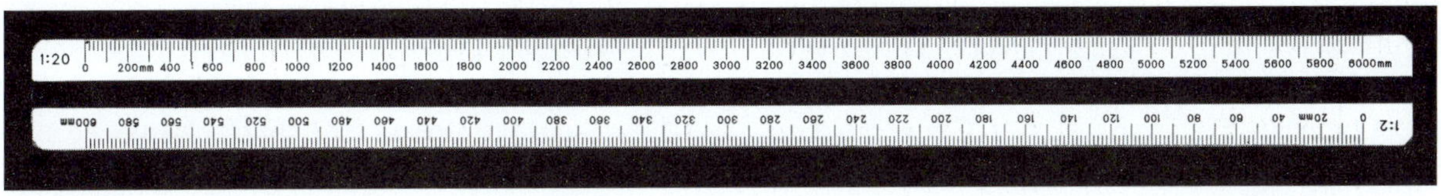

division (four lines, five spaces) representing 20 mm. Once you get used to working with scales, this process becomes quick and automatic. There is a shortcut that seems to be consistent with most metric scales (note that this shortcut does not work with imperial scales), which is whatever the metric scale is (e.g. 1:20), it is also what the space between each measurement bar equals in mm. For example, the 1:50 scale has a space between bars representing 50 mm. The 1:100 scale has a space between bars representing 100 mm.

FIGURE 2.6

One measurement using different metric scales. As the scale is reduced, the clarity of measurement becomes more difficult to read.

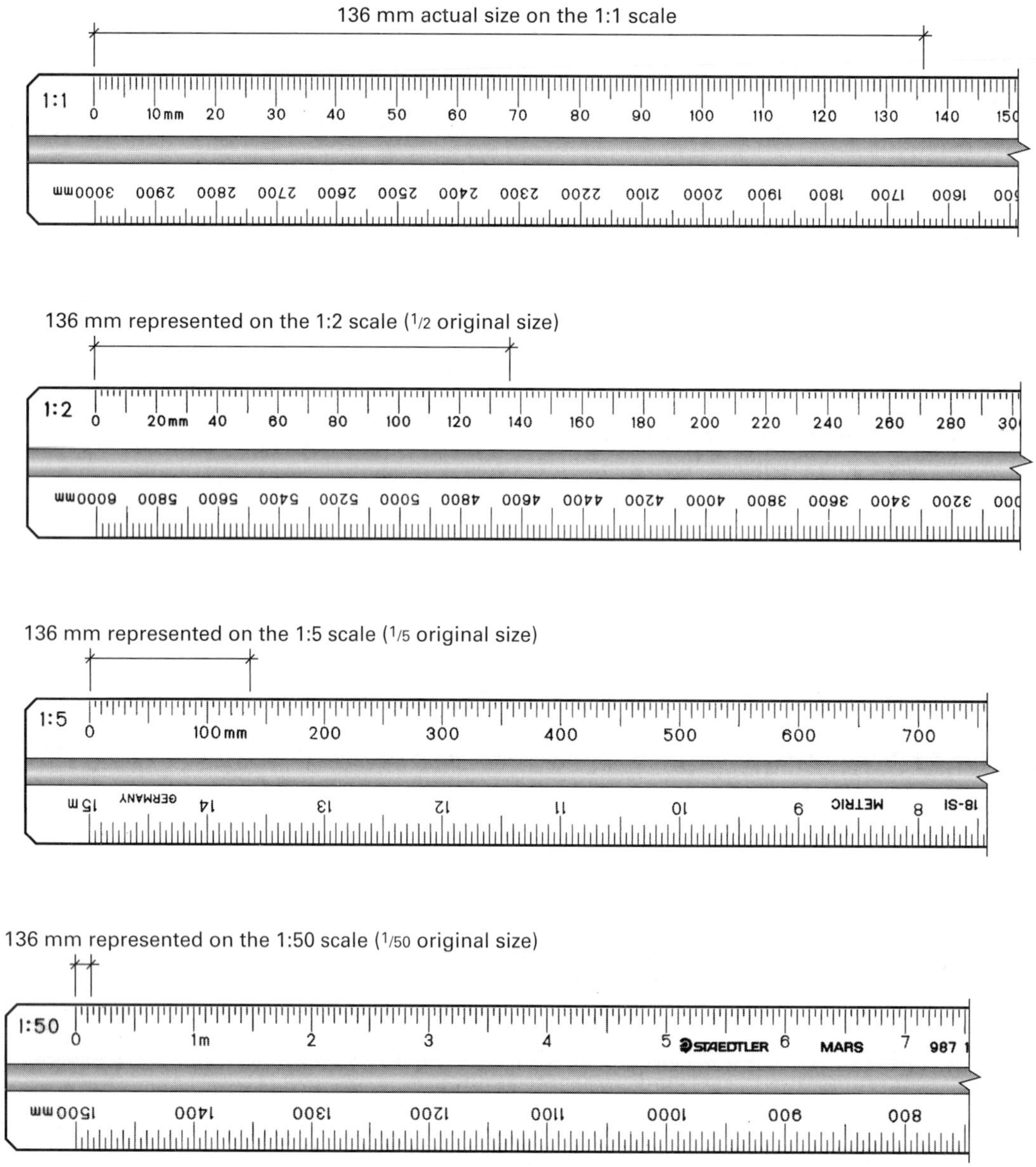

Chapter 2 Assignment: Reading Metric Scales

Assignment Example:

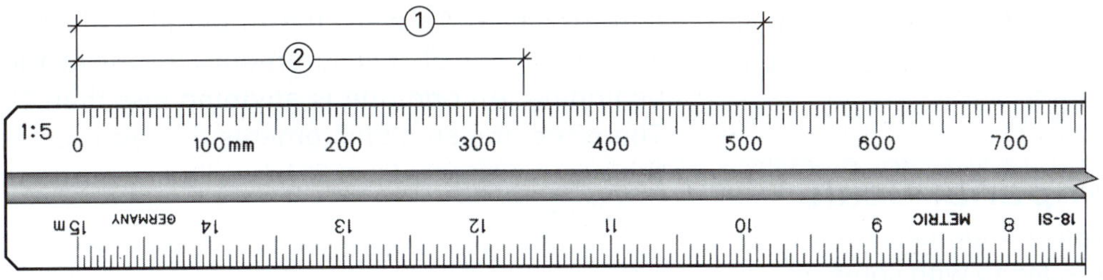

1. 515 mm
2. 335 mm

Refer to the chapter regarding metric scales and write down the correct answer for the following dimensions indicated.

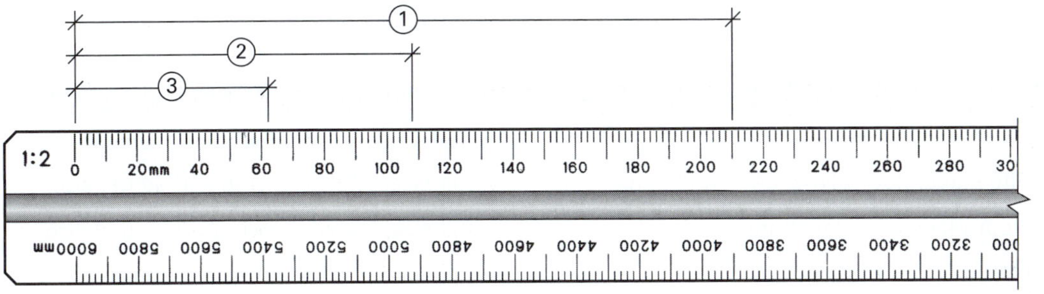

1. _____
2. _____
3. _____

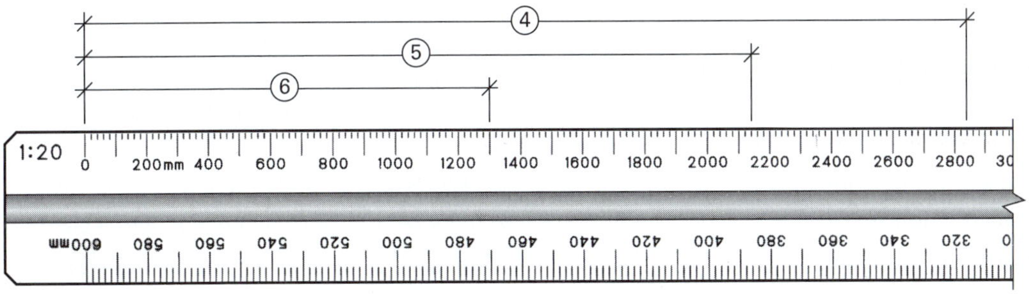

4. _____
5. _____
6. _____

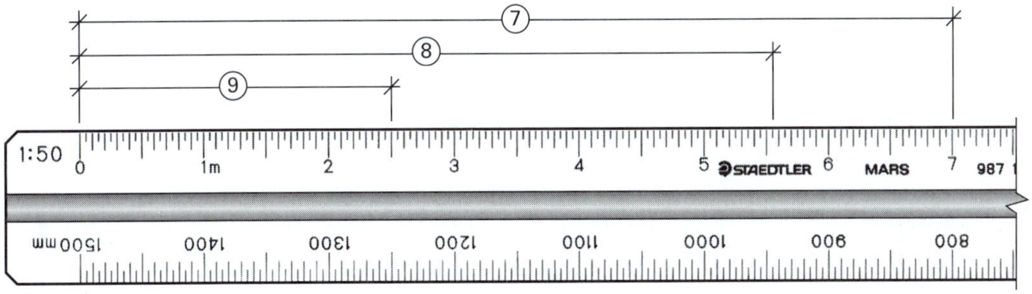

7. _____
8. _____
9. _____

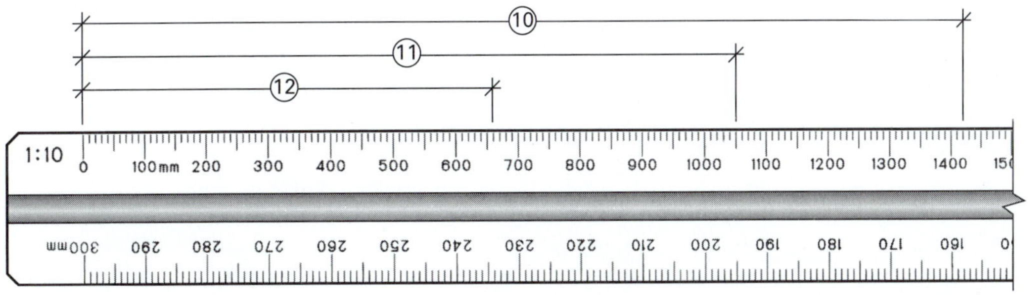

10. _____
11. _____
12. _____

Imperial Scales

Imperial scale rules are often referred to as an architectural scale. They differ from metric scales in that they accommodate two scales per edge with the exception of the full-size edge. Therefore, one imperial/architectural scale rule has 11 different scales on it.

Reading the Imperial Scale

Table 2.3 is a list of typical imperial scales used for construction projects, along with some examples of where they may be used.

Imperial scales differ from metric scales in that each edge of the scale has two scales on it. One reads from left to right, while the other reads from right to left, so using one takes a little practice. As well, the imperial scale has any odd units of measurement set out at the beginning. An odd unit is anything less than 1'. Therefore, the small increments that you see at the beginning of an imperial scale are portions of 12" or 1'.

TABLE 2.3
Typical Imperial Scales

Imperial scale type	Typical drawing type	Reduction size
3/32"	Site plans	1/128 full size; 3/32" of drawing = 1' actual size
1/8"	Site plans, floor plans, and elevations	1/96 full size; 1/8" of drawing = 1' actual size
3/16"	Floor plans and elevations	1/64 full size; 3/16" of drawing = 1' actual size
1/4"	Floor plans and elevations	1/48 full size; 1/4" of drawing = 1' actual size
3/8"	Interior elevations and cabinet details (e.g., kitchen, study, bathroom)	1/32 full size; 3/8" of drawing = 1' actual size
1/2"	Interior elevations and cabinet details (e.g., kitchen, study, bathroom)	1/24 full size; 1/2" of drawing = 1' actual size
3/4"	Interior elevations and cabinet details (e.g., kitchen, study, bathroom)	1/16 full size; 3/4" of drawing = 1' actual size
1"	Construction details	1/12 full size; 1" of drawing = 1" actual size
1 1/2"	Construction details	1/8 full size; 1 1/2" of drawing = 1' actual size
3"	Construction details	1/4 full size; 3" of drawing = 1' actual size
Full Size	Objects that can be drawn full size, usually detail drawings	Full size; 12" of drawing = 1' actual size

FIGURE 2.7
Imperial scale.

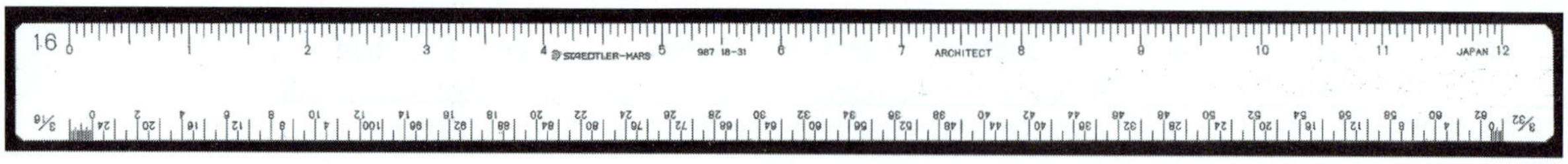

FIGURE 2.8
Reading an imperial scale.

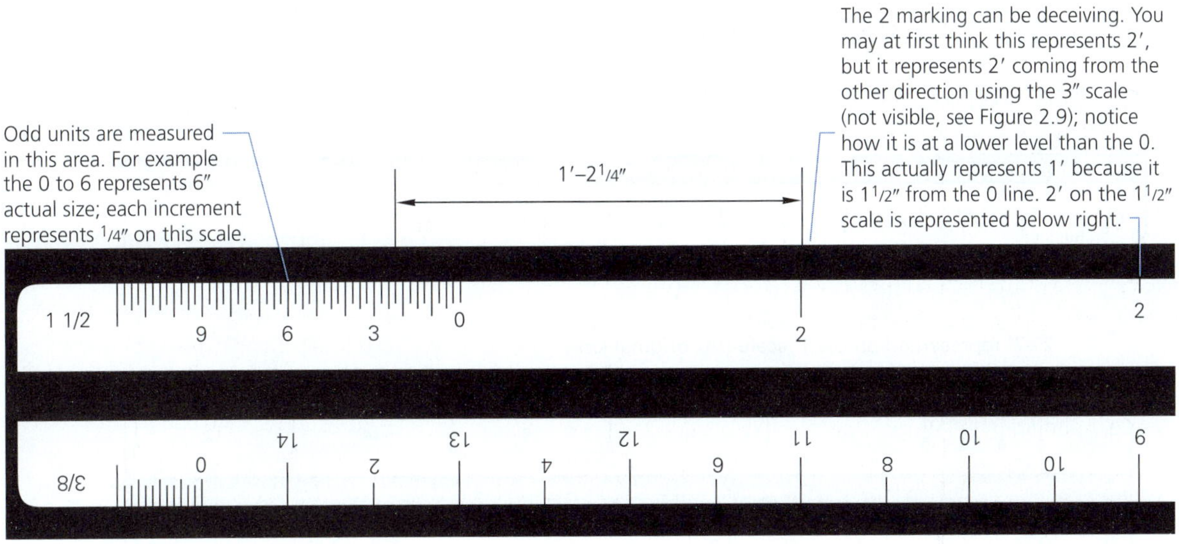

FIGURE 2.9
3″ scale.

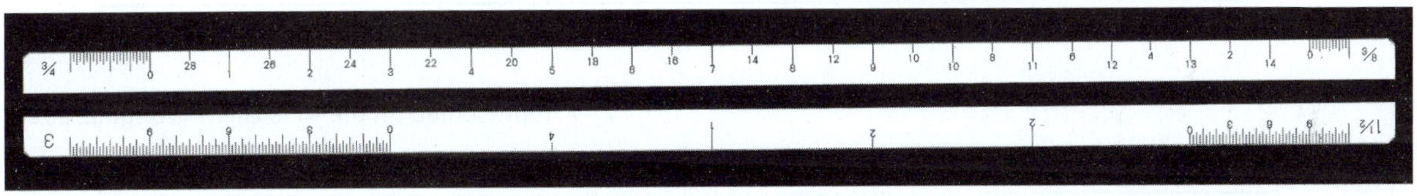

FIGURE 2.10

One measurement using different imperial scales.

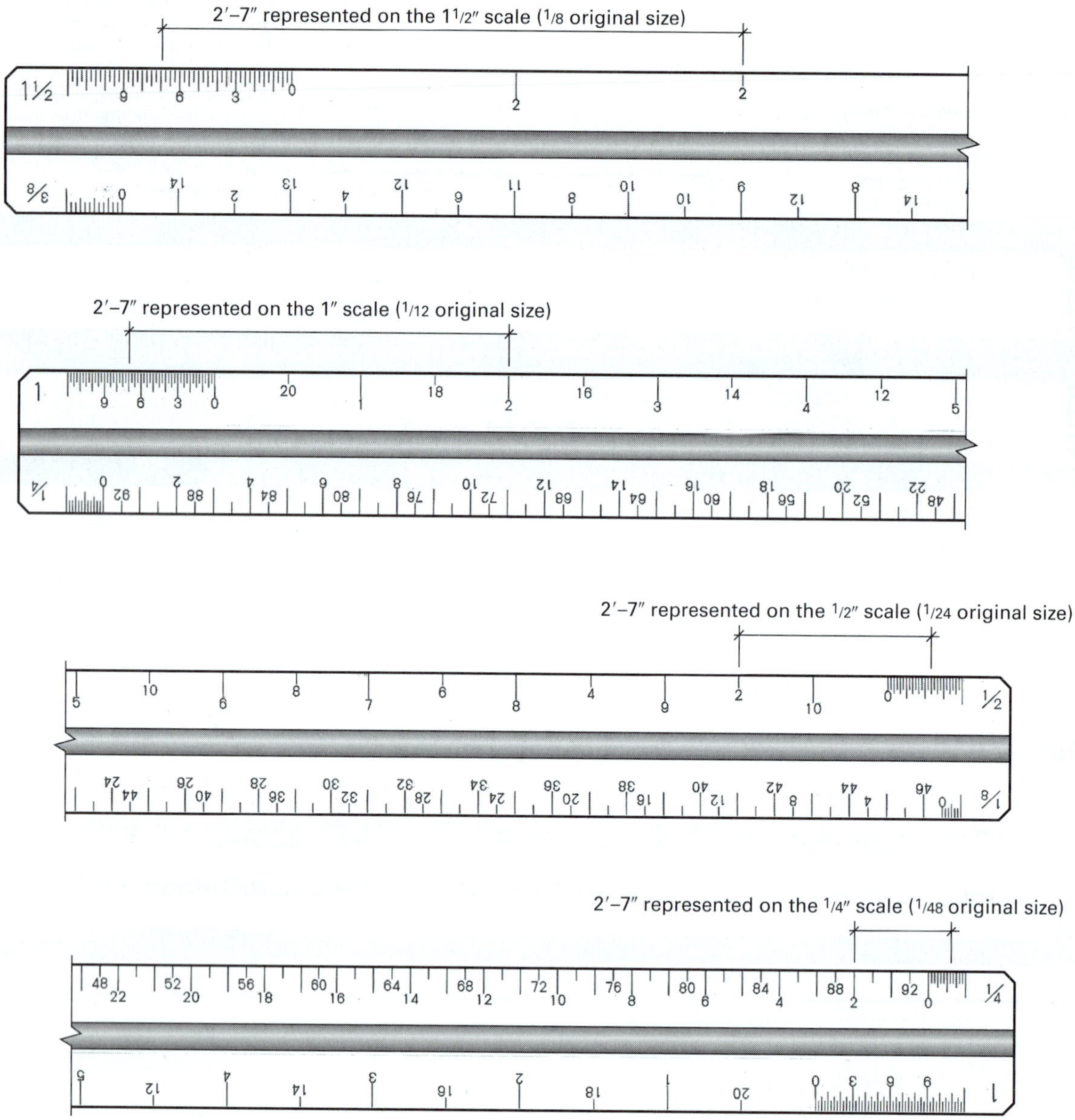

Chapter 2 Assignment: Reading Imperial Scales

Assignment Example:

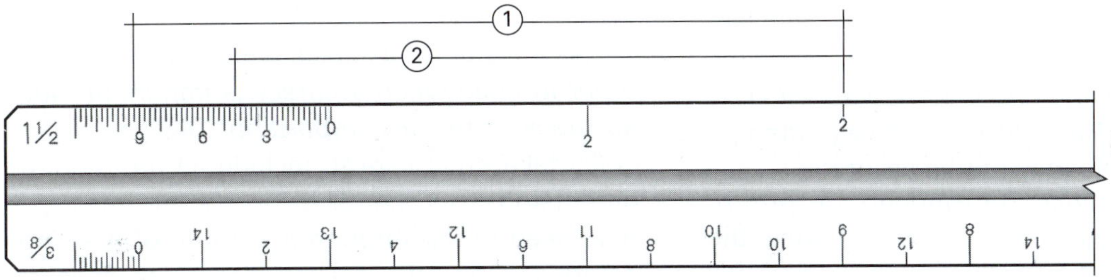

1. 2'–9¼"
2. 2'–4½"

Refer to the chapter regarding imperial scales and write down the correct answer for the following dimensions indicated.

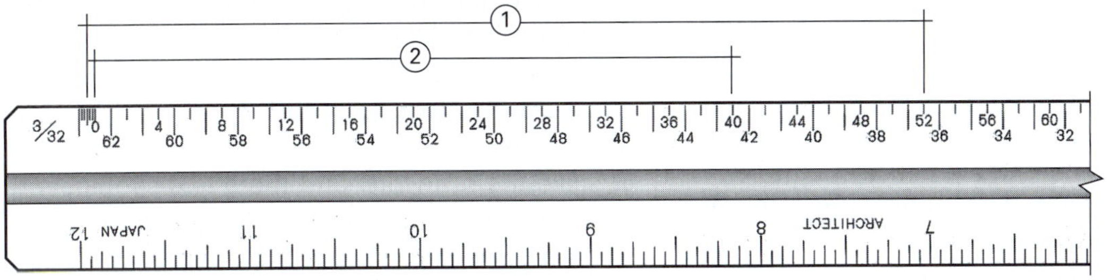

1. _____
2. _____

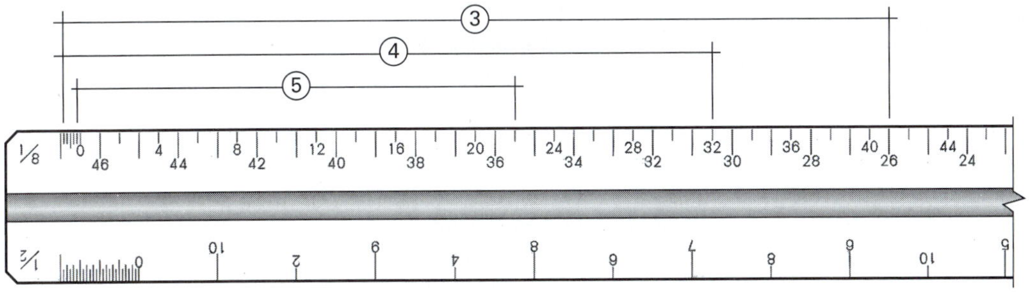

3. _____
4. _____
5. _____

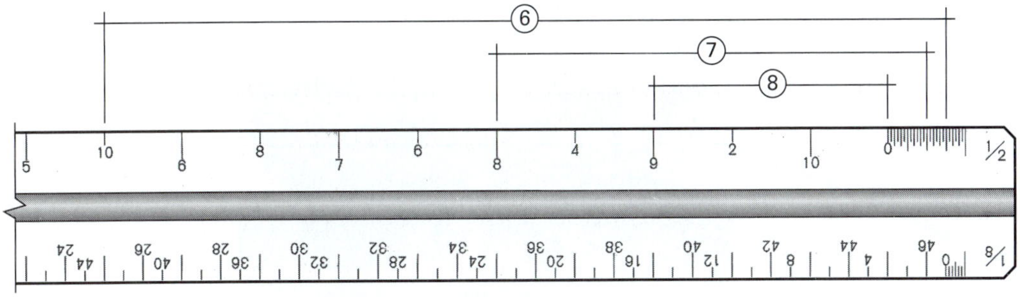

6. _____
7. _____
8. _____

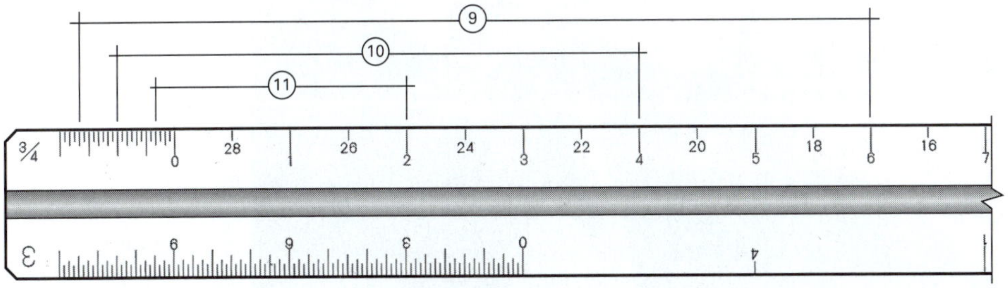

9. _____
10. _____
11. _____

Use of Scales

Scale rules were designed to be used for producing manual drawings. They also can be used in developing material estimates where it is quicker to use the scale than to calculate a length such as a rafter. Remember, however, that although drawings are drawn to a scale, that scale may not be 100% accurate once plotted.

Reasons a scale may not be accurate:

- The drawing was reproduced or photocopied. Copies of copies will reduce the accuracy of the drawing; nevertheless, this practice is commonly used in the reproduction of drawings.
- The drawing was enlarged or reduced from the scale listed. Photocopied and CAD-plotted drawings can easily be enlarged or reduced, with the scale not being changed on the drawings.
- There was a change made to the drawing and instead of redrawing it, the designer changed only the dimension. This practice can happen on any drawing but is most common on manually produced drawings.
- You don't read the scale rule properly, or you don't look at it directly. For example, if you view the scale at an angle you will not be able to read it accurately because your view will be distorted.
- The scale drawing is reduced to such an extent that you can't accurately distinguish the measurement to the accuracy required using the scale rule. In other words, the degree of accuracy depends on the scale used.

As a result, *drawings should not be scaled in order to build when accuracy is important (note: accuracy is usually important)*. Designers generally take great care to include all the required dimensions on the drawing. Sometimes you may need to flip through a number of drawings before you find the desired dimension because, in general, designers include the desired dimension only once to avoid conflict in case a change is made. Also, from a legal perspective, where there are conflicting dimensions, those dimensions provided on detail drawings are generally accepted as being more accurate than those placed on drawings drawn to a smaller scale. Designers usually place a note "**Not to scale**" or an abbreviation, NTS, on the drawings. As a result, if you choose to scale the drawings and there is a conflict with your interpretation, you will be deemed at fault.

If a dimension is not provided or there is a conflict, your first action should always be to get a written confirmation of the correct dimension. This generally requires a call to the designer responsible for the drawings.

FIGURE 2.11

Estimating touch screen monitor uses an interactive pen to scale off the screen in order to assist at getting accurate estimates from electronic plans. This is becoming much more commonplace and can also be done with a mouse and specialized software.

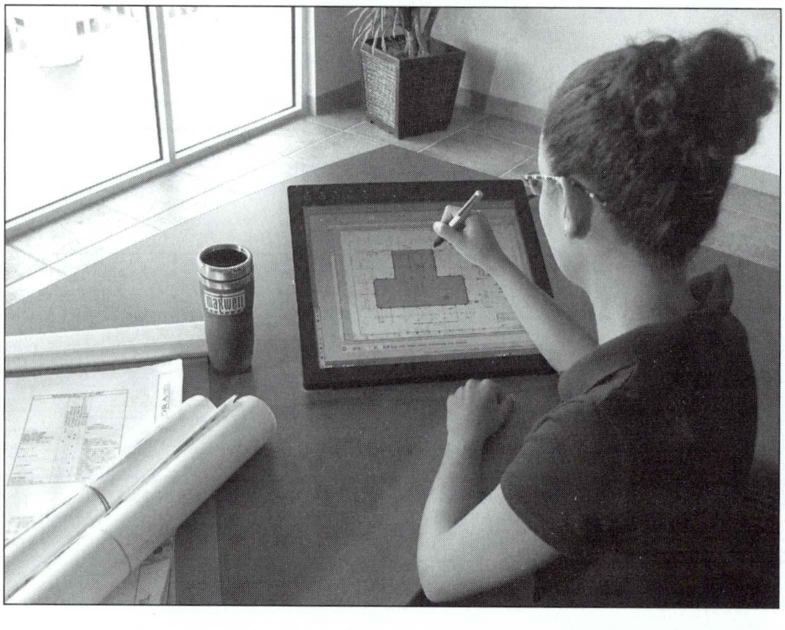

Courtesy of Maxwell Systems

Electronic Rules/Scales

Electronic rules or scales can take dimensions directly from the drawings when estimating. They can be adjusted to the appropriate metric or imperial scale. Using these tools can improve efficiency in estimating takeoffs from drawings, and, when used properly, can provide accuracy within the tolerances accepted for estimating purposes. *(Note: electronic rules or scales should not be used to take measurements for actual building purposes where a higher degree of accuracy is required.)*

DIMENSIONING PRACTICES

Although there are a number of standards for dimensioning drawings, dimensioning practices tend to vary from designer to designer and from region to region.

FIGURE 2.12

Portion of a floor plan displaying swing of door with door dimension indicated by reference number in circle. (Reference number refers to a door schedule where the actual door size is indicated.)

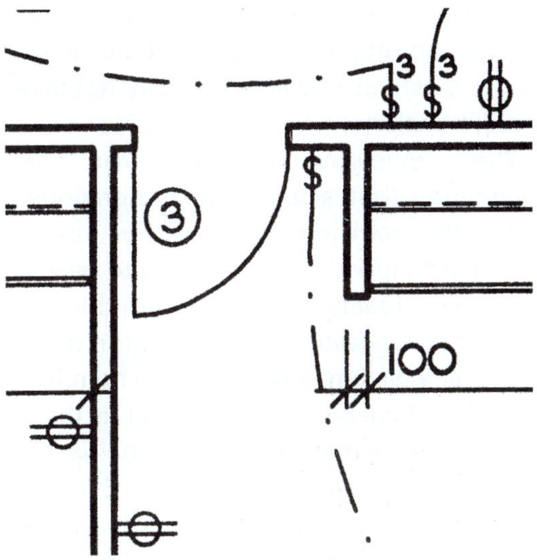

FIGURE 2.13

Examples of how dimensions for frame construction may be indicated. Note the centre line indicator used for the walls, windows and door on the top image and the edge of walls, windows and door dimensions used for the bottom image.

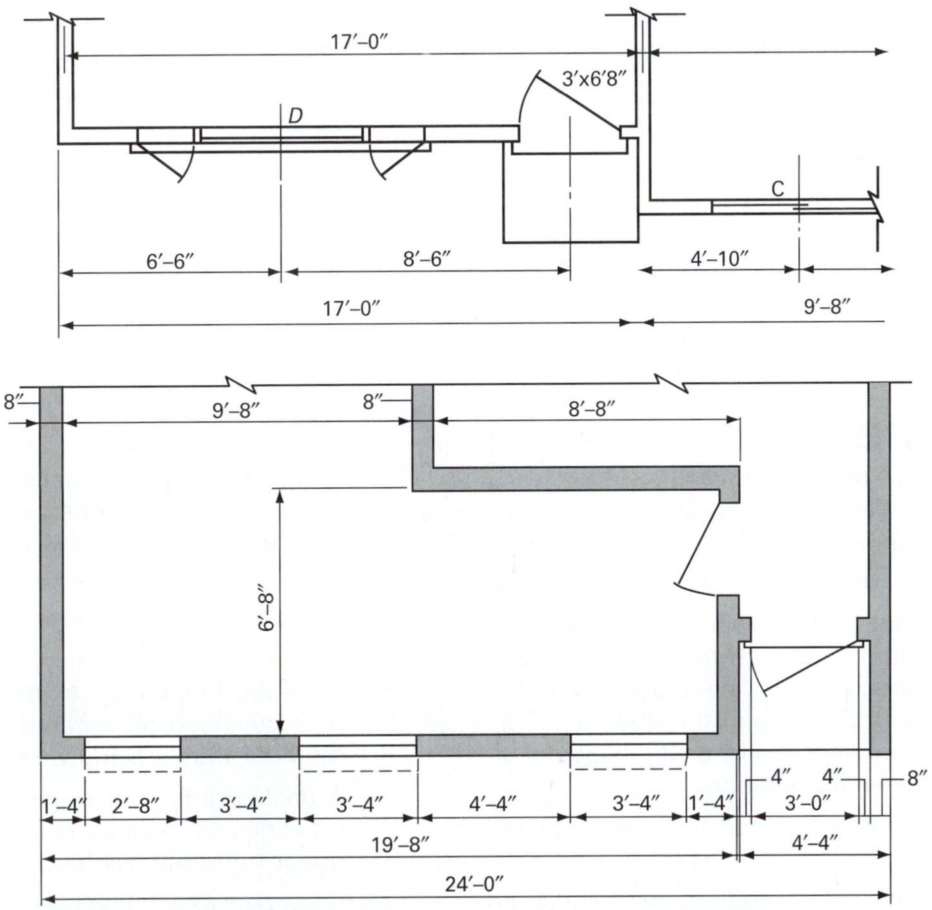

Drawings are produced to scale as accurately as possible; dimensions are then added to describe the size or location of the object. Dimensioning will typically be taken from critical or structural elements. Dimensions must provide the information required by the trades in order to fulfil their job requirements. Finish items such as drywall and sheathing are generally not included in the dimensions.

When reviewing drawings, you need to acknowledge that actual field dimensions may vary slightly from the drawings. For example, a wall may be shown on the drawing to be 100 mm thick, while in reality it is only 89 mm thick as indicated in the Doncaster and Whittington drawings. In imperial, a wall may be shown as being 4" thick when it is actually 3½" as indicated in the Brook drawings. A brick may be shown as being 100 mm thick while actually being only 90 mm thick. *A good site superintendent or tradesperson will look at these variances and think ahead to any possible problems that might occur as a result.* They will make the appropriate adjustments long before the variance becomes an issue, thinking ahead, consulting, taking responsibility, and making these decisions where appropriate to bring the project successfully from paper to reality.

Doors and windows in brick-veneer construction are usually dimensioned to the centre line of the door or window, and the door or window is indicated by its actual size. In other words, if a 2' door is being used, the drawing will indicate 2' either on the plan or in a door table. The rough opening, however, will be 2'–2" as opposed to 2', which would include the 2' door, the door frame on both sides ¾" + ¾", plus a ½" for shimming.

FIGURE 2.14

Example of a cutting-plane line indicating where the section detail is taken and the direction toward which it is viewed.

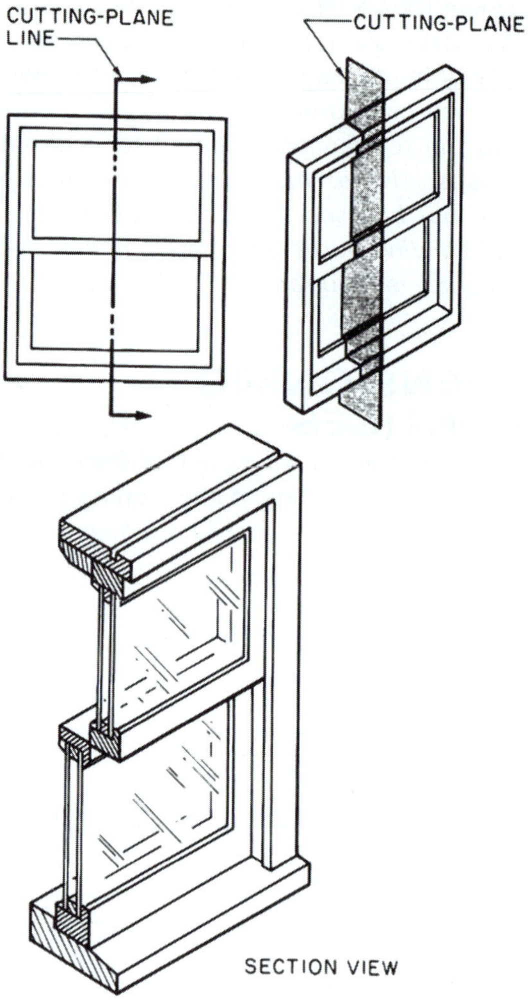

LINE TYPES

A number of line types and line weights are used in construction drawings. Their purpose is to quickly provide information regarding the type of object being described. The list of line types is as follows:

- ***Object line:*** A darker or more heavily weighted line that identifies the object being constructed. This line is used to identify objects (for example, the lines around frame or foundation walls, or the edges of a building shown in elevation).
- ***Dashed line:*** Can be used for a number of reasons:
  - A hidden line indicates an object covered by another object.
  - In renovation work, a dashed line shows the location of existing walls that are to be removed.
  - In new construction, it can show something that may be installed in the future (for example, a basement rough-in, which can be finished with walls in the future).
  - An × pattern using dashed lines shows a change in the elevation of the floor or ceiling overhead.
  - It can indicate upper cabinets in a kitchen, or a bulkhead or coffered ceiling above the floor that is being viewed.
  - A dashed line is also used on elevation drawings to indicate whether a window opens and how it is hinged.
- ***Cutting-plane line:*** A line used to cut through an object in a specific direction so that a

cross-sectional view can be provided to better illustrate the components that comprise the object.

- ***Detail designation:*** Provides the reference information on where the section detail that the cutting plane line cuts through can be found. On simpler drawings that have a limited number of section details, there will usually be a letter referencing which drawing to look for (for example, letter A would be detail A). On more extensive drawings where there are a large number of drawings and drawing

FIGURE 2.15

Example of a cutting-plane line used to provide a section through a house. A detail designation is used to describe section A-A shown below. Section A-A is a cut-through of the building taken between the two As facing the direction of the arrow (the arrow is the directional point at the top of the detail designation), and is referenced as detail A-A on another page of the drawings.

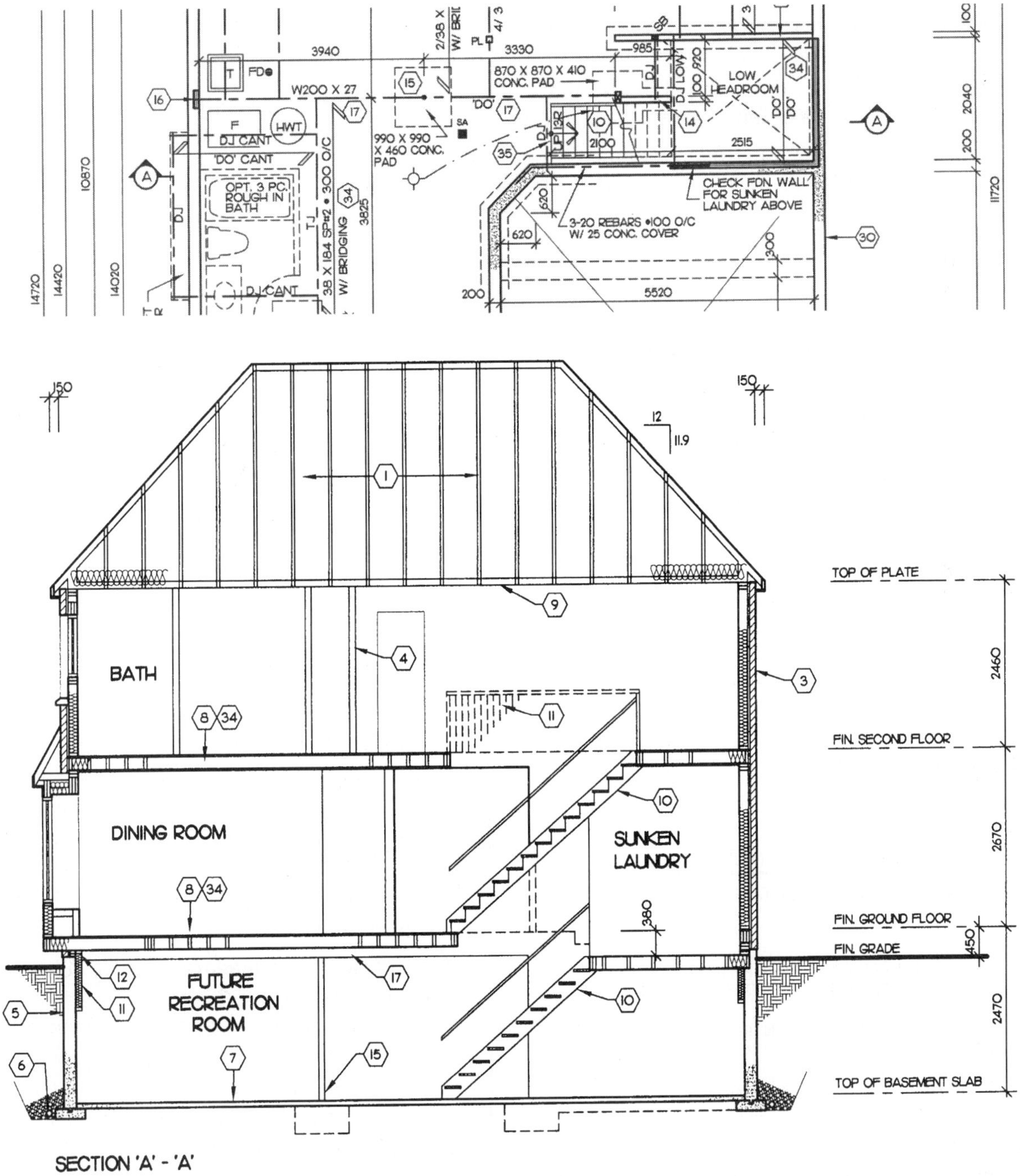

Source: Cassidy & Co. Architectural Technologists

sheets, the top half of the circle indicates the section number while the bottom half indicates the drawing sheet where the detail can be found (refer to Chapter 16 for more advanced examples of this in use).

- ***Extension line:*** Used to indicate where a dimension is taken from.
- ***Dimension line:*** Indicates the point-to-point length of the object being dimensioned.
- ***Centre line:*** Indicates where the centre of an object is (for example, the centre of a window, door, beam, or support column). The centre line is formed by a long line with a dash followed by a long line. There may be a CL abbreviation to indicate that it is a centre line.
- ***Leader line:*** Used to point to an object so that a description can be given.
- ***Break line:*** Used to cut off an object that continues but would not fit on the page, or where another object may be located directly over it (as is the case with most stairs).

FIGURE 2.16
Descriptive line to represent beam.

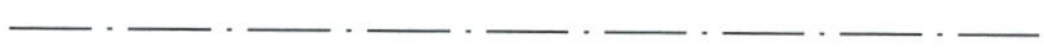

FIGURE 2.17
Solid line to represent direction of floor joists.

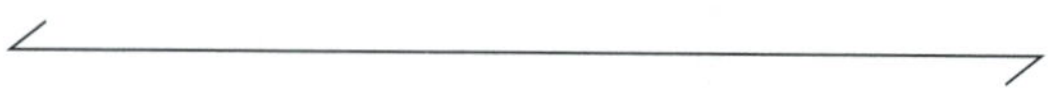

FIGURE 2.18
Line types.

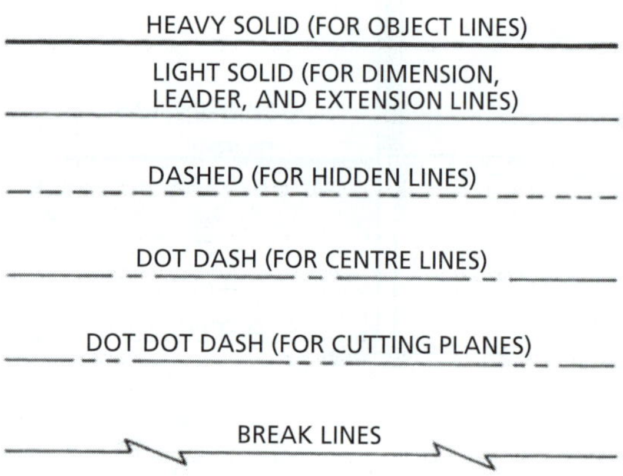

FIGURE 2.19
Dimension and extension line intersections.

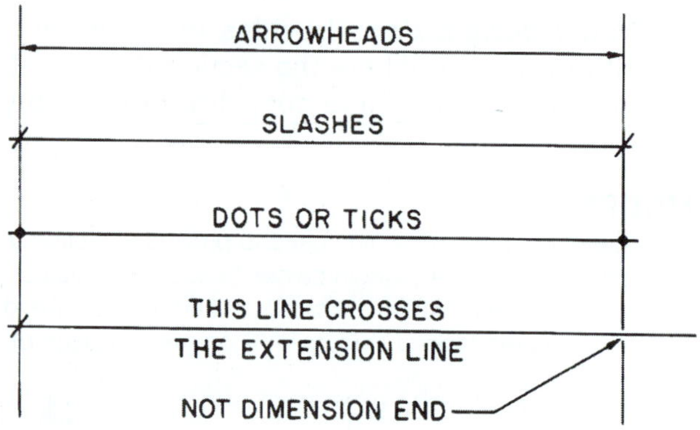

- ***Descriptive line:*** Used to indicate the location and direction of structural elements. For example, a beam may be shown as a long object line with intermittent dashes indicating the centre of the beam (see Figure 2.16), or the direction of floor joists may be indicated by a solid line drawn in the direction of the joists (see Figure 2.17).

KEY TERMS

Definitions for key terms appear in the glossary at the end of the text.

Break line (p. 32)
Centre line (p. 32)
Cutting-plane line (p. 30)
Dashed line (p. 30)
Descriptive line (p. 32)
Detail designation (p. 31)
Dimension line (p. 32)
Extension line (p. 32)
Floor joists (p. 19)
Hard conversion (p. 19)
Imperial system of measurement (p. 19)
Leader line (p. 32)
Metric system of measurement (p. 18)
Not to scale (p. 28)
Object line (p. 30)
Scale (p. 18)
Soft conversion (p. 19)
Studs (p. 19)
Subfloor (p. 19)

FIGURE 2.20

Foundation drawing taken from the Doncaster model house drawings located at the back of the text. Notice the different line types used in the drawing, and cutting plane line B in the garage indicating section detail B at the lower left corner of the page.

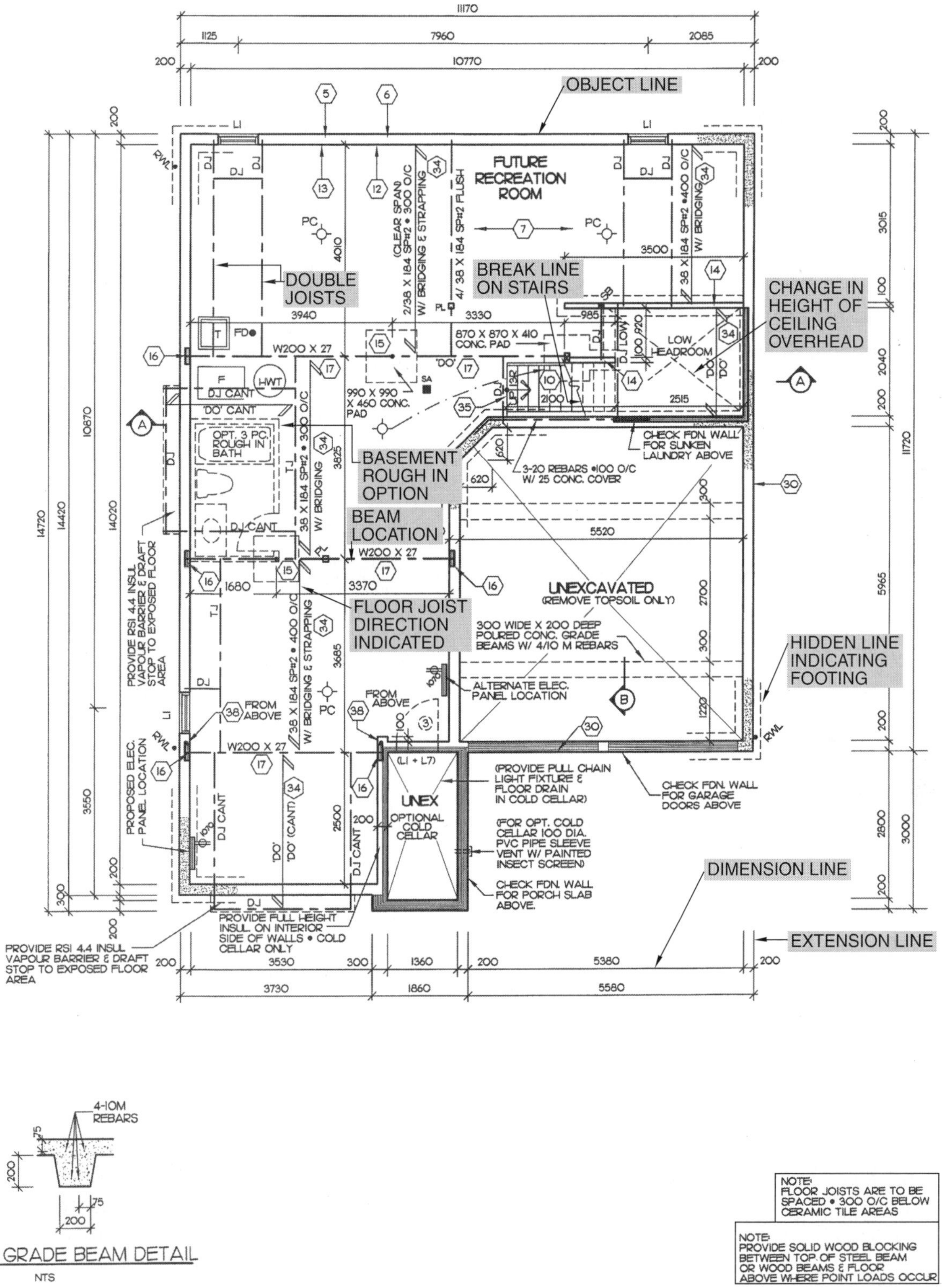

Source: Cassidy & Co. Architectural Technologists

Chapter 2 Quiz

Identify the choice that best answers the question.

_____ 1. The drawing shown below is a corner of a foundation wall shown in plan view with the footings indicated by dashed lines. What type of line is the foundation wall drawn with?

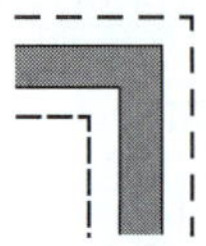

a. dimension line
b. hidden line
c. break line
d. object line

_____ 2. What type of line is shown below?

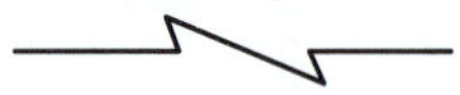

a. extension line
b. dimension line
c. break line
d. centre line

_____ 3. What type of measurement system does the National Building Code of Canada (NBC) use as its standard system of measurement?

a. the imperial system
b. the metric system
c. the foot metre system
d. the ISO system

_____ 4. What is the arrow symbol with the letter B in the circle shown below?

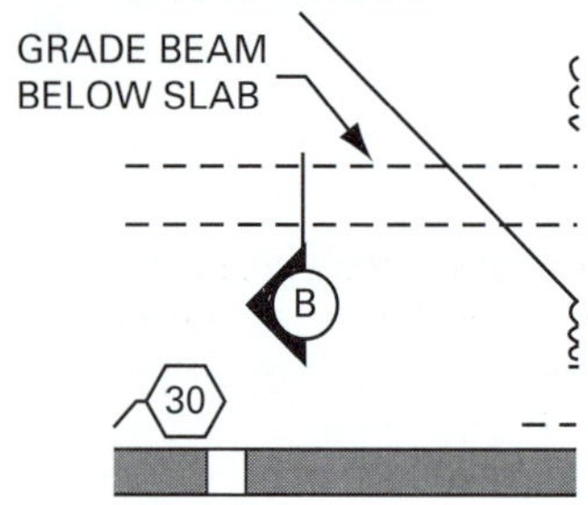

a. an object line
b. a centre line
c. a cutting plane line and detail designation
d. a hidden line

_____ 5. What determines the scale that is being used to produce a drawing?

a. the size of paper the drawing is to be plotted on
b. the type of project, e.g. renovation or house
c. the architectural style of house being built
d. the monetary value of the project

_____ 6. What type of line is the dashed line forming a square shown below which represents a concrete footing under a concrete floor slab?

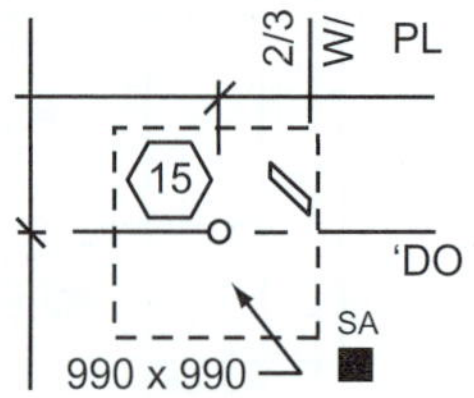

a. object line
b. centre line
c. cutting plane line
d. hidden line

_____ 7. How will a typical 8" concrete block requested at a Canadian building supply have been manufactured?

a. using the metric system of measurement to a size that lays to 200 mm excluding the mortar joint
b. using the metric system of measurement to a size that lays to 200 mm including the mortar joint
c. to a size that lays to exactly 8" including the mortar joint
d. to a size that lays to exactly 8" excluding the mortar joint

_____ 8. What system of measurement does residential construction in Canada use as its standard unit of measurement for construction drawings?

a. only the metric system
b. only the imperial system
c. both the imperial and metric system are used
d. the ISO system of measurement

_____ 9. What kind of conversion is the conversion between 16" on centre spacing and 400mm on centre spacing?

a. hard conversion
b. soft conversion
c. easy conversion
d. mild conversion

_____ 10. What is the conversion factor used as a multiplier to convert inches to mm?

a. 25.4

b. 254

c. 16

d. 12

_____ 11. What is 2000 millimetres in inches?

a. 78¾"

b. 6'-10¾"

c. 48"

d. 6'-0"

_____ 12. What is 14" in millimetres?

a. 3556 mm

b. 355.6 mm

c. 0.5511 mm

d. 400 mm

_____ 13. What type of line is the line with the arrow coming from the hexagon that has the number 38 inside of it?

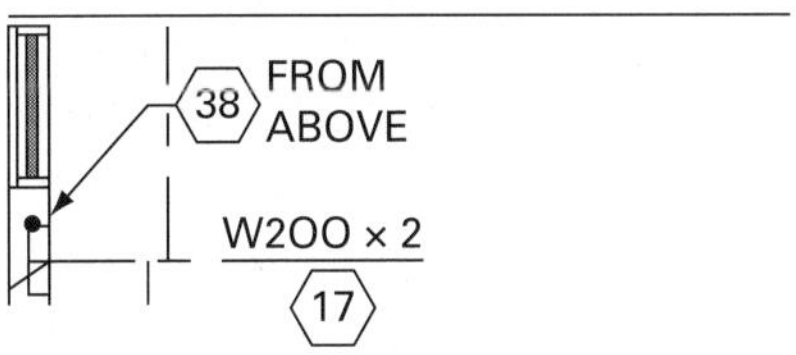

a. object line

b. leader line

c. hidden line

d. extension line

_____ 14. How do construction drawings that are drawn with a ⅛" imperial scale compare to actual size?

a. They are drawn so that ⅛" drawn on paper represents 12" of actual size.

b. They are drawn so that ⅛" drawn on paper represents ⅛ of actual size.

c. They are drawn that ⅛" scale is drawing the object 1/48th of actual size.

d. They are drawn so that ⅛" drawn on paper represents 16" of actual size.

_____ 15. Construction drawings using a ¼" scale are ______________________.

a. drawn at 1/96th actual size

b. drawn at 1/48th actual size

c. drawn at ¼th actual size

d. drawn at 1/12th actual size

_____ 16. What does the line below represent?

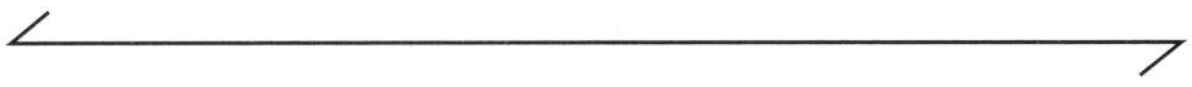

a. dimension line

b. extension line

c. the direction of joists

d. the direction of studs

_____ 17. What does the line below represent?

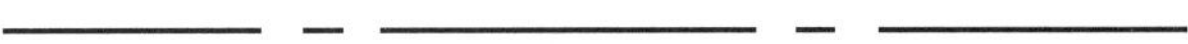

a. dimension line

b. object line

c. centre line

d. hidden line

_____ 18. What is an exact conversion of 16" to millimetres?

a. 400 mm

b. 394.6 mm

c. 16 mm

d. 406.4 mm

_____ 19. What is an exact conversion of 300 millimetres to inches?

a. 300"

b. 12.2"

c. 12"

d. 11.8"

_____ 20. What is the actual length of a line that is drawn to represent 5'-0" using the ¼" architect's scale?

a. 5'-0"

b. 1¼"

c. ¼"

d. ⅝"

3 Elevation and Floor-Plan Drawings

LEARNING OBJECTIVES

After completing this chapter, you will be able to:

- Interpret basic information found on elevation and floor-plan drawings.
- Visualize in three dimensions the exterior of a building based on viewing its exterior elevations.
- Visualize the interior of the building based on the floor-plan drawings.
- Determine the finish materials used for the building based on the elevation drawings.
- Determine the various elevations (height) and layout dimensions required to construct the building based on the elevation and floor-plan drawings.
- Determine the orientation of the elevations based on compass directions.

INTRODUCTION

This chapter covers the most commonly used types of drawings, floor plans, and elevations. These drawings provide most of the information required to build a structure. The text case house, the "Doncaster," is introduced in this chapter to help readers visualize and analyze information found on the drawings. The drawings for the Doncaster came attached to this text and are referred to by the word "Doncaster" located in the title block at the bottom of each page. (There are seven pages in total; for ease of use, staple the sets of drawings together. You can also staple the other drawings together based on the names provided in the title blocks: Whittington, the Brook, Mini-mart. ***Organizing by stapling the drawings together now will make referring to them and doing the assignments and quizzes that follow much easier.***)

ELEVATION DRAWINGS

Elevation drawings show the height and width of a single face of the object. For example, a front elevation of a house provides a direct front view of the house. In this type of orthographic drawing you cannot see any three-dimensional detail or depth of the object from any one view. It is necessary to refer to more than one elevation view to visualize the object three-dimensionally.

The builder or site superintendent refers to these drawings as the building elevations. Other separate elevations may be drawn for interior rooms requiring more detail than a floor plan can provide (for example, the kitchen to display cabinets; the bathroom to display vanities and special tile layouts; and the study to display custom built-in shelving).

In this chapter we focus on the building's exterior elevations. Compare the elevations on the following pages to the photographs of the actual house. Observe the accuracy of the drawings and also note some of the differences that occur between the drawings and the actual house.

Elevation Orientation

Most construction drawings will have four elevations: one each for the front, right, rear, and left sides. For work unique to an individual site, such as a custom home, addition, or industrial project, the elevations will be oriented according to compass direction. For a wall facing north, the elevation will be called the north elevation. So if you are standing with your back against the outside wall of the building looking northward, the wall against your back is the north elevation. Similarly, if you are standing with your back to the wall looking eastward, the wall that your back is against is the east elevation of the building. Refer to Figure 3.10.

A single set of construction drawings may be used more than once. For example, in

FIGURE 3.1

Front elevation from Doncaster drawing (refer to the attached drawings with Doncaster in the title block for the complete set of drawings for this house).

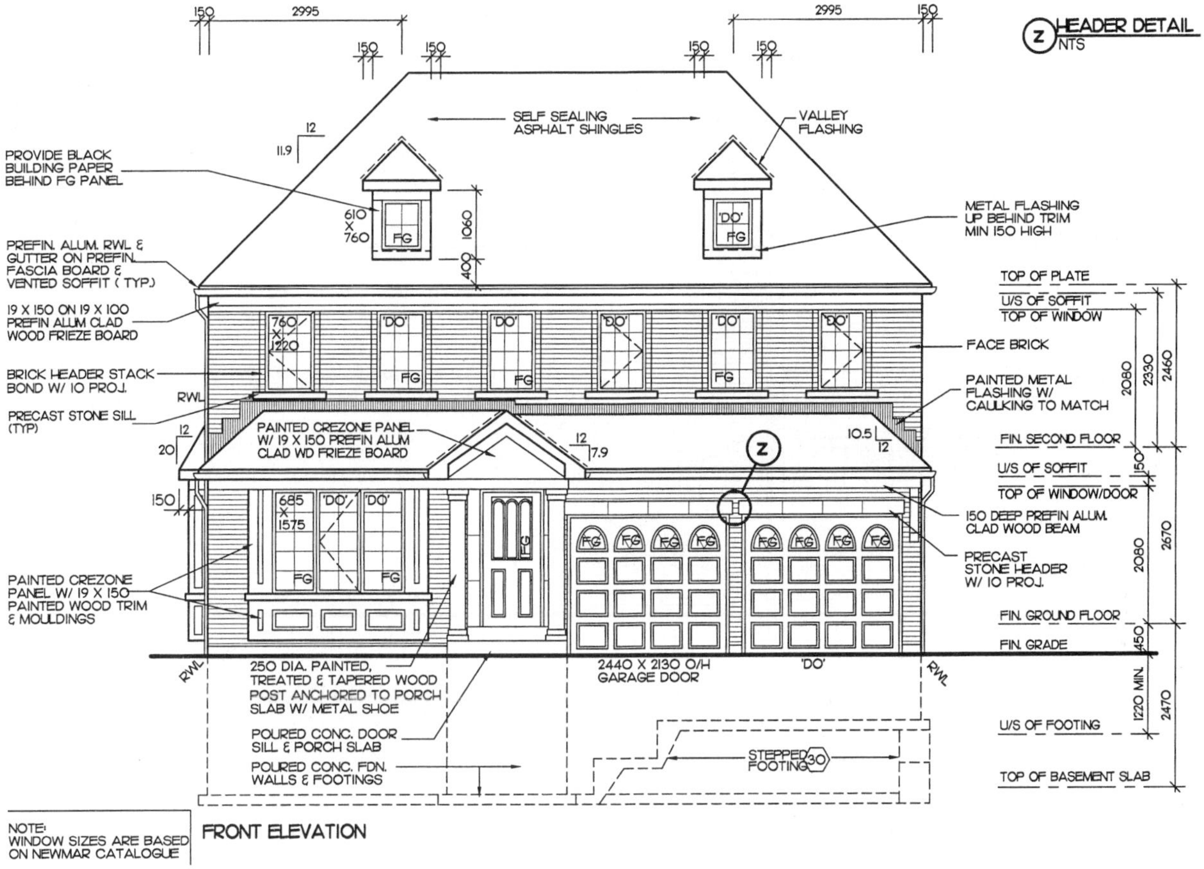

Source: Cassidy & Co. Architectural Technologists

FIGURE 3.2

Photo of actual Doncaster house, Lot 61. Take note of similarities and minor differences.

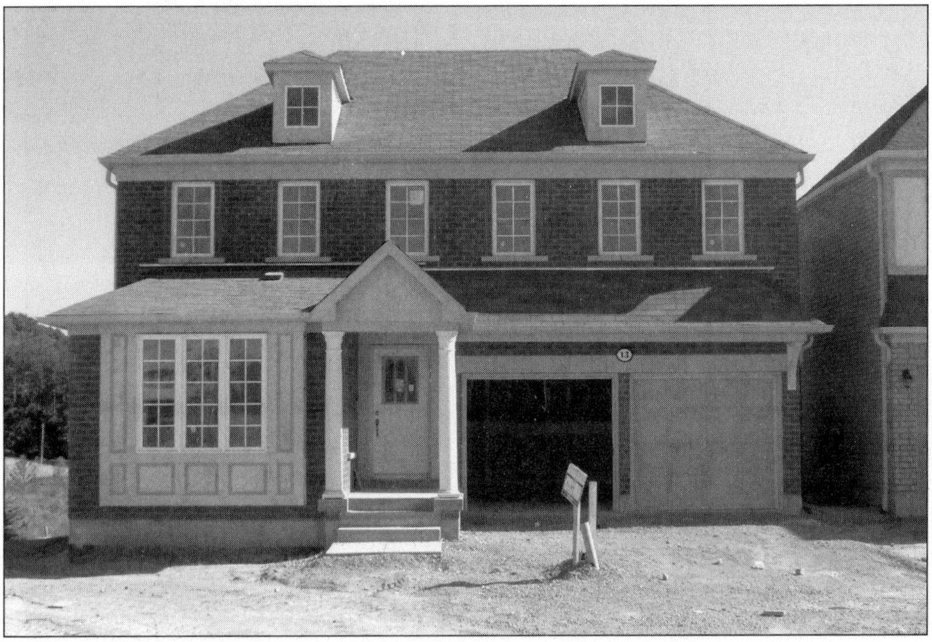

Courtesy of Tom Stephenson

FIGURE 3.3

Right elevation drawing, Doncaster.

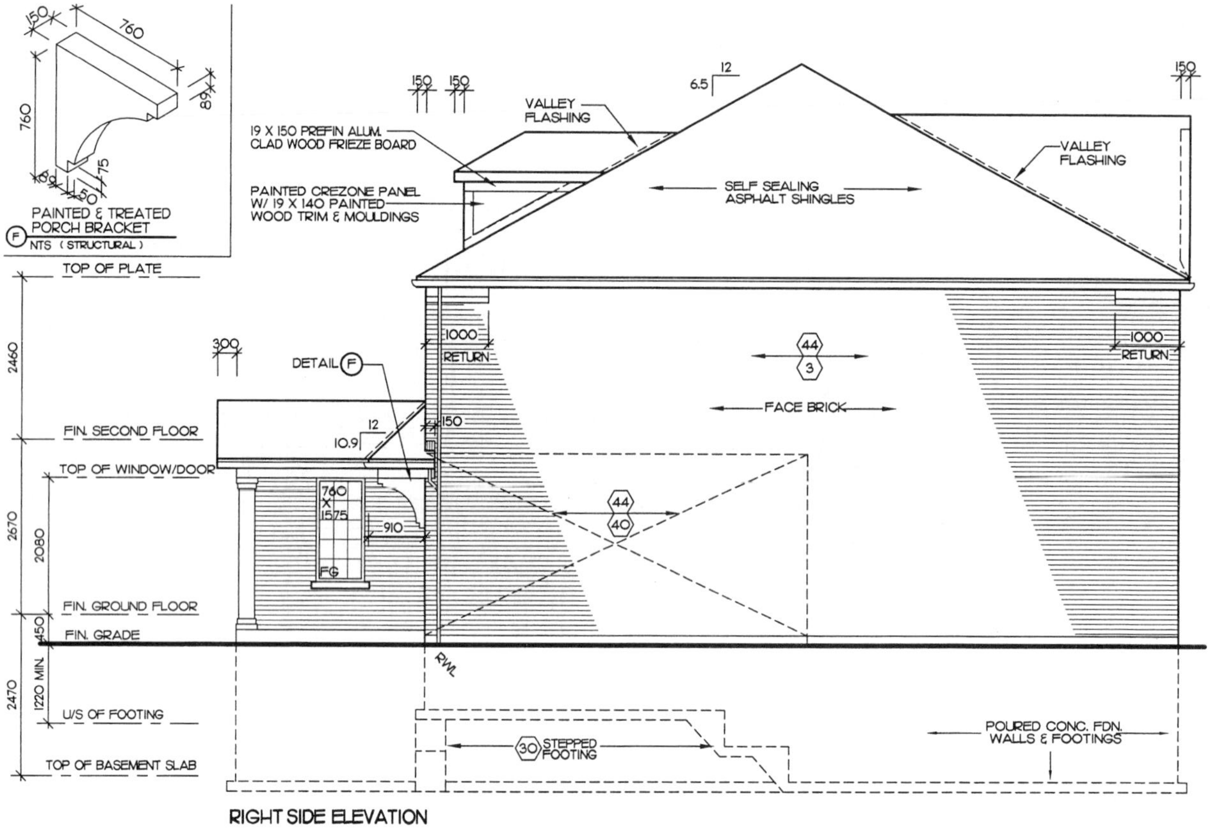

Source: Cassidy & Co. Architectural Technologists

FIGURE 3.4

Partial photo of right elevation at front of Doncaster.

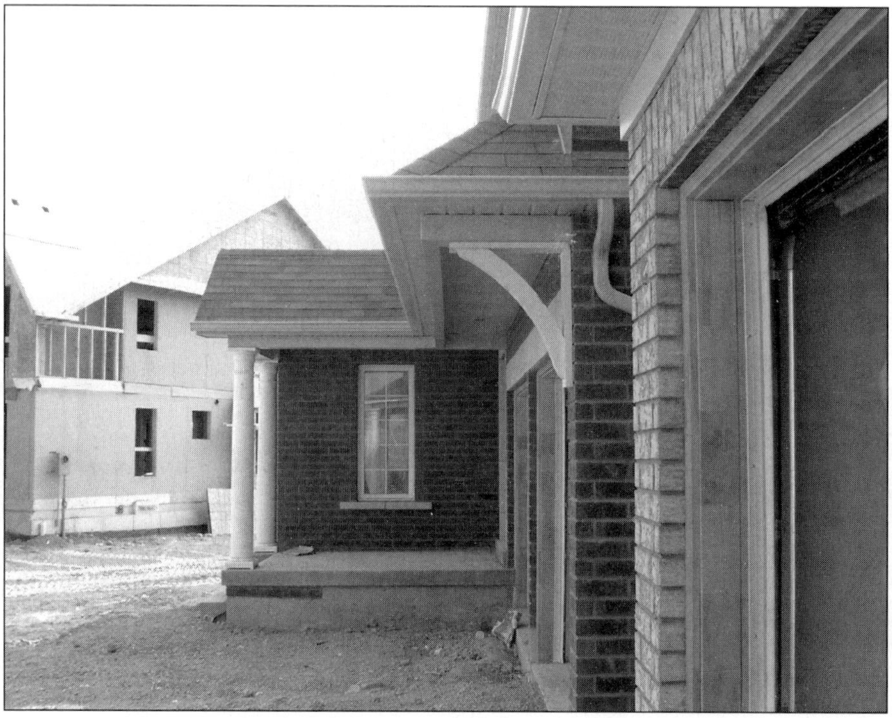

Courtesy of Tom Stephenson

FIGURE 3.5

Left elevation drawing, Doncaster.

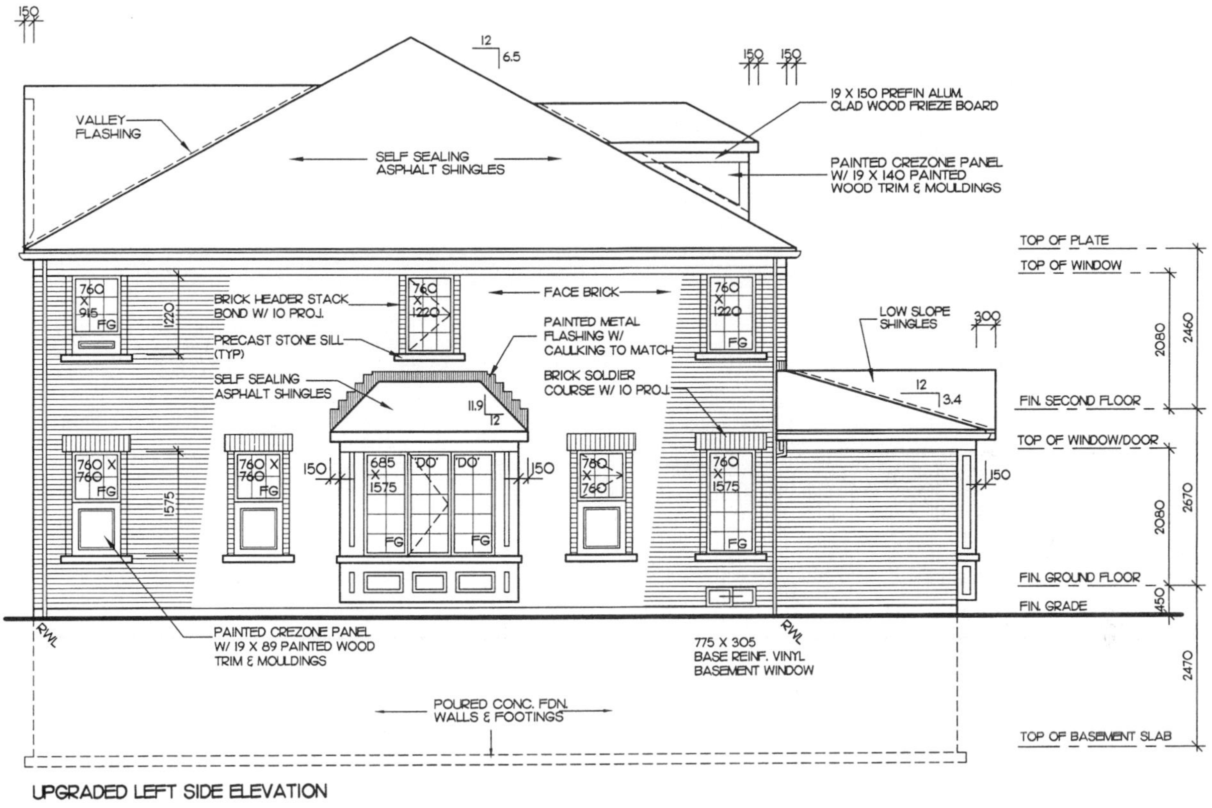

Source: Cassidy & Co. Architectural Technologists

FIGURE 3.6

Photo of left side of Doncaster. Note: the lot has a drop in site elevation toward the rear of the house. This is unique to this particular lot. In construction, every lot has unique qualities that must be considered in both design and construction.

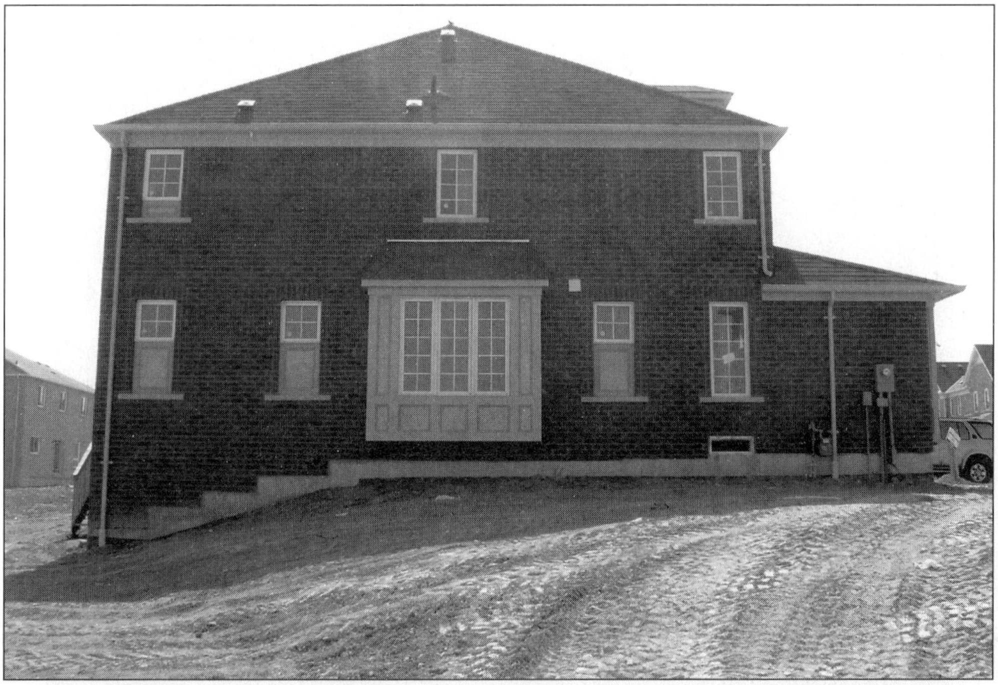

Courtesy of Tom Stephenson

FIGURE 3.7

Rear elevation drawing, Doncaster.

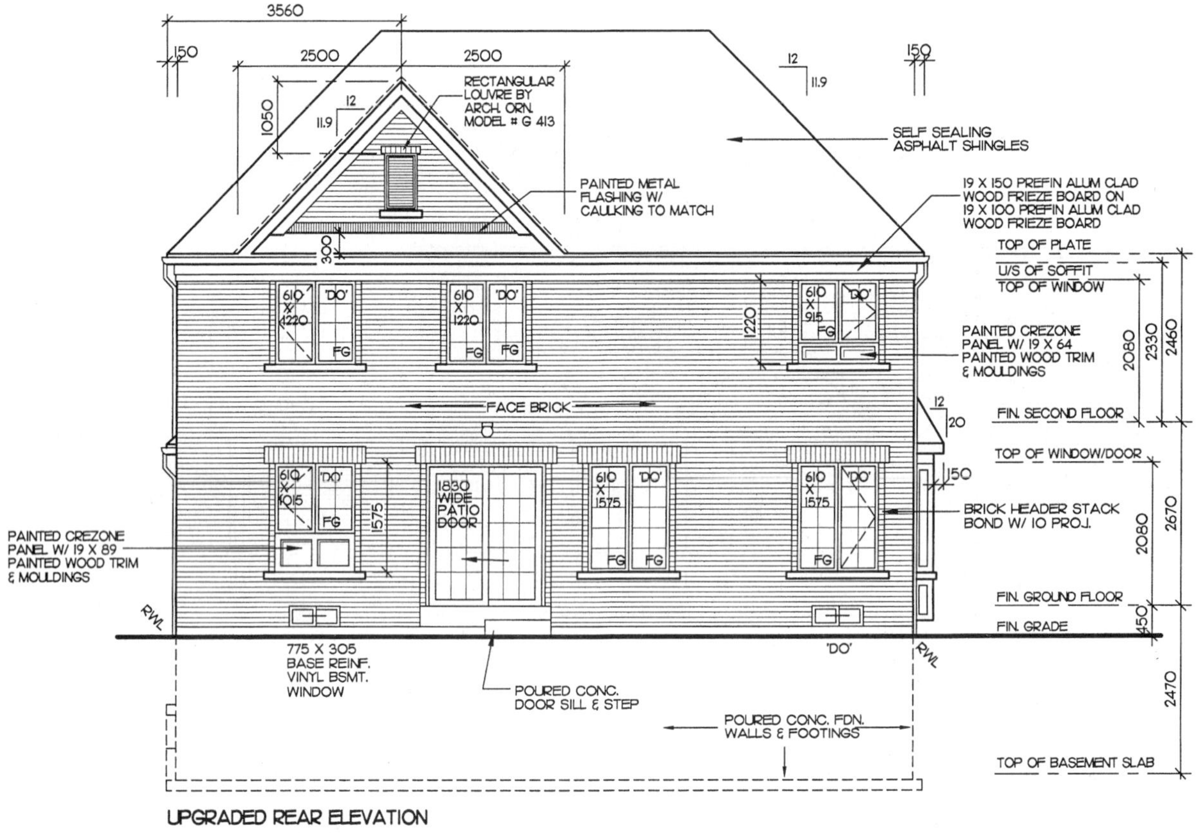

Source: Cassidy & Co. Architectural Technologists

FIGURE 3.8

Photo of rear elevation of Doncaster house. Note the extra depth for this example based on the grading of the lot dropping at the rear of the house.

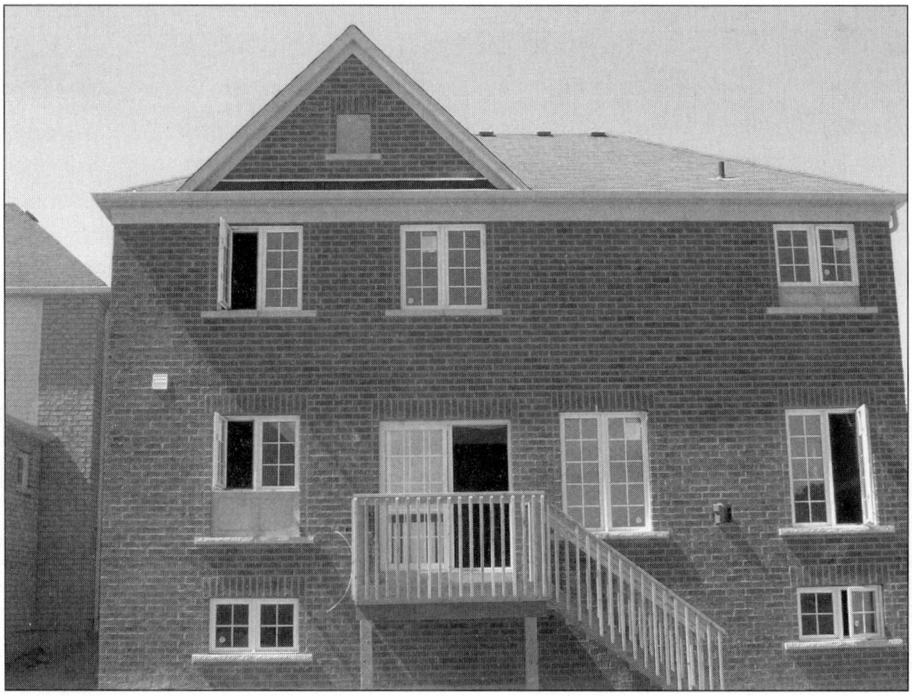

Courtesy of Tom Stephenson

FIGURE 3.9

Photo of rear and left elevation of Doncaster. Refer to and view the rear and left elevation drawings to visualize them in three dimensions.

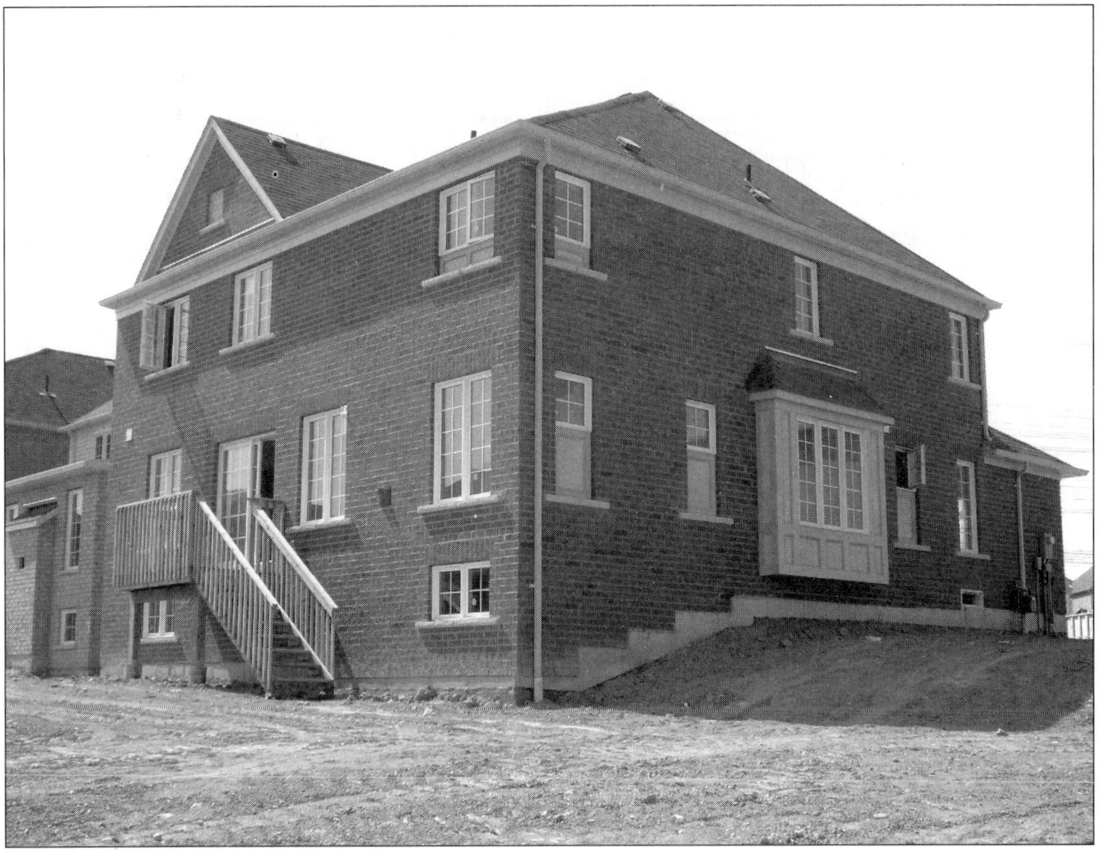

Courtesy of Tom Stephenson

FIGURE 3.10

Elevations are usually named according to their compass directions.

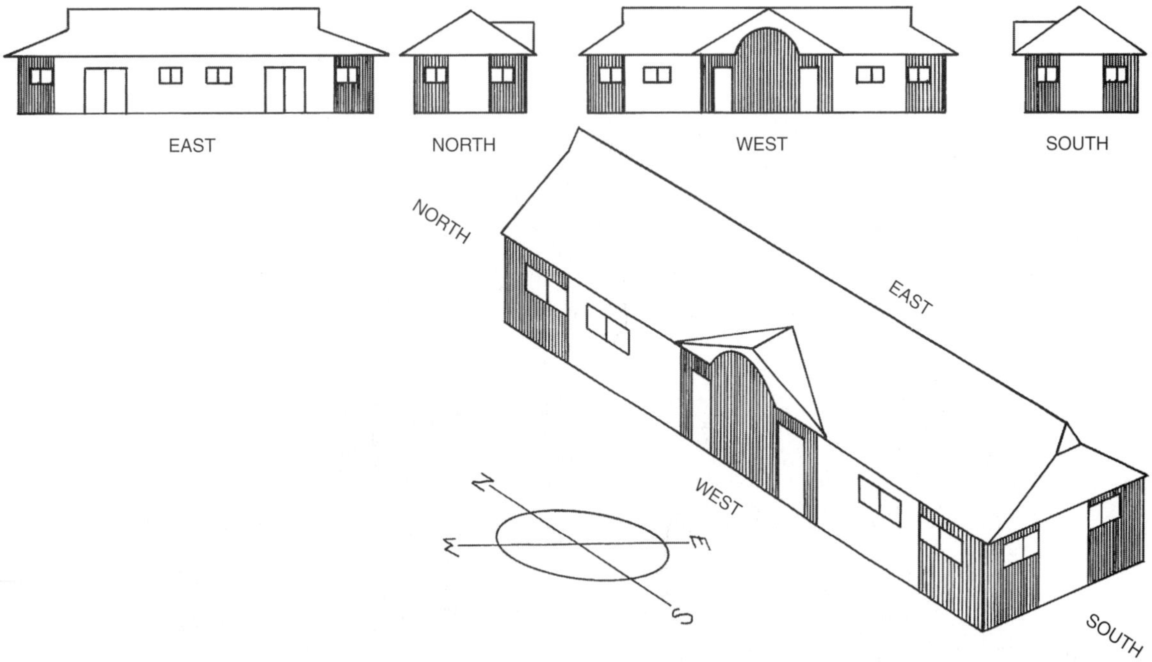

FIGURE 3.11

Plan labelled to help orientation to north arrow. This floor plan is for a semi-detached house.

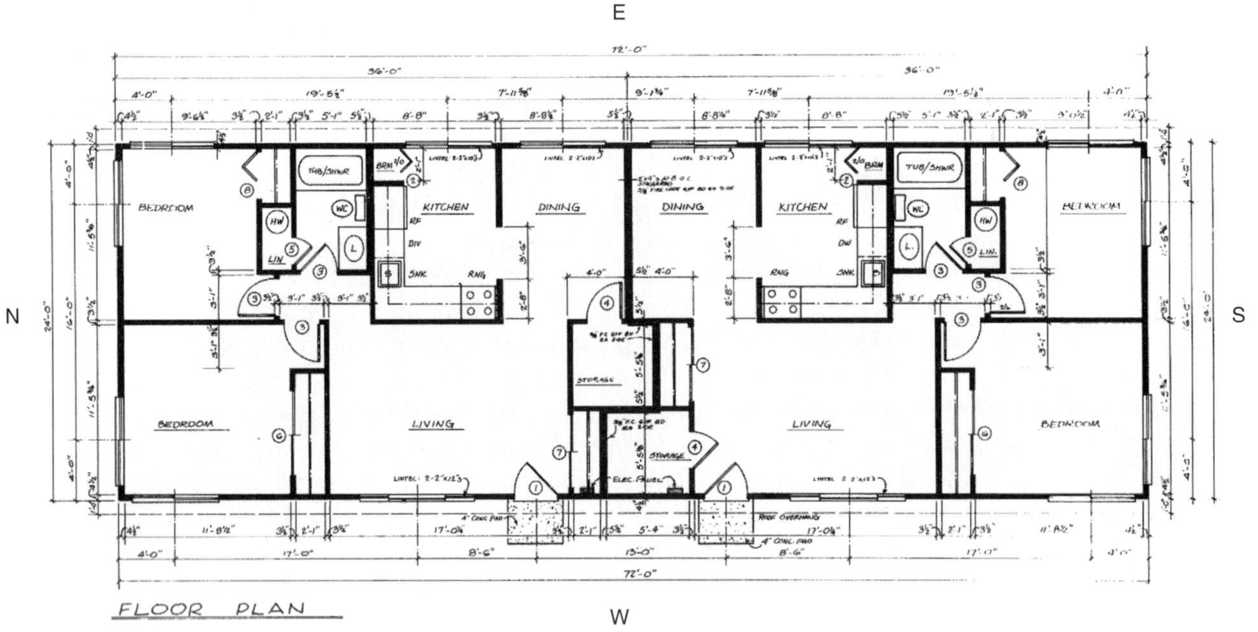

Source: Robert C. Kurzon

production home building, a builder may use one set of plans many times in a single subdivision and perhaps in more than one subdivision. A high-rise building may be repeated with several towers facing different directions. In these cases the drawings will be oriented by front, rear, right, or left sides because the front of one house may be facing south on one street while the front of another house, using the same design, may face north on another street.

The front elevation is the key indicator for naming the other elevations. The right side is

FIGURE 3.12

Simple plan view showing different views in relation to the front elevation; right, left, rear.

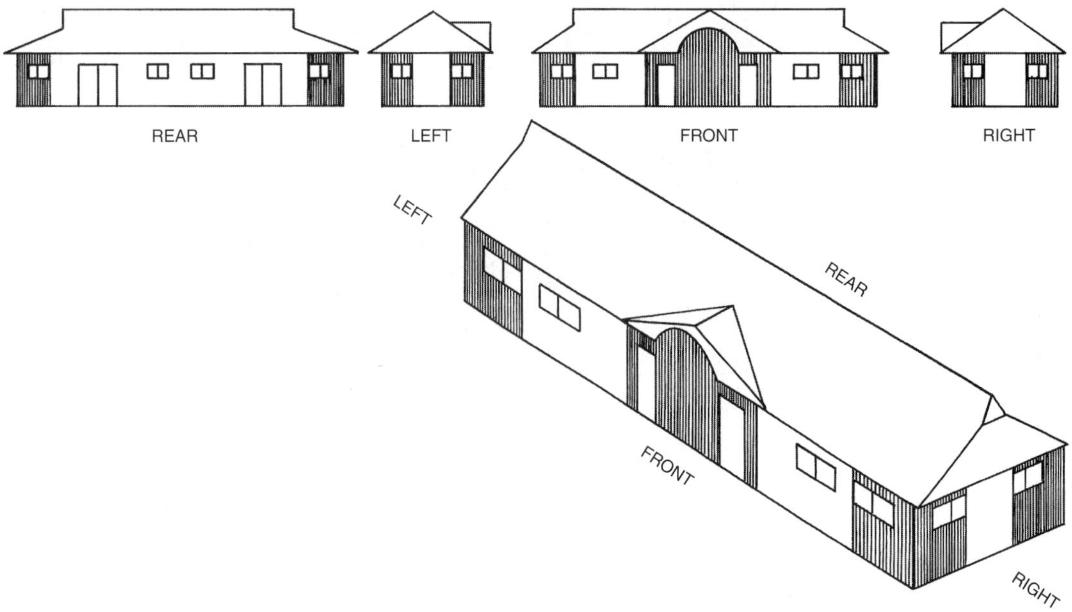

based on the side located to the right of the front elevation while looking at the front of the building. Therefore, the left elevation is the side located to the left of the front elevation.

Production Home Elevation Drawings

Aside from having one set of drawings used many times in a subdivision, a builder may also have alternative front elevations and upgrades for a single set of house plans. This means that a builder can use a set of plans that are basically the same, except for some aesthetic changes made to the front elevation. For example, the shape, size and number of windows may change, as well as the roof type and slope. There may be subtle changes to the room sizes, such as a second-floor bedroom extending out over a living room in one elevation and not in another. There may be two, three, or even four different front elevations. They are usually referred to by letter, for example, front elevation A, front elevation B, and front elevation C. These changes will generally also affect the front portions of the side elevations, and the front portions of the various floor plans and roof plan.

When the homes are complete, they may appear to be three completely different home designs. However, 80% or more of the layout of the houses is the same. A production builder is thus able to build a community from as few as seven standard plans and have 21 different front elevations. The builder will also usually have mirror or reverse images drawn of the plans to further change the look of the street while better orienting each house to a particular lot or meeting zoning requirements.

The changes to the front elevations provide a unique streetscape to the community, thus making

FIGURE 3.13

Elevation C of Doncaster house: Alternate front elevation for the Doncaster elevation shown in Figure 3.1. Can you visualize how this alternate front elevation will affect the side and rear elevations of the Doncaster house shown in Figures 3.3, 3.5, and 3.7?

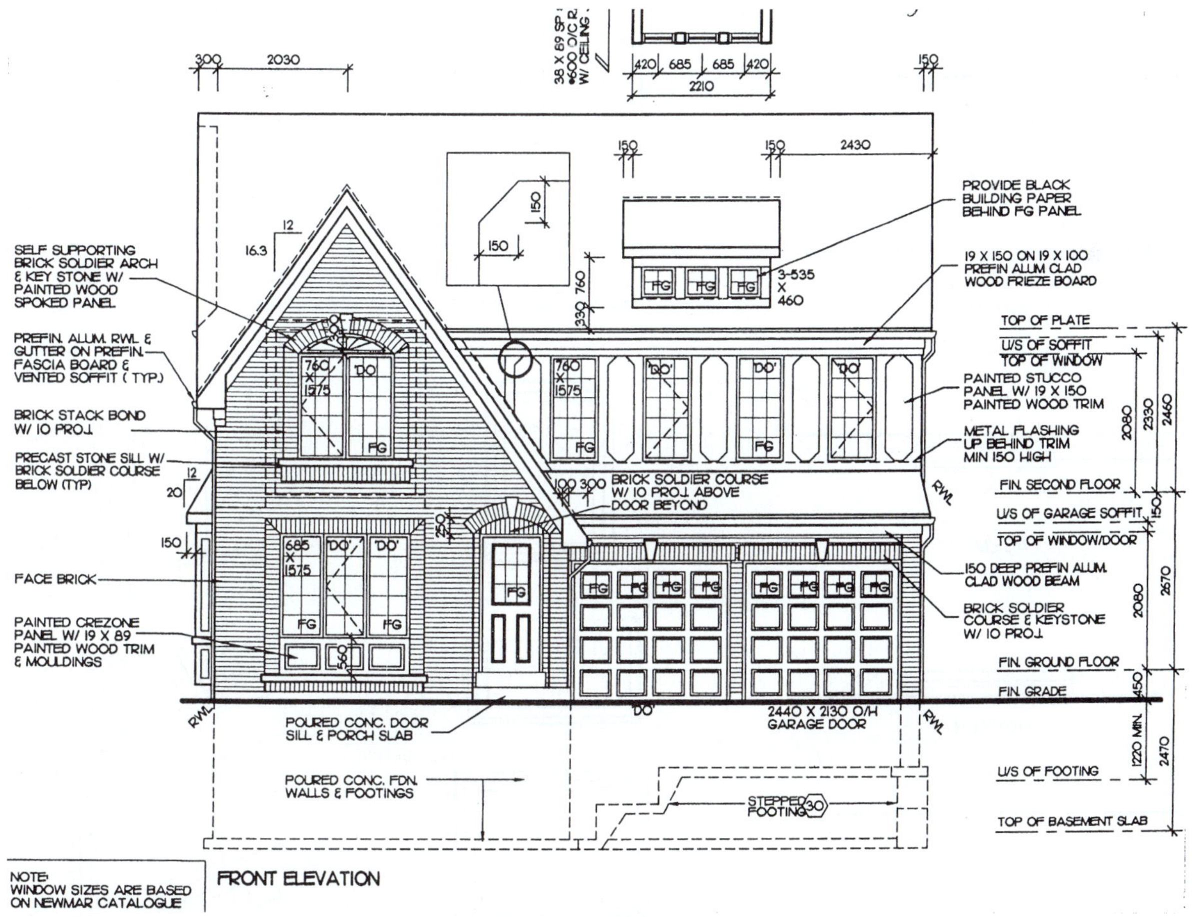

Source: Cassidy & Co. Architectural Technologists

the homes more attractive to purchasers. In many cases it is a requirement of the city/town and/or developer to have **architectural control(s)** over the design of the homes being built there. This helps to promote a high quality of new development and results in increased quality of the built environment where the sum or total of homes built in an area is larger than the parts. Architectural control helps to prevent so-called "cookie cutter" subdivisions, while providing opportunities to streamline production with designs that have similar side and rear elevations, along with similar floor plans. The builder still benefits from the advantages that come with repeat production. When building or manufacturing anything, savings are incurred when components of a project are the same or similar. For this reason, the time to construct a production home is generally much less than for a custom home. Other benefits of using similar plans include **economies of scale** such as scheduling, organization, and savings on procurement and design costs.

The Brook Drawings: two different front elevations The Brook drawings that come attached to your textbook provide an excellent example of production drawings that utilize two

FIGURE 3.14
Window types as shown in elevation and plan views.

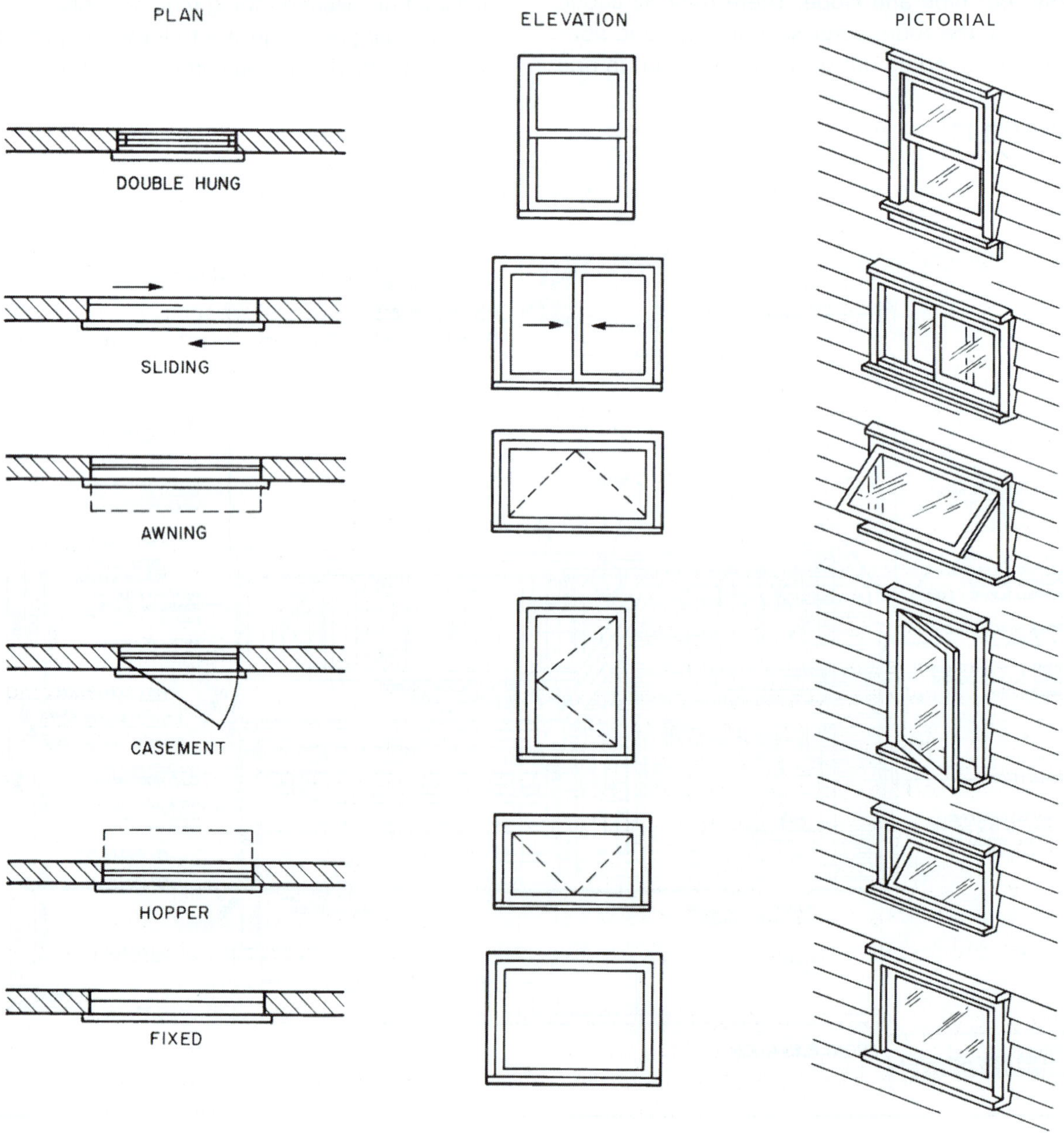

different front elevations. When you review the attached Brook drawings you will note several things. First, the drawings are drawn in imperial measurement. Second, there are two different front elevations contained in these drawings, Front Elevation A (page 8 of the Brook Drawings) and Front Elevation B (page 11 of the Brook Drawings). Flip back and forth between these drawings so that you can compare and contrast the differences. You will note that having two different front elevations also affects the side elevations, and even the rear elevation is slightly different. That is why this set of drawings has corresponding side elevations based on whether the house will use Front Elevation A or Front Elevation B, and the rear elevation indicates the difference in roof with a dashed line with a leader line stating "LINE OF ROOF ELEV. 'B' SHOWN DOTTED" (page 14 of the Brook Drawings).

Information Found on Building Elevations

Building elevations provide a great amount of detail regarding how the building will look. The overall shape and design of the exterior of the building comes together when the exterior elevations are viewed in unison. You will need to refer to and view all the elevations to gain a true understanding of the building. Is the building L-shaped, square, or U-shaped? Does the garage come out beyond the second-floor line? Are any parts **cantilevered** out from the building? (Refer to Chapter 7, Figure 7.50 for more detail on cantilevers.)

The elevation view provides information regarding the window and door styles, and whether the windows are fixed or operating. Window and door sizes may be indicated if a window schedule is not used (for more detailed information on window schedules, refer to the Khalmur renovation drawings in Chapter 15). A **schedule** is a table of information about a type of component that is found throughout the drawings and located away from the graphic image of the drawing. Schedules organize information and reduce the clutter of text that would otherwise be required directly on the drawing itself.

Other information provided in the elevation views includes the following:

- **Window wells**: these are curved or square inserts that come out from the building, often made of corrugated steel, used to isolate basement windows from moisture if the window is located partially below grade
- The type of exterior finish used:
  - Brick
  - Stone
  - Wood or vinyl siding
  - **Exterior insulated finish system** (EIFS, a synthetic insulated wall cladding that emulates stucco)
  - Type of sills used below windows and doors, e.g., concrete or stone
- Style and type of roof, with the following related information:
  - Type and location of eavestroughs and rainwater leader/downspouts
  - Type of roof finishes, e.g., asphalt, cedar, concrete tile
  - Location of photovoltaic cells (solar panels) would be indicated if they are being specified
  - Location and type of flashings for the roof, chimney, and material intersections such as at a dormer and roof
  - Roof vents
  - Detail of fascia and soffit
  - Slope of the roof, usually given as a ratio, with 12", 250 mm, or 300 mm used as the standard **unit run** while another number represents the **unit rise** (e.g., 6"). The slope may also use some standard ratios such as 1:2, 1:3, 1:4, 1:6. For example, a 1:3 roof would have a unit rise of 1 unit and a unit run of 3 units—the slope equivalent to a 4:12 roof expressed differently. The angle symbol shown in Figure 3.19 is drawn against the roof slope of the house to provide the slope information. Sometimes a triangle is used instead of the angle symbol. For an example of this in practice, refer to Figure 3.5. Note that odd units of unit rise are sometimes used instead of standard even numbers such as 11", in order to maximize slope and lower costs due to union contract requirements. For example, a 10.9" rise might be indicated because if the slope were 11" the framing costs would be higher due to a union collective agreement that recognizes the increased time to complete the roof due to its steepness.
  - Slope of vaulted or cathedral ceilings, shown as a dashed line in elevation view; note that when using roof trusses, a cathedral ceiling uses what is known as a scissor truss (refer

to Figure 3.20), which has a steeper roof slope than the ceiling slope (refer to Chapter 9 for more details on roof trusses).

- Special detailing using any of the exterior finishes (refer to Figure 3.15 for examples detailed below).
    - Arches
    - Stone lintels and surrounds
    - Brick details
    - Siding and eaves (cornice) details
    - Shutters
    - Columns
- Foundation walls and footings shown with dashed lines below grade; any stepped footings will be indicated on the elevation drawings. The type of material used for the foundation walls will also be indicated; for example, concrete block or poured concrete.
- The elevation of the grade, known as the grade line.
- Important information regarding both vertical and horizontal dimensions, including:
    - Roof projection: taken from finished fascia to finished wall surface
    - Thickness of footing (sometimes this information is also located in the construction notes)
    - Top of basement slab or top of footing to finished ground floor, where finished floor refers to the top of the subfloor (the plywood or oriented strand board fastened on top of the joists that forms the base for finished flooring material) and does not include floor finishes such as tile, carpet, or hardwood
    - Top of basement slab to top of foundation wall
    - Minimum depth of footing in relation to grade may be provided to ensure footing depth is below the frost line and meets applicable codes (measured from the grade line to underside of the footing)
    - Finished ground floor to top of window or underside of lintel
    - Finished ground floor to finished second floor
    - Finished second floor to top of plate (the top of the framed wall)
    - Finished second floor to underside of soffit

FIGURE 3.15
Example of exterior finishes found on elevation drawings: brick and brick detailing, precast stone around arched window, and staccato board panels with Tudor wood detailing on second floor.

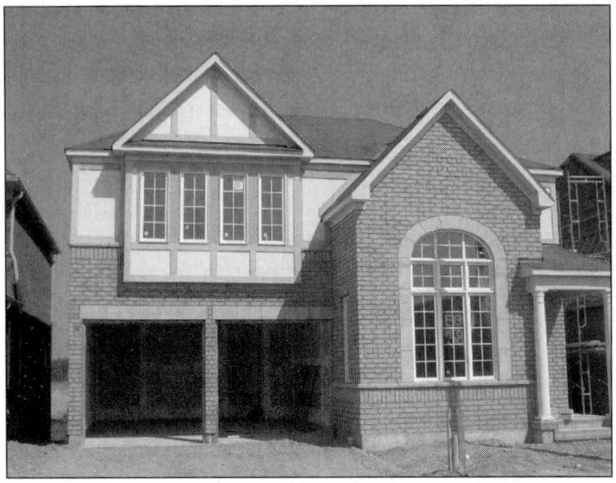

Courtesy of Tom Stephenson

FIGURE 3.16
Basic roof types.

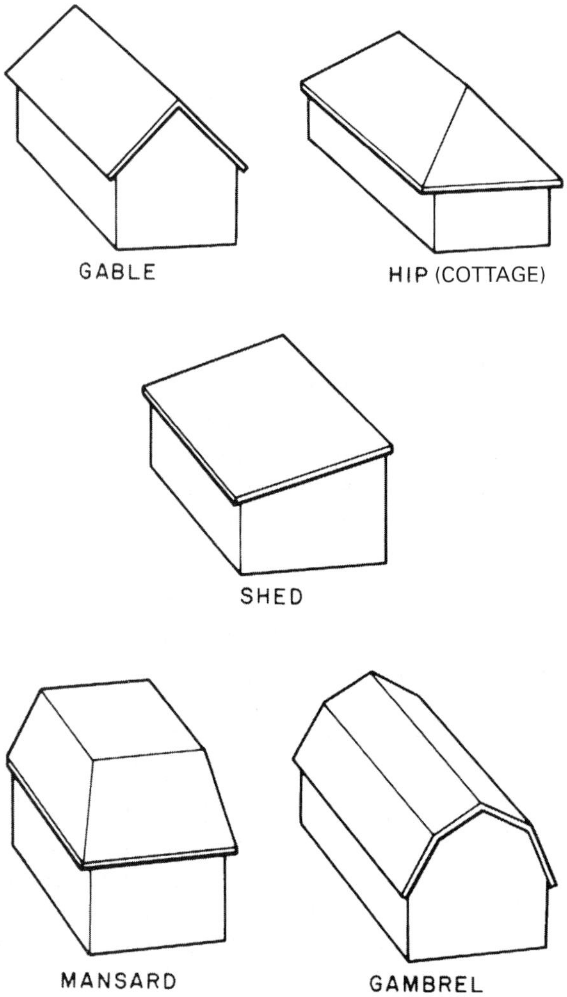

FIGURE 3.17

Intersecting roofs and mixed roof types. Carefully review the names of rafters used to construct the roof (refer to Chapter 9 for more information on roofs).

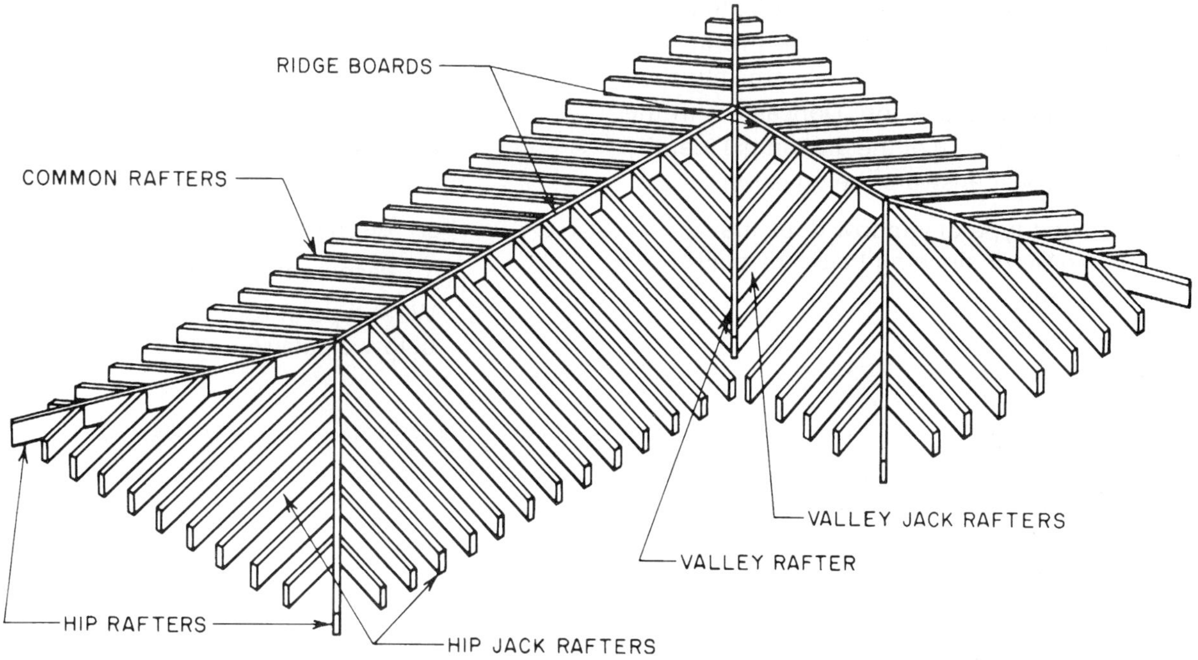

FIGURE 3.18

Dormer types include shed dormer, gable dormer, and hip/dog house dormer. Names are indicative of roof type used for the dormer.

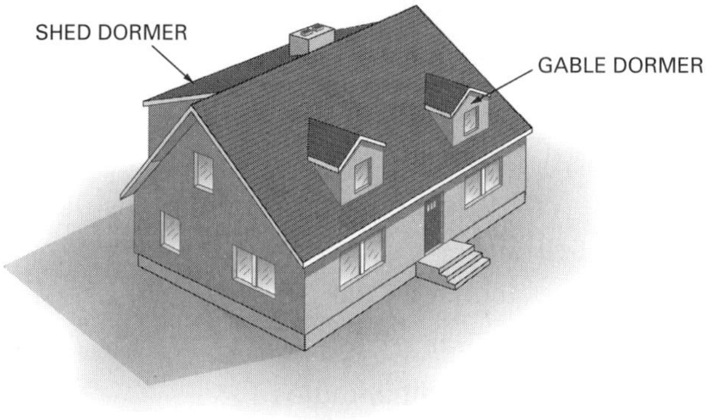

FIGURE 3.19

Slope angle.

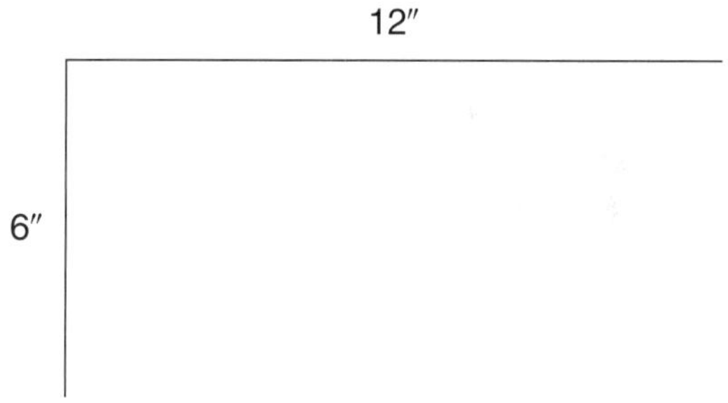

FIGURE 3.20

Scissor truss; notice the exterior slope is steeper than the interior slope.

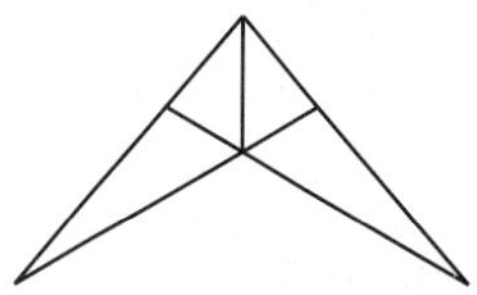

- Information regarding the limiting distance: restrictions regarding allowable openings and proposed openings due to proximity of adjoining properties (refer to Chapter 5 regarding building code and zoning).
- Location of exterior lights that appear in elevation. Note, only the lights that are visible in elevation, such as wall sconces, will appear. Pot lights located in the soffit and electrical outlets will appear in the floor plans; wall lights will appear in the floor plan views. Note: designers often forget to show the lights on the elevations, so review the floor plans carefully; there may be a light shown on one and not the other. Review the elevation drawings for the Doncaster and the Brook drawings and see if you can spot many of the features listed in the bullets above.

Chapter 3 Assignment: Elevation Drawings Using Doncaster Drawings

To answer questions 6 through 20, refer to pages 5 and 6 of the Doncaster drawings (front, rear, left, and right side elevations) that came attached with this textbook. (Look for "Doncaster" in the title block at the bottom of the drawings.) Identify the letter that best answers the question.

_____ 1. What type of window does the isometric drawing below display?

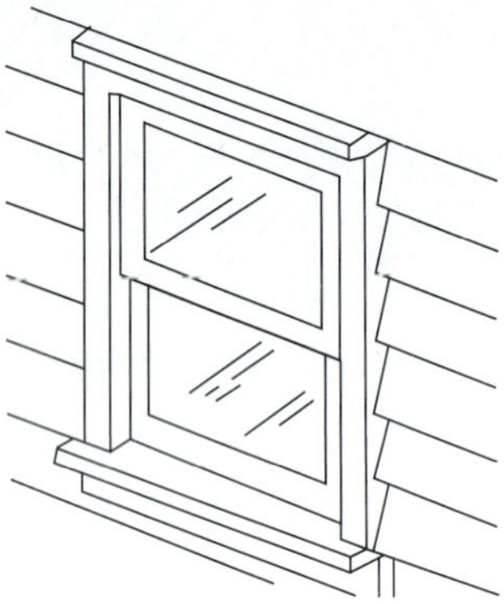

a. double hung
b. awning
c. hopper
d. sliding

_____ 2. What type of window does the elevation drawing below display?

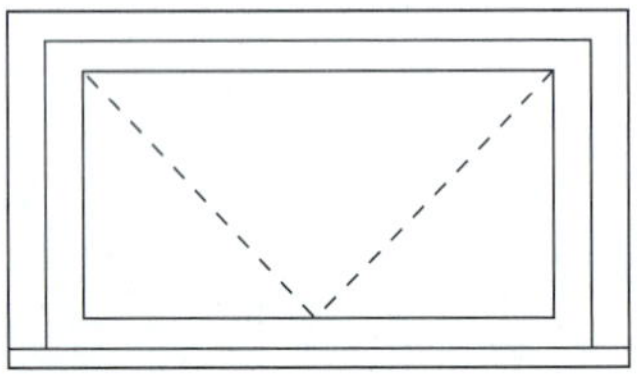

a. double hung
b. awning
c. hopper
d. sliding

_____ 3. What type of window does the plan view below display?

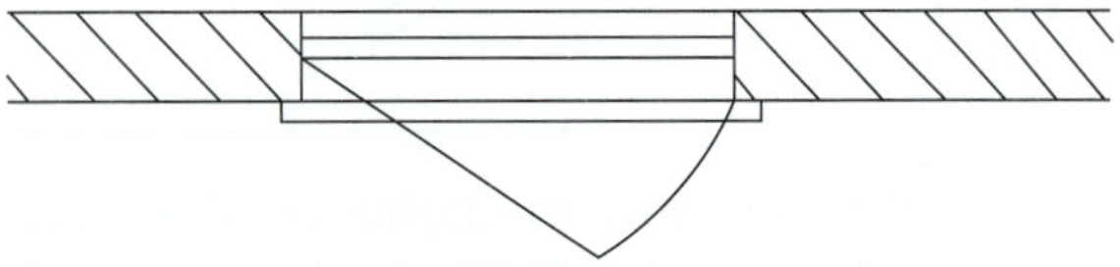

a. double hung
b. awning
c. hopper
d. casement

_____ 4. What type of window does the elevation drawing below display?

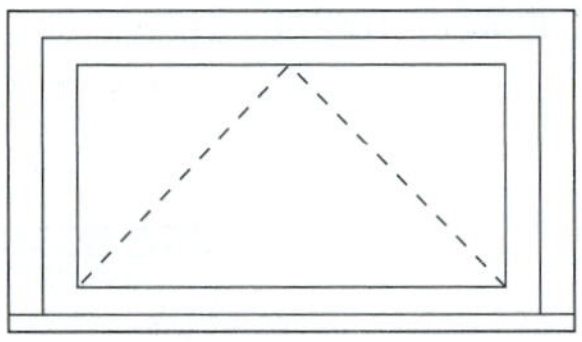

a. double hung
b. awning
c. hopper
d. casement

_____ 5. What type of roof does the image below display?

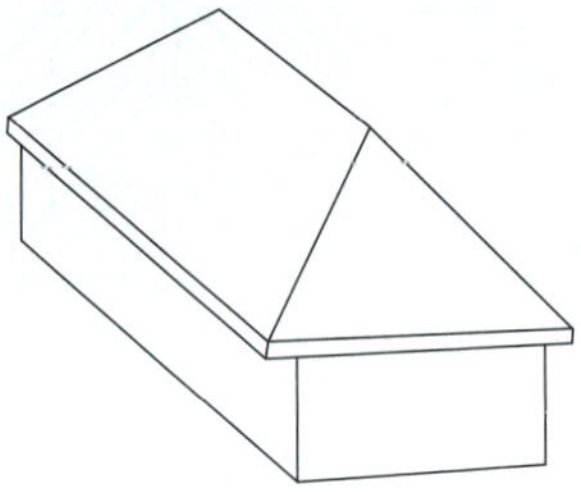

a. hip (cottage)
b. mansard
c. gable
d. shed

_____ 6. What type of roof is the main roof over the second floor of the house?

a. hip (cottage roof)
b. gable roof
c. shed roof
d. mansard roof

_____ 7. What type of roofing material is used on the roof?

a. concrete tile
b. cedar shingles
c. asphalt shingles
d. metal roof

_____ 8. What is the slope of the main (second floor) roof, as viewed from the front elevation drawing?

a. 11.9 : 12
b. 20 : 12
c. 10.5: 12
d. 6.5:12

_____ 9. What kind of roof do the decorative dormers over the front of the house have?

a. gable
b. hip (cottage)

c. shed

d. none of the above

_____ 10. Which elevation displays only one window?

a. the front elevation

b. the left elevation

c. the right elevation

d. the rear elevation

_____ 11. What type of windows are found on this house?

a. double hung, awning, and fixed

b. fixed, sliding, and double hung

c. hopper, awning, and casement

d. casement, fixed, and sliding

_____ 12. What is the slope of the main roof, as viewed from the left side elevation?

a. 11.9 : 12

b. 20 : 12

c. 10.5 :12

d. 6.5 : 12

_____ 13. How many light fixtures are visible on the exterior elevations?

a. 1

b. 2

c. 3

d. 4

_____ 14. What is the measurement from finished ground (first) floor to finished second floor?

a. 2080 mm

b. 2460 mm

c. 2470 mm

d. 2670 mm

_____ 15. What is the size of the overhead (OH) garage door?

a. 2220 mm × 2230 mm

b. 2440 mm × 2130 mm

c. 2340 mm × 2340 mm

d. 2000 mm × 3000 mm

_____ 16. Where on the elevation drawings is the brick header stack bond?

a. There is a brick header stack bond course running along the bottom of the windows that have brick under them.

b. There is a brick header stack bond course running overtop of the windows that have brick overtop of them.

c. There is a brick header stack bond course at the sides of the second floor windows that have brick at their sides.

d. There is a brick header stack bond course at the sides of the basement windows.

_____ 17. What is the purpose of the dormer at the front of the house?

a. to add space to the inside of the house by providing extra area on the third floor

b. to have a gable roof on top of them

c. to add to the aesthetic look of the house, but not to add livable space

d. to have the windows darkened to block out the sun to the interior bedrooms

_____ 18. After reviewing the four elevations, what can you determine about the roof?

a. There are two dormers on the rear elevation.

b. There are three dormers on the front elevation.

c. There is one gable roof intersecting the main (second floor) roof on the rear elevation.

d. There is one gable roof intersecting the main (second floor) roof on the front elevation.

_____ 19. Which of the following statements is true?

a. The soffit (located at the eaves) is vented.

b. The soffit (located at the eaves) is not vented.

c. The top of the windows on the first floor is 2460 mm from the finished first floor.

d. The top of the windows on the first floor is 2670 mm from the finished first floor.

_____ 20. What is the distance from the finished ground (first) floor to the underside of the second-floor soffit?

a. 2330 mm

b. 5000 mm

c. 2230 mm

d. 5430 mm

FLOOR PLANS

While being able to read all types of construction drawings is important, the ability to read and interpret floor plans is integral to understanding the structural requirements of a building. The floor plans are usually drawn first because they establish the layout for the rooms. In a custom home, the designer will gather information about the client and his/her requirements, and will make decisions regarding the layout based on the client's input. In a production home, the builder will determine what the target market needs most and offer a design that responds to those needs. For example, move-up buyers' needs may be quite different than those of first-time buyers who may want to start a family and perhaps can't afford all the bells and whistles. Homes targeted at seniors may have a **barrier-free access** design, including a single storey with the first floor easily accessible from grade to avoid stair climbing and providing wheelchair accessibility into the house and its various rooms including the bathrooms.

A typical two-storey house will require a foundation plan, and first- and second-floor plans. Each plan will establish the size of each room and hallway.

When working with construction drawings, take time to familiarize yourself with the plans. While reviewing the plans it is good practice to imagine yourself walking up to the front door (front elevation) and into the front hallway (first-floor plan). Now walk around the first floor taking special note of any features: for example, kitchen cabinets, bathrooms, closets, changes in floor levels. Imagine yourself walking up to the second floor. Look around the bedrooms, walk into the ensuite master bathroom; note the location of the tub, the basin, and the toilet.

If you flip to the elevation drawings you will notice that the location of the windows on the elevations match the location of the windows in the floor plan. You see the location of all the doors and windows as you walk from room to room. Continue the walk-through for the basement; note the location for the electrical panel, and the location of the furnace. Exit the house and walk around the outside looking at the elevations.

Once you have completed your imaginary walk, you will have gained an overall sense of the completed building. You will now begin to visualize the finished building in your mind. From this point forward no matter what your role or position, whether you are a builder, supervisor, or tradesperson, you will be working to take this project from design to reality.

Doncaster Walk-Through

The floor plans viewed in Figures 3.21 and 3.22 are simplified plans used in home builders' sales and promotions offices to convey the use of space for the home purchaser. These plans omit all detailed structural and related information so as not to clutter the drawing. From these drawings, potential purchasers get a sense of

FIGURE 3.21

Conceptual first-floor plan of the Doncaster house.

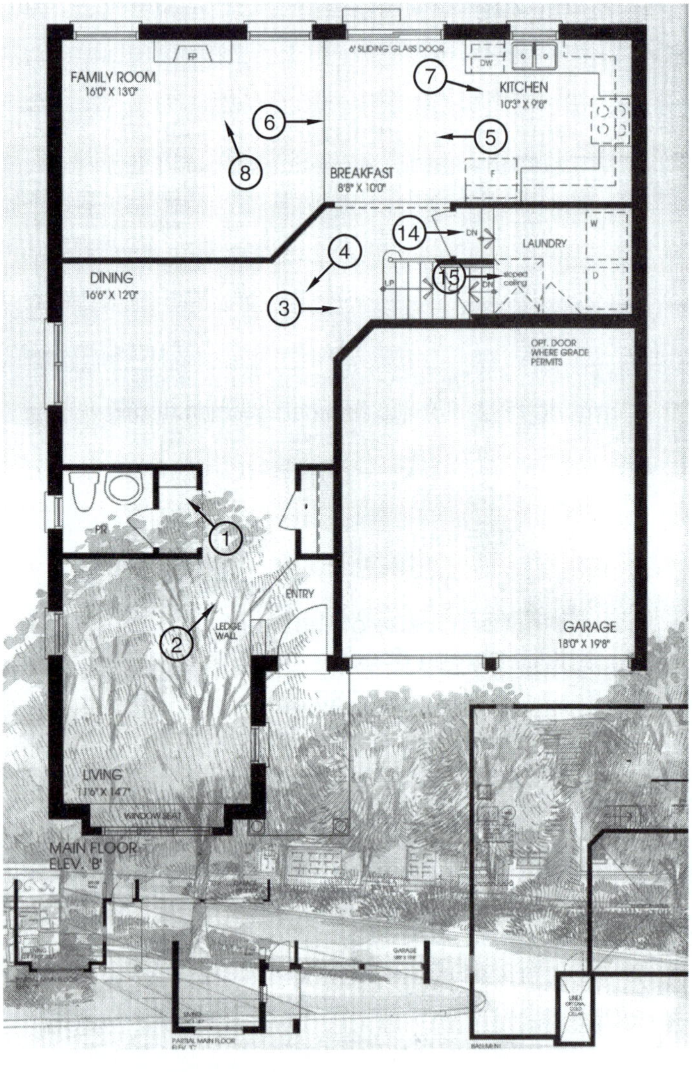

Source: Cassidy & Co. Architectural Technologists

FIGURE 3.22

Conceptual second-floor plan of the Doncaster house.

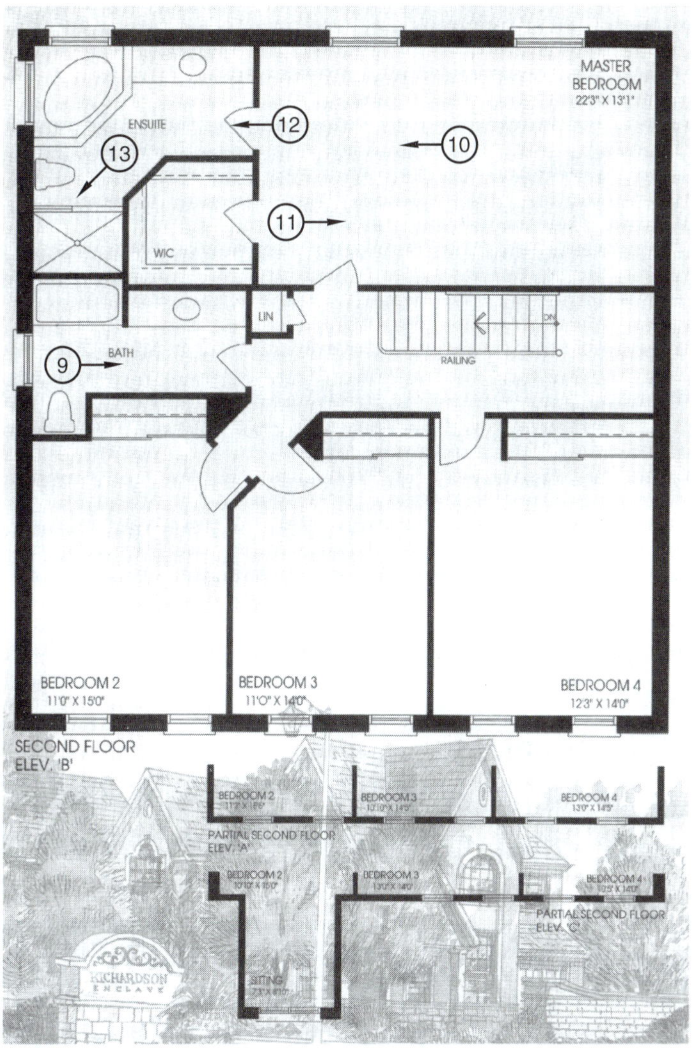

Source: Cassidy & Co. Architectural Technologists

the use of space and how the finished house will look. Figures 3.23 to 3.37 are photos taken of the finished model house based on the Doncaster drawings. This house was built as a model home and is thus furnished. Full construction drawings are included with this text, and are marked "Doncaster" in the title block at the bottom of the drawing page. Figures 3.21 and 3.22 display directional number labels on the floor plans that show where the photograph was taken. To gain the most understanding, view the photographs, simplified plans, and Doncaster construction plans that come with this text. Note that there are some minor differences between the photos and plans because the interior photographs are of a model house with a number of extras.

FIGURE 3.23

Photo position 1 (from Figure 3.21), taken from front entrance at first floor, looking toward powder room and built-in shelves.

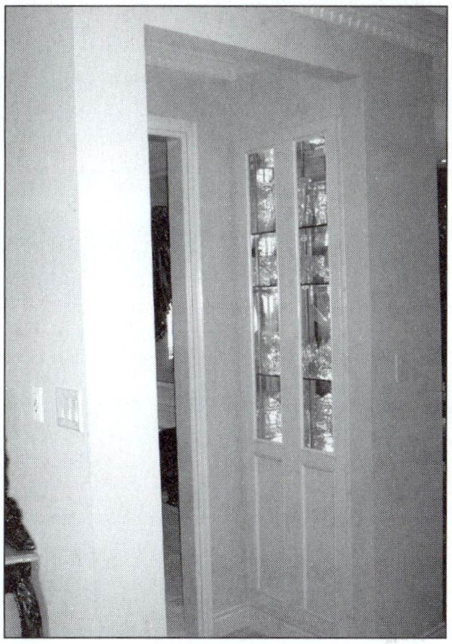

Courtesy of Tom Stephenson

Floor Plan Information

- Floor plans are a horizontal cut taken through the walls (horizontal section) at a varying height so as to maximize the information provided. For example, the cut will display the location of all the windows and doors, so if a 600 mm (2') high window was located high in the wall, the slice would move up so the window would be displayed on the plan.

FIGURE 3.24

Photo position 2 (from Figure 3.21), taken from just to the left of the front door in the living room, looking through the hall and dining room to the sliding doors at the rear of the house in the breakfast area.

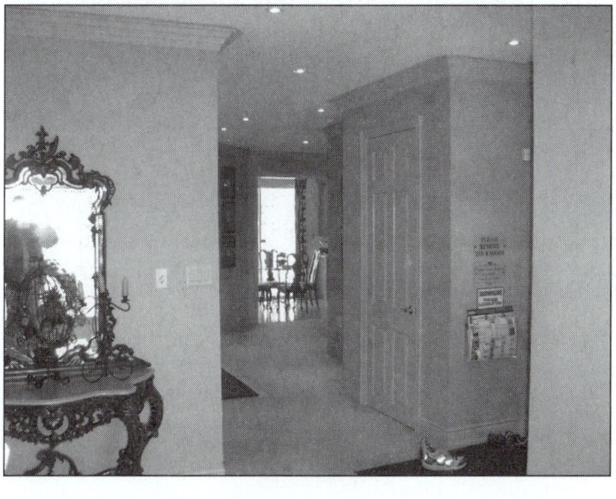

Courtesy of Tom Stephenson

FIGURE 3.25

Photo position 3 (from Figure 3.21), taken at base of first-floor stairs looking toward stairs and laundry room at left.

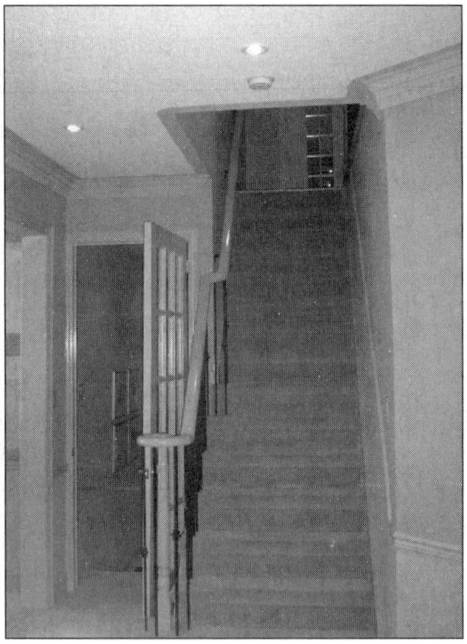

Courtesy of Tom Stephenson

FIGURE 3.27

Photo position 5 (from Figure 3.21), taken from the breakfast area looking toward the family room; notice the entrance with angled wall on the left, and the built-in cabinet that houses the gas fireplace coming out from the wall (compare to the Doncaster plans).

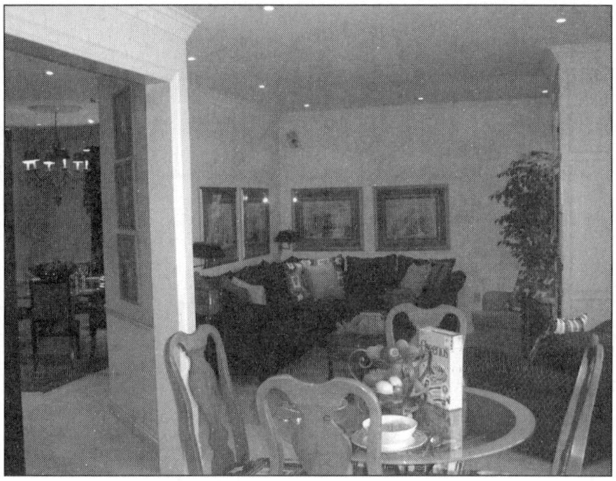

Courtesy of Tom Stephenson

FIGURE 3.26

Photo position 4 (from Figure 3.21), taken looking toward dining room. In this example, the window is not boxed/cantilevered out as the Doncaster drawings that come with the text indicate.

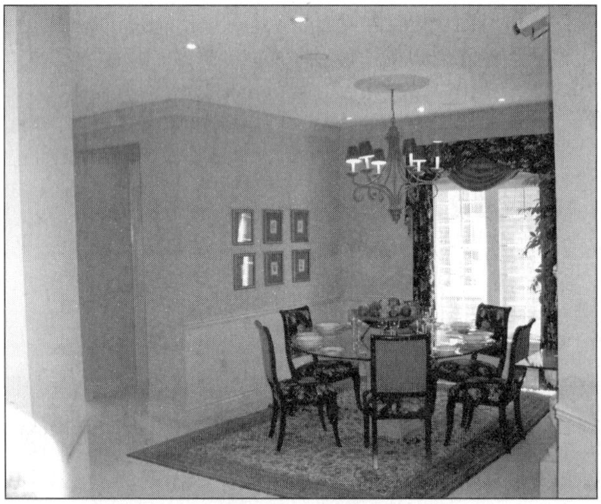

Courtesy of Tom Stephenson

FIGURE 3.28

Photo position 6 (from Figure 3.21), taken from the family room looking toward the kitchen; notice the location of the sliding door, dishwasher, stove, and fridge (compare to the first-floor plan).

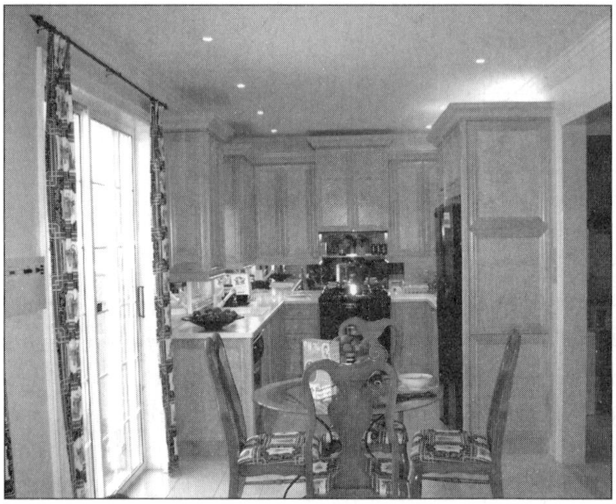

Courtesy of Tom Stephenson

FIGURE 3.29

Photo position 7 (from Figure 3.21), taken from the sliding door looking toward right side of kitchen; notice the microwave built into the cabinets (check the construction drawings to see how this is indicated).

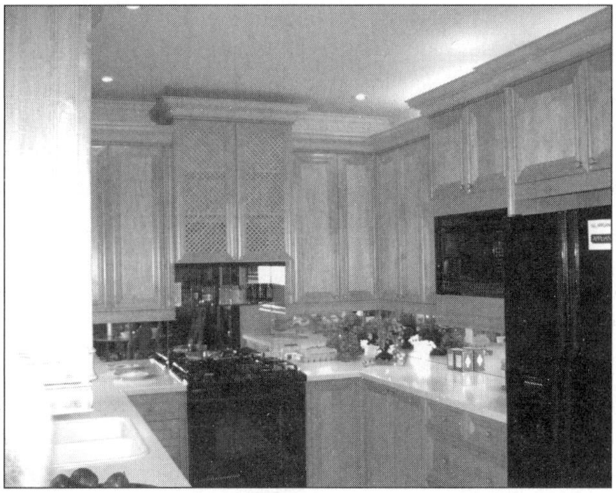

Courtesy of Tom Stephenson

FIGURE 3.30

Photo position 8 (from Figure 3.21), taken from entrance to the family room looking toward gas fireplace. This photo was taken during construction before the final finishes were completed. Note the fireplace mantel surrounding the gas fireplace. This finish would have been chosen as a finish option by the homeowner at the builder's decor centre.

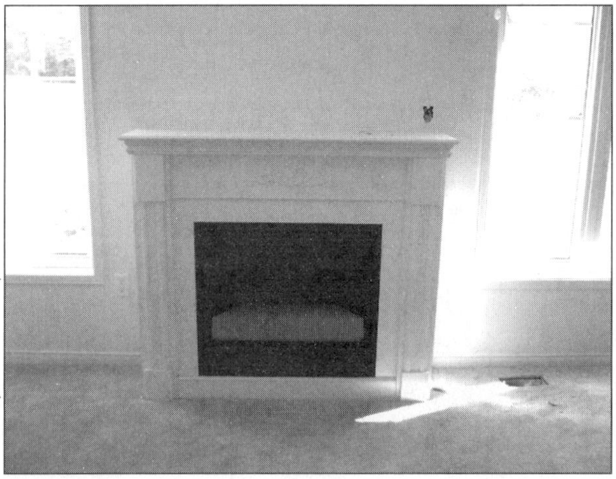

Courtesy of Tom Stephenson

FIGURE 3.31

Photo position 9 (from Figure 3.22), taken from inside the second-floor bathroom looking toward the stairs.

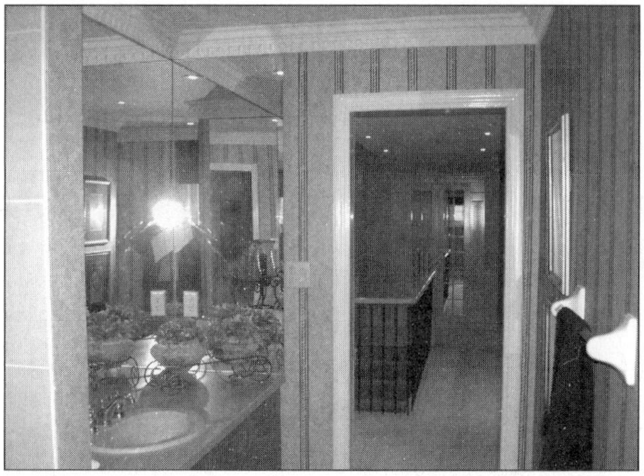

Courtesy of Tom Stephenson

FIGURE 3.32

Photo position 10 (from Figure 3.22), taken in the master bedroom looking toward the walk-in closet at left, and the ensuite bathroom at right.

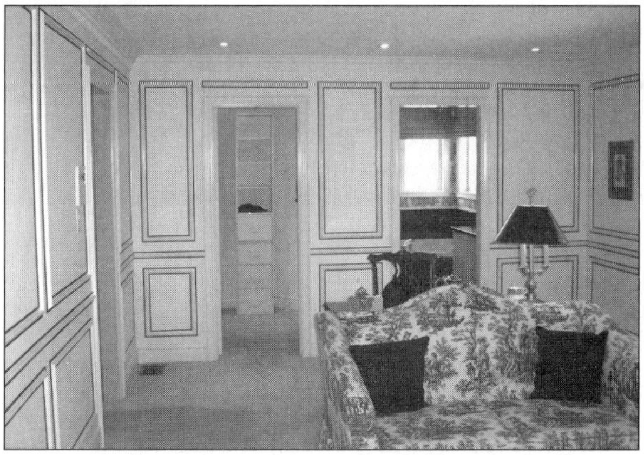

Courtesy of Tom Stephenson

FIGURE 3.33

Photo position 11 (from Figure 3.22), taken looking toward the opposite wall from the ensuite; notice the location of the window at left.

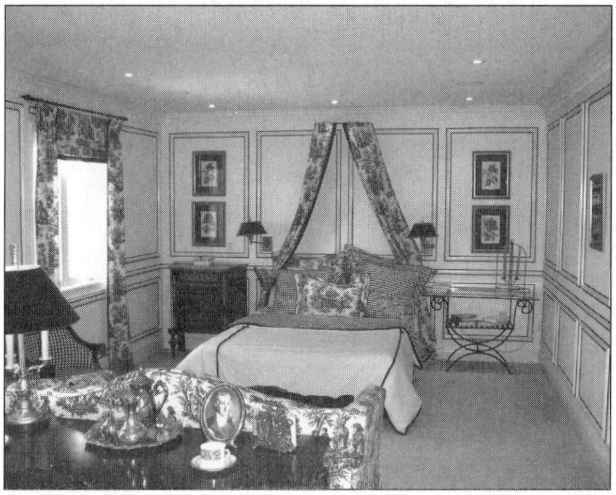

Courtesy of Tom Stephenson

FIGURE 3.34

Photo position 12 (from Figure 3.22), taken from the entrance doorway looking to the rear of the ensuite; notice and compare the location of windows, tub, and vanity to the drawings. Also refer to the rear and left elevations of the Doncaster house or refer to Figures 3.5 and 3.7. This explains why the second-floor windows at the rear left corner do not drop down as much as the other windows.

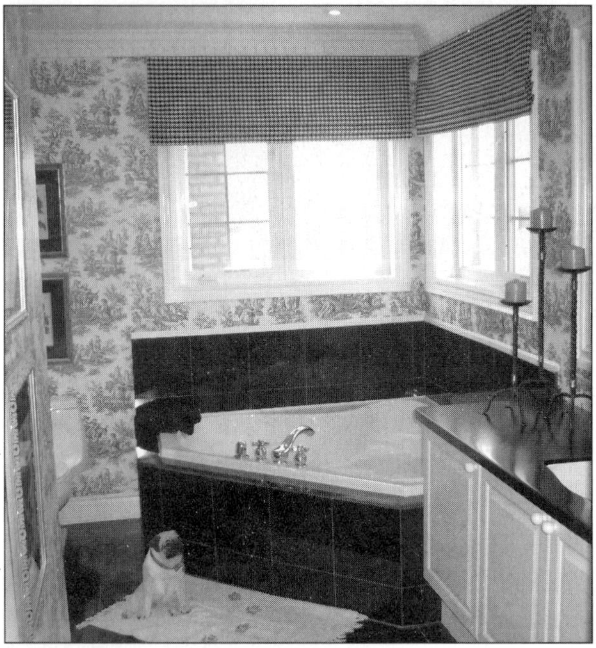

Courtesy of Tom Stephenson

FIGURE 3.35

Photo position 13 (from Figure 3.22), taken looking toward the shower in the ensuite bathroom; notice the half wall and compare to the construction drawings.

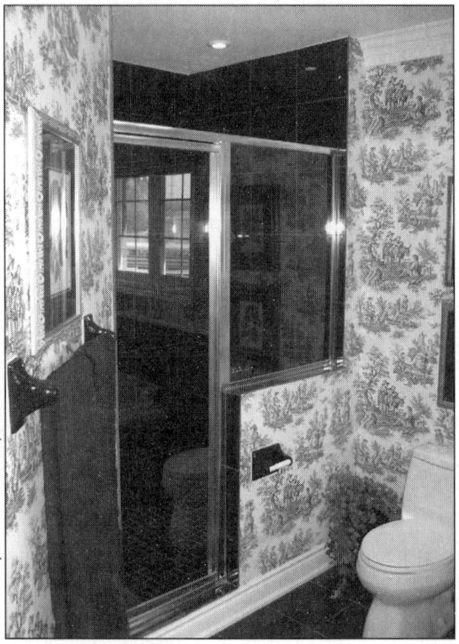

Courtesy of Tom Stephenson

FIGURE 3.36

Photo position 14 (from Figure 3.21), taken from doorway to the left of the stairs looking down toward the sunken laundry room; notice the angled wall at the top right of the photo (these are the stairs from the first floor extending up to the second floor). Compare with the drawings and note that the floor of the laundry room is at a different elevation than the first floor.

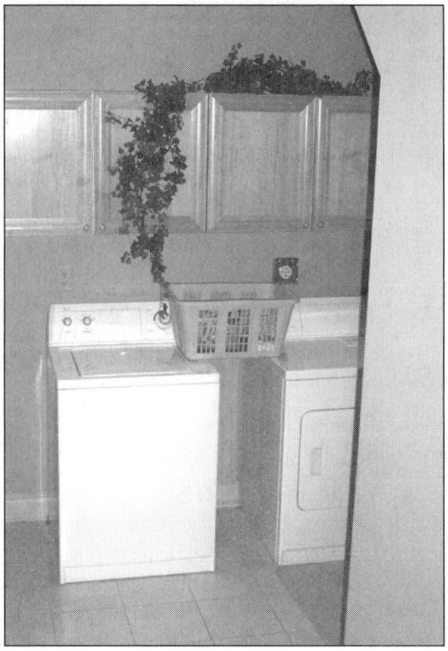

Courtesy of Tom Stephenson

FIGURE 3.37

Photo position 15 (from Figure 3.21), taken from base of basement stairs looking up toward the laundry room; notice the door at right, which exits to the garage.

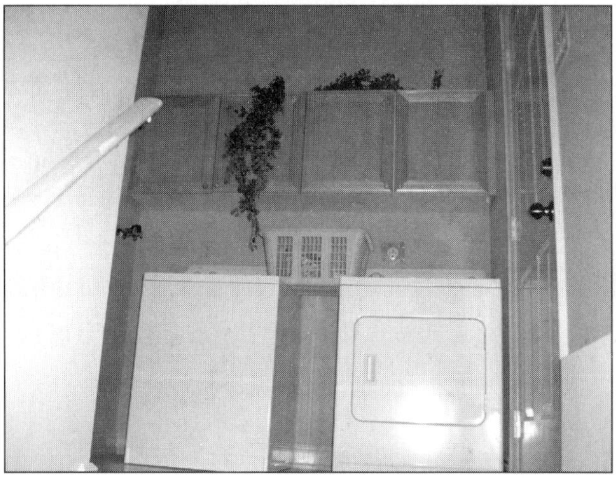

Courtesy of Tom Stephenson

FIGURE 3.38

3D view of first- and second-floor plans for Doncaster house with finishes indicated. Review the Doncaster first-floor plan (page 3 of Doncaster drawings). Compare and contrast, and note any differences or missing finish items. This is a good way to practice your observation skill, which is a valuable skill to have during the construction process, particularly before the finishes are in place.

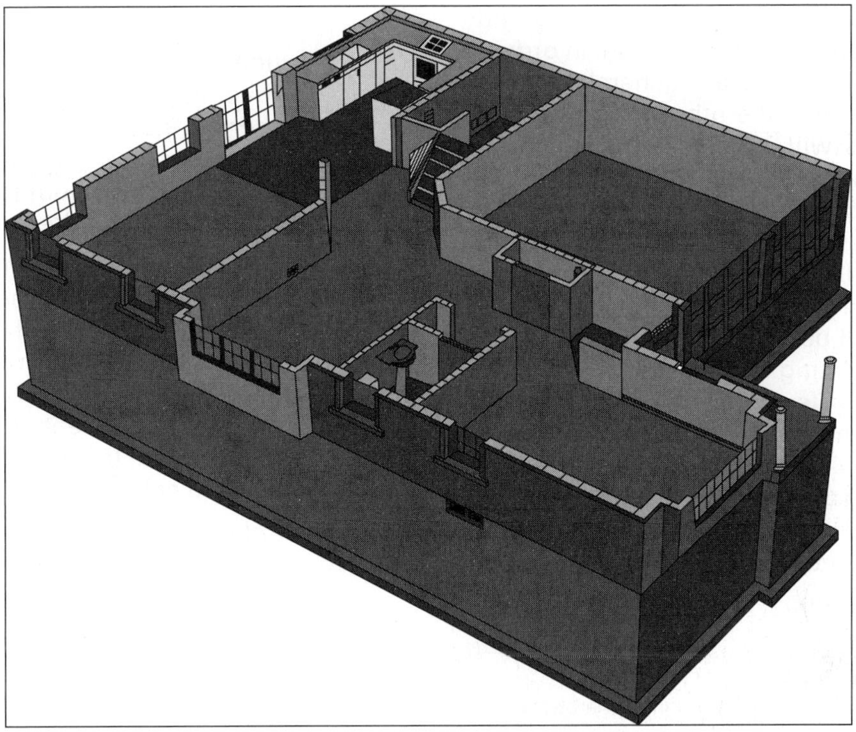

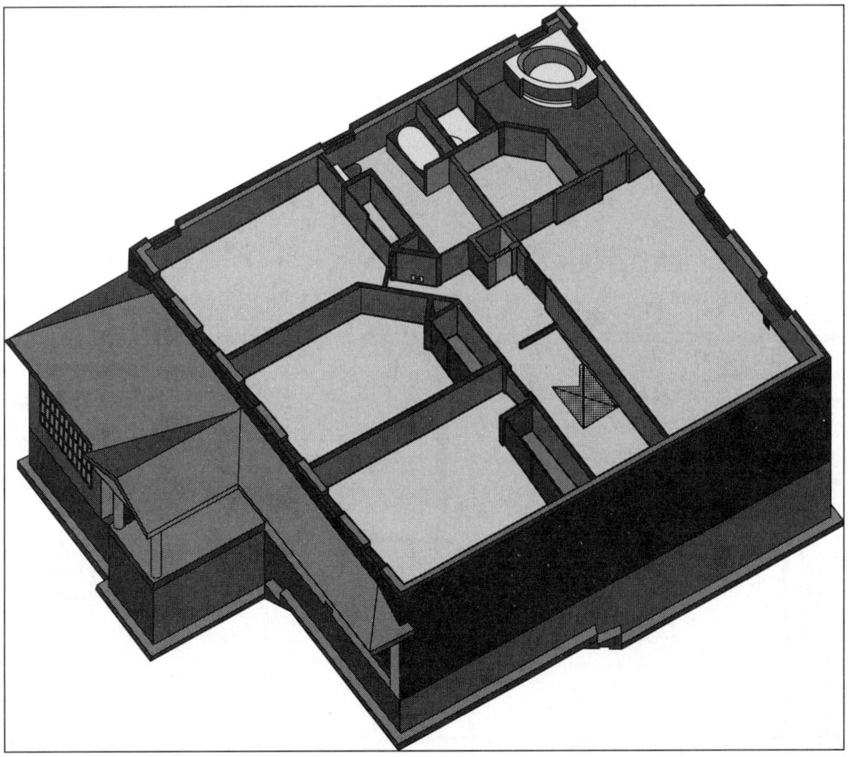

SOURCE: Source: Courtesy of Clint Kissoon, Chair, George Brown College, School of Architectural Studies.

For general purposes, the cut is made at approximately half the height of the wall.

- Individual **floor-plan drawings** must be co-ordinated with each other:
  - Where the stairs start in the foundation plan (basement), the first floor must indicate exactly where they finish. In order to save space, stairs are generally laid out one floor over the other; as a result, the drawings will indicate the stairs going up about halfway, e.g., first to second floor, and then there will be a break line. On the other side of the break line will be an indicator of the stairs going down, e.g., from the first floor to the basement. There will be an arrow with an abbreviation indicating UP or DN, and the floor plan will also indicate the number of risers from one floor to the other (review Figure 3.40 for an example of this). A **riser** is measured vertically from the top of one step, referred to as a tread, to the top of the next tread. The **tread** is the part of the stairs that your foot rests on when climbing stairs, measured horizontally from the face of riser to the face of the riser on the next step plus the **nosing** (the lip formed at the edge of the tread). Remember that the number of risers may differ on the same floor plan depending whether you are going up to the floor above or down to the floor below. If there is a difference, it is probably due to the difference in floor-to-ceiling height between the two floors. (For more information on stairs, refer to Chapter 13).
  - Load-bearing walls must be co-ordinated and relate to each other from floor to floor so that the load from one bearing wall will transfer to the floor and load-bearing walls or beams below it. These loads are then eventually transferred to the footings at the base of the foundation walls and columns.
- Floor plans indicate the location, direction, size, type, or grade of joists, beams,

FIGURE 3.39

A floor plan is a horizontal cross-section of the building.

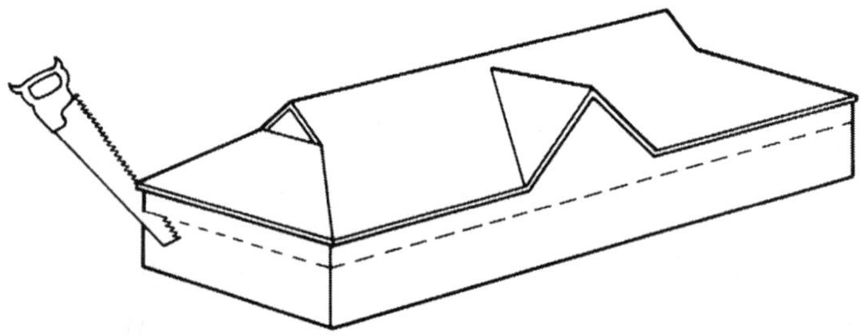

(a) An imaginary cut is made at a level that passes through all windows and doors.

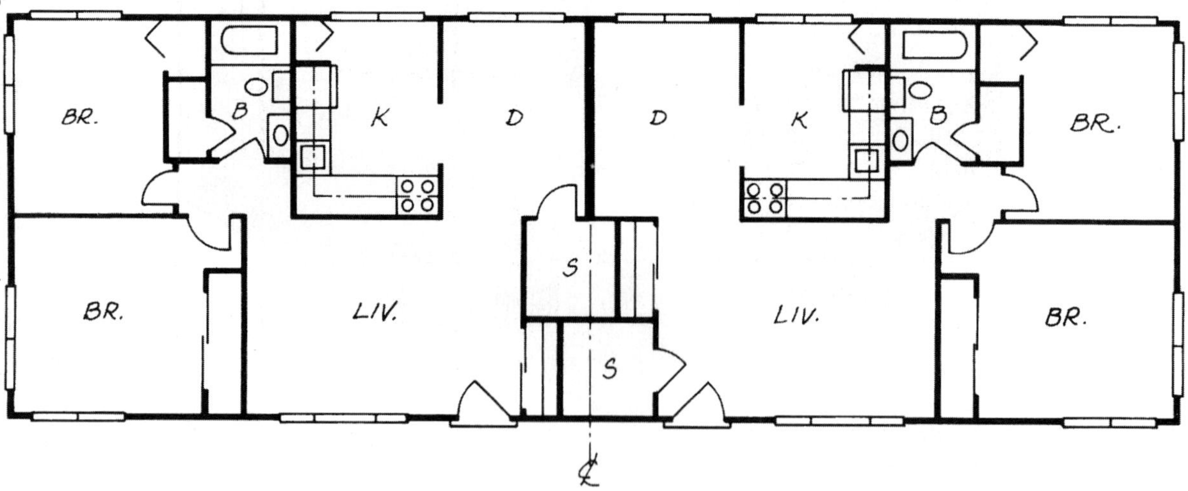

(b) The floor plan shows what is left when the top is removed.

Source: Cassidy & Co. Architectural Technologists

FIGURE 3.40

First-floor plan detail of stairs from the Doncaster house (refer to Doncaster first-floor plan); notice that there are two risers down from the first floor to the laundry room marked DN 2R, and that there are 11 risers down from the laundry room to the basement marked DN 11R. The break line indicates the separation between the stairs going down and the stairs going up. There are 14 risers going up from the first floor to the second floor. Also, notice the sloped ceiling indication at the top of the basement stairs (marked with a dashed X pattern); this is the angled wall/ceiling that you see at the top right of Figure 3.36.

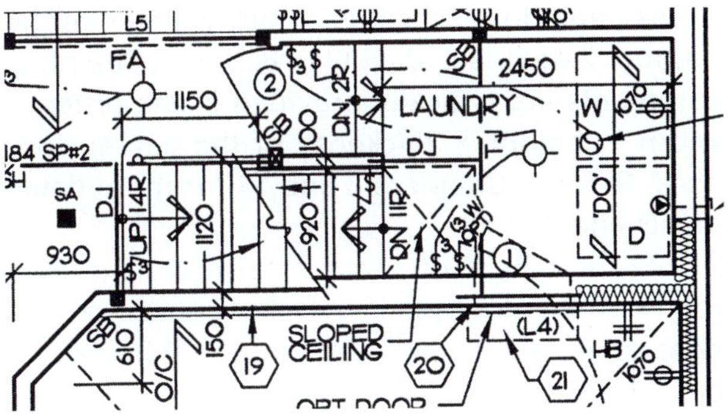

Source: Cassidy & Co. Architectural Technologists

and lintels. For example, the foundation plan (basement) will indicate the joists and beams that support the first floor. The first-floor plans will indicate the joists and beams used to support the second floor. The second or top floor will indicate the direction of the ceiling joists and rafters, or roof trusses, as located directly above the second floor. Note: the previously mentioned point regarding which floor plan displays the floor joist sizes for that floor is an area that often confuses students. If in doubt, start with the basement plan/foundation plan and work your way up floor by floor.

- The layout of the rooms indicates the location of fireplaces, cabinets, fixtures (water closets/toilets, tubs, basins, etc.), and appliances. Upper cabinets, bulkheads, and coffered ceilings (rooms with a decorative ceiling, typically a bulkhead around the edge of the room) are generally shown with a dashed line.
- As mentioned in Chapter 1, dashed lines will also be used as hidden lines displaying the location of footings and grade beams, as well as showing the location of potential rooms, such as a rough-in for a bathroom in the basement. Refer to Figure 3.41 for an example of an optional three-piece rough-in bath. There are other uses for dashed lines as will be described further in the textbook.
- Changes in the elevation (height) of the ceiling are indicated on the floor plan by a dashed line appearing as an × through the area affected by the change in elevation. For example, on the Doncaster basement plan that comes with this text, the garage floor is at a different elevation than the basement ceiling, and the laundry area is sunk down two risers at the top of the basement stairs. These changes in elevation are brought to your attention by the × pattern of dashed lines, or in some cases solid lines.
- The electrical layout is shown on the floor plans. The electrical layout displays the location of all lights, switches, and outlets. If the electrical layout is complicated or extensive, it may be addressed in its own separate floor plan. ICI and high-rise

FIGURE 3.41

Direction of floor joists and beams are indicated on this blowup of a basement plan. "38 × 184 SP" indicates the floor joists while "W200 × 27" indicates the steel beam (refer to Chapter 7 for more detail on floor construction).

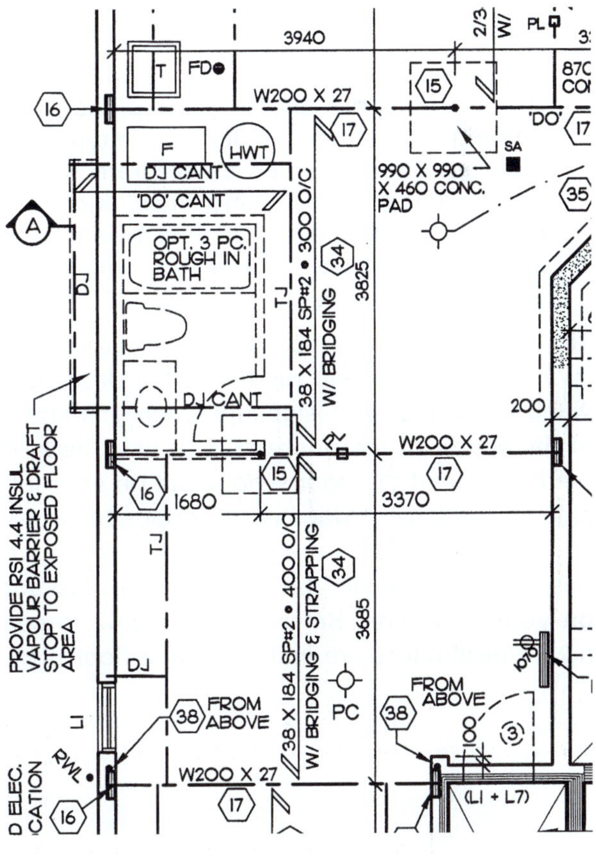

Source: Cassidy & Co. Architectural Technologists

FIGURE 3.42

Typical door types as shown in plan view. The type of door used and its swing will be based on style, function, and space. For example, a pocket door takes little space from the room. Also check that there is adequate space between the door frame and adjoining walls to ensure adequate spacing for trim (designers sometimes miss this or clients change to a larger casing trim after drawings are completed).

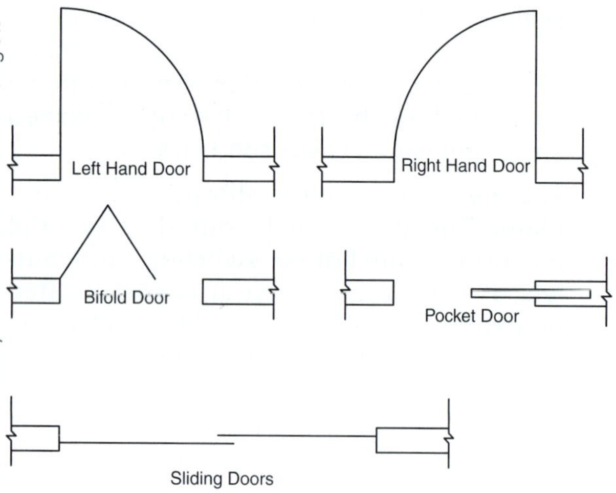

Source: Cassidy & Co. Architectural Technologists

residential projects have separate electrical and mechanical drawings laid out in individual floor plans.

- The type of door, and the way a door swings is also indicated on the floor plans. It is important to be able to identify the type and swing of door based on an accurate review of the plans because, many times, doors are ordered pre-hung based on their swing; if the swing is identified incorrectly, it will cause delays and add cost to the project.

The following is a simple way to interpret the swing of a door when reviewing the drawings. A left hand door is designated when you stand in the centre of the doorway with your back to the hinges and the door swings out on your left side. A right hand door is the opposite; when you stand in the centre of the doorway with your back to the hinges, the door swings out on your right side. Refer to Figure 3.42 to see how typical doors are indicated on a floor plan. An example of a left hand swing and right hand swing are shown. The size of door will also be indicated directly on the plan, or indirectly by a number referencing a door schedule. In the case of the Doncaster drawings, the door is referenced by a number to the door schedule found on the Construction Notes page. For example, Figure 3.40, a close-up of the Doncaster drawings' first-floor plan, has a left hand door with the number 2 inside a circle. To determine the size of door, you then reference the Construction Notes page door schedule, which states that the number 2 door is 2'-8" × 6' to 8' × 1⅜". The measurements are typically provided in the order of width × height × thickness of the actual door itself. Refer to Chapters 8 and 13 for more detail on schedules.

A great deal of other information is also found on the floor plans, which will become evident as you work your way through this text.

KEY TERMS

Definitions for key terms appear in the glossary at the end of the text.

Architectural control (p. 44)
Barrier-free access (p. 50)
Cantilever (p. 45)
Economies of scale (p. 44)
Elevation drawing (p. 36)
Exterior insulated finish system (EIFS) (p. 45)
Floor-plan drawing (p. 56)
Nosing (p. 56)
Production home building (p. 42)
Riser (p. 56)
Schedule (p. 45)
Tread (p. 56)
Unit rise (p. 45)
Unit run (p. 45)
Window well (p. 45)

Chapter 3 Quiz Review: Floor Plans Using Doncaster Drawings

To answer the following questions, refer to the basement (foundation), first- and second-floor plans of the Doncaster drawings that came attached to this textbook (pages 2, 3, and 4 of the Doncaster drawings). Identify the letter that best answers the question.

_____ 1. What swing of door is indicated in the plan drawing below?

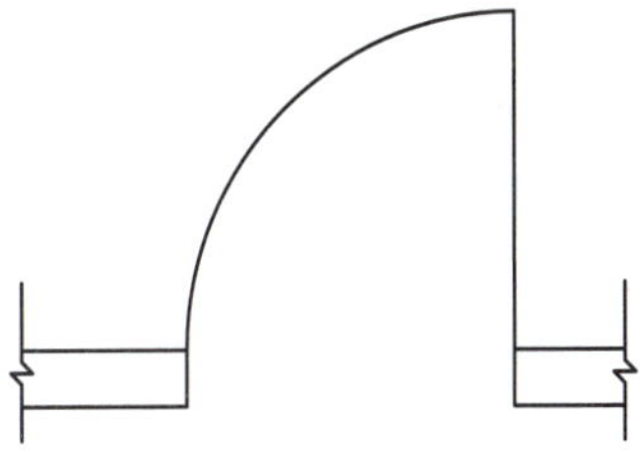

a. right hand door

b. left hand door

c. pocket door

d. sliding door

_____ 2. From the first floor plan, how thick are the exterior above ground walls of the garage?

a. 230 mm

b. 250 mm

c. 280 mm

d. 300 mm

_____ 3. With the exception of the garage, how thick are the exterior above-ground walls of the house?

a. 230 mm

b. 250 mm

c. 280 mm

d. 300 mm

_____ 4. What are the interior dimensions of the master bedroom (width × length)?

a. 3990 mm × 6780 mm

b. 4090 mm × 6880 mm

c. 4370 mm × 7160 mm

d. 3700 mm × 4200 mm

_____ 5. What is the wall thickness between bedrooms 3 and 4, as indicated on the drawings?

a. 150 mm

b. 100 mm

c. 75 mm

d. 125 mm

_____ 6. Where is the oval-shaped bathtub located?

a. in the basement bathroom

b. in the second floor bathroom that you enter from the hallway

c. in the powder room

d. in the master ensuite bathroom

_____ 7. Where does the foundation plan show a change in elevation from the main ceiling of the basement?

a. where the optional bathroom can be roughed in

b. below the laundry room

c. above the floor drain

d. near the proposed electrical panel

_____ 8. How many risers are there from the basement to the laundry room? (Review both the first floor and basement plans carefully.)

a. 11 risers

b. 12 risers

c. 13 risers

d. 14 risers

_____ 9. How many risers are there from the basement to the first floor (not laundry room)?

a. 11 risers

b. 12 risers

c. 13 risers

d. 14 risers

_____ 10. What do the dashed lines indicate in the close-up snapshot taken from the Doncaster drawing basement plan at the right front corner of the garage?

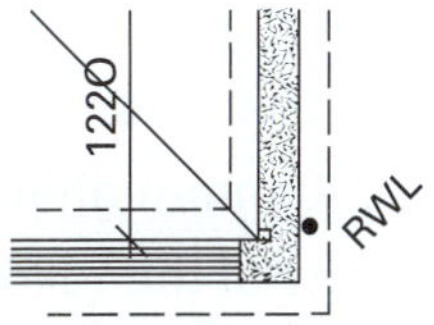

a. the location of joists

b. the location of the footings

c. the location of grade beams

d. the location of the steel beam

_____ 11. What do the dashed lines shown under the basement floor slab indicate?

a. the location of the foundation wall footings and column footings

b. the direction of floor joists

c. the location of bridging

d. the stairs to the basement from the first floor

_____ 12. Where on the basement plan are the foundation walls to be insulated from floor to ceiling?

a. between the garage and basement walls

b. around the exterior foundation walls, excluding the garage and cold room

c. on the interior side of the cold cellar walls (only if this option is chosen)

d. around the exterior foundation walls underneath the laundry room

_____ 13. What is the size and spacing of the first-floor floor joists in at least one area of the first floor?

a. 2/38 × 235 @ 300 O/C

b. 38 × 89 @ 300 O/C

c. 38 × 184 @ 400 O/C

d. 38 × 184 @ 600 O/C

_____ 14. How many column footings/concrete pads are indicated on the basement plan?

a. one

b. two

c. three

d. four

_____ 15. What is located to the left of the front entrance door as you enter the house?

a. a wall outlet

b. light switches

c. the closet door

d. an 810-mm-high half wall

_____ 16. Which one of the following is indicated on the first-floor plan?

a. a family room that is 18'-0" × 12'-0"

b. a window seat in the breakfast area

c. a dropped ceiling in the garage

d. an L-shaped layout for the kitchen

_____ 17. How many direct vent fireplaces are indicated on the drawings?

a. 0

b. 1

c. 2

d. 3

_____ 18. Which of the following can you determine is true after reviewing the second floor plan?

a. There is a walk-in closet in bedroom 4.

b. There is a linen closet in the second floor hallway.

c. There is a shower but no bathtub in the second floor bathroom leading from the hallway.

d. There are three windows in the master bedroom.

_____ 19. From the floor plans, what do you know about the interior walls?

a. The interior wall between bedroom 2 and bedroom 3 is 150 mm thick.

b. The interior wall between the family room and the dining room is 200 mm thick.

c. The interior wall between the powder room and the dining room is 150 mm thick.

d. The interior wall between the ensuite and second floor bathrooms is 100 mm thick.

_____ 20. What do you know from reviewing the floor plans for the first and second floors?

a. Bedroom 2 is directly over the living room and is the exact same size as the living room.

b. The second-floor roof has Girder Trusses (GT) shown in two locations.

c. Bedroom 3 partially covers the first-floor porch.

d. The ensuite bathroom is directly over the kitchen.

4 Abbreviations, Hatchings, Symbols, Construction Notes, and Detail Drawings

LEARNING OBJECTIVES

After completing this chapter, you will be able to:

- Identify and interpret abbreviations commonly used in construction drawings.
- Identify and interpret hatchings commonly used in construction drawings.
- Locate information on the construction notes based on identification symbols used in the construction drawings.
- Locate and interpret section details as indicated by a cutting-plane line in the construction drawings.
- Visualize the construction and assembly of various building materials by reviewing detail drawings.

INTRODUCTION

This chapter covers abbreviations, hatchings, symbols, construction notes, and detail drawings, which designers use to communicate detailed information on construction drawings in a uniform, easy-to-interpret way that does not clutter the drawings. The chapter provides lists of abbreviations, symbols, and hatchings that you can use throughout the text for some of the quizzes and assignments, and ***keep as a reference for when you are working in the field***.

ABBREVIATIONS

Construction drawings require a lot of information to be placed in a small area. **Abbreviations** are used in order to supply sufficient information about the structural and material requirements without overcrowding the drawings with unnecessary text.

Abbreviations usually consist of a few letters that best describe the word or words being abbreviated. For example, FG is used as an abbreviation for fixed glass, and AFF is used to describe an above-finished floor elevation. Abbreviations use a single capitalized letter or series of letters to represent the word they describe. Lowercase letters are typically used for measurements (e.g., mm for millimetres) except measurements named after a person, such as MPa for megapascal. MPa is typically used in construction to indicate the compressive strength of concrete.

Not all abbreviations need to be memorized; once you gain a better understanding of the materials and processes used in construction, you will generally be able to figure out what the abbreviation means when looking at the drawing. However, note that sometimes an abbreviation can mean two different things; for example, BM can be an abbreviation for a beam on floor plans, and an abbreviation for **benchmark** on site plans. With some experience, you will be able to distinguish the difference by viewing the plans and recognizing what an abbreviation means in a particular drawing.

While most abbreviations are standardized, designers will sometimes improvise and create their own. Therefore, it is advisable to always first check the drawings for a legend where the designer will list the more common abbreviations and symbols used on that particular set of drawings. For an example, check the **construction notes** on the Doncaster drawings that came attached to the text and review the legend listed at the bottom right-hand side of the construction notes. A close-up of the **legend** is illustrated in Figure 4.1.

Table 4.1 is a helpful reference of commonly used abbreviations for construction drawings for housing and small buildings. The drawings may or may not indicate periods after abbreviations, and abbreviations may take more than one meaning depending on the circumstances.

FIGURE 4.1
Legend from Doncaster construction notes.

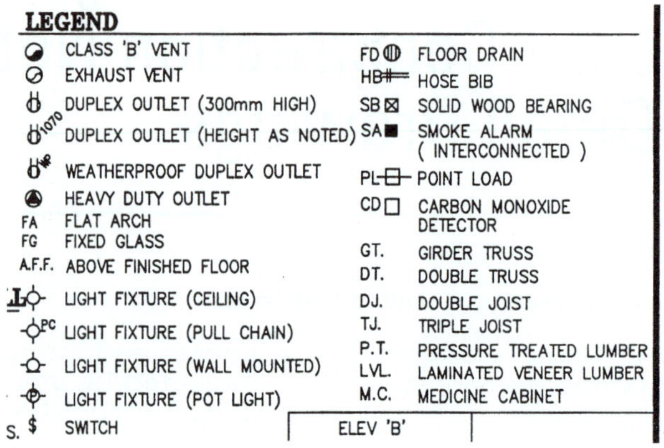

Source: Cassidy & Co. Architectural Technologists

TABLE 4.1
Common Abbreviations

AB	Anchor Bolt
AC or A/C	Air Conditioning
AP	Access Panel
AL. or ALUM	Aluminum
ACST	Acoustic
ACT	Acoustic Tile
ABS	Acrylonitrile Butadiene Styrene (Drain Pipe)
ADJT	Adjustable
ADH	Adhesive
AFF	Above Finished Floor (above subfloor)
AGG	Aggregate
ASPH	Asphalt
APPROX.	Approximate
ARCH	Architectural
@	At
AUTO	Automatic
BA	Bathroom
BD	Board
BS	Backset, or Backsight
BLDG.	Building
BT	Bathtub
BRG	Bearing
BPL or B.PL	Bearing Plate
BR	Bedroom
BLK or BLKG.	Block or Blocking
BLR	Boiler
BRS	Brass
BRDG	Bridging
BRZ	Bronze
BC or BRM	Broom Closet
BOT	Bottom
BSMT	Basement
BF	Basement Floor Elevation
BL	Building Line
BVL	Bevelled
CAB.	Cabinet
CI	Cast Iron
CL	Centre Line
CLG.	Ceiling
CLKG	Caulking
CNTR or CTR.	Centre
CMU.	Concrete Masonry Unit
CAP.	Capacity
CANT	Cantilever
CSMT.	Casement Window
CB	Catch Basin
CER	Ceramic
CO	Cleanout
CA	Cold Air
CD	Carbon Monoxide Detector
CW	Cold Water
CL, C, or CLOS.	Closet
COL	Column
COMB.	Combination
COM	Common
C/W	Compete with
COMP	Composition
CONC.	Concrete
COND	Conduit
CONST	Construction
CONT	Continuous
CONTR	Contract or Contractor
CORRUG.	Corrugated
CU	Copper
CJT	Control Joint
DBL.	Double
DC	Drip Cap
DET.	Detail
DIA	Diameter
DIAG	Diagonal
DT	Double Truss
DJ	Double Joist
DN	Down
DIM	Dimension or Dimmer
DR	Dining Room or Door
DW	Dishwasher
DA	Double acting (door)
DH or DHW	Double Hung Window
DS	Downspout
DP	Dampproof

TABLE 4.1 ***(continued)***

DR	Door
DWLS	Dowels
D or DR	Drain
DWV	Drain Waste Vent System
DWG	Drawing
D	Dryer
DO	Ditto (repeats, same as)
DW	Drywall
DUP	Duplicate
E	East
EA	Each
EE	Each End
EF	Each Face
EL or ELEV	Elevation
ES	Each Side
EW	Each Way
ELEC	Electrical or Electric
EMT	Electric Metallic Tubing
ENT	Entrance
EP	Electrical Panel
EX or EXC	Excavate
EXP JT	Expansion Joint
EXST	Existing
EXT	Extension or Exterior
F	Fridge
F to F	Face to Face
FF	Finished Floor Elevation
FAM RM	Family Room
FBRK	Firebrick
FP or FPL	Fireplace
FD	Floor Drain
FIN	Finish or Finished
FLG or FL	Flashing, Flooring
FLR	Flooring
FT	Foot or Feet
FTG.	Footing
FDN. or FOUND.	Foundation
FLUR or FLUOR	Fluorescent
FL	Flush
FR	Frame
F.R.R.	Fire Resistance Rating
FRWK	Frame work
FRM	Framing
FUT	Future
GAR	Garage
G	Gas
GV or GALV	Galvanized
GI	Galvanized Iron
GL	Grade Line, Glass
GLB or GL BL	Glass Block
GS	Galvanized Steel
GFI or GFCI	Ground Fault Circuit Interrupter
GYP BD	Gypsum Board
G1S	Good One Side
G2S or GS	Good Two Sides or Good Solid
GT	Girder Truss
HB	Hose Bib
HDB	Hardboard
HDW	Hardware
HDWD	Hardwood
HTR	Heater
HEX	Hexagonal
HA	Hot Air
HVAC	Heating Ventilation and Air Conditioning
HW	Hot Water
HWH	Hot Water Heater
HWT	Hot Water Tank
HT	Height
HPT	High Point
HORZ	Horizontal
ID	Inside Diameter
IF	Inside Face
IN	Inch or Inches
INSUL or INS	Insulation
INT	Interior
I	Iron
JB	Jamb
JT	Joint
JST or JSTS	Joist or Joists
KD	Kiln Dried or Knock Down
KIT	Kitchen
L	Lintel, light, length
LG	Long
LAM	Laminated or Laminate
LDG	Landing
LAU	Laundry Room
LT	Laundry Tub or Laundry Tray
LAV	Lavatory
LIN	Linen Closet
LH	Left Hand
LHR	Left Hand Reverse
L CL	Linen Closet
LR	Living Room
LBR	Lumber
LVL	Laminated Veneer Lumber (Manufactured Beam)
LVR	Louvre
MRB	Marble
MAS	Masonry
MATL	Material
MAX	Maximum
MST	Measurement
MECH	Mechanical
MC	Medicine Cabinet
MEMB	Membrane
MIN	Minimum
MISC	Miscellaneous
MIX	Mixture
MLDG	Moulding
MULL	Mullion
MOD	Modular
MPa	Megapascal
N	North
NOS	Nosing
NIC	Not in Contract
NTS	Not to Scale
NBC	National Building Code (first letter may change to indicate province, e.g., OBC—Ontario Building Code)
OBSC GL	Obscure Glass
OC or O/C	On Centre
OWSJ	Open Web Steel Joist
OSB	Oriented Strand Board
OD	Outside Diameter
OH or OVHD or O	Overheador
OPT	Optional

TABLE 4.1 *(continued)*

OPNG	Opening
PERIM	Perimetre
PL	Plate or Plaster or Point Load
PLMG	Plumbing
PLY or PLY WD	Plywood
PVC	Polyvinyl Chloride (used for drainage pipes)
PRCST	Precast
PR	Powder Room
PFB or PREFAB	Prefabricated
PNT	Painted
PNL	Panel
PREFIN	Prefinished
PSI	Pounds per square inch
PSF	Pounds per square foot
PWF	Preserved Wood Foundation
PC or PS	Pull Chain or Pull Switch
PT	Pressure Treated
RAD	Radius or Radiator
R	Radius, Riser, Right, Range, or R-value
RD	Roof Drain
REBAR	Reinforcing Bar
REINF	Reinforcement
REQD or REQ	Required or Requirement
REF	Refrigerator
REG	Register
REV	Reverse
RGH	Rough
RH	Right Hand
RHR	Right Hand Reverse
RM	Room
RO	Rough Opening
ROW	Right of Way
RD	Round
RWL	Rain Water Leader
S	S-Beam, or Shelf
SA	Smoke Alarm
SAN	Sanitary
SB	Solid Wood Bearing
SCN or SCR	Screen
SDL	Saddle
SECT	Section or Sectional
SEL	Select
SERV	Service
SHLP	Shiplap
SHT or SH	Sheet
SHTG	Sheathing
SHWR	Shower
SIM	Similar
SL	Sliding
S&P	Shelf and Pole
SPF	Spruce, Pine, Fir (grading)
SPEC	Specifications
SP or SPR	Spruce
S&R	Shelf and Rod
SST	Stainless Steel
STL	Steel
STN	Stone
SQ	Square
STR	Storm
STY	Storey
T&G	Tongue and Groove
THK	Thick
TEL	Telephone
TER	Terrazzo
T or THERMO	Thermostat
T	Toilet
T or TR	Tread
TBW	Top of Basement Wall Elevation
TJ	Triple Joist
TYP	Typical
U/S	Underside
UF	Underside of Footing Elevation
UNEXCA or UNEX	Unexcavated
UNFIN	Unfinished
VB	Vapour Barrier
VERT	Vertical
VS	Vent Stack
VIN	Vinyl
VT	Vinyl Tile
W/	Width
W	Width, West, Water, Welded, Wide Flange I-Beam
WAIN or WSCT	Wainscot
WC	Water Closet
WD	Wood
WG or WGL	Wire Glass
WH	Water Heater or Weephole
WIC	Walk-in Closet
WP	Waterproof or Weatherproof
WR	Water Resistant
WWF	Welded Wide Flange
WWM	Welded Wire Mesh
WF	Wide Flange I-Beam
WDW	Window
WI	Wrought Iron
WOB	Walkout basement
WOD	Walkout deck

Source: Reproduced with the permission of Tom Stephenson

HATCHINGS AND SYMBOLS

Hatchings and **symbols** are a simple way to identify materials and products used in the project. Their use and application will become more evident as you read through the text and do the assignments and quizzes.

Hatchings

Standardized hatchings are used to identify materials such as concrete, brick, wood, steel, and insulation. The hatching may differ depending on the type of drawing and may not be shown for the whole object. As well, some designers may vary the appearance of the hatching. Standardized hatchings are shown in Figure 4.3. Refer to them as often as needed.

Symbols

Symbols indicate objects on the drawings and are used extensively by the mechanical and electrical subtrades to indicate plumbing fixtures, ducts, fixtures, lights, switches, outlets, etc. Some common symbols are identified in Figure 4.4. Refer to them as often as needed.

FIGURE 4.2

Review the upper right corner section of floor-plan drawing indicating brick with the diagonal lines, insulation with the curved lines, while the squares on the floor indicate floor tile. Also, note the symbol used for hose bib (lawn hose shut-off connection) with the abbreviation HB, and the waterproof electrical outlet symbol with the abbreviation WP. You can also see electrical outlet and switch symbols, along with the height of outlets, when they are not the usual standard 300 mm, e.g. 1070 mm for over-counter outlets in the kitchen.

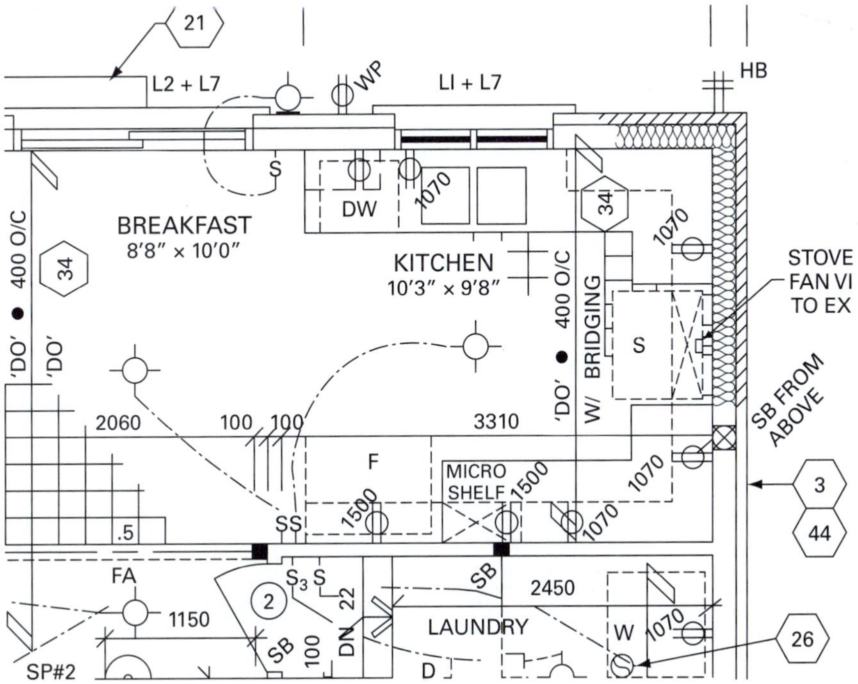

Source: Cassidy & Co. Architectural Technologists

FIGURE 4.3
Hatchings.

	GENERAL LOCATION DRAWINGS SCALE 1:50 OR SMALLER		ASSEMBLY DRAWINGS SCALE 1:10 OR LARGER
	PLAN & SECTION	ELEVATION	PLAN & SECTION
BRICK MASONRY	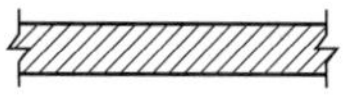	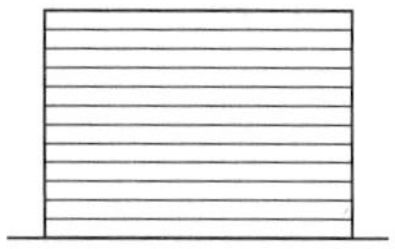 SPACING TO REPRESENT COURSING	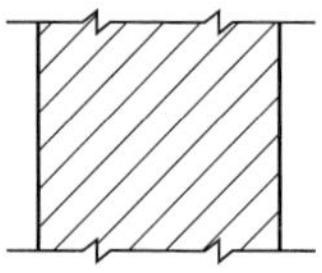
CUT STONE MASONRY	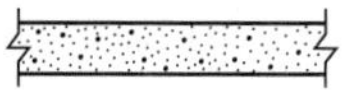	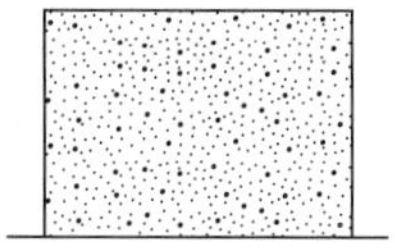	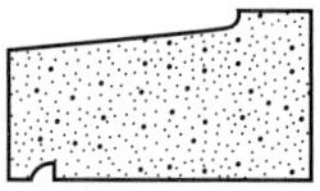
ARTIFICIAL STONE MASONRY	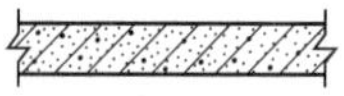 LINE SPACING TWICE AS WIDE AS FOR BRICK	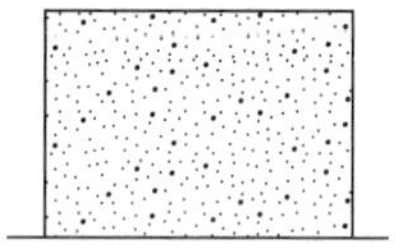 STIPPLE DENSITY ONE-HALF OF THAT FOR CUT STONE	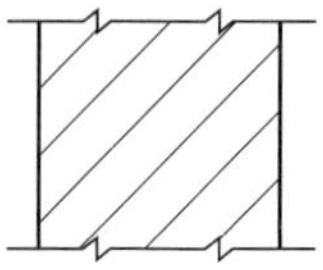
MARBLE	TOO FINE TO HATCH	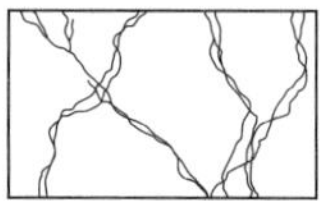	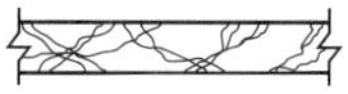
CONCRETE	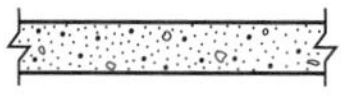	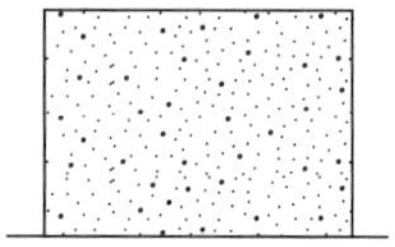 STIPPLE DENSITY ONE-THIRD OF THAT FOR CUT STONE	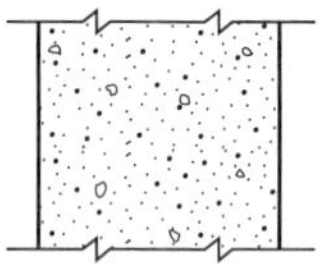
REINFORCED CONCRETE	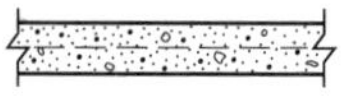	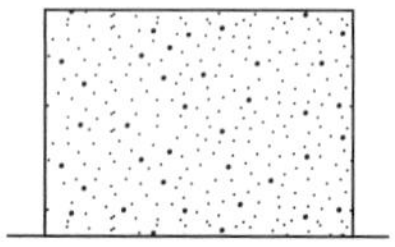 STIPPLE DENSITY ONE-THIRD OF THAT FOR CUT STONE	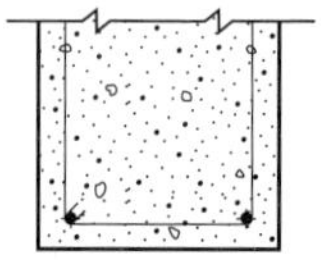
CONCRETE OR CINDER BLOCK	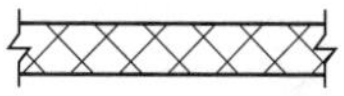 LINE SPACING THREE TIMES AS WIDE AS FOR BRICK	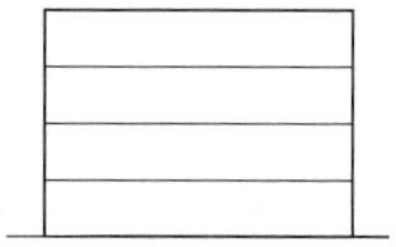 SPACING TO REPRESENT COURSING	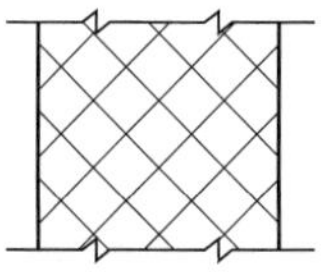
CERAMIC TILE	TOO FINE TO HATCH	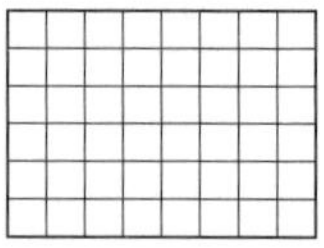	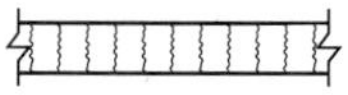

Source: With the permission of Canadian Standards Association (operating as CSA Group), material is reproduced from CSA Group Standard, CAN3 B78.3-M77 entitled "Building Drawings - Basic Engineering" which is copyrighted by CSA Group, 5060 Spectrum Way, Suite 100, Mississauga ON, L4W 5N6. This material is not the complete and official position of CSA Group on the referenced subject, which is represented solely by the standard in its entirety. While use of the material has been authorized, CSA is not responsible for the manner in which the data is presented, nor for any interpretations thereof. For more information or to purchase standards from CSA Group, please visit http://shop.csa.ca/ or call 1-800-463-6727.

FIGURE 4.3 *(continued)*

	GENERAL LOCATION DRAWINGS SCALE 1:50 OR SMALLER		ASSEMBLY DRAWINGS SCALE 1:10 OR LARGER
	PLAN & SECTION	ELEVATION	PLAN & SECTION
TERRAZZO	TOO FINE TO HATCH	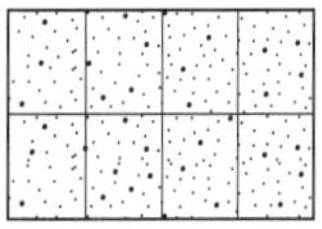 SHOW PATTERN OF STRIPS	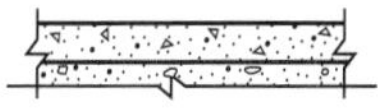
GLASS	—	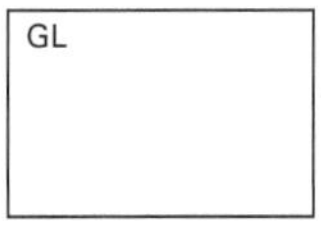 INCLUDE GL WHERE REQUIRED FOR CLARITY	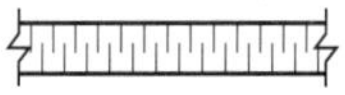
WOOD FRAMING	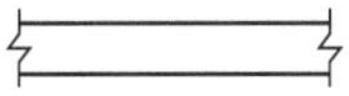 FOR NEW WORK 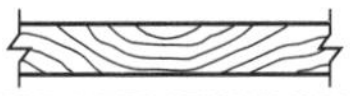 FOR ALTERATION WORK	—	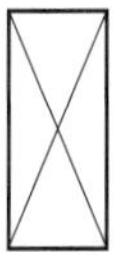
WOOD (FINISHED)	TOO FINE TO HATCH	—	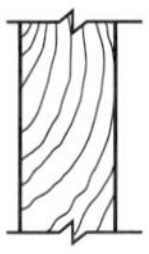
RIGID INSULATION	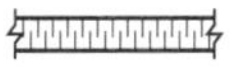	—	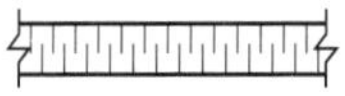
BATT INSULATION	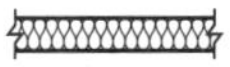	—	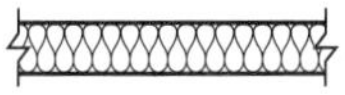
STRUCTURAL STEEL	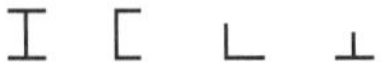	—	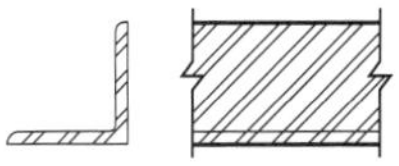
BRONZE, BRASS, COPPER, & ASSOCIATED ALLOYS	AS FOR STRUCTURAL STEEL	—	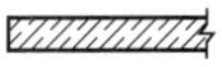
ALUMINUM	AS FOR STRUCTURAL STEEL	—	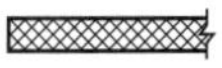

FIGURE 4.3 *(continued)*

	GENERAL LOCATION DRAWINGS SCALE 1:50 OR SMALLER		ASSEMBLY DRAWINGS SCALE 1:10 OR LARGER
	PLAN & SECTION	ELEVATION	PLAN & SECTION
EARTH	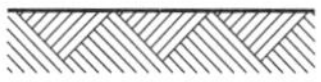	—	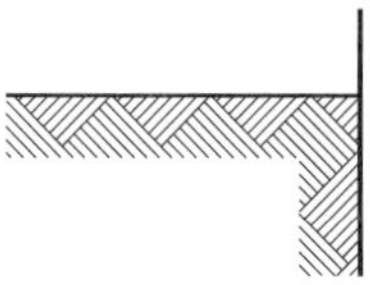
ROCK	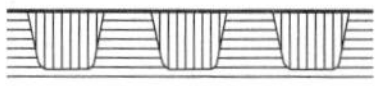	—	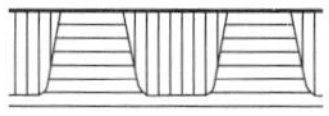
GRAVEL FILL	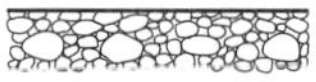	—	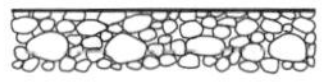
CINDER OR SLAG FILL	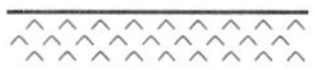	—	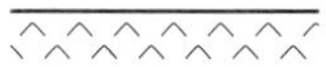
FIRE BRICK	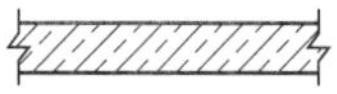	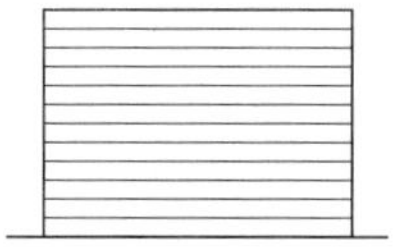	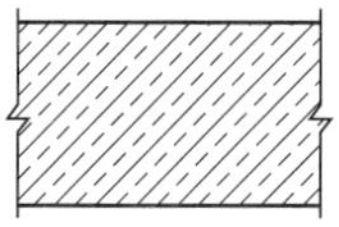
PLYWOOD		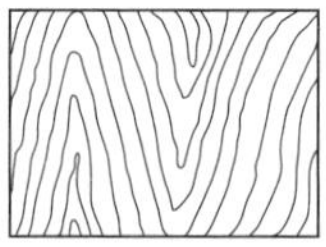	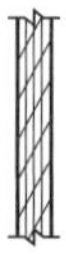
GLASS BLOCK	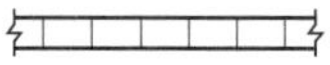	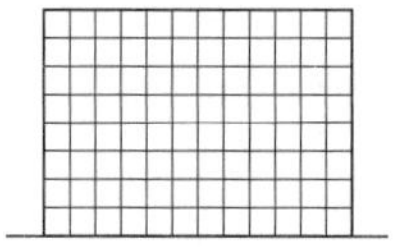	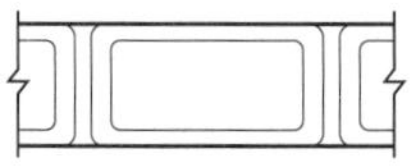
NATURAL STONE – RIP, RAP, FIELD STONE, ETC.	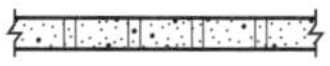	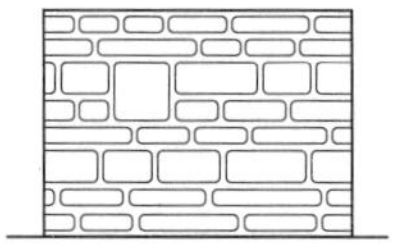	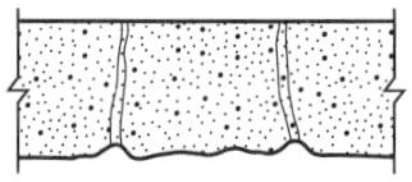
SAND, FILL, PLASTER, GYPSUM BOARD, & CEMENT	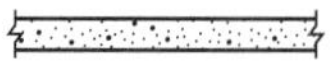	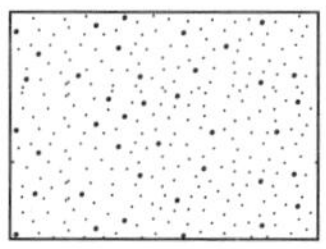	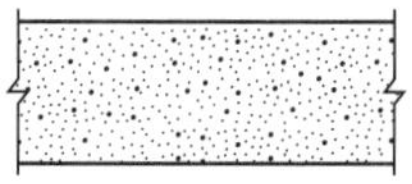

FIGURE 4.4
Symbols.

CEILING	WALL	
○	-○	OUTLET
Ⓑ	-Ⓑ	BLANKED OUTLET
Ⓓ		DROP CORD
Ⓔ	-Ⓔ	ELECTRIC OUTLET IF CONFUSION WITH OTHER SYMBOLS POSSIBLE
Ⓕ	-Ⓕ	FAN OUTLET
Ⓙ	-Ⓙ	JUNCTION BOX
Ⓛ	-Ⓛ	LAMPHOLDER
Ⓛ$_{PS}$	-Ⓛ$_{PS}$	LAMPHOLDER WITH PULL SWITCH
Ⓢ	-Ⓢ	PULL SWITCH
Ⓥ	-Ⓥ	OUTLET FOR VAPOUR DISCHARGE LAMP
Ⓧ	-Ⓧ	EXIT LIGHT OUTLET
Ⓒ	-Ⓒ	CLOCK OUTLET
		DUPLEX RECEPTACLE
	$_{1, 3}$	SINGLE, TRIPLEX RECEPTACLE
		SPLIT-SWITCHED-DUPLEX RECEPTACLE
		THREE-CONDUCTOR SPLIT-DUPLEX RECEPTACLE
		THREE-CONDUCTOR SPLIT-SWITCHED-DUPLEX RECEPTACLE
	$_{WP}$	WEATHERPROOF RECEPTACLE
	$_{R}$	RANGE RECEPTACLE
	$_{S}$	SWITCH & RECEPTACLE
	$_{R}$	RADIO & RECEPTACLE
		SPECIAL-PURPOSE OUTLET, UNDESIGNATED
		FLOOR RECEPTACLE
○$_{a, b, c, etc.}$		DESIGNATED SPECIAL-PURPOSE OUTLET
	$_{a, b, c, etc.}$	DESIGNATED SPECIAL-PURPOSE RECEPTACLE
		FLUORESCENT FIXTURE

S	SINGLE POLE SWITCH
S_2	DOUBLE POLE SWITCH
S_3	THREE-WAY SWITCH
S_4	FOUR-WAY SWITCH
S_D	AUTOMATIC DOOR SWITCH
S_E	ELECTROLIER SWITCH
S_K	KEY-OPERATED SWITCH
S_P	SWITCH & PILOT LAMP
S_{CB}	CIRCUIT BREAKER
S_{WCB}	WEATHERPROOF CIRCUIT BREAKER
S_{MC}	MOMENTARY CONTACT SWITCH
S_{RC}	REMOTE-CONTROL SWITCH
S_{WP}	WEATHERPROOF SWITCH
S_F	FUSED SWITCH
S_{WF}	WEATHERPROOF FUSED SWITCH
$S_{a, b, c, etc.}$	SPECIAL SWITCH OR CIRCUIT BREAKER
	LIGHTING PANEL
	POWER PANEL
———	BRANCH CIRCUIT IN CEILING OR WALL
—·—	BRANCH CIRCUIT IN FLOOR
- - - -	EXPOSED BRANCH CIRCUIT
	HOME RUN TO PANELBOARD NUMBER OF CIRCUITS INDICATED BY NUMBER OF ARROWS
	UNDERFLOOR DUCT & JUNCTION BOX
Ⓖ	GENERATOR
Ⓜ	MOTOR
Ⓘ	INSTRUMENT
Ⓣ	POWER TRANSFORMER
	CONTROLLER
	ISOLATING SWITCH

Source: With the permission of Canadian Standards Association (operating as CSA Group), material is reproduced from CSA Group Standard, CAN3 B78.3-M77 entitled "Building Drawings - Basic Engineering" which is copyrighted by CSA Group, 5060 Spectrum Way, Suite 100, Mississauga ON, L4W 5N6. This material is not the complete and official position of CSA Group on the referenced subject, which is represented solely by the standard in its entirety. While use of the material has been authorized, CSA is not responsible for the manner in which the data is presented, nor for any interpretations thereof. For more information or to purchase standards from CSA Group, please visit http://shop.csa.ca/ or call 1-800-463-6727.

FIGURE 4.4 *(continued)*

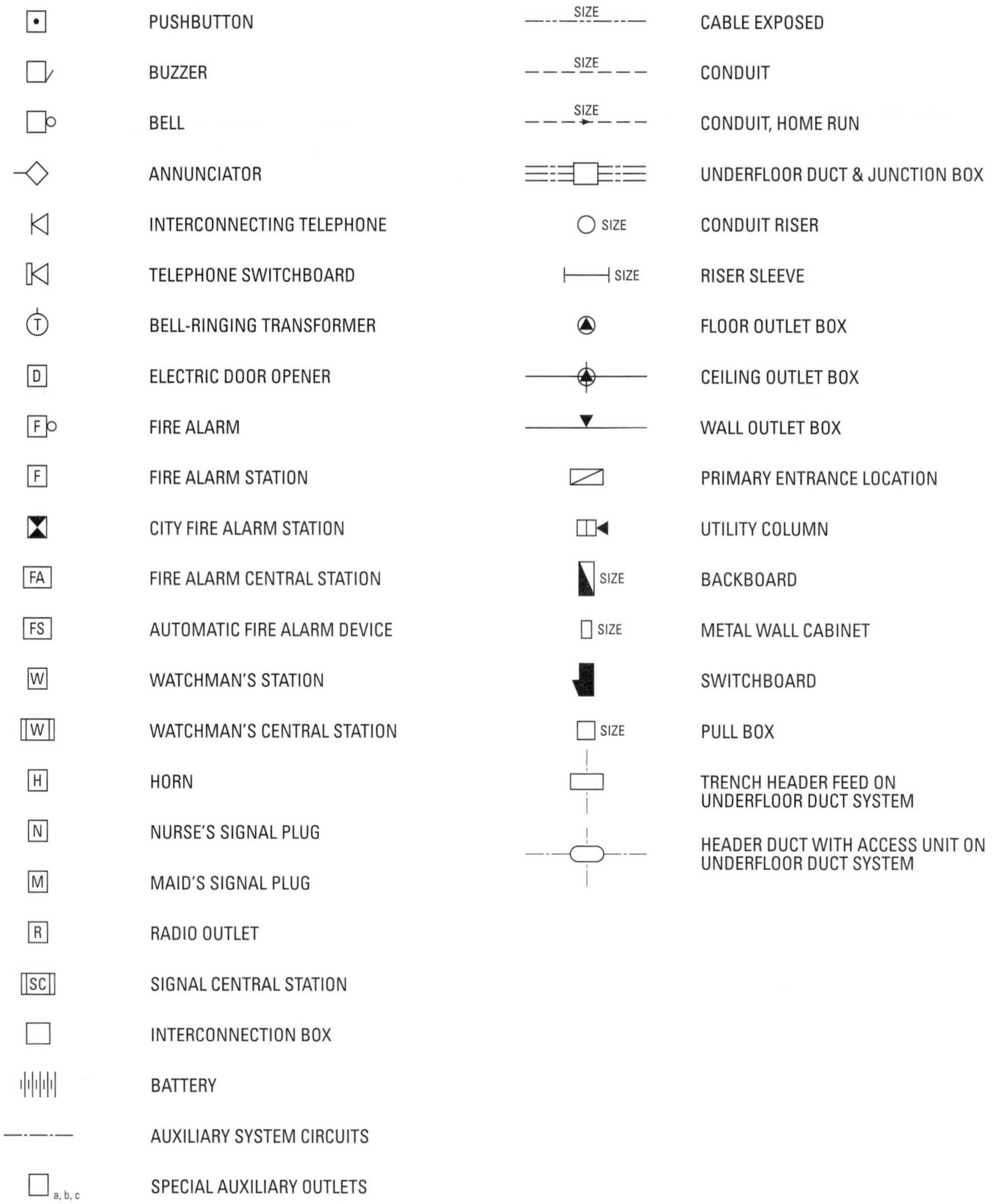

FIGURE 4.4 *(continued)*

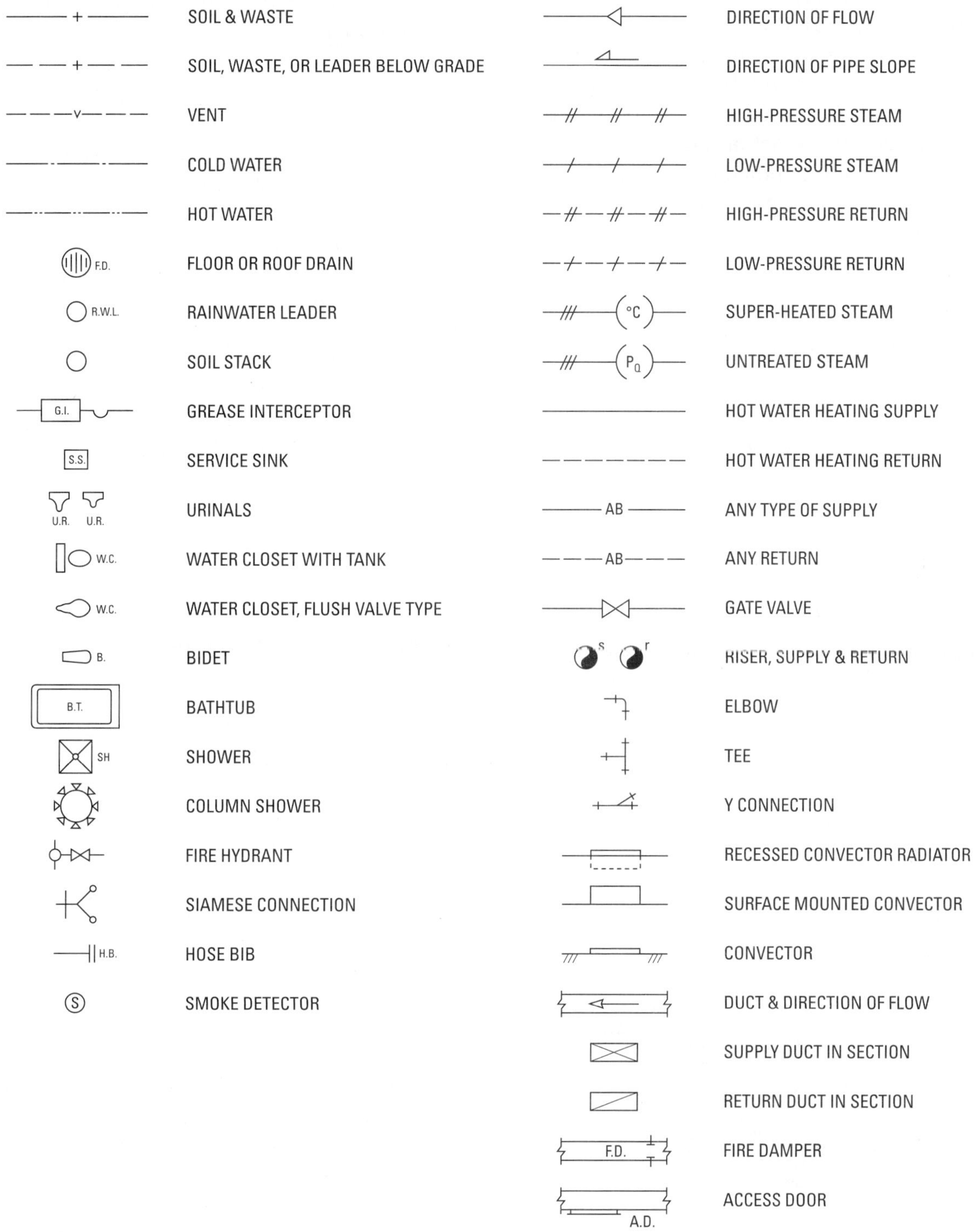

CONSTRUCTION NOTES

The notes are usually found on the first or last page of the drawings. They denote more specific information about the structural requirements for the materials used and their placement in the project. This information would be too cumbersome to place on the actual drawings; instead, numbered reference indicators are shown on the drawings at the most descriptive location. A **hexagon and reference number** is commonly used for this purpose.

FIGURE 4.5
Hexagon and reference number.

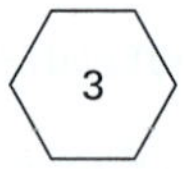

If the reference symbol cannot be placed close to the object it is representing, a leader line may be used to indicate the area or object. When interpreting the drawings, readers can view the symbol on the drawings and then refer to the appropriate number on the construction notes. See Figure 4.6 for an example.

FIGURE 4.6
Close-up of floor plan indicating hex figure with leader line taken from Doncaster first-floor plan kitchen area.

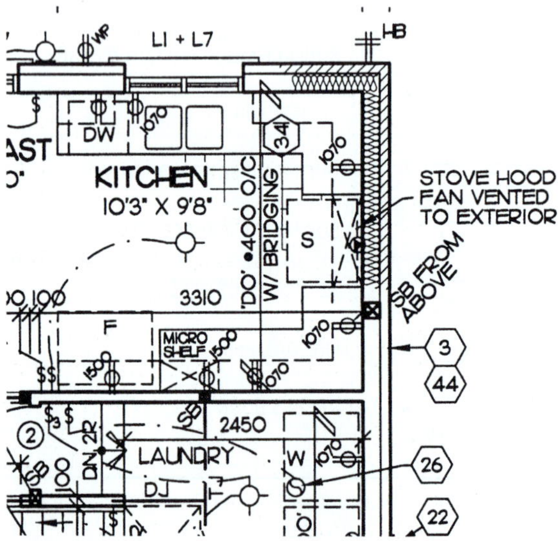

Source: Cassidy & Co. Architectural Technologists

In Figure 4.6, readers would note the reference number 3 and the reference number 44 inside the hexagon symbol and refer to the construction notes under numbers 3 and 44, which state:

3 **BRICK VENEER CONSTRUCTION (2 × 6 Studs)**

90 mm (4") OR 75 mm (3") FACE BRICK, 25 mm (1" AIR SPACE, 22 × 180 × 0.76 mm (7/8" × 7" × 22 ga)
METAL TIES @ 400 mm (16") O.C. HORIZ.
600 mm (24") O.C. VERTICAL, 0.7 kg/m² (No. 15) BUILDING PAPER, 12.7 MM (1/2") EXTERIOR TYPE SHEATHING
(UNLESS OTHERWISE SPECIFIED), 38 × 140 (2 × 6) STUDS @ 400 mm (16") O.C., RSI 3.25 (R19) BATT INSULATION AND 6 Mil AIR/VAPOUR BARRIER, 12.7 mm (1/2") INT. DRYWALL FINISH. PROVIDE WEEP HOLES @ 800 mm (30") O.C. HORIZ. @ BOTTOM COURSE ONLY & OVER OPENINGS, PROVIDE BASE FLASHING UP MIN. 150 mm (6") BEHIND SHEATHING PAPER.

44 **EXTERIOR WALL LESS THAN 1.2 M TO PROPERTY LINE (45 MIN. F.R.R.)**

BRICK VENEER WALL OR FRAME WALL CONSTRUCTION
CONSTRUCTION OF WALLS AS PER NOTES 2, 3, 40, 42 EXCEPT AS PER THE FOLLOWING NOTES.
INSULATION WITH A MASS OF 0.032 kg/m² PER 1 mm OF THICKNESS AND 12.7 mm (1/2") TYPE 'X' INT. DRYWALL FINISH.

The above references provide all the detailed information regarding the wall, listing the wall components beginning from the outside face, and naming the components in order from outside to inside face in reference 3.

Recall in Chapter 3 that production builders use the same drawings a number of times, which results in varying directional locations and setbacks for the buildings. Reference 44 provides important information regarding fire-resistance ratings and requires changes to the wall specification when the wall is located closer than 1.2 m to the property line (part of limiting distance requirements, refer to Division B Part 3.2.3.1 of the NBC). The wall will look similar to Figure 4.7 when viewed in section. Notice the hatchings used to indicate the face brick, building paper, sheathing, insulation, vapour/air barrier, and drywall. Also notice the air space is shown in this section view but not shown in the plan view in Figure 4.6. This is common practice due to the

FIGURE 4.7

Example of hatchings used for a brick veneer wall section; these help the reader to quickly and graphically visualize information from the drawings.

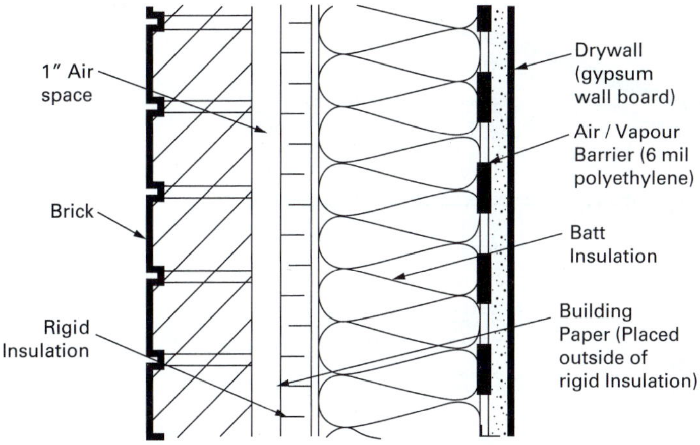

Source: Toronto Area Chief Building Officials Committee (TACBOC)

small scale of the floor-plan drawings and not being able to effectively display a small space.

DETAIL DRAWINGS

Detail drawings display more detailed information about the object they represent. They are typically, but not always, drawn at a larger scale than floor plans and elevation drawings. The following is a brief list of types of detail drawings you may encounter.

1. Section details include:
   a) Wall sections (Figures 4.8 and 4.9)
      - A wall section is a section view cut through the wall in order to display the materials used, and their placement within the wall. There is usually a cutting-plane line drawn on the floor plan that indicates where the section of the wall is being viewed from and in which direction the view is facing.
      - Figure 4.8 illustrates a wall section of a wood frame house, and Figure 4.9 illustrates a wall section of a wood frame house with brick veneer cladding. While some of the materials and finishes used may vary, these sections illustrate the materials and their placement for the two most common methods of constructing wood frame houses in Canada.
      - In the examples viewed in Figures 4.8 and 4.9, you can determine that this is a wall section of a single-storey house with a full basement. The basement in Figure 4.8 is insulated from floor to ceiling, and the foundation consists of a poured concrete footing with blocks being used for the foundation wall. The wall consists of 2 × 6 construction with siding used as the exterior finish. The roof is constructed with rafters and ceiling joists. Figure 4.9 is similar except that it uses a brick veneer for the exterior finish.
      - From Figures 4.8 and 4.9, note the hatchings that are used for the:
      - Insulation, both rigid and batt types
      - Block
      - Concrete footing
      - Wood plates
      - Fascia
      - Sub-floor
      - Gravel (crushed stone)
      - Grade
      - Brick veneer
      - Continuous air and vapour barrier
   b) Insulated sheathing section and heated crawlspace section (Figure 4.10)
      - The section drawings in Figures 4.8, 4.9, and 4.10 provide information about a framing method that uses insulated sheathing, usually extruded polystyrene (XPS) placed outside of the studs, to reduce thermal bridging and thus increase the energy efficiency of the wall. In some cases the thickness of both the stud wall and polystyrene are increased to maximize the insulating qualities of the wall. Refer to Chapter 12 for more detail.
      - The section at the bottom of Figure 4.10 provides information regarding the construction of an addition to a house that does not have a basement but has a heated crawlspace instead. Compare this with Figures 4.8 and 4.9 that display full basements.
   c) Porch and stair section (Figure 4.12)
      - The section used in Figure 4.12 displays cross-sectional views of an exterior solid masonry porch (brick and block construction).

FIGURE 4.8

Wood frame wall section detail.

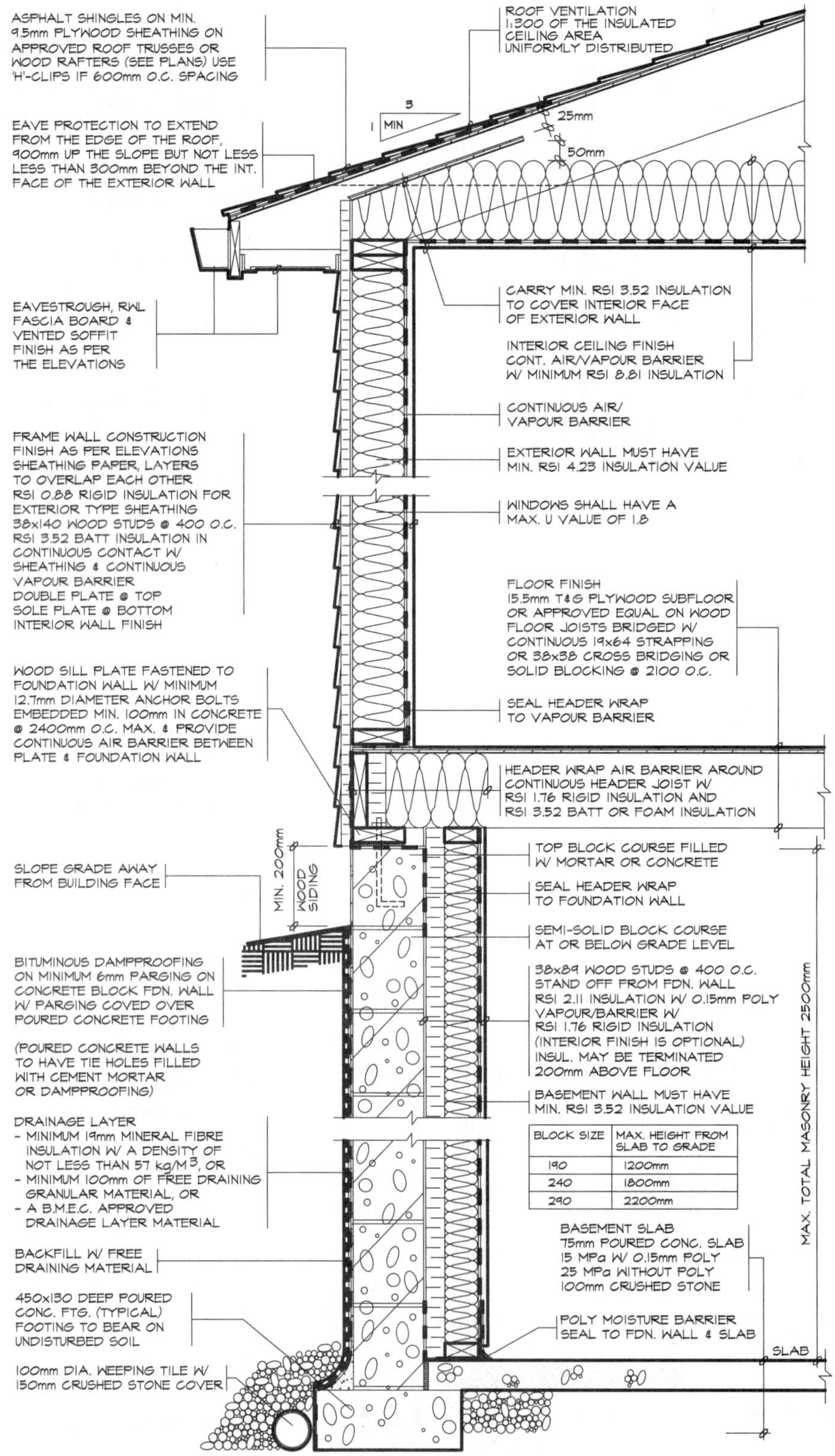

Source: Toronto Area Chief Building Officials Committee (TACBOC)

FIGURE 4.9

Wood frame wall section with non-structural brick veneer cladding. Review detail carefully, reviewing all the notes and observing the various hatchings and abbreviations.

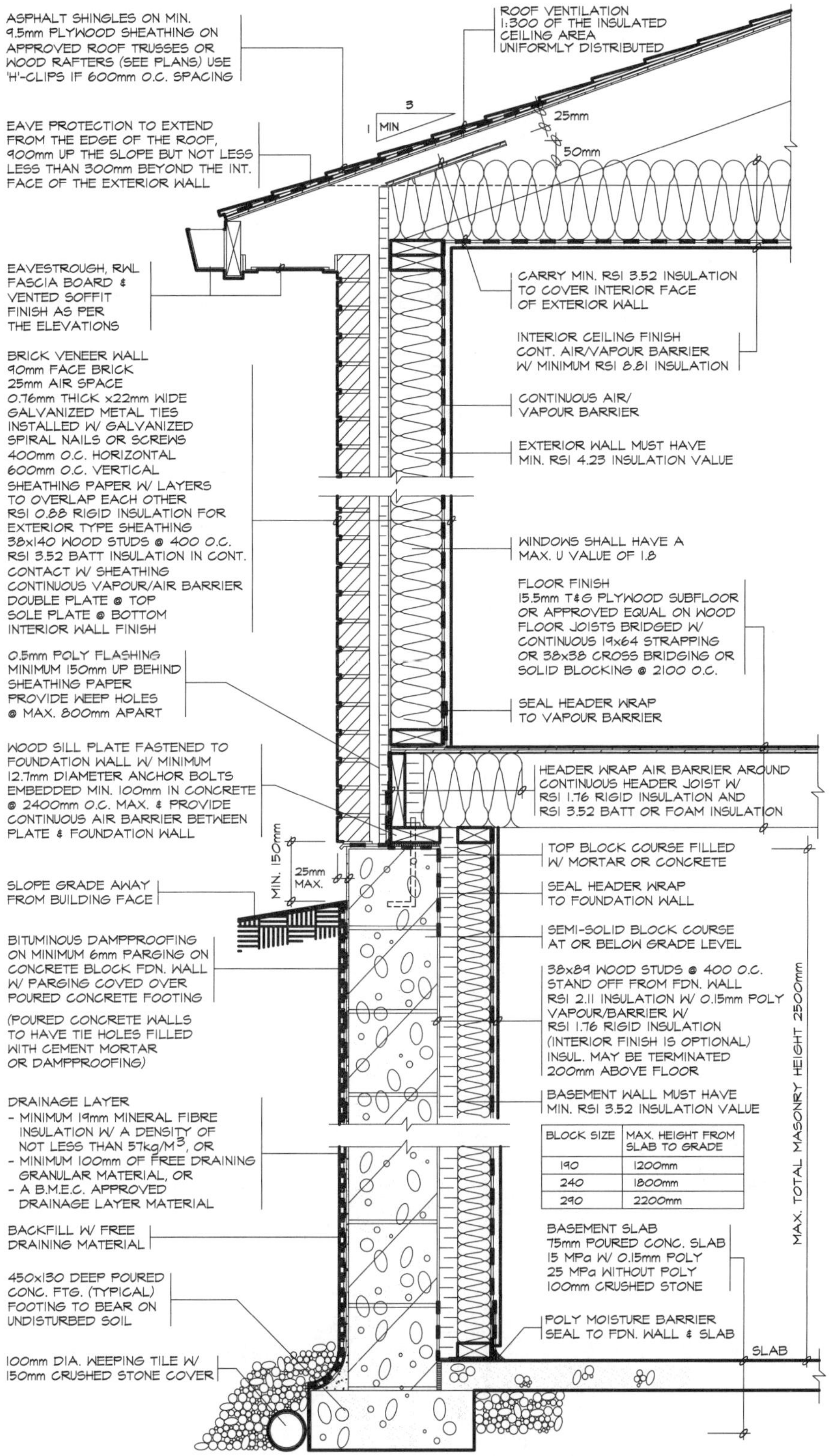

Source: Toronto Area Chief Building Officials Committee (TACBOC)

FIGURE 4.10

Top of page section detail indicates a wood frame wall with insulated sheathing and no basement with a unheated crawl space. (Note: Green Building practices may increase sheathing thickness and/or stud width to increase energy efficiency of wall. Refer to Chapters 7 and 12 for more detail.) Bottom of page section detail indicates a heated crawlspace (no basement). Compare and contrast the difference between the two details. Once you learn more about building science you will see that there can be a number of advantages to having a heatedcrawl space (conditioned space) rather than an unheated crawlspace.

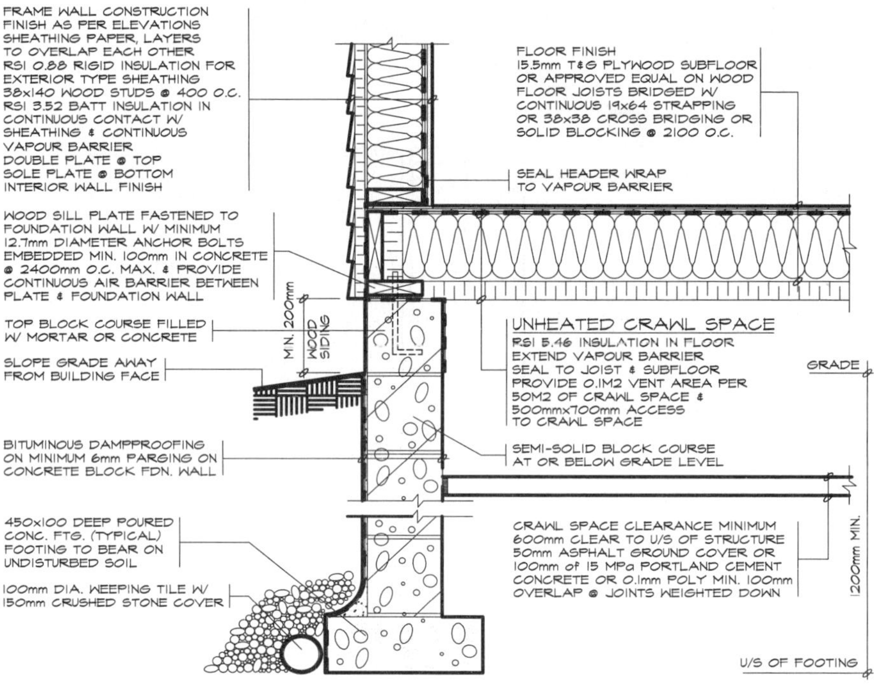

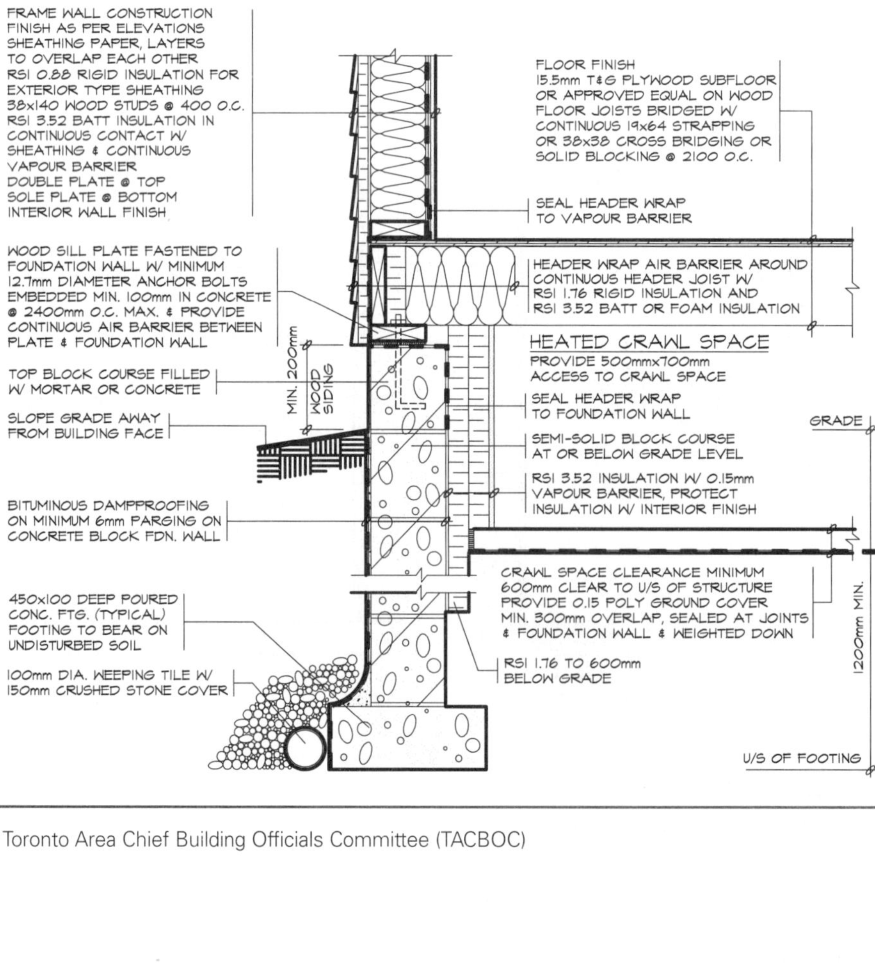

Source: Courtesy of Toronto Area Chief Building Officials Committee (TACBOC)

FIGURE 4.11

Slab on grade section detail for both brick veneer and frame buildings. This type of detail is used for residential buildings that don't have basements; similar details can also be used for basement walkouts.

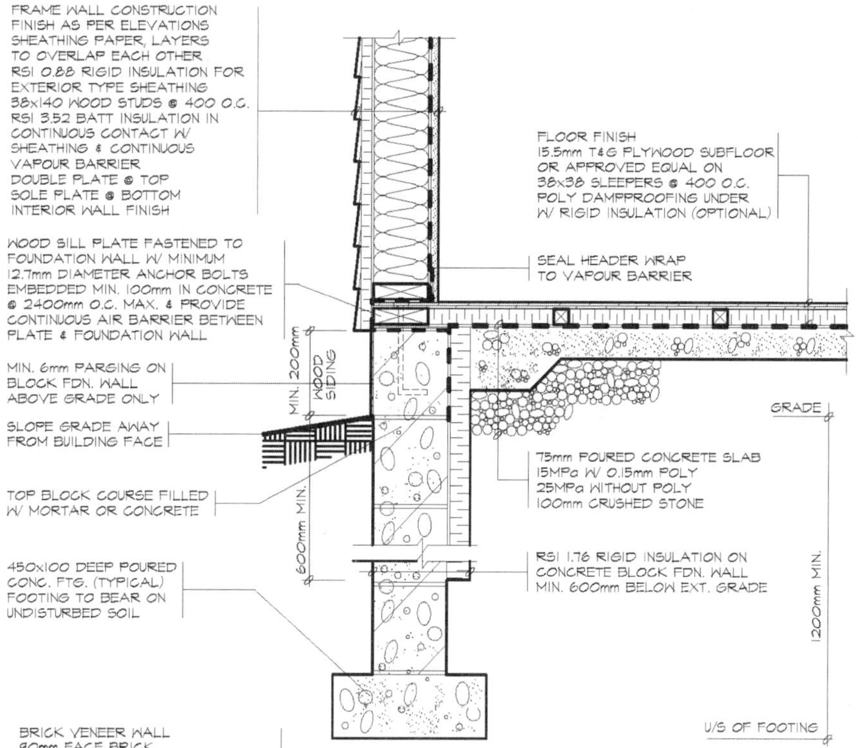

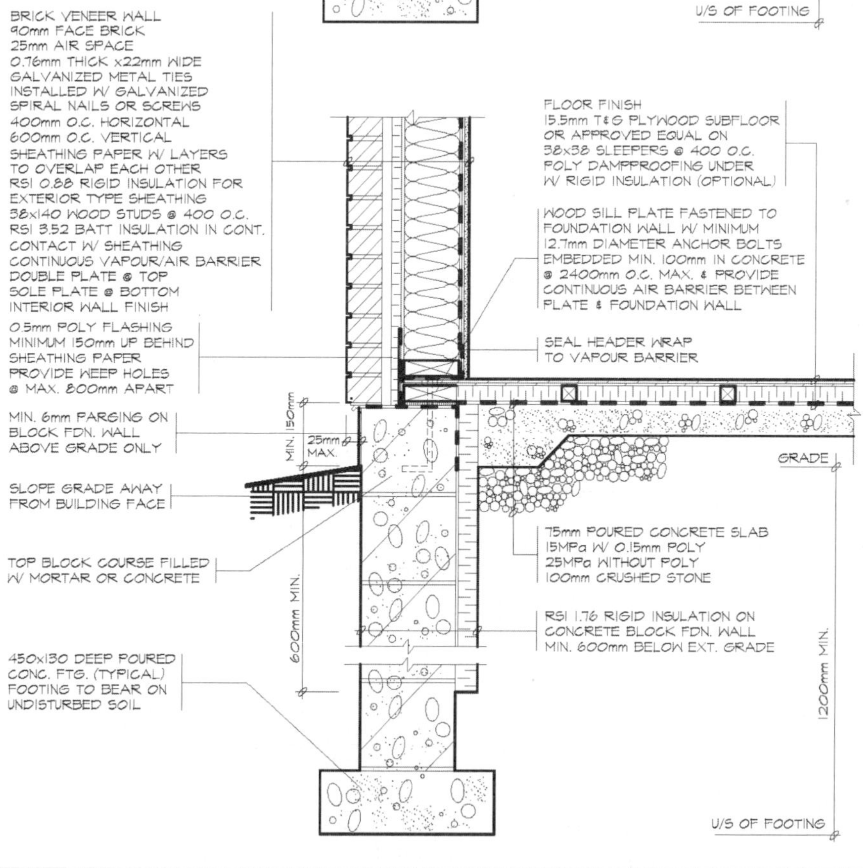

Source: Toronto Area Chief Building Officials Committee (TACBOC)

FIGURE 4.12

Concrete porch and cold cellar plans detail sections and notes. Review carefully the use of cutting-plane lines and detail designation references A and B, their direction, and how the section details display the information. Don't forget that the detail sections are based on the direction of the point of the detail designation.

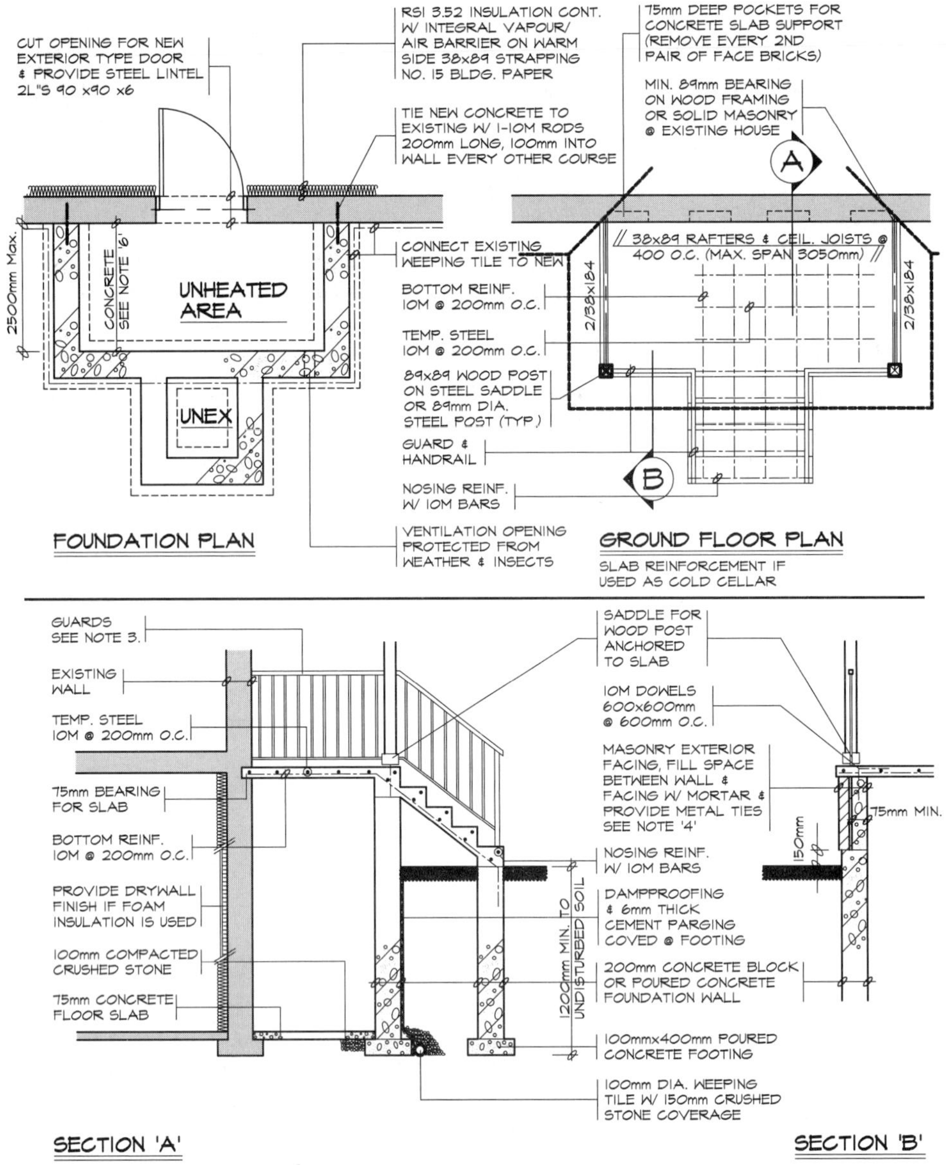

GENERAL NOTES

1. EXTERIOR STAIRS

   125mm - 200mm RISE
   210mm - 355mm RUN
   235mm - 355mm TREAD
   STEPS ARE TO BE UNIFORM THROUGHOUT FLIGHT

2. HANDRAILS

   ARE REQUIRED WHERE STEPS HAVE MORE THAN 3 RISERS. HANDRAIL HEIGHT 800mm - 965mm

3. GUARDS

   ARE REQUIRED AROUND CONCRETE SLAB IF MORE THAN 600mm ABOVE GRADE & ON BOTH SIDES OF STAIRS
   MINIMUM 900mm HIGH FOR STAIRS
   MINIMUM 900mm HIGH FOR PORCHES UP TO 1800mm ABOVE GRADE.
   MINIMUM 1070mm HIGH FOR GREATER HTS.
   MAXIMUM 100mm BETWEEN PICKETS AND NO MEMBER DESIGNED TO FACILITATE CLIMBING BETWEEN 140mm & 900mm

4. MASONRY TIES

   WHEN BRICK FACING IS USED ABOVE GROUND LEVEL, PROVIDE 0.76mm THICK & 22mm WIDE CORROSION RESISTANT METAL TIES @ 600mm HORIZ. & 500mm VERTICAL

5. FOUNDATION WALLS

   THICKNESS OF UNREINFORCED FOUNDATION WALLS LATERALLY SUPPORTED AT THE TOP ARE DEPENDANT UPON HEIGHT OF FINISH GRADE ABOVE BASEMENT FLOOR

   UNIT MASONRY THICKNESS 190mm - MAX. HEIGHT 1200mm
   UNIT MASONRY THICKNESS 240mm - MAX. HEIGHT 1800mm
   UNIT MASONRY THICKNESS 290mm - MAX. HEIGHT 2200mm

6. CONCRETE

   MINIMUM CONCRETE STRENGTH SHALL BE 32Mpa W/ 5%-8% AIR ENTRAINMENT
   CONCRETE SLAB THICKNESS 125mm
   PROVIDE MIN. 30mm CLEAR CONCRETE COVER TO REINFORCING BARS

Source: Toronto Area Chief Building Officials Committee (TACBOC)

- The floor plan on top indicates two cutting-plane lines A and B. Section A, below, is the detailed section of cutting-plane A shown above; notice the direction that the cutting-plane line is pointing in, and how it matches Section A shown below.
- Section A displays a cross-sectional view taken right through the top slab and stairs of the porch.
- Section B indicates the composition of the walls of the porch.
- The reinforcing bars are displayed as a grid in the plan view and as dark circles and dark lines in the section view. (Steel that is placed in concrete strengthens the concrete and is often called reinforcing steel or reinforcing bars. Concrete that has reinforcing in it is called reinforced concrete.) More information regarding the placement and coverage of the reinforcing bars can be found in the general notes at the bottom of the drawing page.

d) Whole building section
   - Figure 4.13 displays a cross-sectional view of the Doncaster house. The floor plans found at the back of the book have cutting-plane lines that indicate where the section is taken and in which direction the section is facing. The designer has made extensive use of reference notes to indicate that more information can be found by referring to these numbers on the construction notes.

2. Detailed sections may be drawn for a variety of structural elements such as beams, lintels, and roof intersections. Figure 4.14 shows three section details indicating how a solid masonry garage addition can tie into an existing two-storey house.
   - Section view 1 displays the intersection into a wood frame house. The roof is fastened to the house using a ledger nailed to the existing stud wall.
   - Section view 2 displays the intersection into a brick veneer house. This detail

FIGURE 4.13
Doncaster house section.

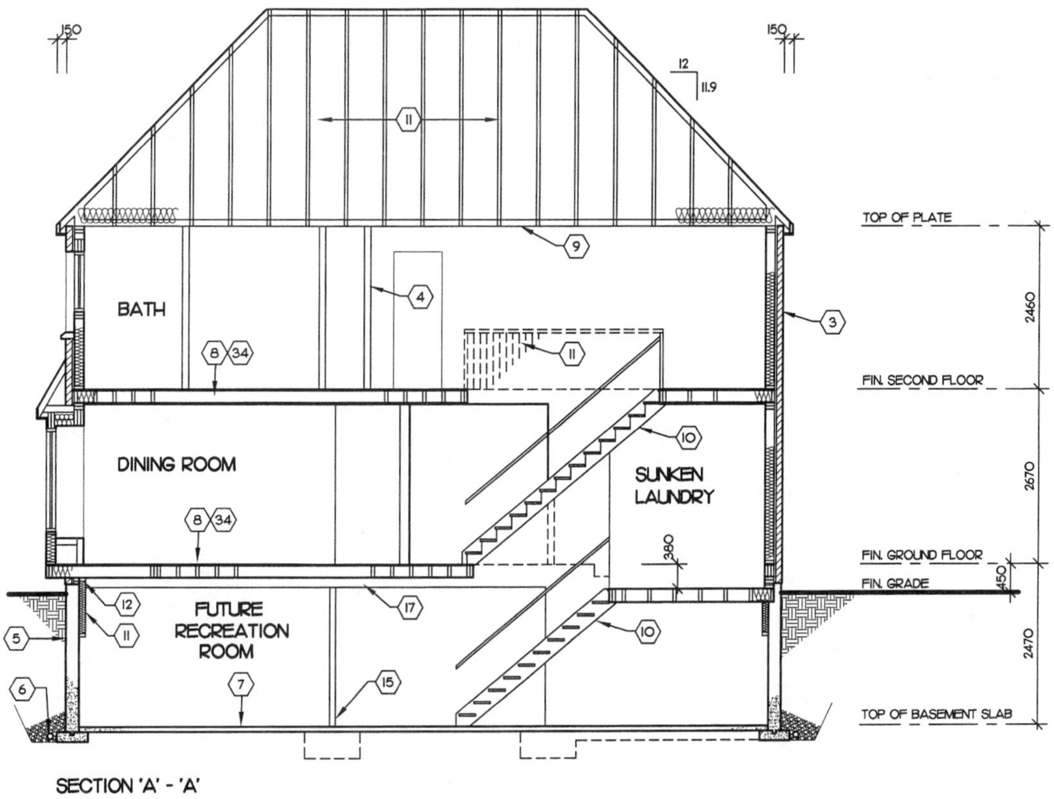

Source: Cassidy & Co. Architectural Technologists

FIGURE 4.14

Solid masonry outer wall for garage addition with three examples of attaching roof to an existing house depending on house construction (frame, brick veneer, or solid masonry; solid masonry refers to a wall that is built from both brick and block and is carrying a structural load).

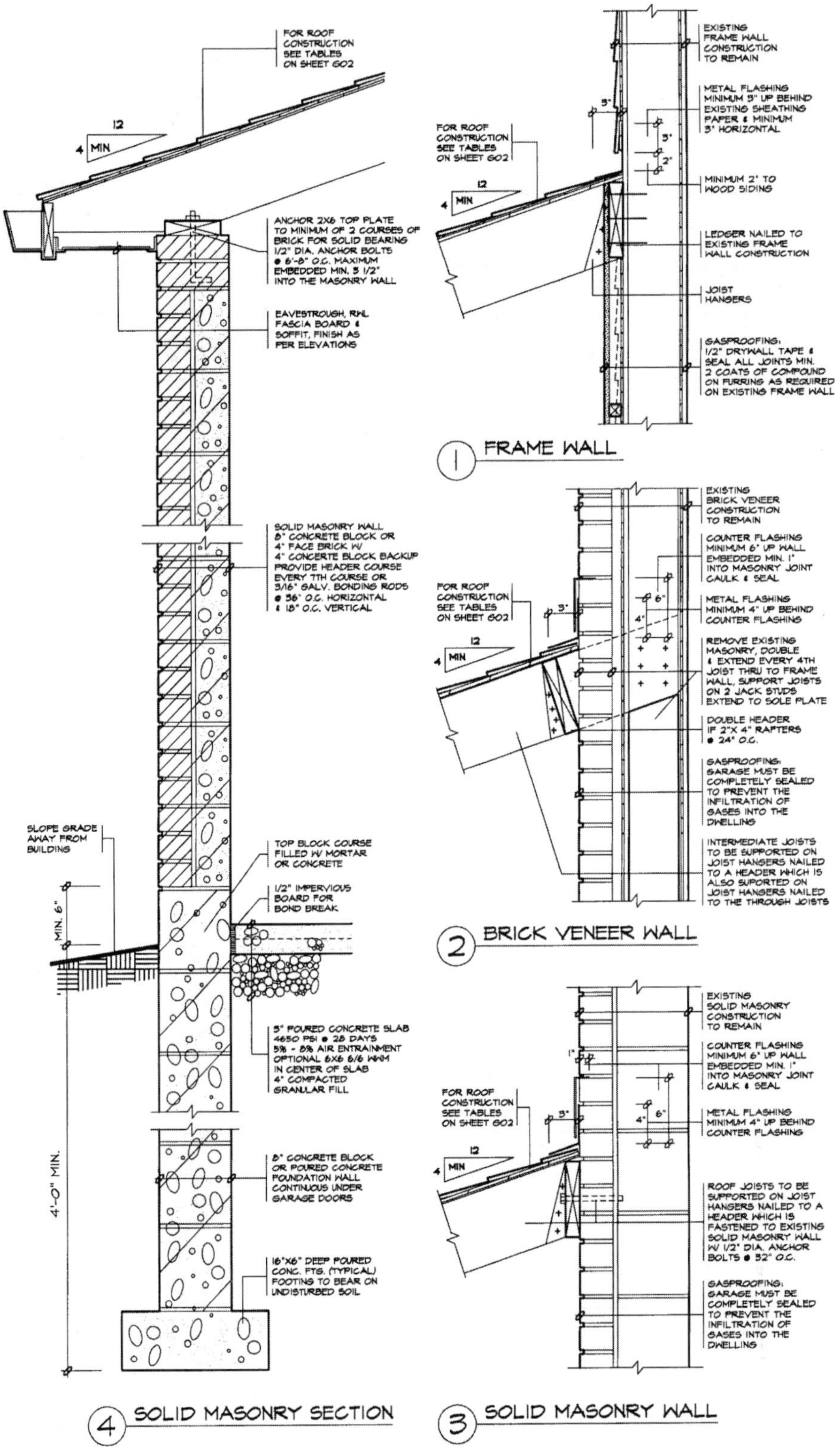

Source: Toronto Area Chief Building Officials Committee (TACBOC)

requires a close look because the brick on a brick veneer house is **non-load-bearing.** Non-load-bearing means that the brick is designed to support only its own weight, not the weight of other parts of the structure, such as the roof. In this case, the rafters are cut through the veneer and tied into the existing wall at 4′ intervals. The extended rafters are nailed to the existing studs and supported on jack studs. The rest of the rafters run between the extended rafters, supported by headers that run between the extended rafters. The above terminology will become more clear to you once you study Chapters 7, 8, and 9 on floor, wall, and roof framing.

- Section view 3 displays the intersection of the roof into a solid masonry house. A solid masonry house is generally built from either solid brick (usually homes/garages that were built 60 or more years ago) (2 courses or **wythes** in width), or currently a combination of brick and block as indicated in Section view 3 (brick on the outside with block used as a backup). These walls are considered load-bearing, so in this case the roof can be fastened directly to the wall using a wood ledger fastened by lag bolts and shields.
- Section view 4 illustrates the construction of the exterior wall of the garage, which is to be constructed of solid masonry consisting of brick and block. Information such as the height of the brick from grade, location of bond break for concrete slab, and connection of rafters to wall are provided.

3. Detailed elevation views may also be used to provide information that cannot be found on floor plans alone. For example:
   - Kitchen elevations can display information regarding the number of drawers in a cabinet, the swing of cabinet doors, and the height of cabinets (Refer to Chapter 13 for examples).
   - Bathroom elevations can display information regarding the layout of ceramic tile and borders, as well as cabinet and shower details.
   - Elevations for rooms requiring special details such as a study or dining room indicate where shelving or panelling is located. Refer to page 6 of the Brook drawings that came attached with this textbook for an example of this, called the Bookshelf Niche Detail. Can you see where on page 6 the detail is located on the plan?

KEY TERMS

Definitions for key terms appear in the glossary at the end of the text.

Abbreviation (p. 61)
Benchmark (p. 61)
Construction notes (p. 61)
Detail drawings (p. 73)
Hatching (p. 65)
Hexagon and reference number (p. 72)
Legend (p. 61)
Non-load-bearing (p. 81)
Symbol (p. 65)
Wythe (p. 81)

FIGURE 4.15

Carefully review the deck plan and section drawings in Figures 4.15 and 4.16. Note how the cutting-plane line is indicated in the plan view and review the resulting sections A and B.

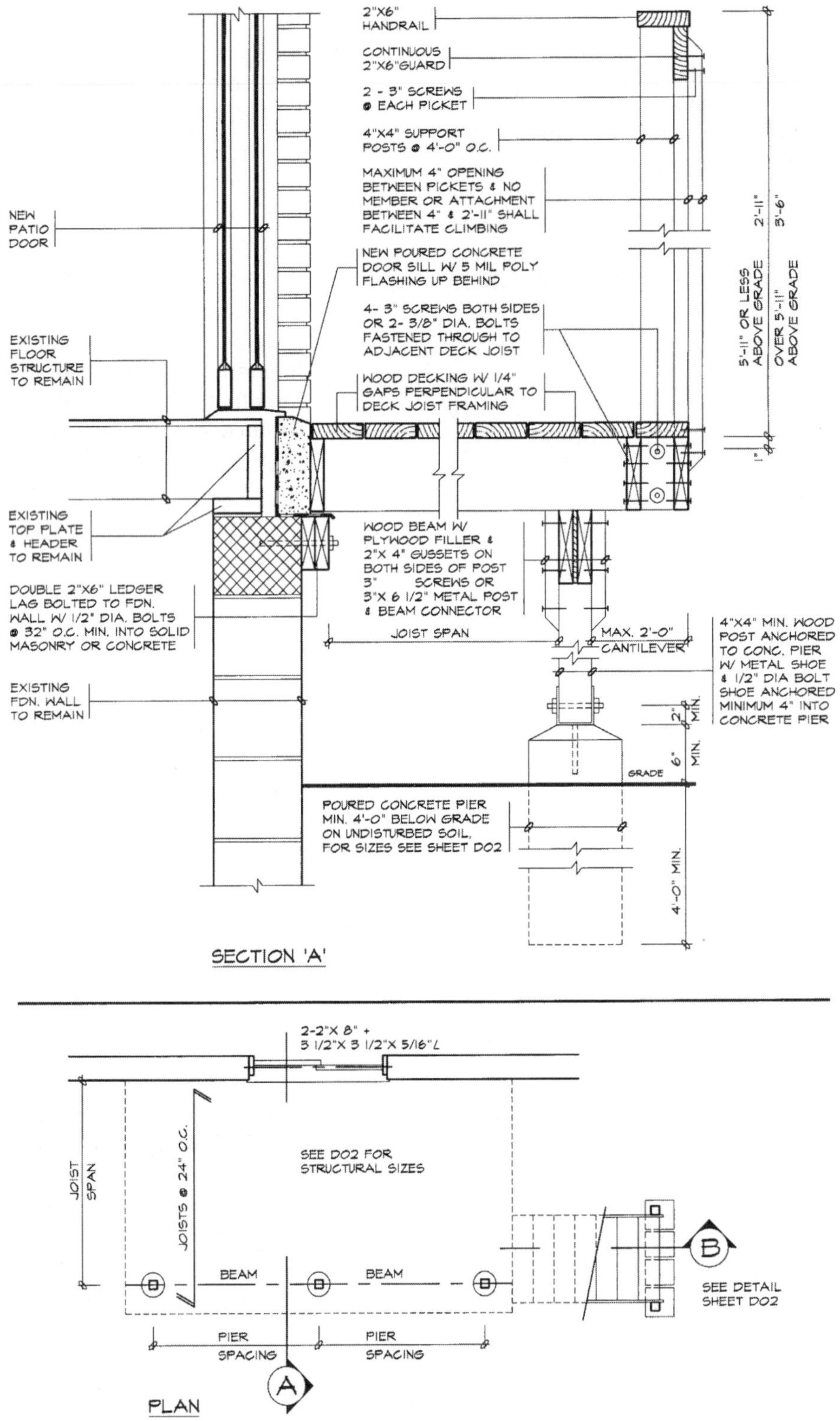

Source: Toronto Area Chief Building Officials Committee (TACBOC)

FIGURE 4.16

Deck stair section notes and table.

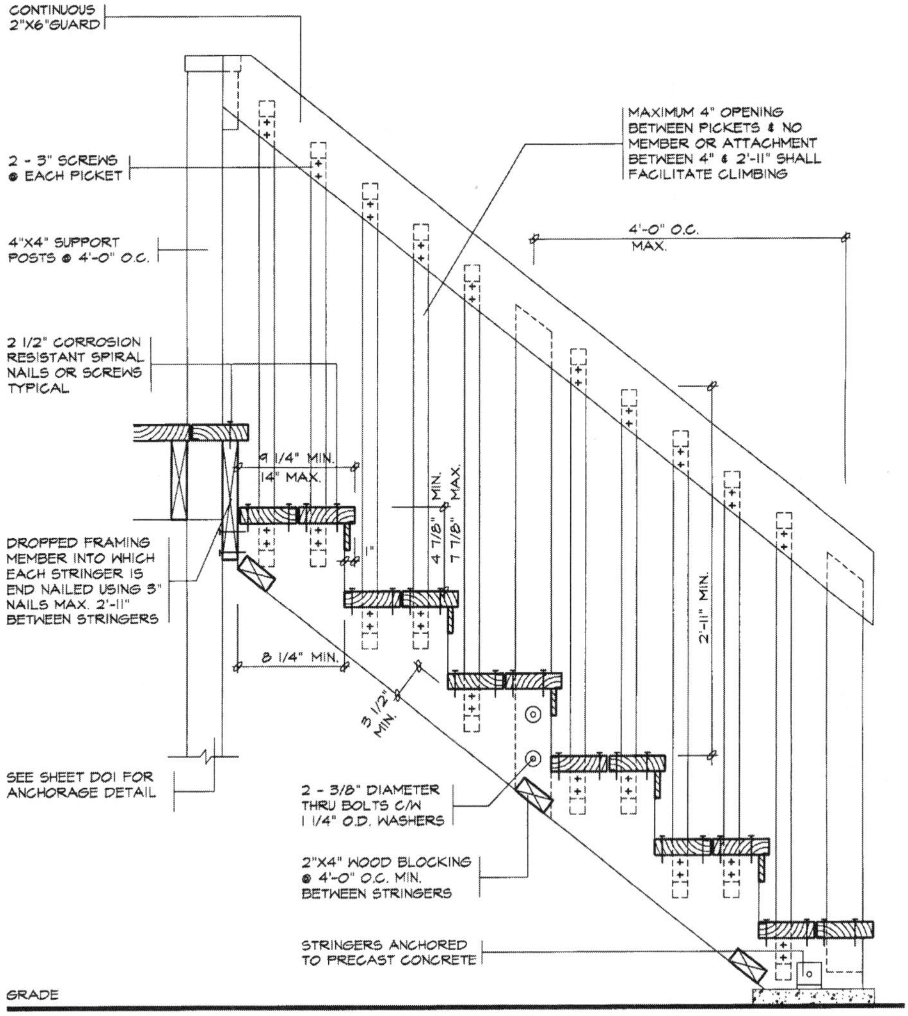

SECTION 'B'

GENERAL NOTES

1. SITE PLAN OR SURVEY IS REQUIRED SHOWING ALL LOT LINES & DIMENSIONS SIZE & LOCATION OF ALL EXISTING BUILDINGS, LOCATION & SIZE OF DECK

2. LUMBER NO. 2 SPF OR BETTER, WOOD POSTS MIN. 4"X4" (SOLID). USE CORROSION RESISTANT SPIRAL NAILS OR SCREWS.

3. DECK IS NOT PERMITTED TO BE SUPPORTED ON BRICK VENEER

4. CONCRETE PIERS SHALL BEAR ON UNDISTURBED SOIL. THE BEARING CAPACITY OF THE SOIL SHALL BE DETERMINED PRIOR TO CONSTRUCTION

5. PROVIDE A HANDRAIL 31"-38" HIGH ON STAIRS IF MORE THAN THREE RISERS. PROVIDE A GUARD ON BOTH SIDES IN ACCORDANCE W/ THE DETAIL ABOVE WHERE THE STAIR EXCEEDS 6 RISERS.

		PIER SIZE				BEAM SIZE				
	JOIST SPAN	PIER SPACING				PIER SPACING				JOIST SIZE
		4'	6'	8'	10'	4'	6'	8'	10'	
1500 PSF	6'	8"Φ	10"Φ	12"Φ	14"Φ	2/2"X6"	2/2"X6"	2/2"X8"	2/2"X10"	2"X6"
	8'	10"Φ	12"Φ	14"Φ	16"Φ	2/2"X6"	2/2"X8"	2/2"X10"	2/2"X12"	2"X6"
	10'	12"Φ	14"Φ	16"Φ	18"Φ	2/2"X6"	2/2"X8"	2/2"X10"	2/2"X12"	2"X8"
	12'	12"Φ	14"Φ	16"Φ	18"Φ	2/2"X6"	2/2"X8"	2/2"X10"	2/2"X12"	2"X10"
2500 PSF	6'	8"Φ	8"Φ	10"Φ	10"Φ	2/2"X6"	2/2"X6"	2/2"X8"	2/2"X10"	2"X6"
	8'	8"Φ	10"Φ	10"Φ	12"Φ	2/2"X6"	2/2"X8"	2/2"X10"	2/2"X12"	2"X6"
	10'	8"Φ	10"Φ	12"Φ	14"Φ	2/2"X6"	2/2"X8"	2/2"X10"	2/2"X12"	2"X8"
	12'	10"Φ	12"Φ	14"Φ	14"Φ	2/2"X6"	2/2"X8"	2/2"X10"	2/2"X12"	2"X10"
4000 PSF	6'	8"Φ	8"Φ	8"Φ	8"Φ	2/2"X6"	2/2"X6"	2/2"X8"	2/2"X10"	2"X6"
	8'	8"Φ	8"Φ	8"Φ	10"Φ	2/2"X6"	2/2"X8"	2/2"X10"	2/2"X12"	2"X6"
	10'	8"Φ	8"Φ	10"Φ	12"Φ	2/2"X6"	2/2"X8"	2/2"X10"	2/2"X12"	2"X8"
	12'	8"Φ	10"Φ	12"Φ	12"Φ	2/2"X6"	2/2"X8"	2/2"X10"	2/2"X12"	2"X10"

Source: Toronto Area Chief Building Officials Committee (TACBOC)

Chapter 4 Assignment: Abbreviations, Hatchings, Symbols, and Construction Notes

Refer to the Doncaster drawings to assist you in answering questions 1 through 5. Refer to the information found in Chapter 4 to assist you in answering questions 6 through 20. Identify the letter that best answers the question.

_____ 1. Referring to the construction notes, what does the hexagon symbol 4, shown below, refer to?

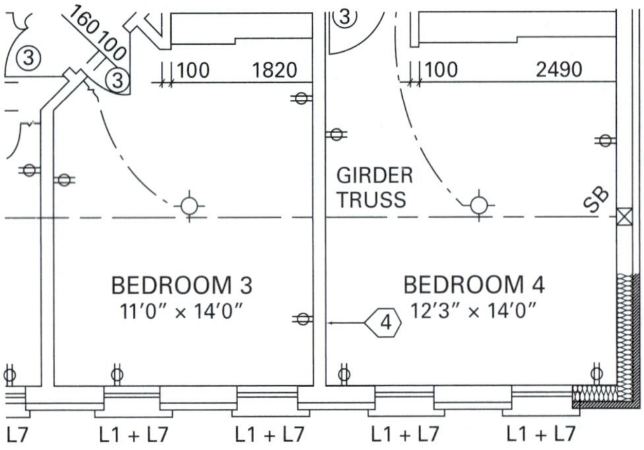

Source: Cassidy & Co. Architectural Technologists

a. interior walls having 5/8"drywall on each side

b. interior walls having 3/8"drywall on each side

c. interior walls having a double top plate and a single bottom plate

d. interior walls having a single top plate and a single bottom plate

_____ 2. The drawing below refers to a water closet (toilet); what else is it referring to?

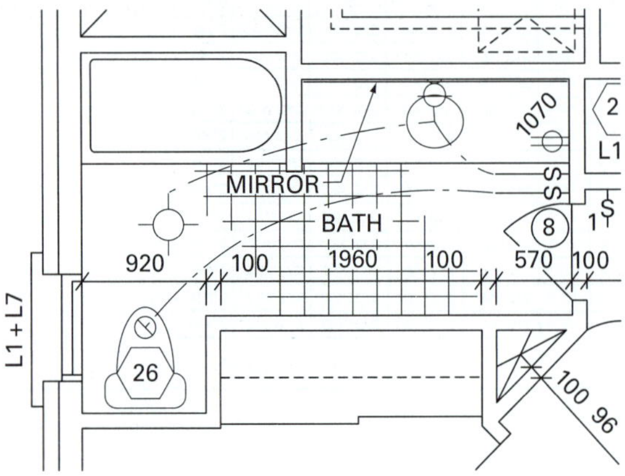

Source: Cassidy & Co. Architectural Technologists

a. desk

b. exhaust fan

c. kitchen cabinet

d. a right hand door

_____ 3. How many shelves is the linen closet supposed to have?

a. three

b. four

c. five

d. six

_____ 4. Where can the abbreviation FD be found?

a. in the laundry room on the second floor

b. near the stove in the kitchen

c. near the furnace and laundry tub in the basement

d. in the powder room on the first floor

_____ 5. In the construction notes, hand-rail information is indicated by the hexagon with what number?

a. 9

b. 21

c. 32

d. 11

_____ 6. The letter R, when used as an abbreviation, can refer to range for a stove and R-value for effectiveness of insulation. What else might it refer to?

a. rating

b. risk

c. riser

d. rhomboid

_____ 7. In construction drawings, what is the abbreviation for ditto (meaning "repeat the information here")?

a. DT

b. DI

c. DO

d. DA

_____ 8. What is the abbreviation used in the Doncaster drawings to identify a double floor joist?

a. DFJ

b. DJ

c. Dble Jsts

d. Doble Jts

_____ 9. What does the abbreviation PR stand for?

a. powder room
b. principle riser
c. place to the right
d. power to range

_____ 10. What does this hatching represent?

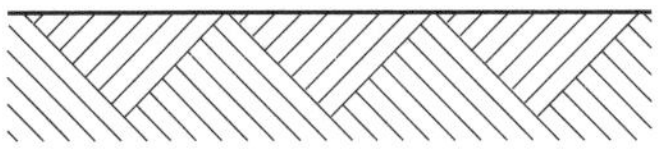

a. insulation
b. grade
c. dimensional lumber
d. concrete

_____ 11. What does this hatching represent?

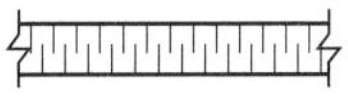

a. fibreglass insulation
b. rigid insulation
c. plywood
d. oriented strand board

_____ 12. What does this hatching represent?

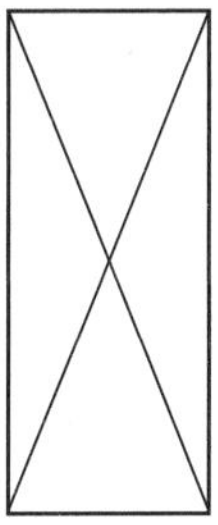

a. a piece of dimensional lumber seen in plan or section view
b. a casement window shown in elevation
c. a cross-section of tubular steel
d. a cross-section of rigid insulation

_____ 13. What does this hatching represent?

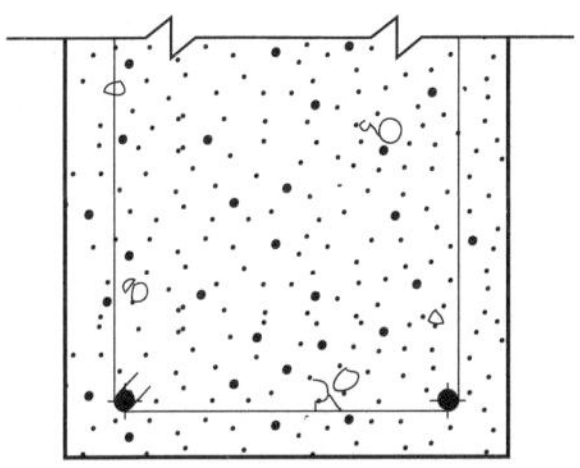

a. sand and gravel
b. gravel
c. reinforced concrete
d. unreinforced concrete

_____ 14. What does this hatching represent?

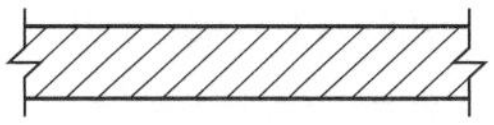

a. a wood stud wall
b. brick in a wall
c. a steel iron pipe
d. batt insulation in a wall

_____ 15. What does this hatching represent?

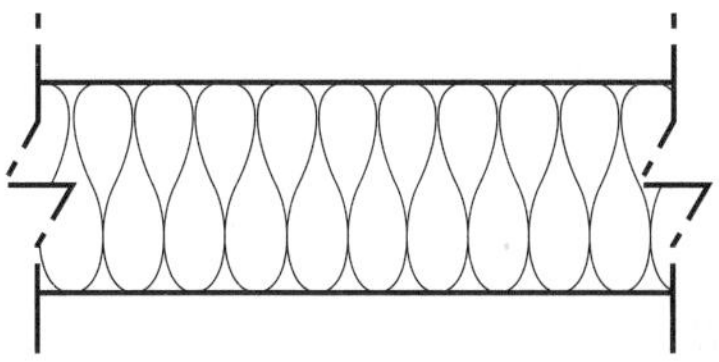

a. rigid insulation
b. dimension lumber viewed in elevation
c. batt insulation
d. concrete

_____ 16. What does the symbol shown below refer to?

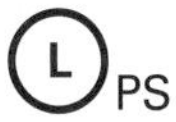

a. faucet and pull valve switch
b. fire alarm pull station
c. lampholder with a pull switch
d. light with a plug and a switch

_____ 17. What does the symbol shown below refer to?

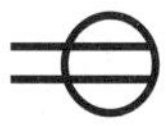

a. refrigerator outlet
b. duplex receptacle
c. split receptacle
d. three-way switch

_____ 18. What does the symbol shown below refer to?

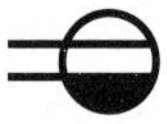

a. refrigerator outlet
b. duplex receptacle
c. stove receptacle
d. split-switched duplex receptacle

_____ 19. What does the symbol shown below refer to?

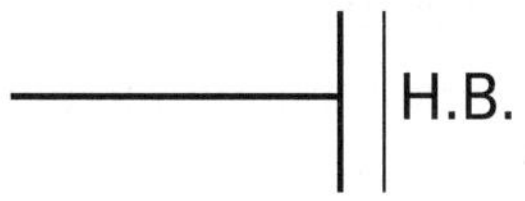

a. hose bib
b. gas hook-up connection
c. high bench mark
d. horizontal beam

_____ 20. What does the symbol shown below refer to? a _________.

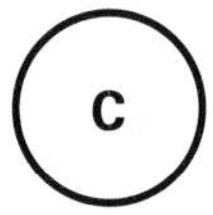

a. shielded outlet
b. clock outlet
c. sump pump
d. stationary switch

Chapter 4 Quiz Review: Plans, Section Details, and Deck Drawings

Answer questions 1 through 16 from the deck plan and section detail drawings illustrated in Figures 4.15 and 4.16, and answer questions 17 through 20 based on information found throughout the chapter. Identify the letter that best answers the question.

_____ 1. What is each picket on the railing fastened with?

a. 3 - 2" screws

b. 2 - 3" screws

c. 2 - 3' screws

d. 2 - 3½" galvanized nails

_____ 2. What space is left between the decking surface boards?

a. 1/8" gap between the boards

b. 1/4" gap between the boards

c. 3/8" gap between the boards

d. 5/16" gap between the boards

_____ 3. What are the bottoms of the stair stringers anchored to?

a. a built-up wood frame

b. precast concrete

c. poured in place concrete sidewalk

d. asphalt driveway

_____ 4. What should the concrete piers rest on?

a. disturbed soil

b. undisturbed soil

c. engineered fill

d. surface fill

_____ 5. Which one of the following answers correctly describes the placement of the pickets and railing system?

a. must have pickets with no more than a 3" space between them

b. must have pickets with no more than a 3½" space between them

c. must have pickets with no more than a 4" space between them

d. must have pickets with no more than a 4½" space between them

_____ 6. How many section details do the cutting plane lines display?

a. two

b. three

c. four

d. five

_____ 7. What type of line is drawn to represent the concrete pier below grade?

a. centre line

b. leader line

c. object line

d. hidden line

_____ 8. What is placed directly under and behind the concrete door sill?

a. floor joists

b. 5 mil poly flashing up behind the concrete door sill

c. 5/8" tongue-and-groove plywood placed vertically directly behind the concrete door sill

d. building paper

_____ 9. What is the maximum amount that the deck can cantilever out from the wood post?

a. The floor joists can cantilever out 1' past the wood beam.

b. The floor joists can cantilever out 2' past the wood beam.

c. The floor joists can cantilever out 3' past the wood beam.

d. The floor joists can cantilever out 3'-6" past the wood beam

_____10. What is the required spacing for the floor joists that support the deck?

a. The joists are spaced at 10" OC.

b. The joists are spaced at 12" OC.

c. The joists are spaced at 16" OC.

d. The joists are spaced at 24" OC.

_____11. What is the minimum riser height for the stairs?

a. 4⅞"

b. 5⅞"

c. 7⅞"

d. 8"

_____ 12. What does the abbreviation SPF refer to that is indicated in the deck drawings?

a. support for fir floor

b. support for pine floor

c. specific pounds per floor

d. spruce, pine, fir

_____ 13. What is the deck *not* allowed to be supported on?

a. solid masonry

b. wood frame

c. concrete

d. brick veneer

_____ 14. How is the deck supported against the wall?

a. by bearing directly on the block foundation wall

b. by two ledgers bolted to the top course of the concrete block foundation wall

c. by resting on concrete columns placed near the wall

d. by the brick veneer

_____ 15. What does the diamond-shaped (criss-cross) hatching in the drawing indicate?

a. the end grain of the wood beam that supports the weight of the entire deck

b. the brick veneer that supports the deck

c. the end grain of the decking on top of the deck

d. the top course of block-filled solid (solid masonry) or concrete that supports the deck

_____ 16. What size and type of fasteners are listed for fastening the deck boards to the surface of the deck?

a. 1½" corrosion-resistant nails or screws

b. 2" corrosion-resistant nails or screws

c. 2½" corrosion-resistant nails or screws

d. 3" corrosion-resistant nails or screws

_____ 17. What does the symbol below represent?

S_3

a. switch for three rooms

b. sump pump

c. three-way switch

d. sump-pump switch

_____ 18. What does the symbol below represent?

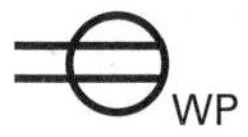

a. a weatherproof receptacle

b. a weatherproof switch

c. a lampholder for outside use

d. a window plug

_____ 19. What does the hatching below represent?

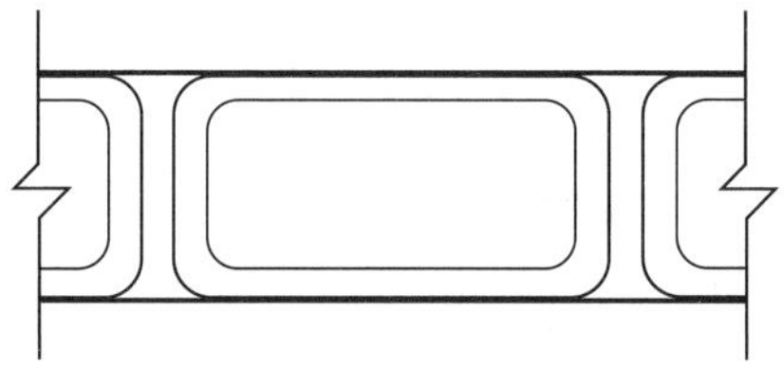

a. chimney flues

b. glass block viewed in plan view

c. brick laid end to end

d. concrete block

_____ 20. What does the hatching shown below indicate?

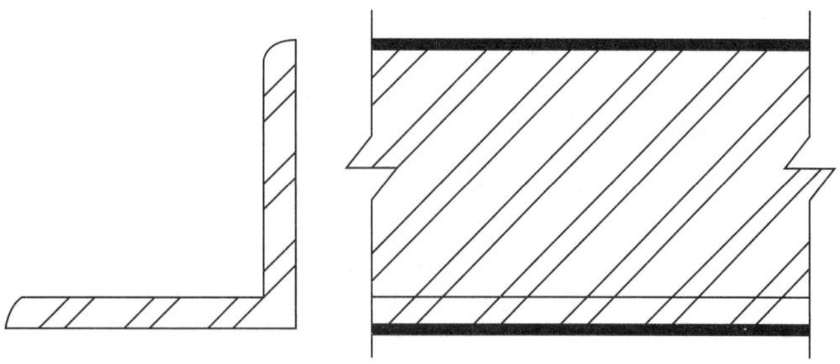

Source: Tom Stephenson

a. wood corner

b. aluminum structure

c. structural steel

d. sectional timber

5 Reading Site Plans, Zoning, the Permit Process, and Building Code Enforcement

LEARNING OBJECTIVES

After completing this chapter, you will be able to:

- Interpret and apply the information found on a site plan, including:

  Locating the setbacks for the building from established property lines

  Establishing grade heights at various points on the lot

  Establishing the depth of excavation for the footings

  Locating storm and sanitary drain connections, and water service

  Locating utility services.
- Interpret and understand the purpose of zoning requirements.
- Gather the appropriate drawings and documentation to apply for a building permit.

INTRODUCTION

This chapter reviews how to read and interpret information on a site plan (also referred to as a plot plan) and provides an overview of zoning and its potential effect on the development of a property. The chapter also discusses the building permit and inspection process and explains when a permit and inspections are required.

SITE AND PLAT PLANS

A plat plan displays many or multiple lots like a subdivision. These drawings are a high-level view of the overall development project situated on a block of land that is then subdivided into numerous lots based on zoning requirements and draft plan approvals. Each individual lot on the plat map is assigned an identifier, usually a number. The plat plan is then officially recorded with the applicable government land registry and becomes the legal description of all the lots in the subdivision. The individual lot site plans are then derived and detailed from the subdivision plat plan.

Engineers develop a series of infrastructure and grading plans in conjunction with the developed plat plans to ensure that grading is appropriate to provide adequate runoff for storm water, under the severest conditions, to prevent flooding. Engineers also make sure that the requirements for all utilities and infrastructure have been planned with drawings submitted to meet all regulatory requirements. Refer to Figure 5.11 for an example of a grading plan.

A key plan is used to orient the view as to where the building or structure is located based on the overall phase of the project. It is shown in the title block at the bottom or side of the drawing page. In Figure 5.10 the key map is shown as part of the site engineering drawings for lot grading.

The **site plan,** also called the plot plan, provides a plan view of the property. The plan is developed from the registered survey of the property and must conform to local zoning requirements. The site plan provides the following information:

- Size, shape, and location of property lines
- Setback, size, and location of building from property lines
- **Contour lines** that connect the same elevations of land along lines that are referenced from a benchmark or starting point
- Elevations of land and buildings
- Location of building services
- Location of existing trees

FIGURE 5.1
Batter boards.

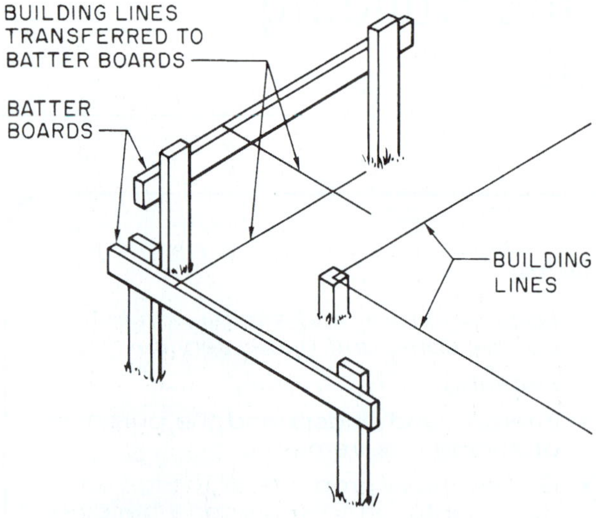

- Location of any special features such as **easements,** the right to use the property for a specific purpose, for example, as a right of way for utilities.
- Location of property lines and setback for the building.

FIGURE 5.2
Two blocks and line to lay out addition along a wall.

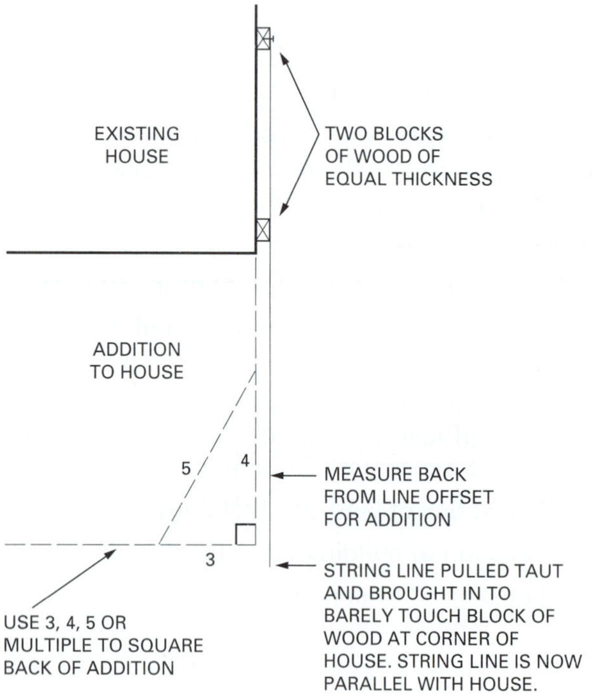

The site plan indicates the location and boundaries of the property lines. A registered land surveyor will have marked these locations with rods indicating the corners of the lots. The building location is then located based on setbacks from these markers. The layout person or land surveyor will then use a transit/total station instrument to locate the building corners. The area to be excavated is always larger than the building to allow space to work. The excavation area is marked with batter boards that indicate the location of the excavation, footings, and foundation walls. The layout for simple buildings or for additions onto existing buildings can be performed using a tape measure and applied mathematics. For example, an addition may be measured from an existing wall using a line and two blocks of wood. Once the location of a wall is complete, then the Pythagorean formula can be used to square the corners (refer to Figure 5.2 and Appendix C for more information on the Pythagorean formula).

Production Home Building

For production home building, the houses are excavated as a strip that is removed for all the houses in a row (see Figures 5.3 and 5.4). The footings and foundations are then located using reference pins or dowels driven into the ground as building corner locators. The location of the footings is referenced from these pins, while the walls are placed in line with the pins (see Figures 5.5 and 5.6).

FIGURE 5.3
Strip excavation for a row of houses in progress.

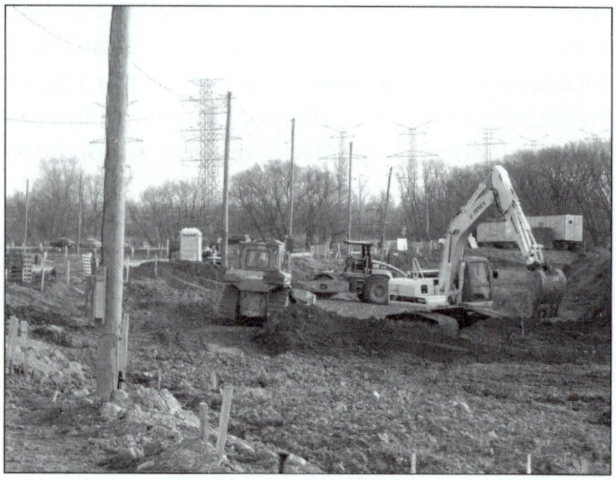

Courtesy of Tom Stephenson

FIGURE 5.4
Strip excavation completed; notice protection skids placed around services.

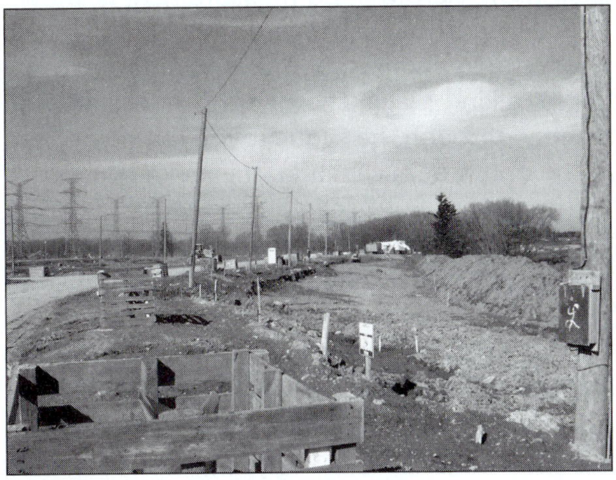

Courtesy of Tom Stephenson

FIGURE 5.5
Footings for multiple houses constructed based on references from foundation wall location pins.

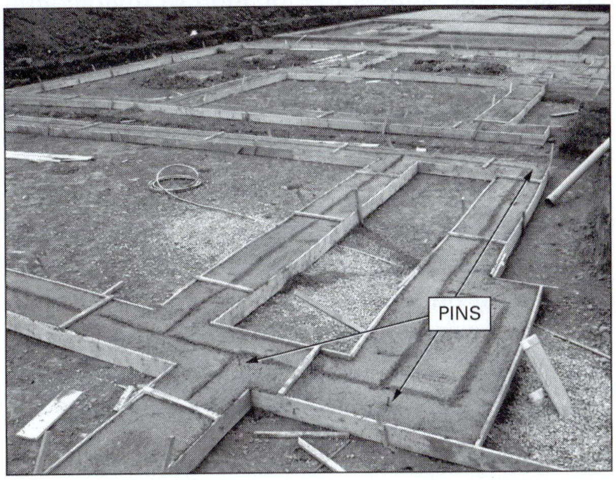

Courtesy of Tom Stephenson

FIGURE 5.6
Wall location pins with nylon line indicating exact foundation wall location.

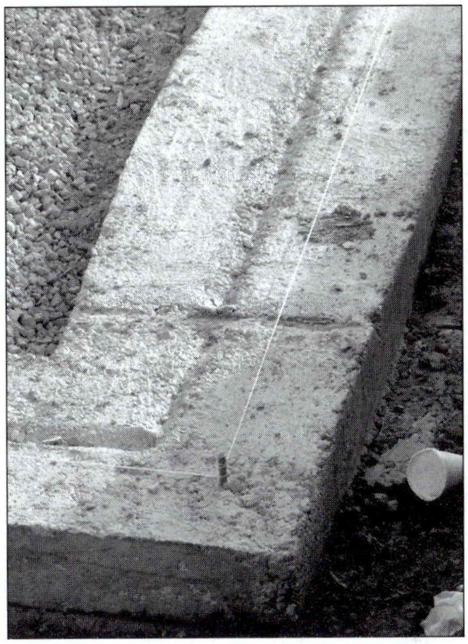

Courtesy of Tom Stephenson

FIGURE 5.7
The top image indicates that the driveway would slope from the house to the road 9.71' on the left side (61.20–51.49) and 9' on the right side (61.20–52.20). The bottom image has contour lines indicating original grade elevation and the required new grade for the project (N.G. = natural grade; F.G. = finished grade).

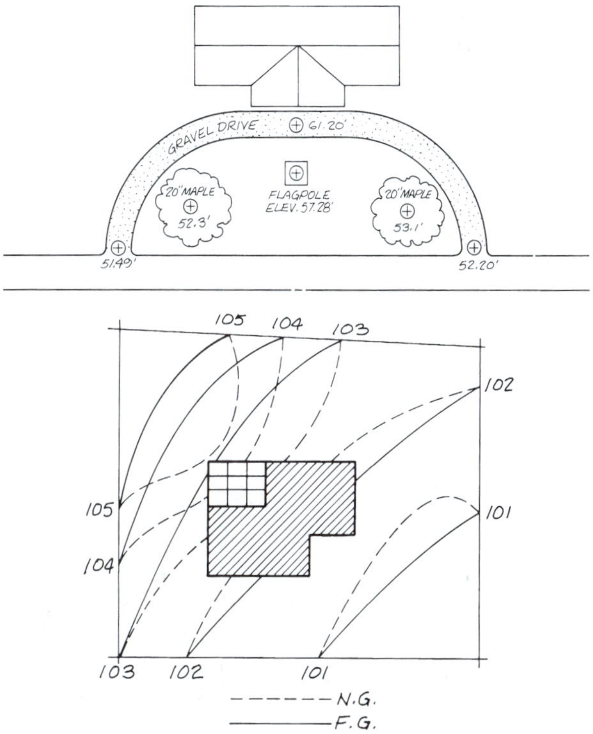

Contour Lines

Contour lines are drawn on site plans to indicate an elevation that falls on the line. For example, a contour line may be given an elevation of 101 m. This means that, at any point on the line, the elevation is 101 m in relation to a **benchmark** (BM) elevation (reference or starting point); if the BM is 100, then it is 1 metre higher than the BM.

When reading a site plan with contour lines, consider the following:

- Contour lines placed close together indicate a steep slope.

- Contour lines spaced farther apart indicate a much more moderate slope; the farther the distance, the smaller the relative slope.
- At some point along the line an elevation must be indicated; sometimes for steep slopes where the lines are very close, the elevation may be indicated on every fifth or sixth line.
- Contours that form concentric circles or shapes indicate a hill or a depression.
- Contour lines that run parallel and are evenly spaced indicate a uniform and even slope.
- If contour lines are not used, elevations will be indicated at key locations, usually with an × or cross-type symbol.
- A **swale** is indicated on a site plan by a line with an arrow pointing in a specific direction with an elevation given at each end. Swales are designed to slope and move water off the property so that it can drain in a controlled fashion to storm sewers, collection ponds, or catchment areas (refer to Figure 5.15).

Elevations of Property and Buildings

The elevation of the depth of footing is determined based on an established benchmark. The benchmark can be referenced from a monument placed by the city or township, usually on the side of a well-established structure (usually an easily identifiable and accessible location) that has footings below the frost line. The monument has been carefully established and measured by provincial surveyors to accurately indicate the elevation at the marker point above mean sea level; more commonly on construction projects it is simply a point located near or on the property that will not move during the construction process. For example, the top of a fire hydrant can be a suitable benchmark. For construction purposes, the benchmark is usually given a starting number of 10 or 100. The logic for starting at 10 or 100 instead of 0 is that you will not run into negative numbers once the grade falls off or goes down considerably, as is the case when excavating.

The depth of excavation can be done on site with the use of a builder's level, transit or total station instrument. The builder's level provides a level line to be sighted through a scope to a measuring rod while it is held on top of the benchmark; this sighting is known as a back sight (BS). The BM + BS measurement is known as the height of instrument (HI). The builder's level is then spun around to where the excavation is taking place. The depth can be taken by sighting through the scope to the rod being held at the depth of excavation; this is known as the FS (foresight). Subtracting the FS elevation from the HI provides the elevation of the bottom of the footing, referenced from the original BM. This point is checked repeatedly as the excavation proceeds and ensures that the bottom of the footing is at the proper elevation.

The calculations for this process become cumbersome to follow on larger sites where the grade may change elevation considerably. In these cases, a standardized format is followed to track elevations at various turning points or at various intermediate points (refer to Figure 5.8).

Location of Building Services

The site plan shows the location of utilities and services, such as:

- Storm and sanitary sewers, and their connection points
- Cable junctions (pedestal location when run underground) or hydro poles when run above ground
- Utility junctions (pedestal location when run underground) or cable/telephone poles when run above ground
- Water service (shutoff valve location)
- Fire hydrant
- Hydro transformer (when run underground) or hydro/telephone poles when run above ground.

Location of Trees

Many municipalities now have requirements regarding the protection of existing trees with trunks over a specified diameter. In these cases it may be a permit requirement to show the location of these trees on the site plan as well as indicating that a tree protective zone will be maintained around the drip line base of the tree during construction.

FIGURE 5.8

Determining the height of the instrument and calculating grade rod.

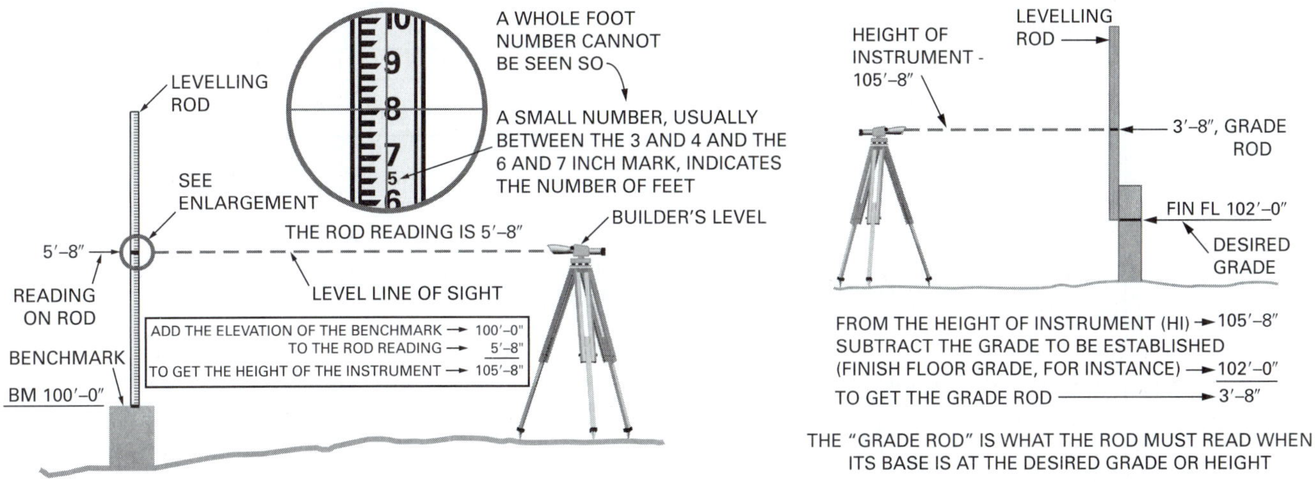

Source: From VOGT, *Carpentry*, 4E, © 2006 Delmar Learning, a part of Cengage Learning, Inc. Reproduced by permission. www.cengage.com/permissions

Easement and Right of Way

Locations of established easements and rights of way will be indicated on the site plan to ensure that no encroachments are made. An easement is a right given to a third party to use part of the property for special purposes such as running a hydro line overhead or having a storm sewer run underneath the ground. Another common form of easement is a right of way that provides the legal right for a person or persons of adjacent land to walk or drive across a certain property to access another property that might otherwise be inaccessible.

Commercial Site Plan Example

The site plan in Figure 5.9 is for a small commercial building. Note that the elevations, lot line, and building measurements are in metres instead of millimetres, as is typical for site plans based on a registered land survey. Often measurements need to be converted from metres to millimetres, which is a relatively simple process when using the metric system of measurement. For example, if 16.800 is indicated on the site plan, it refers to 16.8 metres. This can easily be converted to millimetres by moving the decimal place three places to the right, which in this case means that 16.800 m is equivalent to 16,800 mm. Move the decimal place three places to the left to convert back from millimetres to metres. Review the site plan and elevations while carefully observing the different abbreviations and legend.

FIGURE 5.9

Site plan example.

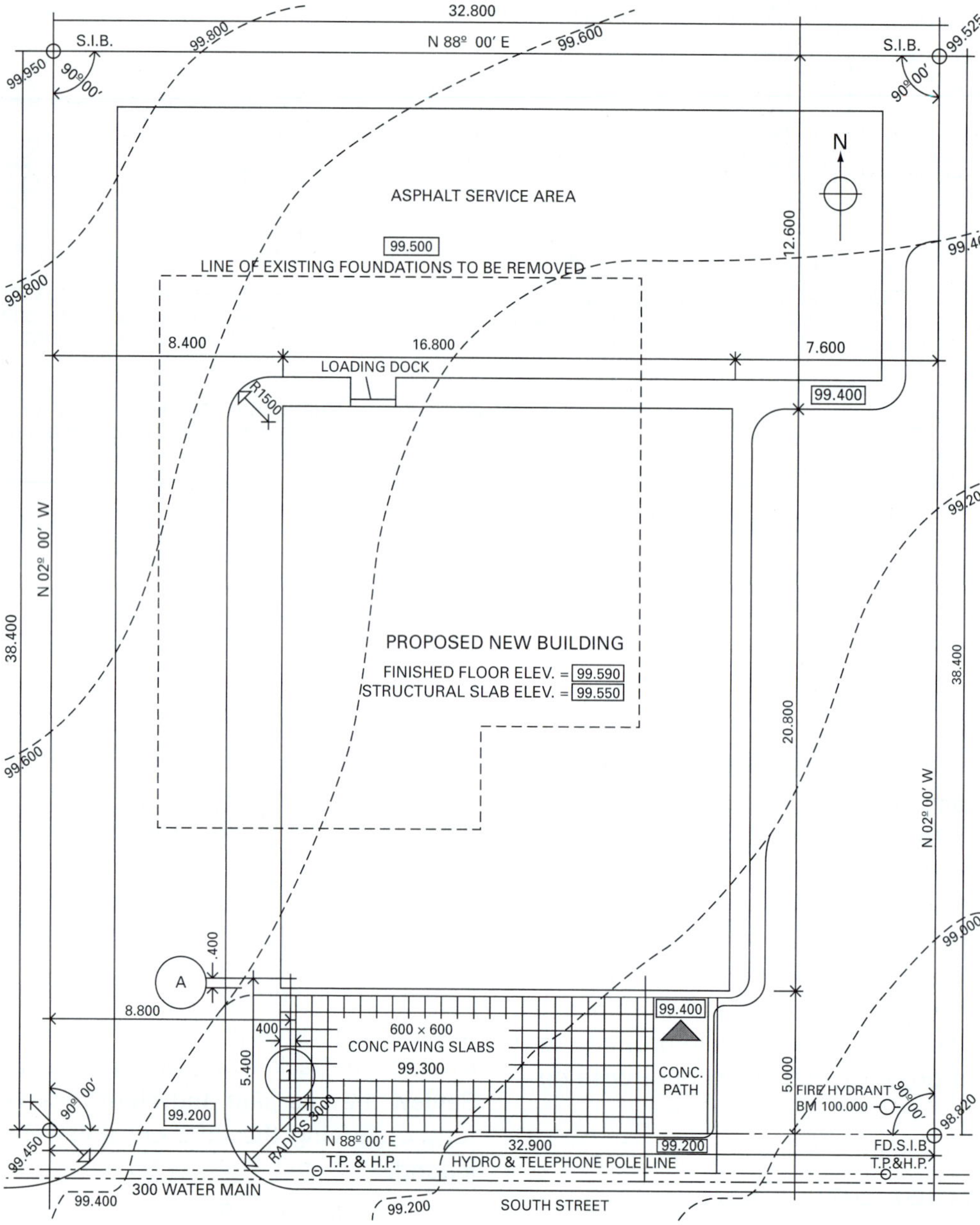

SITE PLAN
LOCATION DRAWING
SCALE 1:200

BM	BENCH MARK
TP	TELEPHONE POLE
HP	HYDRO LINE POLE
SIB	STEEL IRON BAR
NG -----	NATURAL GRADE CONTOUR LINES
99.000	FINISHED GRADE INDICATED BY BOX

Chapter 5 Assignment: Reading Site Plan Drawings

The site plan in Figure 5.9 lists all measurements in metric; dimensions are listed in metres. Based on the site plan in Figure 5.9, identify the letter that best answers the question.

_____ 1. What is the size of the water main that is shown on the site plan?

a. 200 mm
b. 300 mm
c. 400 mm
d. 600 mm

_____ 2. What size are the concrete paving slabs?

a. 600 mm × 600 mm
b. 300 mm × 300 mm
c. 400 mm × 400 mm
d. 250 mm × 800 mm

_____ 3. What compass direction does the natural grade of the property elevation slope down from?

a. northwest to southeast
b. southeast to northwest
c. northeast to southwest
d. southwest to northeast

_____ 4. What is being used as the benchmark for this site plan?

a. an elevation monument
b. a fire hydrant
c. the top of a steel iron rod
d. the finished first floor of the building

_____ 5. What is the finished floor elevation of the new building?

a. 410 mm higher than the BM
b. the same height as the BM
c. 99.590 mm higher than the BM
d. 410 mm lower than the BM

_____ 6. Once calculated, what is the area of the new building?

a. 1259.52 m^2
b. 500 m^2
c. 349.44 m^2
d. 400 m^2

_____ 7. Is there an existing building on this site that needs to be removed?

a. Yes.
b. No.
c. Yes, but the building stays and the new building is to be attached to the existing one.
d. No, there is not an existing building, but there is a small existing shed that is to remain.

_____ 8. What is the finished floor elevation for the new building?

a. 99.200 m
b. 99.550 m
c. 99.590 m
d. 100.000 m

_____ 9. What is the elevation of the structural slab?

a. 99.200 m
b. 99.300 m
c. 99.550 m
d. 100 m

_____ 10. What is the elevation of the benchmark?

a. 99.500 m
b. 99.550 m
c. 99.590 m
d. 100.000 m

_____ 11. What is the difference in elevation (height) of the concrete path that leads from the front property line to the south elevation?

a. 150 mm
b. 200 mm
c. 300 mm
d. 400 mm

_____ 12. What is the highest elevation (point) of finished grade on the property?

a. 99.400 m
b. 99.600 m
c. 99.950 m
d. 99.990 m

_____13. How far is the west elevation of the building set back from the west property line?

a. 8,200 mm

b. 8,400 mm

c. 8,600 mm

d. 8,845 mm

_____14. How far is the south elevation of the building set back from the south property line?

a. 5,400 mm

b. 5,000 mm

c. 4,800 mm

d. 4,600 mm

_____15. What do the abbreviations TP & HP stand for?

a. turning point and high point

b. turning point and hydro pole

c. telephone pole and hydro line pole

d. telecommunications vault and high point

_____16. What are the overall dimensions of the new building?

a. 16,100 mm × 21,000 mm

b. 16,800 mm × 20,800 mm

c. 32,800 mm × 38,400 mm

d. 30,000 mm × 40,000 mm

_____17. What is the difference in elevation between the southeast corner of the property and the northeast corner of the property?

a. 1.180 m

b. 1.130 m

c. .830 m

d. .705 m

_____18. What is the setback of the east elevation of the building from the east property line?

a. 7.600 m

b. 8.400 m

c. 12.600 m

d. 8.800 m

_____19. If the underside of the footing for the building is to be 2000 mm below the structural slab, what elevation would the underside of the footing be placed at?

a. 99.550 m

b. 98.550 m

c. 97.590 m

d. 97.550 m

_____20. What is the difference in elevation between the structural slab and the finished floor?

a. 99.590 m

b. 99.550 m

c. .040 m

d. 40 m

FIGURE 5.10

Key map/block plan of the Tribute Nottingham subdivision. The arrow points to Lot 61, which is where the Doncaster house is situated and was photographed for this textbook.

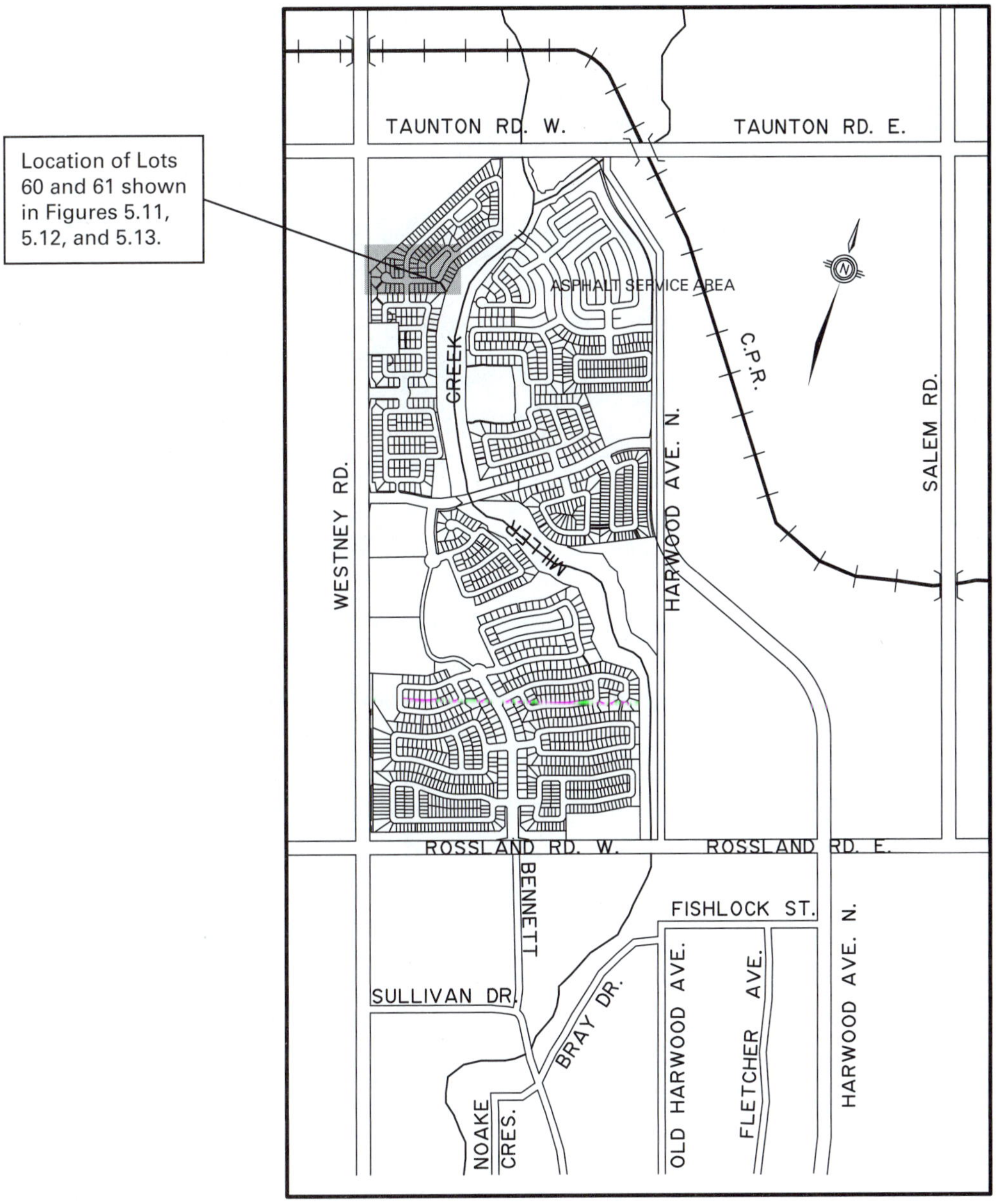

Source: Sabourin Kimble & Associates Ltd.

FIGURE 5.11

Portion of lot grading plan. Note details of Lots 60 and 61 (Doncaster house) and Lot 67 (Whittington house).

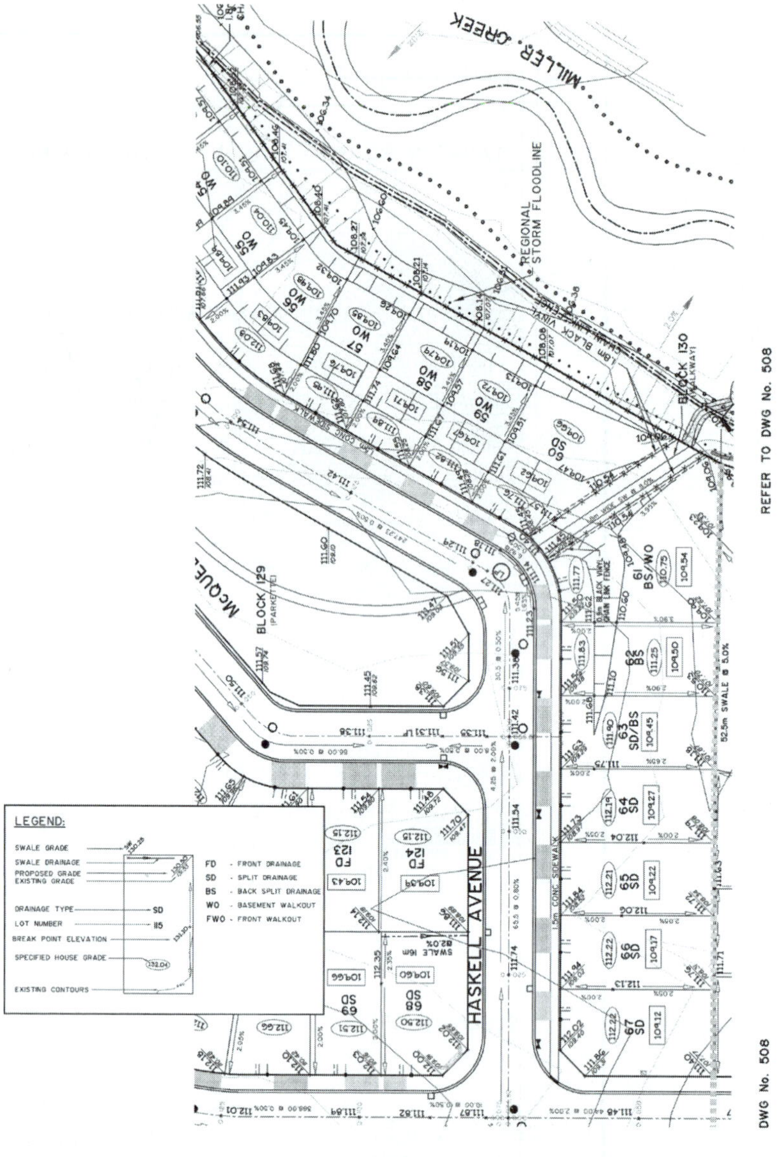

Source: Sabourin Kimble & Associates Ltd.

Site Plan Examples

The two site plans seen in Figures 5.12 and 5.13 are for new production homes being built side by side in the Tribute Nottingham subdivision. The houses are separated by a walkway/storm easement. Lot 60 has a walkout basement (abbreviation WOB) while Lot 61 slopes to the rear but not enough to have a basement walkout; instead it has a semi walkout (abbreviation SWO), which is a walkout from the breakfast area on the first floor and requires a deck/landing and 10 risers to get down to the grade (see Figure 5.24). The pictures in Figures 5.14–5.27 depict various features and elevations shown on the site plan, and will help you to visualize the site. When interpreting any construction drawing, always pay close attention to the legend; in this case the legend is at the bottom left corner of the site plan. The legend provides information about abbreviations and symbols used on the drawing and may differ from designer to designer, as stated in Chapter 4.

FIGURE 5.12

Site plan of Doncaster C reverse house: this house uses a similar Doncaster layout but is a mirror image with a different façade than the drawings included with the text. Lot 60. Note that both Lot 60 and Lot 61 (Figure 5.13) adjoin the storm easement.

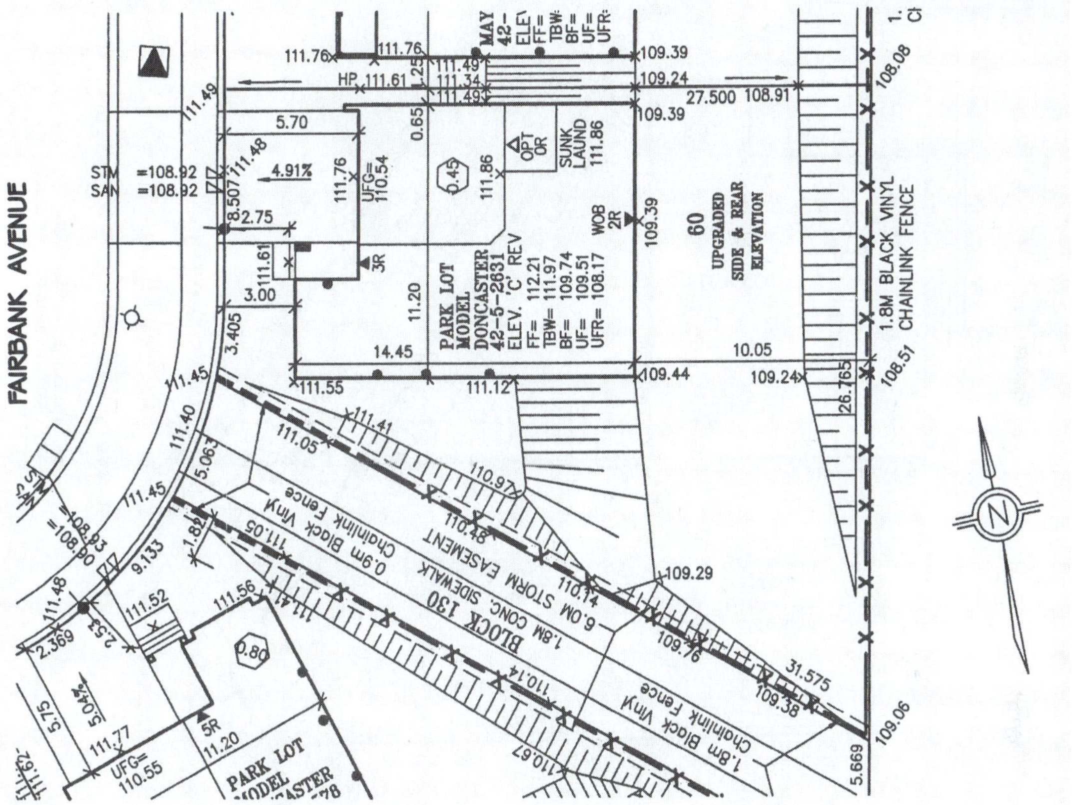

WALL AREA	= 80.63 SM
LIMITING DISTANCE	= 1.2M @7%
MAX ALLOWABLE OPENINGS	= 5.64 SM
PROPOSED OPENINGS	= 5.22 SM

	STREET LIGHT	—H—	HYDRO
	HYDRANT	FF	FINISHED FLOOR ELEVATION
	TRANSFORMER	TBW	TOP OF BASEMENT WALL
	VALVE CHAMBER	BF	BASEMENT FLOOR ELEVATION
	WATER SERVICE	UF	UNDERSIDE OF FOOTING
	CATCH BASIN	UFG	UNDERSIDE OF FTG.-GAR
	STM & SAN CONNECT	UFR	UNDERSIDE OF FTG.-REAR
	SINGLE STM & SAN	WOD	WALK OUT DECK
	SWALE DIRECTION	WOB	WALK OUT BASEMENT
	CABLE TV PEDESTAL	SWO	SEMI WALK OUT
	BELL PEDISTAL	REV	REVERSE PLANS
	SUPER MAIL BOX	2R,3R	NUMBER OF RISERS
	EXTERIOR DOOR LOCATION	000.00	PROPOSED ELEVATION
	SIDE WINDOW LOCATION	(000.00)	EXISTING ELEVATION
	EMBANKMENT 3:1 MAX SLOPE		

LOT AREA	= 534.50 SM
BUILDING COVERAGE	= 142.56 SM
LOT COVERAGE	= 26.67%

Source: Cassidy & Co. Architectural Technologists

FIGURE 5.13

Site plan of Doncaster house, Lot 61. This is the lot for the Doncaster house that is used throughout the text. The full size site plan for Lot 61 also came attached to the textbook.

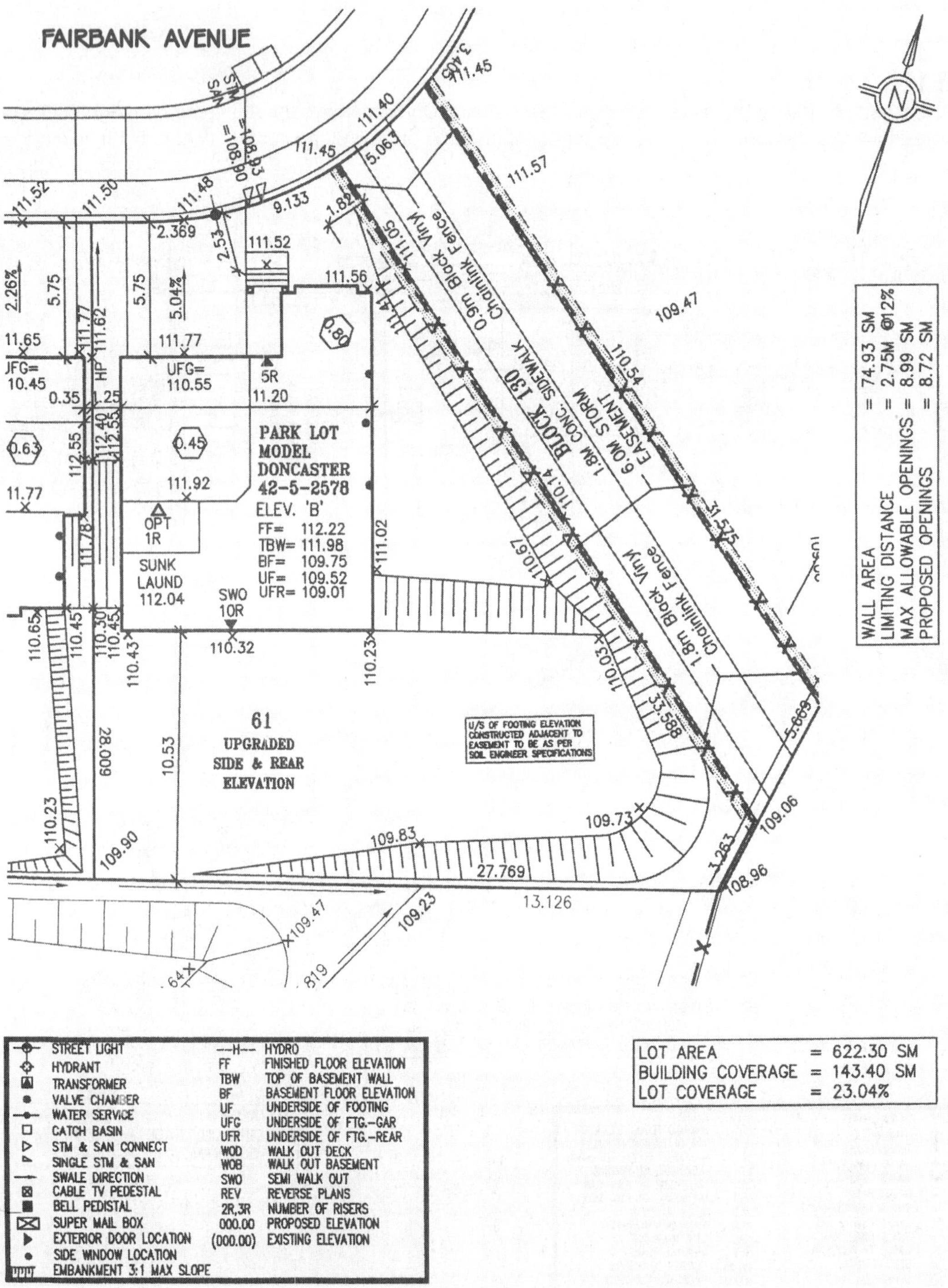

Source: Cassidy & Co. Architectural Technologists

FIGURE 5.14

As indicated on site plans (see Figures 5.12 and 5.13), catch basins are located on the street in front of Lot 61. Refer to legend for catch basin symbols.

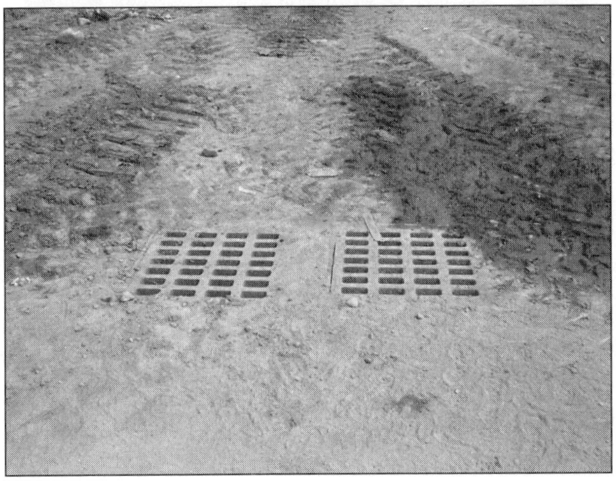

Courtesy of Tom Stephenson

FIGURE 5.16

Storm easement and walkway running between lots 60 and 61; notice the drop in elevation from front to back and compare to elevation points provided on the site plan (see Figures 5.12 and 5.13). Note that as the elevation number gets smaller, the grade falls or slopes downward.

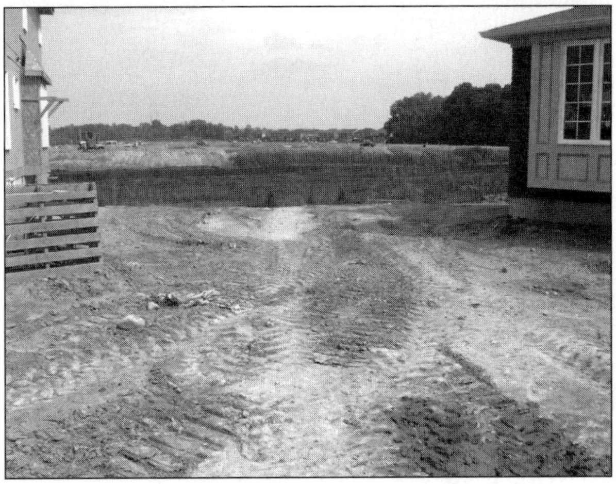

Courtesy of Tom Stephenson

FIGURE 5.15

Grading at rear of property Lot 61 at immediate right. Note the development of the swale (indentation to direct water flow) at the rear property line (see back of lot Figure 5.13).

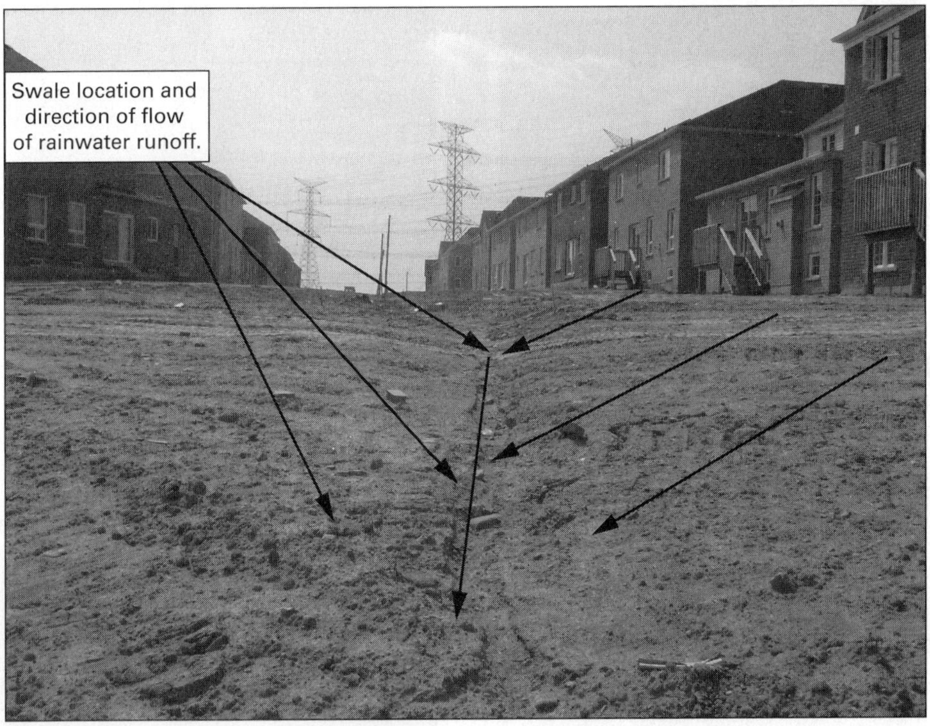

Courtesy of Tom Stephenson

FIGURE 5.17
East elevation of house looking south on Lot 61 (see Figure 5.13); notice the change in elevation toward the rear of the house. How is this shown on the site plan?

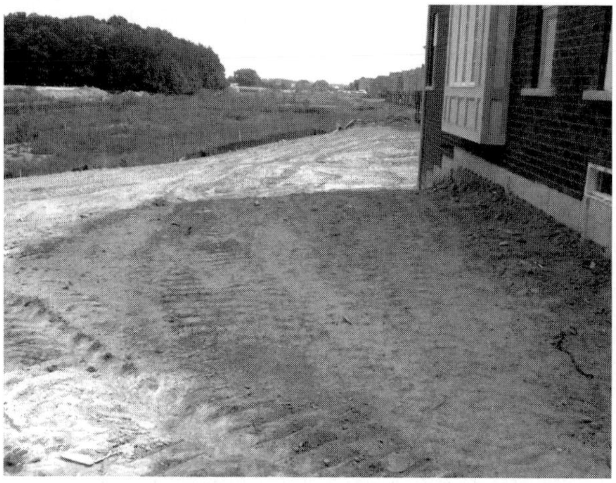

Courtesy of Tom Stephenson

FIGURE 5.18
Another view of the east elevation of house on Lot 61 (see Figure 5.13).

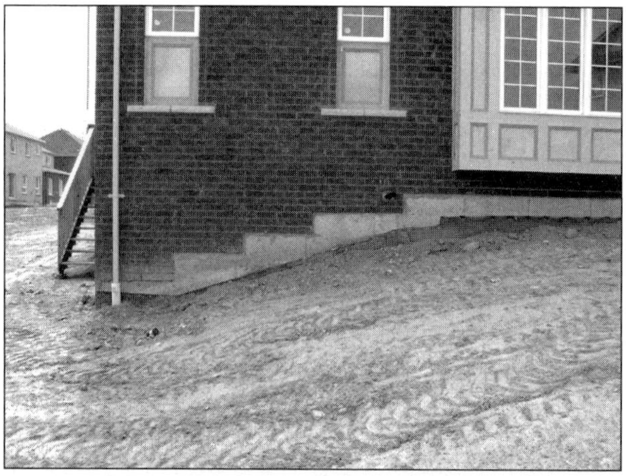

Courtesy of Tom Stephenson

FIGURE 5.19
Steel iron bar installed by land surveyors to mark and identify corner of property line.

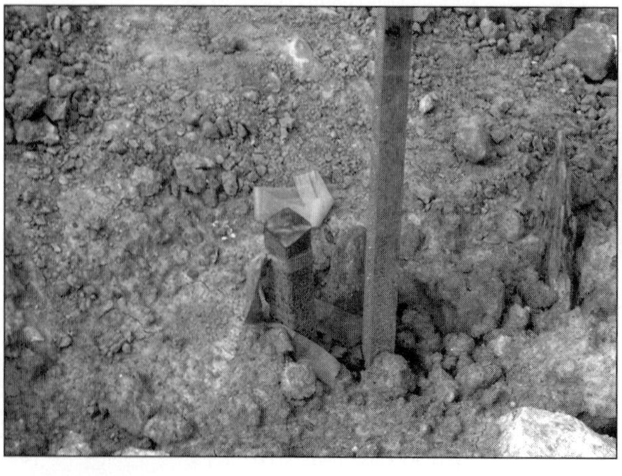

Courtesy of Tom Stephenson

FIGURE 5.20
Grade elevation at southwest corner of house on Lot 61 (see Figure 5.13) sloping to rear.

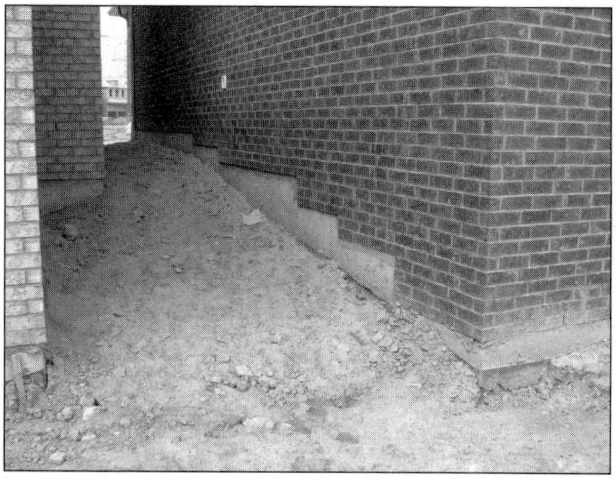

Courtesy of Tom Stephenson

FIGURE 5.21
Location of water service curb box, which provides ability to shut off water supply to house on Lot 61 (see Figure 5.13); observe location relative to garage and front porch and compare to site plan (refer to legend on site plan for symbol identification).

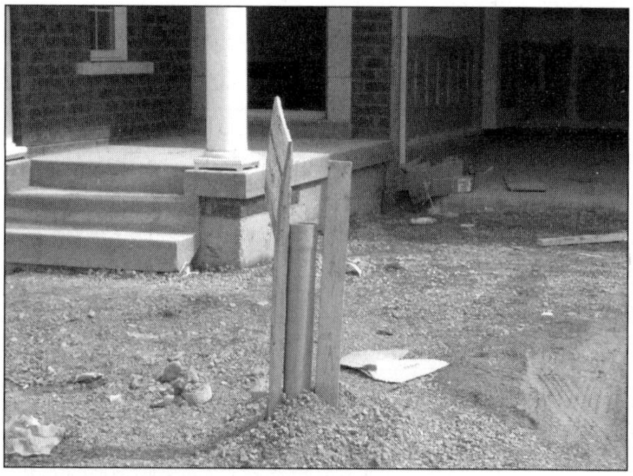

Courtesy of Tom Stephenson

FIGURE 5.22
Location of transformer noted on site plan for Lot 60 (see Figure 5.12).

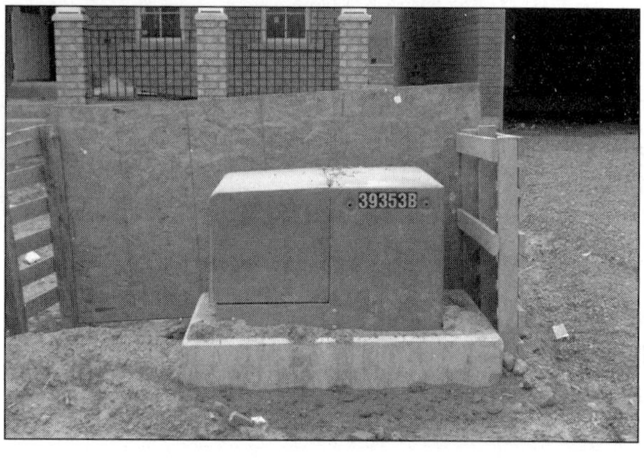

Courtesy of Tom Stephenson

FIGURE 5.23

Location of fire hydrant shown on Lot 60 (see Figure 5.12).

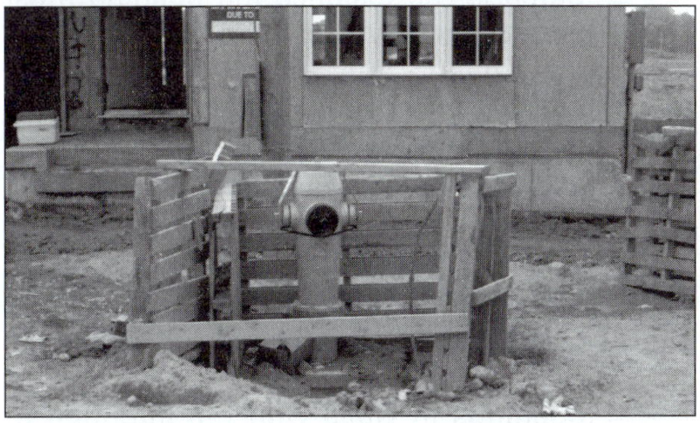

Courtesy of Tom Stephenson

FIGURE 5.24

Semi walkout from first floor, Lot 61; sloping grade at side of house provides opportunity for larger windows at rear (see Figure 5.13).

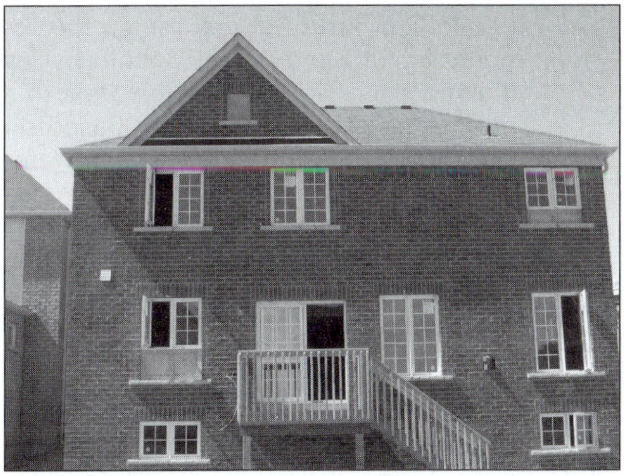

Courtesy of Tom Stephenson

FIGURE 5.25

View of southwest/right elevation, Lot 60 (see Figure 5.13); notice the slope down the side where the storm easement and walkway are. Also notice how grade falls off again at rear of lot next to grassy area; compare this with the site plan elevations and embankment symbols on the site plan for Lot 60 (refer to legend on site plan for embankment symbol). Also review Figures 5.10 and 5.11 to get a sense of where in the subdivision these lots are found, and their surroundings.

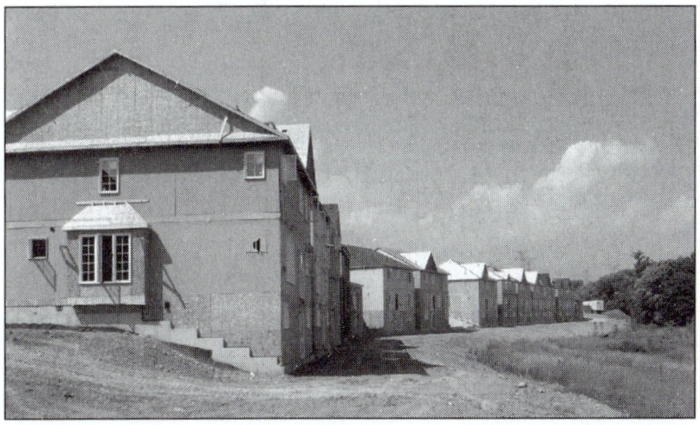

Courtesy of Tom Stephenson

FIGURE 5.26

Storm sewer; refer to site plans for Lots 60 and 61 (see Figures 5.12 and 5.13) and notice the location for the storm easement that runs between the properties. This is the exit point for the storm water from the storm sewer located to the northwest between the lots.

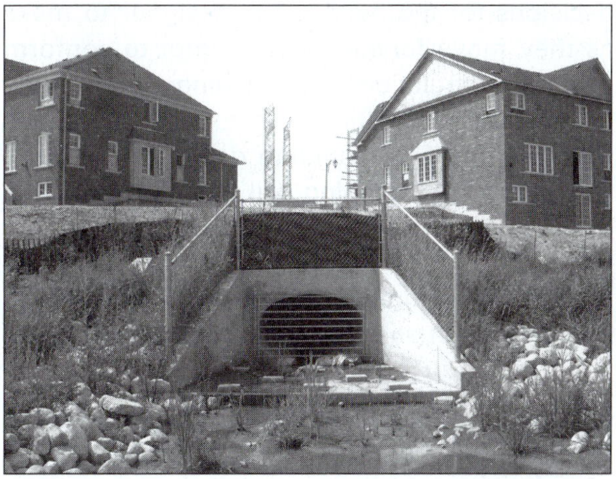

Courtesy of Tom Stephenson

FIGURE 5.27

Other services and utilities may also be shown on a site plan, such as telephone and cable connection boxes.

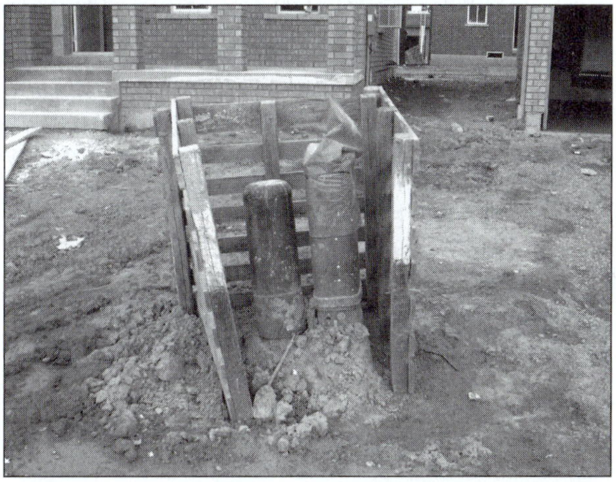

Courtesy of Tom Stephenson

ZONING

Towns and urban centres have zoning requirements that outline the type of use and occupancy permitted for individual properties. For example, properties may be zoned agricultural, residential, commercial, or industrial, with many subcategories and combinations used to designate land use.

Zoning designates a system of land use regulation. It separates areas known as zones and designates uses for the land within these zones. The process generally follows an **official plan** set out by the municipality. The official plan looks forward over a long period of time (10 to 20 years) and establishes how the town or city is to grow in the most beneficial and prosperous way for the community as a whole. There are many decisions for the builder and designer to make as they move forward and attempt to conform to zoning regulations. The building permit application process requires that zoning restrictions be met, and if the building does not conform, then a change to zoning will be required or the drawings will need to be revised to conform to the zoning.

Zoning requirements and changes vary from street to street and from municipality to municipality and may include the following:

- Ensuring that the land use of one property does not negatively affect the use of an adjoining property and that the land use is compatible with neighbouring land use. This is why residential neighbourhoods are separated from industrial areas and why height, lot coverage, and setback restrictions are placed on properties.
- Defining what affects the use of an adjoining property can be a challenging task that may involve the community, politicians, the developer, and various levels of government. When a major zoning change is requested, there will be notification of a community meeting either by mail, signage, or both (see Figure 5.28). Smaller changes known as a **minor variance**, will go through a more streamlined process that involves only those properties that are more directly affected.

FIGURE 5.28

Example of community meeting invitation.

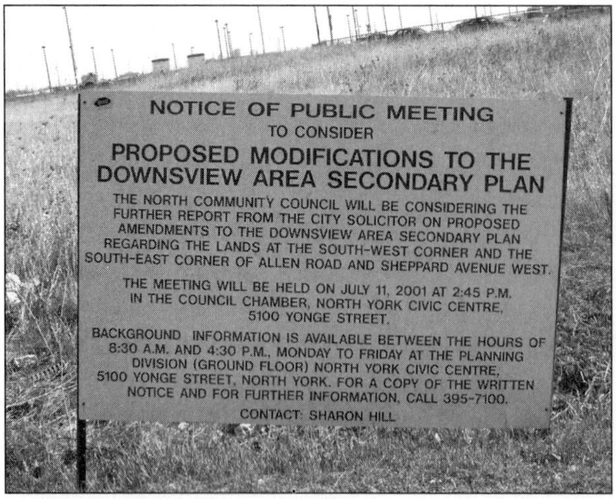

Courtesy of Etobicoke York Planning Office, City Planning.

- Restrictions that may be placed on properties include:
  - *Setback:* The distance the building must be set back from the property line. Among other things, this restriction ensures that buildings line up along the street and do not unnecessarily block sunlight from adjoining properties.
  - ***Limiting distance:*** Although not directly a part of zoning, the limiting distance requirements of the building code affect setbacks. The setback distance affects the choice and placement of materials and openings in walls with the intention of limiting the potential of fire spreading from building to building and is a requirement for buildings placed in close proximity. The limiting distance requirement is part of the NBC and information will be placed on the site and elevation plans. The limiting distance requirements include maximum opening sizes allowed on a wall based on its proximity to another building and based on a percentage of wall surface area (refer to Figures. 5.12 and 5.13 for examples of limiting distance restrictions for between the houses). There are also specific code requirements regarding the flame resistance of materials and wall coverings in close proximity to property lines. Fire-resistant rating requirements will be listed on the drawings, or in the construction notes (NBC Reference Division 3.2.3).
  - *Building coverage:* This restriction limits the area the building can occupy in relation to the area of the lot. There may also be other restrictions on overall square footage (square footage (SF) or square metres (SM) may include the area of each floor, excluding an unfinished basement) in relation to the lot size. This information is usually provided on the drawings in the construction notes page.
  - *Building height:* This may be an overall height of building, height of building to middle of roof, or an angular height maximum from a specific lot line. For example, a condominium building may be restricted from overhanging adjoining low-rise residential properties. This is accomplished by an angular plane requirement of X degrees from the rear property line; the building height cannot cross the angular plane.

- The municipality may request an environmental impact study to ensure that a change to zoning does not negatively affect the environment and neighbouring properties.
- The municipality may request a retail impact study, which analyzes how the development may affect retail shopping in the area, either positively or negatively.
- The municipality may request a traffic impact study, which considers how the development may affect local traffic patterns. The traffic impact study responds to questions such as:
  - Are the roads capable of handling the additional traffic?
  - Are the traffic lights adequate?
  - Is public transit suitable?
  - Will there be adequate parking available?
  - Is the speed limit appropriate?
  - Will this project affect crossing patterns of children going to school?
- A heritage buildings impact determines whether a zoning change will affect any heritage buildings in close proximity to the property.
- A storm and waste water management plan asks the following:
  - Will a zoning change fit with municipal capacity to manage storm and waste water?
  - Will this change alter the drainage patterns in the area and, if so, what measures will need to be addressed to prevent flooding?

The designer must assess whether the project meets existing zoning requirements and/or whether a request for a change to zoning is worth the time and effort involved in getting it passed. The time period required to change or adjust the zoning can vary from a few months to years and can become an embroiled political process when communities get involved in attempting to block a change to the zoning that they think will negatively impact the community. Changes to zoning and the official plan do not always align with local communities' beliefs about what projects are or are not appropriate.

The provinces set out guidelines in their planning acts, which the municipalities use to develop their official plan and zoning requirements.

THE PERMIT PROCESS

In order to receive permission to build, renovate, or perform alterations, a request for building permit must be submitted and approved. The request must show that the project meets the requirements of all applicable codes and regulations. Governments have a responsibility to protect their citizens, and accomplish this through the enactment of codes and regulations that ensure public health and safety. The National Building Code of Canada 2010 (NBC) is the most current building code and is accepted as being the model code for all regulations pertaining to the building, renovating, or altering of buildings in Canada.

The NBC sets out the minimum requirements and standards for fire protection, structural design, environmental separation, heating ventilation and air conditioning, plumbing, and housing and small building construction. Note that the code lists the *minimum* acceptable standards for building; you can build to higher standards, but you cannot build to lesser standards.

As a model code, the NBC has no enforcement power on its own. As a result, each province has adopted its own provincial code based on the NBC, or in some cases just adopted the NBC directly with some minor changes. Individual provincial codes address special regional needs and requirements. Provincial codes are law. The power to grant permits and enforce the building code is subsequently granted by the province to the municipal government.

An application for a permit can usually be sought at the local city hall or township office building department.

General Submission Requirements

In order to receive a **building permit** you must not be in contravention of any zoning or building code regulations. You must fill out the appropriate forms and include the number of full sets of drawings as required by the building department, including:

- Survey of the property
  - Usually a relatively recent survey (15 years or less) because an older survey may not show additions or alterations that changed the size of the building on the lot. The lot may also be vacant and a recent survey will indicate

any geographical changes that may have occurred (refer to NBC Division C 2.2.2).

- Site plan
- Foundation plan
- Floor plans (one for each floor and roof)
- Elevation drawings (one for each face of the building)
- Section and detail drawings
- Construction notes as required.

Permits are generally required for any kind of structural work, and when the work may affect the health and safety of the general public, including the occupants.

The following are examples of when a permit is required:

- Building a new home
- Adding an addition
- Removal or alteration of load-bearing walls
- Decks built above ground level, e.g., 600 mm or more above grade
- Altering or adding to the plumbing system
- Demolition of a building
- Masonry fireplace installation
- Finishing a basement and basement apartments.

The following are examples of when a permit is *generally* not required:

- Removal of non-load-bearing walls
- Installation of non-load-bearing walls
- Changing a plumbing fixture in the same location
- Sheds built to less than 10 sq. metres
- Replacement of a furnace
- Decks less than 600 mm above grade.

To clarify whether a permit is required, always check with the local municipal office; other situations may require a building permit.

Once you receive approval for your permit, you must follow your municipality's guidelines regarding the posting of the building permit. The permit is usually posted at eye level on or at the front of the building or property (see Figure 5.29).

You are also required to keep an approved copy of the plans at the site, so the inspector can review them during his/her inspections.

FIGURE 5.29

Building permit location.

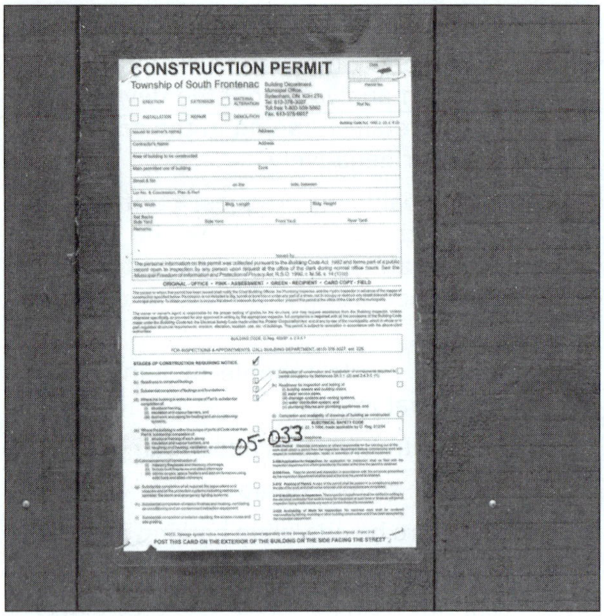

Courtesy of the Ontario Ministry of Municipal Affairs and Housing

As part of the permit requirements, the project generally will also require other permits, depending on the scope of construction. These may include:

- Demolition
- Plumbing
- HVAC
- Electrical

Also refer to your provincial requirements regarding builder licensing and warranty requirements. Some provinces have mandatory home warranty and licensing programs and these vary significantly in requirements from province to province.

Inspections

Inspections can occur at any time; however, the permit will designate when certain inspections **must occur**. The builder or contractor must make sure that the building department is notified with appropriate lead time (usually 48 hours) regarding the need for an inspection. The following is a typical list of inspections that a building department may require for the construction of a new house:

- Excavation inspection
- Footing foundation wall and drainage inspection (before backfilling)

FIGURE 5.30
Flow chart of zoning and permit process.

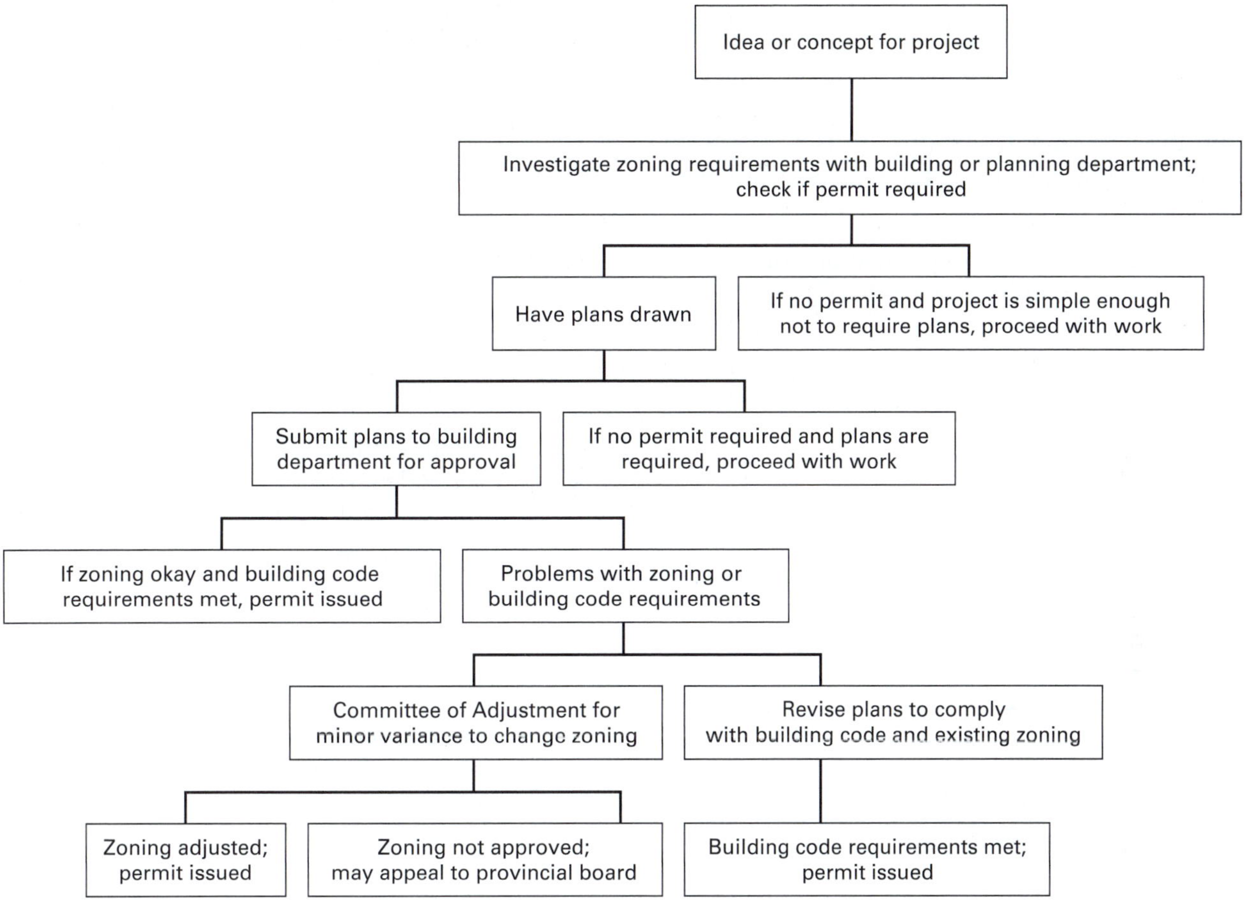

Courtesy of Tom Stephenson

- HVAC rough-in inspection (separate HVAC permit required, and gas permit if applicable)
- Plumbing rough-in inspection (separate plumbing permit required)
- Electrical rough-in inspection (separate electrical permit required)
- Framing inspection (taken after mechanical subtrades are finished, this ensures that the framing has not been compromised by the subtrades)
- Envelope inspection (insulation and vapour retarder/barrier)
- Occupancy inspection and final HVAC, plumbing, and electrical inspections
- Final inspection and certificate of completion.

Inspections are mandatory, and the inspector has the power to place a stop-work order on the project if it is not being carried out according to the approved plans. The building inspector has the right to have the work opened up if it has been enclosed without the appropriate inspection taking place.

KEY TERMS

Definitions for key terms appear in the glossary at the end of the text.

Benchmark (p. 91)
Building permit (p. 105)
Contour lines (p. 89)
Easement (p. 90)
Limiting distance (p. 104)
Minor variance (p. 104)
Official plan (p. 104)
Site plan/plot plan (p. 89)
Swale (p. 92)
Zoning (p. 103)

Chapter 5 Quiz Review: Site Plans, Zoning, and Permits

Use the site plan (siting and grading) for Lot 67 Nottingham Community, Whittington Drawings, attached to the textbook, which lists all measurements in metric. Answer questions 1 through 15 from the site plan. Questions 16 to 20 are from Chapter 5. Identify the letter that best answers the question.

____ 1. How much is the house on Lot 67 set back from the east side of the lot?

a. .35 m
b. 1.25 m
c. 2.15 m
d. 3.95 m

____ 2. How much is the house on Lot 67 set back from the south side of the lot?

a. 2.15 m
b. 3.87 m
c. 10.05 m
d. 27.5 m

____ 3. How many square metres is the Lot area for Lot 67?

a. 138.86 m^2
b. 142.72 m^2
c. 407.90 m^2
d. 489.34 m^2

____ 4. What is the difference in elevation between the spot elevation marking at the front of the garage door (at driveway intersection) and the sunken landing (referred to as sunk land on the drawings)?

a. .15 m
b. .32 m
c. .40 m
d. 1.54 m

____ 5. How many metres long is the lot at its longest point?

a. 15.2 m
b. 27 m
c. 27.5 m
d. 28 m

____ 6. What is the elevation of the finished first floor?

a. 109.43 m
b. 110.02 m
c. 110.25 m
d. 112.72 m

____ 7. What is the underside of footing elevation indicated on the site plan?

a. 112.00 m
b. 111.25 m
c. 110.02 m
d. 108.25 m

____ 8. What is the width of the house at its widest point from east to west, excluding the porch?

a. 111.92 m
b. 10.70 m
c. 11.20 m
d. .35 m

____ 9. How many risers are there from the house to grade using the deck at the south elevation?

a. 1 riser
b. 2 risers
c. 3 risers
d. 4 risers

____ 10. Where is a swale located on the site plan?

a. There is a swale located at the north property line.
b. There is a swale located along the east property line.
c. There is a swale located along the west property line.
d. There is a swale located right where the walk-out deck will be placed.

____ 11. Where is the hydro transformer located?

a. There is a hydro transformer located near the southwest corner of the property.
b. There is a hydro transformer located near the northeast corner of the property.

c. There is a hydro transformer located near the southeast corner of the property.

d. There is a hydro transformer located near the northwest corner of the property.

____ 12. How much is the house on Lot 67 set back from the north lot line?

a. .35 m

b. 3.65 m

c. 3.87 m

d. 10.05 m

____ 13. On which elevation is a walk-out deck indicated for Lot 67?

a. south elevation

b. north elevation

c. east elevation

d. west elevation

____ 14. What is the change in elevation from the underside of footing to the finished first floor?

a. 2.23 m

b. 2.46 m

c. 2.52 m

d. 2.70 m

____ 15. What is the elevation of the high point of the grade (highest point outside the house) on Lot 67?

a. 112.27 m

b. 111.91 m

c. 112.21 m

d. 112.91 m

____ 16. Which of the following is a building code requirement?

a. setback for house

b. limiting distance requirements

c. building coverage

d. building height

____ 17. What is the name of the building code(s) that all building codes are modelled after in Canada?

a. Canadian Building Code (CBC)

b. National Building Code of Canada (NBC)

c. National Research Code (NRC)

d. International Building Code (IBC)

____ 18. When would a building permit be required?

a. to remove a non-load-bearing interior wall

b. to remove a small load-bearing wall

c. to build a shed 10'-0" × 7'-0"

d. to build a shed 8'-0" × 8'-0"

____ 19. Which one of the following options requires a building permit?

a. replacement of a furnace

b. a wood deck built 300 mm from the ground

c. installation of non-load-bearing walls

d. construction of an addition to a house

____ 20. Who produces and regulates the zoning requirements for a particular town or city?

a. the National Building Code of Canada

b. the Provincial Building Code

c. the local government of the town or city

d. the federal government

6 Footings, Foundations, Dampproofing, and Waterproofing

LEARNING OBJECTIVES

After completing this chapter, you will be able to:

- Identify the different methods and materials used to construct footings and foundations.
- Read, interpret, and apply information regarding the construction of footings and foundations.
- Identify the dimensions for the footing and foundation wall layout.
- Interpret basic NBC requirements as applied to housing and small buildings regarding footings and foundations.
- Identify the methods and materials used to dampproof and waterproof a foundation wall.

INTRODUCTION

The text now begins to integrate drawing theory and building practices together, starting with the foundation. When you are reviewing construction drawings for information regarding the foundation, it is important to fully understand the principles and materials behind the information you are reviewing. This chapter provides a background for the types of footings and foundations, code references, materials, and methods of construction. The chapter also includes important points to watch for when interpreting and applying information from the drawings to the actual construction of the house or addition. Errors made on the foundation can be catastrophic, and carry through the entire project, directly affecting its success; therefore, review foundation (basement) drawings and relevant specifications carefully, keeping in mind site conditions and time of year, while planning ahead to how the rest of the building will integrate with the foundation.

FOOTINGS

Footings provide the base for the building. All of the loads imposed on the building are distributed to the footings. The two types of loads imposed on a structure are:

- ***Live loads:*** These are loads that are not part of the building structure and that change; for example, people, furniture, and composite snow loads (combination of snow and rain) are examples of live loads.
- ***Dead loads:*** These loads are part of the building structure and do not change. Walls; floors; roof; brick; and block and concrete foundation walls are examples of dead loads. The weight of both dead and live loads is in turn distributed through the footings to the soil/ground (bearing surface).

You will improve your understanding of structures and your ability to interpret construction drawings if you think of all of the weight (load) from the building, both live and dead loads, being distributed through the structure and eventually being transferred through the footings to the soil. Any load-bearing component that is not properly supported or is not transferring its weight through the house to the footings will cause settlement problems to the structure. Settlement problems include the shifting or movement of parts of the building structure, which can cause unsightly cracking and even structural failure.

For the building to be structurally sound, the bearing capacity of the soil must be strong enough to absorb the load being placed on it without moving. To make sure of this, the NBC requires that the bearing capacity be 75 kPa (10.9 psi) for footings used in housing and small buildings and that the footing rest on undisturbed soil, rock, or compacted granular fill (refer to Division B Part 9.15.1). **Undisturbed soil** refers to soil that has not been moved or altered from its original condition; it is thus suitably compacted to carry the load of the structure, provided it has a minimum bearing capacity of 75 kPa.

A couple of examples of disturbed soil (soil that would not be suitable to build on directly) are:

- An excavation that was dug too deep and then backfilled to the correct elevation. The backfilled portion is loose fill and will compress and cause settlement problems as the house or building is built. The problem can be resolved by re-engineering the area to meet specific compaction criteria or by filling with clean gravel that does not compact; this process must be designed and certified by an engineer.
- When adding an addition to a house and the footings for the addition are higher than the footings for the existing house. These footings would be resting on soil close to the house. When the house was originally built, the area close to the foundation wall would have been excavated to provide room to construct the original foundation. Therefore, the soil close to the house, even if the house was built 25 years previously, could cause settlement of the addition. This problem is resolved by stepping the footings down to the existing house foundation or undisturbed soil.

Footings must be placed below the **frost line** because, as temperatures reach below freezing, the ground expands and moves whatever is resting on it. This is known as heaving and can cause extensive damage to the structure (see Figure 6.1). Local municipalities will say how deep the footings must be placed below grade in order to ensure that the frost level does not go deeper than the footings. Areas in warm climates such as the southern United States may require the footings to be dug only a minimal distance below grade because frost is not an issue. However, in most parts of Canada the footings are required to be dug much deeper. Table 9.12.2.2 of the NBC (see Table 6.1) provides information regarding the minimum depth of foundations for buildings containing a heated basement and for foundations containing no heated space. The standard depth of a foundation for clay or soils not clearly defined with a heated basement, where the foundation is not insulated to reduce heat loss through the footings, is 1.2 m or 3'–11"(most regions round upto 4'). For foundations containing no heated space, the depth of footing is below the frost line as set out by the local municipality.

Foundations can also be designed to allow for the movement caused by frost. In these cases, a reinforced slab is designed to move when the ground freezes. The slab is reinforced and rests on a bed of gravel to drain excess water away from the foundation. In northerly climates with **permafrost,** the ground does not thaw; in this case the foundation may be engineered and constructed without excavating. This type of foundation requires careful design consideration, including the potential for interior building heat loss to thaw the ground. As well, global warming and climate change is a matter of concern.

TABLE 6.1

Minimum Depths of Foundations (NBC Table 9.12.2.2)

	Minimum Depth of Foundation Containing Heated Basement or Crawl Space[1]		Minimum Depth of Foundation Containing No Heated Space[2]	
Type of Soil	Good Soil Drainage[3]	Poor Soil Drainage	Good Soil Drainage[3]	Poor Soil Drainage
Rock	No limit	No limit	No limit	No limit
Coarse grained soils	No limit	No limit	No limit	Below the depth of frost penetration
Silt	No limit	No limit	Below the depth of frost penetration	Below the depth of frost penetration
Clay or soils not clearly defined	1.2 m	1.2 m	1.2 m but not less than the depth of frost penetration	1.2 m but not less than the depth of frost penetration

Notes to Table 6.1
1 Foundation not insulated to reduce heat loss through the footings.
2 Including foundations insulated to reduce heat loss through the footings.
3 Good soil drainage to not less than the depth of frost penetration.

Source: Volume 1, Division B, of the National Building Code of Canada 2005, published by the National Research Council of Canada (NRC). Reproduced with the permission of the National Research Council of Canada, copyright holder.

FIGURE 6.1
Foundation heaving.

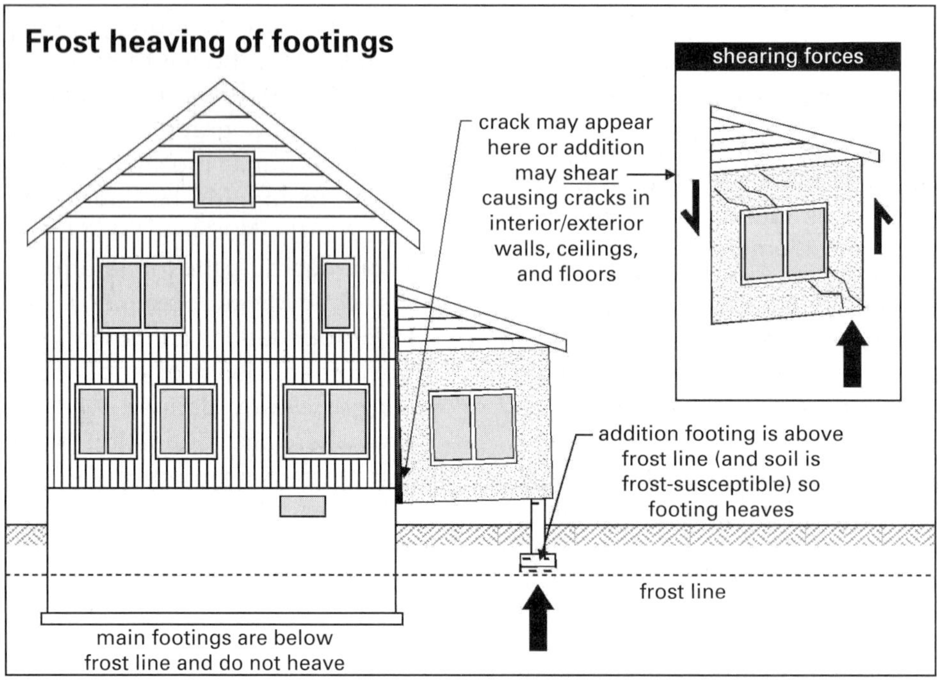

Source: Image excerpted from The Illustrated Home by Carson Dunlop www.carsondunlop.com

Foundations may also be constructed on top of rock, if the size of the rock is extensive as in the Canadian Shield; there, any topsoil is removed and the footings placed directly on top of the rock (see Figure 6.2). This type of situation will need to be verified by an engineer.

Concrete is the most common material used for footings; forms are constructed to hold the concrete while it is poured into place. The two basic types of footings found in most housing and small buildings are strip footings and column footings.

Strip footings also called spread footings because they spread out to distribute the loads. Strip footings are continuous and found under exterior foundation walls and interior load-bearing walls. Column footings are found under columns and support loads transferred from beams.

Footings cover a larger surface area than the walls or columns they support so the load is distributed over a larger area, thus lessening the load at any specific point, which could cause settlement. The relationship between the width of an un reinforced footing, the projection of the footing from the foundation wall, and the thickness of the footing must be maintained. The size of the footings is determined by the loads they

FIGURE 6.2
Footing of deck placed directly on rock. Footing is stabilized with steel dowel drilled to rock; note 4" block incorrectly placed on side under wood beam. Block should always be placed upright when supporting a load.

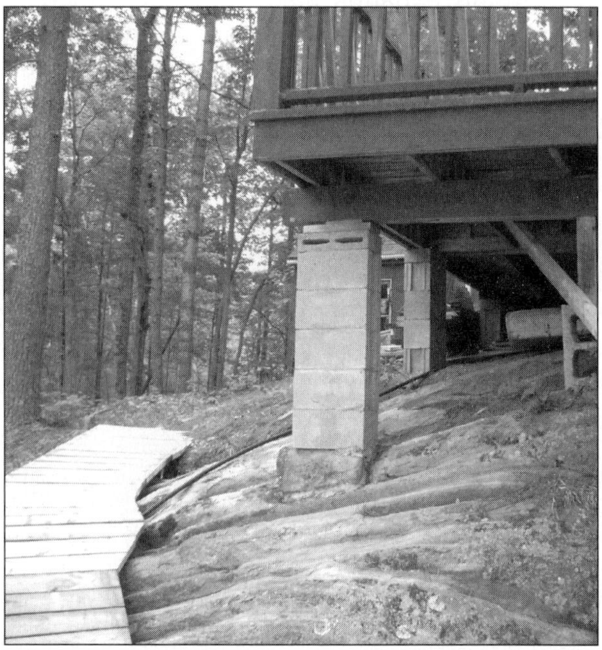

Courtesy of Tom Stephenson

FIGURE 6.3

Strip footings for a house; notice the gravel is placed where the concrete floor for the basement will be. The area without the gravel is the garage area, which has been excavated to the same depth as the house footings. On individual custom homes the garage footing would likely be stepped to the house footing because the basement is deeper than the minimum depth below the frost line. For production homes it is easier and more economical to trench excavate all the homes in a row at once to the same elevation depth (note, production drawings such as the Doncaster still may show the garage footing stepped, thus giving the builder the option).

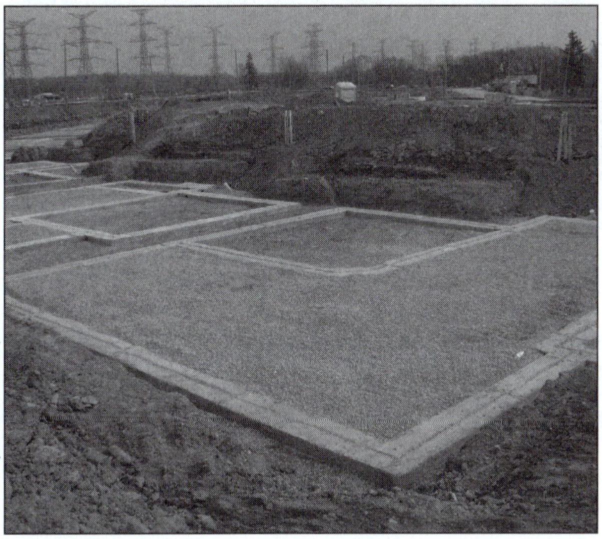

Courtesy of Tom Stephenson

FIGURE 6.4

Forming for interior column footing pad to be placed on undisturbed soil; additional stakes will be added to stiffen the form.

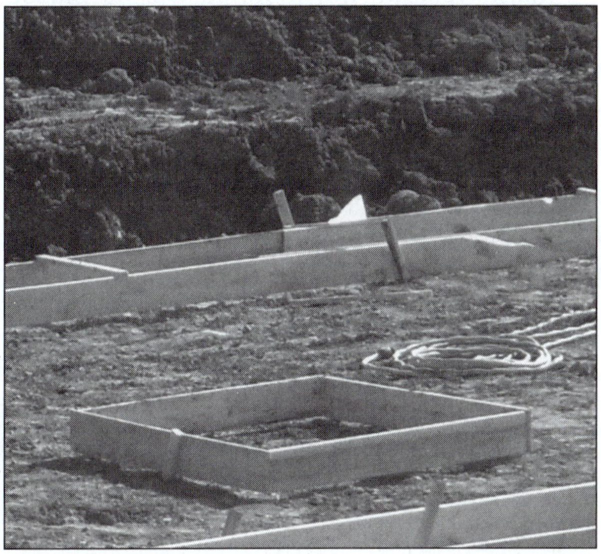

Courtesy of Tom Stephenson

carry. For example, footings supporting a one-storey wood frame house will be smaller than footings supporting a three-storey wood frame house. Also, because of the increase in dead-load weight, footings supporting a two-storey solid masonry house will need to be larger than footings supporting a two-storey wood frame house. Part 9.15.3 and Table 9.15.3.4 of the NBC list footing sizes for housing and small buildings up to three storeys, and 600 sq. metres or less in size. In order to calculate the footing size, the designer needs to know the following:

- The thickness of the foundation wall that is to be used
- The number of storeys the building is to be
- The type of construction for the building and roof system
    - Wood frame construction
    - Brick veneer construction
    - Solid masonry construction
    - Flat insulating concrete form wall construction (ICF)
- The supported joist span (the length of floor supported by the wall).

If any of the requirements do not conform to Part 9, then a geotechnical/structural engineer will be employed to design a suitable base and footing that conforms with the specific soil conditions on site and that meets the requirements of other particular sections of the Building Code, in particular Part 4 as specified in the NBC, which sets out the procedures that engineers must follow in calculating structural load requirements.

Footing Construction

Footings are constructed by building forms from dimensional lumber boards, usually 1" × 6" or 1" × 8", depending on the size requirements. The forms are placed the proper distance from the location pins or by dropping plumb lines from lines placed on the batter boards (refer to Chapter 5 for site plan information). The edges of the forms are nailed to stakes that are driven into the ground, and concrete is poured into the forms to construct the footing.

FIGURE 6.5
Strip and column forms for footings.

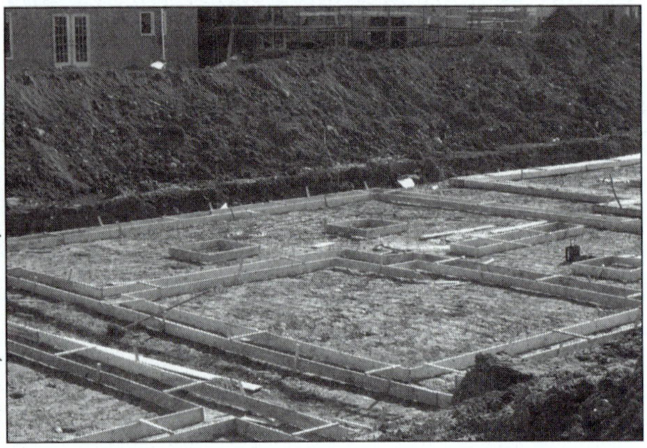

Courtesy of Tom Stephenson

FIGURE 6.6
Footings being poured using a concrete pump truck with boom for easier and more economical placement.

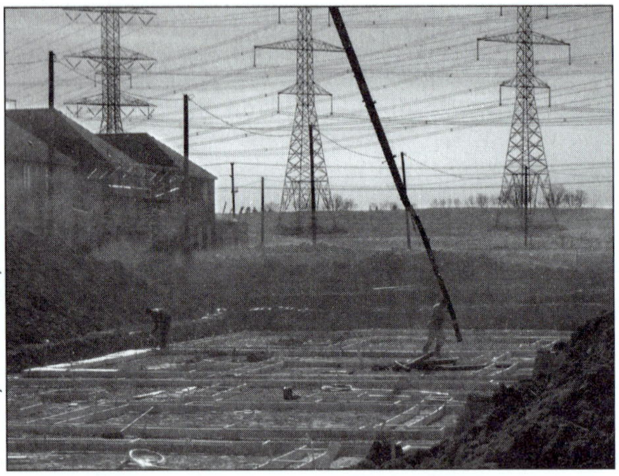

Courtesy of Tom Stephenson

FIGURE 6.7
Concrete footings in place for four homes.

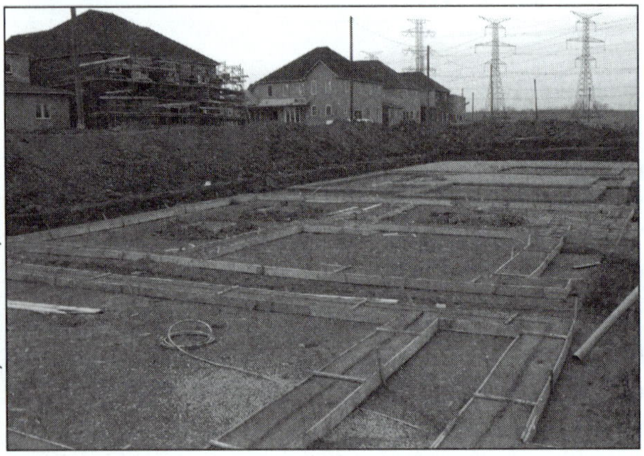

Courtesy of Tom Stephenson

FIGURE 6.8
Concrete footing is keyed with a V notch in the middle to tie foundation wall to footing and prevent lateral movement; notice corner wall locator pin. These pins are placed at the exact corners of the building according to the site plan and foundation/basement plan.

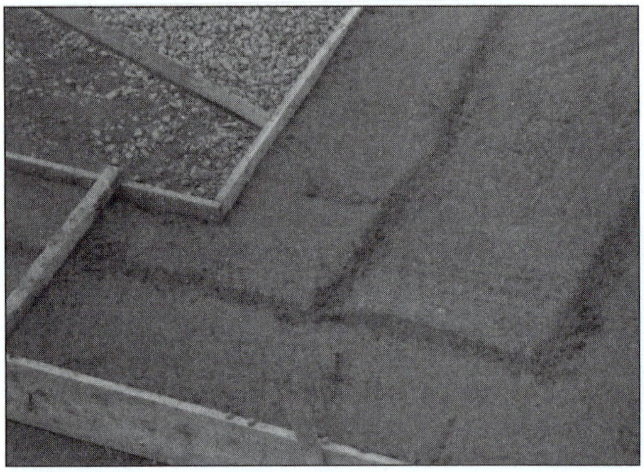

Courtesy of Tom Stephenson

Stepped Footings

Stepped footings are required when there is a change in the elevation of grade or in the requirements for the depth of footing. Examples of when a footing may be stepped include the following:

- If the grade level slopes toward the rear of the house, the footings will be stepped in order to ensure that the underside of the footing is maintained at a level below the frost line at all points. For example, if there is a basement walkout, the footings will be placed below the frost line at the walkout point.
- For construction of single homes, or in some cases production homes, the footing will be stepped if there is an attached garage and the basement of the house is at a depth well below the minimum for the frost line.
- If there is an attached porch that is unexcavated inside the porch walls, the footing of the house and the porch may be at different levels, requiring the footing to be stepped.
- Additions with a crawlspace attached to a house with a full basement will require the footings to be stepped.

FIGURE 6.9
Stepped footings being formed for one house on the left and already poured for house on the right. Both houses have a walkout basement. Notice the concrete consolidation problem with the second step on the right; can you spot it?

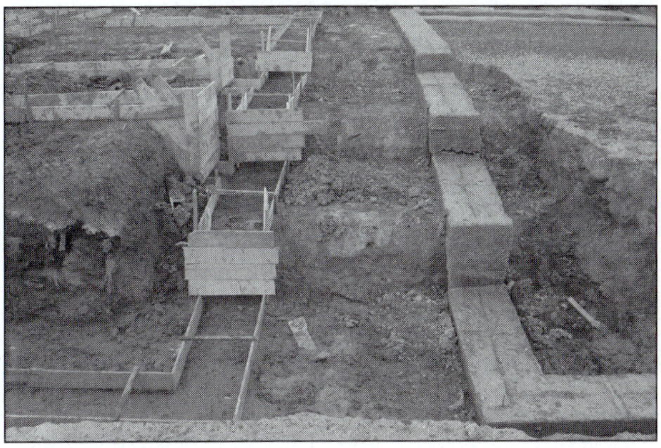

Courtesy of Tom Stephenson

FIGURE 6.10
Notice the level of gravel; it will form the base for the entire basement floor, which will be at one elevation. The bottom of the footings at all points must be at least 1.2 m below grade as per building code requirement (Review Table 6.1).

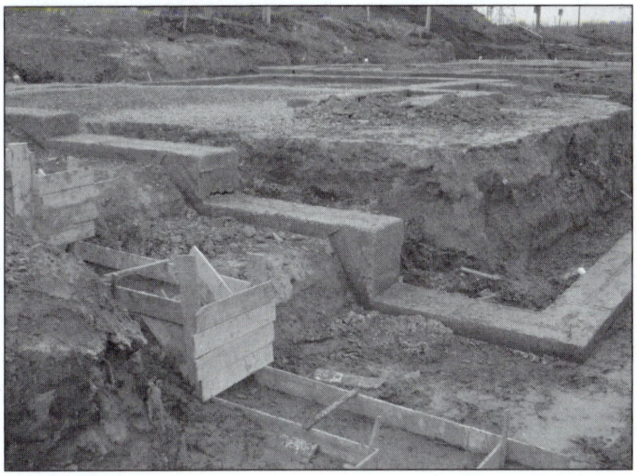

Courtesy of Tom Stephenson

FIGURE 6.11
Same house as Figure 6.10; photo taken from front of lot. The basement floor will be installed at the same elevation. The footings and foundation will be backfilled with a sloping grade, maintaining a minimum allowable distance below the frost line.

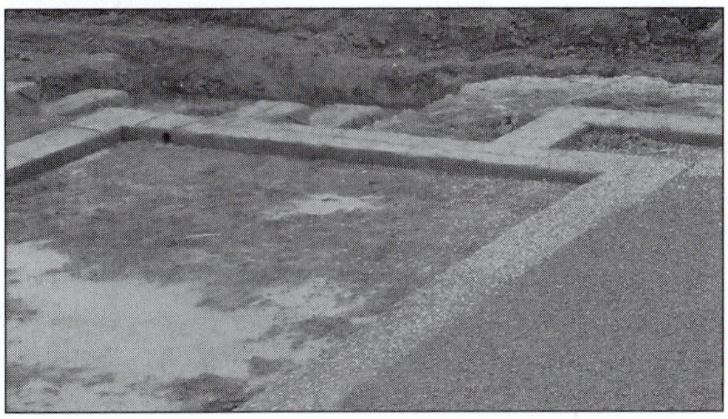

Courtesy of Tom Stephenson

FOUNDATION WALLS

There are several methods of constructing foundation walls, including poured concrete, concrete block, and permanent wood foundations (also known as preserved wood foundations or PWF). The type of foundation chosen will usually be based on:

- Economics; lowest-cost or best value solution
- Project requirements including site conditions (bearing capacity, soil conditions, water table, etc.) and structural requirements
- Regional preferences
- Availability of materials
- Weather conditions during placement
- Availability of skilled trades for the method selected.

To determine the type of material used for the foundation, refer to the elevations and view the foundation below grade shown as a dashed line, or refer to the construction notes and specifications.

Poured Concrete Foundations

Poured concrete foundations use forms that are constructed to contain the concrete while it sets up and cures. The forms are then removed and reused on other foundations. These walls may be reinforced with steel reinforcing rod or be unreinforced, depending on the project requirements. Check the construction notes or specifications to find information regarding the strength of concrete used in the foundation wall. Notes regarding reinforcing may also be found on the foundation plan, particularly if the reinforcing is localized due to a particular condition. Further review of plat and site plans may be required to identify lots that have engineered fill requiring specific reinforcing requirements in the footing and foundation walls.

Insulated Concrete Forms

Insulated concrete forms (ICF) is another method of construction that has been growing in popularity over the last few years. Numerous manufacturers produce ICF systems. The system may be used to construct concrete foundations or whole buildings, if desired. ICF construction uses forms that are similar to children's building blocks; they stack, interlock, and use reinforcing rods that clip into place. The outer walls of the

FIGURE 6.12
Forming for foundation wall; notice stepped footing and top of wall being formed to step. Note, in this example, rebar lying on top of footing is to be placed at top of wall below window openings following engineered fill requirements for this particular lot and its unique soil requirements. Site inspections would be carried out to ensure that the rebar is correctly placed in the wall before pouring of concrete.

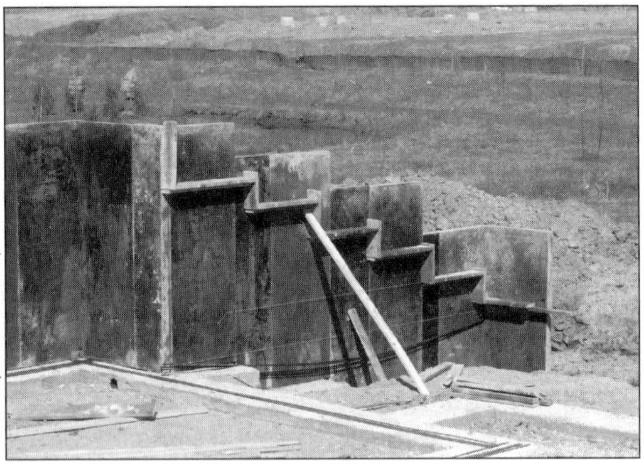

Courtesy of Tom Stephenson

FIGURE 6.13
Concrete forms in place.

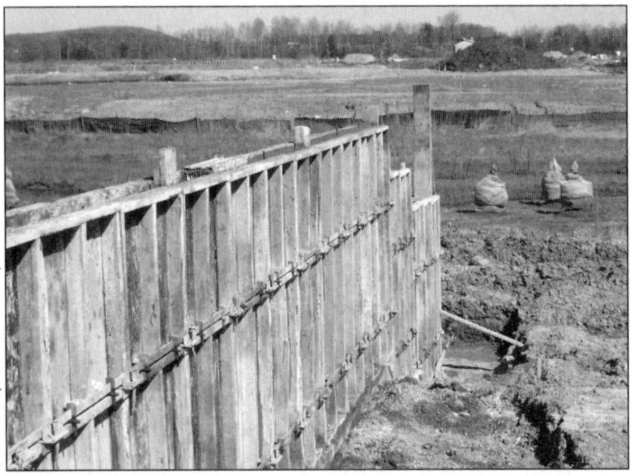

Courtesy of Tom Stephenson

FIGURE 6.14
Concrete forms set up and ready for concrete.

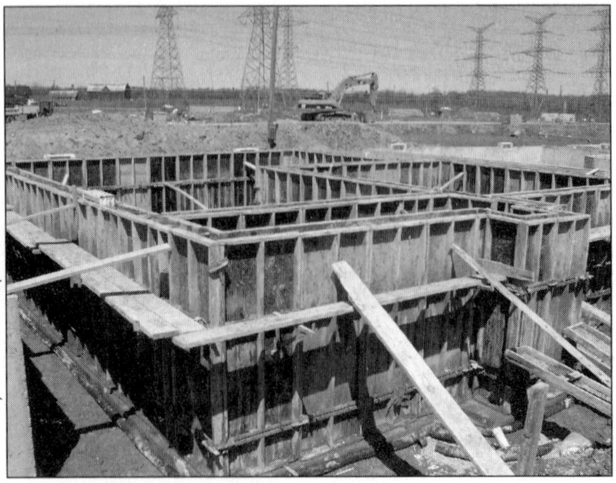

Courtesy of Tom Stephenson

FIGURE 6.15
Concrete being poured in forms using a concrete pump with boom.

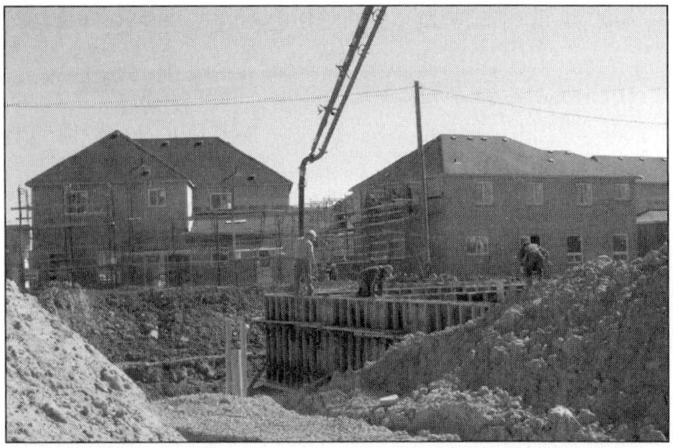

Courtesy of Tom Stephenson

FIGURE 6.16
Concrete in place and curing; notice the steel threaded rods sticking up from the concrete. These are called anchor bolts; they are L shaped and embedded into poured concrete and concrete block foundation walls. They tie the wood frame structure to the foundation wall once the sill plate is bolted down with washers and nuts.

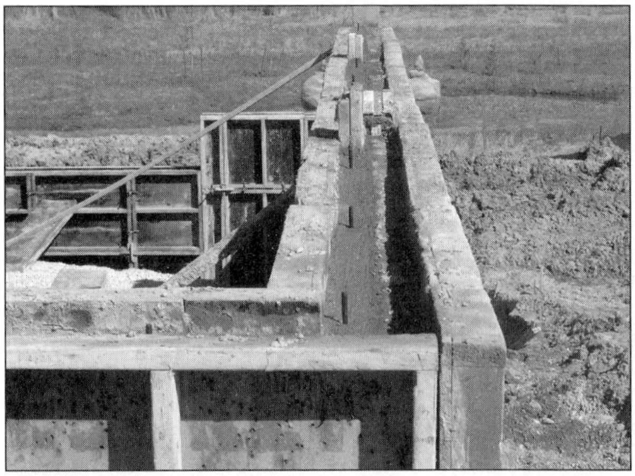

Courtesy of Tom Stephenson

FIGURE 6.17
Poured concrete foundation walls with most of the forms removed. Garage area would be backfilled and compacted and have gravel added before concrete slab is poured.

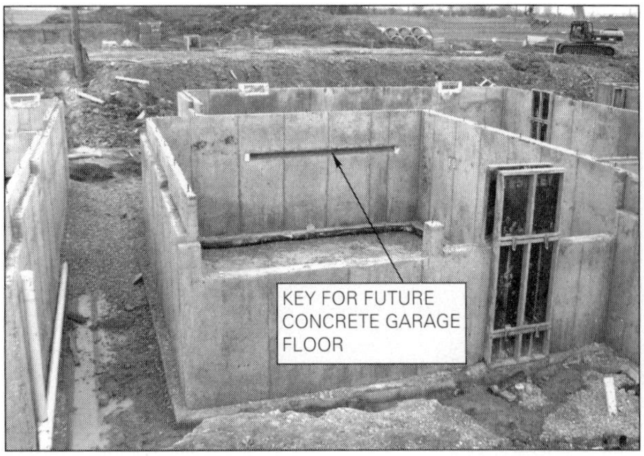

Courtesy of Tom Stephenson

forms are usually constructed of extruded polystyrene, providing the foundation wall with excellent insulating value afforded by a **thermal break** that greatly reduces heat flow paths through the wall. These paths are called a **thermal bridge**. The panels have lateral members that flatten out at the ends, providing surfaces that drywall and any other required finishes can be attached to.

FIGURE 6.18

ICF detail drawing indicating requirements for an ICF foundation wall with window well, drawn by the manufacturer Quad-Lock.

1. REINFORCED CONCRETE CORE
2. QUAD-LOCK REGULAR 2.25" PANEL
3. QUAD-LOCK TIE (8" YELLOW SHOWN)
4. VERTICAL REINFORCEMENT AS SPECIFIED
5. HORIZONTAL REINFORCEMENT AS SPECIFIED
6. LINTEL REINFORCEMENT AND STIRRUPS AS SPECIFIED
7. BOMAN KEMP 4040" 7.5" THICK EASY-BUCK"
8. FASTEN BUCK TO QL PANELS WITH WIND-LOCK SCREWS @ 6"O.C.
9. BOMAN KEMP CLEAR LEXAN GRATE COVER
10. BOMAN KEMP STEEL SAFETY GRATE
11. BOMAN KEMP 4'-0 ESCAPE LADDER (WHEN EXIT HEIGHT EXCEEDS 44")
12. FINISHED GRADE LINE APPROX.3" BELOW WELL SLOPE GRADE AWAY FROM WELL
13. GRAVEL BASE
14. PEEL&STICK WATER PROOFING OR SIM.
15. OPTIONAL BACKFILL PROTECTION - DIMPLEBOARD OR SIM.
16. WELL DRAIN SYSTEM
17. PERIMETER DRAIN
18. GEO-TEXTILE FABRIC (OPTIONAL)
19. FOOTING STRIP REINFORCEMENT AS SPECIFIED
20. ALTERNATE DOWELS AND EXTEND INTO WALL AS SPECIFIED
21. FOOTING STRIP SEE PLANS FOR SIZE AND REINF.
22. QUAD-LOCK TRACK
23. ZERO COMPACTION SAND OR GRAVEL
24. UNDER SLAB INSULATION - IF REQUIRED BY CODE
25. WWM OR LONG. & TRANS. REBAR FOR SLAB REINFORCEMENT
26. SLAB ON GRADE AS SPECIFIED
27. BOMAN KEMP 7.5" THICK 4040DOUBLE INSULATED VINYL WINDOW

SEE http://www.boman-kemp.com/ FOR MORE INFORMATION

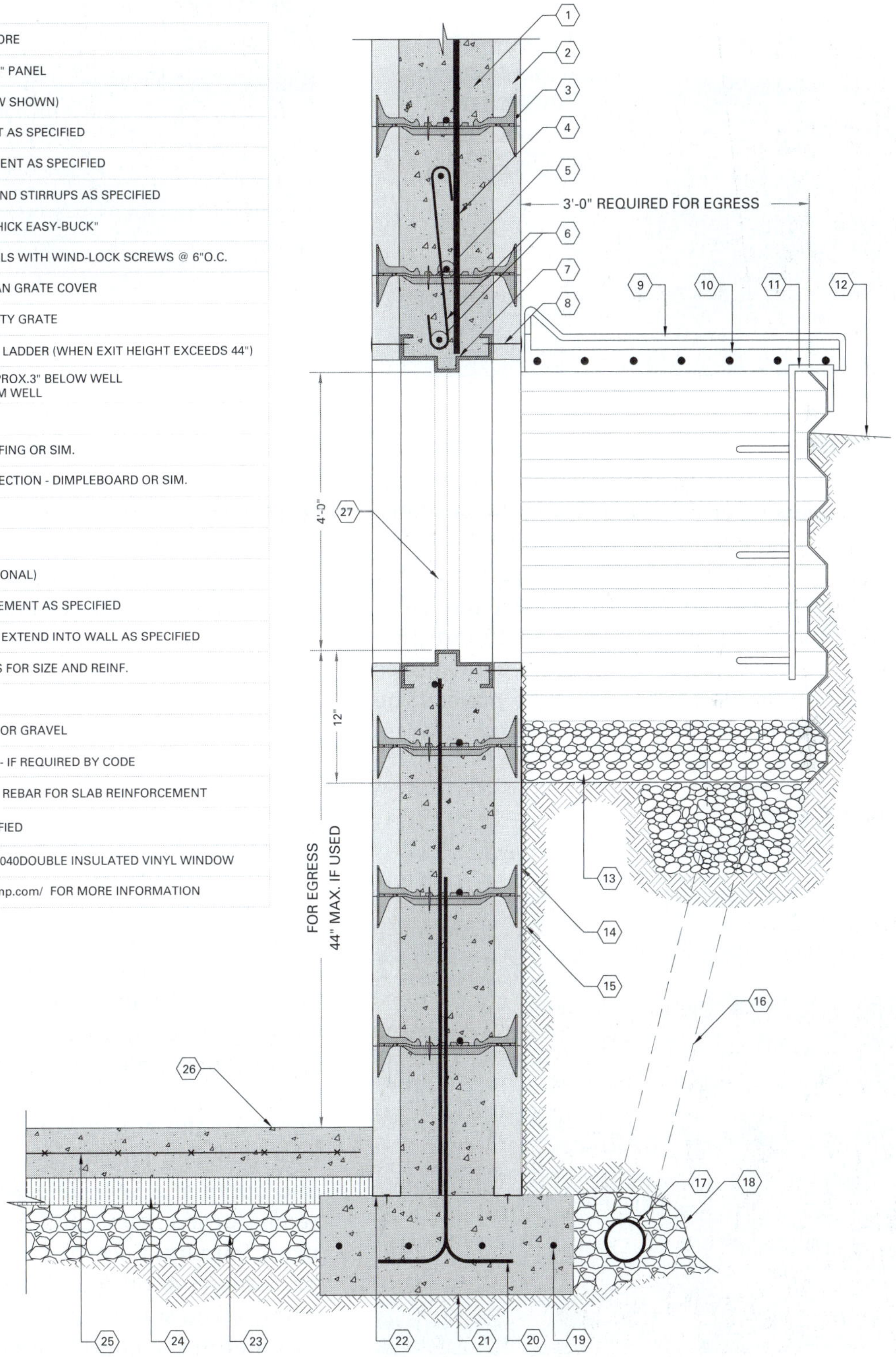

Source: Quad-Lock Building Systems Ltd.

FIGURE 6.19

Example of an ICF foundation that has been poured. Notice the ledge stepping out from the wall; it is for brick veneer that will be added later. In this case, the whole house and foundation will be constructed from ICF forms. Notice the rebar extending out above the forms and the forms on deck waiting to be placed.

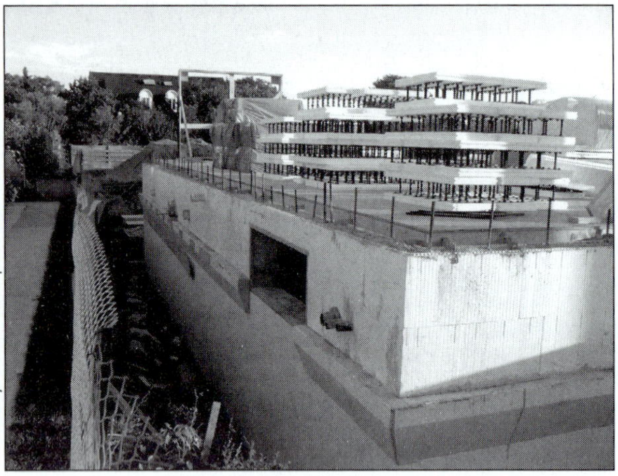

Courtesy of Tom Stephenson

Concrete Block Foundations

Foundations constructed using concrete block are long lasting and durable. Blocks are available in size increments of 50 mm starting from 100 mm (4") to 300 mm (12"). The foundation wall is constructed of either 200 mm (8"), 250 mm (10"), or 300 mm (12") blocks (including 10 mm [$\frac{3}{8}$"] mortar joint). The size of concrete block varies depending on the conditions of the project. The blocks are placed half lapped where successive courses lap over each other on mortar that provides the structural bond for the blocks. The top course of the block is filled solid with concrete (filler material is stuffed in the core holes to prevent the mortar from falling down to the bottom, or semi-solid blocks are used for the top course), and **anchor bolts** are inserted. The outer surface of the blocks is coated with a thin coat of mortar (approximately $\frac{1}{4}$" thick), which is known as parging or rendering (refer to Part 9.13 of the NBC). The parge coat fills any gaps or voids and is the first step in sealing the wall against moisture penetration.

FIGURE 6.20

Concrete block foundation wall.

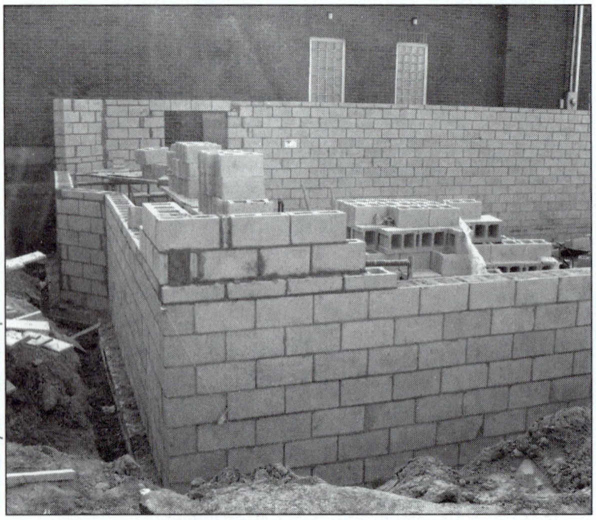

Courtesy of Tom Stephenson

FIGURE 6.21

Concrete block foundation wall with parge coat applied to 200 mm minimum above grade.

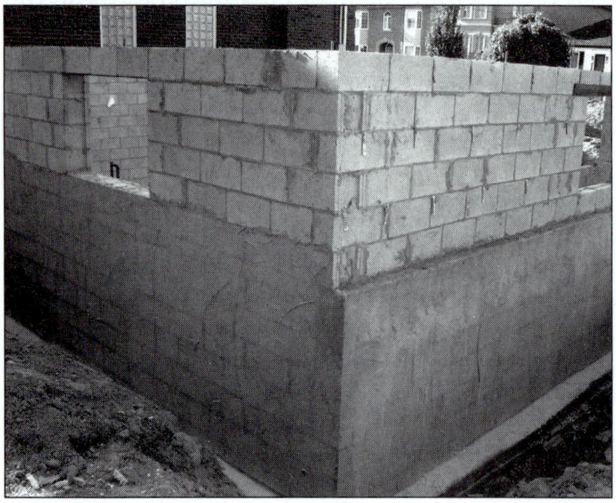

Courtesy of Tom Stephenson

Permanent Wood Foundations

This form of construction is more commonly used in some regions of Canada than others. Permanent wood foundations (PWF or preserved wood foundations) are constructed with pressure-treated wood that meets CSA 0322 qualification (refer to Figure 6.22). Advantages include requiring only a single framing crew to construct the house from footings through the roof, allowing for the easy addition of windows and doors at a later date. Also, the basement is ready for insulation and finishing, requiring no extra framing. Each piece of lumber used for a permanent wood foundation must carry a specific grading stamp that indicates that the lumber has met the applicable treatment requirements of permanent wood foundations under Standard CSA 0322 (refer to Figure 6.22) for a sample of the PWF grading stamp. A complete guide to designing and specifying is found in *Preserved Wood Foundations*, published by the Canadian Wood Council and CAN/CSA-S406 *Construction of Preserved Wood Foundations*.

Other Foundation Conditions and Designs

Piles

Where soil conditions are considered to be inadequate to provide the proper bearing capacity, piles may be used by either driving or drilling

FIGURE 6.22
PWF Grading Stamp.

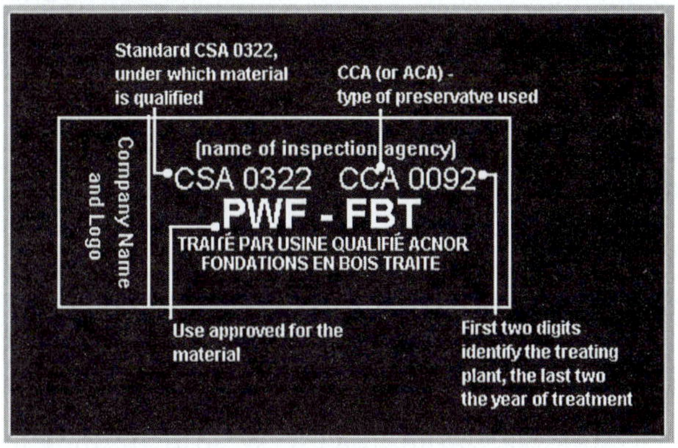

Source: Canadian Wood Council

FIGURE 6.23
Detail of a preserved wood foundation; note that there is no drainage (weeping) tile with this type of foundation, but there is the requirement to have a sump pump.

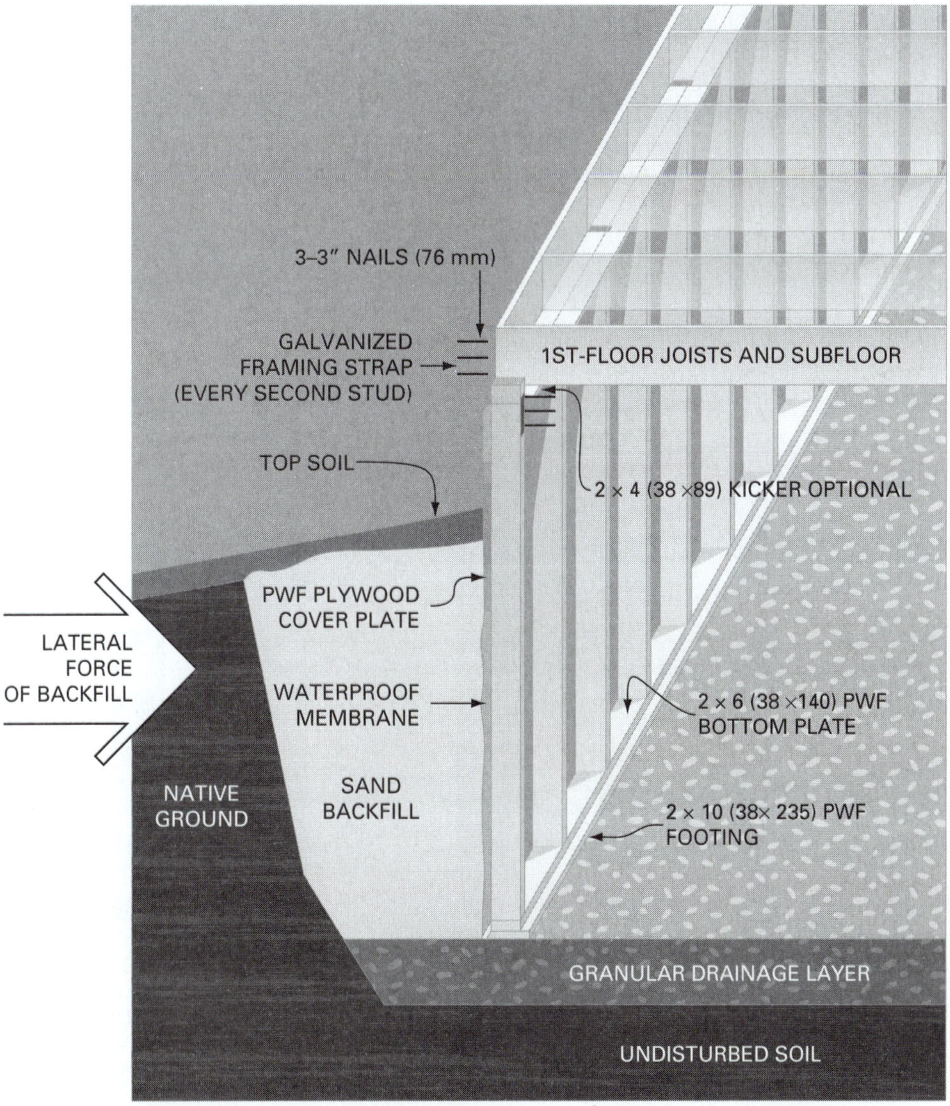

Source: From VOGT. Carpentry, 1E. © 2009 Nelson Education Ltd. Reproduced by permission. www.cengage.com/permissions

them into the ground. **Piles** can be constructed out of wood, steel, or concrete. Piles are designed by an engineer, who determines the depth and type of pile.

There are many types of piles and foundation designs that the engineer can use in designing a suitable foundation or retaining wall that will facilitate construction and/or support of the foundation.

Pile types:

- Bearing piles: These piles are driven until they hit a suitable bearing surface capable of supporting the foundation loads placed on them. This surface is often bedrock. A reinforced grade beam or strip footing is then placed on top of the bearing piles, which then support their load.
- Friction piles: These piles are driven into the ground and are designed to form a suitable foundation base as a result of the surface area of the pile, the type of surrounding soil, and the friction that is created from the soil placing pressure on the pile. A reinforced grade beam or strip footing is then placed on top of the bearing piles, which then support their load.

FIGURE 6.24

Example of soldier piles being used to act as a retaining wall while construction proceeds on building. Steel piles are driven into the ground prior to excavation and lagging is placed between piles as excavation proceeds. Tiebacks are used as specified for deep excavations.

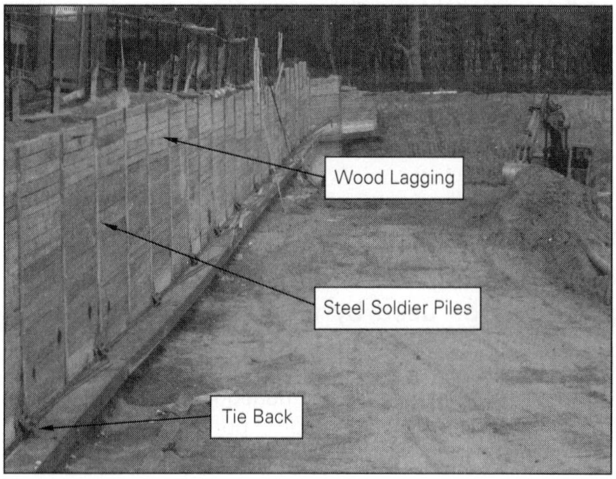

Courtesy of Tom Stephenson

FIGURE 6.25

Commonly used pile methods and materials.

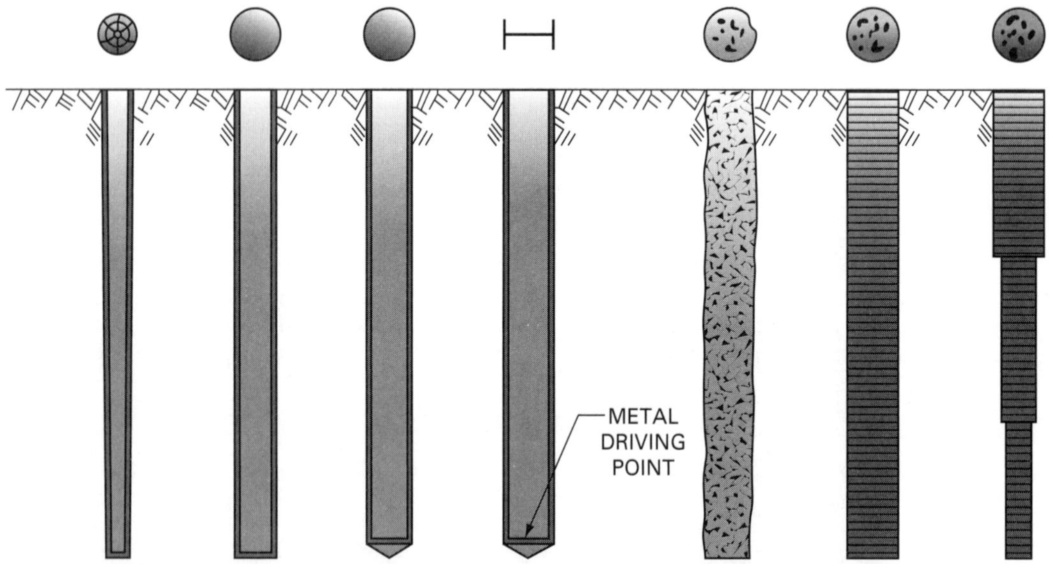

Source: From SPENCE, Construction Materials, Methods, and Techniques, 2E. © 2007 Delmar Learning, a part of Cengage Learning, Inc. Reproduced by permission. www.cengage.com/permissions

- Soldier piles: The purpose of these piles is to act as a retaining wall that facilitates ongoing construction of a foundation without risk of collapse of adjoining properties. An example is when steel piles are driven into the ground and wood laggings are placed between the steel piles to retain the wall of earth beside them. Tiebacks may be placed on deep excavations. (Review Figure 6.24)

Helical pier foundations:

- This type of foundation uses a helical (screw) steel pier that is installed by screwing the steel into the soil. The engineer can design the depth of the pier based on the torque, diameter of the screw, and the type of foundation that will be bearing on the pier. This foundation system can also be used to support existing foundations that are experiencing settlement problems.

Slab on Grade

Slab on grade design uses a reinforced concrete slab placed on undisturbed soil (the level of reinforcing will vary with purpose of use and site conditions). This type of foundation can be used in conditions where there is a high water table and where waterproofing of a basement does not therefore make economical sense. It can also be used as an economical solution for building simplified structures such as garages. The slab is designed to float up and down evenly during temperature changes caused by frost. Most jurisdictions will require an engineer's stamp when using this type of design.

Window Well

Window wells are used where basement windows are placed below grade. A window well provides an opportunity for extra light to be brought into the basement when the first floor is not placed very much above grade. If not properly installed and maintained, window wells are prone to leakage problems.

FIGURE 6.26

Grade beam on piles/piers and column footing.

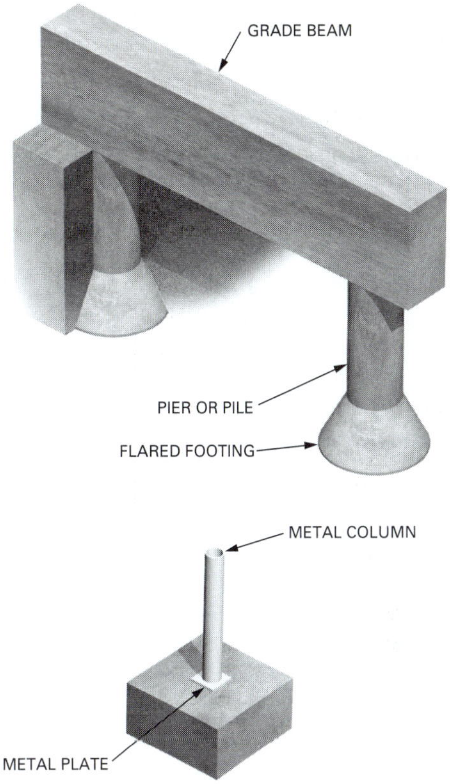

Source: From VOGT. Carpentry, 1E. © 2009 Nelson Education Ltd. Reproduced by permission. www.cengage.com/permissions

FIGURE 6.27

Section example of slab on grade for a one-storey garage not to exceed 538 square feet; notice use of lines and hatching to indicate materials and components, e.g., anchor bolt. All organic material to be excavated and removed, e.g. top soil.

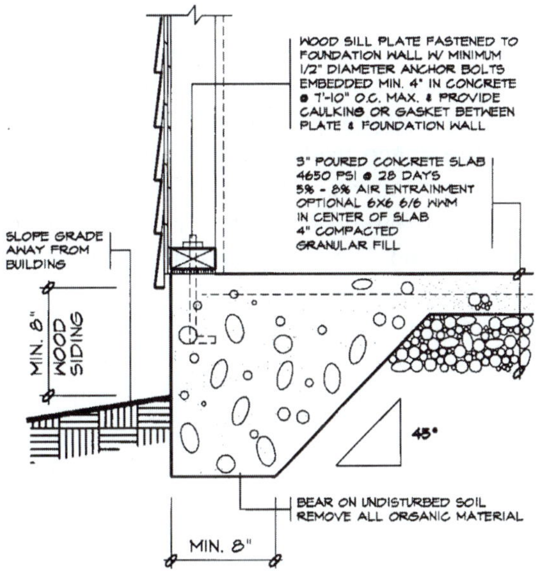

Source: Toronto Area Chief Building Officials Committee (TACBOC)

FIGURE 6.28

Example of how a window well is indicated on a basement floor plan drawing. This example is taken from the Brook drawings that came attached to this text. Review the basement plans on pages 3 and 4 to provide an overall perspective of window wells in use.

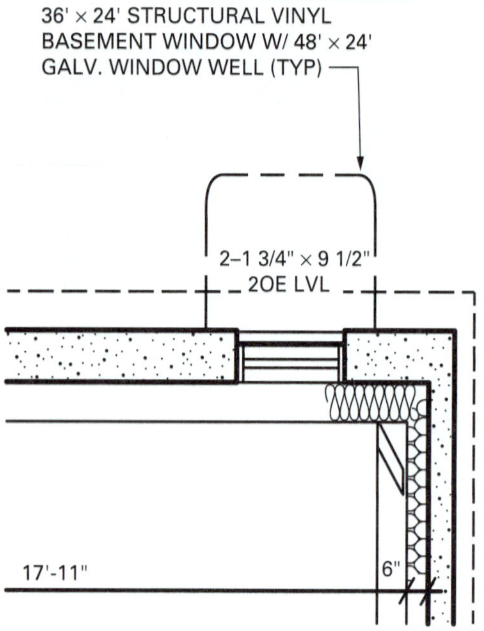

Source: Cassidy & Co. Architectural Technologists

Underpinning: Commonly Found on Renovation Drawings

This section provides background information that will help you review and visualize the walkout basement underpinning case drawings in Figures 6.33 and 6.34.

Underpinning is required when an existing foundation requires lowering. For example, a client may wish to excavate and finish an existing basement that does not have sufficient height to make the space useful. This process is quite different than new construction. The existing building must remain supported while the footings are lowered. The designer will supply a set of foundation drawings that indicate the process of sequencing for the underpinning.

Using the designer's process, the contractor divides the perimeter of the existing foundation into small sections, usually between 3' and 4' long, and numbers the sections in sequences of 3 or 4 depending on the design, e.g., 1, 2, 3, 4, 1, 2, 3, 4. This process is repeated for the length of foundation wall being lowered. The contractor then excavates one section at a time; for example, the workers will break the existing basement floor and dig down under the existing foundation for all the section 1s. The section 2s, 3s, and 4s will support the existing house. The section 1s will then be formed and have the concrete poured underneath, leaving a small space for grouting. Once the concrete is properly cured, a nonshrink grout is packed in between the underpinning and the existing foundation to ensure there is no settlement. When section 1 reaches its design strength, the method is then carried out for section 2 and so on.

Underpinning Case Drawings

The details provided in Figures 6.33 and 6.34 depict underpinning required for a foundation where a basement walkout is added to an existing house. In this example, the area below and to each side of the doorway requires that the footing be extended lower, as indicated in Section A Figure 6.31. This requirement ensures that all of the footings for the house remain below the frost line. If, as in our example, we open up and excavate an area to install a door, the existing footing at the opening and to the sides of the opening is now at grade and exposed to the cold and potential frost. This is corrected in an existing building by underpinning the foundation to below the frost line. In a new building the footing would be stepped to below the frost line. **From the drawings, observe carefully the use of cutting-plane lines, elevations, and sections to orient and visualize the house. You will need to view all of the details to fully understand the process involved in reviewing drawings.**

FIGURE 6.29

Section detail of window well. Note that footing drain is also referred to as weeping tile or drainage tile.

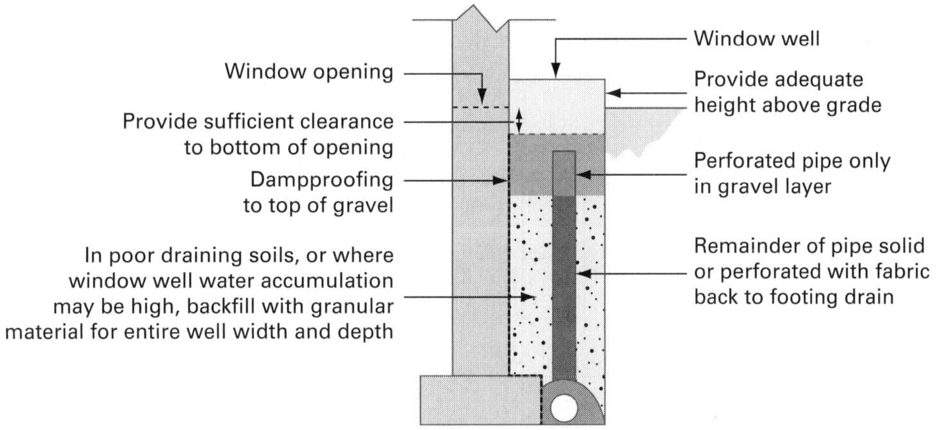

Source: NRC Performance Guidelines for Basement envelope systems and material final research report, Pg 98, Ted Kesik and Michael Swinton Institute for Research in Construction, National Research Council of Canada, 2005. Reproduced with the permission of the Minister of Public Words and Government Services Canada, 2010.

FIGURE 6.30

Isometric drawing of window well.

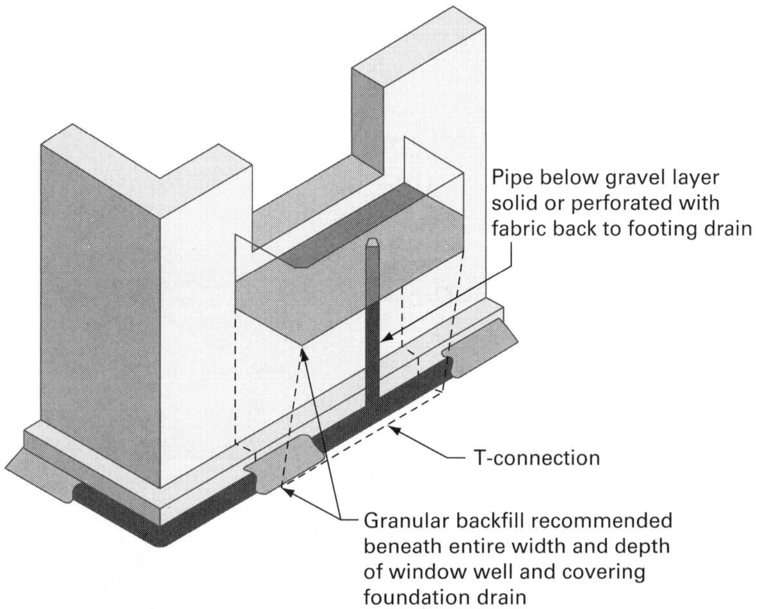

Source: NRC Performance Guidelines for Basement envelope systems and material final research report, Pg 99, Ted Kesik and Michael Swinton Institute for Research in Construction, National Research Council of Canada, 2005. Reproduced with the permission of the Minister of Public Words and Government Services Canada, 2010.

FIGURE 6.31

Example of underpinning a basement foundation wall; notice the drainage tile added under the basement slab. A nonshrink grout would normally be placed between the existing footing and the underpinning.

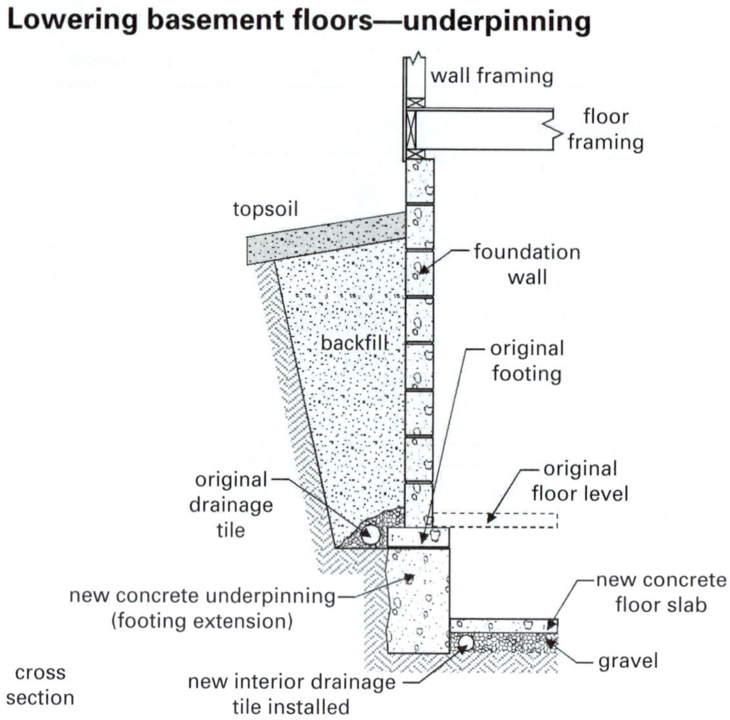

Source: Image excerpted from The Illustrated Home by Carson Dunlop www.carsondunlop.com

FIGURE 6.32

The secret to the underpinning process is the timing of the pours (as indicated below), which allows the concrete to set up so that the existing building structure is always supported as the footings are lowered.

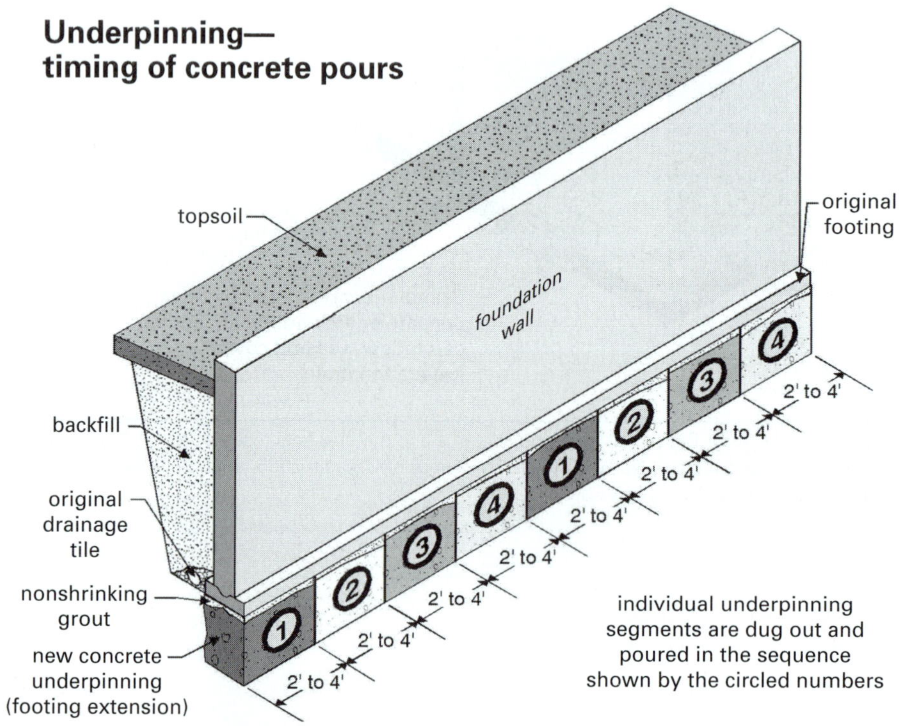

Source: Image excerpted from The Illustrated Home by Carson Dunlop www.carsondunlop.com

FIGURE 6.33

Basement walkout underpinning drawings; walkout plan, section, and notes.

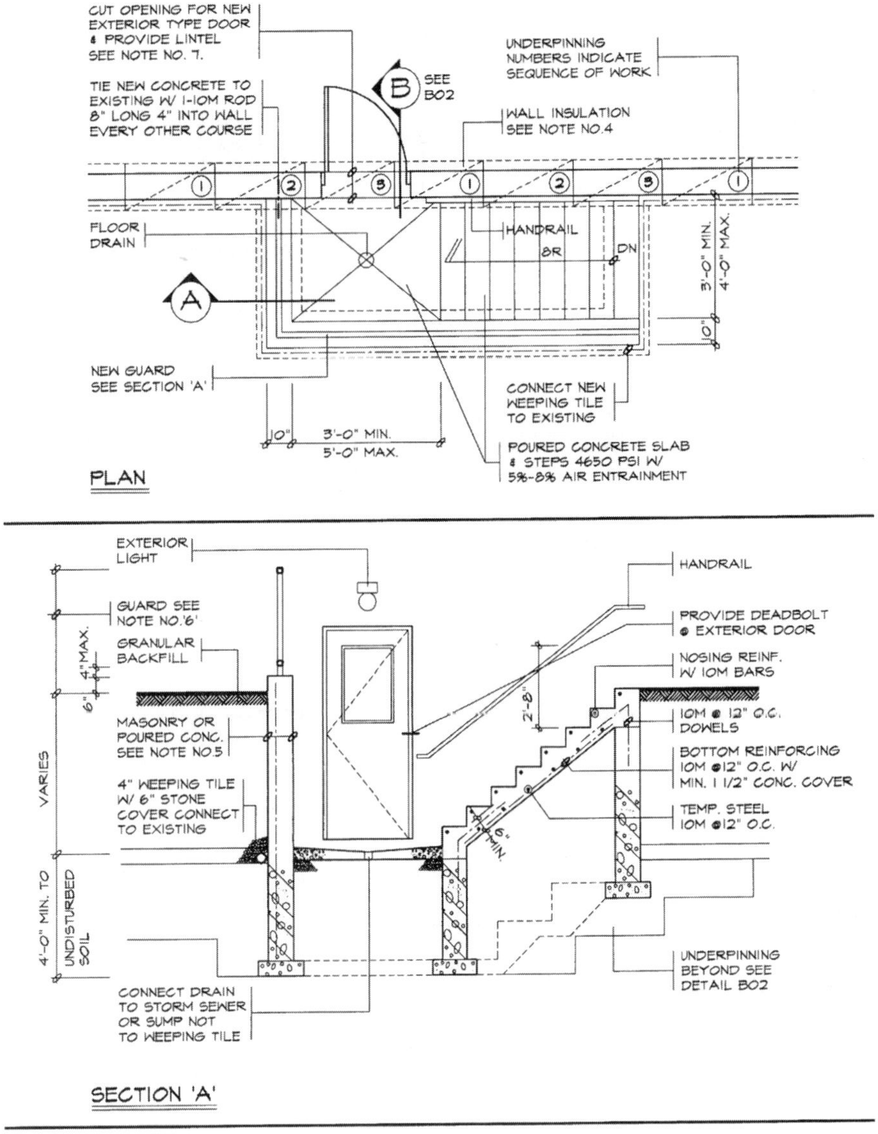

GENERAL NOTES

1. FOOTINGS

   16"X 6" POURED CONC. FOOTING ALL FOOTINGS SHALL REST ON NATURAL UNDISTURBED SOIL OR COMPACTED GRANULAR FILL

2. CONCRETE

   MINIMUM COMPRESSIVE STRENGTH OF 4650 PSI @ 28 DAYS W/ 5% TO 8% AIR ENTRAINMENT

3. EXTERIOR STAIRS

   7 7/8" RISE MAXIMUM 4 7/8" MINIMUM
   8 1/4" RUN MINIMUM 14" MAXIMUM
   9 1/4" TREAD MINIMUM 14" MAXIMUM

4. INSULATION

   MINIMUM R8 INSULATION & VAPOUR BARRIER ON THE INSIDE FACE OF THE EXPOSED FOUNDATION WALL

5. RETAINING WALL

   10" MASONRY OR POURED CONCRETE W/ NO REINFORCING REQUIRED FOR WALL HEIGHTS TO A MAX. OF 3'-11" PROVIDE 15M VERTICAL REINFORCEMENT @ 16" O.C. FOR WALL HEIGHTS TO 7'-0"

6. GUARDS

   3'-6" HIGH WHERE DISTANCE FROM GRADE TO BOTTOM OF WALKOUT EXCEEDS 5'-11"; 2'-11" FOR LESSER HEIGHTS. MAXIMUM 4" BETWEEN VERTICAL PICKETS

7. LINTELS

   1. SOLID MASONRY: 2- 3 1/2"X3 1/2"X1/4" ANGLES
   2. BRICK VENEER: 1- 3 1/2"X3 1/2"X1/4" L + 2-2X8
   3. WOOD FRAME/SIDING: 2-2"X8"

Source: Toronto Area Chief Building Officials Committee (TACBOC)

FIGURE 6.34

Basement walkout underpinning drawings; underpinning section B taken from cutting plane line B in Figure 6.31.

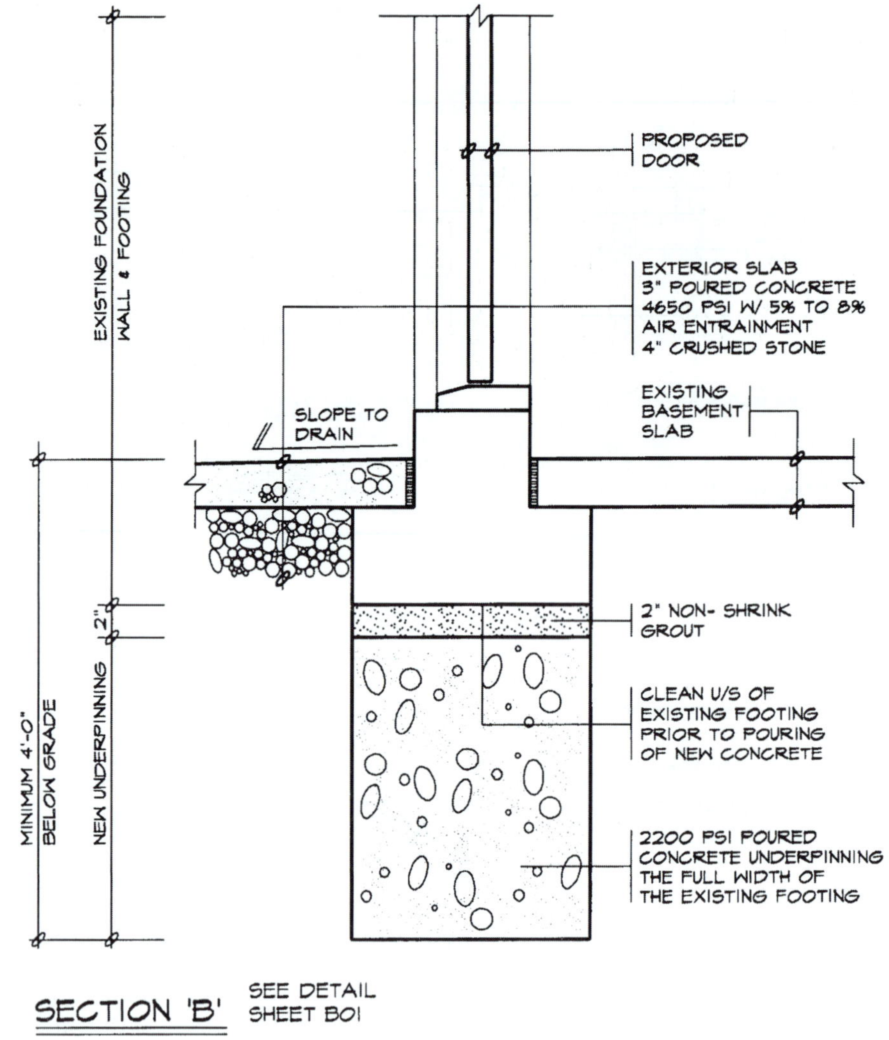

Source: Toronto Area Chief Building Officials Committee (TACBOC)

Foundation Wall Thickness

A designer will determine the thickness of a foundation wall by using tables found in Part 9.15.4 of the NBC. The requirements involve:

- Determining the type of material used: concrete, concrete block, or ICF
- If using concrete, determining the strength: 15 MPa (2200 psi) or 20 MPa (2900 psi)
- Measuring the overall height of wall from the top of the concrete floor slab to grade
- Determining whether the wall is considered **laterally supported** or **laterally unsupported.** For example, a sill plate anchoring a wall to the floor joists provides a great deal of lateral strength. However, walls that have large window openings or stairwell openings will be considered laterally unsupported because they are weak in these areas. In most cases where walls are laterally unsupported, they may need to be strengthened using reinforcing. The drawings will usually indicate the additional reinforcing requirements on the foundation plan or in a detail drawing. **To read drawings, you need to become knowledgeable regarding structural requirements. That means you will be less likely to omit something when reading drawings for estimating or construction purposes because you will be looking for it.** (Refer to the NBC Division B Part 9.15.4.2.)

FIGURE 6.35

The drawing below is a close-up of page 2 of the Doncaster basement plan, indicating an example of a laterally unsupported wall at the side of a stairwell opening and the addition of rebar to provide extra strength. The note 3-20 REBARS @100 O/C W/25 CONC. COVER refers to the requirement for three pieces of 20M (20 mm diameter) steel reinforcing bars to be placed horizontally near the top of the foundation wall 25 mm in from the inside face of the foundation wall (25 mm Conc. Cover) at a spacing of 100 mm on centre. This strengthens the wall where there is no lateral support provided by the floor due to the stairwell opening.

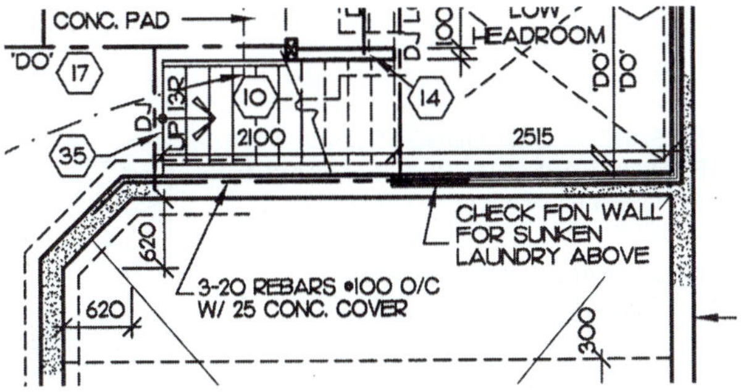

Source: Cassidy & Co. Architectural Technologists

FIGURE 6.36

The image below and the succeeding two figures are from the unfinished basement plan of the Brook drawings, pages 3, 8, and 17, and comprise an example of a laterally unsupported wall at stairs. Note the darkened centre line inside the concrete wall with the leader line pointing to it stating "reinforced foundation wall (see detail E)," which is based on cutting plane line E. In this particular case, the reinforcing is carried around the wall due to the dropped floor, which also requires a reverse veneer cut in the foundation wall.

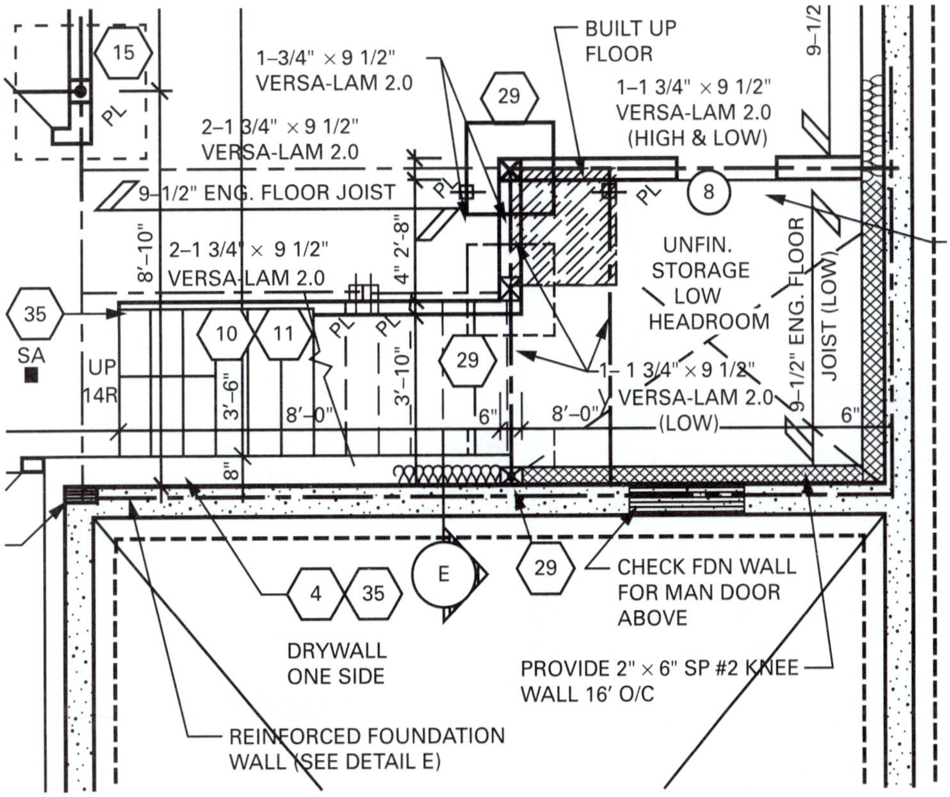

Source: Cassidy & Co. Architectural Technologists

FIGURE 6.37

Section detail indicating reinforcing to be placed in a laterally unsupported wall. The reinforcing rods are to be placed to the basement side of the foundation wall.

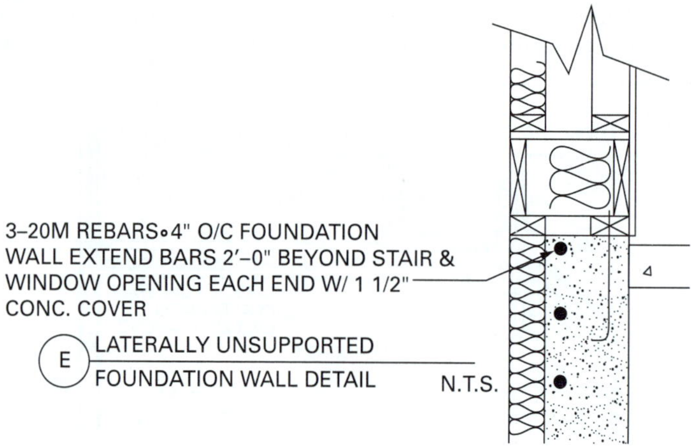

Source: Cassidy & Co. Architectural Technologists

FIGURE 6.38

This section view is a close-up of the whole house section A-A, taken from page 17 of the Brook drawings, indicating the reverse concrete reveal/veneer in the concrete foundation wall taken at the mud room. Note that the finished main level is 8" higher than the finished mud room (technically by building code this must be kept to 200 mm or 7⅞), and that the finished grade is 10" to the mud room floor. The actual top of the reveal would need to be at least 6" above grade to meet NBC requirements. The reinforcing indicated in Figure 6.36 would be placed on the inside of the concrete wall below the sill plate as indicated by the leader line inserted below.

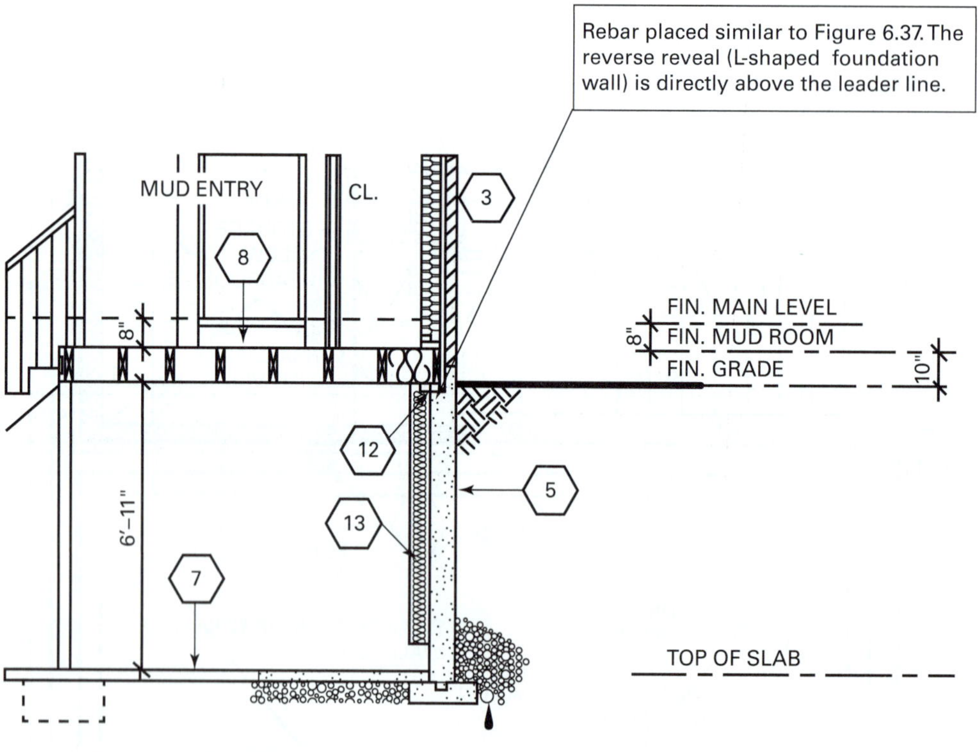

Source: Cassidy & Co. Architectural Technologists

WATERPROOFING, DAMPPROOFING, AND DRAINAGE

Foundations need to be protected from moisture leaking or penetrating through the foundation walls into the basement. Basements in residential construction are usually dampproofed (e.g. concrete and block foundations usually have a bituminous or other approved dampproof coating applied unless hydrostatic pressure is present). However, in the presence of **hydrostatic pressure,** they are required to be waterproofed (refer to Division B Part 9.13.3.1 of the NBC). **Waterproofing** provides a higher standard of protection in more severe conditions.

Dampproofing involves sealing all penetrations in the foundation wall left by form ties; concrete block walls are parged with a 6 mm (¼") sand and cement-based coating. The wall is then coated with a bituminous or other approved coating (refer to Division B Part 9.13.2 of the NBC). Many builders also add a drainage layer or fabric to the outside of the foundation wall to help prevent a buildup of hydrostatic pressure and allow the water to flow freely down to the footings. To ensure that no water gathers at the foundation, a **drainage pipe** (also referred to as **weeping tile,** drainage tile, or foundation drain) is placed along the side of the footing. Gravel is then placed over the pipe at the base of the footing and next to the foundation wall. Gravel is also placed under the concrete floor slab to ensure that water and or moisture does not travel upward through the slab by capillary action. See Figures 6.41 to 6.46 for examples of dampproofing a foundation wall.

FIGURE 6.39

Many approved waterproofing materials and membranes are available and must be installed in accordance with manufacturer's installation guidelines.

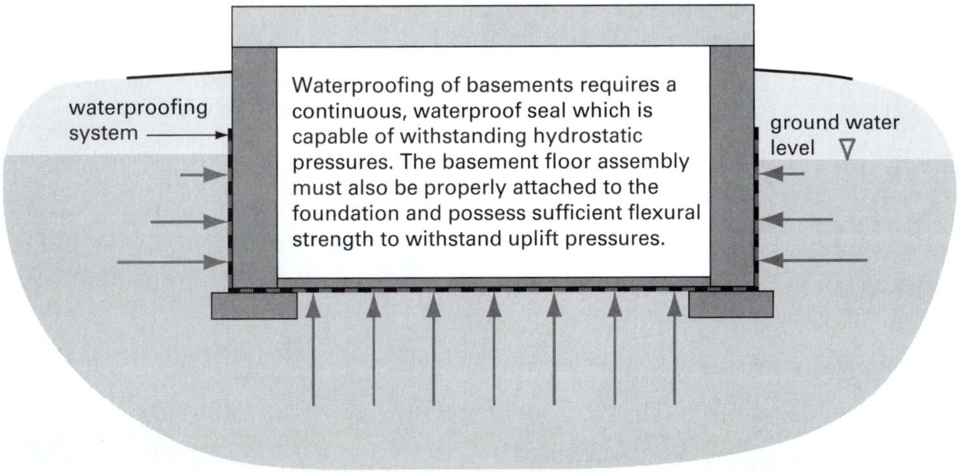

Source: NRC Performance Guidelines for Basement envelope systems and material final research report, Pg 50, Ted Kesik and Michael Swinton Institute for Research in Construction, National Research Council of Canada, 2005. Reproduced with the permission of the Minister of Public Words and Government Services Canada, 2010.

FIGURE 6.40

The photo depicts the foundations of two side by side houses. The foundation wall on the right illustrates snap tie form holes patched to prevent moisture ingress and drainage tile with nylon sock (sock prevents drainage tile from clogging over time) placed at side of footing. Snap ties are used in the construction of the formwork to hold the walls of the form at a preset distance, for example, 200 mm for a 200 mm foundation wall. The ties snap off below the surface of the concrete and then are patched to seal and prevent moisture migration to the basement. The finished grade line is marked on walls based on the elevations taken from the site plan, as discussed in Chapter 5.

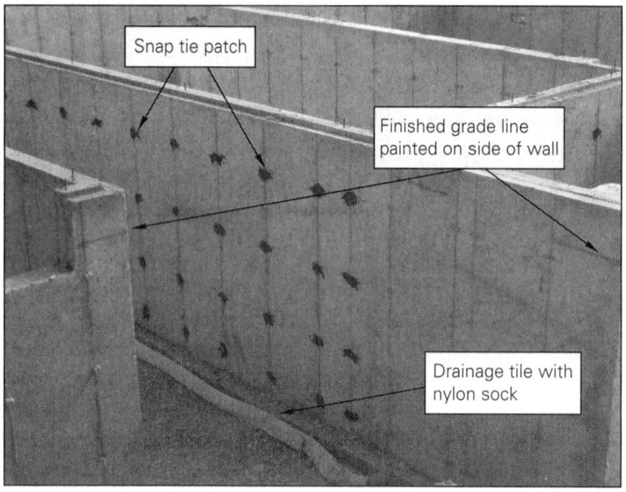

Courtesy Tom Stephenson

FIGURE 6.41

Foundation wall being dampproofed with bituminous coating applied and drainage tile placed around foundation; notice at the right side the coating does not extend around garage but does extend inside garage area where the foundation wall is connected to the basement.

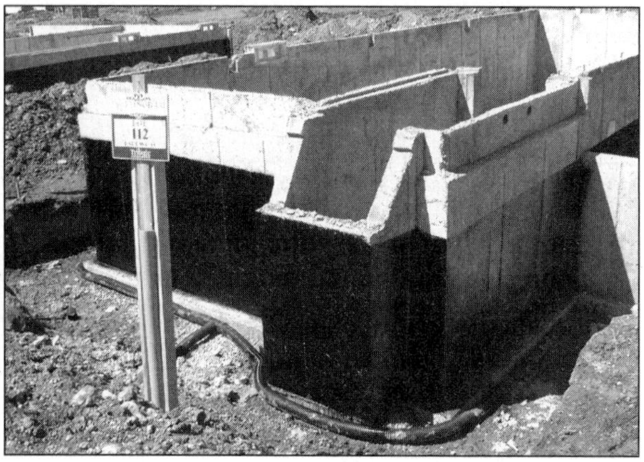

Courtesy Tom Stephenson

FIGURE 6.42

Drainage tile draining into storm drain at T joint; notice connection pipe for rain water leaders (RWL). Often soil conditions will dictate the use of a nylon sock to prevent clogging of the drainage tile over time.

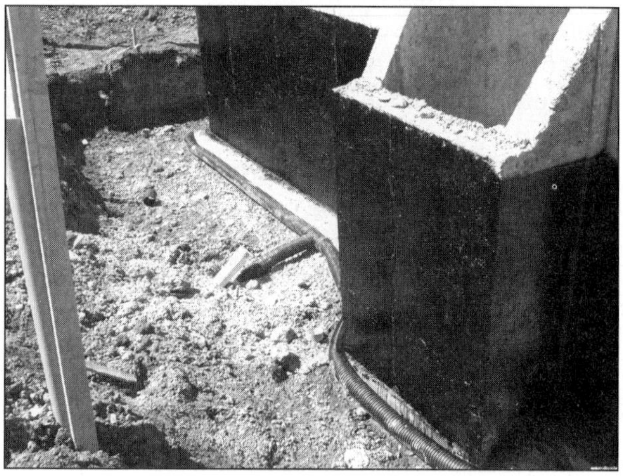

Courtesy Tom Stephenson

FIGURE 6.43

Example of a drainage pipe (weeping tile, drainage tile); serrations are placed down so that tile does not become clogged with sediment. As water drains to footings, the drainage tile allows the water to rise in the tile and flow freely to the storm drain and then to the storm sewer. A nylon sock is placed over the drainage pipe to prevent clogging in sandy soil conditions.

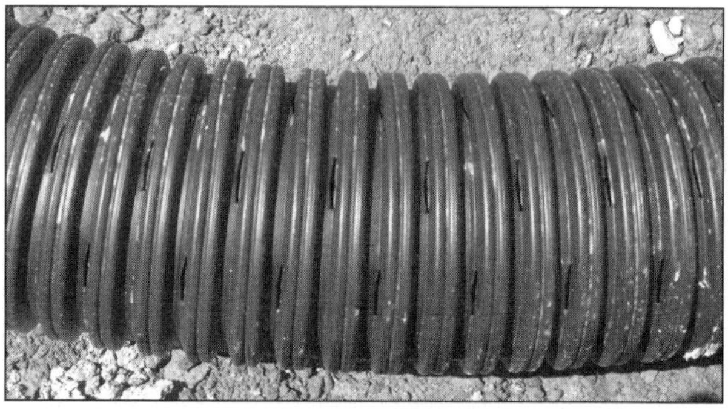

Courtesy Tom Stephenson

FIGURE 6.44

RWL drain pipes installed and hooked up to storm drain as a separate system that is not connected to drain tile. The RWLs' purpose is to remove surface water as part of a water management system; this practice will vary depending on local water management regulations. Notice the gravel at the rear right of the foundation wall and on the inside of the garage. The gravel is covering the drainage tile, which is placed along the exterior walls of the basement to protect it from moisture ingress.

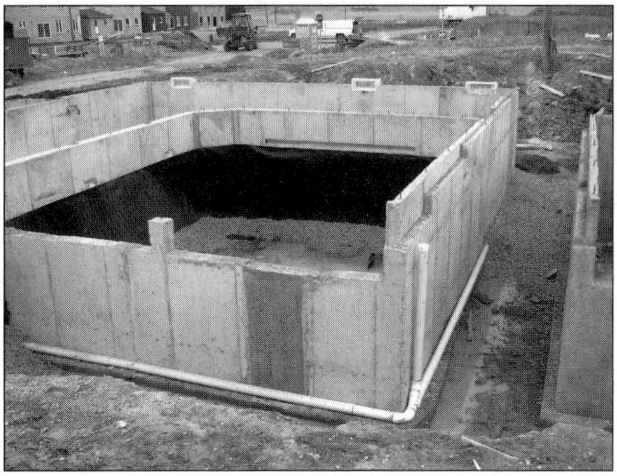

Courtesy Tom Stephenson

FIGURE 6.46

Closeup view of drainage layer and drain for RWL; notice the bubbles in the drainage layer, which provide a space between the wall to permit water to flow freely to the drain pipe at the base of the footing.

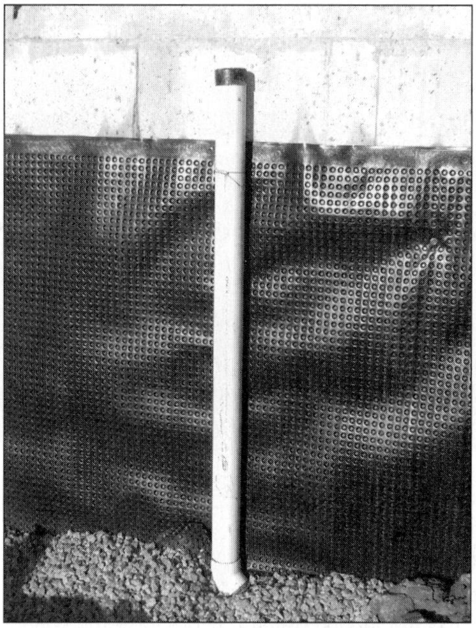

Courtesy Tom Stephenson

FIGURE 6.45

Drainage layer applied against foundation wall after bituminous coating has been applied; notice the gravel covering over the drainage tile and backfill placed over gravel.

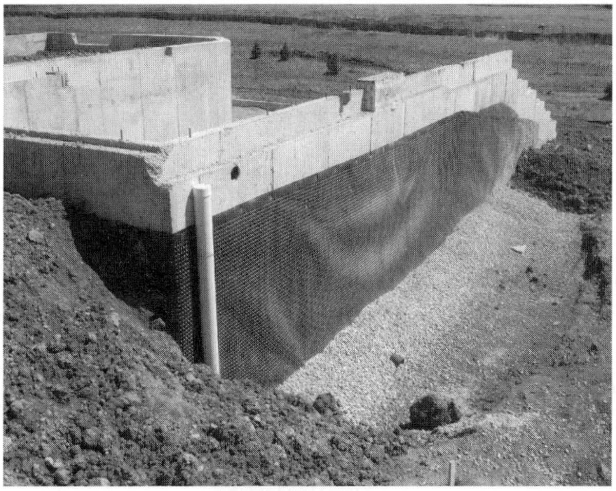

Courtesy Tom Stephenson

FIGURE 6.47

Stone slinger installing gravel base for basement concrete floor slab; notice concrete pump and boom in the background.

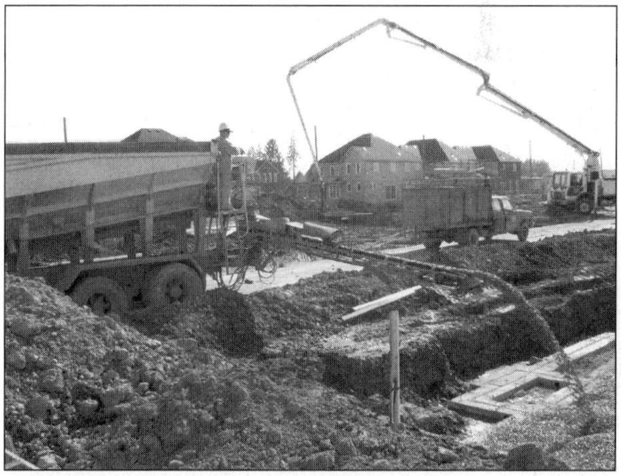

Courtesy Tom Stephenson

FIGURE 6.48

Doncaster house foundation: Pay attention to the outline of the footing and the location of garage, porch, and front step area. How are the footings shown on the Doncaster basement plan? And on the elevations?

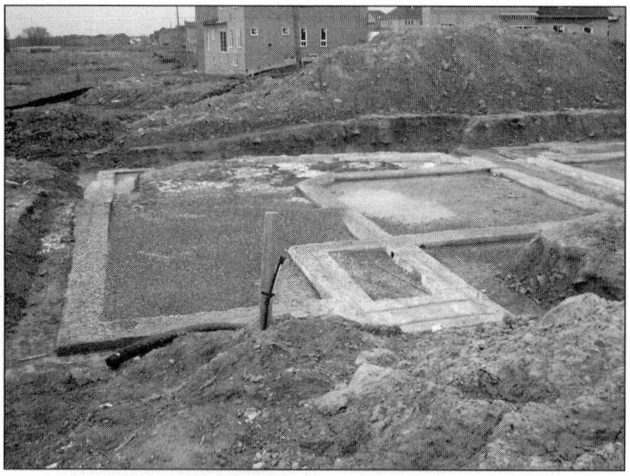

Courtesy Tom Stephenson

FIGURE 6.49

When reviewing the drawings, you will see that the rear of the foundation steps down due to the change in grade; the drawings do not indicate the stepping of the footing and foundation wall because they were drawn for a standard-level lot. Notice the indentation in the rear garage wall. The indentation is used to support the dropped floor for the laundry room; check the foundation and first-floor plan of the Doncaster house to see this more clearly. (Refer to Figure 6.50.)

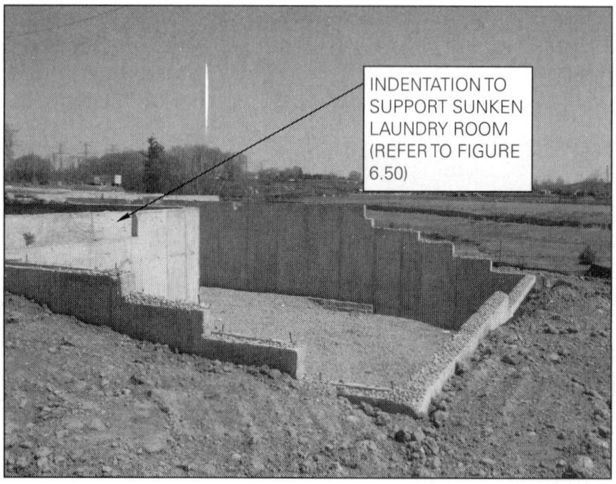

Courtesy Tom Stephenson

FIGURE 6.50

Compare floor-plan indentation referred to as "Check FDN wall for sunken laundry above" found in this figure. with the indentation in the foundation wall shown in the photo in Figure 6.49. This indentation will support the sunken laundry room floor later when the floor is framed.

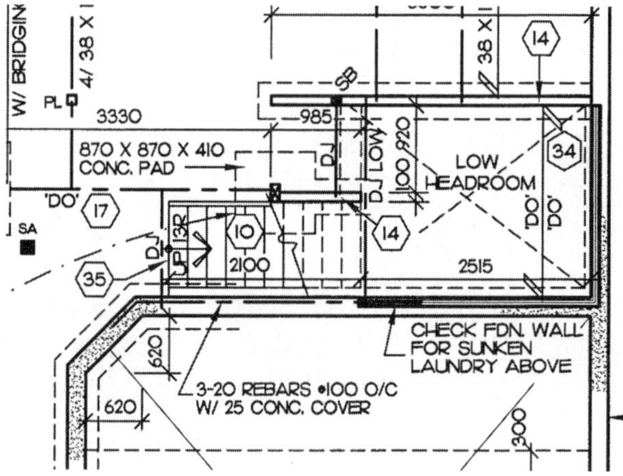

Source: Cassidy & Co. Architectural Technologists

FIGURE 6.51

3D view of foundation for Doncaster house. This view assumes a level grade: that is why the foundation wall does not step down as per Figure 6.49. Compare with Figure 6.50 and the basement floor plan (page 2) of the Doncaster drawings. Note the location of the column pads (footings) and the location of the indentation for the sunken landing.

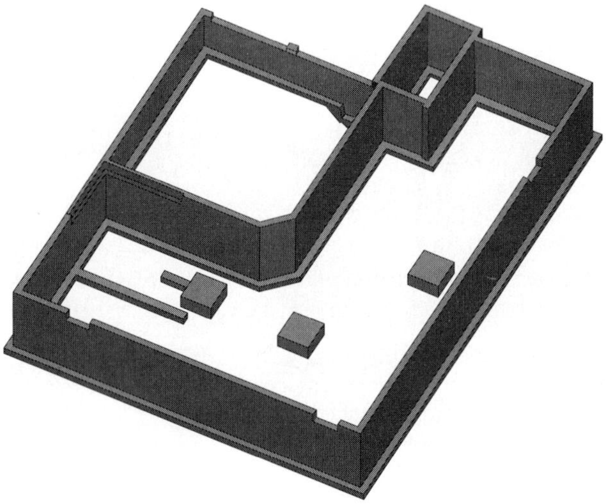

Source: Courtesy of Clint Kissoon, Chair, George Brown College, School of Architectural Studies.

FIGURE 6.52

Construction note taken from Doncaster drawings for foundation including wall and basement slab. Reference 5 provides information regarding the size of footing, thickness and strength of wall, dampproofing, and backfill requirements; reference 6 provides information regarding dampproofing including the wall coating, the weeping tile (drainage tile), and gravel covering; and reference 7 provides information regarding the strength of concrete slab and dampproofing requirements.

5 **FOUNDATION WALL**

200mm (8") POURED CONC. FOUNDATION WALL WITH 15 Mpa (2200 psi) CONC. STRG. AND BITUMINOUS DAMPPROOFING, 19mm (3/4") MINERAL FIBRE INSULATION WITH DENSITY OF 57 KG/M3 MIN. OR EQUIVALENT DRAINAGE LAYER. 480mm x 155mm (19" x 6") OR 600mm x 200mm (24" x 8") FOR PARTY WALLS CONTINUOUS KEYED CONC. FTG OR AS PER SOIL REPORT. BACKFILL WITH NON-FROST SUSCEPTIBLE SOIL.

6 100mm (4") DIA. WEEPING TILE, 150mm (6") CRUSHED STONE OVER AND AROUND ALL WEEPING TILES.

7 80mm (3") 15Mpa (2200 PSI) CONC. SLAB ON 100mm (4") CRUSHED STONE WITH DAMPPROOFING, 25 Mpa (3600 psi) CONC. WITHOUT DAMPPROOFING

Source: Cassidy & Co. Architectural Technologists

KEY TERMS

Definitions for key terms appear in the glossary at the end of the text.

Anchor bolt (p. 118)
Dampproofing (p. 129)
Dead load (p. 110)
Drainage pipe/weeping tile (p. 129)
Frost line (p. 111)
Hydrostatic pressure (p.129)
Insulated concrete forms (ICF) (p. 115)
Laterally supported (p. 126)
Laterally unsupported (p. 126)
Live load (p. 110)
Permafrost (p. 111)
Piles (p. 120)
Stepped footing (p. 114)
Strip footing (p. 112)
Thermal break (p. 117)
Thermal bridge (p. 117)
Underpinning (p. 122)
Undisturbed soil (p. 110)
Waterproofing (p. 129)
Window well (p. 121)

Chapter 6 Assignment: Foundations

Answer the questions below using the material in this chapter. Identify the letter that best answers the question.

_____ 1. What must occur in order to let more light into a basement that has a first-floor elevation that is placed relatively close to grade?

a. An underpinning condition is required.
b. A window well must be used.
c. A stepped footing is used to facilitate extra light.
d. An alternate solution requiring raising the first floor is the only solution.

_____ 2. Where will the greatest amount of information regarding the size of the footing for the walls, the thickness of the walls, and dampproofing be indicated?

a. the foundation plan
b. the first floor plan
c. the elevations
d. the construction notes

_____ 3. When the basement of an existing house requires additional head room, what is the technical name of the construction method used to accomplish this known as?

a. installing friction piles
b. installing bearing piles
c. underpinning the foundation
d. lowering the foundation

_____ 4. Which of the following responses is an example of a dead load?

a. snow on a roof
b. ice on a roof
c. sofa sitting in the living room
d. drywall installed on a wall

_____ 5. Which of the following responses is an example of a live load?

a. the floor joists that make up the floor system
b. the studs in the wall
c. the snow on the roof
d. the brick on the foundation wall

_____ 6. Which of the following is one of the criteria in the definition for "housing and small buildings" in Part 9 of the National Building Code of Canada?

a. at least 200 sq. ft. in size
b. 600 sq. metres or less in size
c. 4 storeys or less in height
d. 600 sq. metres or more in size

_____ 7. What other technical type of load do footings transfer to the ground besides live loads?

a. motion loads
b. dead loads
c. snow loads
d. joist loads

_____ 8. What type of soil is a foundation on when it is resting on soil that was excavated and then casually filled in?

a. undisturbed soil
b. compacted soil
c. disturbed soil
d. solid ground for building

_____ 9. According to the NBC, what is the minimum depth of footing for a heated basement placed on clay soil, where the basement wall is not insulated so as to prevent heat loss to the footing?

a. no limit
b. 1 metre
c. 1.2 metres
d. 2.4 metres

_____ 10. What does the abbreviation ICF refer to?

a. industrial concrete foundation
b. insulated concrete form
c. institutional concrete foundation
d. important concrete formation

_____ 11. For a poured concrete or block foundation wall, what item makes the connection for the framing to the foundation by fastening the sill plate to the foundation wall?

a. anchor bolts
b. the rim joists
c. screws and shields
d. construction adhesive

_____ 12. What foundation unit is designed wider than the walls that they support so they can better distribute the load over a greater surface area to the ground?

a. strip footings and column footings
b. concrete block foundations
c. poured concrete foundations
d. preserved wood foundation walls

_____ 13. When are stepped footings usually required?

a. when there is a large difference in elevation of the finished grade (sloping grade)
b. when stairs are used on a deck that leads from the rear of the house
c. when stairs are used on a front porch that leads from the front porch to the driveway
d. when there is a side entrance to the house's main floor

_____ 14. What special steps may be needed if the bearing surface of the soil falls below 75 kpa?

a. the floor joist system needs to be increased so that it can handle the deflection of the foundation
b. using judgement to increase the size of the footings to make up for the issue with the bearing surface of the soil
c. hiring a geotechnical/structural engineer to design the foundation of the house
d. designing according to the predefined tables in Part 9 of the building code

_____ 15. Which of the following does a permanent wood foundation use?

a. the same pressure-treated wood that is used for decks and fences
b. wood that carries a special grading stamp signifying that it is approved for permanent wood foundations
c. poured concrete on the outside wall to handle all of the lateral pressures that take place on the foundation
d. concrete block as a secondary barrier to moisture

_____ 16. Are there any circumstances where a footing may be placed above the frost line?

a. Where there is permafrost it is possible when carefully engineered.
b. Under no circumstances should a footing be placed above the frost line.
c. Whenever reinforced concrete is used, the footing can be placed above the frost line.
d. Whenever a preserved wood foundation is used, the footing can be placed above the frost line.

_____ 17. What steps must be done to the foundation when hydrostatic pressure is present?

a. The foundation must be dampproofed.
b. The foundation must be constructed with insulated concrete forms.
c. The foundation must be waterproofed.
d. The foundation must be constructed out of concrete.

_____ 18. What do foundations constructed out of concrete block require?

a. a parge coat on the outside of the foundation wall
b. patching of the snap ties
c. the cores of all the courses of blocks to be filled solid with concrete
d. waterproofing under all conditions

_____ 19. How will the footings on the foundation/basement plans be indicated in most cases?

a. heavy object lines
b. dashed hidden lines
c. cutting plane lines
d. centre lines

_____ 20. How are the foundation walls below grade displayed on elevation drawings?

a. by heavy object lines
b. by a blank space
c. by dashed hidden lines
d. by centre lines

Chapter 6 Assignment 2: Foundations Using the Brook Drawings

Use the Brook drawings that came attached to this text. Use the Elevation A drawings (note that these drawings include two different front elevations for the house, A and B), and focus in particular on Lower Level Plan Elevation A on page 3, while referring to the construction notes and appropriate elevation drawings as required by the question. Identify the letter that best answers the question.

_____ 1. Review the drawings and construction notes; what is the size of the strip footings for the house?

a. 19" × 6"
b. 21" × 4"
c. 22" × 6"
d. 24" × 8"

_____ 2. Where are the step footings indicated on the drawings?

a. under the kitchen area
b. under the laundry room
c. under the garage
d. under the living room area

_____ 3. What can you determine about the basement floor by reviewing the basement plan?

a. The basement floor steps down under the sunken mud room.
b. The basement floor is all one level.
c. The basement floor steps down under the garage.
d. The basement floor steps up at the optional rough-in bathroom.

_____ 4. How thick are the foundation walls for the house?

a. 4"
b. 6"
c. 8"
d. 10"

_____ 5. What is the size of the column footing (concrete pad) that is underneath the support column placed at the junction between the great room, dining room, and breakfast area above?

a. 34" × 34" × 16"
b. 36" × 36" × 24"
c. 24" × 24" × 12"
d. 20" × 20" × 10"

_____ 6. What reinforcement is required for the laterally unsupported section of the foundation wall?

a. 4-10M rebars @ 4" O/C foundation wall with bars extended 2'-0" beyond opening with 1½" concrete cover
b. 6-20M rebars @ 6" O/C foundation wall with bars extended 2'-0" beyond opening with 1½" concrete cover
c. 3-20M rebars @ 4" O/C foundation wall with bars extended 2'-0" beyond opening with 1½" concrete cover
d. 3-25M rebars @ 4" O/C foundation wall with bars extended 2'-0" beyond opening with 1½" concrete cover

_____ 7. Where is the foundation wall laterally unsupported and as a result, where does it need extra reinforcement?

a. near the front porch
b. at the window well areas for all basement windows
c. directly behind the optional washroom
d. at the side of the stairs and mud room

_____ 8. What does the concrete slab floor in the basement with dampproofing rest directly on?

a. 2" crushed stone
b. 3" crushed stone
c. 4" crushed stone
d. 5" crushed stone

_____ 9. What strength is the concrete specified for the foundation wall?

a. 2500 psi
b. 2900 psi
c. 4650 psi
d. 5000 psi

_____ 10. How many beam pockets are indicated on the foundation plan?

a. one
b. two

c. three

d. four

_____ 11. What size window well is used for the basement windows?

a. 32" × 18"

b. 33" × 12"

c. 35" × 16"

d. 48" × 24"

_____ 12. What is the overall longest length of the foundation (excluding the porch) measured at its longest point measured from outside of wall to outside of wall (front to back)?

a. 33'-9"

b. 48'-0"

c. 43'-2"

d. 53'-0"

_____ 13. What size weeping tile (drainage tile) is specified for the foundation?

a. 3"

b. 4"

c. 5"

d. 6"

_____ 14. What is the interior width (left side to right side of basement plan) of the foundation wall measured from inside wall to inside wall at its widest point?

a. 32'-5"

b. 33'-9"

c. 34'-6"

d. 53'-0"

_____ 15. What strength is the concrete specified for the garage floor?

a. 2500 psi

b. 2900 psi

c. 4650 psi

d. 5000 psi

_____ 16. What are the inside measurements of the garage foundation wall as indicated on the foundation plan?

a. 16'-0" × 20'-6"

b. 18'-5" × 21'-5"

c. 18'-11" × 21'-5"

d. 18'-5" × 22'-9"

_____ 17. What do the drawings indicate is the difference in elevation between the first-floor kitchen and the mud room?

a. 6"

b. 7"

c. 8"

d. 9"

_____ 18. What is the minimum (least) depth below grade indicated on the drawings for the underside of footings?

a. 3'-0"

b. 4'-0"

c. 8'-5"

d. 8'-7"

_____ 19. What is the height from the top of the basement slab to the top of the mud room finished floor?

a. 8'-7"

b. 8'-5"

c. 7'-11"

d. 6'-8"

_____ 20. What is the height from the top of the basement slab to the top of the first floor finished floor?

a. 8'-7"

b. 8'-5"

c. 7'-11"

d. 6'-8"

CHAPTER 6 Quiz Review: Foundations Using Walkout Drawings and Doncaster Drawings

Answer questions 1 through 8 from the basement walkout underpinning drawings in Figures 6.33 and 6.34, and answer questions 9 through 20 from the Doncaster drawings that came attached to this text. Identify the letter that best answers the question.

_____ 1. What is the minimum strength of concrete for the underpinning of the footings?

a. 2000 psi
b. 2200 psi
c. 4000 psi
d. 4650 psi

_____ 2. What is the height of the handrail going down the stairs?

a. 2'-4"
b. 2'-8"
c. 2'-11"
d. 3'-6"

_____ 3. What height is the concrete retaining wall shown to be above the finished grade?

a. 2"
b. 4"
c. 6"
d. 8"

_____ 4. How many risers are shown for the stairs?

a. three
b. six
c. seven
d. eight

_____ 5. How thick is the foundation retaining wall surrounding the stairs?

a. 6"
b. 8"
c. 10"
d. 12"

_____ 6. What must the weeping tile surrounding the retaining wall connect to?

a. the sump pump
b. the existing weeping tile
c. the sanitary drain
d. the floor drain

_____ 7. What size are the footings for the new foundation wall?

a. 16" wide and 6" deep
b. 18" wide and 8" deep
c. 20" wide and 10" deep
d. 24" wide and 12" deep

_____ 8. What needs to be placed between the existing footing of the house and the new footing placed underneath it (underpinned footing)?

a. silicone sealant
b. non-shrink grout
c. construction adhesive
d. concrete with a compressive strength of 4650 psi

_____ 9. Review the construction notes from the Doncaster drawings; what is the size of the footings for the house?

a. 332 mm × 221 mm
b. 480 mm × 155 mm
c. 600 mm × 200 mm
d. 800 mm × 250 mm

_____ 10. Where on the drawings are the step footings indicated?

a. under the kitchen area
b. under the laundry room
c. under the garage
d. under the living room area

_____ 11. What can you determine from reviewing the basement plan regarding the basement floor?

a. The basement floor steps down under the laundry room.
b. The basement floor is all one level.
c. The basement floor steps up at the optional cold room.
d. The basement floor steps up at the optional rough-in bathroom.

_____ 12. How thick are the foundation walls for the house?

a. 150 mm
b. 200 mm
c. 250 mm
d. 300 mm

_____ 13. What is the size of the column footing (concrete pad) nearest the stairs?

a. 400 mm × 400 mm × 600 mm

b. 870 mm × 870 mm × 410 mm

c. 990 mm × 990 mm × 460 mm

d. 1000 mm × 1000 mm × 600 mm

_____ 14. What do the drawings indicate is the difference in elevation between the first floor and the laundry room?

a. 300 mm

b. 380 mm

c. 400 mm

d. 420 mm

_____ 15. What size weeping tile (drainage tile) is specified for the foundation?

a. 60 mm

b. 80 mm

c. 100 mm

d. 120 mm

_____ 16. What reinforcement is required for the foundation wall, and why is it required?

a. The wall by the cold room door is laterally unsupported; therefore, to increase its strength there are 3-20 rebar @100 O/C W/25 CONC. COVER.

b. The wall by the cold room door is laterally unsupported; therefore, to increase its strength there are 3-10 rebar @100 O/C W/25 CONC. COVER.

c. The wall by the basement stairs is laterally unsupported; therefore, to increase its strength there are 6-20 rebar @200 O/C W/25 CONC. COVER.

d. The wall by the basement stairs is laterally unsupported; therefore, to increase its strength there are 3-20 rebar @100 O/C W/25 CONC. COVER.

_____ 17. What type of material is specified on the drawings to prevent water from seeping through the foundation walls into the basement?

a. waterproof coating and drainage layer

b. bituminous dampproofing and drainage layer

c. two inches of extruded polystyrene and peel-and-stick waterproof membrane

d. transparent spray-on sealer

_____ 18. What is the interior width (left side to right side of basement plan) of the foundation wall measured from inside wall to inside wall at its widest point?

a. 10370 mm

b. 10770 mm

c. 11170 mm

d. 14020 mm

_____ 19. What is the overall length of the foundation (excluding porch) measured at its longest point measured from outside of wall to outside of wall?

a. 10770 mm

b. 11170 mm

c. 14020 mm

d. 14420 mm

_____ 20. What is the minimum depth below grade indicated on the drawings for the underside of footings?

a. 1000 mm

b. 1220 mm

c. 1800 mm

d. 2300 mm

7 Methods of Residential Construction and Floor Framing

LEARNING OBJECTIVES

After completing this chapter, you will be able to:

- Identify the methods used in residential construction.
- Identify the structural elements of wood frame construction, and visualize the materials used and their orientation in a wood frame building.
- Read and apply floor framing information found on construction drawings.
- Locate and interpret information about columns and beams as indicated on construction drawings.
- Analyze and interpret information from construction drawings for the purpose of laying out a floor system.
- Identify, analyze, and apply sound framing methods that meet the minimum standards of the NBC.

INTRODUCTION

This chapter introduces you to typical residential construction methods used in Canada and focuses on wood frame construction for floor systems. An overview of the components of both dimensional lumber and engineered (manufactured) floor systems is provided. Understanding the components and how they go together is critical to being able to interpret and analyze information from drawings. As you work your way through the chapter, pay attention to the terminology used and how it is applied in the construction of the building. This will help you to identify and visualize the terms when you see them on the drawings. There are two assignments in this chapter: Assignment 1 is based on the theory of wood frame construction, and Assignment 2 uses the Brook drawings to provide practice at interpreting floor framing and related requirements on a house that uses imperial measurement and an engineered (manufactured) floor system. The chapter quiz is based on theory from the chapter and uses the Doncaster drawings to provide practice at interpreting flooring framing and related requirements.

METHODS OF RESIDENTIAL CONSTRUCTION

The materials and methods commonly used in construction are driven by value—value for the consumer and value for the builder. Consumers are willing to purchase when a building meets not only their budget but also their expectations of performance, energy efficiency, sustainability, and quality. The balance between these factors and budget is **value,** which becomes the driving force for designers, builders, and contractors when choosing materials and construction methods.

The following is a list of the more common construction methods used in Canada:

1. ***Wood frame construction:*** This method (also referred to as light frame, Western, platform, or stick construction) is the most popular and commonly used method of construction in Canada and North America because it is a

FIGURE 7.1
Platform construction wood frame.

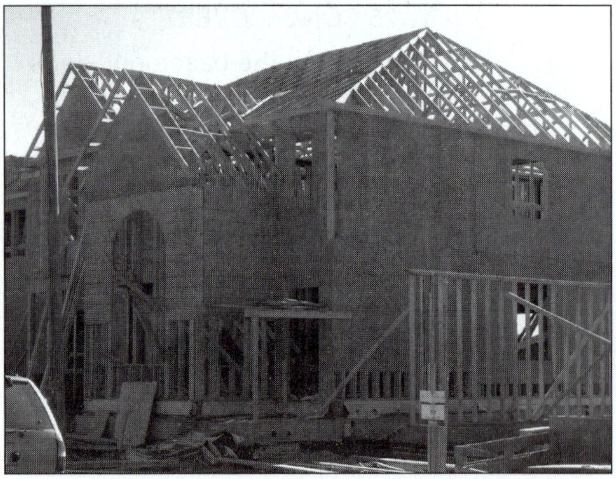

Courtesy Tom Stephenson

very efficient and economical way to build houses. When properly designed and constructed, wood frame construction is very durable and energy efficient. Due to its popularity, this text will focus mainly on wood frame construction. The principles of interpreting the drawings remain similar when different building systems are used, but it does require background research to fully appreciate how different building systems are manufactured and assembled on site.

Wood frame construction consists of walls built out of 2" × 4" or 2" × 6" pieces of wood known as **studs.** The house is built one floor at a time. This means that the first floor, which is constructed using floor **joists** and sheet material such as plywood or oriented strand board (OSB), rests on top of the foundation wall. The floor is then used as a platform on which to build the walls. Once the first-floor walls are completed, the second floor is built, again forming a platform on which to build the second-floor walls, and then the roof. Exterior and interior finishes are fastened directly to the stud walls.

2. **Brick veneer construction** is wood frame construction with a brick veneer (a single **wythe,** one brick in width front to back) on the exterior. The only real difference results from the wood frame being recessed to provide room for the brick to bear on the foundation wall. The brick itself is not considered structural; it is used for aesthetic purposes and acts as an exterior cladding, protecting the structure from the elements. This means that brick veneer is not allowed to directly support things such as floors and roofs and that they must be supported by the structural members of the building. Keep this in mind when reviewing drawings that have brick veneer. Brick veneer is very popular in some regions where bricks and brick masons are readily available to install the veneer. Many house designs will utilize a combination of exterior finishes such as siding,

FIGURE 7.2

Wood frame construction 3D Model of Doncaster house. Note the location of window and door openings and other framing details while comparing to Doncaster drawings. You will learn the terminology for wood framing construction components in Chapters 7, 8, and 9.

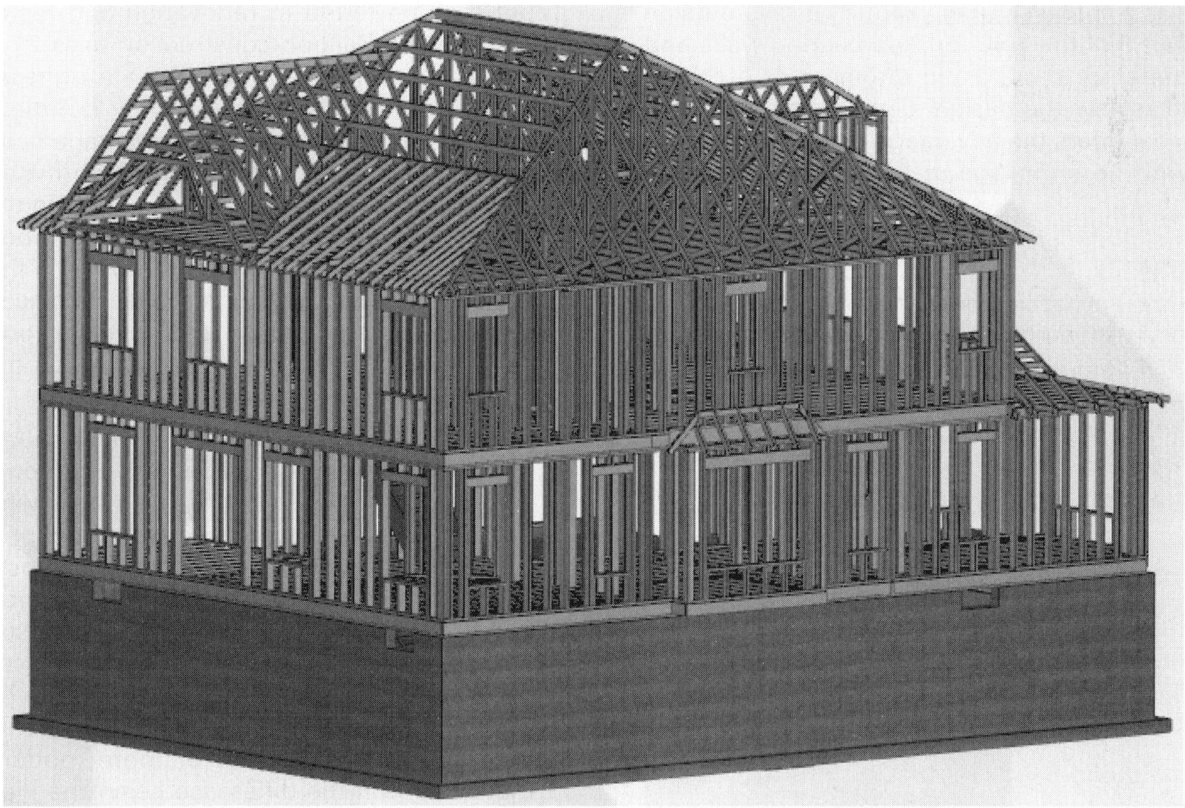

Source: Courtesy of Clint Kissoon, Chair, George Brown College, School of Architectural Studies.

FIGURE 7.3

Platform construction with brick veneer; notice that the bricks are being placed after the walls and roof are constructed, and are not carrying any structural load.

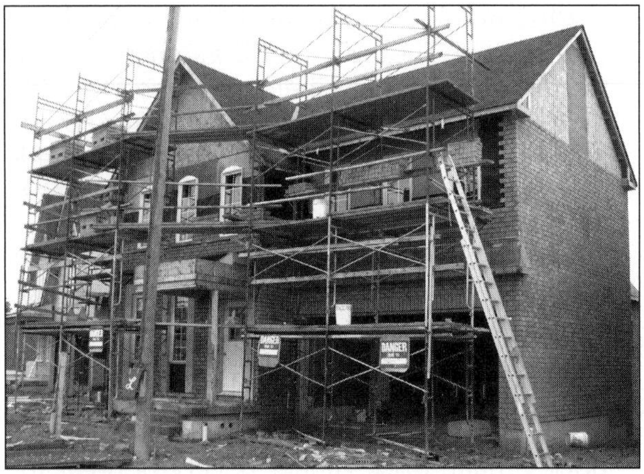

Courtesy Tom Stephenson

exterior insulated finish system (EIFS), brick, and/or stone. Refer to Chapter 10 for more information on exterior finishes.

3. **Lightweight structural steel framing,** used extensively in the ICI sector, has been adopted by some home builders as a material that can substitute for wood when platform framing. Constructed from galvanized steel, it requires the use of different tools. Because it is a manufactured product, it falls outside Part 9 of the NBC for load-bearing walls and therefore requires an engineer or architect to review the design. Lightweight structural steel offers the advantage of staying straight with no shrinkage or expansion problems.

FIGURE 7.4

An example of platform construction using lightweight structural steel; notice the steel roof rafters supported on steel stud walls.

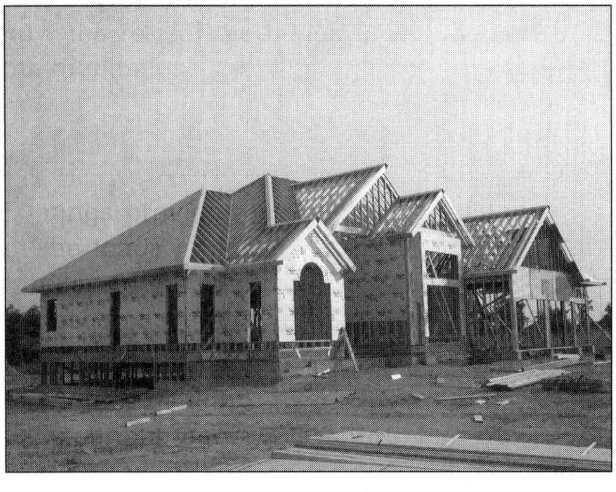

Courtesy ArcelorMittal Dofasco Design Steel Design Magazine

FIGURE 7.5

Lightweight structural steel used to construct a low-rise condominium project; notice the diagonal bracing and open web steel joists (OWSJ).

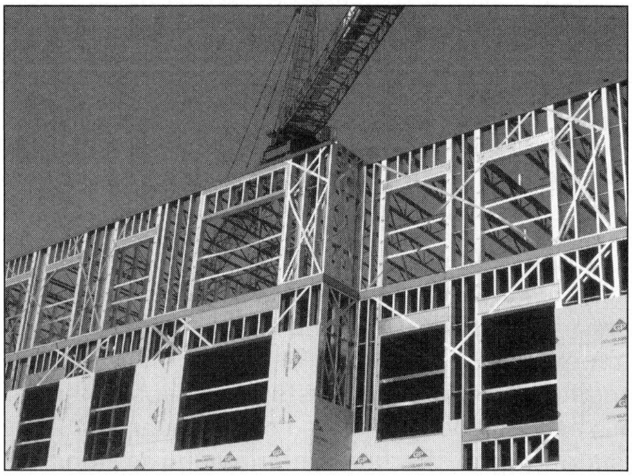

Courtesy Tom Stephenson

This system requires the use of an exterior insulating material (e.g., extruded polystyrene) to provide a continuous thermal break in the wall, because steel is a conductive material (refer to Division B Part 9.24 of the NBC). Joists and studs use the same standard spacing as wood frame construction.

4. ***Balloon construction:*** This method is no longer very common in new construction, but may be found in renovation and restoration work. Balloon construction uses 2" × 4"s that extend from the sill plate right up to the top plate of a two-storey building. Balloon construction's main advantage is its minimal shrinkage. Most lumber shrinking occurs across the grain, with very little in the direction of the grain. Floors rest on ribbon boards that are let (cut) into the walls. A fire stop in the form of blocking must be placed between the studs at all floor levels to prevent fire from quickly travelling through the structure.

5. ***Timber frame construction:*** Also called post-and-beam construction, this traditional method is popular with upscale clients who like the aesthetic that exposed wood provides. Timber frame construction uses posts and **beams** to provide the structural requirements for the building, while thick exposed planking is used to span the greater distances of the floor structure. Variations of this type of construction use either more traditional methods of construction or more modern methods, with the difference being the type of connections made between the posts and beams. For example, mortise and tenon

FIGURE 7.6

Section drawing of balloon framing.

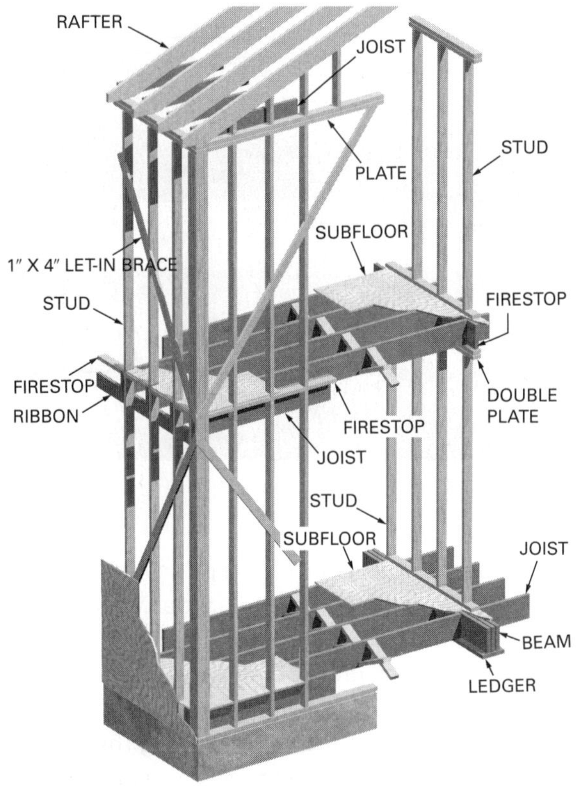

Source: From VOGT. Carpentry, 4E. c 2006 Delmar Learning, a part of Cengage Learning, Inc. Reproduced by permission. www.cengage.com/permissions.

joints may be used with dowels, or specified metal connectors can be used instead.

6. ***Solid masonry construction:*** This is an older method that uses solid masonry walls to form the structure of the building. Masonry walls consist of brick, block, or combinations of both. In some regions of Canada older homes built before the Second World War used solid masonry consisting of two wythes of brick (a wythe is defined as rows of brick placed side by side), or one wythe of brick and one wythe of block held together by a course of header brick, usually placed every sixth course. Solid masonry walls built today no longer use header brick to tie the wythes together; instead they use metal ties. While this method of construction is very sound, it has lost popularity in low-rise residential due to the extra labour and cost requirements needed to construct, insulate, and finish the building. However, solid masonry construction is common for residential garage construction, and is very popular in the ICI sector in, for example, construction of industrial and retail type buildings.
7. ***Insulated concrete forms (ICF):*** This method (discussed in Chapter 6) uses reinforced concrete sandwiched with extruded or expanded polystyrene foam, which acts as both the form and permanent insulation while also providing the structural means to attach

FIGURE 7.7

Timber frame construction.

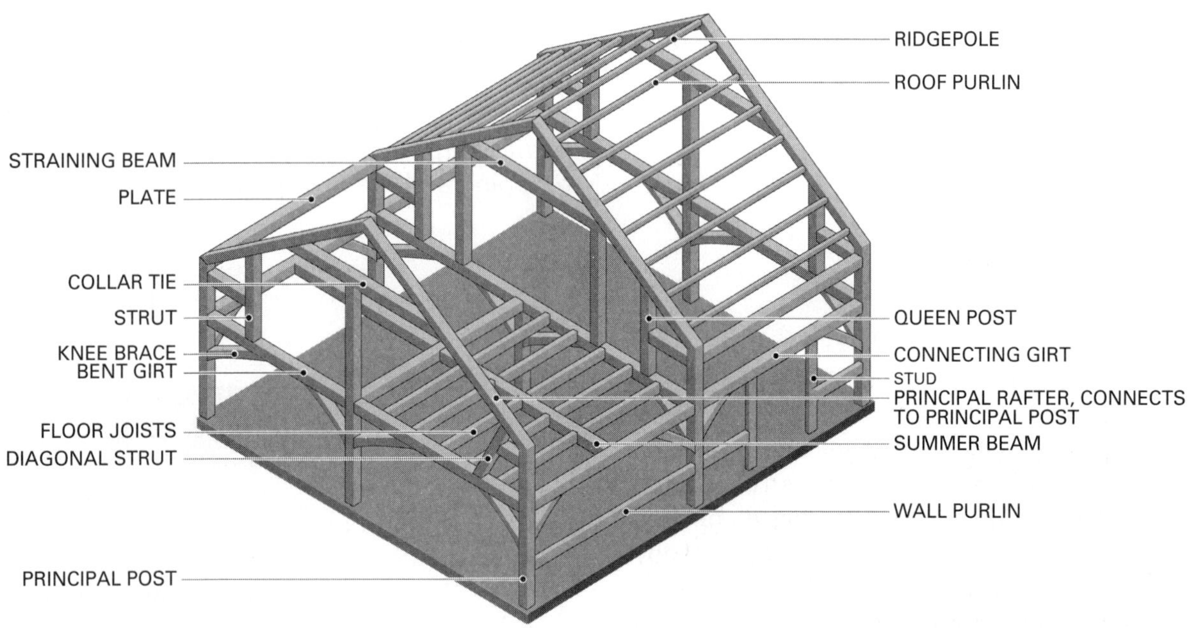

Source: The Canadian Wood Council

FIGURE 7.8

Section detail drawing of existing solid masonry wall using brick with a block backup (two wythes). This wall is structural so the roof can be fastened directly to the wall with 1/2" bolts and shields or sleeves.

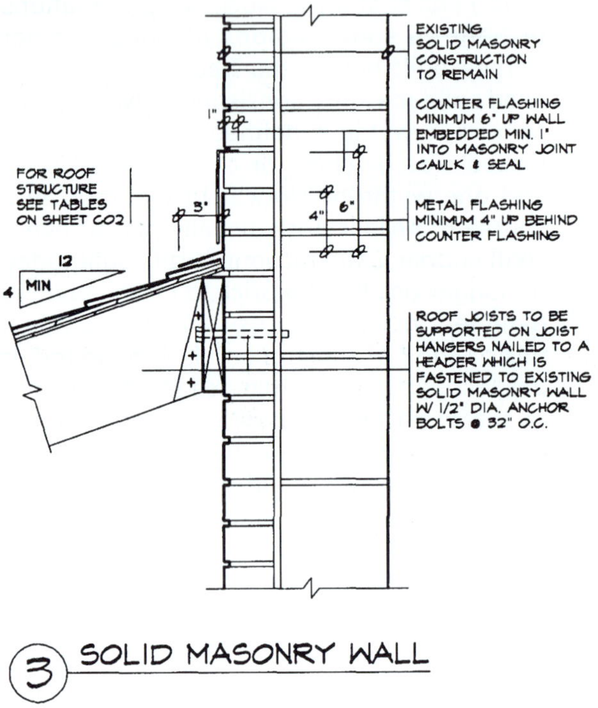

Source: Toronto Area Chief Building Officials Committee (TACBOC)

both exterior and interior finishes. ICFs can be used for construction with either wood or concrete floor systems. This method of construction is growing in popularity and provides a very sound structural, noise resistant, and energy efficient building that is attractive to more upscale clients.

WOOD FRAME CONSTRUCTION

Dimensional Lumber

Dimensional lumber and panel sheets (sheathing or subfloor material usually 4' × 8' in size and of varying thicknesses—typically ⅜" to ¾") are the main materials used to construct wood frame buildings. **Dimensional lumber** comprises standardized lumber that is milled and readily available for purchase from materials suppliers. Dimensional lumber sizes indicated on drawings in metric are actual size, while sizes indicated in imperial are smaller than the size indicated

FIGURE 7.9

ICF house under construction; notice the rebar protruding up from the walls ready to tie first-floor walls to the foundation.

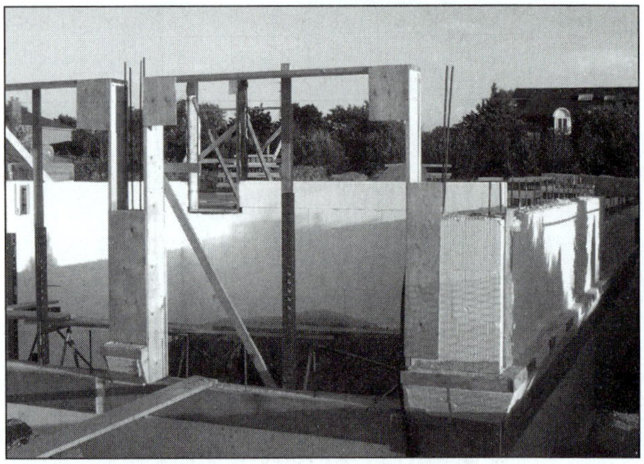

Courtesy Tom Stephenson

on the drawings. This discrepancy dates back 100 years or more to when lumber was cut at a mill to dimensional sizes, and shipped without jointing or smoothing the surface, meaning that a 2" × 4" was actually 2" × 4". Today the lumber is cut, jointed, and smoothed, reducing the size but providing a straighter, smoother, more consistent material that is easier to work with. A 2" × 4" indicated on the drawings is actually 1½" × 3½" but is still referred to as a 2" × 4". Tradespeople involved in renovation and restoration work will commonly come across the older, fuller sizes in existing buildings. Table 7.1 is a list of the more commonly used dimensional lumber sizes with corresponding actual sizes.

Lumber in Canada is graded to the National Lumber and Grading Association (NLGA) Standard Grading Rules. Each piece of lumber carries a grading stamp indicating the species grouping and grade of lumber.

TABLE 7.1

Commonly Used Dimensional Lumber

Drawing Reference Size Imperial	Actual Imperial Size of Material	Drawing and Actual Reference Size Metric
2" × 4"	1½" × 3½"	38 × 89 mm
2" × 6"	1½" × 5½"	38 × 140 mm
2" × 8"	1½" × 7¼"	38 × 184 mm
2" × 10"	1½" × 9¼"	38 × 235 mm
2" × 12"	1½" × 11¼"	38 × 286 mm

FIGURE 7.10

Example of grading stamp on a floor joist. Notice S-P-F refers to the species spruce, pine, fir; the #2 is the grade of lumber; and KD refers to kiln dried, the method by which the lumber was dried to a moisture content of 19% or less as required by the NBC.

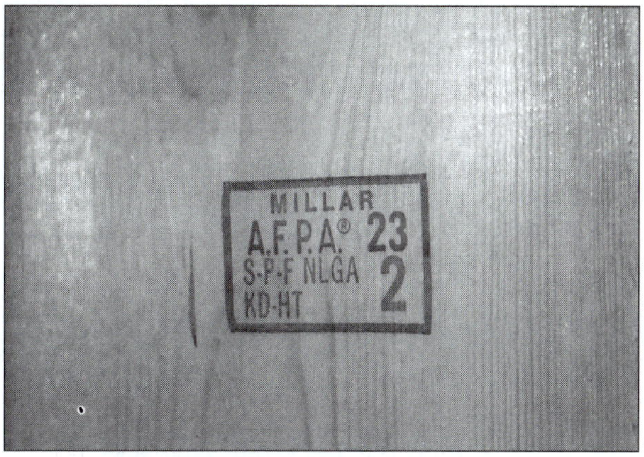

Courtesy Tom Stephenson

The construction drawings will indicate the species group, or species and grade of lumber to be used; for example, SPF refers to spruce, pine, fir, and SP #2 is used to indicate spruce for which the grade is SPF #2 or better. Lumber is graded so that the consistency of the material is maintained. Wood joists and beams are designed using tables that have calculated the spans based on the strength of the materials. When the material is unsuitable or of a low quality it will receive a grading that is not acceptable for construction purposes. Building inspectors may perform a visual check when inspecting a project by referencing the drawings and checking to make sure that the grade stamp matches the approved drawings.

Wood Shrinkage

According to the NBC, lumber used for framing must have a moisture content not exceeding 19% (refer to Division B Part 9.3.2.5 of the NBC) because lumber shrinks as it dries. If the moisture content is too high, excessive shrinkage will take place, causing the material to warp, cup, check, split, and twist. Wood shrinkage results in a number of problems in residential construction, including squeaky floors and drywall screw pops. This problem is most pronounced during the first year of occupancy; after that period the lumber has acclimatized to the surrounding humidity level.

Builders budget for an expected callback to repair the screw pops and other shrinkage problems after the first year. Lumber also expands and contracts differently during the seasons, which is why a wood door may stick during a humid summer and be fine during the winter when the humidity levels in the house are much lower. Good building and design practices can help to reduce the effects of shrinkage on the structure by carefully accounting for shrinkage and ensuring that it does not negatively affect the structure in ways that cause unnecessary cracking, or changing of sill slopes that would facilitate moisture ingress. Wood shrinks most in its width, then thickness. The effect of shrinkage on the length of lumber is negligible.

Wood Frame Construction Components and Spacing

Floor, wall, and roof members consist of joists for floors, studs for walls and trusses, and rafters and ceiling joists for roofs. All of these members have specific spacing designed to meet the structural requirements for the building. The standard on-centre spacing sizes must work with the standard panel sizes used for sheathing, subfloors, and drywall. The standard panel size is 4' × 8' (note that panel sizes are imperial, not metric); as

FIGURE 7.11

On-centre spacing works out to even spacing when using 8' sheets of material (plywood, oriented strand board, drywall, etc.). As noted earlier in the text, the NBC uses 300, 400, 600 standard spacings. Because our sheet materials are in imperial measurement (typically 4' × 8' sheets), we use the slightly different imperial spacing of 12", 16", 19.2", and 24" in their place.

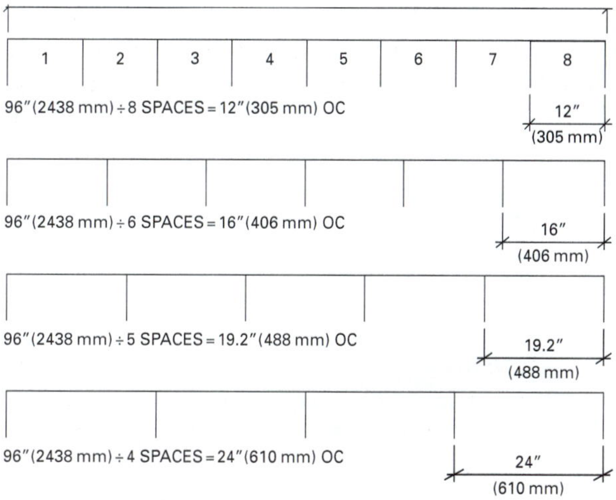

Source: From VOGT. *Carpentry*, 1E. © 2009 Nelson Education Ltd. Reproduced by permission. www.cengage.com/permissions

a result, the four choices for standard on-centre spacing used in wood frame construction are:

- 12" spacing
  - Commonly used for longer spans of floor joists within a dimensional size. For example, a 2" × 8" may not be able to span the required distance at 16" OC but at 12" OC it can. This dimension is also commonly used to reduce deflection under tile floors where movement can cause cracking of the tile.
  - May be required on walls that extend to a height above NBC requirements for 16" OC spaced walls (refer to Division B Table 9.23.10.1 of the NBC).
- 16" spacing
  - Most commonly used OC spacing for floors, walls, and rafters.

FIGURE 7.12

Sill plate laid out at 16" OC, first spacing at 15¼" (16" – ½ of joist thickness (¾") = 15¼") to have sub-floor sheets finish centred over joist at 8' joint the length of the sheets (recall that 16" divides evenly into 96").

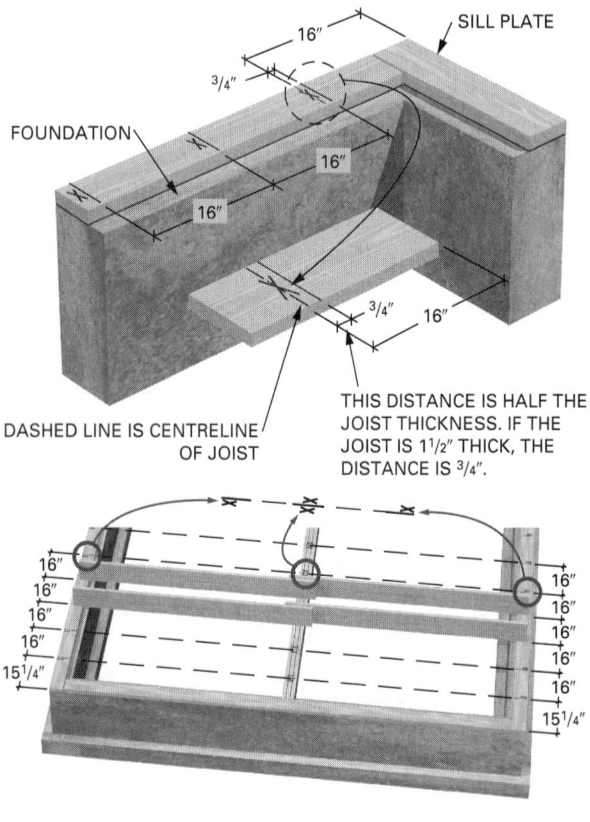

Source: From VOGT. *Carpentry*, 4E. © 2006 Delmar Learning, a part of Cengage Learning, Inc. Reproduced by permission. www.cengage.com/permissions

FIGURE 7.13

Wood frame construction components.

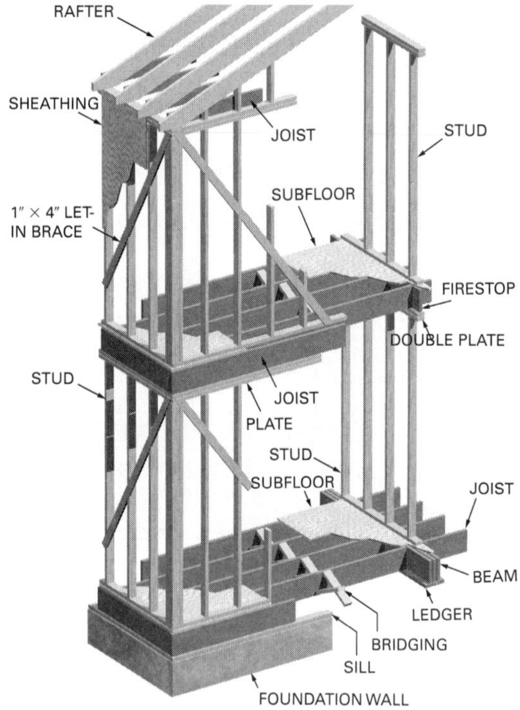

Source: From VOGT. *Carpentry*, 4E. © 2006 Delmar Learning, a part of Cengage Learning, Inc. Reproduced by permission. www.cengage.com/permissions

- 19.2" spacing
  - Gaining popularity with some builders for use in trusses and walls. Saves material and reduces thermal bridging (the path of least thermal resistance, by which heat will pass through a material) by increasing the surface area of the wall that can be insulated.
- 24" spacing
  - More commonly used for roof-truss spacing, can be used for wall and floor spacing but requires thicker sheathing and subflooring.

All of these measurements will divide evenly into a standard 8' or 96" panel. Laying out the floors, walls, and roofs to correspond to the OC spacing greatly reduces the amount of cutting required because the sheet goods will line up on the centre of the structural member (for example, a joist). As well, the building can be designed as a system with a predictable placement of components that react to the spans and loads placed upon them.

Chapter 7 Assignment 1: Methods of Residential Construction

Based on the material in the chapter, identify the letter that best answers the question.

_____ 1. What materials are ICF built houses constructed with?

a. preserved wood

b. galvanized steel

c. reinforced concrete and polystyrene insulation

d. structurally insulated forms made of oriented strand board and polystyrene

_____ 2. What type of construction method uses roof purlins?

a. timber frame

b. western or platform

c. balloon frame

d. ICF

_____ 3. Assuming a consistent subfloor and joist size and thickness, which standard floor spacing would be the most favourable for a ceramic tile floor?

a. 10" OC

b. 12" OC

c. 16" OC

d. 24" OC

_____ 4. Which of the following is currently the most popular method of residential construction used in Canada?

a. insulated concrete form construction

b. timber frame construction

c. solid masonry construction

d. platform construction

_____ 5. What will the edge of the first joist be placed at when laying out a single span floor joist system at 16" OC using standard dimensional joists 2" × 8" (38mm × 184mm)?

a. 16¼" from the edge of the floor to the centre of the first joist

b. 16" from the edge of the floor to the edge of the first joist

c. 15¼" from the edge of the floor to the edge of the first joist

d. 15" from the edge of the floor to the edge of the first joist

_____ 6. What are the following terms used to describe: light frame, platform, and stick construction?

a. insulated concrete form construction

b. solid masonry construction

c. timber frame construction

d. wood frame construction

_____ 7. What is the main difference for wood frame construction when using brick as compared to using siding?

a. the brick rests on the foundation wall, and the sill plate is set back to provide room for the brick and an air space

b. the brick is supported on the floor joists of the first floor

c. the siding rests on the sill plate

d. the siding requires the sill plate to be set back from the edge of the foundation wall to allow for a sloping flashing that rests on top of the sill plate

_____ 8. What is the typical cause for screw pops in drywall?

a. wood shrinkage

b. metal joists and studs

c. shrinkage in drywall

d. structural weakness in the building

_____ 9. What methods used for lightweight structural steel in residential construction are similar to the methods used in wood frame construction?

a. The tools to construct the house are exactly the same.

b. The materials behave the same in terms of expansion and contraction.

c. The design and sizing of members is the same as for dimensional lumber.

d. The spacings for studs and joists are the same standard spacings.

_____ 10. How does brick veneer differ from solid masonry construction?

a. Brick veneer is load-bearing.

b. Brick veneer supports both live and dead loads, such as floors, walls, and roof systems.

c. Solid masonry construction is non-load-bearing.

d. Solid masonry construction is load-bearing, while the brick veneer is a non-load-bearing.

_____ 11. Why is balloon construction typically not used anymore?

a. It has too many issues with shrinkage.

b. It is not structurally sound.

c. It is awkward, more time-consuming, and costly to construct.

d. It requires too high a skill level to construct due to its different spacing methods.

_____ 12. What does the NBC require the moisture content to be for dimensional lumber?

a. a moisture content of 18% or less

b. a moisture content of 19% or less

c. a moisture content of 27% or less

d. a moisture content of 29% or less

_____ 13. What is the actual size of a 2"× 4" piece of dimensional lumber?

a. 1¼" × 3¼"

b. 1½" × 3½"

c. 1¾" × 3¾"

d. 2" × 4"

_____ 14. What is the actual metric size for a 2"× 10"?

a. 38 × 89 mm

b. 38 × 140 mm

c. 38 × 235 mm

d. 38 × 286 mm

_____ 15. In which direction does wood shrink the most?

a. along its length

b. along its width

c. along its thickness

d. equally in all directions

_____ 16. What is the actual size of a 2"× 6" piece of dimensional lumber?

a. 1¼" × 5¼"

b. 1½" × 5½"

c. 1¾" × 5¾"

d. 2" × 6"

_____ 17. What is the actual metric size for a 2"× 6"?

a. 38 × 89 mm

b. 38 × 140 mm

c. 38 × 235 mm

d. 38 × 286 mm

_____ 18. What do trimmer joists directly support?

a. header joists

b. joists

c. tail joists

d. sill joists

_____ 19. How much would a W200 × 27 steel beam weigh if it was 5 metres long?

a. 1000 lbs

b. 1000 kg

c. 135 lbs

d. 135 kg

_____ 20. What is the beam height (bottom flange to top flange measurement) of a W8 × 15 steel beam?

a. 15 inches

b. 15 mm

c. 8 cm

d. 8 inches

FLOOR FRAMING

Beams and Columns to Support Floor Framing

Joists carry the load of the floor area that they span. Floor joists run directly under the subfloor, spanning from supporting wall to supporting wall. In many cases, the distance between supporting walls is too great for a joist to span unsupported. In this case, beams are used to reduce the span distance. Division B Part 9 of the NBC provides span tables that define the minimum clear spans for beams based on the length of joists they are supporting. For example, a beam supporting a joist that spans 3 m on one side and a joist that spans 4 m on the other side would be supporting 3.5 m of joists (floor). This length is called the **supported joist length (SJL).**

A beam must have adequate support (known as **bearing** support). Part 9.23.8 of the NBC states that "beams shall have even and level bearing and shall have not less than 89 mm length of bearing at end supports." For example, a steel beam that spans 3 metres would have to extend 89 mm (3½") on each side to have adequate bearing. Beam length would be 3178 mm (3 m + 89 mm + 89 mm). Alternatively, using imperial measurement, if the span of the steel beam extends 10'-2", the actual length of the beam would need to be at least 10'-9" (10'-2"+3½"+ 3½"). Drawings or detail drawings will indicate if the bearing for a particular beam must be greater than 89 mm or if there are special requirements regarding connections or bearing plates (bearing plates are used to distribute the load to a greater area).

The end bearing for built-up wood beams varies based on the span, the supported joist length (NBC refers to it as the supported length), and the number of floors being supported. It should be noted that, in some cases, that bearing may not be indicated on the drawings and the framers, should consult with their provincial building code, or if unsure, the designer (structural engineer) should be consulted.

Part 9 Tables A8 to A10 of the NBC require that the minimum bearing for a built-up wood beams design be as follows:

1 Beams supporting not more than one floor:

- 3-ply beams that support not more than one floor, with a supported length greater than 4.2 m, require 114 mm bearing. All other beams supporting not more than one floor require 76 mm.

2. Beams supporting not more than two floors:

- 3-ply beams that support not more than two floors require 114 mm bearing. 4 ply and 5 ply beams with supported lengths greater than 3 m require 114 mm bearing. All other beams supporting not more than two floors require 76 mm bearing.

3. Beams supporting not more than three floors:

- 3-ply beams that support not more than three floors with supported lengths greater than 4.2 m require 152 mm bearing. All other beams require 114 mm bearing.

FIGURE 7.14

Doncaster garage beam section detail taken at bearing support for beam. Pay special attention to the bearing requirements for the steel plate under the beam and the width of the built-up support column. Also pay attention to the note requiring solid blocking under the column below the subfloor. The beam transferring the load to the column places a point load on the floor and therefore blocking must be filled in to effectively transfer the load to the foundation.

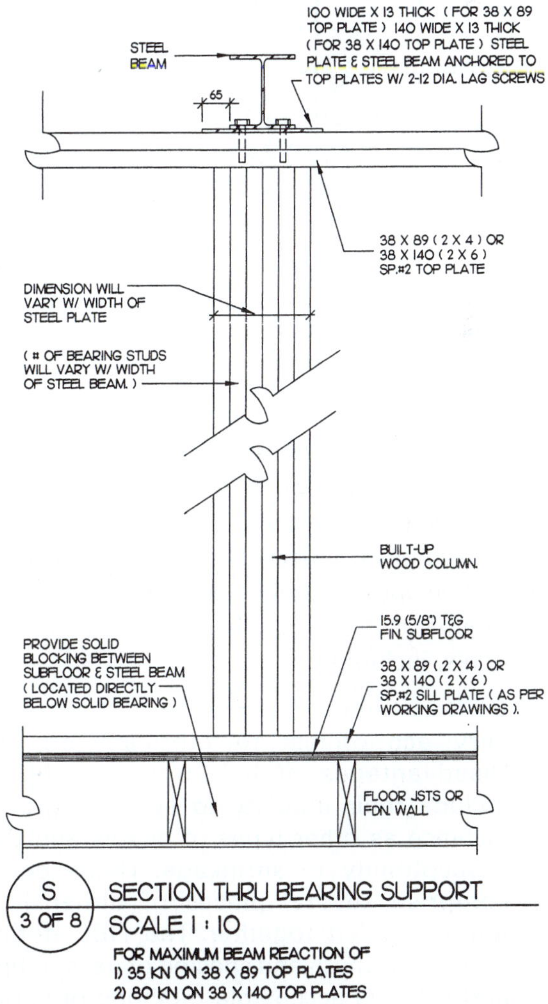

Source: Cassidy & Co. Architectural Technologists

FIGURE 7.15

Doncaster drawing example of a built-up wood beam indicated on a drawing; this beam consists of three pieces of 38 × 184 (2 × 10) #2 spruce and is installed flush with the ceiling as indicated. There is also a flat arch indicated with the abbreviation FA, and the lintel reference L5, for which the size can be found from the construction notes on page one of the Doncaster drawings lintel/beam schedule. The FA flat arch means the beam would drop below the ceiling, forming a more formal separation of rooms, whereas the flush beam will be level with the ceiling so that when the drywall is applied the average person will not know there is a beam there.

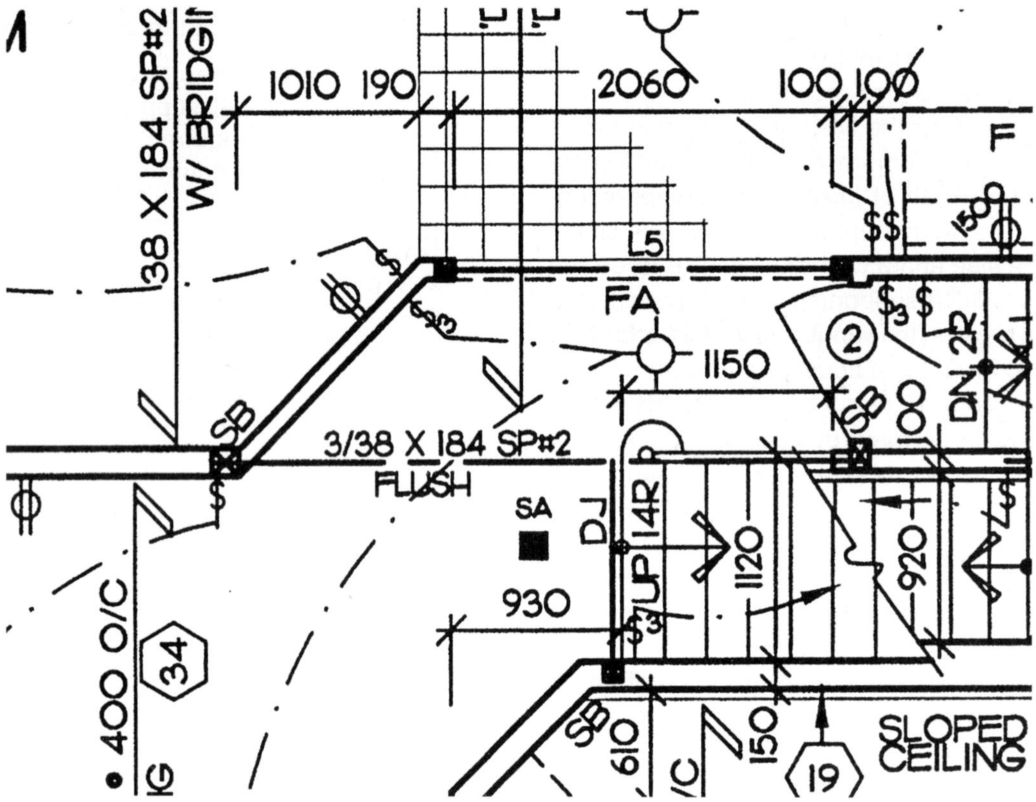

Source: Cassidy & Co. Architectural Technologists

You will note that the number of ply and bearing support corresponds exactly with 38 mm thickness of dimensional lumber. 76 mm equals 2 × 38, 114 mm equals 3 × 38. Refer to Figure 7.14 for an example of a detail drawing for a steel beam installation and its required support.

Beams can be constructed from various materials, including:

1. ***Built-up wood beams:*** These beams are very economical to fabricate on site. Disadvantages of built-up wood beams include their inability to span as great a distance as other types of beams and their susceptibility to shrinkage. These beams comprise several pieces of dimensional lumber nailed together. The NBC defines minimum length and spacing of nails or bolts and lists the acceptable methods of joining the beams where required (refer to Division B Part 9.23.8.3 of the NBC). The drawings will usually note built-up wood beams by stating the number and size of pieces, also referred to as plies, as well as the species and grade of lumber. For example:
   - 4/38 × 235 SP #2 or 4/2 × 10 SPF#2 indicates that four pieces of 2 × 10 spruce or spruce, pine, fir grade #2 are to be used for the built-up wood beam.
2. ***Manufactured (engineered) beam and joist systems:*** Manufactured floor systems are popular and dominate some regional markets because of their ability to both span longer distances than dimensional lumber and avoid the shrinkage problems associated with dimensional lumber. They are also popular with subtrades because there is greater flexibility in the sizes of holes that can be drilled through the joists. The Brook drawings that came attached to this text use manufactured joists.

FIGURE 7.16

Built-up wood beam, supporting floor joists over an opening; notice the joist hangers supporting the floor joists and that the underside of the joists and beam are on the same horizontal plane. Once the drywall is installed the ceiling will be continuous with no indication of a beam. This beam would be indicated as a flush beam on the drawings, similar to the flush beam indicated in Figure 7.15.

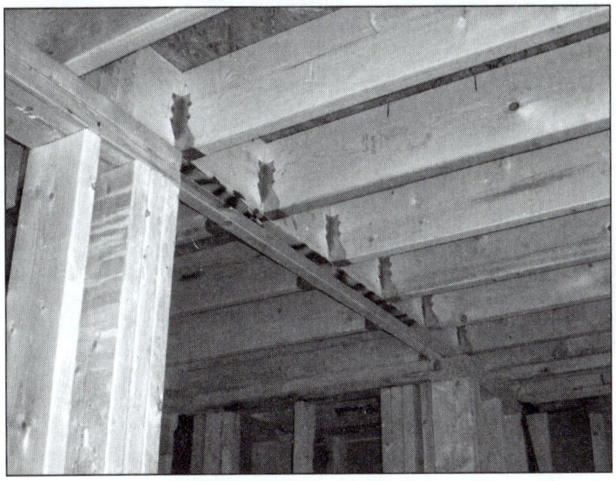

Courtesy Tom Stephenson

Drawings that specify manufactured joists or beams must have a structural drawing (also referred to as a shop drawing) done by the manufacturer and carry an engineer's stamp of approval. This is necessary because, unlike dimensional lumber, which is standardized, manufacturers' products have different design strengths and requirements. It is therefore left to the manufacturer to design their products to satisfy code requirements as set out in Division B Part 4 of the NBC. Generally, the construction drawings are submitted to the local building supply, which then sends them to the manufacturer for the floor-system design drawings and stamp. In some cases the drawings might be done in-house by the supplier using manufacturer's specialized software. The structural drawings are then submitted to the building department for final approval. Framers must follow the manufacturer's specifications when installing the system.

The floor systems comprise a truss-type joist that may be constructed in a truss or I-beam shape.

Laminated veneer lumber (LVL) and parallel strand lumber (PSL) beams are examples of manufactured beams that are used to span long distances. LVL is constructed from consecutive plies of lumber laminated together, while PSL is constructed from strands of lumber

FIGURE 7.17

A drawing specifying Laminated Veneer Lumber (LVL) beams for over the garage doors and roof, and extending out in front of garage. There are two plies or pieces 45 X 235 LVL. Each piece is 45 mm thick by 235 mm high. The "+ L9" refers to a steel lintel size listed in the lintel schedule used to support the brick veneer (refer to Chapter 10).

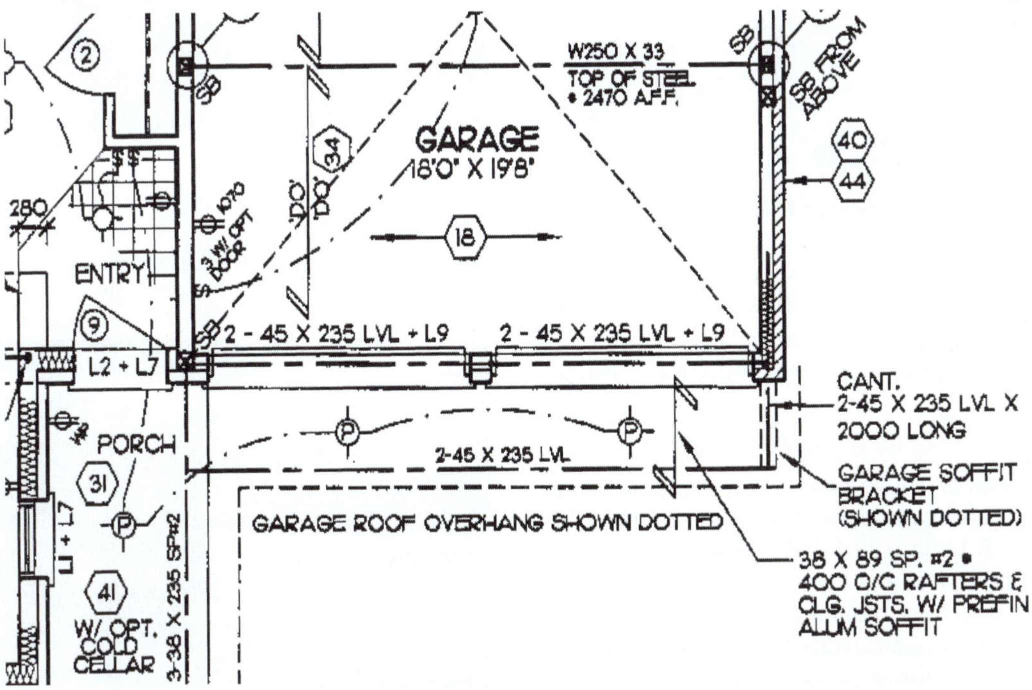

Source: Cassidy & Co. Architectural Technologists

FIGURE 7.18

LVLs nailed together to form a beam that supports roof trusses.

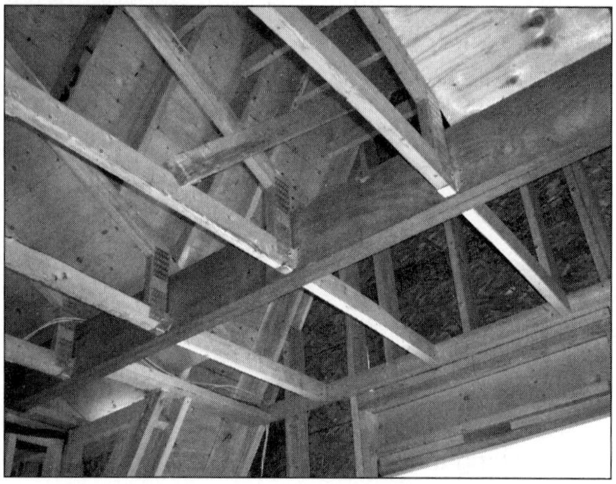

Courtesy Tom Stephenson

FIGURE 7.19

LVLs connected with joist hangers to built-up wood beam; notice that the hanger type and nailing requirements (all nail holes in hangers must be used with the approved nails) must be approved by the LVL manufacturer. Regular hangers are unacceptable.

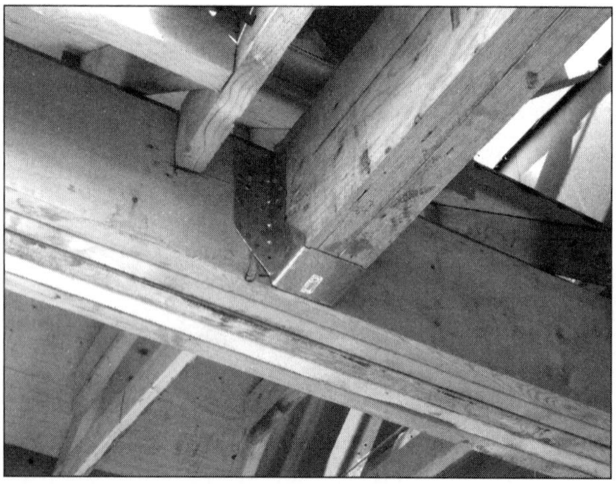

Courtesy Tom Stephenson

oriented parallel to each other and laminated together. These types of beams do not have the shrinkage problems associated with built-up wood beams. LVLs and PSLs have much greater strength than dimensional lumber and can therefore span greater distances.

3. ***Steel beams:*** These beams are comparatively stronger and can span long distances (refer to Division B Table 9.23.4.3 of the NBC). They are very economical, but can be quite heavy and in some cases may require a crane to move into place. They carry structural loads including loads placed by masonry (wood beams do not carry masonry loads). Steel beams are identified on drawings as shown in Figures 7.27 and 7.28.

FIGURE 7.20

Manufactured joist system tied into an LVL built-up beam; notice the joists are resting on hangers approved for this system.

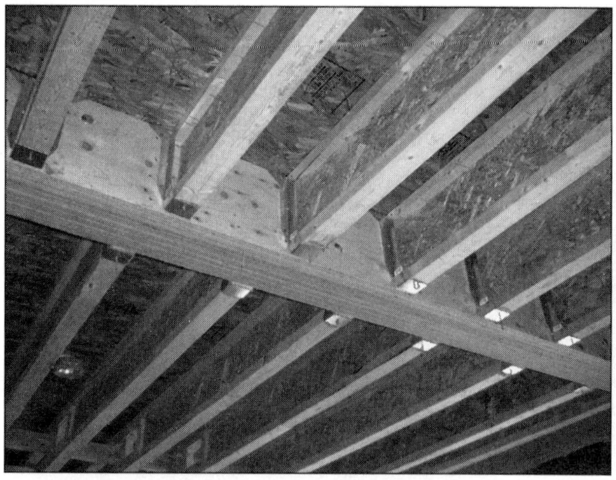

Courtesy Tom Stephenson

FIGURE 7.21

Manufactured joist system resting on a wide flange I-beam.

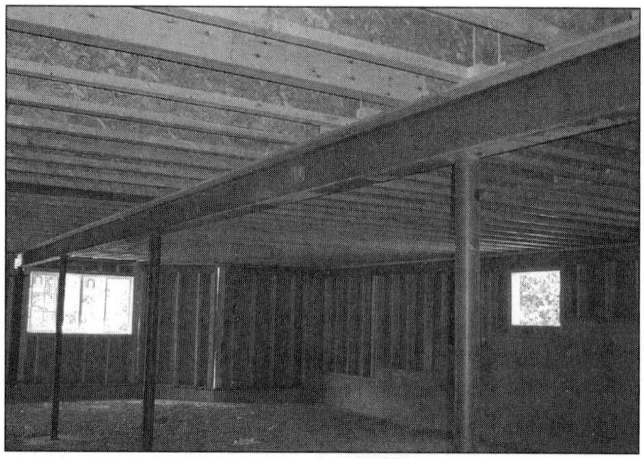

Courtesy Tom Stephenson

FIGURE 7.22

Manufactured joists with designed cutout for installation of mechanical system pipes and ducts.

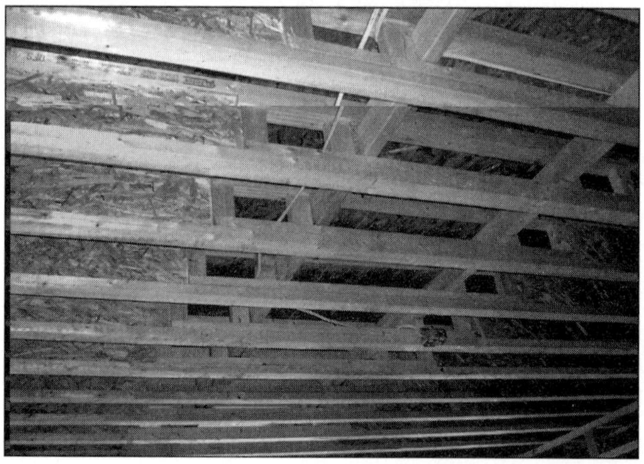

Courtesy Tom Stephenson

FIGURE 7.23

Manufactured joist system framed around a stairwell opening, using LVLs for the headers and trimmers; note how the stairwell is temporarily filled in for safety purposes.

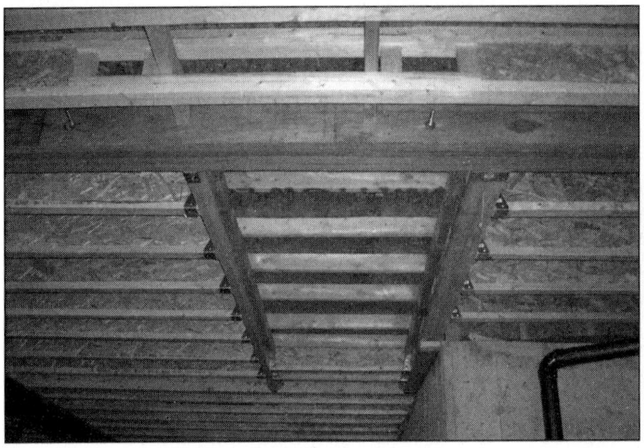

Courtesy Tom Stephenson

The weight of a steel beam can be calculated by multiplying the length by the weight. For example, a 20' long W8 × 15 will weigh 300 lbs (20 × 15 = 300).

The weight of a metric-sized beam can also be calculated, for example, a W200 × 27 that is 5 metres long would weigh 135 kgs (5 × 27 = 135).

The beam height and weight are indicators of strength; the larger the beam height and the more it weighs, the stronger the beam. This means the beam will be able to carry more load and span more distance. Part 9 has predefined tables that are used to design the appropriate beam sizes. For spans beyond those indicated in Part 9, an engineer must be hired to design the correct beam size.

FIGURE 7.24

Typical manufactured joist system details. Remember to always follow the individual manufacturer's requirements when using engineered floor and roof systems. Note the details on squash blocks, which are used to transfer loads from above so the wood I-joists are not negatively affected by these loads.

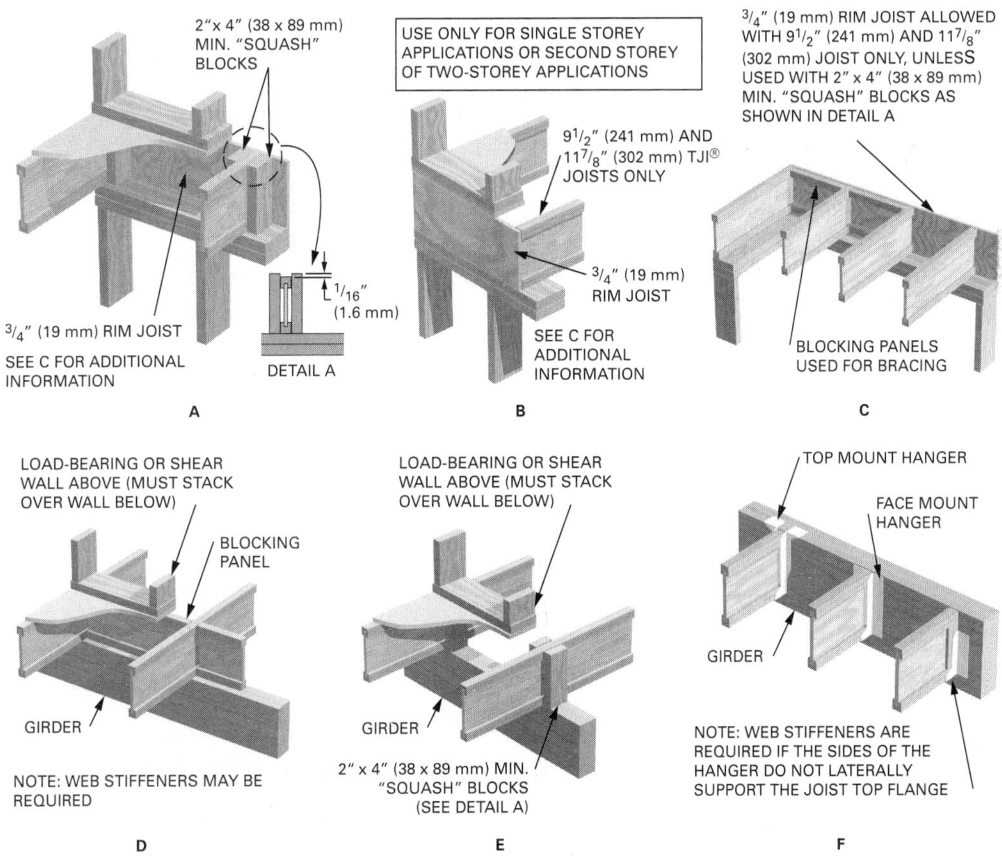

Source: From VOGT. *Carpentry*, 1E. © 2009 Nelson Education Ltd. Reproduced by permission. www.cengage.com/permissions.

FIGURE 7.25

Plan and Section details taken from page 19 of the Brook drawings that came attached with the text. Note the use of squash blocks at all point loads and that there is an outer rim board which is a joist that wraps around the exterior of the floor framed wall also referred to as a perimeter joist, sometimes referred to as a "header joist" (see Plan Section D), although "header" is technically not the correct term for this use as defined in the NBC. Also note that there is a 1⅛" rim board between the joists. This is placed there to ensure that there is adequate bearing for the structural loads from the walls above and that the loads are effectively transferred to the sill plate and foundation wall below (see Elevation Section D1). Review the details carefully and visualize how the manufactured floor system intersects with the exterior walls and foundation.

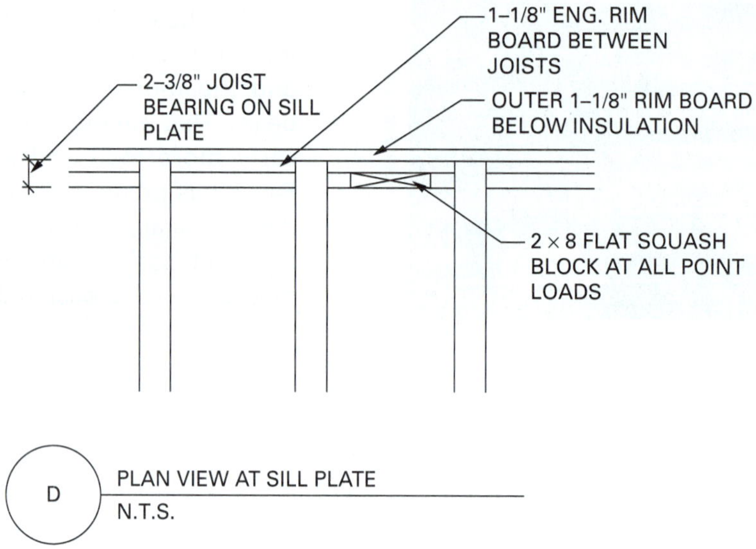

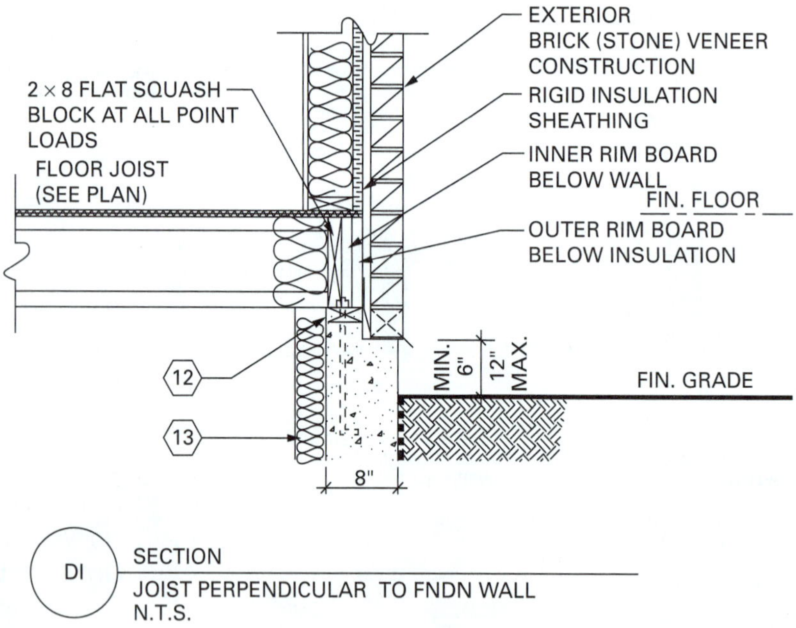

Source: Cassidy & Co. Architectural Technologists

The following photos (Figures 7.29 to 7.36) display various steel beams and their supports.

Floor Frame Layout and Details

Sill plates are fastened to the top of the foundation wall with **anchor bolts** spaced at a minimum 2.4 m OC (refer to Part 9.23.6.1 of the NBC). Sill plates rest on a sill gasket or are sealed to the foundation with adequate protection from dampness to maintain the air barrier and prevent heat loss; a bed of mortar may be used to level the sill when the top of the foundation wall is not straight or level.

FIGURE 7.26

Manufactured joist systems in place, based on manufacturer's requirements as indicated in shop drawings supplied by manufacturer. As in all designed manufactured systems that do not fall under Part 9 of the NBC, the shop drawings will require an engineer's stamp in order to receive municipal approval.

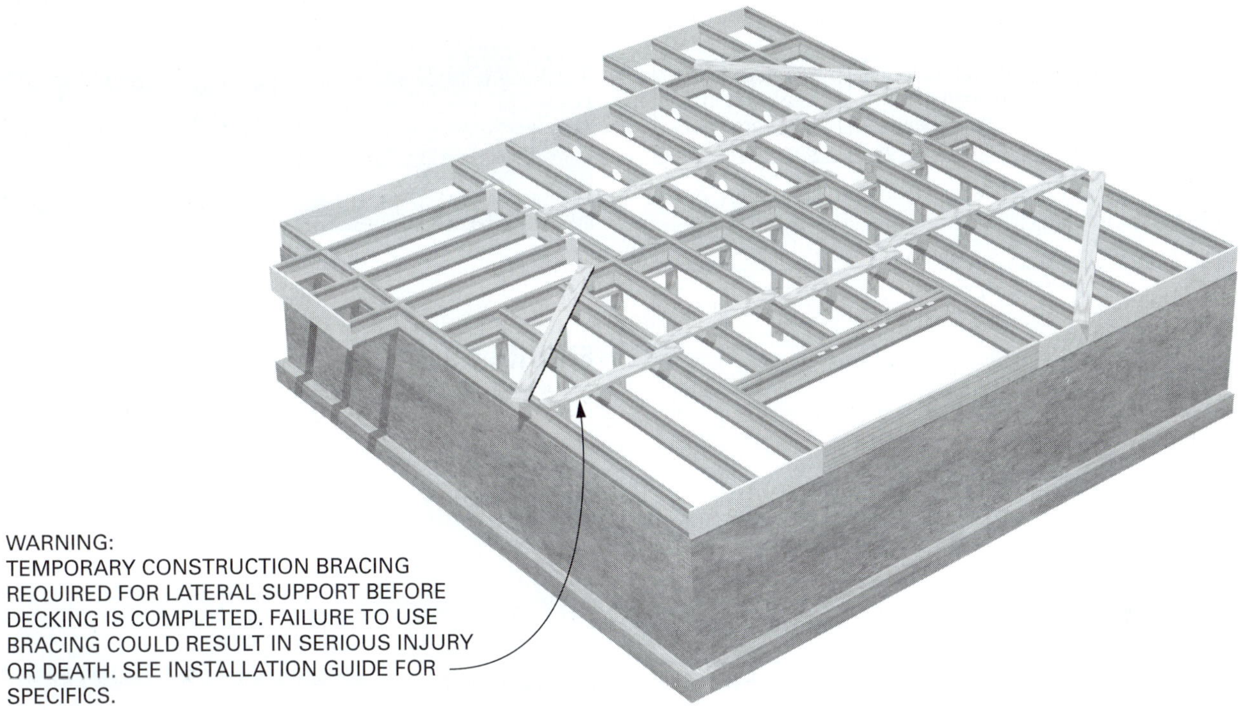

Source: From VOGT. *Carpentry*, 1E. © 2009 Nelson Education Ltd. Reproduced by permission. www.cengage.com/permissions.

Floors are framed with joists at a uniform spacing, usually 16" OC or 12" OC. The joists are fastened to the sill plate and a rim joist (a header joist that goes around the perimeter of the floor as illustrated in Figure 7.37). In most cases, floor systems are designed so that the joists span the shortest distance, which generally reduces the size requirement of the joist, thus reducing the cost. For example, using a 2" × 8" instead of a 2" × 10" results in a lower cost per linear foot.

The design and installation of the floor system requires consideration of the following:

- Type of finished floor being used. For a tile, marble, or granite floor the designer may prefer 12" OC to reduce the deflection. This is one reason you may see the spacing change from 16" to 12" OC under a kitchen or bathroom that has a tile floor. Another reason for a spacing change is that the span of the joist may have changed (refer to Figure 7.40).
- A joist must have adequate bearing support. Part 9.23.9.1 of the NBC states that:

  Except when supported on ribbon boards, floor joists shall have not less than 38 mm (1½") length of end bearing.
- The headroom requirements and aesthetics of the room will determine how the joists transfer their load to any beams. For example, if a flush beam is to be installed, the underside of the joists will align with the bottom of the beam as in Figure 7.16 if the beam is wood, and Figure 7.40 if the beam is steel. If it does not matter whether the joists are flush, the beam layout will be similar to Figure 7.45 where the joists rest on top of the beam. A flush beam is indicated on the floor plan with a note or abbreviation such as FLUSH.
- Any **point loads** and solid bearing requirements that occur will need to be supported and effectively have their loads transferred to beams and support walls. A point load is a load being transferred through the structure at a particular point. For example, a load being placed by a beam that is transferred by a column to the floor will create a point load

FIGURE 7.27

Metric beam size.

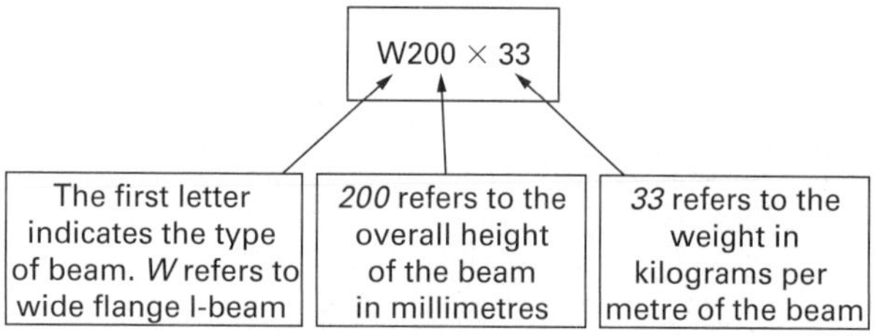

Beam height in millimetres or inches provided after W (wide flange I-beam)

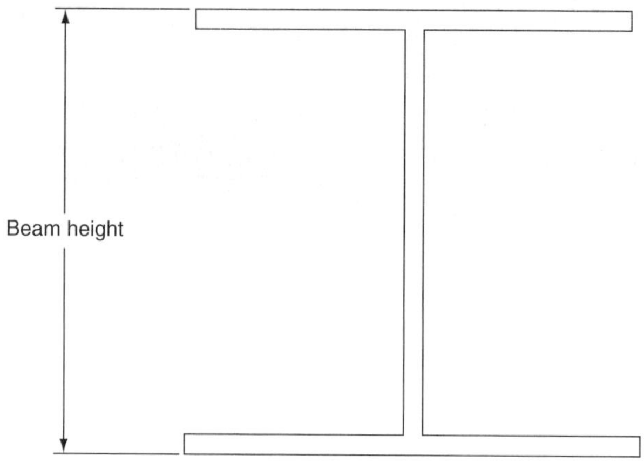

Imperial beam size.

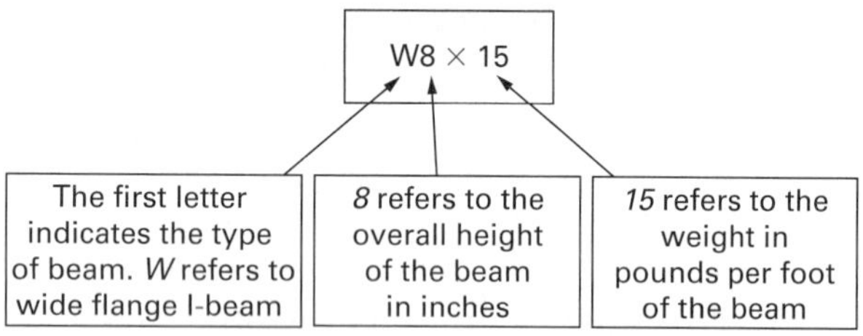

Source: Tom Stephenson

FIGURE 7.28

Steel beam as viewed on a drawing; notice the column footing shown as a dashed line (⑮) and the beam pocket at left (⑯). This is a wide flange I-beam, as indicated: W200 × 27, that is, 200 mm in height and weighs 27 kg per metre (⑰).

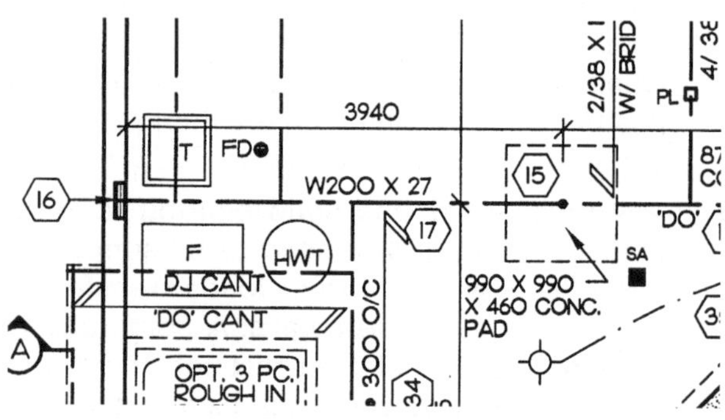

Source: Cassidy & Co. Architectural Technologists

FIGURE 7.29

Beam pocket and beams ready to be installed; similar to drawing in Figure 7.28.

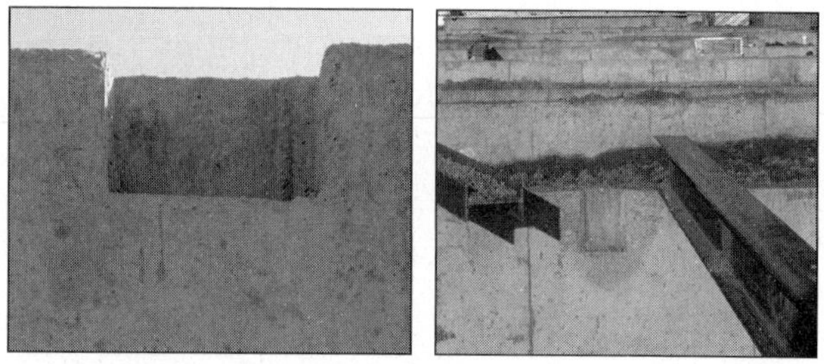

Courtesy Tom Stephenson

FIGURE 7.30

Steel beam in place; notice columns on gravel waiting to be placed on footings and adjusted (two column footings shown).

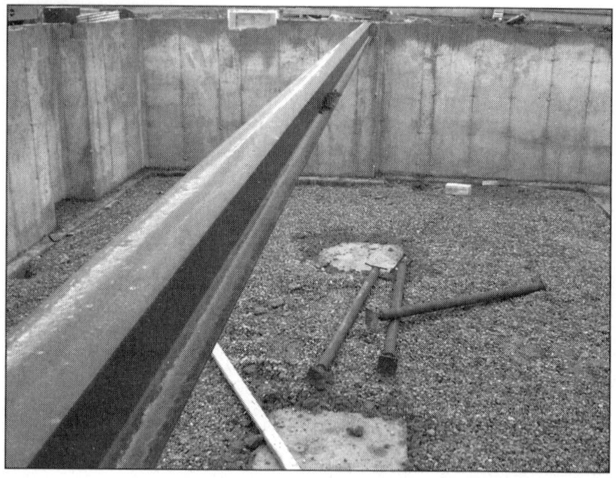

Courtesy Tom Stephenson

FIGURE 7.31

Adjustable steel support column; this column rests on the footing and has the basement floor slab poured around it. It supports either a wood or steel beam. Refer to Figure 7.30 for an example of columns about to be placed under a steel beam. The columns can be adjusted up or down by turning the nut at the base (telescoping).

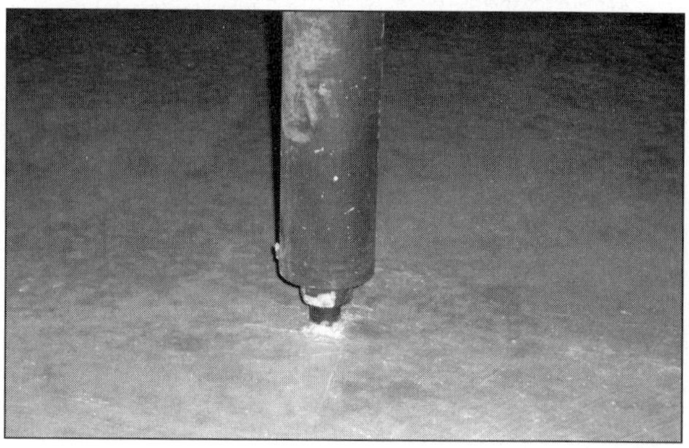

Courtesy Tom Stephenson

FIGURE 7.32
Steel beam supporting floor joists; notice the wood headers bolted through the steel beam. The headers provide a means of fastening the joists to the beam with joist hangers.

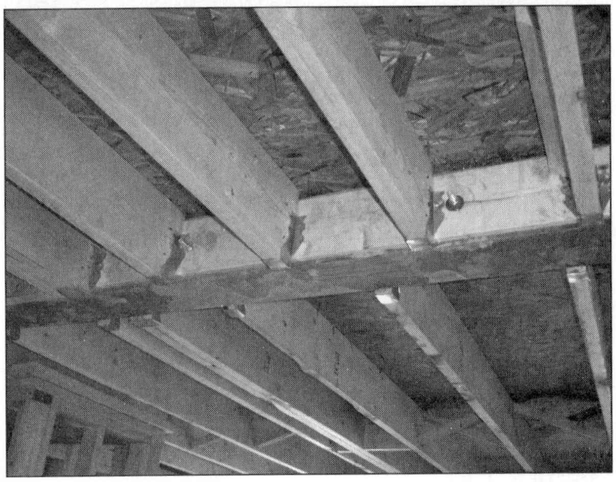

Courtesy Tom Stephenson

FIGURE 7.33
Steel beam supported on steel pipe column; this is a commonly used placement. Bearing plate and connection must meet NBC 9.17.3.2 requirements.

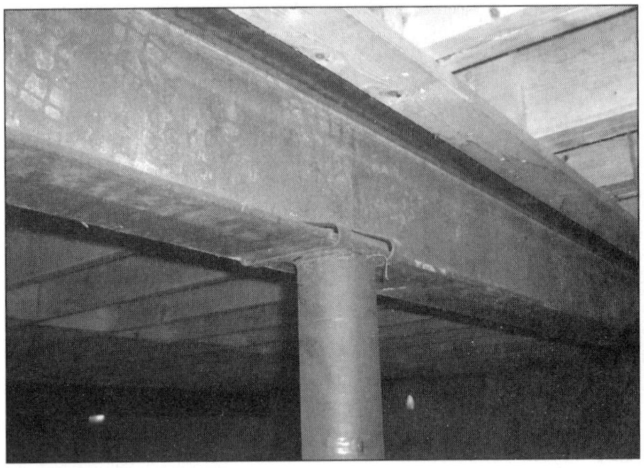

Courtesy Tom Stephenson

FIGURE 7.34
Steel beam with beam pocket filled in; notice that the joists rest on top of the beam, and that the strapping is placed on the side of the beam to hold the joists and beam in place.

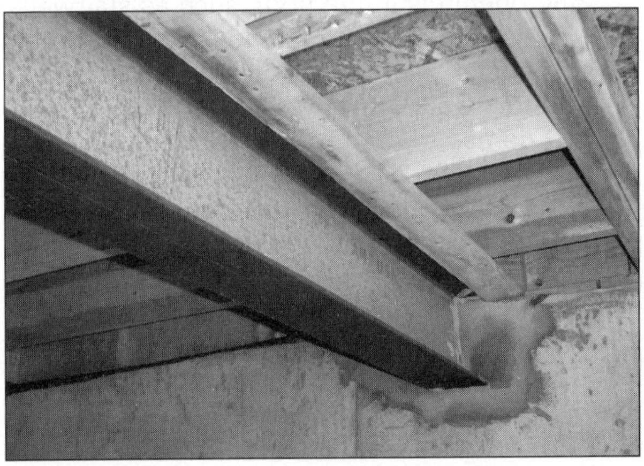

Courtesy Tom Stephenson

FIGURE 7.35
Steel beam supported on wood column; notice column resting on footing wrapped in polyethylene to protect from moisture. Basement slab will be poured around column.

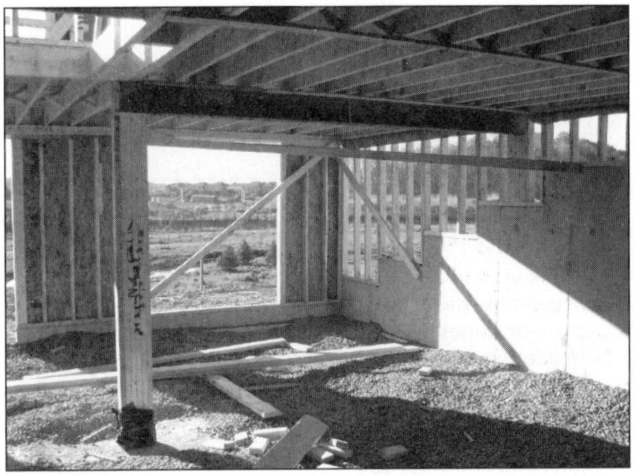

Courtesy Tom Stephenson

FIGURE 7.36
Steel beam with welded plate ready to support masonry; notice how the plate protrudes out to support brick from the wide flange I-beam, and how the I-beam supports the roof over the garage.

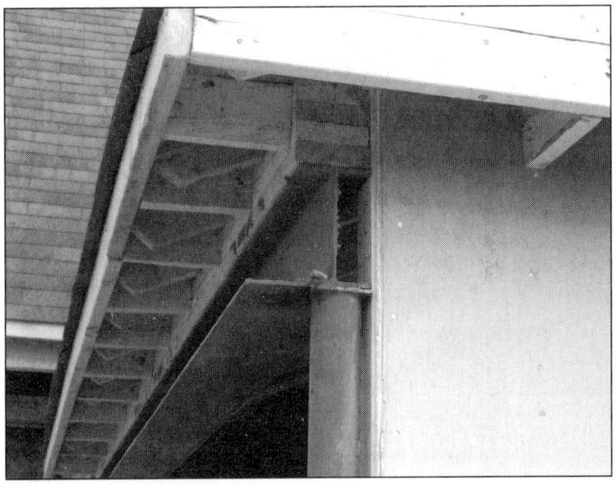

Courtesy Tom Stephenson

at that spot. The load then needs to be transferred through the floor and effectively to the beam below it. Point loads are usually indicated with an abbreviation PL on the floor plan of the floor below where the point load is occurring, for example, the basement plan will indicate the location for the point load on the first-floor framing. See Figure 7.43 for an example of a point load being transferred to a beam through blocking inserted between the floor and the beam. If there was no blocking placed as a support, then over time the subfloor would settle and the walls would crack.

FIGURE 7.37

Section detail of joists and sill plate attached to foundation wall with anchor bolt (refer to 9.23.6 and 9.23.7 of NBC).

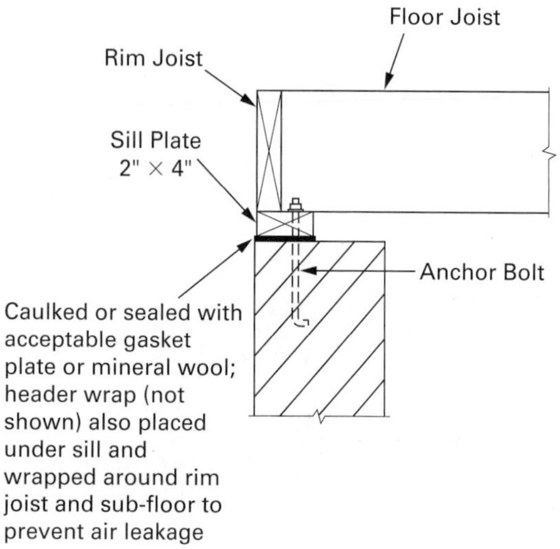

FIGURE 7.38

Sill plate ready to be fastened to foundation wall; notice the header wrap material placed under the sill. The header wrap seals the numerous joints between the foundation, sill, joists, and sub-floor (that otherwise would be very difficult to seal), thus maintaining an air barrier to protect against air leakage (refer to Chapter 12).

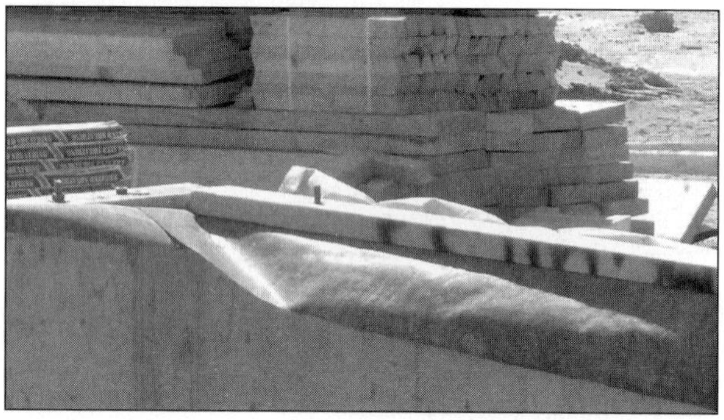

Courtesy Tom Stephenson

FIGURE 7.39

Knee wall (a shortened wall) fastened to foundation with anchor bolt and gasket.

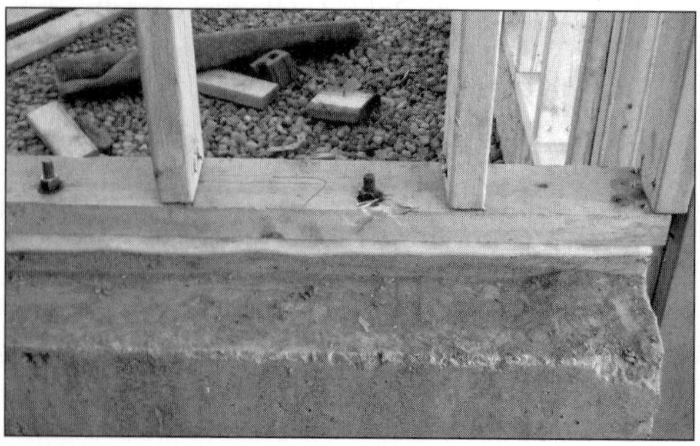

Courtesy Tom Stephenson

FIGURE 7.40

Change of spacing from 16" OC at far side of beam to 12" OC at near side of beam.

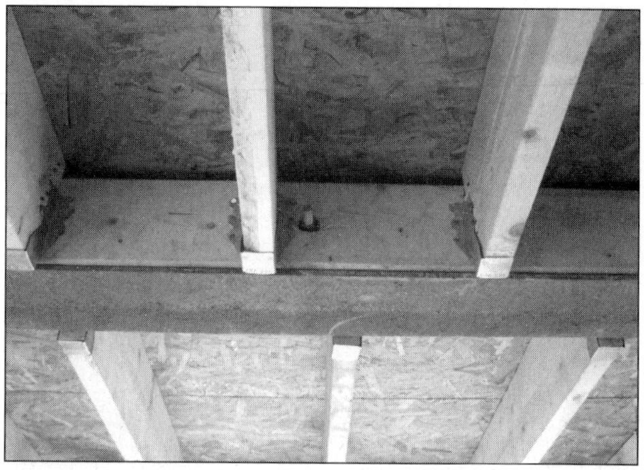

Courtesy Tom Stephenson

FIGURE 7.41

NBC minimum end bearing for a floor joist (refer to Part 9.23.9.1 of the NBC).

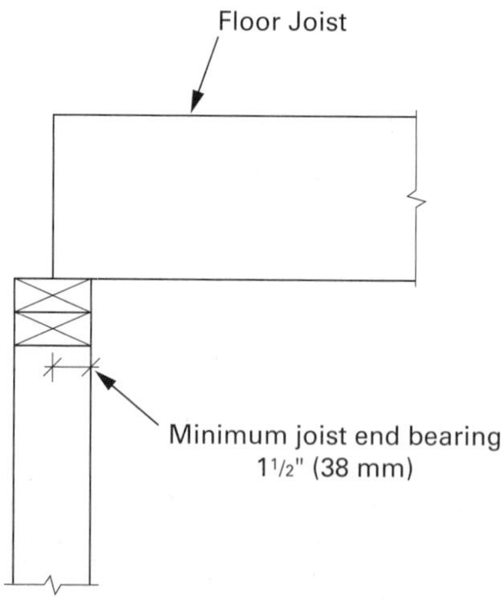

- The location of non-load-bearing walls will require extra blocking or joists for support. See Figure 7.45 for details.
- Any openings in the floor, such as stairwells, will require the use of headers and trimmers to transfer the loads to nearby columns and support walls. Review Figure 7.46 and note the names given to the various floor members known as joists. The key difference between header joists and trimmer joists is that headers can support joists and tail joists, while trimmers can support headers. In simple terms, this means headers are carrying more load than the joists, and trimmers are carrying more load than the headers. The drawings will generally indicate whether headers and trimmers need to be doubled or in some

FIGURE 7.42
Framing details for joists at beam support.

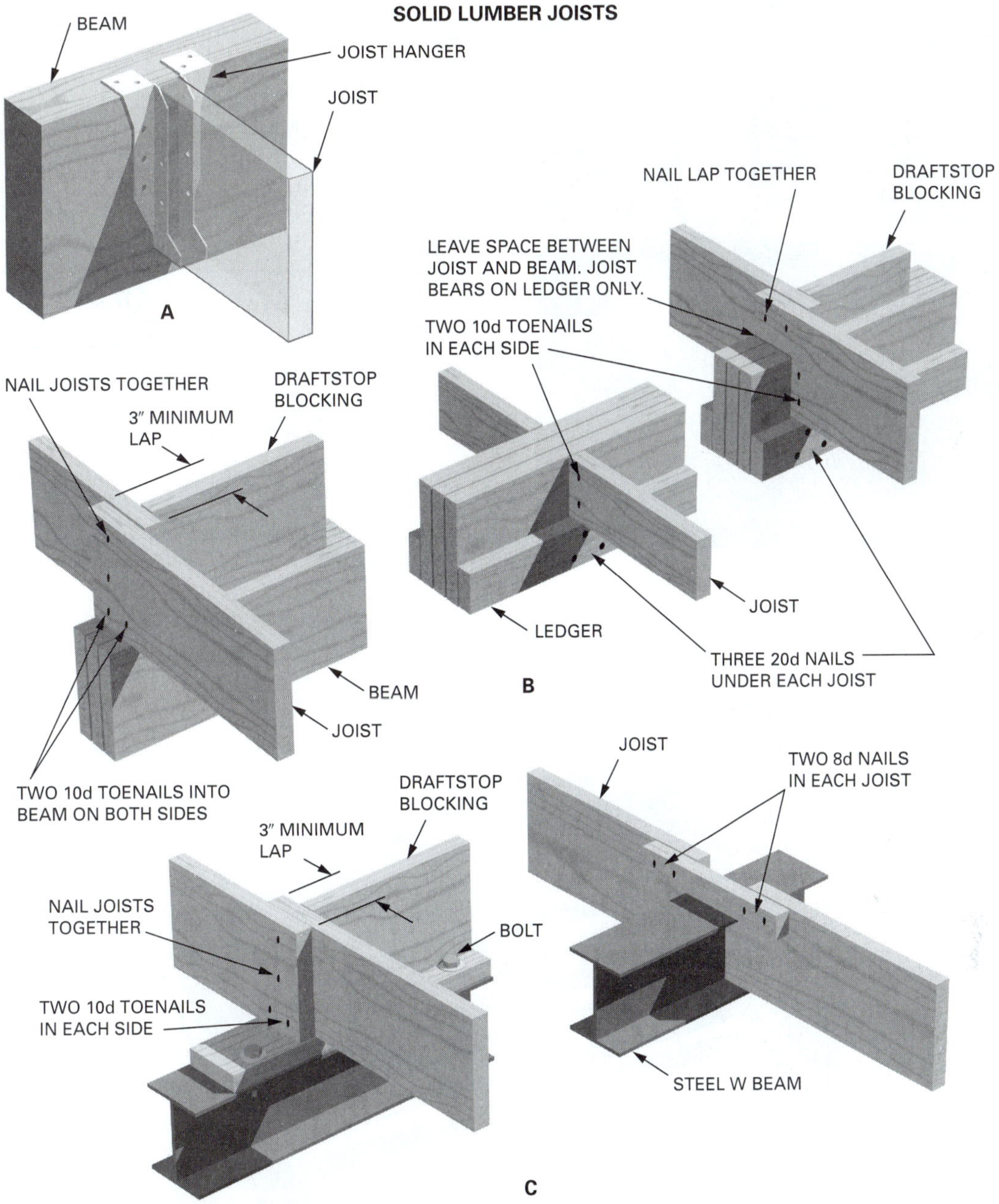

Source: From VOGT, *Carpentry*, 4E, © 2006 Delmar Learning, a part of Cengage Learning, Inc. Reproduced by permission. www.cengage.com/permissions

cases tripled. In all cases, the requirements of the NBC or adopted code must be followed, at minimum. The NBC states in Parts 9.23.9.5 and 9.23.9.6 that:

> Header joists around floor openings shall be doubled when they exceed 1.2 m in length.
> The size of header joists exceeding 3.2 m in length shall be determined by calculations (engineered).
> Trimmer joists around floor openings shall be doubled when the length of the header joist exceeds 800 mm.
> When the header joist exceeds 2 m in length, the size of the trimmer joists shall be determined by calculations (engineered).

- Bridging and strapping are used to reduce deflection and strengthen the floor as a whole. Strapping may be used by itself or in

FIGURE 7.43
Point load blocking in Doncaster house.

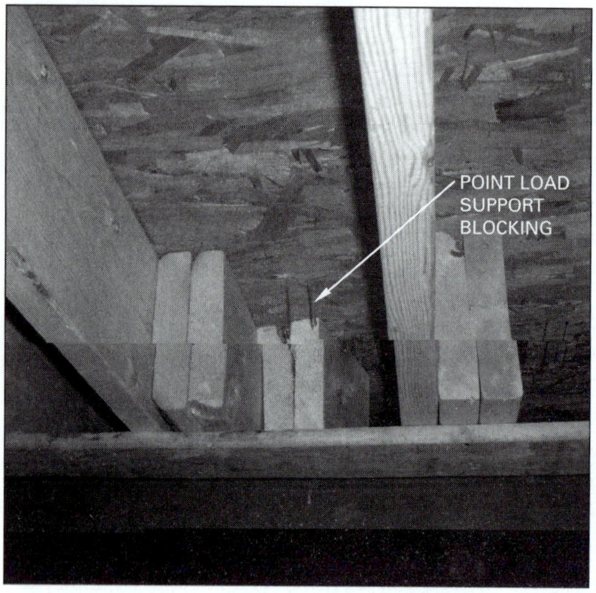

Courtesy Tom Stephenson

FIGURE 7.44
Point load displayed on drawing with the abbreviation PL (Doncaster basement drawing).

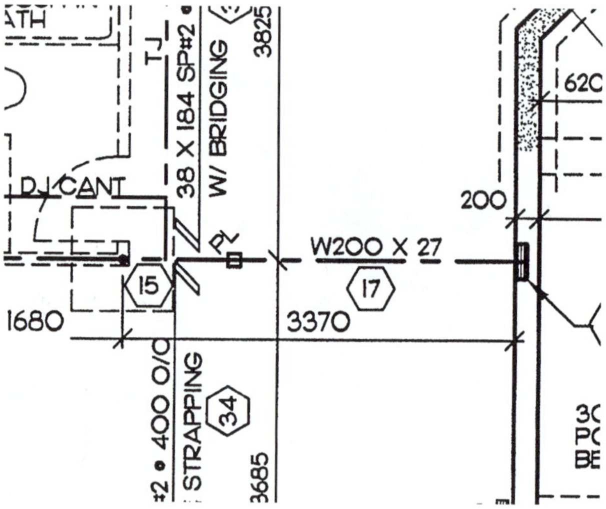

Source: Cassidy & Co. Architectural Technologists

FIGURE 7.45
Floor frame layout.

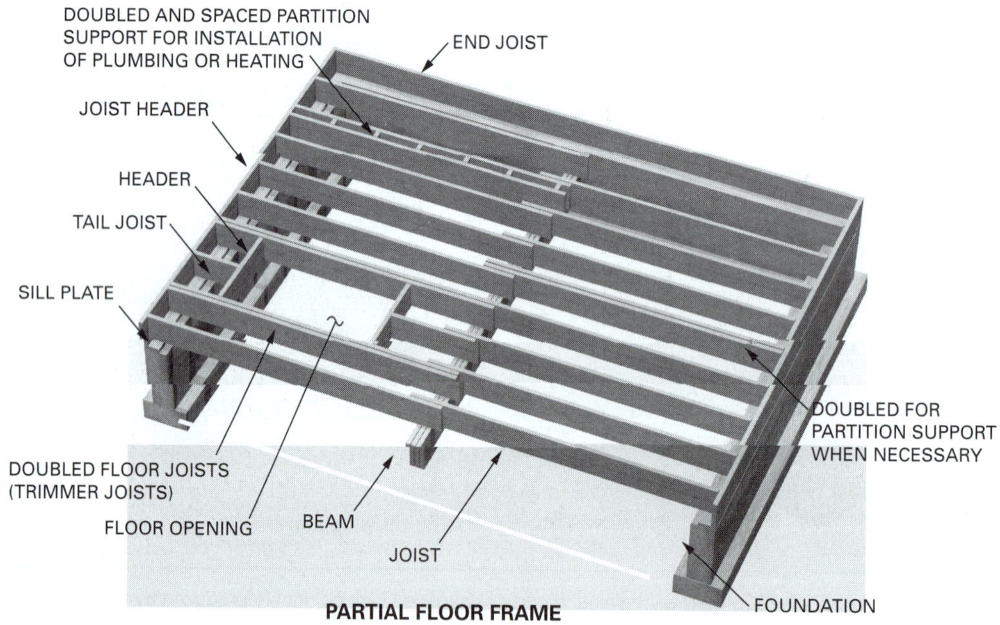

Source: From VOGT, *Carpentry*, 4E, © 2006 Delmar Learning, a part of Cengage Learning, Inc. Reproduced by permission. www.cengage.com/permissions

combination with bridging, and may be used where floor joists are not being finished or enclosed (for example, an unfinished basement ceiling). The span distance of the joist will determine how many rows of bridging are to be installed. Part 9.23.9.4 of the NBC requires that bridging be placed at 2.1 m minimum OC spacing. Bridging can be either cross bridging, solid bridging (blocking), or engineered bridging. Depending on the span being used, the NBC may also require the use of strapping to strengthen the floor system. See Figures 7.47, 7.48, and 7.49 for details.

FIGURE 7.46

Framing layout and floor framing members used for floor openings such as a stairwell opening. Note: Tail and header joists will be connected with joist hangers.

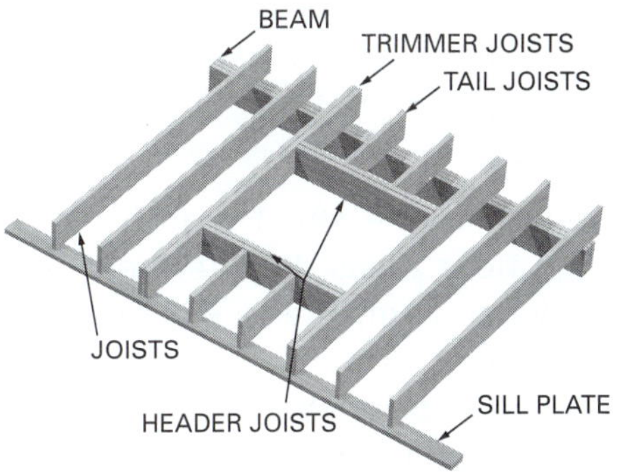

Source: From VOGT, *Carpentry*, 4E, © 2006 Delmar Learning, a part of Cengage Learning, Inc. Reproduced by permission. www.cengage.com/permissions

FIGURE 7.47

Completed strapping and bridging. Strapping is not required in all cases; the span tables in Part 9 of the NBC will specify when strapping is to be used. and the drawings will list this requirement. Because the strapping drops below the joists, it is usually used only in basement conditions where drywall is not being installed.

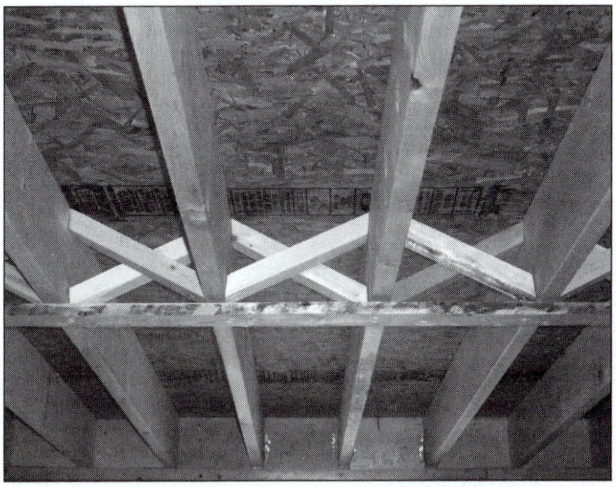

Courtesy Tom Stephenson

FIGURE 7.48

Solid bridging; notice the floor joists are doubled in this area of the floor. Can you find where the doubling of the joists requirement is on the Doncaster drawings? This framing option is not typical and it falls outside of Part 9 of the NBC, but it does allow the builder to span further with a 2" × 8" (38 mm × 184 mm) than would otherwise be possible. This system is designated under Part 4 of the NBC.

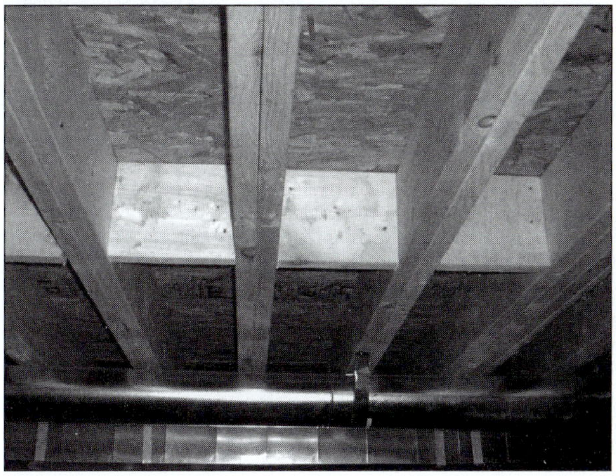

Courtesy Tom Stephenson

FIGURE 7.49

Engineered bridging, a manufactured system designed to increase the strength of floor.

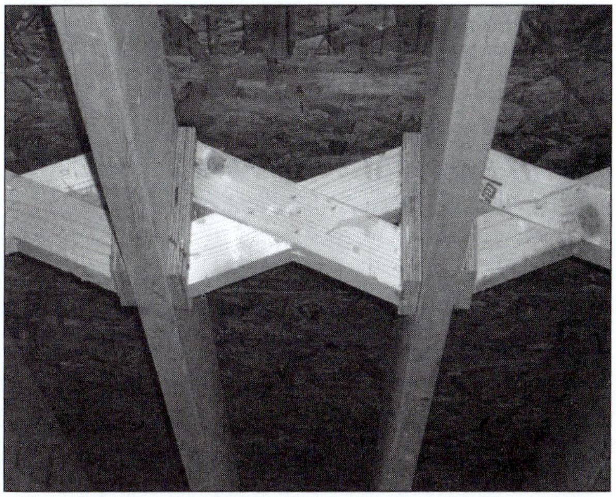

Courtesy Tom Stephenson

Cantilevered Joists

Joists are cantilevered for a number of reasons, including bay, box, or bow windows; closets; and second floors of houses. By cantilevering the joists, the builder avoids having to place a foundation under the cantilevered portion, thus expanding floor space economically. Part 9.23.9.9 of the NBC states that:

> Floor joists supporting roof loads shall not be cantilevered more than 400 mm beyond their supports where 38 mm by 184 mm joists are used and not more than 600 mm beyond their supports where 38 mm by 235 mm or larger joists are used.
>
> The cantilevered portions referred to in Sentence (1) shall not support floor loads from other stories unless calculations are provided to show that the allowable design stresses of the cantilevered joists are not exceeded.
>
> Where cantilevered floor joists described in Sentences (1) and (2) are at right angles to

FIGURE 7.50
Cantilever supporting one storey of house.

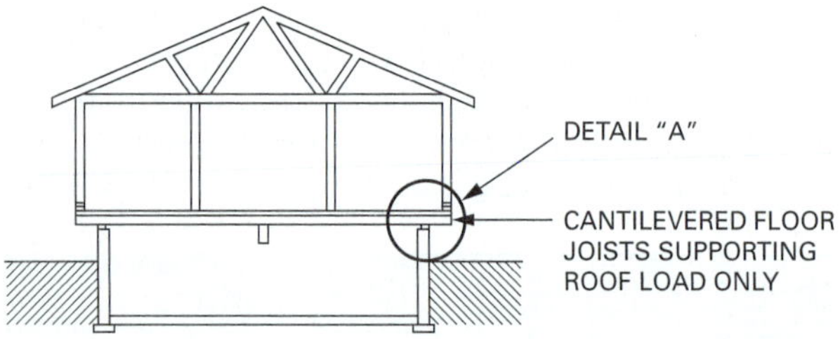

Source: Cassidy & Co. Architectural Technologists

the main floor joists, the tail joists in the cantilevered portion shall extend inward away from the cantilever support a distance equal to not less than 6 times the length of the cantilever, and shall be end nailed to an interior doubled header joist in conformance with NBC Table 9.23.3.4.

Doncaster Cantilever Example

The Doncaster drawings provide an example of a cantilever and how it is to be framed. Figure 7.51 is a close-up of the first-floor framing requirements taken from the basement plan while Figure 7.52 is a section view that displays the box window. Figures 7.53 to 7.58 illustrate the actual framing of the cantilever for the box window on the left elevation of the Doncaster house. View the drawings and photos carefully to fully grasp the transition from drawings to house.

Subfloor

The subfloor is the sheet material placed on top of the floor joists. It is termed "subfloor" because, when complete, it will lie underneath the finished floor covering. Subflooring may consist of plywood or **oriented strand board (OSB)** with a tongue-and-groove edge. The tongue and groove prevents deflection where the sheets span between the joists. The subfloor material is available in 4' × 8' panels and is placed perpendicular (90 degrees) to the run of the joists. The sheets must also be staggered to provide better stiffening to the floor and prevent longitudinal cracking of tile finishes. Subflooring must be nailed or screwed down in accordance with the NBC (refer to Division B Table 9.23.3.5.A). For example, a 20 mm thick plywood subfloor would require minimum 57 mm spiral nails nailed at 150 mm OC along edges and 300 mm OC along intermediate supports.

FIGURE 7.51
Doncaster cantilever close-up; the abbreviation DJ Cant and TJ stands for double joist cantilever and triple joists acting as a header (required because of the cantilever).

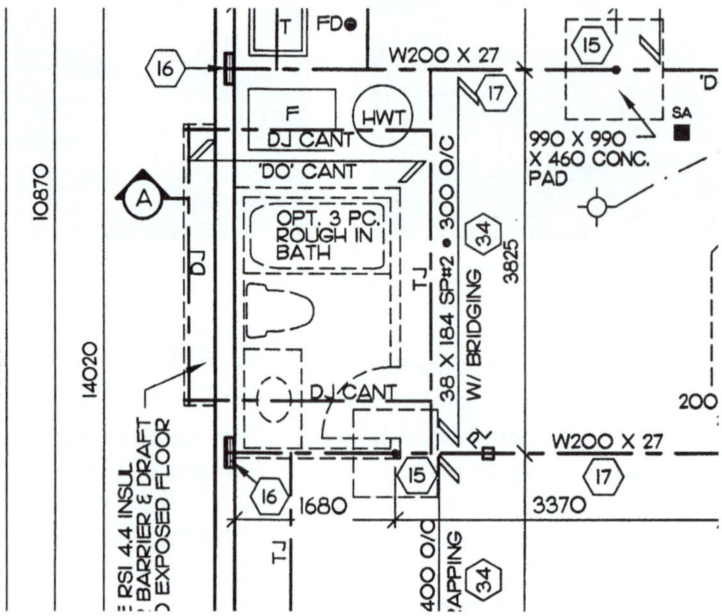

Source: Cassidy & Co. Architectural Technologists

FIGURE 7.52

Section A-A; close-up of cantilevered box window.

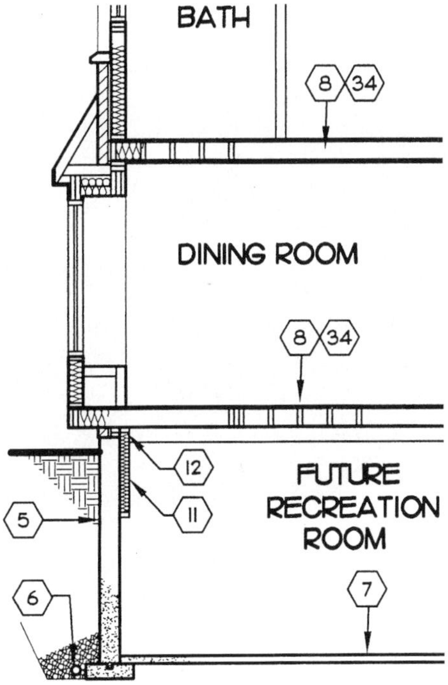

Source: Cassidy & Co. Architectural Technologists

FIGURE 7.53

Doncaster cantilevered box window on left elevation.

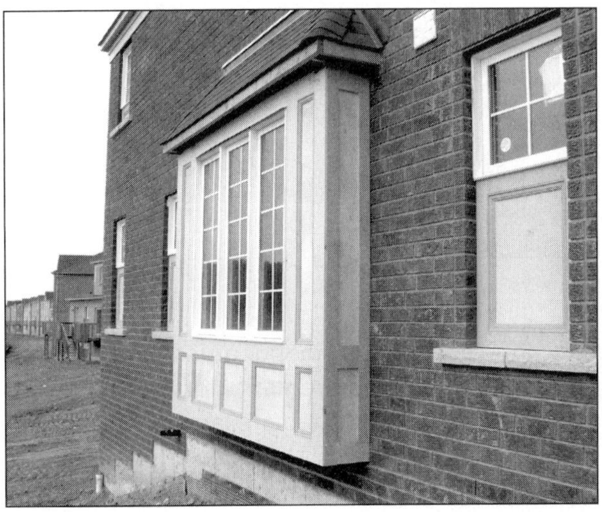

Courtesy Tom Stephenson

FIGURE 7.54

Doncaster inside view of box window opening.

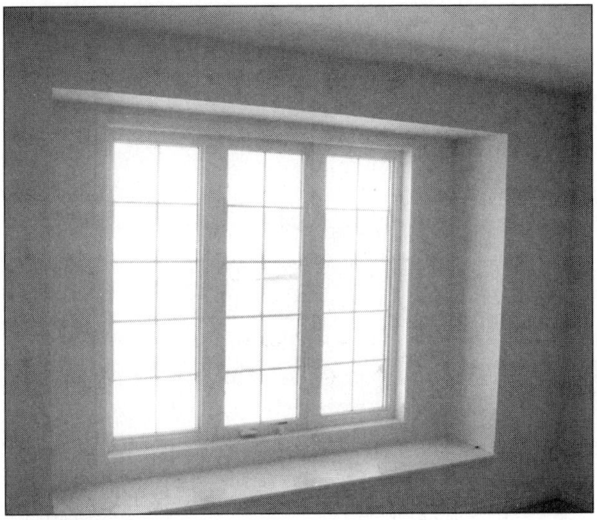

Courtesy Tom Stephenson

FIGURE 7.55

Doncaster outside view of joists cantilevering out from foundation wall.

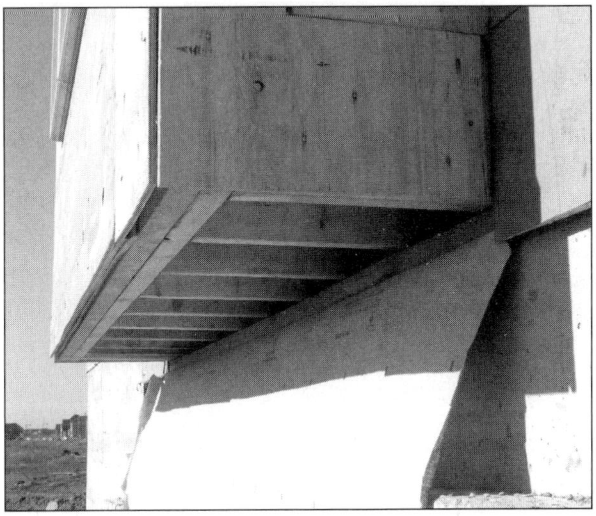

Courtesy Tom Stephenson

FIGURE 7.56

Doncaster inside view of joists cantilevered back more than six times the length of the cantilever; notice how they are fastened to an interior triple header. This exceeds the NBC minimum of an interior double header. You can also see the underside of the OSB subfloor placed above the joists.

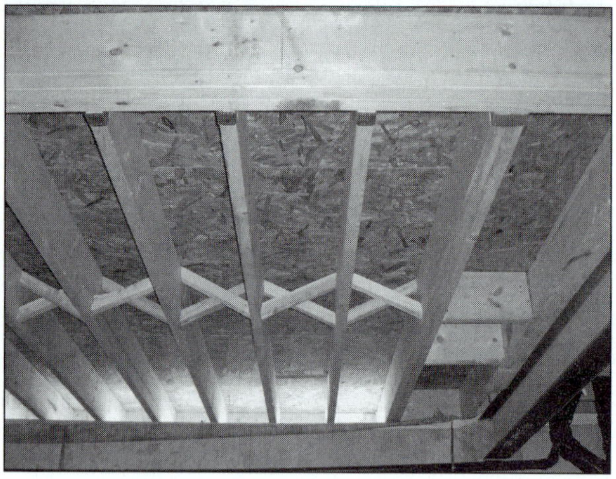

Courtesy Tom Stephenson

FIGURE 7.57

Doncaster cantilever from another viewpoint. Note the triple header (TJ) and how it is supported on the steel beam.

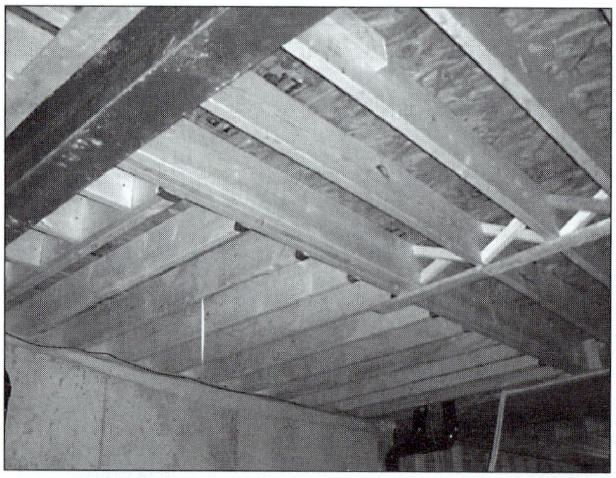

Courtesy Tom Stephenson

FIGURE 7.58

3D illustration representing how the floor for the cantilever is framed. Note where the double and triple joists are located.

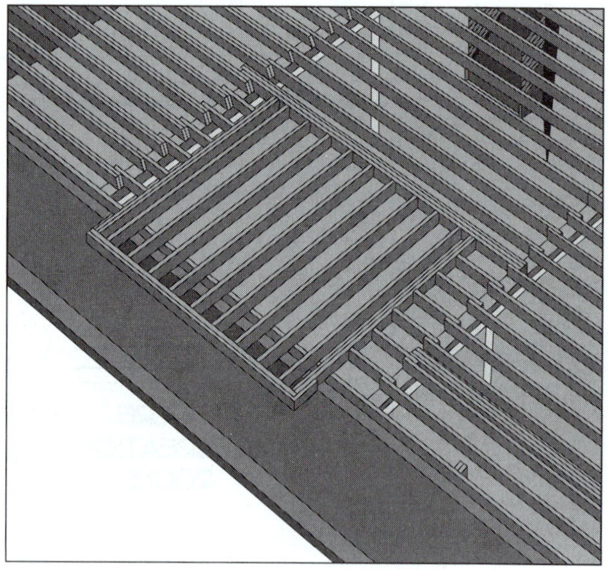

Source: Courtesy of Clint Kissoon, Chair, George Brown College, School of Architectural Studies.

FIGURE 7.59

Plywood subflooring installed perpendicular to joists and staggered.

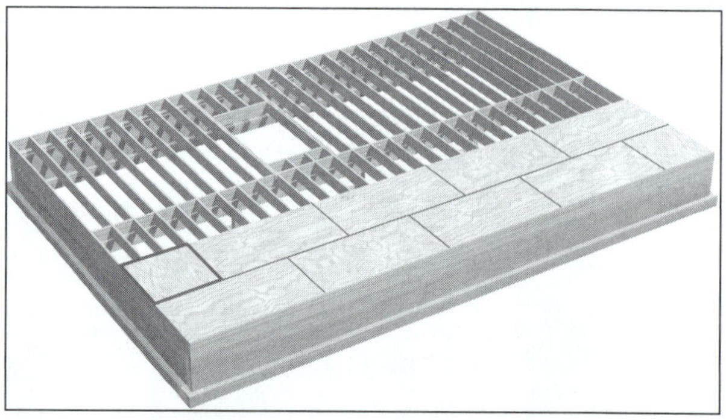

NBC Span Tables

Part 9 of the National Building Code (NBC) provides a number of tables for designing common structural components such as footings, foundation walls, floor joists, steel beams, built up wood beams, ceiling joists, and roof rafters to name just a few. These tables save a lot of time and provide accurate information on what is the minimum acceptable size. Refer to Figure 7.60 for an example of a floor joist span table and how it works to determine the size of floor joist required. A designer will base their choice of on centre spacing and other elements based on the availabilty of material (e.g. No. 2 Spruce is very common in a number of jurisdictions across Canada) and the requirements for their design.

Figure 7.60 A span table example from Part 9 of the NBC (Span Tables Maximum Spans for Floor Joists – General Cases) for how Part 9 is used for designing the correct size floor joist based on span, spacing, species, and use of strapping, bridging or both. Example: Determine the joist size for a joist spanning 3940 mm with Bridging and Strapping and the on centre spacing of 300 mm using Spruce No. 2 (Note: strapping does not neccesarily have to be placed on the underside of joists if a panel type finish such as drywall is to be used but the same span can be used under the bridging and strapping heading).

FIGURE 7.60

A span table example from Part 9 of the NBC (Span Tables Maximum Spans for Floor Joists - General Cases) for how Part 9 is used for designing the correct size floor joist based on span, spacing, species, and use of strapping, bridging or both. Example: Determine the joist size for a joist spanning 3940 mm with Bridging and Strapping and the on centre spacing of 300 mm using Spruce No. 2 (Note: strapping does not neccesarily have to be placed on the underside of joists if a panel type finish such as drywall is to be used but the same span can be used under the bridging and strapping heading).

Span Tables

Table A-1 (Continued)

Commercial Designation	Grade	Joist Size, mm	Maximum Span, m								
			With Strapping[2]			With Bridging			With Strapping[2] and Bridging		
			Joist Spacing, mm			Joist Spacing, mm			Joist Spacing, mm		
			300	400	600	300	400	600	300	400	600
	Select Structural	38 x 89	1.95	1.81	1.64	2.06	1.87	1.64	2.06	1.87	1.64
		38 x 140	3.05	2.85	2.57	3.24	2.95	2.57	3.24	2.95	2.57
		38 x 184	3.66	3.48	3.31	3.94	3.70	3.38	4.12	3.84	3.38
		38 x 235	4.31	4.10	3.90	4.59	4.31	4.05	4.76	4.44	4.14
		38 x 286	4.91	4.67	4.45	5.18	4.87	4.57	5.34	4.98	4.64
Spruce – Pine – Fir (includes Spruce (all species except Coast Sitka Spruce), Jack Pine, Lodgepole Pine, Balsam Fir and Alpine Fir)	No. 1 and No. 2	38 x 89	1.86	1.72	1.58	1.99	1.81	1.58	1.99	1.81	1.58
		38 x 140	2.92	2.71	2.49	3.14	2.85	2.49	3.14	2.85	2.49
		38 x 184	3.54	3.36	3.20	3.81	3.58	3.27	3.99	3.72	3.27
		38 x 235	4.17	3.96	3.77	4.44	4.17	3.92	4.60	4.29	4.00
		38 x 286	4.75	4.52	4.30	5.01	4.71	4.42	5.17	4.82	4.49
	No. 3	38 x 89	1.81	1.68	1.55	1.96	1.78	1.55	1.96	1.78	1.55
		38 x 140	2.84	2.64	2.43	3.08	2.80	2.43	3.08	2.80	2.43
		38 x 184	3.47	3.30	2.95	3.74	3.52	2.95	3.92	3.61	2.95
		38 x 235	4.09	3.89	3.61	4.36	4.09	3.61	4.52	4.22	3.61
		38 x 286	4.67	4.44	4.19	4.92	4.62	4.19	5.08	4.73	4.19
	Construction	38 x 89	1.81	1.68	1.55	1.96	1.78	1.55	1.96	1.78	1.55

KEY TERMS

Anchor bolt (p. 154)
Balloon construction (p. 142)
Beams (p. 142)
Bearing (p. 149)
Brick veneer construction (p. 141)
Built-up wood beam (p. 150)
Dimensional lumber (p. 144)
Insulated concrete forms (ICF) (p. 143)
Joists (p. 141)
Lightweight structural steel framing (p. 142)
Manufactured beam and joist systems (p. 150)
Oriented strand board (OSB) (p. 162)
Point load (p. 155)
Sill plate (p. 154)
Solid masonry construction (p. 143)
Steel beams (p. 152)
Studs (p. 141)
Supported joist length (SJL) (p. 149)
Timber frame construction (p. 142)
Value (p. 140)
Wood frame construction (p. 140)
Wythe (p. 141)

Chapter 7 Assignment 2: Floor Framing Using the Brook Drawings

Use the Brook drawings that came attached to this text. Use the Elevation A drawings (note that the Brook drawings include two different front elevations for the house, Elevation A and Elevation B) and focus in particular on the various floor plan drawings for Elevation A (pages 3, 4, 5, and 6), the construction notes, and the floor framing details on page 19; for some questions you will also need to review the elevation drawings for front Elevation A (pages 8, 9, 10, and 14). Identify the letter that best answers the question.

_____ 1. What size engineered floor joists are used for the first floor?

a. 7½"

b. 8½"

c. 9½"

d. 10½"

_____ 2. How much must the joist bear on the sill plate? (Hint: review the detail drawings on page 19.)

a. 1½"

b. 2¼"

c. 2⅜"

d. 2⅝"

_____ 3. What size is the steel beam that is shown supporting part of the first floor below the breakfast and great rooms?

a. W150 × 20

b. W200 × 27

c. W250 × 33

d. W300 × 40

_____ 4. How far does the steel beam that supports the first floor below the breakfast and great rooms span, from end of wall to centre of column?

a. 10'-11"

b. 8'-10"

c. 19'-1"

d. 19'-9"

_____ 5. Where on the floor plans is a coffered ceiling indicated?

a. rec. room 1

b. master bedroom

c. great room

d. dining room

_____ 6. What height above the finished floor is the coffered ceiling to be placed?

a. 7'-6"

b. 8'-0"

c. 8'-6"

d. 9'-0"

_____ 7. How many point loads are indicated on the lower level plan elevation A, page 3?

a. 4

b. 5

c. 6

d. 7

_____ 8. What size Versa-Lam beams are used at various points throughout the lower level plans?

a. 1¾" × 6½"

b. 1¾" × 7½"

c. 1¾" × 8½"

d. 1¾" × 9½"

_____ 9. From the optional finished lower level plan elevation A (page 4), what is the inside size of bath 3, measured inside frame wall to inside frame wall, on page 4?

a. 4'-0" × 5'-9"

b. 5'-0" × 8'-0"

c. 7'-2" × 8'-0"

d. 5'-9" × 8'-4"

_____ 10. What do the floor framing details on page 19 state regarding when squash blocks need to be installed around the exterior walls?

a. everywhere to protect the floor joists

b. everywhere there is a sill plate

c. wherever there is a point load

d. wherever there is a requirement for an inner rim board

_____ 11. How far does the optional coffered ceiling drop down from the main ceiling? (Hint: remember how cutting plane lines help to find section details.)

a. 6"

b. 8"

c. 10"

d. 12"

_____ 12. If the W250 × 33 steel beam over the garage door must have minimum bearing for a steel beam on each end of the beam, what will the total length of the beam need to be?

a. 18'-4"

b. 18'-7½"

c. 18'-11"

d. 19'-10"

_____ 13. Why would the PAN (pantry/closet) located in the kitchen have a dropped ceiling in it?

a. so that the shelves are easier to reach

b. to compensate for the stairs which slope directly above it

c. to allow room for a structural steel support beam to run through it

d. to help support the second floor

_____ 14. What is the overall span of joist over the basement from the interior garage foundation wall (the wall between the garage and the basement) to the inside of the left exterior foundation wall?

a. 8'-6"

b. 13'-4"

c. 13'-9"

d. 18'-8"

_____ 15. What is the span for the joists over the dining room, taken from the inside of the exterior dining room wall to the inside of the wall that separates the dining room and garage?

a. 11'-9"

b. 12'-11"

c. 13'-3"

d. 13'-11"

_____ 16. How far is the top of the steel beam in the garage to be placed above the finished floor?

a. 8'-0"

b. 8'-7"

c. 9'-1"

d. 10'-6"

_____ 17. What must be placed under the sill plate, between the air barrier and the foundation wall?

a. engineered floor joist

b. continuous caulking or 1" mineral wool

c. 1" rigid extruded polystyrene

d. a course of brick

_____ 18. What size engineered floor joists are used for the second floor?

a. 7½"

b. 8½"

c. 9½"

d. 10½"

_____ 19. How long would the steel beam that runs below and between the breakfast and great rooms supporting the first floor need to be if it is bearing on the outside walls the minimum amount for steel beams?

a. 19'-8"

b. 20'-½"

c. 20'-4"

d. 21'-2"

_____ 20. What is the overall height from the finished concrete slab in the basement to the finished main floor level?

a. 7'-0"

b. 7'-11"

c. 8'-7"

d. 9'-1"

Chapter 7 Quiz Review: Floor Framing Theory and Doncaster Drawings

Answer questions 1 through 6 using information found throughout the chapter, and answer questions 7 to 20 using the Doncaster drawings that came attached with the textbook. Identify the letter that best answers the question.

_____ 1. What is the beam height, measured from top flange to bottom flange, for a W250 × 35 steel beam?

a. 35 mm
b. 35 inches
c. 250 cm
d. 250 mm

_____ 2. What is the term for the total length of floor supported by a beam?

a. the supported joist length
b. the span of the beam
c. the run of the beam
d. the supported beam length

_____ 3. What does a W8 × 22 beam that is 21' long weigh?

a. 220 lbs
b. 220 kg
c. 440 kg
d. 462 lbs

_____ 4. What does the NBC require as the minimum bearing amount for a steel beam?

a. 38 mm
b. 89 mm
c. 140 mm
d. 235 mm

_____ 5. What is an LVL?

a. a type of dimensional lumber beam
b. a type of L-shaped steel support beam known as an angle iron
c. a laminated veneer lumber beam
d. a type of timber frame beam

_____ 6. What does the NBC require as the minimum bearing for a floor joist?

a. 38 mm
b. 89 mm
c. 140 mm
d. 184 mm

_____ 7. Which of the following size, type, grade, and spacing of floor joists (excluding built-up wood beams) are indicated for some part of the first floor?

a. 2/38 × 235 SP #2 @ 300 OC
b. 38 × 235 SP #2 @ 400 OC
c. 38 × 184 SP #2 @ 400 OC
d. 38 × 184 SP #3 @ 300 OC

_____ 8. How many point loads are indicated on the foundation plan?

a. 1
b. 2
c. 3
d. 4

_____ 9. What size and spacing of joists are used over the kitchen?

a. 38 × 184 @ 300 mm OC
b. 38 × 184 @ 400 mm OC
c. 38 × 235 @ 300 mm OC
d. 38 × 235 @ 400 mm OC

_____ 10. What size are all the floor joists for the second floor?

a. 38 × 184
b. 38 × 235
c. 38 × 140
d. 38 × 286

_____ 11. Review the drawing carefully to determine what is the widest clear span from face of support to face of support (e.g. wall to wall, or wall to centre of beam) for any single floor joist on the second floor?

a. 3630
b. 3940
c. 3440
d. 4440

_____ 12. How many locations (areas) on the basement plan require cantilevered joists?

a. one
b. two
c. three
d. four

_____ 13. What type of beam is used to support the garage roof overhang across the front of the house?

a. LVL

b. built-up wood beam

c. steel beam

d. PSL

_____ 14. If the steel beam inside the garage must have the minimum bearing for a steel beam on each end of the beam, what will the total length of the beam be?

a. 5468 mm

b. 5668 mm

c. 5870 mm

d. 7318 mm

_____ 15. What size, spacing, grade, and type of joists are used over the garage?

a. 38 mm × 184 mm SP#2 @ 400 OC

b. 38 mm × 184 mm SP#2 @ 300 OC

c. 38 mm × 235 mm SP#2 @ 300 OC

d. 38 mm × 235 mm SP#2 @ 400 OC

_____ 16. What does the abbreviation DO refer to when used to reference joists on the drawings?

a. double opening

b. double joists

c. ditto, repetition of the size and spacing of joists listed next to or near it

d. do over, the same size and spacing of joists is used for all floors throughout the house when DO is listed once

_____ 17. What is the overall length required for the steel beam that spans over the basement from the interior garage foundation wall (the wall between the garage and the basement) to the left exterior wall, using the minimum NBC bearing requirement for a steel beam?

a. 1680 mm

b. 3370 mm

c. 5050 mm

d. 5228 mm

_____ 18. What is the span for the joists over the dining room?

a. 2800 mm

b. 3630 mm

c. 3825 mm

d. 3940 mm

_____ 19. How far is the top of the steel beam in the garage to be placed above the finished floor?

a. 2080 mm

b. 2200 mm

c. 2470 mm

d. 2670 mm

_____ 20. Where do the drawings state that the floor joists need to be spaced at 300 mm OC?

a. underneath the entire living room

b. underneath bedroom 3

c. underneath bedroom 4

d. underneath ceramic tile areas and oval tub

8 Wall Framing

LEARNING OBJECTIVES

After completing this chapter, you will be able to:

- Identify the structural elements and components of framed walls.
- Analyze and interpret information from construction drawings for the purpose of laying out a wall system.
- Visualize the materials used and their orientation in a framed wall.
- Identify, analyze, and apply sound framing methods that meet the minimum standards of the NBC.
- Distinguish between measurements found on a drawing and actual framing requirements.
- Identify and distinguish between standard framing practices and optimum value engineering (OVE) used in advanced framing techniques.

INTRODUCTION

This chapter introduces you to the components and methods used in the construction of walls. Understanding the components and how they go together is critical to being able to interpret and analyze information from construction drawings. At the end of the chapter there are numerous examples of walls and floors from the Doncaster house to help you visualize what the drawings depict. The end-of-chapter assignments use both the Doncaster drawings (Assignment 1) and the Brook drawings (Assignment 2) so that you can become accustomed to drawings in both metric and imperial, as well as being able to compare and contrast between dimensional lumber floor systems such as those used in the Doncaster drawings and the engineered flooring system used in the Brook drawings. The chapter quiz review introduces a new set of drawings—the Whittington—also found with this text. The quiz review ties together floor framing, addressed in Chapter 7, with wall framing, addressed in this chapter, because both are done in sequence on site and are interconnected.

WALL FRAMING

The methods used to construct walls are similar to those used in floor construction. Walls use the same standard OC spacing as floors to ensure that the exterior sheathing and drywall panels line up to the wall studs with minimal cutting. The most popular spacing is 16"; however, walls may be spaced at any of the standard spacings. The size and spacing of the wall depends on the following factors:

- Height of wall
- Size of stud (in most cases 38 × 89 mm or 38 × 140 mm)
- Method of insulating the walls to reach the required **R value** or **RSI value**, which are the imperial and metric measurements for resistance to heat loss. For example, the walls may use 2" × 4" studs and have fibreglass batts placed inside with 1" or greater thickness extruded Styrofoam on the outside (or other similar insulating material), or be constructed with 2" × 6" studs with fibreglass insulation placed inside the walls. As new codes require higher energy efficiency (higher R or RSI values), the 2" × 4" with Styrofoam insulation is giving way to 2" × 6" walls constructed with Styrofoam insulation on the outside. Other systems may also be introduced in order to further improve energy efficiency.
- Loads being carried, for example, load-bearing or non-load-bearing wall. Is the wall an interior or exterior wall? How many floors are being supported?
- The unsupported height of the wall (this refers to the height of **stud** without the restraint of **blocking**).

An energy-efficient design may opt for the largest OC spacing possible, so as to reduce the surface area that the dimensional lumber uses. For example, if the wall is constructed using 12" OC as compared to 24", almost double the surface area will be taken up by the edge of the studs. If this surface area is reduced by using 24" OC, and replaced with insulation, the energy efficiency of the house is increased.

Division B Table 9.23.10.1 of the NBC defines the minimum size and spacing of studs for interior and exterior walls. Information regarding the stud size or wall thickness is found on the drawings.

- Check the dimensions on the floor plan for the thickness of the wall.
- The hexagon reference indicator located next to the walls on the floor plan will tell you where to find the wall specification in the construction notes.
- Check to see if there is a section detail that provides information regarding the stud size.
- Check elevations to determine floor to ceiling heights.

Wall Layout

Walls are laid out on the wall plates. The plates are laid out from one end to the other at the determined spacing. Once the wall is laid out, the window, door, and intersecting wall locations are marked. The heights of **lintels** and **rough sills** are determined by referencing the elevation drawings and acquiring the window and door sizes from the appropriate schedule or drawing (e.g., elevation drawing). A typical exterior wall uses the components as illustrated in Figure 8.1.

A lintel is a beam placed over the tops of windows and doors in load-bearing walls; on site it may be referred to as a header (an American term for beams in walls). However, it is incorrect to use these words interchangeably, as they have different meanings in the NBC; headers are found in floors as support in floor openings (see Chapter 7), while lintels are found in walls over window, door, and room openings. A lintel is typically comprised of two pieces of dimensional lumber placed on edge. The lintel is supported on **jack studs** and transfers loads

FIGURE 8.1

Components found in a wall, including plates, studs, jack studs, cripple studs, lintels, bracing, and blocking as required. Provide single jack studs 38 mm bearing up to 3 m and double jack studs 76 mm bearing for lintel spans over 3 m in length.

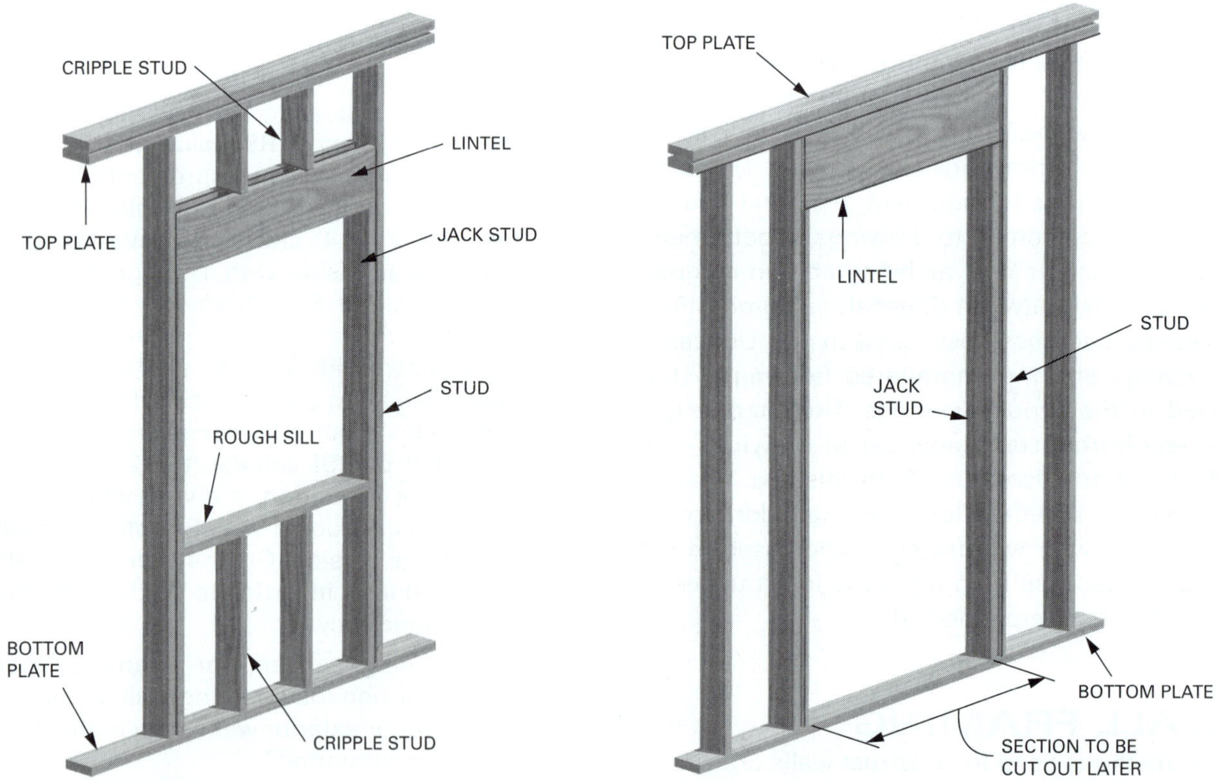

Source: From VOGT, Carpentry, 4E. © 2006 Delmar Learning, a part of Cengage Learning, Inc. Reproduced by permission. www.cengage.com/permissions

(e.g., floor, ceiling, and roof loads) from above the opening over to the jack studs. Supporting jack studs require doubling on openings that span more than 3 m long. There must be an outer stud placed against the jack stud and this stud must run from bottom to top plate (sometimes referred to as a king stud) (Refer to Division B 9.23.10.6 of the NBC.) In an exterior wall, the lintel carries both live loads and dead loads. The live load includes composite snow loads (combined weight of snow and rain), which vary substantially from one jurisdiction to another. Therefore, the designer must size the lintel based on local climatic data regarding composite snow loads. This is one reason generic stock drawings must be adjusted to meet local code requirements.

Lintels are sized based on their span and the number of storeys that they support. (Refer to Division B Part 9 Tables A13–A19.)

Lintels can be framed from the top of the opening up, as in Figure 8.1, or from the plates down, as in Figure 8.2; either is an acceptable framing method.

Walls are typically framed with double top plates. This allows the framer to lay out the joists, rafters, or trusses anywhere along the plate without concern about the plate deflecting. Also, when the top plate is double, the plates can lap over and be nailed to each other, forming strong corners and intersections. See Figure 8.3.

According to the NBC, walls can also be framed with a single top plate, as long as concentrated loads such as joists, rafters, or trusses fall within 50 mm of the stud on the plate. Wall corners and intersections must be suitably tied with metal plates (refer to Division B Parts 9.23.11.3 and 9.23.11.4 of

FIGURE 8.2

Lintel framed from the top down over sliding door; the area framed directly below the window acts as a filler and does not provide support for the lintel. The lintel, however, must be supported on jack studs.

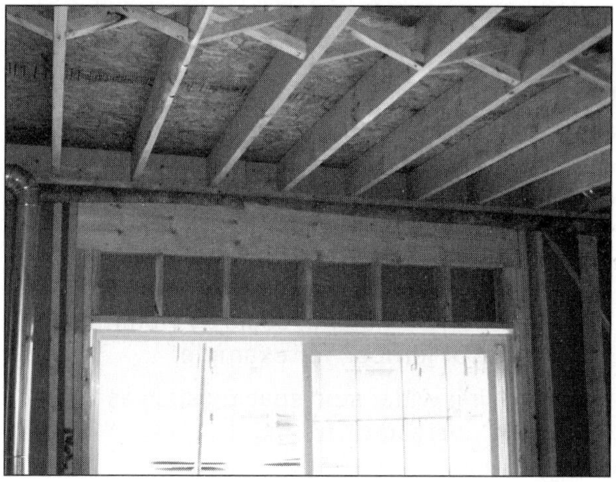

Courtesy Tom Stephenson

FIGURE 8.3

Common framing terminology and lapping of plate detail at wall corner. Generally the cut-in bracing uses a metal type brace that requires only a saw cut and nails. The bracing may be omitted if a structural type sheathing such as plywood or oriented strand board is used.

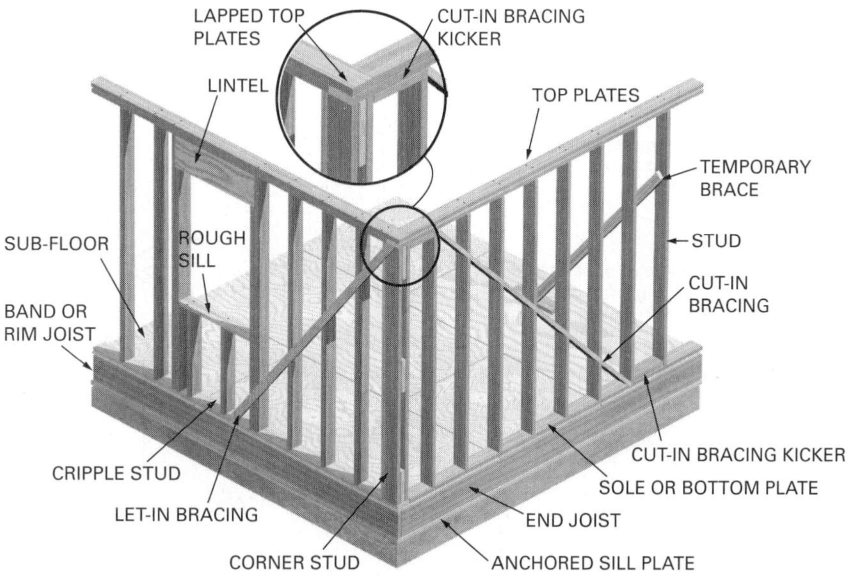

Source: From VOGT, Carpentry, 4E. © 2006 Delmar Learning, a part of Cengage Learning, Inc. Reproduced by permission. www.cengage.com/permissions

the NBC). Regardless of whether a single or double top plate is used, it is always good practice to align the joists and studs as closely as possible. This makes it a lot easier for the subtrades to run their pipes and ducts from one floor to another.

Optimum Value Engineering

Optimum value engineering (OVE), also termed advanced framing techniques, is a method of design and construction that strives to miminize the use of lumber in the building process.

The advantages and drivers for the use of this system include:

1. Reduction of material costs by minimizing the use of lumber, for example:
   - Using wider stud spacing such as 19.2" or 24" instead of 16" OC
   - Using a single top plate instead of a double top plate
   - Reducing lumber at corner and wall intersections
2. The builder not only reduces costs but at the same time wastes less lumber, which translates into a more sustainable building process that requires fewer trees.
3. An added bonus is that when we frame with less lumber, we reduce thermal bridging by increasing the surface area of the wall that is insulated, thus providing greater energy efficiency.

Currently OVE is not that widely adopted, mainly because it does take more initial planning in the design stage, and does require more on-site coordination and education with the trades; also some of the buying public may think that the builder is just trying to cut costs. However, with increasing energy costs, a more educated consumer, and increased use of green certification programs, as well as ongoing research such as Canada Mortgage and Housing Corporation's (CMHC) EQuilibrium™ projects (designing and building test houses that are sustainable and meet net zero energy requirements, where the

FIGURE 8.4

Corner framing and top plate connection details; note that the goal is to reduce the number of studs: the best option uses the metal clips for fastening drywall. The clips also reduce the effects of shrinkage cracks in finished drywall that result when the studs dry out. Using a single top plate instead of a double requires planning (during design and construction phases) at connection points and for the layout of joists and roof trusses that must align with studs to effectively transfer dead and live loads.

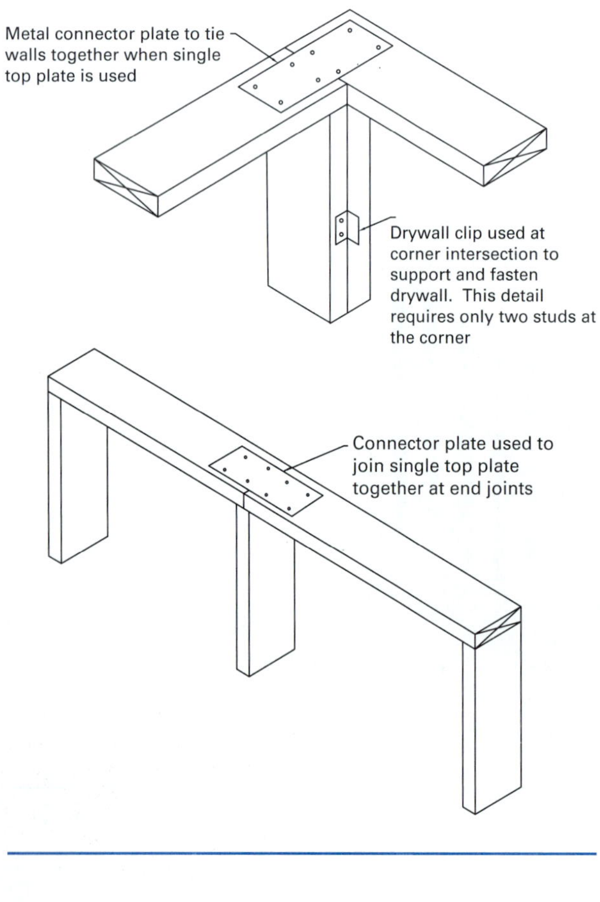

FIGURE 8.5

This detail indicates the use for drywall support required at interior wall intersections. There would be clips used on each side of the intersecting wall or in some cases a space may be left to allow the drywall to slide behind the intersecting wall. A single top plate is used, but does require that the joists align with the studs so that the loads transfer without causing the plate to sag.

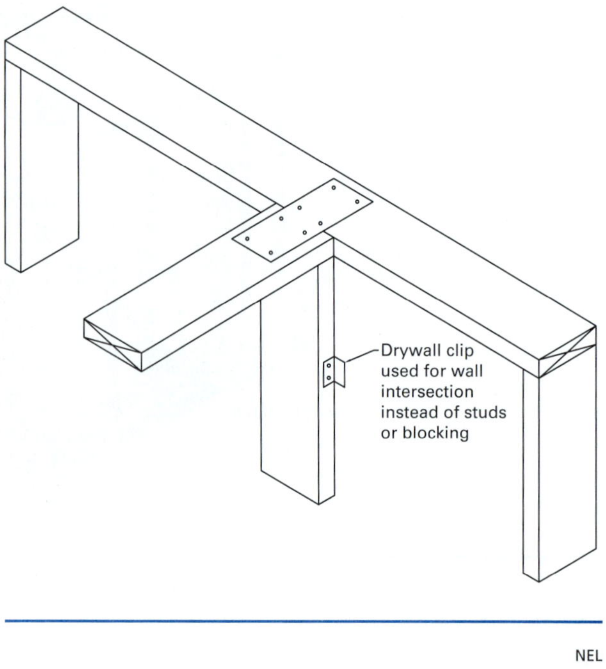

FIGURE 8.6

Framing sketch of rough opening for an exterior door; note that shim spacing in this example is ½" for the rabbeted jamb door, which is an exterior door, providing the necessary space for foam insulation (shim spacing may vary slightly based on site requirements and installer preferences).

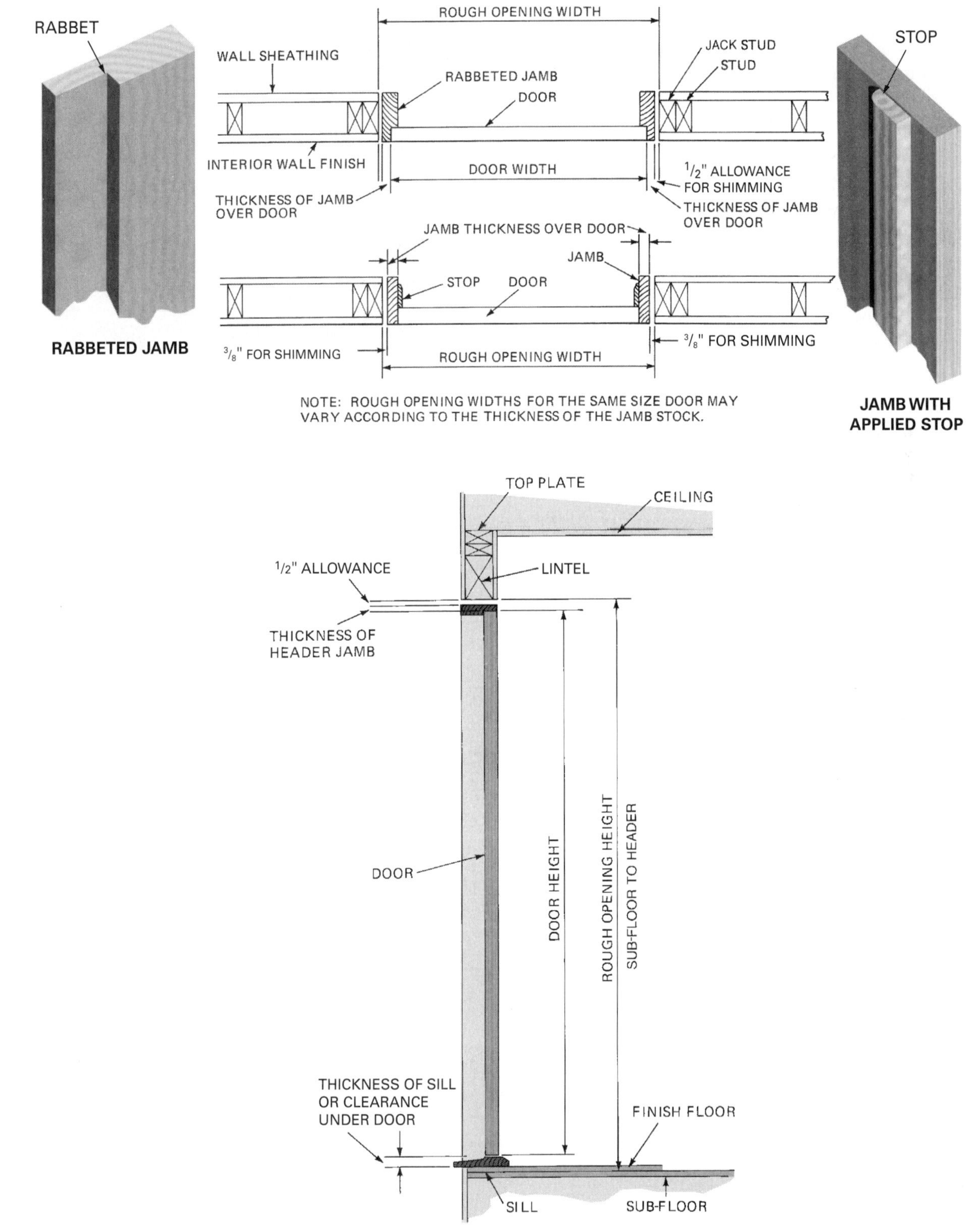

Source: From VOGT, Carpentry, 4E. © 2006 Delmar Learning, a part of Cengage Learning, Inc. Reproduced by permission. www.cengage.com/permissions.

house produces as much energy as it uses), it is expected that the value and use of OVE will increase over time. Other framing methods that use materials in a different way, such as the use of Structural Insulated Panels, known as SIP, are also coming into wider use because of their increased energy efficiency and the ability to pre-manufacture the panels so that they are ready for fast assembly on site. A SIP uses an exterior sheathing material such as OSB that is laminated to thick boards of Styrofoam to form a structural system. Like manufactured joist systems, these systems must be engineered by the manufacturer, who provides a set of shop drawings that are also provided to the inspections office, and are used for the assembly of the pre-manufactured panel system on site.

Rough Openings

The openings framed for windows and doors are called **rough openings**. They are framed larger than the actual size of the windows and doors, which is not always evident on the drawings. The larger opening allows the installer to use shims to install the doors and windows perfectly plumb and level. On exterior walls, the shim space also provides room to insulate around the opening. While the amount of extra rough-opening size varies based on preference, a window will usually be framed ¾" to 1" (25 mm) wider and higher than the window's actual size. For example, a 3'-0" × 4'-0" window (width is provided first, then height) would be framed with a ½" space on all sides, resulting in a rough-opening size of 3'-1" × 4'-1". The ½" space around windows and exterior doors is preferred in order to provide space for foam insulation (See Figure 8.7). An interior door will have similar allowances except generally ⅜" is allowed for shimming as the frame does not have to be insulated. For example, a 2'-6" × 6'-8" interior door using a ¾" thick door frame will have a rough opening of:

- 2'-6" + ¾" door jamb (left side) + ¾" door jamb (right side) + (2 × ⅜" shim space, 1 for each side of jamb) = 2'-8¼"
- The height will be 6'-8"+ ¾" head jamb + 1" space for shimming and space under door = 6'-9¾"
- Therefore, the rough-opening size is 2'-8¼" × 6'-9¾".

The drawings generally indicate the actual door and window sizes; it is expected that the

FIGURE 8.7

Window installed in rough opening; notice the shims at the bottom corner and the space between the window frame and rough framing. The space will be filled later with insulation. Polyurethane foam is a popular choice for insulating around windows and doors.

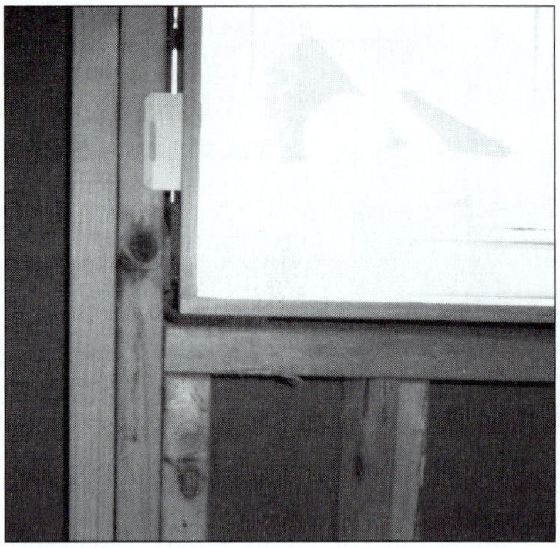

Courtesy Tom Stephenson

FIGURE 8.8

This lintel/beam schedule is found on the construction notes of the Doncaster drawings. The windows indicated in the close-up of the floor plan display the letters L1 and L7. In this case, the Lintel schedule indicates that L1 is 2 – 2" × 8"s used to frame the lintel over the window. The L7 refers to a 3½" × 3½" × ¼" L (steel angle iron) used to support the brick veneer on the outside over the window (for more information on brick veneer, refer to Chapter 10).

LINTEL/BEAM SCHEDULE

L1	2 – 38 x 184	(2 – 2 x 8) SPR.
L2	3 – 38 x 184	(3 – 2 x 8) SPR.
L3	2 – 38 x 235	(2 – 2 x 10) SPR.
L4	3 – 38 x 235	(3 – 2 x 10) SPR.
L5	2 – 38 x 286	(2 – 2 x 12) SPR.
L6	3 – 38 x 286	(3 – 2 x 12) SPR.
L7	90 x 90 x 6.0	(3 1/2 x 3 1/2 x 1/4 L)
L8	90 x 90 x 8.0	(3 1/2 x 3 1/2 x 5/16 L)
L9	100 x 90 x 6.0	(4 x 3 1/2 x 1/4 L)
L10	125 x 90 x 8.0	(5 x 3 1/2 x 5/16 L)
L11	125 x 90 x 10.0	(5 x 3 1/2 x 3/8 L)
L12	150 x 100 x 10.0	(6 x 4 x 3/8 L)

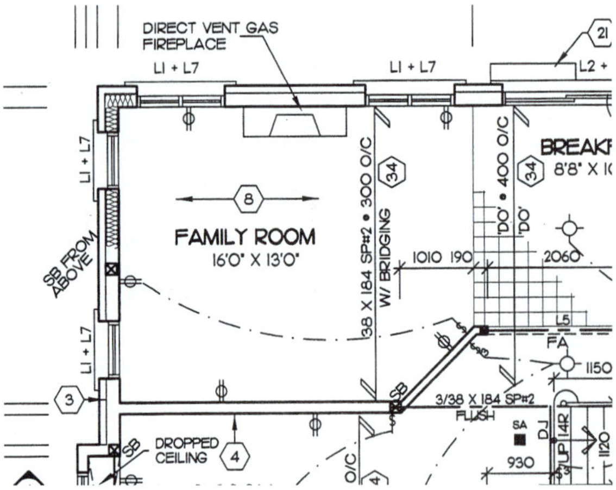

Source: Cassidy & Co. Architectural Technologists

tradesperson will calculate the rough-opening size. **Conflicts can easily occur over the misinterpretation of door and window sizes as indicated on the drawings; therefore, it is highly recommended that you confirm and reconfirm the actual sizes of the windows and doors that have been ordered for the project before any wall framing takes place.** Taking 10 to 15 minutes to confirm this information avoids many needless hours of rework and arguments that occur as a result of openings that are incorrectly framed due to misinformation.

Schedules

The sizes of the windows and doors are found on the plans and elevations, or on a window and door **schedule.** A schedule will number the doors and windows, and provide information regarding their size and type. When a schedule is used, the windows and doors are provided with a reference number found on the floor plans, which corresponds with the number provided in the schedule. Schedules are also used for other components such as lintels and beams. Check the construction notes of the Doncaster drawings to find a door schedule and a lintel/beam schedule; on the floor plans you will find corresponding reference numbers next to the windows and doors. The Doncaster drawings list the actual sizes of the windows on the elevation drawings. Other schedules (e.g., room schedules) may also be used to indicate room finishes, such as colour of paint, type of trim, and type of flooring.

Framing Details

Walls are framed to enclose the space and support the structure. There are a number of ways to frame the walls, particularly at corners and intersections. The framing must take into consideration the transfer of loads, including providing solid bearing under beams and ensuring that **point loads** are properly transferred through the floor system. The following illustrations and photos (Figures 8.9 to 8.13) show some of the different methods commonly used to frame walls.

FIGURE 8.9
Illustrations of corner and wall intersections.

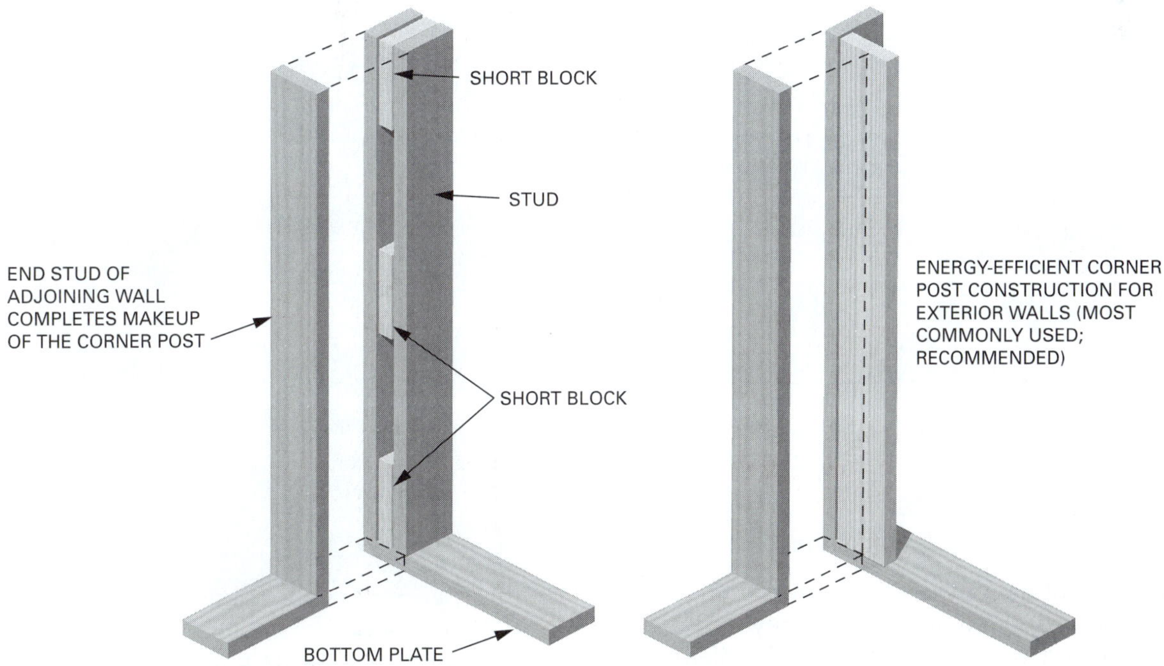

Source: From VOGT, Carpentry, 4E. © 2006 Delmar Learning, a part of Cengage Learning, Inc. Reproduced by permission. www.cengage.com/permissions.

FIGURE 8.9 *(continued)*

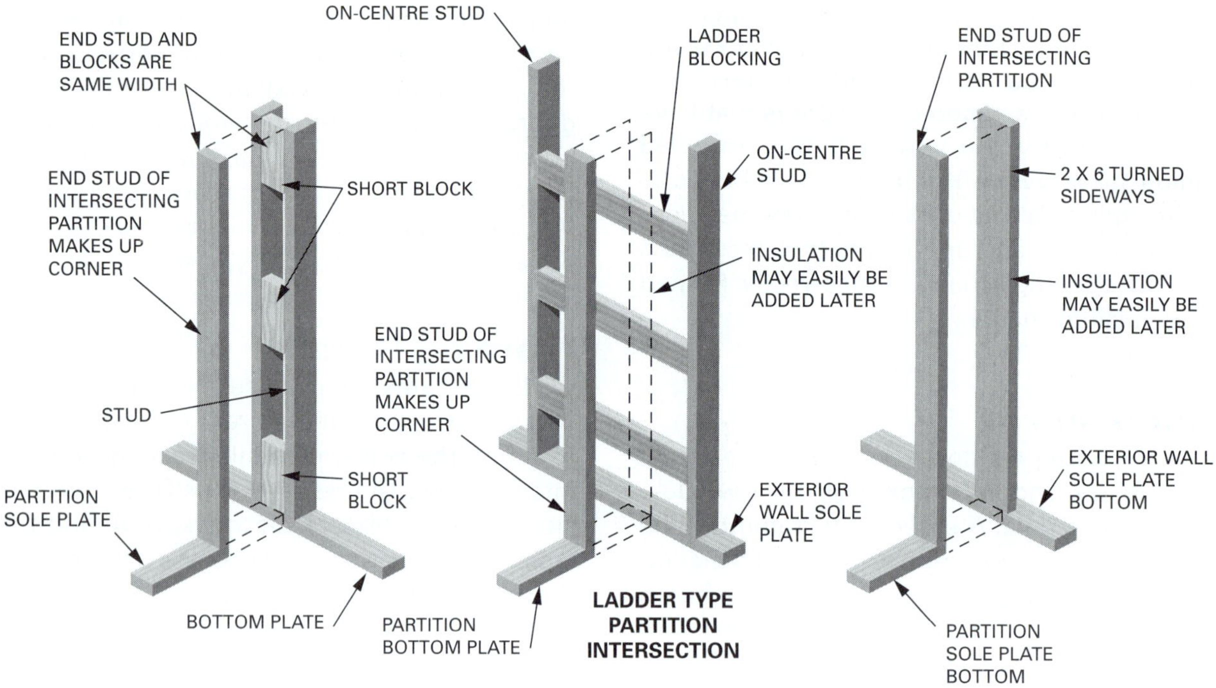

FIGURE 8.10

Example of corner intersection, three-stud corner.

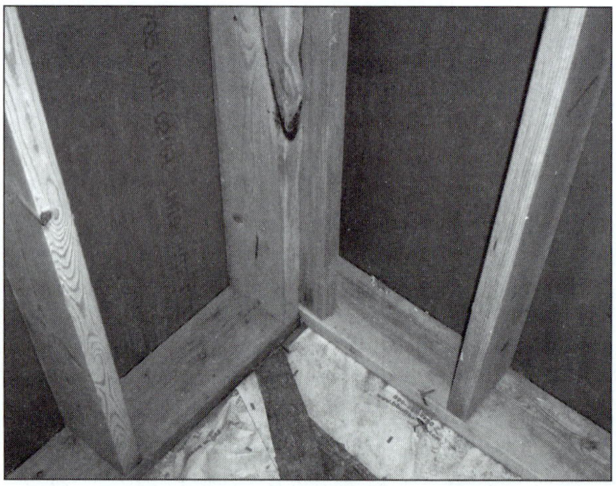

Courtesy Tom Stephenson

Bracing of Walls

Part 9.23.10.2 of the NBC states that:

1. Except as provided in Sentence (2), each exterior wall in each storey shall be braced with not less than one diagonal brace conforming to Article 9.23.18.2.
2. Bracing is not required where walls have an interior finish conforming to the requirements of Section 9.29, or if the walls are clad with panel type siding, diagonal lumber or plywood, OSB, waferboard, gypsum, or fibreboard sheathing.

FIGURE 8.11

Intersecting wall with cross blocking; black poly provides for a continuous air and vapour barrier and is placed between top plates and at intersecting walls (refer to Chapter 12 for more information on the building envelope).

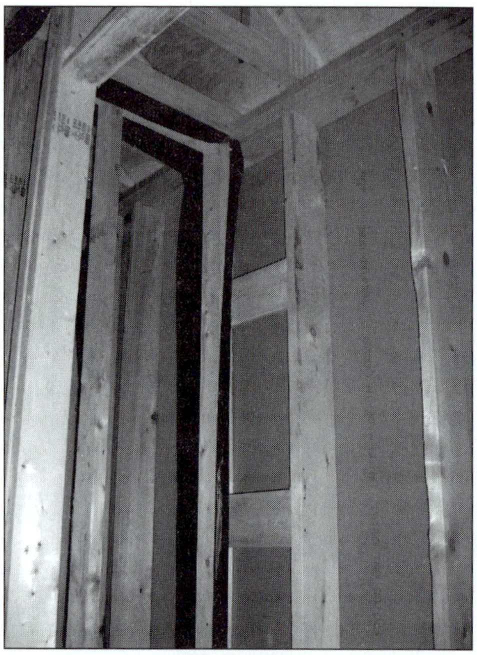

Courtesy Tom Stephenson

FIGURE 8.12
Intersecting wall with stud on each side; notice 2 × 6 on flat at top of wall. The stud on each side with blocking provides a tie between the walls and screw support for drywall, as does the 2 × 6 laid on top of the wall for screwing drywall to the ceiling.

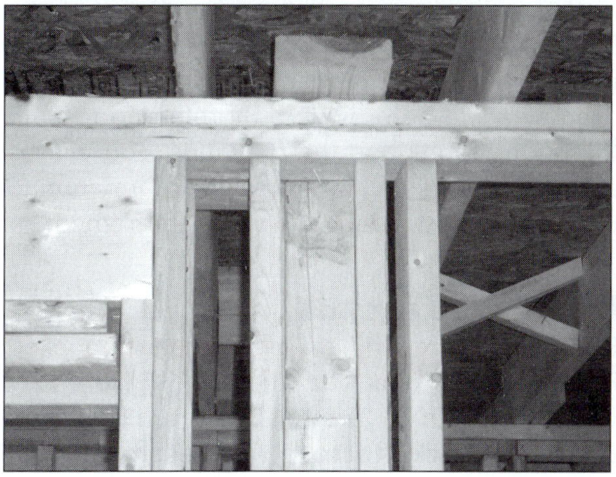

Courtesy Tom Stephenson

FIGURE 8.13
Wood column providing solid bearing for built-up wood beam.

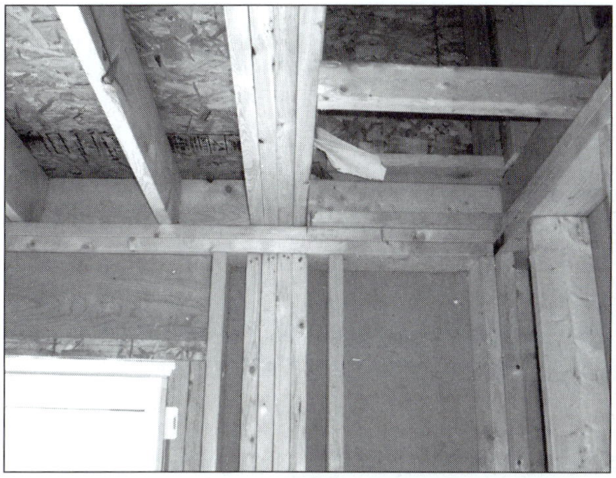

Courtesy Tom Stephenson

Bracing has traditionally been used to strengthen the walls by reducing **racking.** A brace placed on a 45-degree angle fastened to the wall produces a triangle effect that will stabilize a wall and hold it square. Laboratory tests have shown that interior finishes such as those listed in point number 2 above take the place of bracing under normal circumstances. For buildings that have special conditions, bracing will still be required, and many builders prefer to install extra bracing as standard practice. Sometimes builders will install structural **sheathing,** such as oriented strand board (OSB) at the corners of the building while using nonstructural sheathing such as fibreboard for the rest of the walls. Temporary bracing is used to erect the walls and hold them in place until additional floor and roof structures are in place.

Back Framing

Back framing involves going back and making adjustments to the framing, such as straightening any twisted 2" × 4"s so that the drywall goes on straight and shows no humps or inconsistencies. Back framing also includes constructing any necessary bulkheads to conceal exposed ducts or pipes, and decorative bulkheads such as those found in coffered ceilings or on top of kitchen cabinets. Many showers and bathtubs require custom framing as well, which must be coordinated with the plumbing subcontractor. More details on this subject will be included in Chapter 11.

FIGURE 8.14
Example of diagonal metal bracing. The framer snaps a chalk line at 45 degrees once the wall is framed; he/she then cuts a groove about ¾" deep with a circular saw. The metal brace fits in the groove and is nailed at each stud location.

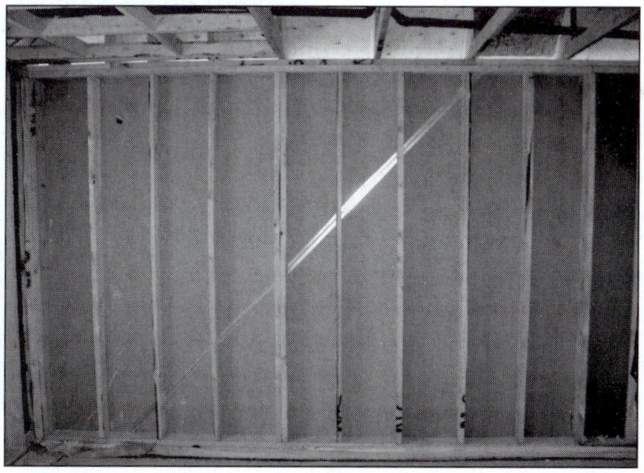

Courtesy Tom Stephenson

FIGURE 8.15

Temporary bracing installed to hold walls plumb and straight while second floor is framed above.

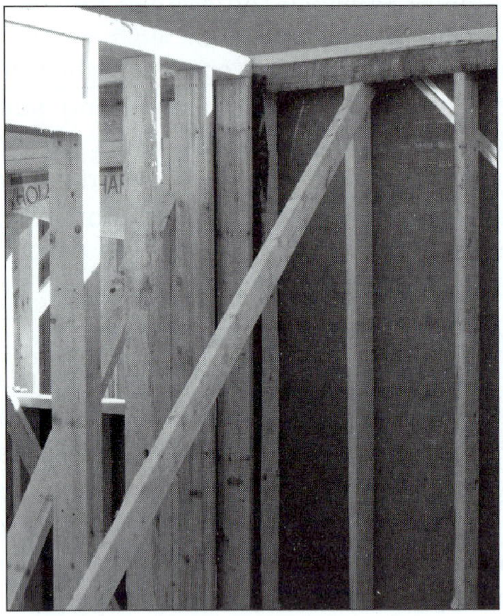

Courtesy Tom Stephenson

DONCASTER DRAWINGS AND EXAMPLES OF FLOOR AND WALL FRAMING

In this section, we will examine some of the framing details from the Doncaster drawings provided with the text. Both floor and wall framing examples will be provided. By referring to both the photographs and the drawings, you will be able to visualize and interpret the information to understand how floors and walls are framed.

Framing Details

Refer to Figure 8.17. (16) is the beam pocket. From the floor plan near the top of Figure 8.17 you see a W200 × 27 steel beam running from the beam pocket across the basement over top of a column and continuing on past the

FIGURE 8.16

Overview of Doncaster floor joist layout. Compare with basement plan of Doncaster drawings (page 2) and observe direction and placement of joists on beams, which are resting on foundation walls and columns. Note that doubling of joists around windows is not indicated.

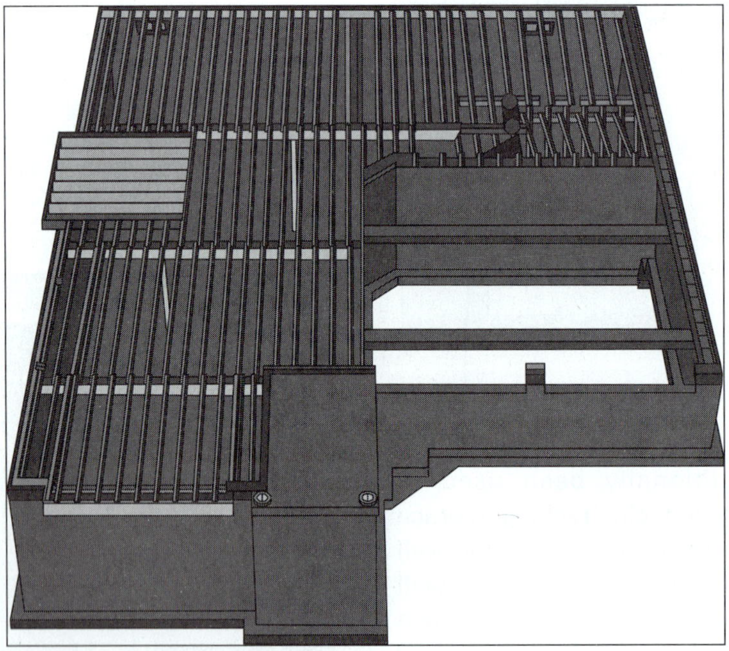

Source: Courtesy of Clint Kissoon, Chair, George Brown College, School of Architectural Studies.

side of the stairs. The Doncaster floor plans that came attached to the text call for 2/38 × 184 SP#2 @ 300 OC toward the rear of the basement and single 38 × 184 SP#2 @ 300 OC on the other side of the beam. They are doubled and at 12" OC to maximize the span of the 38 × 184 mm and reduce deflection of the ceramic tile floor above in the kitchen. See Figures 8.18 and 8.19 for photographs of this beam and joists.

FIGURE 8.17

Close-up of Doncaster basement plan.

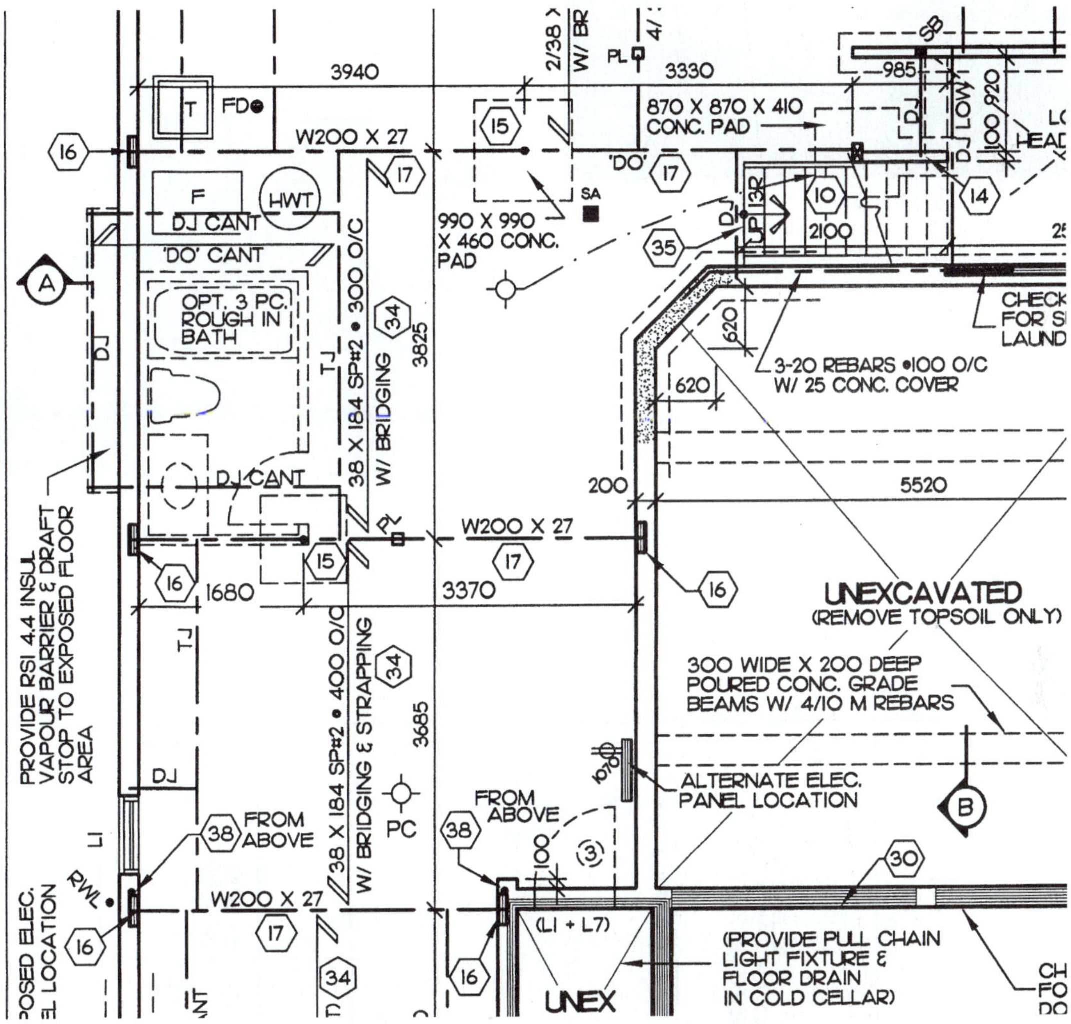

Source: Cassidy & Co. Architectural Technologists

FIGURE 8.18

Double floor joists resting on beam near column in basement, looking toward front of the Doncaster house.

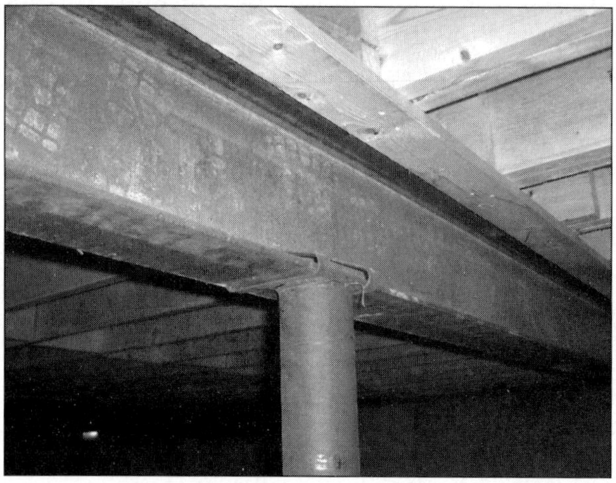

Courtesy Tom Stephenson

FIGURE 8.19

Double joists and built-up wood beam looking to the rear of the house. The wood beam supports a point load, as indicated by PL on the basement plan in Figure 8.20. The point load is from the beam labelled L5 down through the angled wall between the breakfast and family rooms in Figure 8.21. Compare and visualize the Doncaster close-ups of basement and first-floor plans. The end of the foundation wall is framed instead of having concrete under the sill plate because of the SWO (semi-walkout) condition described for the sloping outside grade in the site plan from Chapter 5 for Lot 61.

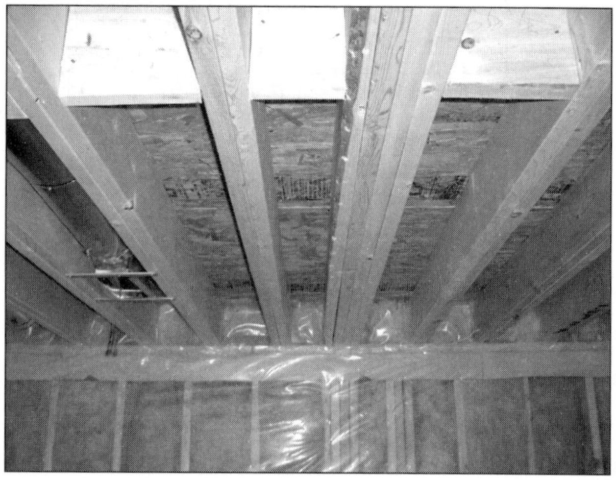

Courtesy Tom Stephenson

FIGURE 8.20

Doncaster basement plan close-up.

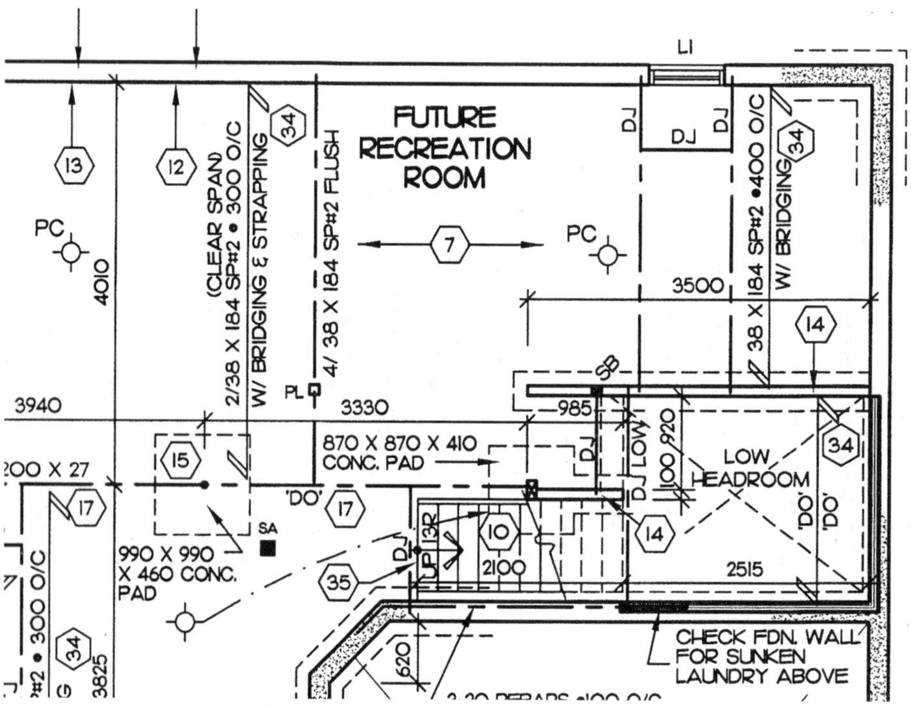

Source: Cassidy & Co. Architectural Technologists

FIGURE 8.21

Doncaster first-floor plan close-up.

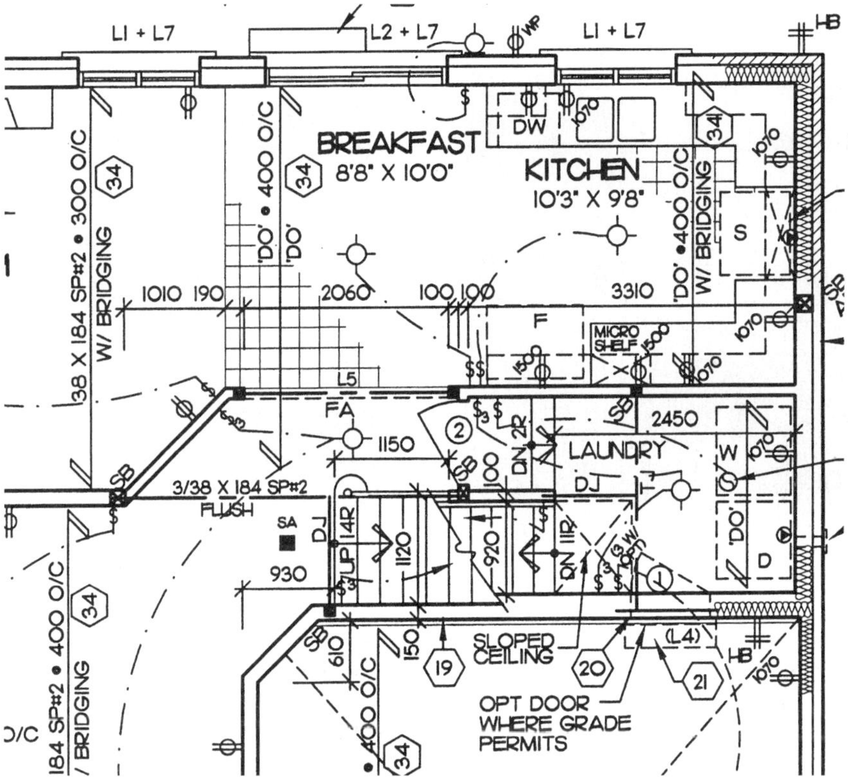

Source: Cassidy & Co. Architectural Technologists

FIGURE 8.22

3D View of sunken laundry room from Doncaster drawing. Review in conjunction with actual photos and Figures 8.20 and 8.21 to better visualize the framing requirements for this construction detail.

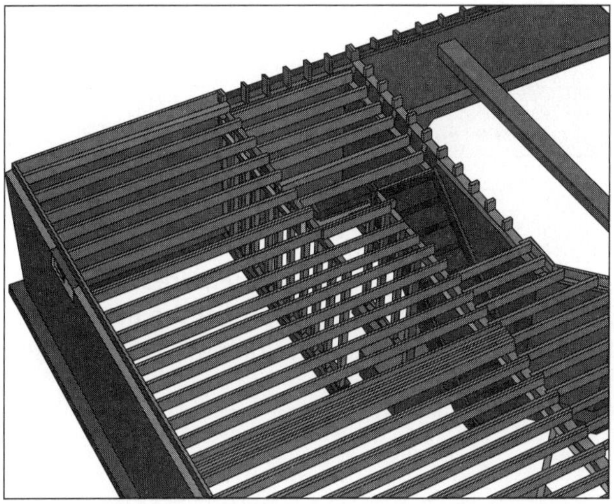

Courtesy Tom Stephenson

Source: Courtesy of Clint Kissoon, Chair, George Brown College, School of Architectural Studies.

FIGURE 8.23

Beam ends beside wall at side of stairs in basement plan; see column and footing indication on basement plan in Figure 8.20.

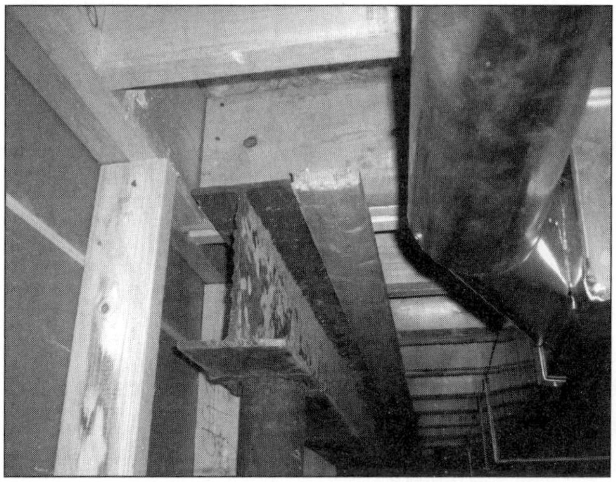

Courtesy Tom Stephenson

FIGURE 8.24

Framing for dropped floor at laundry room at top of basement stairs; joist hangers not yet installed (refer to drawings in Figures 8.20 and 8.21).

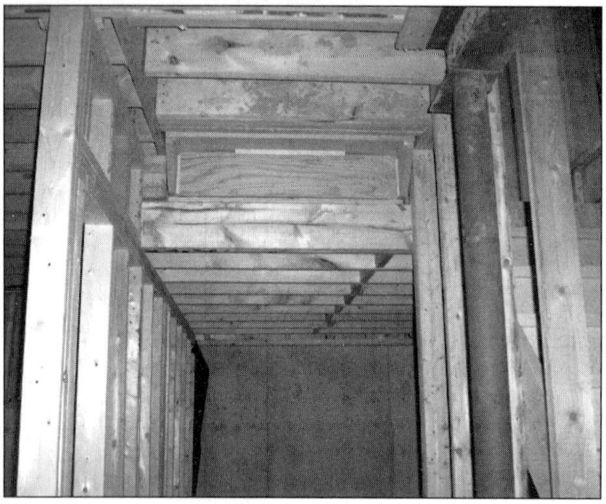

Courtesy Tom Stephenson

FIGURE 8.25

Another view of framing for dropped floor at laundry room. Study the Doncaster drawings and close-ups in Figures 8.20 and 8.21; notice how floor is supported and how elevation of floor changes.

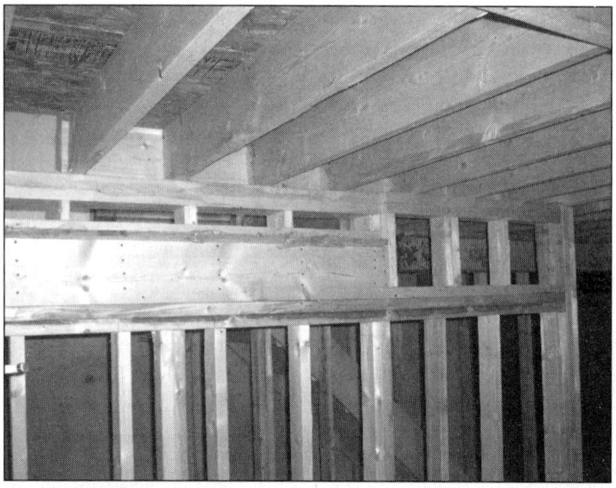

Courtesy Tom Stephenson

FIGURE 8.26

Doncaster basement plan close-up of steel support column and beam intersection. What is the column supporting?

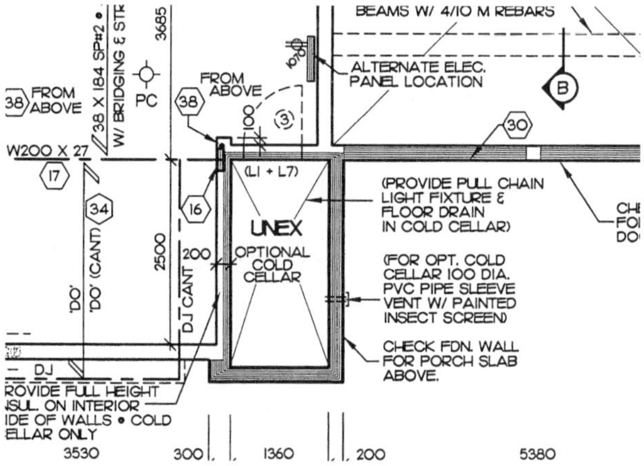

Source: Cassidy & Co. Architectural Technologists

FIGURE 8.27

W200 × 27 beam supports first-floor joists, and steel support column for beam supporting second-floor and masonry veneer.

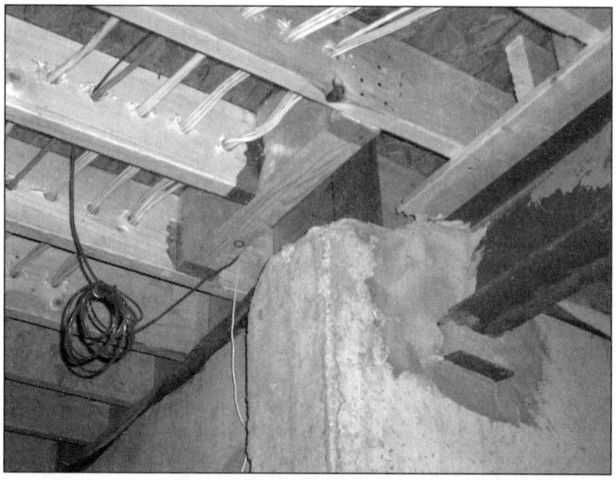

Courtesy Tom Stephenson

Refer to Figures 8.26 and 8.28 for close-ups of the foundation and first-floor plans. Follow the post indicated on the foundation plan with reference number (38) up through the floor to the ceiling of the first floor. See Figures 8.27, 8.29, 8.30, and 8.31 for photographs of the actual post and beam indicated on the plans.

FIGURE 8.28

Doncaster first-floor plan close-up of column and beam supporting second floor and masonry above. The W200 × 27 + 180 × 10 PL means that the wide flange I beam will have a 180 mm steel plate that is 10 mm thick welded to it, as shown in Figure 8.30.

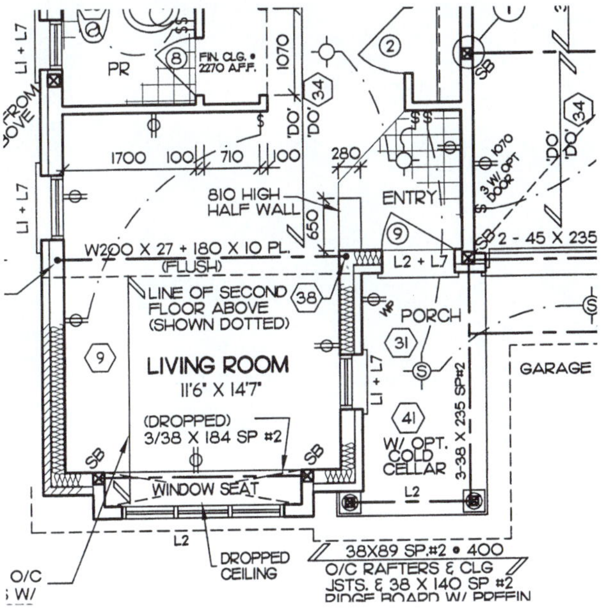

Source: Cassidy & Co. Architectural Technologists

FIGURE 8.29

Steel support column coming up through first floor beside front door; notice partial wall at front door and review first-floor plan in Figure 8.28 to view how this is indicated and what height of wall is.

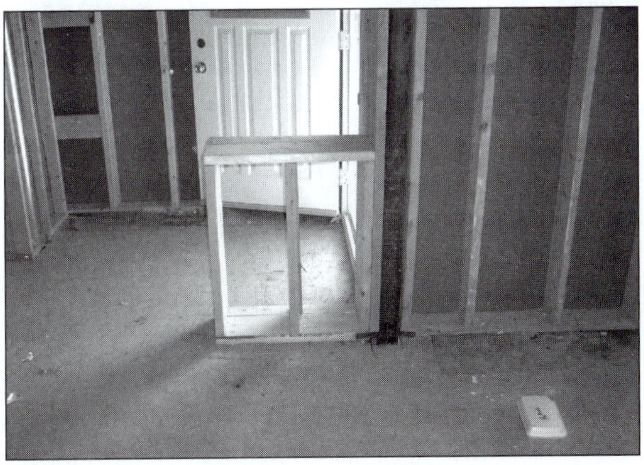

Courtesy Tom Stephenson

FIGURE 8.30

Steel support column providing solid bearing for W200 × 27 with 180 mm × 10 mm steel plate welded onto it to support masonry (review Figure 8.28 to see this beam and wall intersection on drawing).

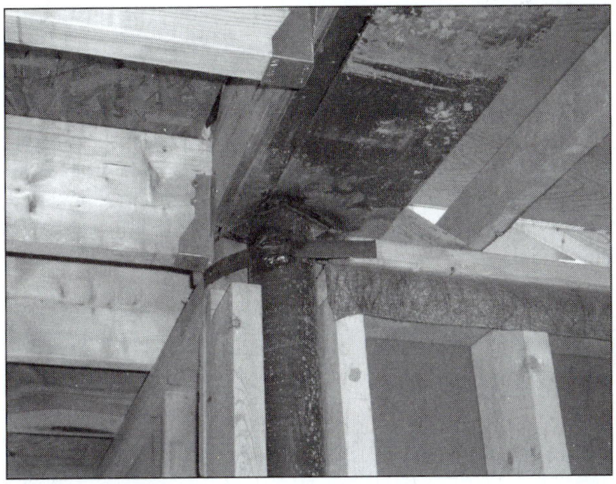

Courtesy Tom Stephenson

FIGURE 8.31

Different view of Figure 8.30 taken from living room looking toward post by front door; notice masonry and second floor supported on steel beam, and roof rafters supported by frame wall behind masonry.

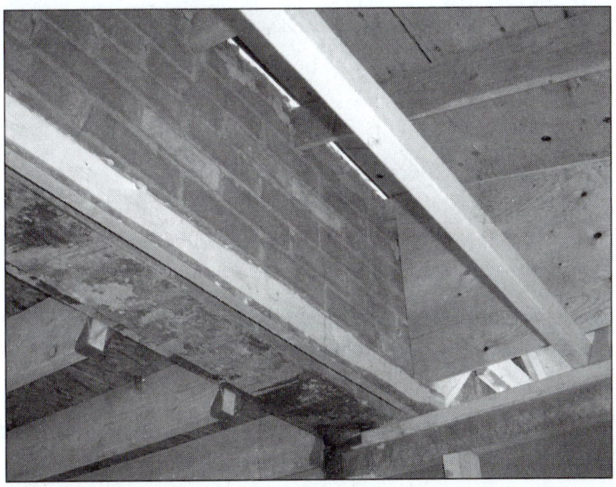

Courtesy Tom Stephenson

The following photographs (Figures 8.32 to 8.34) illustrate the ceiling and wall framing that takes place between the breakfast area and hall leading to the dining room of the Doncaster house.

FIGURE 8.32

Flat arch (FA indicated on Doncaster drawing; refer to Figure 8.20) between hall and breakfast area. Drawing indicates L5 size beam is being used; check the beam schedule for the actual size. Note the 3/38 × 184 mm flush beam (header) running parallel with the stairs and the double joist (DJ) above the stairs. Also note the frame wall and door opening down to the laundry room, using the door schedule. What is the size of the laundry room door (refer to Doncaster drawings)? In the drawing, the number 2 with a circle around it refers to the size of the door on the door schedule of the construction notes, which is 2'-8" × 6'-8" × 1⅜". The rough opening will be larger to accommodate the door frame and shim space.

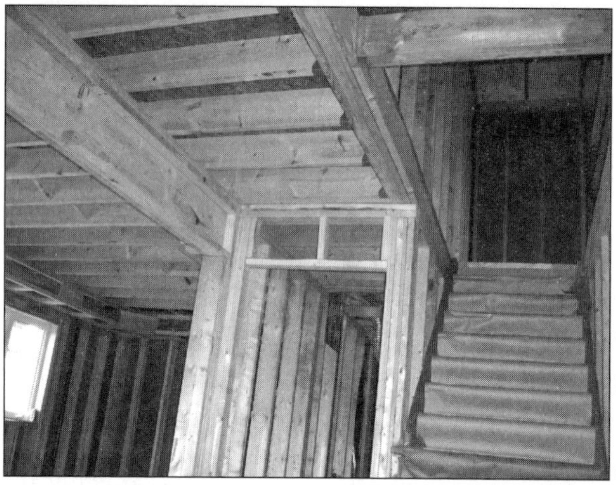

Courtesy Tom Stephenson

FIGURE 8.33

Solid bearing (SB) provided by the wall for 3/38 × 184 flush beam (header) at angled wall; notice how the angled wall has been back framed to ensure proper support for drywall at the corner.

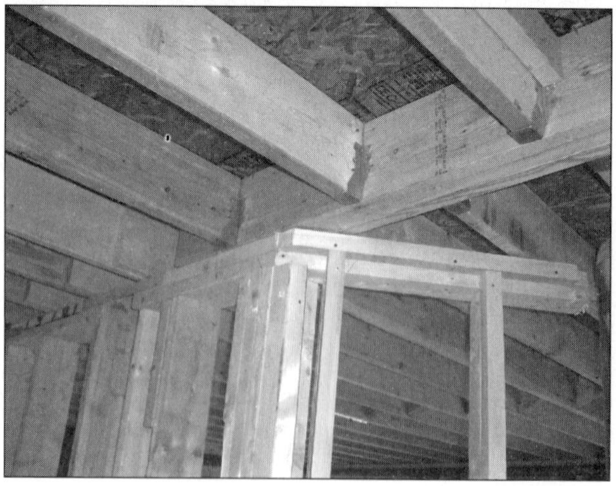

Courtesy Tom Stephenson

FIGURE 8.34

Flat arch, which is a straight dropped archway, marked with the abbreviation FA on the Doncaster first-floor plans; photograph taken looking out from the breakfast/family area toward the stairs (refer to Figure 8.21).

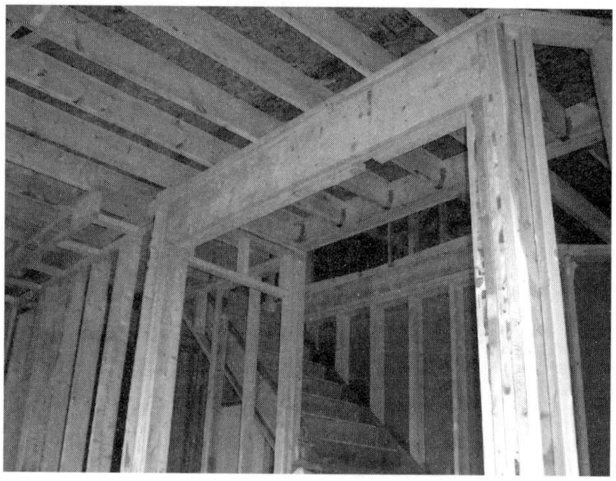

Courtesy Tom Stephenson

Bathrooms require extra framing to accommodate showers and tubs. The following photos (Figures 8.35 and 8.36) illustrate the framing requirements for the ensuite bathroom on the second floor of the Doncaster drawings.

FIGURE 8.35

Shower in ensuite bathroom; refer to Doncaster drawings for information about shower enclosure.

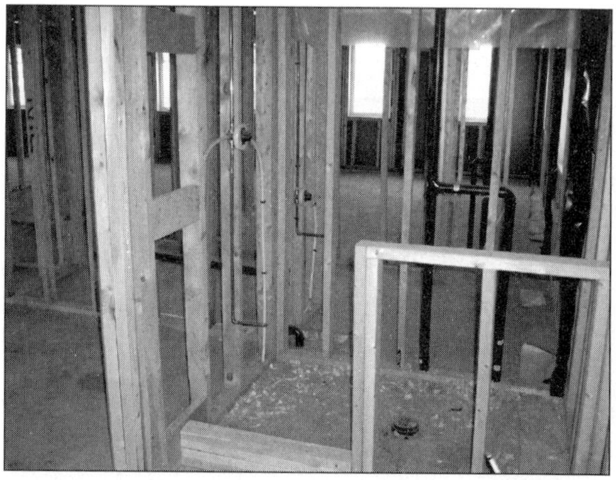

Courtesy Tom Stephenson

FIGURE 8.36

Framed tub surround in ensuite bathroom; refer to Doncaster drawings for information about tub surround. The surround is framed based on a layout template provided by the tub manufacturer (refer to Chapter 11).

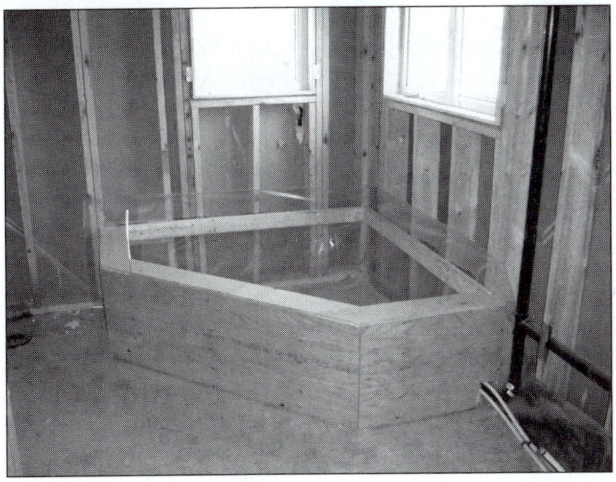

Courtesy Tom Stephenson

Some of the basement windows are raised up into the floor to provide extra light without using window wells, and to allow for a first floor that is not raised too high off the grade level. See Figures 8.39 and 8.40, and compare to close-up of basement floor plan at window in Figures 8.37 and 8.38.

FIGURE 8.37

Close-up of basement window on basement plan; notice how joists are shown doubled around the window, so that the window can be placed higher on the wall. Compare with Figures 8.39 and 8.40.

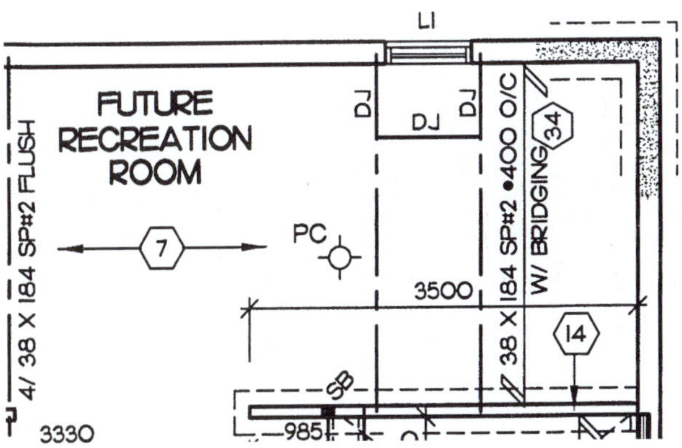

Source: Cassidy & Co. Architectural Technologists

FIGURE 8.38

Section detail of raised window in floor.

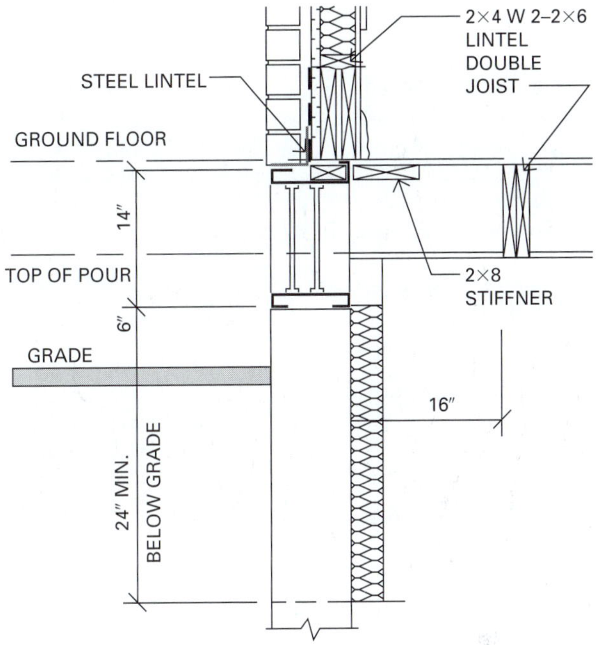

FIGURE 8.39

Framing of double joists around window. The double joists running from left to right are known as header joists, and they carry the loads from the single tail joists and transfer the loads to the double trimmer joists that are shown on each side of the opening.

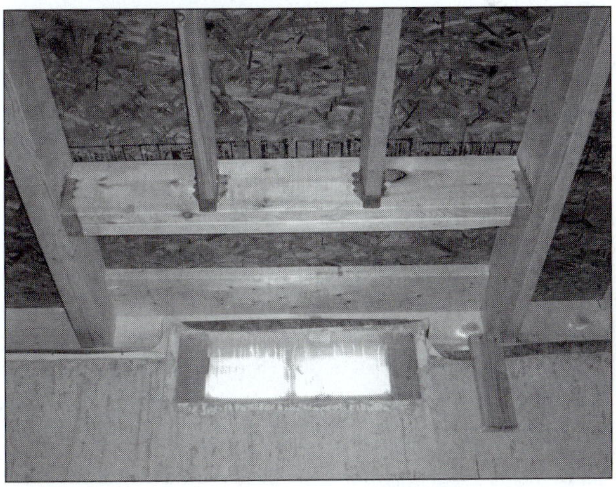

Courtesy Tom Stephenson

FIGURE 8.40
Framing directly above basement window as seen in drawings in Figures 8.37 and 8.38 on first-floor exterior wall; lintel is at plate level to support studs above.

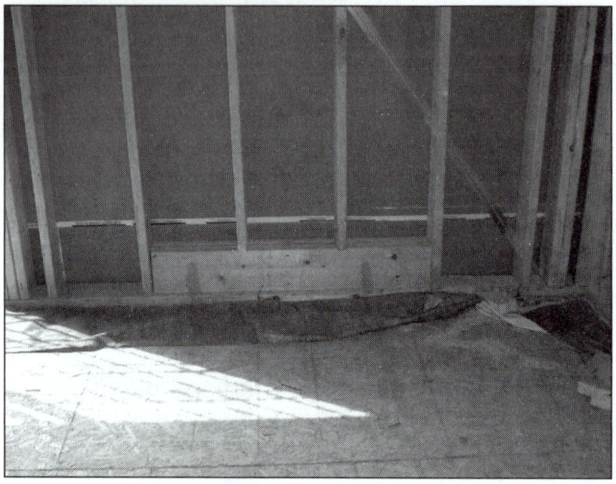

Courtesy Tom Stephenson

FIGURE 8.42
Steel beam supported on steel plate and studs at left side of garage (refer to Figure 7.14 in Chapter 7 for similar drawing detail).

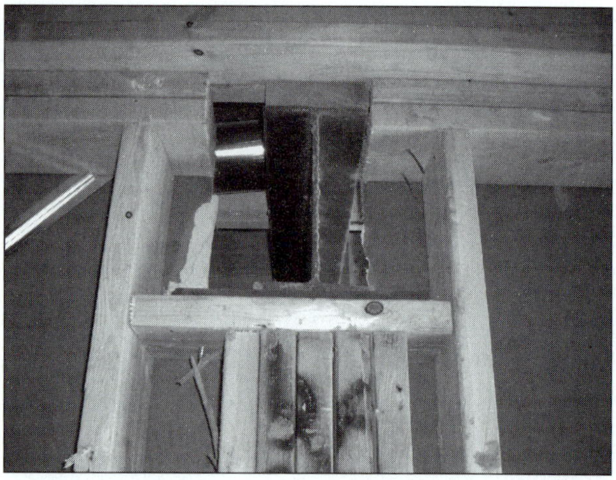

Courtesy Tom Stephenson

The garage ceiling is dropped to provide space for ductwork above. Note the beam location and direction of joists, and how the framing for the dropped ceiling is attached (compare Figures 8.41 and 8.42 to the first-floor plan of the garage from the Doncaster drawings).

FIGURE 8.41
Garage ceiling, beam, and second-floor joists above.

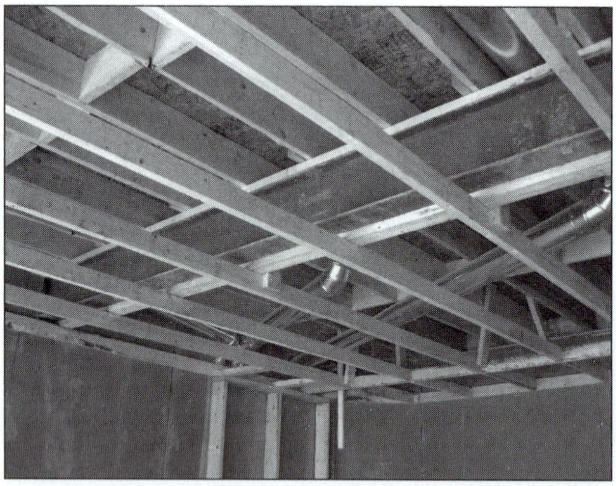

Courtesy Tom Stephenson

KEY TERMS

Back framing (p. 179)
Blocking (p. 171)
Jack stud (p. 172)
Lintel (p. 172)
Optimum value engineering (OVE) (p. 174)
Point load (p. 177)
Racking (p. 179)
Rough opening (p. 176)
Rough sill (p. 172)
RSI value (p. 171)
R value (p. 171)
Schedule (p. 177)
Sheathing (p. 179)
Stud (p. 171)

Chapter 8 Assignment: Wall Framing and Doncaster Drawings

Answer questions 1 through 8 using information found throughout the chapter, and answer questions 9 to 20 using the Doncaster drawings that came attached to the text. Identify the letter that best answers the question.

_____ 1. How does the text define the term optimum value engineering?

a. design changes that increase the strength of the structure to meet NBC requirements

b. changes in framing methodology that improve energy efficiency while utilizing more sustainable building practices

c. decreasing stud spacing and using a double bottom plate for easier fastening of trim

d. using metal studs and joists to increase the efficiency of framing

_____ 2. Which of the following is a feature of optimum value engineering?

a. single top plate

b. double top plate

c. narrower stud spacings

d. three-stud corners

_____ 3. Which of the following is true with regard to the framing of a top plate for a frame wall?

a. A top plate must always be doubled.

b. A single top plate is not as energy efficient as a double top plate.

c. A single top plate is acceptable as long as the stud falls within 50 mm of any supporting member such as a joist.

d. A single top plate is acceptable as long as the stud falls within 100 mm of any supporting member such as a joist.

_____ 4. Which of the following best describes a lintel?

a. a type of beam found in a wall

b. a type of stud found in a wall

c. a type of joist found in a floor

d. a type of header found in a floor

_____ 5. What factor from those listed below will help to determine the size of lintel used in a wall?

a. the length of studs from bottom to top plate

b. the span of joists from inside support to inside support

c. the span of the header joists that are used in the floor opening

d. the local climatic data regarding composite snow loads

_____ 6. Why is black poly (polyethylene) placed between interior wall plates and exterior wall intersections?

a. to prevent methane gas from entering the house

b. to soundproof the walls and ceiling

c. to provide a surface to glue the drywall to

d. to form a continuous seal for the air and vapour barrier

_____ 7. What is the rough opening for a 2'-8" wide interior door, if it uses ¾" thick wood for the jambs of the frame?

a. 2'-8"

b. 2'-9½"

c. 2'-10¼"

d. 2'-11¼"

_____ 8. What size rough opening will a window require if it has an actual outside frame size of 3'-6" × 4'-2"?

a. 3'-6" × 4'-2"

b. 3'-6¼" × 4'-2¼"

c. 3'-7" × 4'-3"

d. 3'-8¼" × 4'-5¼"

_____ 9. What is the size and number of pieces required for the wood lintel over the window where the box seat is in the living room?

a. 2 - 2" × 8"

b. 3 - 2" × 8"

c. 2 - 2" × 10"

d. 2 - 2" × 12"

_____ 10. What is the size of the wood lintels over the garage doors?

a. 3 - 2" × 12"

b. 3 - 2" × 10"

c. 2 - 45 mm × 235 mm LVL

d. 3 - 45 mm × 235 mm LVL

_____ 11. What is the thickness of the main wall on the first floor that separates the family room and dining room?

a. 38 mm

b. 89 mm

c. 100 mm

d. 150 mm

_____ 12. What is the inside size of the Powder Room (PR)?

a. 1600 mm × 1430 mm

b. 1700 mm × 1530 mm

c. 1800 mm × 1530 mm

d. 1800 mm × 1730 mm

_____ 13. What is the length of the studs used for the second floor if there is a double top plate and a single bottom plate used for the wall?

a. 2346 mm

b. 2384 mm

c. 2422 mm

d. 2460 mm

_____ 14. What is the length of the studs used for the first floor if the wall has a double top plate and single bottom plate? (Don't forget to factor in deductions for the thickness of subfloor and joists.)

a. 2356.1 mm

b. 2422 mm

c. 2460.1 mm

d. 2670 mm

_____ 15. What is the height of the wall to the left of the front entry door?

a. 950 mm

b. 810 mm

c. 650 mm

d. 280 mm

_____ 16. What is the height of the half wall at the shower of the ensuite?

a. 810 mm

b. 980 mm

c. 1200mm

d. 1530 mm

_____ 17. What is the inside size of the master bedroom?

a. 2060 mm × 2100 mm

b. 3990 mm × 6780 mm

c. 3890 mm × 6780 mm

d. 4090 mm × 6880 mm

_____ 18. What is the area of the shower for the ensuite in square metres?

a. 1.02 square metres

b. 1.45 square metres

c. 1.5606 square metres

d. 1,560,600 square metres

_____ 19. What is the interior size of bedroom 4 (taken at its longest points)?

a. 3740 mm × 4270 mm

b. 3740 mm × 4980 mm

c. 3550 mm × 6520 mm

d. 3840 mm × 5080 mm

_____ 20. What size diameter are the steel columns that support the steel beam in the living room, and what size are the steel base plates that are welded to the columns?

a. 4" columns with 3½" × 3½" steel plates

b. 6" columns with 6" × 6" steel columns

c. 3½" columns with 6"× 6"× ⅜" steel plates

d. 5" columns with 4" × 4" × ½" steel plates

Chapter 8 Assignment 2: Floors and Walls, Brook Drawings

Use the Brook drawings that came attached to the textbook to identify the letter that best answers the question.

_____ 1. What is the size and number of plys for the built-up engineered beam that runs over and between the great room and breakfast area?

a. 1 - 1¾" × 9½" Versa Lam 2.0

b. 2 - 1¾" × 9½" Versa Lam 2.0

c. 3 - 1¾" × 9½" Versa Lam 2.0

d. 4 - 1¾" × 9½" Versa Lam 2.0

_____ 2. Where is a 42" half wall located?

a. kitchen

b. dining room

c. great room

d. living room

_____ 3. Using the page 4 floor plan, what is the widest clear span (from edge of structural support to edge of structural support, note the concrete foundation wall is a structural support) of first-floor joists shown on the plan?

a. 13'-0"

b. 13'-3"

c. 13'-6"

d. 14'-10"

_____ 4. If the patio door is the size indicated on the drawings, including the frame, what should the rough opening be framed at?

a. 6'- 0" × 8'- 0"

b. 6'- 1" × 8'- 0"

c. 6'- 1" × 8'- 1"

d. 6'- 2" × 8'- 3"

_____ 5. How many risers are suggested for the drop from the kitchen down to the mud room?

a. 1

b. 2

c. 3

d. 4

_____ 6. What is the size of the front entrance door from the porch?

a. 2'-6" × 6'-8" × 1⅜"

b. 2'-8" × 6'-8" × 1⅜"

c. 2'-10" × 6'-8" × 1⅜"

d. 3'-0" × 6'-8" × 1¾"

_____ 7. What is the inside size of the laundry room/closet?

a. 3'-7" × 3'-8"

b. 3'-11" × 4'-2"

c. 4'-10" × 7'-8"

d. 5'-1" × 8'-1"

_____ 8. What is the inside size of the living room from widest point to widest point?

a. 9'- 0" × 12'- 0"

b. 9'- 0" × 12'- 6"

c. 9'- 6" × 12'- 0"

d. 13'- 3" × 12'- 0"

_____ 9. What is the height of the technology centre from finished floor to top plate of wall?

a. 8'-1"

b. 8'-4"

c. 9'-7"

d. 10'-4"

_____ 10. What is the difference in elevation between the first floor and the mudroom?

a. 6"

b. 7"

c. 8"

d. 16"

_____ 11. Calculate the exact floor area of the master bedroom so that you can purchase carpeting which is sold by the square yard. What is the area?

a. 227.5 sq. yds

b. 25.28 sq. yds

c. 163.33 sq. yds

d. 18.15 sq. yds

_____ 12. What size lumber is used to frame the four walls surrounding the laundry room?

a. the four walls are 2" × 4"

b. the four walls are 2" × 6"

c. two walls are 2" × 4" and two walls are 2" × 6"

d. three walls are 2" × 4" and one wall is 2" × 6"

_____ 13. What is the size of the rough opening for the entrance to the master bedroom using 3/4" thick material for the frames, assuming the door is 6'-8" in height?

a. 3'-0" × 6'-8"

b. 3'-2¼" × 6'-9¾"

c. 3'-3¼" × 6'-9¾"

d. 3'-4¼" × 6'-9¾"

_____ 14. What is the length and width of the ensuite bathroom?

a. 8'-13" × 10'-2"

b. 8'-13" × 13'-2"

c. 8'-3" × 10'-2"

d. 8'-3" × 13'-6"

_____ 15. What is the wood lintel size for over the window in the ensuite bathroom?

a. 2 - 2" × 8"

b. 3 - 2" × 10"

c. 2 - 2" × 12"

d. 3 - 2" × 12"

_____ 16. What is the height of the first floor walls from top of first floor subfloor to underside of second floor joists?

a. 8'-11"

b. 9'-⅞"

c. 9'-1½"

d. 9'-11"

_____ 17. What is the length of studs required to frame the first floor (assume double top plate and single bottom plate)?

a. 8'-8⅜"

b. 8'-9"

c. 9'-⅞"

d. 9'-11"

_____ 18. What is the height of the second floor from the top of the subfloor to the top of the plate?

a. 8'-1"

b. 9'-0"

c. 9'-11"

d. 10'-0"

_____ 19. Where can an interior wall be found that is 12" thick?

a. between the master bedroom and walk-in closet

b. between bedrooms 2 and 3

c. between the walk-in closet and the stairs

d. between the dining room and the great room

_____ 20. What size and number of pieces are used for the wood lintel over the sliding door from the breakfast area?

a. 2 - 2" × 6"

b. 3 - 2" × 6"

c. 2 - 2" × 8"

d. 3 - 2" × 8"

Chapter 8 Quiz Review: Floors and Walls Using Whittington Drawings

Use the Whittington drawings that came attached to the textbook to identify the letter that best answers the question.

_____ 1. What is the size of the steel beam that runs under the kitchen and dining room area?

a. W150 × 27
b. W200 × 27
c. W225 × 35
d. W250 × 35

_____ 2. What is the size of the steel lintel that is over the doorway to the optional cold room?

a. 90 mm × 90 mm × 6 mm
b. 125 mm × 90 mm × 10 mm
c. 2 - 2" × 8"
d. 3 - 2" × 8"

_____ 3. What is the widest clear span of first floor joists shown on the plan?

a. 2100 mm
b. 3450 mm
c. 3800 mm
d. 3935 mm

_____ 4. What is the actual length, including the minimum bearing from the NBC, for the W200 × 27 steel beam closest to the optional cold cellar door?

a. 4390 mm
b. 4468 mm
c. 4750 mm
d. 4928 mm

_____ 5. What is the clear span of the steel beam in the garage?

a. 4880 mm
b. 5310 mm
c. 5490 mm
d. 5790 mm

_____ 6. What is the size of the front entrance door from the verandah?

a. 2'-6" × 6'-8" × 1⅜"
b. 2'-8" × 6'-8" × 1⅜"
c. 2'-10" × 6'-10" × 1¾"
d. 3'-0" × 6'-8" × 1¾"

_____ 7. What is the length of the steel beam over the living room, including the minimum bearing amount required by the NBC?

a. 4260 mm
b. 4438 mm
c. 4620 mm
d. 4798 mm

_____ 8. What is the inside size of the living room?

a. 4260 mm × 3520 mm
b. 1810 mm × 4620 mm
c. 4620 mm × 3660 mm
d. 5180 mm × 4040 mm

_____ 9. What is the height/elevation that the top of the steel beam in the garage is placed above the finished floor?

a. 2470 mm
b. 2670 mm
c. 3780 mm
d. 4880 mm

_____ 10. What is longest inside dimension of bedroom 2?

a. 4260 mm
b. 4970 mm
c. 5070 mm
d. 5350 mm

_____ 11. What is the exact floor area of bedroom 3, excluding the area of the wall between the closet and the door?

a. 15,158,500 sq. mm or 15.159 sq. metres
b. 15,087,500 sq. mm or 15.088 sq. metres
c. 16,000,000 sq. mm or 16 sq. metres
d. 17,200,000 sq. mm or 17.2 sq. metres

_____ 12. What size lumber is used to frame the four walls surrounding the laundry room?

a. all are 2" × 4"
b. all are 2" × 6"
c. two walls are 2" × 4" and two walls are 2" × 6"
d. three walls are 2" × 4" and one wall is 2" × 6"

_____ 13. What size should the rough opening for the entrance to the master bedroom be, when using ¾" thick material for the frames?

a. 5'-0" × 6'-8"

b. 5'-2¼" × 6'-9¾"

c. 5'-3¼" × 6'-9¾"

d. 5'-4¼" × 6'-9¾"

_____ 14. What is the inside length and width of the ensuite bathroom?

a. 2650 mm × 2600 mm

b. 2750 mm × 1530 mm

c. 2750 mm × 1630 mm

d. 2840 mm × 1480 mm

_____ 15. What size is the wood lintel for over the window in the ensuite bathroom?

a. 2 - 38 mm × 184 mm

b. 3 - 38 mm × 184 mm

c. 2 - 38 mm × 150 mm

d. 2 - 38 mm × 186 mm

_____ 16. What is the height of the first floor walls from the top of first floor subfloor to the underside of the second floor joists?

a. 2670 mm

b. 2470 mm

c. 2530 mm

d. 2760 mm

_____ 17. What is the length of studs required to frame the first floor (assuming double top plate and single bottom plate)?

a. 2470 mm

b. 2670 mm

c. 2356 mm

d. 2394 mm

_____ 18. What is the height of the second floor from top of subfloor to top of plate?

a. 2080 mm

b. 2330 mm

c. 2460 mm

d. 2670 mm

_____ 19. Where can an interior wall be found that is 2" × 6", other than beside the garage?

a. between the master bedroom and ensuite

b. between bedrooms 2 and 3

c. between the walk-in closet and the stairs

d. between the kitchen and dining room

_____ 20. What size is the wood lintel over the window in bedroom three?

a. 2" × 6"

b. 2" × 8"

c. 2" × 10"

d. 2" × 12"

9 Roof Framing

LEARNING OBJECTIVES

After completing this chapter, you will be able to:

- Identify the different components and terminology used in roof construction.
- Read, interpret, and analyze truss drawings and apply that knowledge to roof construction.
- Read, interpret, and analyze construction drawings for constructing traditional cut and trussroofs.

INTRODUCTION

This chapter provides an overview of how roofs are constructed and installed. The chapter covers both conventional and truss-type roofs. A set of truss drawings is included from the Doncaster house to assist you in reading and analyzing drawings provided by a manufacturer (for example, **shop drawings**). Review a carpentry text for more detail on the process of laying out, cutting, and installing a conventional cut roof or one constructed with roof trusses. Also refer back to Chapter 3 for a refresher on basic roof types (Figures 3.16 to 3.20). Included in this chapter are two assignments: Assignment 1 is based on the Doncaster roof truss drawings included in the chapter, and the Doncaster drawings that came attached with the text. Assignment 2 is based on the theory in the chapter and on the the Brook drawings that came attached with the text. The quiz is based on the theory in the chapter and on the Whittington drawings that came attached with the text.

TYPES OF ROOF CONSTRUCTION

Roofs are constructed using two common methods: roof trusses and cut roofs. The **cut roof** is the traditional method of constructing roofs. It involves laying out and cutting roof rafters on site. A skilled framer is able to cut all the roof components on the ground and pass them up without having to measure and cut each piece individually using trial and error. To lay out a cut roof requires a lot of skill and a thorough understanding of applied geometry and mathematics.

A **truss roof** is a roof that is engineered and built from pre-manufactured trusses. The builder or contractor submits the drawings to the lumber or truss supplier who then designs a truss roof that is engineered to satisfy the drawing's design and applicable code requirements for the jurisdiction that the house is being constructed in. This is essentially making sure that the structure is designed to meet all local weather requirements, such as composite snow loads (heavy wet snow). The supplier delivers the trusses and provides the builder with a copy of the truss drawings. The truss drawings are submitted to the building department and must carry an engineer's stamp of approval which provides the municipality with assurance that the design has been carried out by a qualified professional and that the design and product is protected by appropriate insurance. A copy of the truss drawings is kept by the builder to act as a guide for the framer. The truss drawings display the layout of the trusses and provide information on fastening and bracing requirements.

Roof Terminology

Roof construction is the most technically complicated portion of the framing process and as a result there is a lot of terminology used to identify different components and measurements required to construct a roof. The following is a listing of the more commonly used terminology:

1. ***Total span:*** The span is defined in two ways:
   - For designing rafter sizes using the rafter span tables in the NBC (Division B Part 9). For roof design purposes the rafter span is the distance from support to support, not necessarily wall to wall.
   - For calculating the lengths of rafters. To calculate roof span, the distance from outside supporting wall to outside supporting wall is measured (see Figure 9.1).
2. ***Total run:*** Equal to half of the span and is important for calculating the length of rafter because it forms the bottom side of a right-angled triangle where the length of the rafter is the hypotenuse (the longest side). The other side is the height of the roof. On a set of drawings, the run can be calculated by dividing the span by two or on site by measuring outside of frame wall to outside of frame wall and dividing this measurement by two.
3. ***Theoretical rise:*** The rise from top of wall plate to the outside peak of the triangle (see Figure 9.1). Note: this measurement does not include the extra thickness of material to the peak.
4. ***Actual rise:*** The rise from the top of the wall plate to the actual peak of the roof. This can be an important measurement for calculating overall height of the building to determine whether it meets zoning height requirements.
5. ***Line length (point length):*** The length of rafter excluding the overhang before deductions are made for intersecting materials such as the ridge board. For estimation purposes the rough length of a rafter can be quickly taken by scaling the elevation drawings. In this case, accuracy just needs to be within the next largest 2' increment, as framing lumber is ordered in 2' increments, from 8' to 16' being the most common.
6. ***Unit run:*** A standard unit of run that corresponds with a given unit rise to provide the slope for the roof. The most commonly used standard is 12"; however, 250 mm and 300 mm are also sometimes used when a metric slope is provided. Recall that the slope of the roof is found on the elevation drawings and is simply a ratio between rise and run. Unit run is always a level/horizontal measurement (see Figure 9.2).
7. ***Unit rise:*** The rise of the roof in comparison to the unit run. This is always a vertical/plumb measurement. (see Figure 9.2).

FIGURE 9.1
Roofing terminology.

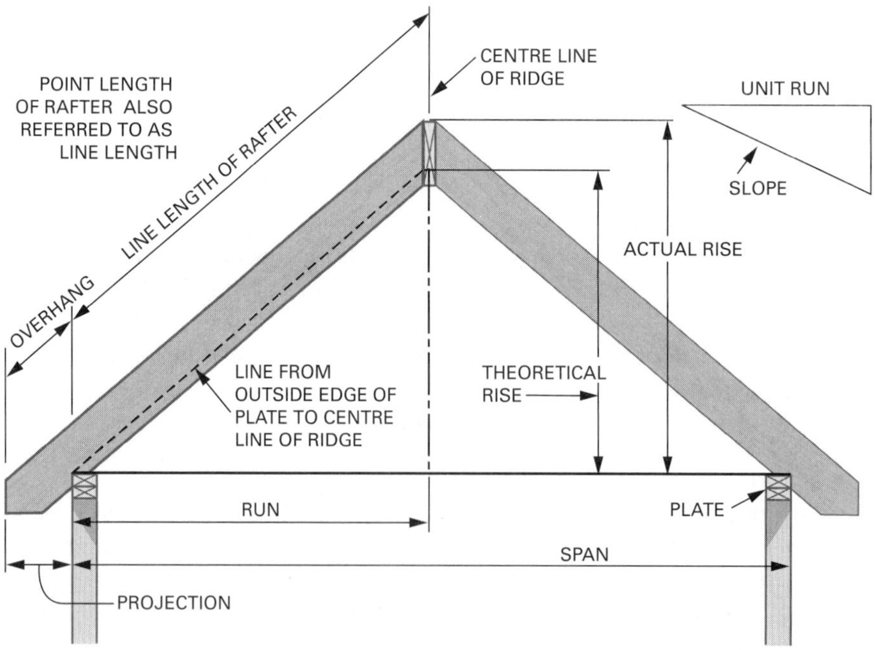

Source: From VOGT, Carpentry, 4E. © 2006 Delmar Learning, a part of Cengage Learning, Inc. Reproduced by permission. www.cengage.com/permissions

8. ***Unit line length (unit point length):*** The hypotenuse of the unit rise and unit run. If the unit rise is 7" and the unit run is 12" then using the Pythagorean Theorem we can compute the unit line length (see Figure 9.2).
   - Pythagorean Theorem: $A^2 + B^2 = C^2$ (to find C all we need to do is find the square root of $A^2 + B^2$).
   - $\sqrt{5^2 + 12^2} = 13$. This number can also be found already calculated on a framing square rafter table. Look under the inch marking along the top. This represents the rise based on a standard unit of run of 12". (See Figure 9.3.)
9. ***Roof rafters:*** Defined by the NBC as the structural members that support the cut roof on slopes greater than 1:4 (greater than 3:12 slope) (refer to Division B Part 9.23.14.1 and Tables A6 and A7 of the NBC).
10. ***Roof joists:*** Defined by the NBC as the structural members that support the cut roof on slopes 1:4 (equals 3:12 as usually listed on drawings) or less and for flat roofs (refer to Division B Part 9.23.14.10 and Tables A4 and A5 of the NBC). Roof joists are larger than rafters of the same span due to the extra loading requirements.
11. *Slope:* The ratio of unit rise to unit run. For example, a 7:12 slope has a unit rise of 7" and a unit run of 12".
12. *Pitch:* Historically, the ratio of unit rise to span. For example, a roof with a 4:12 slope would have a pitch of 4:24, or more accurately expressed as a fraction in lowest terms, 1/6 pitch. Don't be surprised, however, if in some geographic areas this term is sometimes used instead of slope, using the definition of slope.
13. *Projection:* The horizontal distance that the roof projects out from the wall of the house. Take care defining this measurement because problems can occur if the projection extends too far over property lines. For example, if the

FIGURE 9.2

Simple example showing the relationship between unit rise, unit run, and unit line length and how it fits into an overall rafter layout based on a slope given on the drawings. (Refer to a Carpentry text for further detail on rafter layouts). The slope in this example is 5 to 12 with a unit line length of 13. The total run is 4'. The framing square would be held at the 5" marking on the one outer edge (thinner side of square, known as the tongue of the square) and on the number 12" marking on the other outer edge (wider side of square, known as the body of the square). When marking a plumb line, the line is drawn on the 5" or unit rise edge, and when marking a level line, the line is marked along the 12" unit run edge. The total line length is calculated by multiplying the units of run, which is four in this case (this is taken from the 4' total run measurement) by the unit line length of 13" (the hypotenuse of 5 and 12 equals 13). 4 × 13 = 52" line length of rafter.

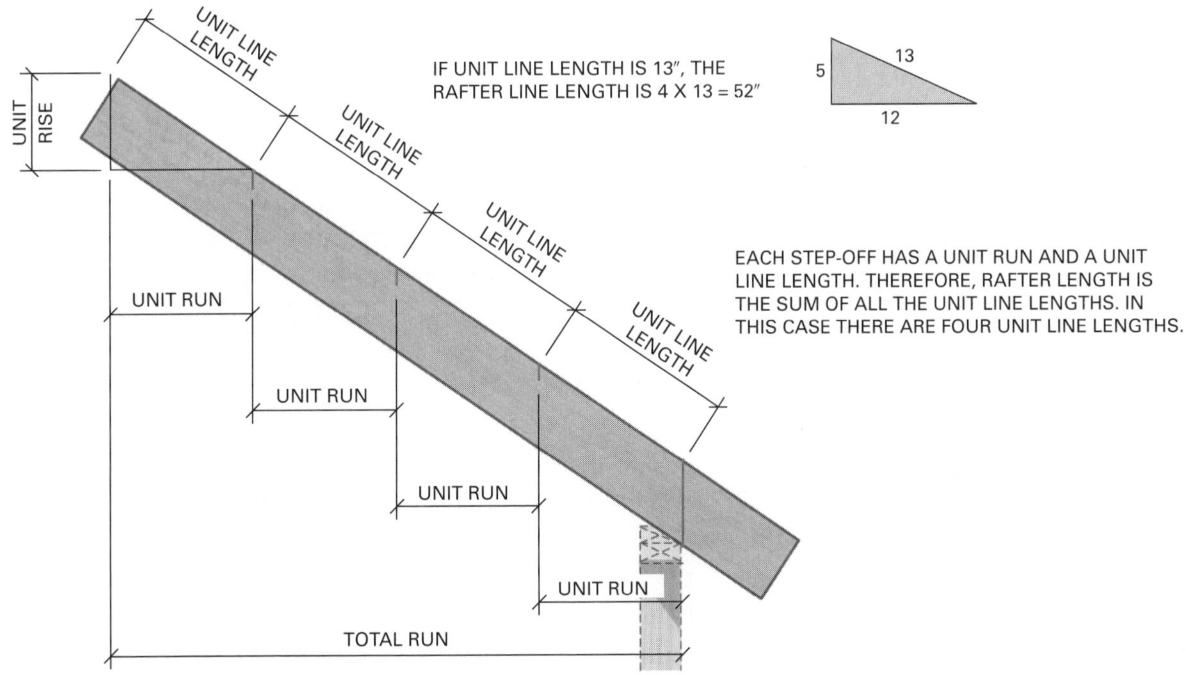

FIGURE 9.3

Framing square rafter tables. The information under the inch marking relates to the layout of a roof with a standard unit of run of 12". This example relates to 5:12 slope where 13 is the hypotenuse of 5 and 12.

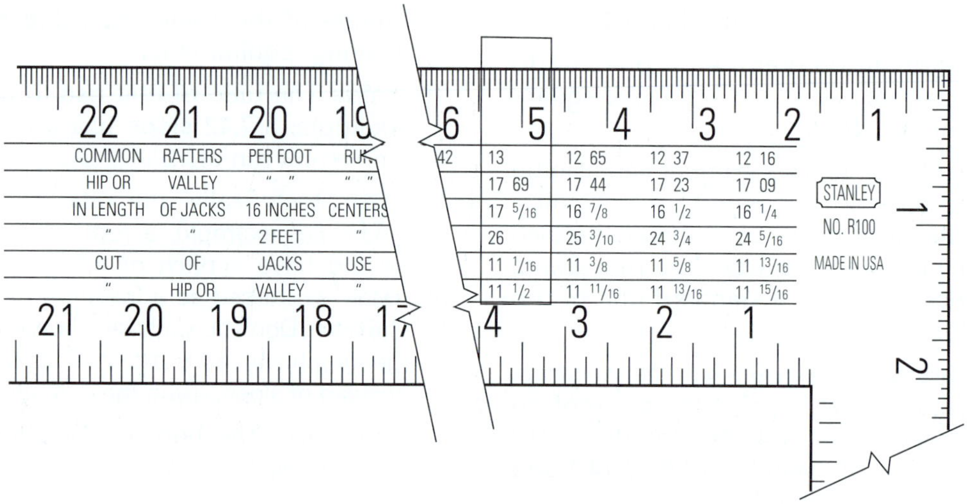

Source: From VOGT/NAUTH, Carpentry, 2E. (c) 2013 Nelson Education Ltd. Reproduced by permission. www.cengage.com/permissions

drawings call for a 10" projection excluding the eavestrough, then the rafter must be shortened to accommodate the fascia board that is added later on.

14. *Overhang:* Technically, the sloping length of the projection. In some areas it may be used with the same meaning as projection.

FIGURE 9.4

Roof joists versus roof rafters.

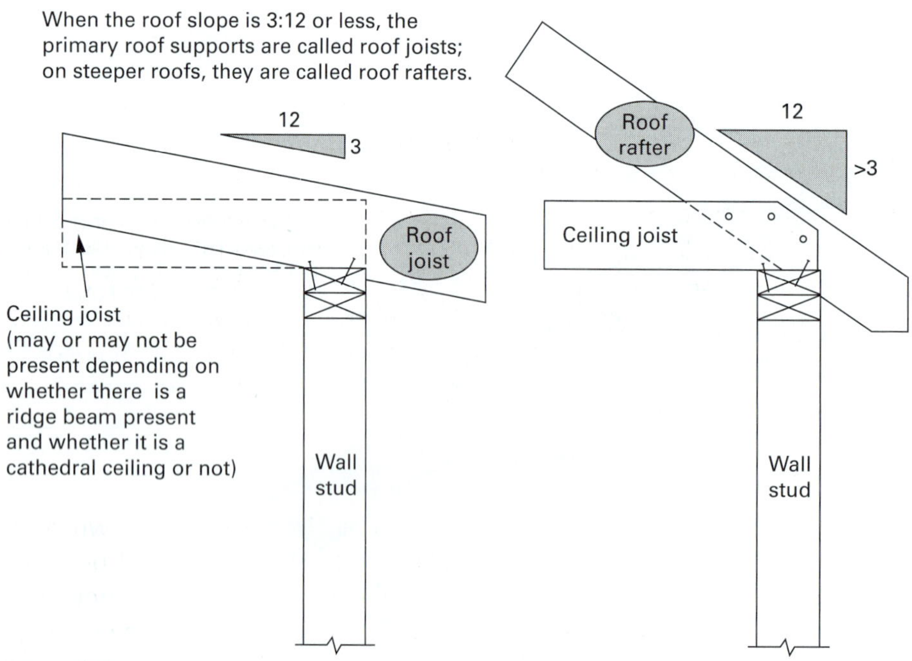

Source: Image excerpted from The Illustrated Home by Carson Dunlop www.carsondunlop.com

FIGURE 9.5
Rafter parts. Note ceiling joist and potential collar tie not indicated on this drawing.

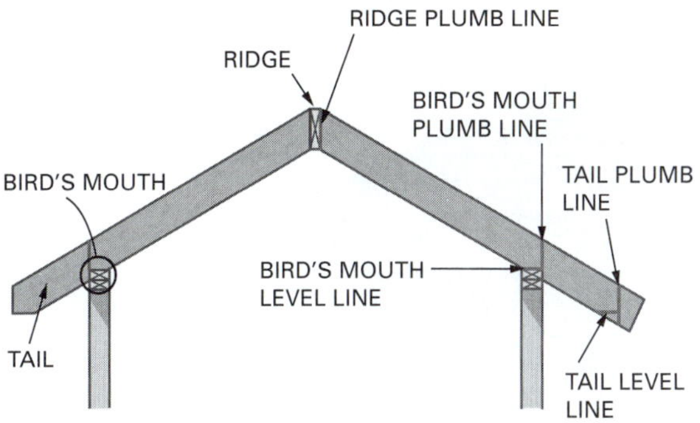

Source: From VOGT, Carpentry, 4E. © 2006 Delmar Learning, a part of Cengage Learning, Inc. Reproduced by permission. www.cengage.com/permissions

FIGURE 9.6
Minimum bearing for rafters and ceiling joists is 1½" (refer to Division B Part 9.23.14.3 of the NBC).

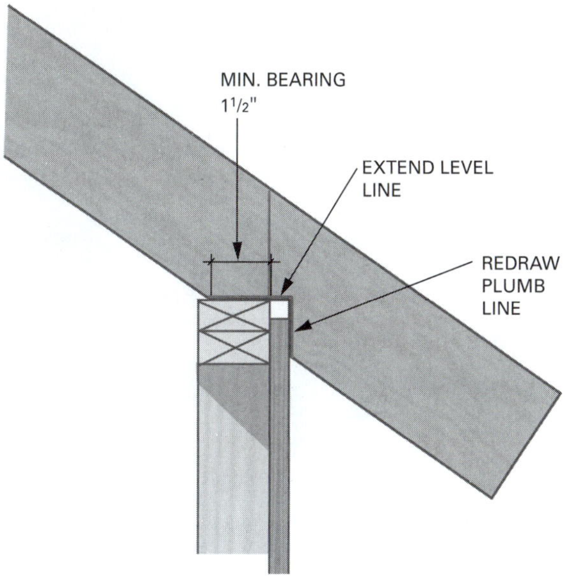

Source: From VOGT, Carpentry, 4E. © 2006 Delmar Learning, a part of Cengage Learning, Inc. Reproduced by permission. www.cengage.com/permissions

Conventional Cut Roof

As previously stated, the cut roof is the conventional method of constructing roofs; the roof is laid out and cut on site. This method is most commonly used when building additions or custom homes. It is also commonly used on smaller sections of truss roofs or in combination with truss roofs.

Advantages:

- Relatively easy to move pieces into place and does not require a crane or forklift to install
- Easy to design space for an attic
- Can be cut on site using dimensional lumber with no special orders or wait time required
- Easy to tie into an existing roof (renovations).

Disadvantages:

- Requires a high level of skill to lay out and generally costs more to construct
- Cannot span as far as a truss roof
- Usually requires load-bearing interior walls to support ceiling joists
- Requires heavier dimensional lumber (for example, 2" × 8"s instead of 2" × 4"s typically used in truss construction).

Gable Roof

A gable roof is the simplest form of pitched roof. It consists of one type of rafter, or for truss roofs, one kind of truss. The rafter extends from the top plate to the ridge board. The ridge board is always at least one size larger than the common rafter. For example, if the drawings state the rafter is 2" × 4" then the ridge board will be at least 2" × 6", to ensure full contact of the rafter against the ridge based on the rafter being cut on an angle. The ceiling joists provide resistance to stop the roof from splaying the walls outward. At the end of a gable roof, a wall is framed vertically to enclose the space between the rafters and the wall below. The gable end wall is considered non-load bearing because the rafters support the roof and direct the loads to the plates at the side walls. Refer to Figure 9.7.

Ceiling Joists

Ceiling joist sizes are shown on the floor plan of the floor that they will be placed above. For example, on a two-storey house the ceiling joist and rafter sizes will be shown on the second-floor plan; if there is a roof covering part of the first floor, its size will be shown on the first-floor plan.

FIGURE 9.7
Gable roof.

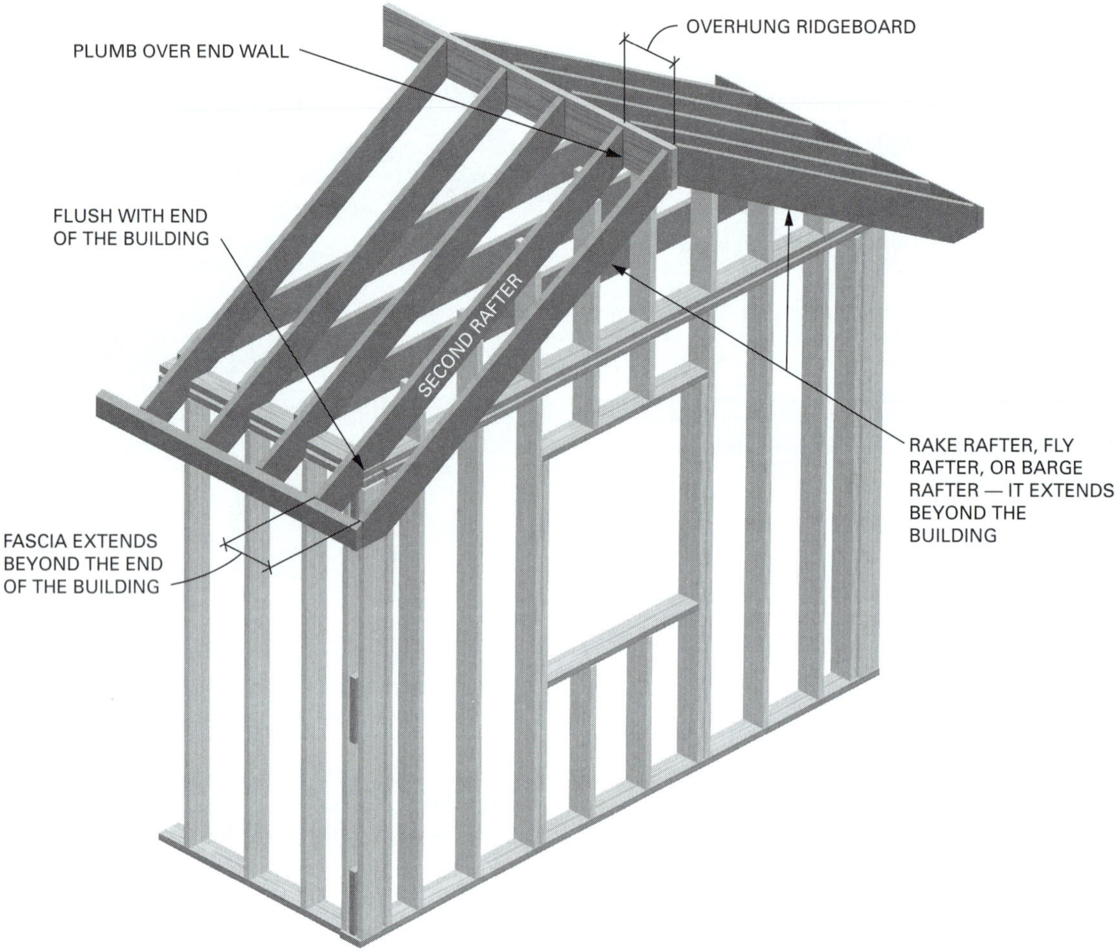

Source: From VOGT, Carpentry, 4E. © 2006 Delmar Learning, a part of Cengage Learning, Inc. Reproduced by permission. www.cengage.com/permissions

The ceiling joist size is based on span, similar to floor joists, except the function of the ceiling joist is to carry dead loads, and limited live loads. Therefore, smaller sizes of dimensional lumber are used for ceilings than for floors, based on the NBC span tables for ceiling joists (refer to span tables in Division B Part 9 of the NBC). Ceiling joists are laid out similarly to floor joists with the same standard on centre (OC) spacings. However, the layout must also consider the location of rafters, depending on the design of roof.

Rafter Spans

Rafter sizes are determined by their spans and the specified snow loads that they are exposed to regionally. Designers may reduce spans through the use of **collar ties** and knee walls that help support them. Refer to Figures 9.8 and 9.9.

Hip or Cottage Roof

A **hip or cottage roof** has its sides sloping to the ridge. The hip roof has hip rafters that run at 45 degrees in plan view on an equal sloped roof (all sides with the same slope). It has additional rafters besides the common rafter, including the hip and hip jack rafters. The hip rafter is always at least one size larger than the common rafter, as per NBC. For example, if the drawings state that the common rafter is 2" × 4" then the hip rafter will be at least 2" × 6" to ensure full contact of the jack rafters against the hip rafter, based on the hip rafter being on a 45-degree angle in plan view for an equal slope roof. A hip roof constructed from roof trusses will require a number of different sizes of trusses, as will be discussed further in the truss section of this chapter.

FIGURE 9.8

Collar tie requirements (refer to Division B Part 9.23.14.7 of the NBC).

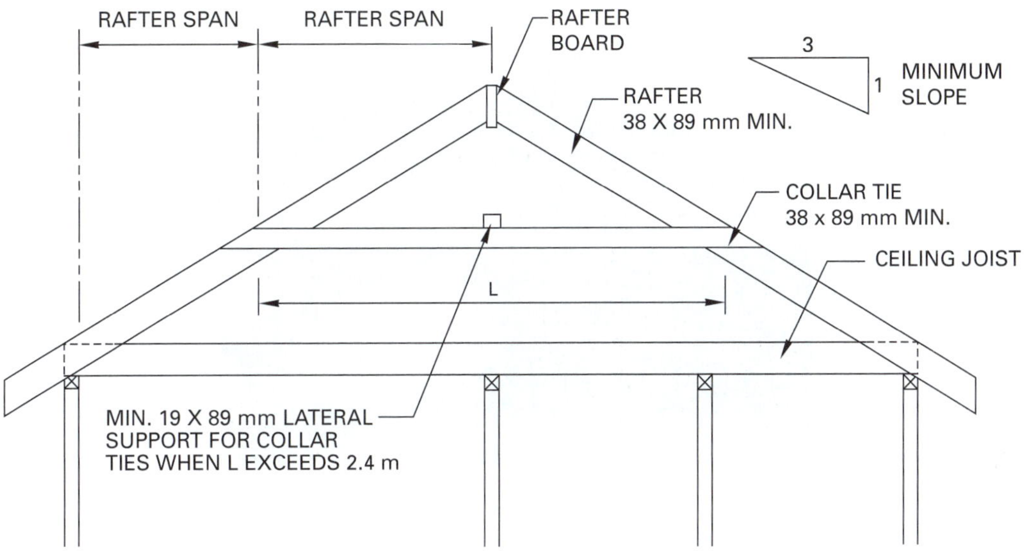

Source: National Building Code of Canada

FIGURE 9.9

Knee walls and how they can shorten rafter spans.

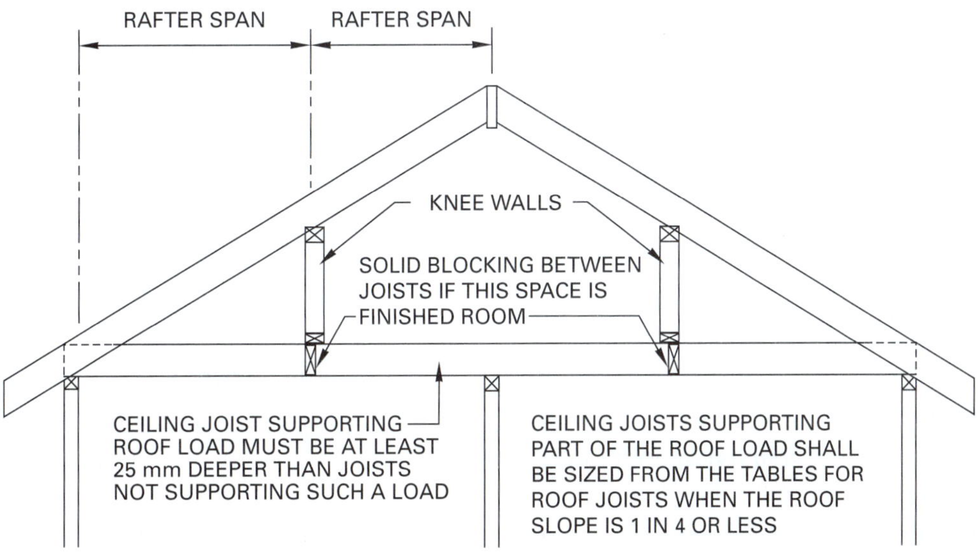

Intersecting Roofs

An intersecting roof occurs when the building or structure is something other than just a square or rectangle. The largest roof is called the major roof and the smaller roof is referred to as the minor roof. Examples of an intersecting roof are L-, T-, and U-shaped buildings (see Figure 9.12).

A traditional cut roof will add valley and valley jack rafters when an intersecting roof occurs.

FIGURE 9.10

Example of a hip roof on both upper and lower roofs.

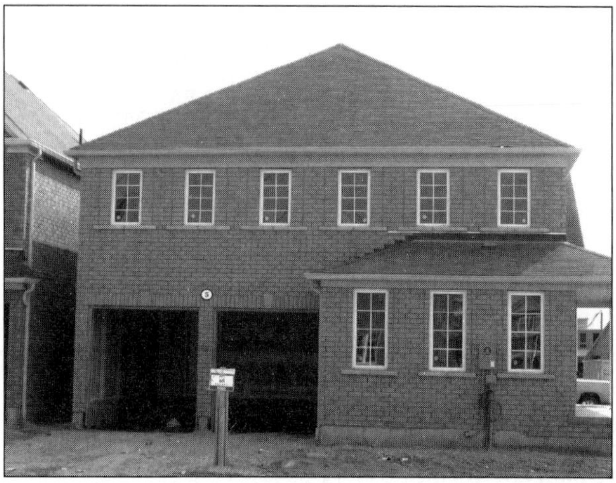

Courtesy Tom Stephenson

FIGURE 9.11

Section detail of a flat roof over the porch taken from the Brook drawings. Note that the flat roof has a minimum two percent slope requirement. This would be achieved by tapering the purlins so they slope to the drain. Also note the flashing details and the use of cant strips at edge of roof and wall (4" × 4" material that is cut on a diagonal).

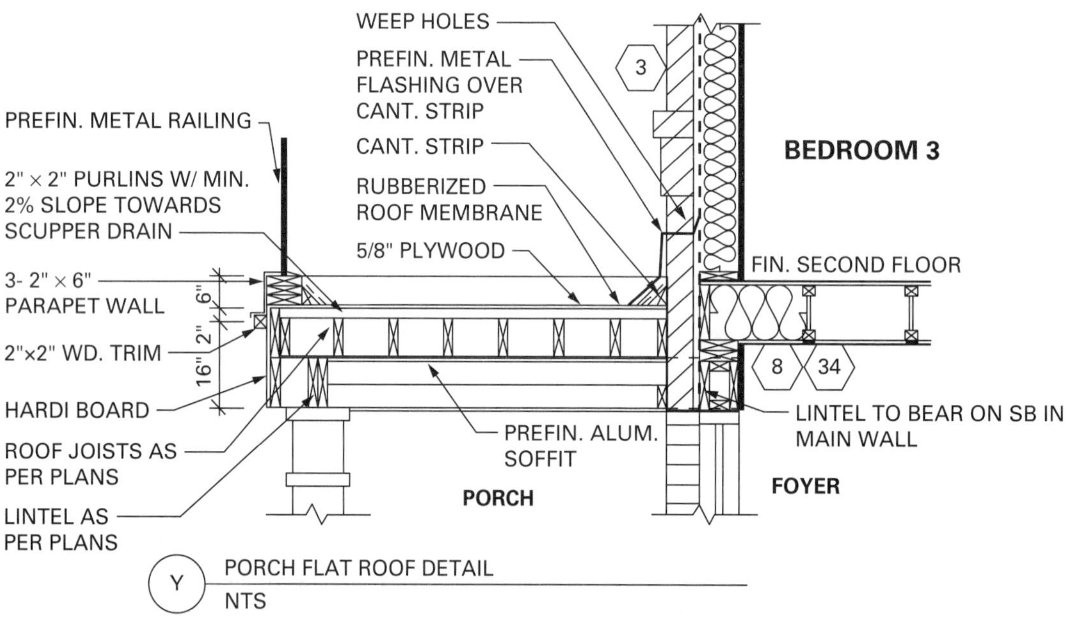

Source: Cassidy & Co. Architectural Technologists

FIGURE 9.12

L-, T-, and U-shaped buildings.

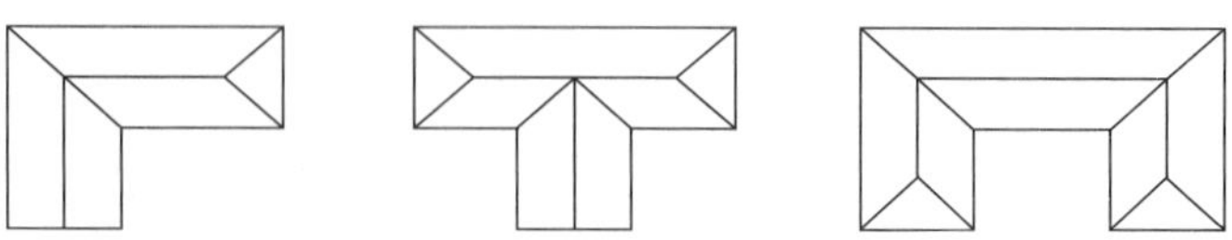

FIGURE 9.13

The illustration below depicts the assembly process and location of the various types of rafters required to construct a hip or cottage roof. In more complex assemblies (for example, multiple intersections with unequal slopes) a roof framing plan similar to the illustration below may be provided to ensure that the roof is constructed as designed. As an alternative layout the common rafters are often laid out to intersect at the exact junction of the hip rafters and ridge board instead of being offset as indicated below.

STEP 1 ERECT THE RIDGE WITH ONLY AS MANY COMMON RAFTERS AS NEEDED.

STEP 2 INSTALL THE HIP RAFTERS, THE HIP RAFTERS EFFECTIVELY BRACE THE ASSEMBLY.

NOTE: THIS PROCEDURE SHOWS, THE HIPS FRAMED TO THE RIDGE. IF HIPS ARE FRAMED TO THE COMMON RAFTERS, INSTALL THE COMMONS TO THE END OF THE RIDGE BEFORE THE HIPS.

STEP 3 INSTALL THE REMAINING COMMON RAFTERS IN PAIRS OPPOSING EACH OTHER. SIGHT THE TOP EDGE OF THE RIDGE FOR STRAIGHTNESS AS FRAMING PROGRESSES.

STEP 4 INSTALL THE HIP JACK RAFTERS IN PAIRS OPPOSING EACH OTHER. SIGHT THE HIP FOR STRAIGHTNESS AS JACKS ARE INSTALLED.

NOTE: IT IS BEST TO INSTALL A PAIR OF JACKS ABOUT HALFWAY UP THE HIP TO STRAIGHTEN IT. THE HIP CAN BE STRAIGHTENED BY DRIVING THE JACK, ON THE CROWN SIDE OF THE BOW, DOWN TIGHTER.

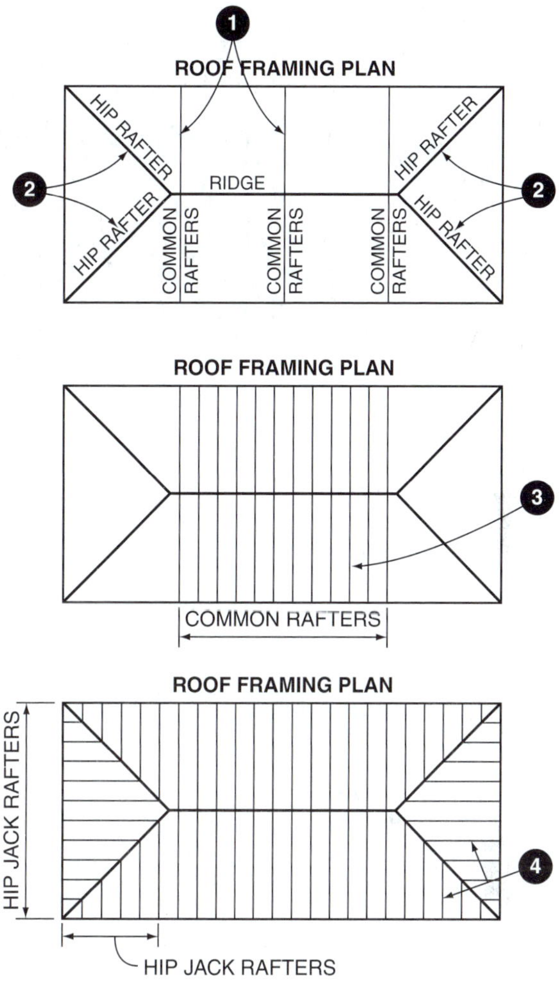

Roof Trusses

A trussed roof is engineered and manufactured to be constructed as a system. To meet this requirement, the installation must follow the truss manufacturer's/engineer's instructions. Care must be taken to ensure that the trusses are maintained in good condition and not damaged during installation. The truss consists of a bottom and top chord with individual webs that are fastened together with **gusset plates.** The trusses are fastened to the top plates and are braced according to the truss drawings. The bottom chord of the truss can clear span without support. This alleviates the need for load-bearing interior walls on the top floor. Long clear spans offer an advantage over a cut roof, which usually requires load-bearing walls to support the ceiling joists. The designer therefore has more flexibility in laying out the walls on the floor below the trusses. The roof sheathing acts as a brace and forms part of the overall design. There are a wide variety of trusses that can be used in countless designs. Trusses are the most popular type of roof used in residential construction today.

Advantages:

- Can span greater distances than a conventional cut roof

FIGURE 9.14

Intersecting roof displaying the various types of rafters.

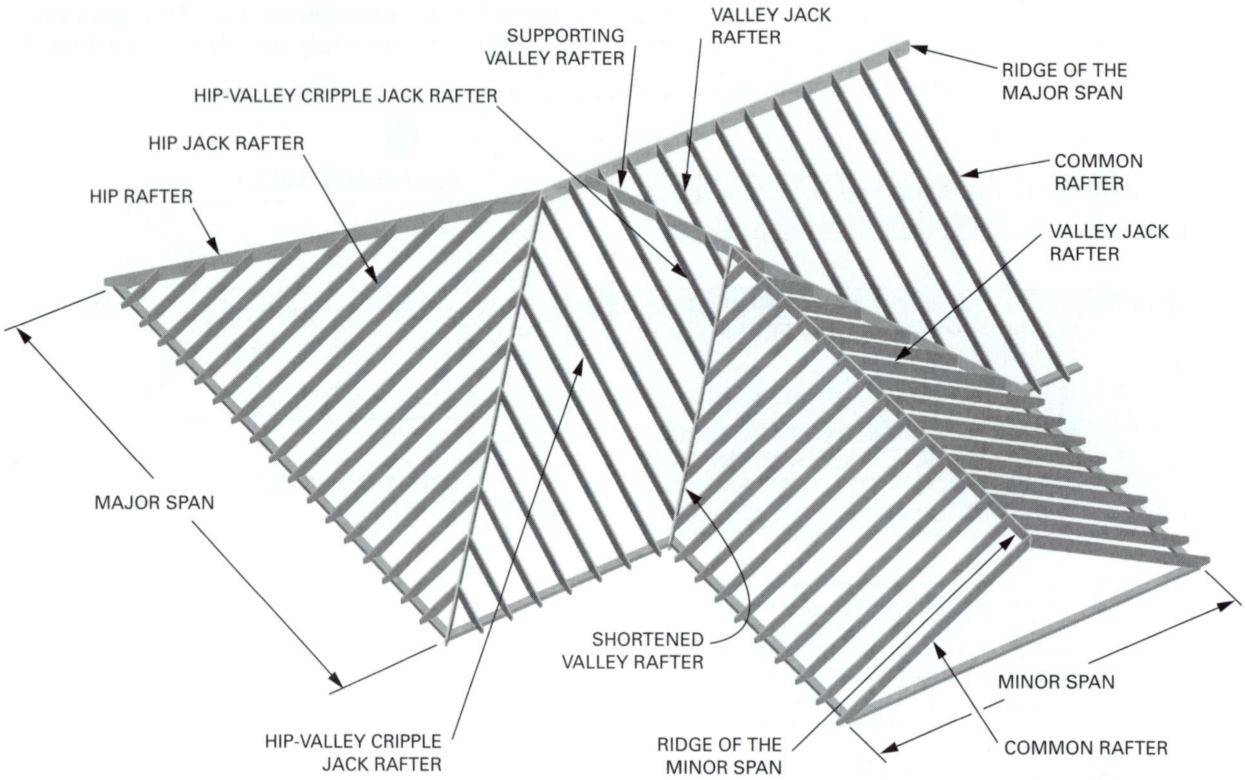

- Intermediate load-bearing walls are not required to support bottom chord. Open-concept designs are easier to achieve as a result.
- Generally constructed of 2" × 4" or 2" × 6" material, they use smaller sizes of dimensional lumber.
- More economical to use
- Can be installed faster
- Requires less expertise to construct than a cut roof. However, thorough knowledge of reading truss drawings and installation are still required. As well, many truss roofs still require portions of the roof to be constructed conventionally.

Disadvantages:

- General loss of attic space, although trusses can be designed to accommodate an attic
- Acts as a system and is only as strong as its weakest link. Therefore, extra care needs to be taken to ensure that trusses are not damaged and are installed according to the manufacturer's requirements.
- In many cases requires a forklift, zoom boom, or crane to lift trusses into place. This can be a disadvantage when constructing custom homes or additions where trees and other obstructions may make access difficult, if not impossible.
- Must be ordered well in advance of the installation date.
- If precautions are not taken during installation, large trusses may, as a result of differential shrinkage, have an uplift of the bottom chord at mid-span. This results in a cracked ceiling where it meets the wall near mid span. This condition is known as truss uplift.

Flat Roofs

Flat roofs are often used in residential construction. They are framed in place similar to how a floor is framed. Flat roofs must be carefully detailed to ensure that the roof does not leak. Contrary to what their name implies, a flat roof does have a slope; generally the minimal recommended amount is two percent, approximately

FIGURE 9.15
Roof truss members.

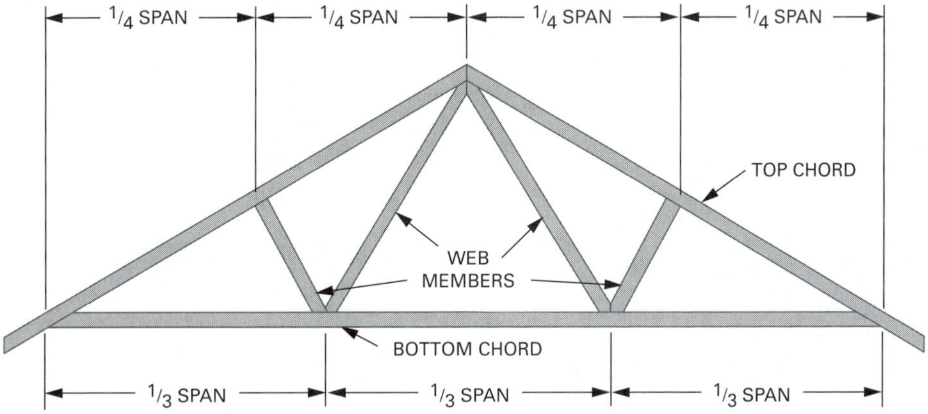

Source: From VOGT, Carpentry, 4E. © 2006 Delmar Learning, a part of Cengage Learning, Inc. Reproduced by permission. www.cengage.com/permissions

1/4" to 1' or greater, in order to have the water run off the roof or to a roof drain/scupper (a scupper is a roof drain). The best way to visualize a two percent roof is to imagine a length of 2" × 8" lumber that is 100" long and placed 2" out of level. That is a two percent slope.

Bracing

Trusses are required to be braced to each other both temporarily until the roof sheathing is applied, and permanently to meet design requirements. The truss drawings will indicate the location and type of bracing, which include 1" × 4"s

FIGURE 9.16
Roof truss types.

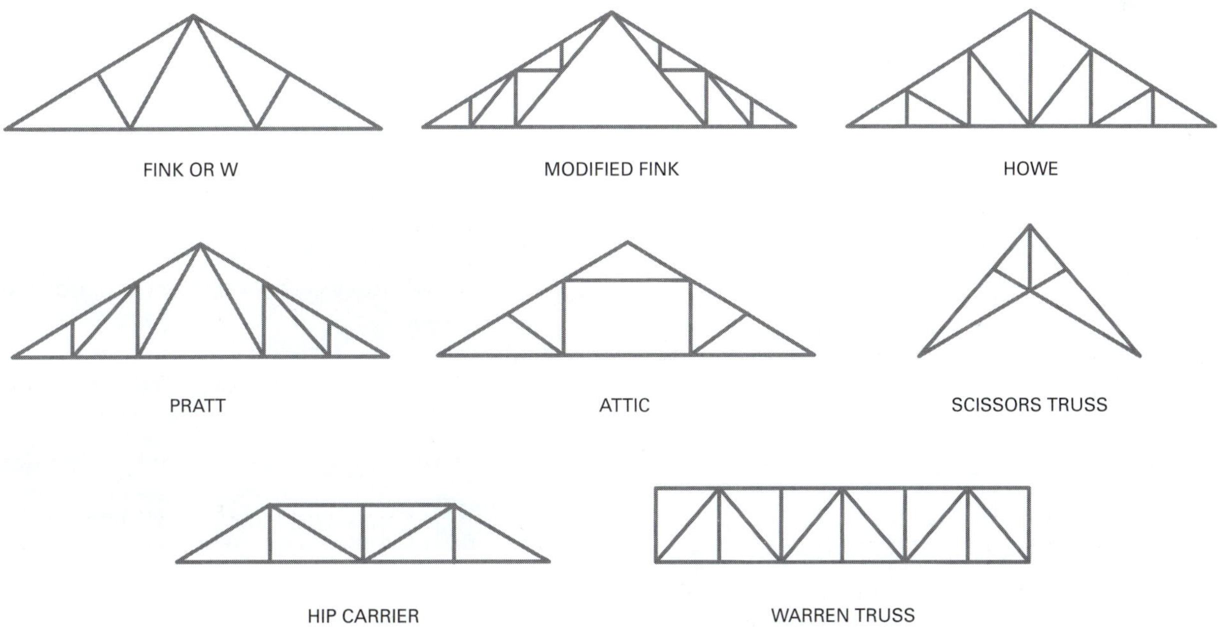

Source: From VOGT, Carpentry, 4E. © 2006 Delmar Learning, a part of Cengage Learning, Inc. Reproduced by permission. www.cengage.com/permissions

FIGURE 9.17
Examples of temporary bracing.

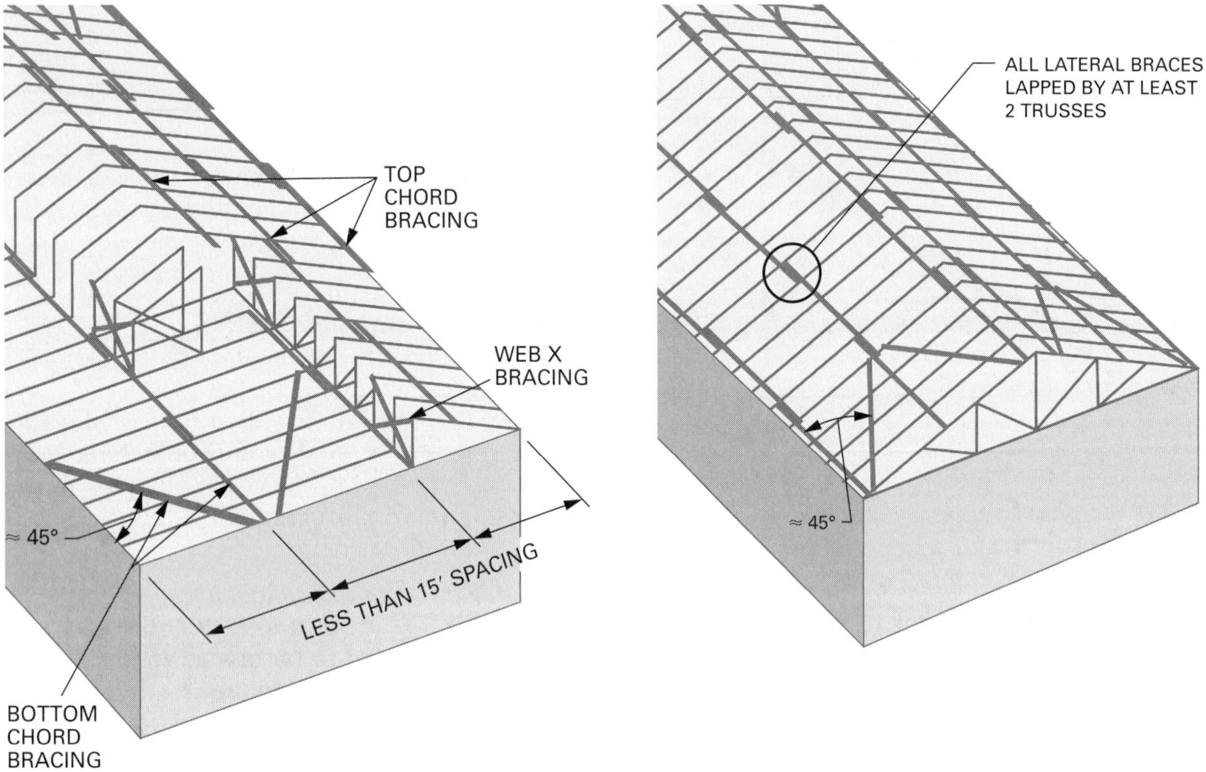

Source: From VOGT, Carpentry, 4E. © 2006 Delmar Learning, a part of Cengage Learning, Inc. Reproduced by permission. www.cengage.com/permissions

FIGURE 9.18
Example of permanent bracing.

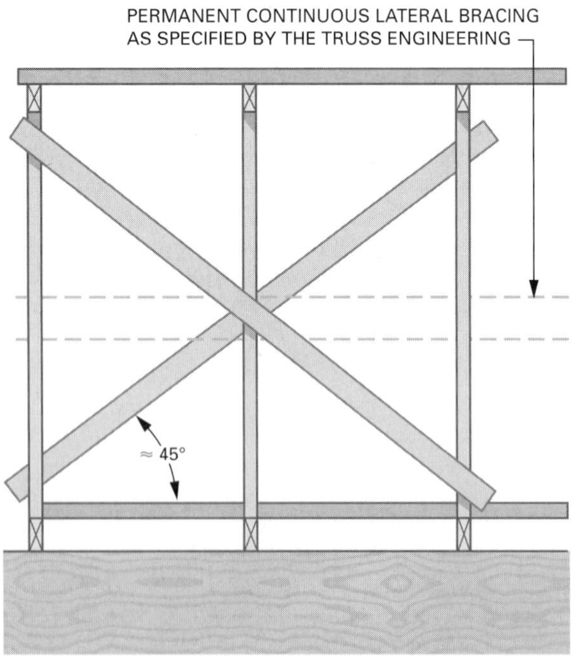

Source: From VOGT, Carpentry, 4E. © 2006 Delmar Learning, a part of Cengage Learning, Inc. Reproduced by permission. www.cengage.com/permissions

FIGURE 9.19
Roof trusses being installed for a hip roof; cut hips not yet in place at lower corner. Notice that the girder truss is the truss that is supporting the common jacks. The Girder Truss is indicated on the drawings by GT and requires solid bearing SB down to the foundation. The solid bearing is required because a girder truss is carrying the weight of other trusses. Review Doncaster floor-plan drawings for these abbreviations.

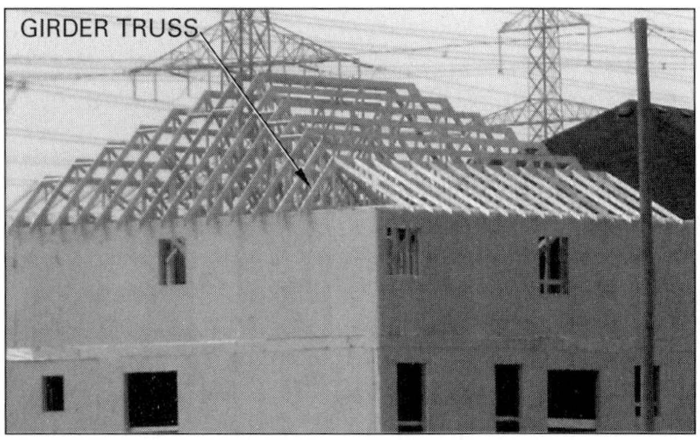

Courtesy Tom Stephenson

fastened as lateral and diagonal braces to webs, roof sheathing placed on the top chord, and strapping or drywall placed on the bottom chord. Bracing is indicated on individual truss drawings supplied by the manufacturer; study these drawings carefully and watch for bracing indicators.

READING TRUSS PLANS AND INSTALLATION PRACTICES

Truss drawings provide the building department with the assurance that the trusses have been designed and manufactured to meet applicable code requirements. The drawings also provide the builder and framer with the information that they require to install the trusses correctly. A set of manufacturer-supplied truss drawings will generally contain the following:

- A roof plan that indicates the location of each individual truss and the building dimensions (refer to Figure 9.20, Doncaster truss drawings, page 1)
- A list of responsibilities and general specifications (refer to Figure 9.20, Doncaster truss drawings, page 2)
- A coding or identification marking that indicates which truss goes where (refer to Figure 9.20, Doncaster truss drawings, page 1; numbering listed beside the trusses)
- An indication if any parts of the roof are to be conventionally framed to the trusses or to be placed on top of the trusses (refer to Figure 9.20, Doncaster truss drawings, page 1; diagonal lines indicate portion of roof requiring a conventional or cut roof)
- A listing of required hangers, if required, including the type, load capacity, and make of hanger (hangers not required for Doncaster truss drawings)
- Any detail drawings that are required to better illustrate truss intersections such as **girder truss** and jack intersections (refer to Figure 9.20, Doncaster truss drawings, page 3)
- The detail drawings will indicate size, type, spacing, and method of connection for these trusses. The elevation drawing will provide design information regarding the loading and criteria used to design the trusses, the permanent lateral bracing required, and required bracing for the top and bottom chords. All bracing requirements should be verified with the truss manufacturer including requirements for diagonal and temporary bracing.
- The elevation drawings also indicate the live load that was used in the design of the trusses. It is important that the truss manufacturer understand the type of finished roof covering to be used and any special weight requirements. The designer should confirm this at the drawings stage, and if there are any changes, the contractor must make sure that they are reflected in the order for the trusses. For example, a client may not be able to change from an asphalt roof to a concrete tile roof after the trusses are ordered because the truss may not be designed to carry the extra weight (dead load) of this type of roof. The contractor therefore must ensure that the client is satisfied with the chosen roof covering before the trusses are ordered because choices are limited after that.

FIGURE 9.20

Doncaster truss drawings; these drawings are an example of the drawings provided by the truss manufacturer and are required as part of the building inspection process. Construction members are not shown, except for the lower roof in front of the garage and over the living room. The dormers and gable at rear of house, not shown; refer to Doncaster construction drawings for this information.

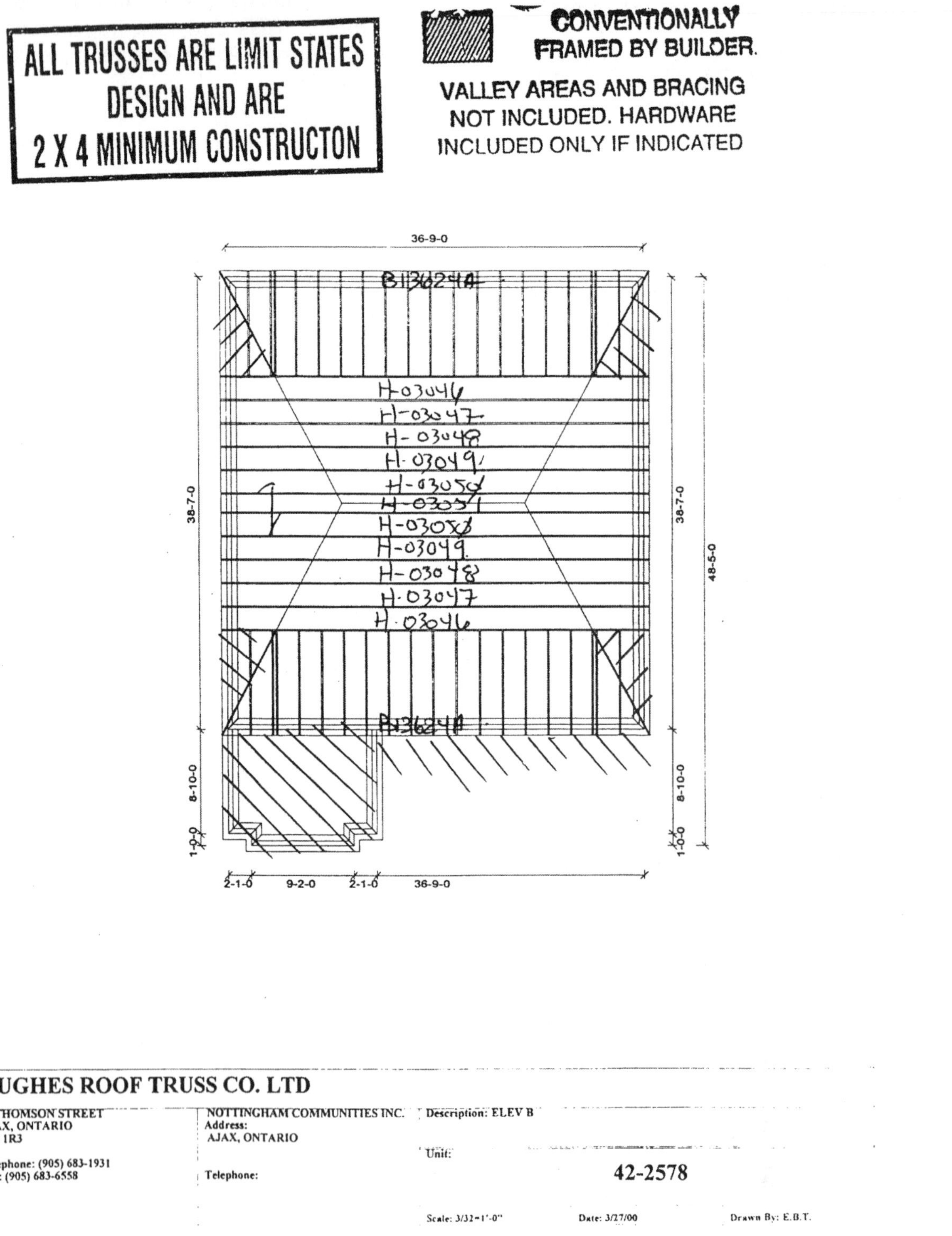

Source: Hughes Roof Truss Co. Ltd.

FIGURE 9.20 (*continued*)

MITEK CANADA INC.

GENERAL SPECIFICATIONS

RESPONSIBILITIES

MITEK CANADA INC. IS RESPONSIBLE FOR THE DESIGN OF TRUSSES AS INDIVIDUAL COMPONENTS AND FOR THE QUALITY OF METAL PRODUCTS (SUCH AS CONNECTORS, HARDWARE, ETC.) MANUFACTURED BY MITEK CANADA AND USED FOR THE ASSEMBLY OF SAID TRUSSES.

IT IS THE TRUSS MANUFACTURER'S RESPONSIBILITY TO ENSURE THAT TRUSSES ARE MANUFACTURED IN CONFORMANCE WITH MITEK CANADA DESIGNS AND WITH THE SPECIFICATIONS OUTLINED BELOW. MITEK CANADA INC. IS NOT RESPONSIBLE FOR QUALITY CONTROL IN THE TRUSS MANUFACTURING PLANT.

THE TRUSS MANUFACTURER IS RESPONSIBLE FOR PROVIDING THE USER OF THE TRUSSES WITH INSTRUCTIONS ABOUT THE PROPER ON-SITE STORAGE, HANDLING AND INSTALLATION, INCLUDING BRACINGS REQUIRED FOR THE STRENGTH AND STABILITY OF THE TRUSS AS SHOWN ON THE DRAWING. FURTHERMORE, IT IS THE TRUSS MANUFACTURER'S OBLIGATION TO WARN THE USER ABOUT THE NEED FOR TEMPORARY AND PERMANENT BRACING DURING THE CONSTRUCTION AS WELL AS FOR THE FINISHED ROOF, THE DESIGN OF WHICH REQUIRES THE GUIDANCE OF A PROFESSIONAL ENGINEER, A SERVICE WHICH MITEK CANADA DOES NOT OFFER.

ALL DRAWINGS, REPORTS AND SPECIFICATIONS PREPARED AND APPROVED BY MITEK CANADA MUST BE USED IN CONJUNCTION WITH THIS DOCUMENT

SPECIFICATIONS

TRUSSES DESIGNED BY MITEK CANADA CONFORM TO THE RELEVANT SECTIONS OF THE NATIONAL BUILDING CODE OF CANADA OR TO THE CANADIAN CODE FOR FARM BUILDINGS, WHICHEVER APPLIES TO THE BUILDING TYPE INDICATED ON THE DRAWING AS WELL AS TO THE PROCEDURES ESTABLISHED BY THE TRUSS PLATE INSTITUTE OF CANADA BEING IN FORCE AT THE DATE OF THE DESIGN. MITEK CANADA GUARANTEES THAT TRUSSES FABRICATED TO OUR DESIGN ARE SUITABLE FOR THE USE INDICATED THEREON PROVIDED THE FOLLOWING CONDITIONS ARE MET:

1. BUILDING TYPE: THE BUILDING IS OF THE TYPE REQUESTED BY THE TRUSS MANUFACTURER AND SO INDICATED ON THE DRAWING.

2. LOADING: THE TRUSS LOADING INTENSITY AND DISTRIBUTION AS WELL AS LOAD TRANSFER MECHANISM IS THAT INDICATED ON THE DRAWING. NO BUILDINGS, TREES, PARAPETS OR OTHER PROJECTIONS HIGHER THAN THE ROOF FOR WHICH THE TRUSSES ARE USED ARE LOCATED WITHIN A DISTANCE LESS THAN TEN (10) TIMES THE DIFFERENCE IN HEIGHT, OR FIVE METERS (16FT.), WHICHEVER IS GREATER, UNLESS THE DRAWING INDICATES THAT THE SNOW DRIFTING HAS BEEN TAKEN INTO ACCOUNT.

3. PERMANENT BRACING: THE COMPRESSION CHORDS ARE LATERALLY BRACED AT INTERVALS NOT EXCEEDING 10 TIMES THEIR THICKNESS AND THE TENSION CHORDS ARE LATERALLY BRACED AT INTERVALS NOT EXCEEDING 80 TIMES THEIR THICKNESS OR BY CONTINUOUS RIGID DIAPHRAGM SHEATHING OR AS SPECIFIED ON THE DRAWING. ALL OTHER MEMBERS ARE LATERALLY BRACED AT THE INTERVALS INDICATED. BRACING REFERRED TO HEREIN IS ITSELF BRACED TO A SECURE ANCHORAGE TO PREVENT MOVEMENT OF THE OVERALL STRUCTURE.

NOTE: LATERAL BRACES AND PURLINS INDICATED FOR TRUSS MEMBERS ARE REQUIRED TO REDUCE BUCKLING LENGTH OF MEMBER, AND SHOULD BE NAILED TO TRUSS MEMBERS WITH MINIMUM OF 2-3" COMMON WIRE NAILS. PROVISIONS MUST BE MADE AT ENDS OR SPECIFIED INTERVALS TO RESTRAIN OR ANCHOR LATERAL BRACING (BY OTHERS).

4. OVERALL BRACING: THE TRUSSES ARE INSTALLED IN A PLUMB POSITION AND MAINTAINED THERE BY A PROPERLY DESIGNED BRACING SYSTEM. MITEK CANADA DOES NOT DESIGN OVERALL WIND AND SWAY BRACING SYSTEMS. BRACING SHOWN ON MITEK CANADA DRAWINGS IS SPECIFIED FOR THE TRUSS AS A COMPONENT ONLY AND FORMS AN INTEGRAL PART OF THE TRUSS DESIGN. TRUSSES ARE DESIGNED FOR VERTICAL LOADS ONLY. LATERAL STABILITY OF TRUSS IS TO BE PROVIDED BY PROJECT ENGINEER OR ARCHITECT.

5. HANDLING AND ERECTION: FIELD ERECTION OF THIS STRUCTURAL COMPONENT, INCLUDING ITEMS SUCH AS PROPER HANDLING, SAFETY PRECAUTIONS, TEMPORARY AND PERMANENT BRACING, AND OTHER SAFEGUARDS OR PROCEDURES CONSISTENT WITH GOOD WORKMANSHIP AND GOOD BUILDING ERECTION PRACTICES IS THE RESPONSIBILITY OF THE ERECTION CONTRACTOR. ERECTION BRACING MUST BE INSTALLED TO HOLD THE TRUSS TRUE AND PLUMB AND IN A SAFE CONDITION UNTIL OTHER PERMANENT COMPONENTS SUCH AS PURLINS AND SHEATHING, CONTRIBUTING TO THE RIGIDITY OF THE COMPLETE ROOF STRUCTURE ARE IN PLACE. IT IS RECOMMENDED THAT PROFESSIONAL ENGINEER'S ADVICE BE OBTAINED FOR THE BRACING OF TRUSSES SPANNING MORE THAN 12.37 METERS (40'-7").

6. SUPPORTS: THE TRUSSES ARE SUPPORTED AT THE BEARING POINTS INDICATED AND ANCHORED TO THE SUPPORTS WHERE CONSIDERED NECESSARY BY THE DESIGN ER OF THE OVERALL STRUCTURE. MITEK CANADA DOES NOT DESIGN SUPPORTING STRUCTURES. BEARING SIZES SHOWN ARE THE MINIMUM REQUIRED TO PREVENT CRUSHING OF THE TRUSS MEMBERS AND DO NOT NECESSARILY TAKE INTO ACCOUNT STABILITY OF THE OVERALL BUILDING STRUCTURE. ELEVATION OF BEARINGS MUST BE CAREFULLY CHECKED AND SHIMMED TO ALIGNMENT FOR SOLID BEARINGS. ADEQUATE WOOD TRUSS BEARING IS THE RESPONSIBILITY OF THE BUILDING DESIGNER.

7. LUMBER: THE LUMBER IS OF THE SPECIES, GRADE AND SIZE INDICATED ON THE DRAWING, PROPERLY GRADED IN ACCORDANCE WITH THE LATEST REVISION OF NLGA STANDARD GRADING RULES FOR CANADIAN LUMBER, UNLESS INDICATED OTHERWISE ON THE DRAWING. MOISTURE CONTENT OF THE LUMBER AT THE TIME OF FABRICATION SHALL NOT EXCEED 19% UNLESS INDICATED OTHERWISE ON THE DRAWING. THE TRUSSES ARE USED IN AN ENVIRONMENT IN WHICH THE EQUILIBRIUM MOISTURE CONTENT OF THE LUMBER WILL AVERAGE 15% OR LESS OVER A YEAR AND WILL NOT EXCEED 19% UNLESS THE DRAWING INDICATES THAT WET SERVICE CONDITION HAS BEEN TAKEN INTO ACCOUNT.

8. PLATES: THE CONNECTOR PLATES ARE SUPPLIED BY MITEK AND ARE OF THE TYPE, GAUGE AND SIZE SHOWN IN THE PLATING SCHEDULE. THE TRUSS IS MANUFACTURED BY AN AUTHORIZED MITEK TRUSS MANUFACTURER AND IN ACCORDANCE WITH A DRAWING APPROVED BY A PROFESSIONAL ENGINEER AUTHORIZED TO DO SO BY MITEK CANADA.

9. DECAY AND CHEMICAL TREATMENT: THE TRUSSES ARE NOT USED IN AN ENVIRONMENT LIKELY TO CAUSE DECAY OF THE LUMBER OR CORROSION OF THE CONNECTOR PLATES UNLESS APPROPRIATE MEASURES ARE TAKEN TO PREVENT SUCH DECAY OR CORROSION. THE LUMBER HAS NOT BEEN TREATED WITH CHEMICALS LIKELY TO IMPAIR ITS STRENGTH OR STIFFNESS.

10. DIMENSIONS: GEOMETRY OF THE TRUSS AND DIMENSIONS INDICATED ON THE DRAWING IS IDENTICAL TO THOSE OF THE INSTALLED TRUSS. TRUSS MEMBERS ARE ACCURATELY CUT AND THE PLATES ARE ACCURATELY PLACED IN ACCORDANCE WITH MITEK CANADA DRAWING AND JOINT DETAILS.

ALTHOUGH THESE GENERAL SPECIFICATIONS BECOME SPECIFIC IN SOME OF THE FOREGOING, NEVERTHELESS, THEY ARE INTENDED TO BE GENERAL IN NATURE AND MUST BE USED AND INTERPRETED CORRECTLY BY THE "AUTHORITY IN JURISDICTION" WHO IS QUALIFIED TO SPECIFY MITEK TRUSSES FOR THE PARTICULAR END USE.

FIGURE 9.20 (*continued*)

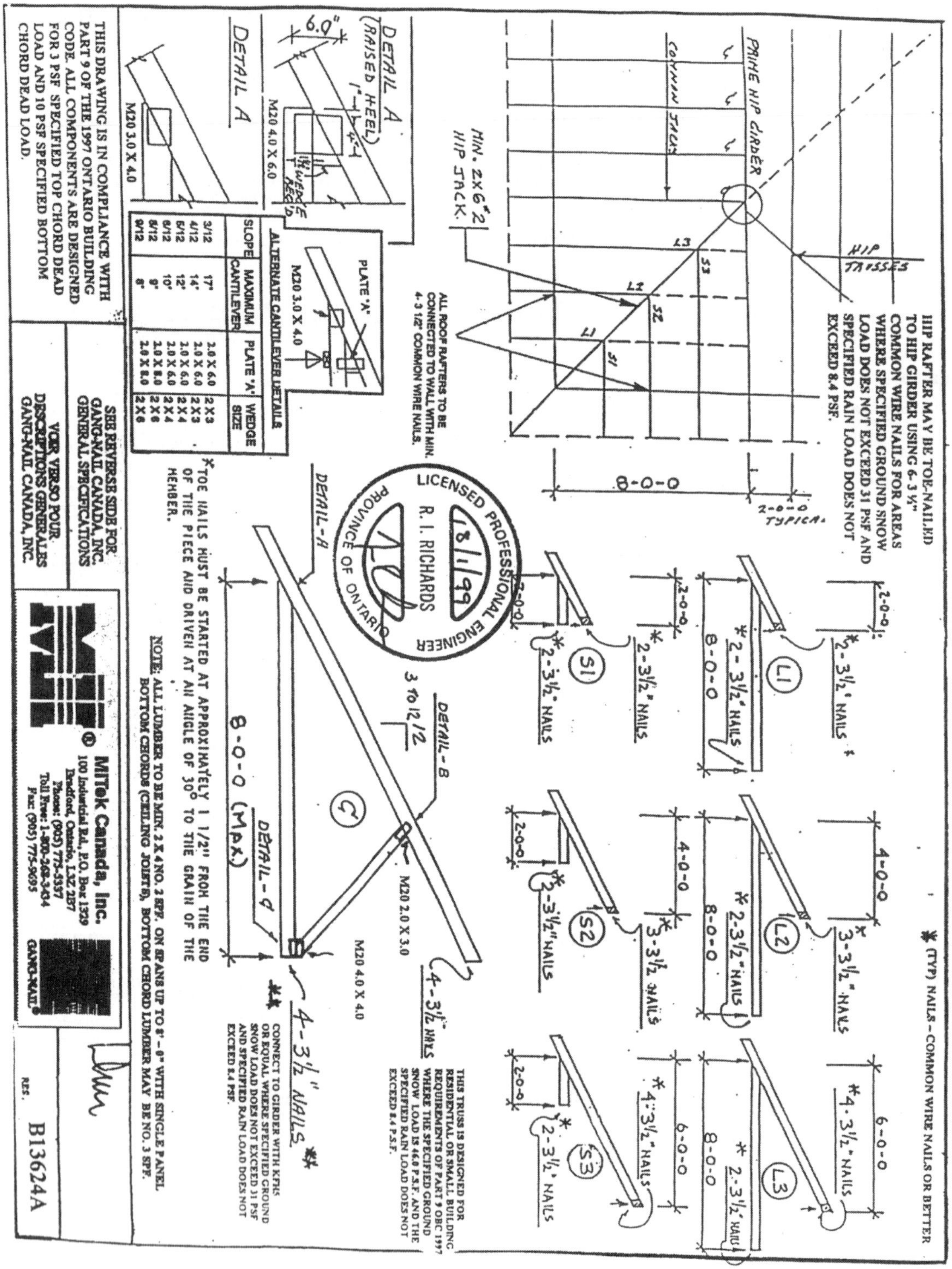

FIGURE 9.20 (*continued*)

3 NAME	TRUSS NAME	QUANTITY	PLY	JOB DESC.	DRWG NO. **H-03046**
9B_ENG	G1	1	2	TRUSS DESC.	

GHES ROOF TRUSS CO. LTD. — Version 4.100 S Mar 2 2000 MiTek Industries, Inc. Wed Mar 29 10:18:45 2000 Page 1 of 1

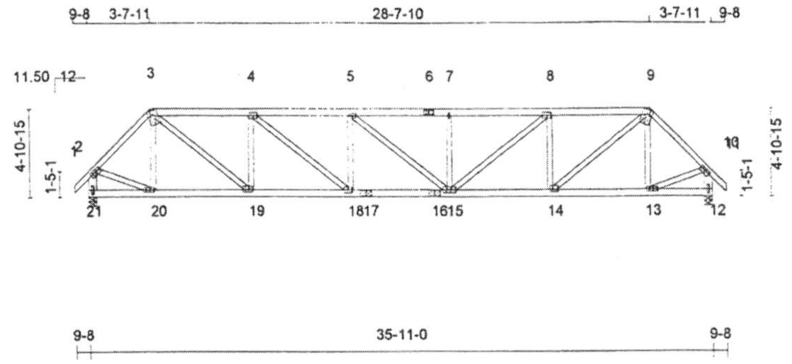

TOTAL WEIGHT = 2 X 236 = 472 lbs

UMBER
L. G. A. RULES

HORDS	SIZE	LUMBER	DESCR.
- 3	2 X 4	No.2	SPF
- 6	2 X 5	No.2	SPF
- 9	2 X 5	No.2	SPF
- 11	2 X 4	No.2	SPF
1 - 2	2 X 6	No.2	SPF
2 - 10	2 X 6	No.2	SPF
1 - 17	2 X 5	No.2	SPF
7 - 16	2 X 5	No.2	SPF
6 - 12	2 X 5	No.2	SPF
LL WEBS	2 X 4	No.2	SPF

ESIGN CONSIST OF 2 TRUSSES BUILT SEPARATELY THEN AILED TOGETHER WITH 3" SPIRAL WIRE NAILS STAGGERED HROUGHOUT BOTH FACES AS PER NAILING PATTERN

AILING MAY BE DONE FROM ONE SIDE ONLY

HORDS	#ROWS	SPACING(IN)	LOAD(PLF)
- 3	1	12	TOP
- 6	1	12	SIDE(39.2)
- 9	1	12	SIDE(39.2)
- 11	1	12	TOP
1- 2	2	12	TOP
2- 10	2	12	TOP
1- 17	1	12	SIDE(125.2)
7- 16	1	12	SIDE(125.2)
6- 12	1	12	SIDE(125.2)
VEBS			
x4	1	6	

RDER NAILING ASSUMES NAILED HANGERS ARE FASTENED ITH MIN. 3" NAILS.

OP - COMPONENTS ARE LOADED FROM THE TOP AND MUST BE LACED ON TOP EDGE OF ALL PLIES FOR THE LOAD TO BE RANSFERRED TO EACH PLY.

DE - PLF SHOWN IS THE MAX. LOAD FROM ONE SIDE THAT THE AILING PATTERN CAN TRANSFER. REMAINING PLF MUST BE PPLIED ON THE OPPOSITE SIDE OR ON THE TOP.

LATES (table is in inches)

T	TYPE	PLATES	W	LEN	Y	X
	CMWW-t	M-20	4.0	8.0	1.50	3.00
	CCWW+m	M-20	7.0	10.0	Edge	1.75
	CMWW-t	M-20	4.0	6.0		
	CMWW-t	M-20	3.0	5.0		
	CS-t	M-20	3.0	7.0		
	CMW+w	M-20	1.5	4.0		
	CMWW-t	M-20	4.0	6.0		
	CCWW+m	M-20	7.0	10.0	Edge	1.75
)	CMWW-t	M-20	4.0	8.0	1.50	3.00
!	BMV+p	M-20	2.0	6.0		
)	CMWW-t	M-20	3.0	8.0	1.50	2.75
!	CMWW-t	M-20	4.0	6.0	1.50	2.00
5	CMWWW-t	M-20	4.0	9.0		
5	CS-t	M-20	4.0	8.0		
'	CS-t	M-20	4.0	8.0		
3	CMWW-t	M-20	4.0	6.0		
)	CMWW-t	M-20	4.0	7.0	1.50	2.75
)	CMWW-t	M-20	3.0	8.0	1.50	2.50
:	BMV+p	M-20	2.0	6.0		

ge - INDICATES REFERENCE CORNER OF PLATE TOUCHES DGE OF CHORD.

DIMENSIONS, SUPPORTS AND LOADINGS SPECIFIED BY FABRICATOR TO BE VERIFIED BY BUILDING DESIGNER

PROVIDE ADEQUATE DRAINAGE TO PREVENT PONDING

BEARINGS

	FACTORED GROSS REACTION		MAX. FACTORED GROSS REACTION		INPUT BRG	REQRD BRG	HEEL
JT	VERT	HORZ	VERT	HORZ	IN-SX	IN-SX	WEDGE
21	4968	0	4968	0	5- 8	2-11	
12	4968	0	4968	0	5- 8	2-11	

MAX. UNFACTORED GROSS REACTIONS

	VERTICAL		HORIZONTAL	
JT	LIVE	DEAD	LIVE	DEAD
21	2595	860	0	0
12	2595	860	0	0

BEARING MATERIAL TO BE OF THE SAME SPECIES AS CHORD MEMBER AND OF GRADE NO. 2 OR BETTER.

BRACING
TOP CHORD TO BE SHEATHED OR MAX. PURLIN SPACING = 3.02FT.
MAX. UNBRACED BOTTOM CHORD LENGTH = 10.00FT.

LOADING (FACTORED)
LOAD CASE (1) OF (2)

CHORDS MEMB. FR-TO	FACT FORCE (LBS)	FACTORED VERT. LOAD (PLF) FROM	TO	FACTORED HORIZ. LOAD (PLF) FROM	TO	MAX UNBRAC LENGTH	WEBS MEMB. FR-TO	FACTORED FORCE (LBS)
1 - 2	27T	-88.50	-88.50	0.00	0.00	10.00	20 - 3	422C
2 - 3	5275C	-88.50	-88.50	0.00	0.00	3.97	13 - 9	423C
3 - 4	7974C	-127.70	-127.70	0.00	0.00	3.46	19 - 4	2364C
4 - 5	9922C	-127.70	-127.70	0.00	0.00	3.02	18 - 5	702C
5 - 6	9919C	-127.70	-127.70	0.00	0.00	3.16	15 - 7	701C
6 - 7	9919C	-127.70	-127.70	0.00	0.00	3.16	14 - 8	2363C
7 - 8	9919C	-127.70	-127.70	0.00	0.00	3.02	3 - 19	5296T
8 - 9	7974C	-127.70	-127.70	0.00	0.00	3.46	4 - 18	2493T
9 - 10	5275C	-88.50	-88.50	0.00	0.00	3.97	5 - 15	4C
10 - 11	27T	-88.50	-88.50	0.00	0.00	10.00	15 - 8	2489T
21 - 2	4921C	0.00	0.00	0.00	0.00	6.55	14 - 9	5297T
12 - 10	4921C	0.00	0.00	0.00	0.00	6.55	2 - 20	3942T
							13 - 10	3942T
21 - 20	0C	-63.20	-63.20	0.00	0.00	10.00		
20 - 19	3791T	-151.70	-151.70	0.00	0.00	10.00		
19 - 18	7974T	-151.70	-151.70	0.00	0.00	10.00		
18 - 17	9922T	-151.70	-151.70	0.00	0.00	10.00		
17 - 16	9922T	-151.70	-151.70	0.00	0.00	10.00		
16 - 15	9922T	-151.70	-151.70	0.00	0.00	10.00		
15 - 14	7974T	-151.70	-151.70	0.00	0.00	10.00		
14 - 13	3791T	-151.70	-151.70	0.00	0.00	10.00		
13 - 12	0C	-63.20	-63.20	0.00	0.00	10.00		

SPECIAL HANGER REQUIRED TO SUPPORT FACTORED CONCENTRATED LOADS

FACTORED CONCENTRATED LOADS

JT	VERTICAL (LBS)	HORIZONTAL (LBS)	MOMENT (FT.LB)
3	-349	0	0.0
9	-349	0	0.0

ROOF TRUSSES
FABRICATED BY
HUGHES ROOF TRUSS CO. LTD.
AJAX: (905) 683-1931 TOR: (416) 213-5665

[M]

DESIGN CRITERIA

UNFACTORED LOADS:
TOP CH. LL = 27.0 PSF
DL = 3.0 PSF
BOT CH. LL = 3.0 PSF
DL = 7.0 PSF
TOTAL LOAD = 40.0 PSF

SPACING = 24.0 IN. C/C

GIRDER TYPE: CPrimeHip
SIDE SETBACK = 3- 7-11
END SETBACK = 8- 0- 0
CORNER FRAMING TYPE: CONVENTIONAL
END JACK TYPE: PARTIAL
APPLIED TO FRONT SIDE
- ADDT'L LOADS BASED ON 60 % OF GSL.

THIS TRUSS IS DESIGNED FOR RESIDENTIAL OR SMALL BUILDING REQUIREMENTS OF PART 9, NBCC 1995 WHERE THE SPECIFIED GROUND SNOW LOAD IS 31.0 P.S.F. OR LESS AND SPECIFIED RAIN LOAD DOES NOT EXCEED 8.4 P.S.F.

THIS DESIGN COMPLIES WITH:
- CSA 086.1-94 (LSD)
- TPIC 1996 (LSD)
- C.C.M.C. ACCEPTANCE NO.:
11996-L(M-20)
10319-L(GNA-16)

60 % OF 31.0 P.S.F. G.S.L. PLUS
8.4 P.S.F. RAIN LOAD EQUALS
27.0 P.S.F. SPECIFIED ROOF LIVE LOAD

ALLOWABLE DEFL. = L/360
CALCULATED VERT. DEFL.(TL) = L/ 935

CSI: TC=0.72 (7-8:1) , BC=0.90 (15-18:1) , WB=0.47 (9-14:1) , SSI=0.25 (13-14:1)

DOL LUMBER= 0% NAIL= 0% LS BEND= 0% COMP= 0% SHEAR= 0% TENS= 0%

TRUSS PLATE MANUFACTURER IS NOT RESPONSIBLE FOR QUALITY CONTROL IN THE TRUSS MANUFACTURING PLANT .

NAIL VALUES

PLATE	GRIP(DRY) (PSI) MAX	MIN	SHEAR (PLI) MAX	MIN	SECTION (PLI) MAX	MIN
M-20	627	380	2010	857	2478	1702

PLATE PLACEMENT TOL. = 0.25 inches

DO NOT USE THIS DESIGN IF THE DESIGN CRITERIA LISTED ABOVE DO NOT MEET PROJECT OR LOCAL BUILDING CODE REQUIREMENTS.

WHEN THIS DESIGN IS SEALED AND SIGNED, MITEK CANADA INC.APPROVES ONLY THE STRUCTURAL DESIGN OF THE TRUSS AS AN INDIVIDUAL COMPONENT BASED ON THE DATA PROVIDED BY THE CUSTOMER AND SHOWN ON THIS DRAWING.

SEE REVERSE SIDE FOR MITEK CANADA, INC GENERAL SPECIFICATIONS

® MiTek Canada, Inc.
100 Industrial Rd., P.O. Box 1329
Bradford, Ontario, L3Z 2B7

GANG-NAIL®

FIGURE 9.20 (*continued*)

·B NAME	TRUSS NAME	QUANTITY	PLY	JOB DESC.	DRWG NO.
·29B_ENG	T1	1	1	TRUSS DESC.	**H-03047**

·GHES ROOF TRUSS CO. LTD.

Version 4.100 S Mar 2 2000 MiTek Industries, Inc. Wed Mar 29 10:18:56 2000 Page 1 of 1

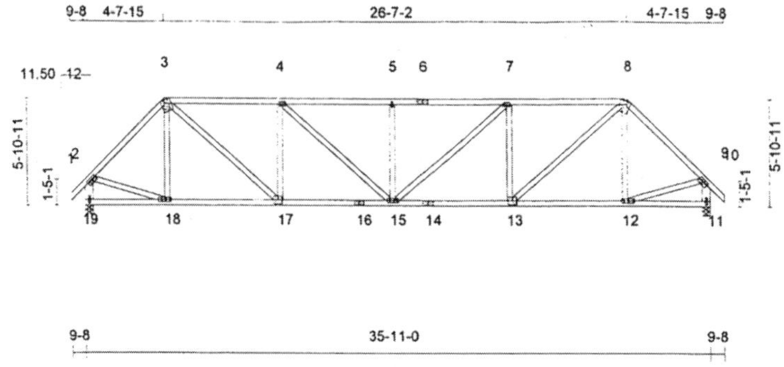

TOTAL WEIGHT = 214 lbs

·UMBER

·L. G. A. RULES

·HORDS	SIZE	LUMBER	DESCR.
· -3	2 X 4	No.2	SPF
· -6	2 X 4	No.2	SPF
· -8	2 X 4	No.2	SPF
· -10	2 X 4	No.2	SPF
·9 - 2	2 X 6	No.2	SPF
·1 - 9	2 X 6	No.2	SPF
·9 - 16	2 X 4	No.2	SPF
·6 - 14	2 X 4	No.2	SPF
·4 - 11	2 X 4	No.2	SPF
·LL WEBS	2 X 4	No.2	SPF

·LATES (table is in inches)

·	TYPE	PLATES	W	LEN	Y	X
	CMVW-t	M-20	4.0	7.0	1.50	2.00
	CCWW+m	M-20	7.0	8.0	Edge	2.25
	CMWW-t	M-20	3.0	5.0		
	CMW+w	M-20	1.5	4.0		
	CS-t	M-20	3.0	7.0		
	CMWW-t	M-20	3.0	5.0		
	CCWW+m	M-20	7.0	8.0	Edge	2.25
	CMVW-t	M-20	4.0	7.0	1.50	2.00
·1	BMV+p	M-20	2.0	5.0		
·2	CMWW-t	M-20	3.0	9.0		
·3	CMWW-t	M-20	5.0	6.0		
·4	CS-t	M-20	3.0	7.0		
·5	CMWWW-t	M-20	3.0	8.0		
·6	CS-t	M-20	3.0	7.0		
·7	CMWW-t	M-20	5.0	6.0		
·8	CMWW-t	M-20	3.0	9.0		
·9	BMV+p	M-20	2.0	5.0		

·dge - INDICATES REFERENCE CORNER OF PLATE TOUCHES ·DGE OF CHORD.

DIMENSIONS, SUPPORTS AND LOADINGS SPECIFIED BY FABRICATOR TO BE VERIFIED BY BUILDING DESIGNER

PROVIDE ADEQUATE DRAINAGE TO PREVENT PONDING

BEARINGS

JT	FACTORED GROSS REACTION VERT	FACTORED GROSS REACTION HORZ	MAX. FACTORED GROSS REACTION VERT	MAX. FACTORED GROSS REACTION HORZ	INPUT BRG IN-SX	REQRD BRG IN-SX	HEEL WEDGE
19	2144	-0	2144	0	5- 8	2- 5	
11	2144	0	2144	0	5- 8	2- 5	

MAX. UNFACTORED GROSS REACTIONS

JT	VERTICAL LIVE	VERTICAL DEAD	HORIZONTAL LIVE	HORIZONTAL DEAD
19	1126	365	0	0
11	1126	365	0	0

BEARING MATERIAL TO BE OF THE SAME SPECIES AS CHORD MEMBER AND OF GRADE NO. 2 OR BETTER.

BRACING

TOP CHORD TO BE SHEATHED OR MAX. PURLIN SPACING = 3.23FT.
MAX. UNBRACED BOTTOM CHORD LENGTH = 10.00FT.

LOADING (FACTORED)

LOAD CASE (1) OF (2)

CHORDS MEMB. FR-TO	FACT FORCE (LBS)	FACTORED VERT. LOAD (PLF) FROM	FACTORED VERT. LOAD (PLF) TO	FACTORED HORIZ. LOAD (PLF) FROM	FACTORED HORIZ. LOAD (PLF) TO	MAX UNBRAC LENGTH	WEBS MEMB. FR-TO	FACTORED FORCE (LBS)
1 - 2	27T	-88.50	-88.50	0.00	0.00	10.00	18 - 3	171C
2 - 3	2123C	-88.50	-88.50	0.00	0.00	4.38	12 - 8	171C
3 - 4	2960C	-88.50	-88.50	0.00	0.00	3.44	17 - 4	1014C
4 - 5	3376C	-88.50	-88.50	0.00	0.00	3.23	15 - 5	535C
5 - 6	3376C	-88.50	-88.50	0.00	0.00	3.23	13 - 7	1014C
6 - 7	3376C	-88.50	-88.50	0.00	0.00	3.23	3 - 17	1865T
7 - 8	2960C	-88.50	-88.50	0.00	0.00	3.44	4 - 15	545T
8 - 9	2123C	-88.50	-88.50	0.00	0.00	4.38	15 - 7	545T
9 - 10	27T	-88.50	-88.50	0.00	0.00	10.00	13 - 8	1865T
19 - 2	2098C	0.00	0.00	0.00	0.00	7.00	2 - 18	1565T
11 - 9	2098C	0.00	0.00	0.00	0.00	7.00	12 - 9	1565T
19 - 18	0C	-26.50	-26.50	0.00	0.00	10.00		
18 - 17	1523T	-26.50	-26.50	0.00	0.00	10.00		
17 - 16	2960T	-26.50	-26.50	0.00	0.00	10.00		
16 - 15	2960T	-26.50	-26.50	0.00	0.00	10.00		
15 - 14	2960T	-26.50	-26.50	0.00	0.00	10.00		
14 - 13	2960T	-26.50	-26.50	0.00	0.00	10.00		
13 - 12	1523T	-26.50	-26.50	0.00	0.00	10.00		
12 - 11	0C	-26.50	-26.50	0.00	0.00	10.00		

[M][F]

DESIGN CRITERIA

UNFACTORED LOADS:

TOP CH.	LL	=	27.0	PSF
	DL	=	3.0	PSF
BOT CH.	LL	=	3.0	PSF
	DL	=	7.0	PSF
TOTAL LOAD		=	40.0	PSF

SPACING = 24.0 IN. C/C

THIS TRUSS IS DESIGNED FOR RESIDENTIAL OR SMALL BUILDING REQUIREMENTS OF PART 9, NBCC 1995 WHERE THE SPECIFIED GROUND SNOW LOAD IS 31.0 P.S.F. OR LESS AND SPECIFIED RAIN LOAD DOES NOT EXCEED 8.4 P.S.F.

THIS DESIGN COMPLIES WITH:
- CSA 086.1-94 (LSD)
- TPIC 1996 (LSD)
- C.C.M.C. ACCEPTANCE NO.:
  11996-L(M-20)
  10319-L(GNA-16)

60 % OF 31.0 P.S.F. G.S.L. PLUS 8.4 P.S.F. RAIN LOAD EQUALS 27.0 P.S.F. SPECIFIED ROOF LIVE LOAD

ALLOWABLE DEFL. = L/360
CALCULATED VERT. DEFL.(TL) = L/ 999

CSI: TC=0.69 (5-7:1) , BC=0.59 (13-15:1) , WB=0.39 (7-13:1) , SSI=0.42 (3-4:1)

DOL LUMBER= 0% NAIL= 0% LS BEND=10% COMP=10% SHEAR=10% TENS= 10%

TRUSS PLATE MANUFACTURER IS NOT RESPONSIBLE FOR QUALITY CONTROL IN THE TRUSS MANUFACTURING PLANT .

NAIL VALUES

PLATE	GRIP(DRY) (PSI) MAX	GRIP(DRY) (PSI) MIN	SHEAR (PLI) MAX	SHEAR (PLI) MIN	SECTION (PLI) MAX	SECTION (PLI) MIN
M-20	627	380	2010	857	2478	1702

PLATE PLACEMENT TOL. = 0.25 inches

DO NOT USE THIS DESIGN IF THE DESIGN CRITERIA LISTED ABOVE DO NOT MEET PROJECT OR LOCAL BUILDING CODE REQUIREMENTS.

WHEN THIS DESIGN IS SEALED AND SIGNED, MITEK CANADA INC.APPROVES ONLY THE STRUCTURAL DESIGN OF THE TRUSS AS AN INDIVIDUAL COMPONENT BASED ON THE DATA PROVIDED BY THE CUSTOMER AND SHOWN ON THIS DRAWING.

SEE REVERSE SIDE FOR MITEK CANADA, INC GENERAL SPECIFICATIONS

ROOF TRUSSES
FABRICATED BY
HUGHES ROOF TRUSS CO. LTD.
AJAX: (905) 683-1931 TOR: (416) 213-5665

MiTek Canada, Inc.
100 Industrial Rd., P.O. Box 1329
Bradford, Ontario, L3Z 2B7

GANG-NAIL®

FIGURE 9.20 (*continued*)

B NAME	TRUSS NAME	QUANTITY	PLY	JOB DESC.	DRWG NO.
:98_ENG	T2	1	1	TRUSS DESC.	**H-03048**

GHES ROOF TRUSS CO. LTD.

Version 4.100 S Mar 2 2000 MiTek Industries, Inc. Wed Mar 29 10:19:49 2000 Page 1 of 1

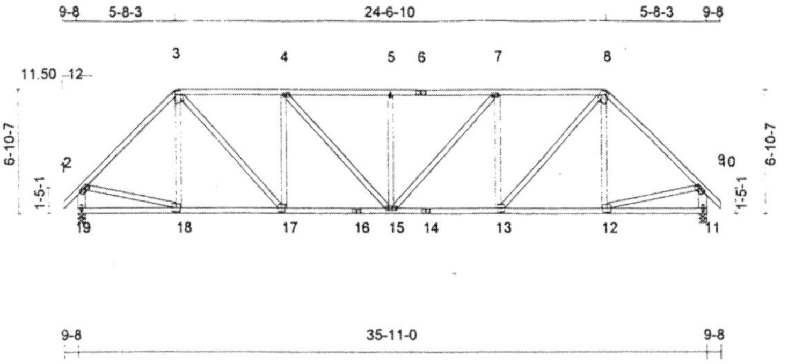

TOTAL WEIGHT = 227 lbs

JMBER

L. G. A. RULES

HORDS	SIZE	LUMBER	DESCR.
- 3	2 X 4	No.2	SPF
- 6	2 X 4	No.2	SPF
- 8	2 X 4	No.2	SPF
- 10	2 X 4	No.2	SPF
9 - 2	2 X 6	No.2	SPF
1 - 9	2 X 6	No.2	SPF
9 - 16	2 X 4	No.2	SPF
5 - 14	2 X 4	No.2	SPF
4 - 11	2 X 4	No.2	SPF
LL WEBS	2 X 4	No.2	SPF

LATES (table is in inches)

TYPE	PLATES	W	LEN	Y	X
CMVW-t	M-20	4.0	7.0	1.50	2.00
CCWW+m	M-20	6.0	8.0	Edge	1.75
CMWW-t	M-20	3.0	5.0		
CMW+w	M-20	1.5	4.0		
CS-t	M-20	3.0	7.0		
CMWW-t	M-20	3.0	5.0		
CCWW+m	M-20	6.0	8.0	Edge	1.75
CMVW-t	M-20	4.0	7.0	1.50	2.00
BMV+p	M-20	2.0	5.0		
CMWW-t	M-20	5.0	6.0		
CMWW+t	M-20	5.0	5.0		
CS-t	M-20	3.0	6.0		
CMWWW-t	M-20	3.0	8.0		
CS-t	M-20	3.0	6.0		
CMWW+t	M-20	5.0	5.0		
CMWW-t	M-20	5.0	6.0		
BMV+p	M-20	2.0	5.0		

)ge - INDICATES REFERENCE CORNER OF PLATE TOUCHES)GE OF CHORD.

DIMENSIONS, SUPPORTS AND LOADINGS SPECIFIED BY FABRICATOR TO BE VERIFIED BY BUILDING DESIGNER

PROVIDE ADEQUATE DRAINAGE TO PREVENT PONDING

BEARINGS

JT	FACTORED GROSS REACTION VERT	HORZ	MAX. FACTORED GROSS REACTION VERT	HORZ	INPUT BRG IN-SX	REQRD BRG IN-SX	HEEL WEDGE
19	2144	0	2144	0	5- 8	2- 5	
11	2144	0	2144	0	5- 8	2- 5	

MAX. UNFACTORED GROSS REACTIONS

JT	VERTICAL LIVE	DEAD	HORIZONTAL LIVE	DEAD
19	1126	365	0	0
11	1126	365	0	0

BEARING MATERIAL TO BE OF THE SAME SPECIES AS CHORD MEMBER AND OF GRADE NO. 2 OR BETTER.

BRACING

TOP CHORD TO BE SHEATHED OR MAX. PURLIN SPACING = 3.65FT.
MAX. UNBRACED BOTTOM CHORD LENGTH = 10.00FT.

LOADING (FACTORED)

LOAD CASE (1) OF (2)

CHORDS MEMB. FR-TO	FACT FORCE (LBS)	FACTORED VERT. LOAD (PLF) FROM	TO	FACTORED HORIZ. LOAD (PLF) FROM	TO	MAX UNBRAC LENGTH	WEBS MEMB. FR-TO	FACTORED FORCE (LBS)
1 - 2	27T	-88.50	-88.50	0.00	0.00	10.00	18 - 3	103C
2 - 3	2122C	-88.50	-88.50	0.00	0.00	4.22	12 - 8	103C
3 - 4	2564C	-88.50	-88.50	0.00	0.00	3.83	17 - 4	937C
4 - 5	2866C	-88.50	-88.50	0.00	0.00	3.65	15 - 5	493C
5 - 6	2866C	-88.50	-88.50	0.00	0.00	3.65	13 - 7	937C
6 - 7	2866C	-88.50	-88.50	0.00	0.00	3.65	3 - 17	1506T
7 - 8	2564C	-88.50	-88.50	0.00	0.00	3.83	4 - 15	445T
8 - 9	2122C	-88.50	-88.50	0.00	0.00	4.22	15 - 7	445T
9 - 10	27T	-88.50	-88.50	0.00	0.00	10.00	13 - 8	1506T
19 - 2	2083C	0.00	0.00	0.00	0.00	7.03	2 - 18	1554T
11 - 9	2083C	0.00	0.00	0.00	0.00	7.03	12 - 9	1554T
19 - 18	0C	-26.50	-26.50	0.00	0.00	10.00		
18 - 17	1526T	-26.50	-26.50	0.00	0.00	10.00		
17 - 16	2564T	-26.50	-26.50	0.00	0.00	10.00		
16 - 15	2564T	-26.50	-26.50	0.00	0.00	10.00		
15 - 14	2564T	-26.50	-26.50	0.00	0.00	10.00		
14 - 13	2564T	-26.50	-26.50	0.00	0.00	10.00		
13 - 12	1526T	-26.50	-26.50	0.00	0.00	10.00		
12 - 11	0C	-26.50	-26.50	0.00	0.00	10.00		

[M][F]

DESIGN CRITERIA

UNFACTORED LOADS:

TOP CH.	LL =	27.0	PSF
	DL =	3.0	PSF
BOT CH.	LL =	3.0	PSF
	DL =	7.0	PSF
TOTAL LOAD	=	40.0	PSF

SPACING = 24.0 IN. C/C

THIS TRUSS IS DESIGNED FOR RESIDENTIAL OR SMALL BUILDING REQUIREMENTS OF PART 9, NBCC 1995 WHERE THE SPECIFIED GROUND SNOW LOAD IS 31.0 P.S.F. OR LESS AND SPECIFIED RAIN LOAD DOES NOT EXCEED 8.4 P.S.F.

THIS DESIGN COMPLIES WITH:
- CSA 086.1-94 (LSD)
- TPIC 1996 (LSD)
- C.C.M.C. ACCEPTANCE NO.:
  11996-L(M-20)
  10319-L(GNA-16)

80 % OF 31.0 P.S.F. G.S.L. PLUS 8.4 P.S.F. RAIN LOAD EQUALS 27.0 P.S.F. SPECIFIED ROOF LIVE LOAD

ALLOWABLE DEFL. = L/360
CALCULATED VERT. DEFL.(TL) = L/ 999

CSI: TC=0.53 (4-5:1) , BC=0.51 (15-17:1) , WB=0.53 (4-17:1) , SSI=0.39 (7-8:1)

DOL LUMBER= 0% NAIL= 0% LS BEND=10% COMP=10% SHEAR=10% TENS= 10%

TRUSS PLATE MANUFACTURER IS NOT RESPONSIBLE FOR QUALITY CONTROL IN THE TRUSS MANUFACTURING PLANT .

NAIL VALUES

PLATE	GRIP(DRY) (PSI) MAX	MIN	SHEAR (PLI) MAX	MIN	SECTION (PLI) MAX	MIN
M-20	627	380	2010	857	2478	1702

PLATE PLACEMENT TOL. = 0.25 inches

DO NOT USE THIS DESIGN IF THE DESIGN CRITERIA LISTED ABOVE DO NOT MEET PROJECT OR LOCAL BUILDING CODE REQUIREMENTS.

WHEN THIS DESIGN IS SEALED AND SIGNED, MITEK CANADA INC.APPROVES ONLY THE STRUCTURAL DESIGN OF THE TRUSS AS AN INDIVIDUAL COMPONENT BASED ON THE DATA PROVIDED BY THE CUSTOMER AND SHOWN ON THIS DRAWING.

SEE REVERSE SIDE FOR MITEK CANADA, INC GENERAL SPECIFICATIONS

ROOF TRUSSES
FABRICATED BY
HUGHES ROOF TRUSS CO. LTD.
AJAX: (905) 683-1931 TOR: (416) 213-5665

MiTek Canada, Inc.
100 Industrial Rd., P.O. Box 1329
Bradford, Ontario, L3Z 2B7

GANG-NAIL®

FIGURE 9.20 (*continued*)

B NAME	TRUSS NAME	QUANTITY	PLY	JOB DESC.	DRWG NO. H-03049
29B_ENG	T3	1	1	TRUSS DESC.	

GHES ROOF TRUSS CO. LTD.

Version 4.100 S Mar 2 2000 MiTek Industries, Inc. Wed Mar 29 10:20:38 2000 Page 1 of 1

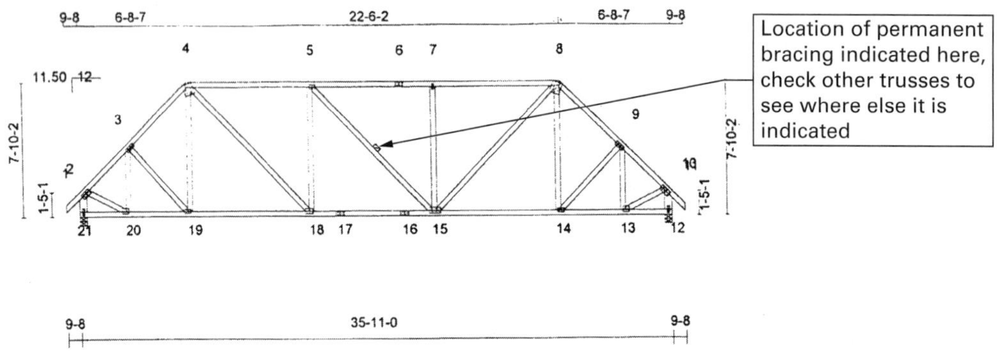

TOTAL WEIGHT = 237 lbs

UMBER

L. G. A. RULES

HORDS	SIZE	LUMBER	DESCR.
- 4	2 X 4	No.2	SPF
- 6	2 X 4	No.2	SPF
- 8	2 X 4	No.2	SPF
- 11	2 X 4	No.2	SPF
1 - 2	2 X 6	No.2	SPF
2 - 10	2 X 6	No.2	SPF
1 - 17	2 X 4	No.2	SPF
7 - 16	2 X 4	No.2	SPF
6 - 12	2 X 4	No.2	SPF
LL WEBS	2 X 4	No.2	SPF

LATES (table is in inches)

TYPE	PLATES	W	LEN	Y	X
CMVW-t	M-20	4.0	7.0	1.50	2.00
CMWW-t	M-20	3.0	7.0		
CCWW+m	M-20	7.0	8.0	Edge	2.25
CMWW-t	M-20	3.0	5.0		
CS-t	M-20	3.0	5.0		
CMW+w	M-20	1.5	4.0		
CCWW+m	M-20	7.0	8.0	Edge	2.25
CMWW-t	M-20	3.0	7.0		
CMVW-t	M-20	4.0	8.0	1.75	3.00
BMV+p	M-20	2.0	5.0		
CMWW-t	M-20	4.0	5.0	1.75	2.00
CMWW-t	M-20	3.0	5.0		
CMWWW-t	M-20	4.0	8.0	1.75	2.75
CS-t	M-20	3.0	6.0		
CS-t	M-20	3.0	6.0		
CMWW-t	M-20	4.0	6.0		
CMWW-t	M-20	3.0	5.0		
CMWW-t	M-20	4.0	5.0	1.50	2.00
BMV+p	M-20	2.0	5.0		

dge - INDICATES REFERENCE CORNER OF PLATE TOUCHES DGE OF CHORD.

DIMENSIONS, SUPPORTS AND LOADINGS SPECIFIED BY FABRICATOR TO BE VERIFIED BY BUILDING DESIGNER

PROVIDE ADEQUATE DRAINAGE TO PREVENT PONDING

BEARINGS

JT	FACTORED GROSS REACTION VERT	HORZ	MAX. FACTORED GROSS REACTION VERT	HORZ	INPUT BRG IN-SX	REQRD BRG IN-SX	HEEL WEDGE
21	2144	0	2144	0	5- 8	2- 5	
12	2144	-0	2144	0	5- 8	2- 5	

MAX. UNFACTORED GROSS REACTIONS

JT	VERTICAL LIVE	DEAD	HORIZONTAL LIVE	DEAD
21	1126	365	0	0
12	1126	365	0	0

BEARING MATERIAL TO BE OF THE SAME SPECIES AS CHORD MEMBER AND OF GRADE NO. 2 OR BETTER.

BRACING

TOP CHORD TO BE SHEATHED OR MAX. PURLIN SPACING = 3.56FT.
MAX. UNBRACED BOTTOM CHORD LENGTH = 10.00FT.
1-1X4 LATERAL BRACE REQUIRED AT 1/ 2LENGTH OF 5-15,

END VERTICAL(S) MUST BE SHEATHED OR HAVE BRACES AS INDICATED IN THE MAX. UNBRACED LENGTH COLUMN OF THE TABLE BELOW

LOADING (FACTORED)
LOAD CASE (1) OF (2)

CHORDS MEMB. FR-TO	FACT FORCE (LBS)	FACTORED VERT. LOAD (PLF) FROM	TO	FACTORED HORIZ. LOAD (PLF) FROM	TO	MAX UNBRAC LENGTH	WEBS MEMB. FR-TO	FACTORED FORCE (LBS)
1 - 2	27T	-88.50	-88.50	0.00	0.00	10.00	19 - 4	128T
2 - 3	2005C	-88.50	-88.50	0.00	0.00	4.64	14 - 8	129T
3 - 4	2120C	-88.50	-88.50	0.00	0.00	4.52	20 - 3	461C
4 - 5	2432C	-88.50	-88.50	0.00	0.00	3.56	18 - 5	717C
5 - 6	2432C	-88.50	-88.50	0.00	0.00	3.72	15 - 7	717C
6 - 7	2432C	-88.50	-88.50	0.00	0.00	3.72	13 - 9	461C
7 - 8	2432C	-88.50	-88.50	0.00	0.00	3.56	3 - 19	60T
8 - 9	2120C	-88.50	-88.50	0.00	0.00	4.52	4 - 18	1300T
9 - 10	2005C	-88.50	-88.50	0.00	0.00	4.64	5 - 15	1C
10 - 11	27T	-88.50	-88.50	0.00	0.00	10.00	15 - 8	1298T
21 - 2	2104C	0.00	0.00	0.00	0.00	7.00	14 - 9	60T
12 - 10	2104C	0.00	0.00	0.00	0.00	7.00	2 - 20	1559T
							13 - 10	1558T
21 - 20	0C	-26.50	-26.50	0.00	0.00	10.00		
20 - 19	1467T	-26.50	-26.50	0.00	0.00	10.00		
19 - 18	1512T	-26.50	-26.50	0.00	0.00	10.00		
18 - 17	2433T	-26.50	-26.50	0.00	0.00	10.00		
17 - 16	2433T	-26.50	-26.50	0.00	0.00	10.00		
16 - 15	2433T	-26.50	-26.50	0.00	0.00	10.00		
15 - 14	1512T	-26.50	-26.50	0.00	0.00	10.00		
14 - 13	1467T	-26.50	-26.50	0.00	0.00	10.00		
13 - 12	0C	-26.50	-26.50	0.00	0.00	10.00		

ROOF TRUSSES
FABRICATED BY
HUGHES ROOF TRUSS CO. LTD.
AJAX: (905) 683-1931 TOR: (416) 213-5665

MiTek Canada, Inc.
100 Industrial Rd., P.O. Box 1329
Bradford, Ontario, L3Z 2B7
GANG-NAIL®

[M][F]

DESIGN CRITERIA

UNFACTORED LOADS:

TOP CH.	LL =	27.0	PSF
	DL =	3.0	PSF
BOT CH.	LL =	3.0	PSF
	DL =	7.0	PSF
TOTAL LOAD	=	40.0	PSF

SPACING = 24.0 IN. C/C

THIS TRUSS IS DESIGNED FOR RESIDENTIAL OR SMALL BUILDING REQUIREMENTS OF PART 9, NBCC 1995 WHERE THE SPECIFIED GROUND SNOW LOAD IS 31.0 P.S.F. OR LESS AND SPECIFIED RAIN LOAD DOES NOT EXCEED 8.4 P.S.F.

THIS DESIGN COMPLIES WITH:
- CSA 086.1-94 (LSD)
- TPIC 1996 (LSD)
- C.C.M.C. ACCEPTANCE NO.:
11996-L(M-20)
10319-L(GNA-16)

60 % OF 31.0 P.S.F. G.S.L. PLUS 8.4 P.S.F. RAIN LOAD EQUALS 27.0 P.S.F. SPECIFIED ROOF LIVE LOAD

ALLOWABLE DEFL. = L/360
CALCULATED VERT. DEFL.(TL) = L/ 999

CSI: TC=0.76 (4-5:1) , BC=0.55 (15-18:1) , WB=0.59 (5-18:1) , SSI=0.47 (4-5:1)

DOL LUMBER= 0% NAIL= 0% LS BEND=10% COMP=10% SHEAR=10% TENS= 10%

TRUSS PLATE MANUFACTURER IS NOT RESPONSIBLE FOR QUALITY CONTROL IN THE TRUSS MANUFACTURING PLANT .

NAIL VALUES

PLATE	GRIP(DRY) (PSI) MAX	MIN	SHEAR (PLI) MAX	MIN	SECTION (PLI) MAX	MIN
M-20	627	380	2010	857	2478	1702

PLATE PLACEMENT TOL. = 0.25 inches

DO NOT USE THIS DESIGN IF THE DESIGN CRITERIA LISTED ABOVE DO NOT MEET PROJECT OR LOCAL BUILDING CODE REQUIREMENTS.

WHEN THIS DESIGN IS SEALED AND SIGNED, MITEK CANADA INC.APPROVES ONLY THE STRUCTURAL DESIGN OF THE TRUSS AS AN INDIVIDUAL COMPONENT BASED ON THE DATA PROVIDED BY THE CUSTOMER AND SHOWN ON THIS DRAWING.

SEE REVERSE SIDE FOR MITEK CANADA, INC GENERAL SPECIFICATIONS

FIGURE 9.20 (*continued*)

B NAME	TRUSS NAME	QUANTITY	PLY	JOB DESC.	DRWG NO	H-03050
98_ENG	T4	1	1	TRUSS DESC.		

GHES ROOF TRUSS CO. LTD.

Version 4.100 S Mar 2 2000 MiTek Industries, Inc. Wed Mar 29 10:21:15 2000 Page 1 of 1

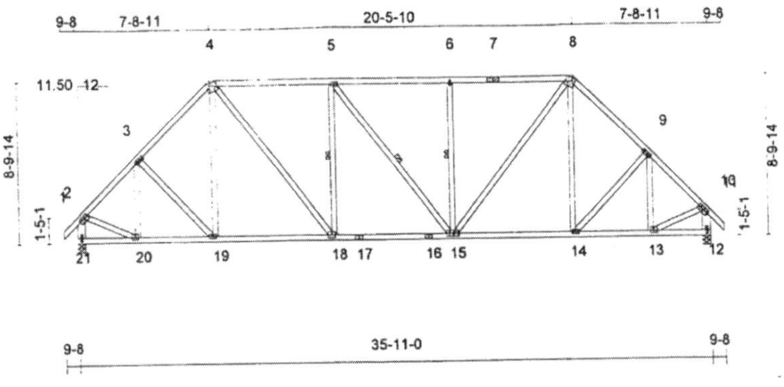

TOTAL WEIGHT = 249 lbs

UMBER

G. A. RULES

HORDS	SIZE	LUMBER	DESCR.
- 4	2 X 4	No.2	SPF
- 7	2 X 4	No.2	SPF
- 8	2 X 4	No.2	SPF
- 11	2 X 4	No.2	SPF
1 - 2	2 X 6	No.2	SPF
2 - 10	2 X 6	No.2	SPF
1 - 17	2 X 4	No.2	SPF
7 - 16	2 X 4	No.2	SPF
5 - 12	2 X 4	No.2	SPF
LL WEBS	2 X 4	No.2	SPF

LATES (table is in inches)

TYPE	PLATES	W	LEN	Y	X
CMVW-t	M-20	4.0	7.0	1.50	2.00
CMWW-t	M-20	3.0	7.0		
CCWW-m	M-20	6.0	7.0	1.75	2.25
CMWW-t	M-20	3.0	5.0		
CMW+w	M-20	1.5	4.0		
CS-t	M-20	3.0	8.0		
CCWW-m	M-20	6.0	7.0	1.75	2.00
CMWW-t	M-20	3.0	7.0		
CMVW-t	M-20	4.0	8.0	1.75	3.00
BMV+p	M-20	2.0	5.0		
CMWW-t	M-20	4.0	5.0	1.75	2.00
CMWW-t	M-20	3.0	5.0		
CMWWW-t	M-20	4.0	9.0		
CS-t	M-20	3.0	5.0		
CS-t	M-20	3.0	5.0		
CMWW-t	M-20	4.0	5.0		
CMWW-t	M-20	3.0	5.0		
CMWW-t	M-20	4.0	5.0	1.50	2.00
BMV+p	M-20	2.0	5.0		

DIMENSIONS, SUPPORTS AND LOADINGS SPECIFIED BY FABRICATOR TO BE VERIFIED BY BUILDING DESIGNER

PROVIDE ADEQUATE DRAINAGE TO PREVENT PONDING

BEARINGS

	FACTORED GROSS REACTION		MAX. FACTORED GROSS REACTION		INPUT BRG	REQRD BRG	HEEL WEDGE
JT	VERT	HORZ	VERT	HORZ	IN-SX	IN-SX	
21	2144	0	2144	0	5- 8	2- 5	
12	2144	0	2144	0	5- 8	2- 5	

MAX. UNFACTORED GROSS REACTIONS

	VERTICAL		HORIZONTAL	
JT	LIVE	DEAD	LIVE	DEAD
21	1126	365	0	0
12	1126	365	0	0

BEARING MATERIAL TO BE OF THE SAME SPECIES AS CHORD MEMBER AND OF GRADE NO. 2 OR BETTER.

BRACING

TOP CHORD TO BE SHEATHED OR MAX. PURLIN SPACING = 4.00FT.
MAX. UNBRACED BOTTOM CHORD LENGTH = 10.00FT.
1-1X4 LATERAL BRACE REQUIRED AT 1/ 2LENGTH OF 18-5, 16-6, 5-15,

END VERTICAL(S) MUST BE SHEATHED OR HAVE BRACES AS INDICATED IN THE MAX. UNBRACED LENGTH COLUMN OF THE TABLE BELOW

LOADING (FACTORED)

LOAD CASE (1) OF (2)

CHORDS

MEMB. FR-TO	FACT FORCE (LBS)	FACTORED VERT. LOAD (PLF) FROM	TO	FACTORED HORIZ. LOAD (PLF) FROM	TO	MAX UNBRAC LENGTH
1 - 2	27T	-88.50	-88.50	0.00	0.00	10.00
2 - 3	2065C	-88.50	-88.50	0.00	0.00	4.53
3 - 4	2084C	-88.50	-88.50	0.00	0.00	4.50
4 - 5	2157C	-88.50	-88.50	0.00	0.00	4.00
5 - 6	2155C	-88.50	-88.50	0.00	0.00	4.10
6 - 7	2156C	-88.50	-88.50	0.00	0.00	4.03
7 - 8	2156C	-88.50	-88.50	0.00	0.00	4.03
8 - 9	2084C	-88.50	-88.50	0.00	0.00	4.50
9 - 10	2065C	-88.50	-88.50	0.00	0.00	4.53
10 - 11	27T	-88.50	-88.50	0.00	0.00	10.00
21 - 2	2103C	0.00	0.00	0.00	0.00	7.00
12 - 10	2102C	0.00	0.00	0.00	0.00	7.00
21 - 20	0C	-26.50	-26.50	0.00	0.00	10.00
20 - 19	1514T	-26.50	-26.50	0.00	0.00	10.00
19 - 18	1484T	-26.50	-26.50	0.00	0.00	10.00
18 - 17	2158T	-26.50	-26.50	0.00	0.00	10.00
17 - 16	2158T	-26.50	-26.50	0.00	0.00	10.00
16 - 15	2158T	-26.50	-26.50	0.00	0.00	10.00
15 - 14	1484T	-26.50	-26.50	0.00	0.00	10.00
14 - 13	1514T	-26.50	-26.50	0.00	0.00	10.00
13 - 12	0C	-26.50	-26.50	0.00	0.00	10.00

WEBS

MEMB. FR-TO	FACTORED FORCE (LBS)
19 - 4	192T
14 - 8	193T
20 - 3	369C
18 - 5	651C
15 - 6	647C
13 - 9	370C
3 - 19	47C
4 - 18	1071T
5 - 15	4C
15 - 8	1067T
14 - 9	46C
2 - 20	1583T
13 - 10	1583T

[M][F]

DESIGN CRITERIA

UNFACTORED LOADS:

TOP	CH.	LL	=	27.0	PSF
		DL	=	3.0	PSF
BOT	CH.	LL	=	3.0	PSF
		DL	=	7.0	PSF
TOTAL	LOAD		=	40.0	PSF

SPACING = 24.0 IN. C/C

THIS TRUSS IS DESIGNED FOR RESIDENTIAL OR SMALL BUILDING REQUIREMENTS OF PART 9, NBCC 1995 WHERE THE SPECIFIED GROUND SNOW LOAD IS 31.0 P.S.F. OR LESS AND SPECIFIED RAIN LOAD DOES NOT EXCEED 8.4 P.S.F.

THIS DESIGN COMPLIES WITH:
- CSA 086.1-94 (LSD)
- TPIC 1996 (LSD)
- C.C.M.C. ACCEPTANCE NO.:
  11996-L(M-20)
  10319-L(GNA-16)

60 % OF 31.0 P.S.F. G.S.L. PLUS
8.4 P.S.F. RAIN LOAD EQUALS
27.0 P.S.F. SPECIFIED ROOF LIVE LOAD

ALLOWABLE DEFL. = L/360
CALCULATED VERT. DEFL.(TL) = L/ 999

CSI: TC=0.59 (4-5:1) , BC=0.48 (15-18:1) , WB=0.25 (2-20:1) , SSI=0.42 (4-5:1)

DOL LUMBER= 0% NAIL= 0% LS BEND=10% COMP=10% SHEAR=10% TENS= 10%

TRUSS PLATE MANUFACTURER IS NOT RESPONSIBLE FOR QUALITY CONTROL IN THE TRUSS MANUFACTURING PLANT .

NAIL VALUES

	PLATE GRIP(DRY) (PSI)		SHEAR (PLI)		SECTION (PLI)	
	MAX	MIN	MAX	MIN	MAX	MIN
M-20	627	380	2010	857	2478	1702

PLATE PLACEMENT TOL. = 0.25 inches

DO NOT USE THIS DESIGN IF THE DESIGN CRITERIA LISTED ABOVE DO NOT MEET PROJECT OR LOCAL BUILDING CODE REQUIREMENTS.

WHEN THIS DESIGN IS SEALED AND SIGNED, MITEK CANADA INC.APPROVES ONLY THE STRUCTURAL DESIGN OF THE TRUSS AS AN INDIVIDUAL COMPONENT BASED ON THE DATA PROVIDED BY THE CUSTOMER AND SHOWN ON THIS DRAWING.

SEE REVERSE SIDE FOR MITEK CANADA, INC GENERAL SPECIFICATIONS

ROOF TRUSSES
FABRICATED BY
HUGHES ROOF TRUSS CO. LTD.
AJAX: (905) 683-1931 TOR: (416) 213-5665

MiTek Canada, Inc.
100 Industrial Rd., P.O. Box 1329
Bradford, Ontario, L3Z 2B7
GANG-NAIL®

FIGURE 9.20 (*continued*)

B NAME	TRUSS NAME	QUANTITY	PLY	JOB DESC.	DRWG NO.
98_ENG	T5	1	1	TRUSS DESC.	H-03051

GHES ROOF TRUSS CO. LTD.

Version 4.100 S Mar 2 2000 MiTek Industries, Inc. Wed Mar 29 10:22:05 2000 Page 1 of 1

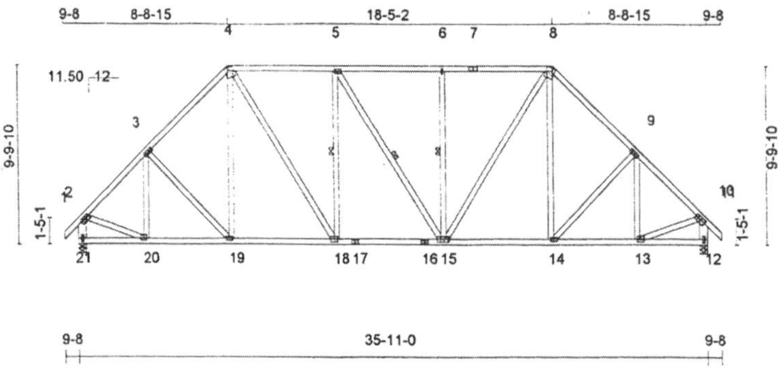

TOTAL WEIGHT = 263 lbs

MBER

L.G.A. RULES

HORDS	SIZE	LUMBER	DESCR.
- 4	2 X 4	No.2	SPF
- 7	2 X 4	No.2	SPF
- 8	2 X 4	No.2	SPF
- 11	2 X 4	No.2	SPF
- 2	2 X 6	No.2	SPF
- 10	2 X 6	No.2	SPF
- 17	2 X 4	No.2	SPF
- 16	2 X 4	No.2	SPF
- 12	2 X 4	No.2	SPF
LL WEBS	2 X 4	No.2	SPF

ATES (table is in inches)

TYPE	PLATES	W	LEN	Y	X
CMVW-t	M-20	4.0	7.0	1.50	2.00
CMWW-t	M-20	3.0	7.0		
CCWW+m	M-20	6.0	7.0	Edge	1.75
CMWW-t	M-20	3.0	5.0		
CMW+w	M-20	1.5	4.0		
CS-t	M-20	3.0	6.0		
CCWW+m	M-20	6.0	7.0	Edge	1.75
CMWW-t	M-20	3.0	7.0		
CMVW-t	M-20	4.0	6.0	1.75	3.00
BMV+p	M-20	2.0	5.0		
CMWW-t	M-20	4.0	5.0	1.75	2.00
CMWW-t	M-20	3.0	5.0		
CMWWW-t	M-20	4.0	8.0		
CS-t	M-20	3.0	5.0		
CS-t	M-20	3.0	5.0		
CMWW-t	M-20	4.0	5.0		
CMWW-t	M-20	3.0	5.0		
CMWW-t	M-20	4.0	5.0	1.50	2.00
BMV+p	M-20	2.0	5.0		

ge - INDICATES REFERENCE CORNER OF PLATE TOUCHES GE OF CHORD.

DIMENSIONS, SUPPORTS AND LOADINGS SPECIFIED BY FABRICATOR TO BE VERIFIED BY BUILDING DESIGNER

PROVIDE ADEQUATE DRAINAGE TO PREVENT PONDING

BEARINGS

	FACTORED GROSS REACTION		MAX. FACTORED GROSS REACTION		INPUT BRG	REQRD BRG	HEEL
JT	VERT	HORZ	VERT	HORZ	IN-SX	IN-SX	WEDGE
21	2144	0	2144	0	5- 8	2- 5	
12	2144	-0	2144	0	5- 8	2- 5	

MAX. UNFACTORED GROSS REACTIONS

	VERTICAL		HORIZONTAL	
JT	LIVE	DEAD	LIVE	DEAD
21	1126	365	0	0
12	1126	365	0	0

BEARING MATERIAL TO BE OF THE SAME SPECIES AS CHORD MEMBER AND OF GRADE NO. 2 OR BETTER.

BRACING
TOP CHORD TO BE SHEATHED OR MAX. PURLIN SPACING = 4.37FT.
MAX. UNBRACED BOTTOM CHORD LENGTH = 10.00FT.
1-1X4 LATERAL BRACE REQUIRED AT 1/ 2LENGTH OF 18-5, 15-6, 5-15,

END VERTICAL(S) MUST BE SHEATHED OR HAVE BRACES AS INDICATED IN THE MAX. UNBRACED LENGTH COLUMN OF THE TABLE BELOW

LOADING (FACTORED)
LOAD CASE (1) OF (2)

CHORDS MEMB. FR-TO	FACT FORCE (LBS)	FACTORED VERT. LOAD (PLF) FROM	TO	FACTORED HORIZ. LOAD (PLF) FROM	TO	MAX UNBRAC LENGTH	WEBS MEMB. FR-TO	FACTORED FORCE (LBS)
1 - 2	27T	-88.50	-88.50	0.00	0.00	10.00	19 - 4	251T
2 - 3	2106C	-88.50	-88.50	0.00	0.00	4.44	14 - 8	252T
3 - 4	2039C	-88.50	-88.50	0.00	0.00	4.48	20 - 3	294C
4 - 5	1938C	-88.50	-88.50	0.00	0.00	4.37	18 - 5	585C
5 - 6	1937C	-88.50	-88.50	0.00	0.00	4.43	15 - 6	581C
6 - 7	1937C	-88.50	-88.50	0.00	0.00	4.38	13 - 9	295C
7 - 8	1937C	-88.50	-88.50	0.00	0.00	4.38	3 - 19	141C
8 - 9	2039C	-88.50	-88.50	0.00	0.00	4.48	4 - 18	893T
9 - 10	2106C	-88.50	-88.50	0.00	0.00	4.44	5 - 15	3C
10 - 11	27T	-88.50	-88.50	0.00	0.00	10.00	15 - 8	869T
21 - 2	2100C	0.00	0.00	0.00	0.00	7.00	14 - 9	141C
12 - 10	2100C	0.00	0.00	0.00	0.00	7.00	2 - 20	1600T
							13 - 10	1600T
21 - 20	0C	-26.50	-26.50	0.00	0.00	10.00		
20 - 19	1546T	-26.50	-26.50	0.00	0.00	10.00		
19 - 18	1450T	-26.50	-26.50	0.00	0.00	10.00		
18 - 17	1938T	-26.50	-26.50	0.00	0.00	10.00		
17 - 16	1938T	-26.50	-26.50	0.00	0.00	10.00		
16 - 15	1938T	-26.50	-26.50	0.00	0.00	10.00		
15 - 14	1450T	-26.50	-26.50	0.00	0.00	10.00		
14 - 13	1546T	-26.50	-26.50	0.00	0.00	10.00		
13 - 12	0C	-26.50	-26.50	0.00	0.00	10.00		

[M][F]

DESIGN CRITERIA

UNFACTORED LOADS:

TOP CH.	LL	=	27.0	PSF
	DL	=	3.0	PSF
BOT CH.	LL	=	3.0	PSF
	DL	=	7.0	PSF
TOTAL LOAD		=	40.0	PSF

SPACING = 24.0 IN. C/C

THIS TRUSS IS DESIGNED FOR RESIDENTIAL OR SMALL BUILDING REQUIREMENTS OF PART 9, NBCC 1995 WHERE THE SPECIFIED GROUND SNOW LOAD IS 31.0 P.S.F. OR LESS AND SPECIFIED RAIN LOAD DOES NOT EXCEED 8.4 P.S.F.

THIS DESIGN COMPLIES WITH:
- CSA 086.1-94 (LSD)
- TPIC 1996 (LSD)
- C.C.M.C. ACCEPTANCE NO.:
11996-L(M-20)
10319-L(GNA-16)

60 % OF 31.0 P.S.F. G.S.L. PLUS
8.4 P.S.F. RAIN LOAD EQUALS
27.0 P.S.F. SPECIFIED ROOF LIVE LOAD

ALLOWABLE DEFL. = L/360
CALCULATED VERT. DEFL.(TL) = L/ 999

CSI: TC=0.45 (4-5:1) , BC=0.42 (15-18:1) , WB=0.27 (5-18:1) , SSI=0.38 (4-5:1)

DOL LUMBER= 0% NAIL= 0% LS BEND=10% COMP=10% SHEAR=10% TENS= 10%

TRUSS PLATE MANUFACTURER IS NOT RESPONSIBLE FOR QUALITY CONTROL IN THE TRUSS MANUFACTURING PLANT .

NAIL VALUES

	PLATE GRIP(DRY) (PSI) MAX	MIN	SHEAR (PLI) MAX	MIN	SECTION (PLI) MAX	MIN
M-20	627	380	2010	857	2478	1702

PLATE PLACEMENT TOL. = 0.25 inches

DO NOT USE THIS DESIGN IF THE DESIGN CRITERIA LISTED ABOVE DO NOT MEET PROJECT OR LOCAL BUILDING CODE REQUIREMENTS.

WHEN THIS DESIGN IS SEALED AND SIGNED, MITEK CANADA INC.APPROVES ONLY THE STRUCTURAL DESIGN OF THE TRUSS AS AN INDIVIDUAL COMPONENT BASED ON THE DATA PROVIDED BY THE CUSTOMER AND SHOWN ON THIS DRAWING.

SEE REVERSE SIDE FOR MITEK CANADA, INC GENERAL SPECIFICATIONS

ROOF TRUSSES
FABRICATED BY
HUGHES ROOF TRUSS CO. LTD.
AJAX: (905) 683-1931 TOR: (416) 213-5665

MiTek Canada, Inc.
100 Industrial Rd., P.O. Box 1329
Bradford, Ontario, L3Z 2B7

GANG-NAIL®

FIGURE 9.21

Truss assembly for a hip roof; a girder truss supports other trusses, in this case the end (common) jacks. A girder truss may be single, double, or triple ply, depending on the load it is carrying.

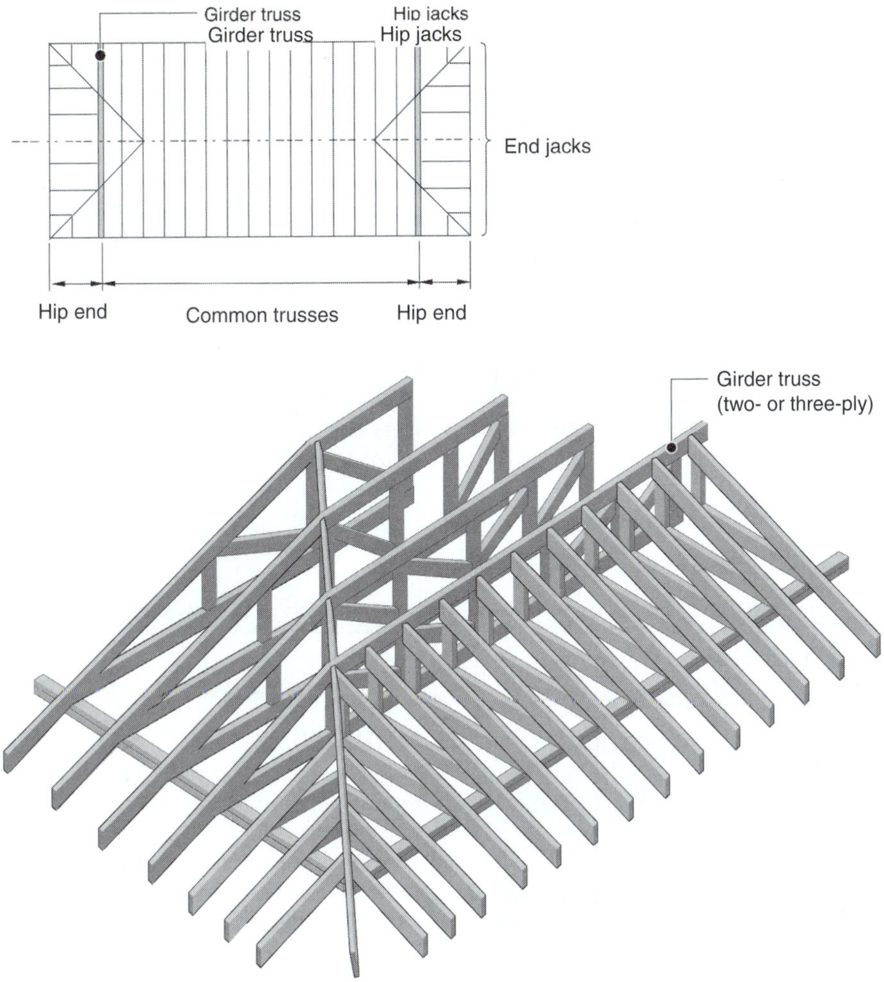

Source: Canadian Wood Council

FIGURE 9.22

Gusset plate structurally fastens truss members together.

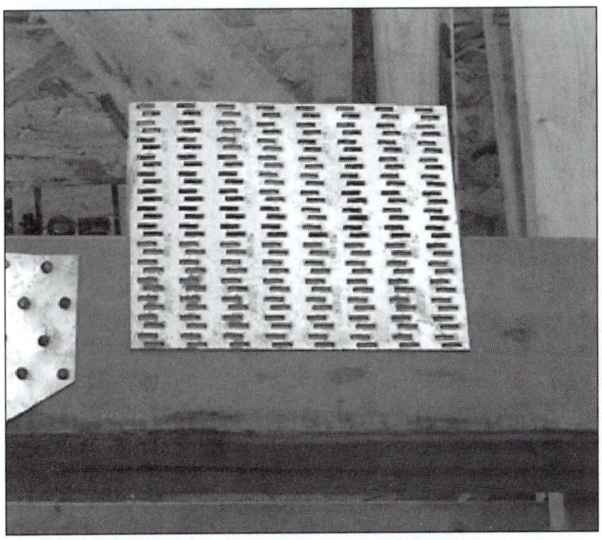

Courtesy Tom Stephenson

FIGURE 9.23

Truss supported on approved truss hanger.

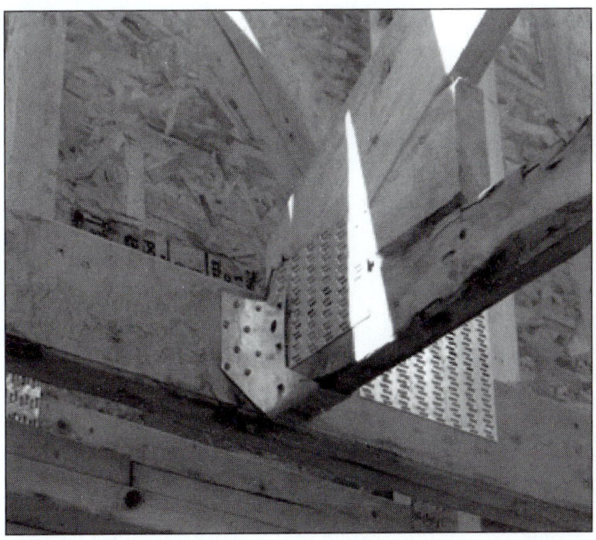

Courtesy Tom Stephenson

FIGURE 9.24

Roof trusses with lateral bracing installed.

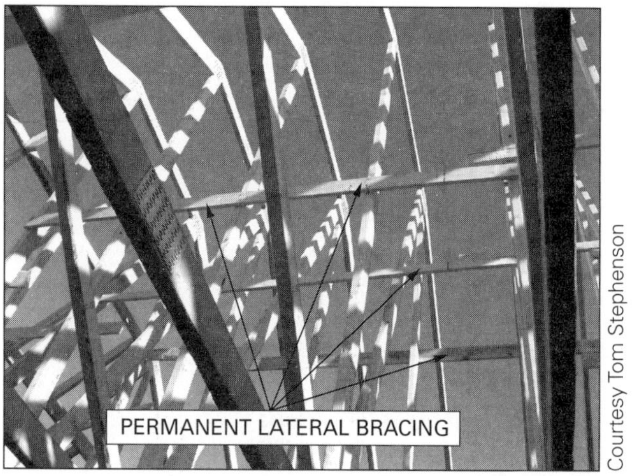

FIGURE 9.25

Raised heel roof truss. The extra height at the plate allows for extra insulation in the ceiling, limiting any cold spots around the perimeter of the house at the ceiling. New codes demanding higher levels of efficiency are making this type of truss much more prevalent.

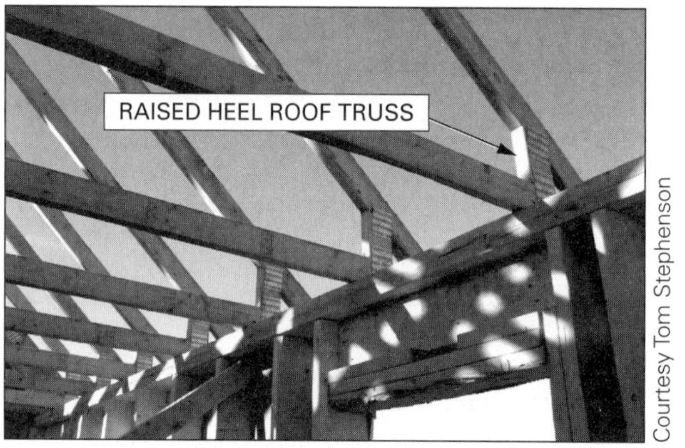

FIGURE 9.26

Girder truss on solid bearing. Notice solid bearing is continuous down to foundation wall, and indicated on the Doncaster construction drawings floor plans as SB (refer to Figure 9.28). The girder truss is placing a point load on the top plate because it is supporting a number of common jack trusses. The double studs distribute the point load down through the walls and eventually through the footings to the undisturbed soil.

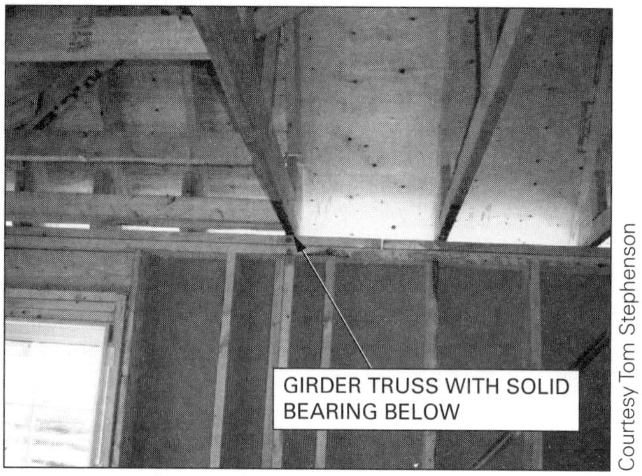

FIGURE 9.27

Below are some examples of connectors that may be a requirement in some municipalities and regional areas that are susceptible to high wind conditions. The truss drawings will indicate fastening requirements for truss to truss members and fastening to plates. Refer to Figure 9.20 page 3 for an example of nailing details for Doncaster house.

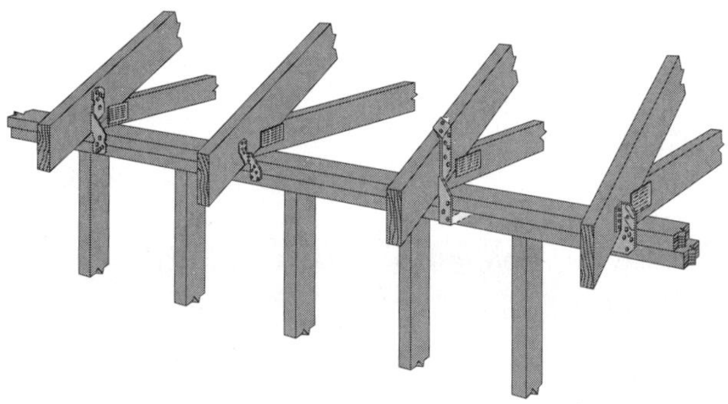

Source: Simpson Strong-Tie Company Inc.

FIGURE 9.28

Close-up of Doncaster floor-plan drawings indicate location of the girder truss (GT) and solid bearing (SB) support moving from second floor down through the first floor. Also note the indication of a cut roof to be placed above the master bedroom. The size of rafters are shown. Refer to the rear elevation of the Doncaster drawings to see the gable cut roof that is resting on top of the hip truss roof.

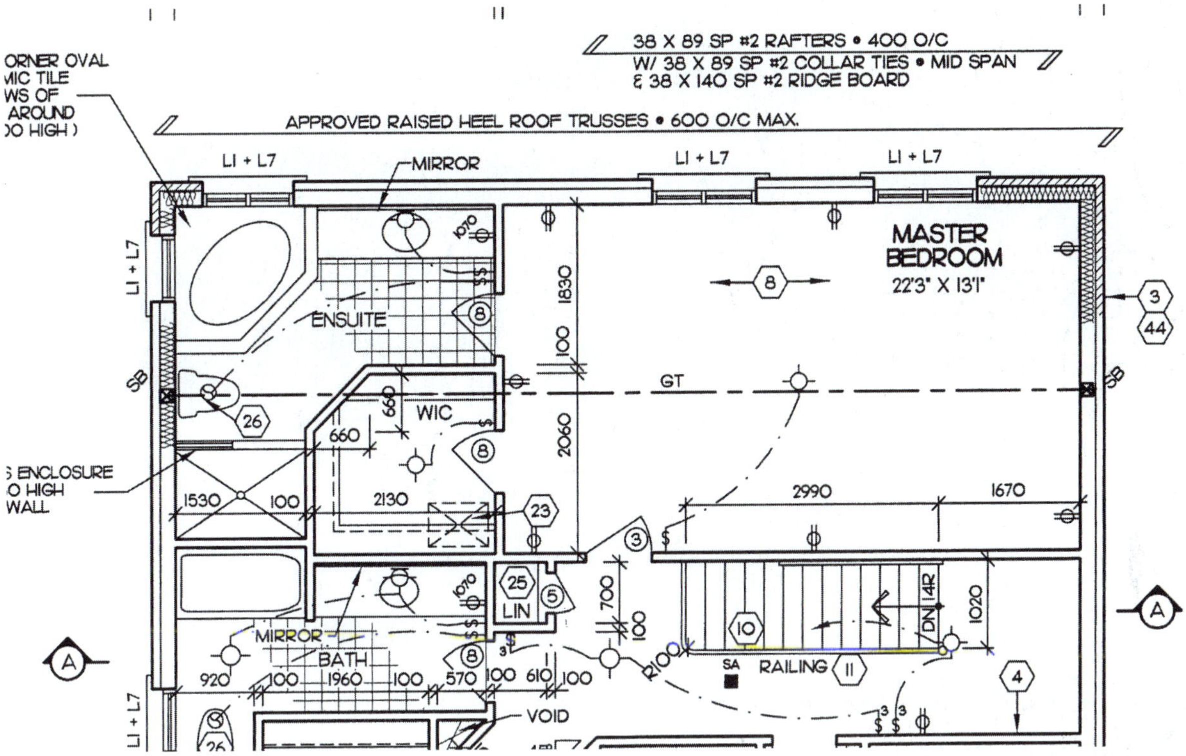

Source: Cassidy & Co. Architectural Technologists

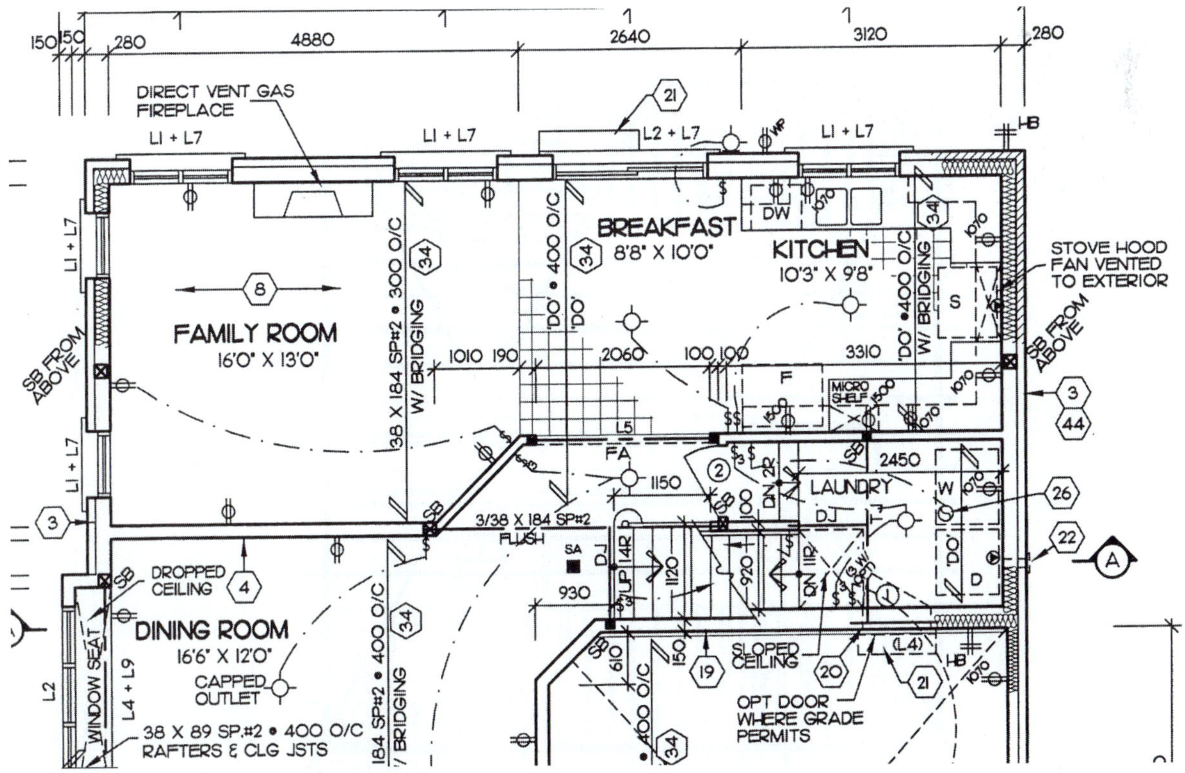

Source: Cassidy & Co. Architectural Technologists

FIGURE 9.29

3D view of roof construction for Doncaster house, including both trusses and cut roof components for two dormers at front and gable roof at rear. Review carefully while looking at Doncaster drawings to help visualize the roof construction.

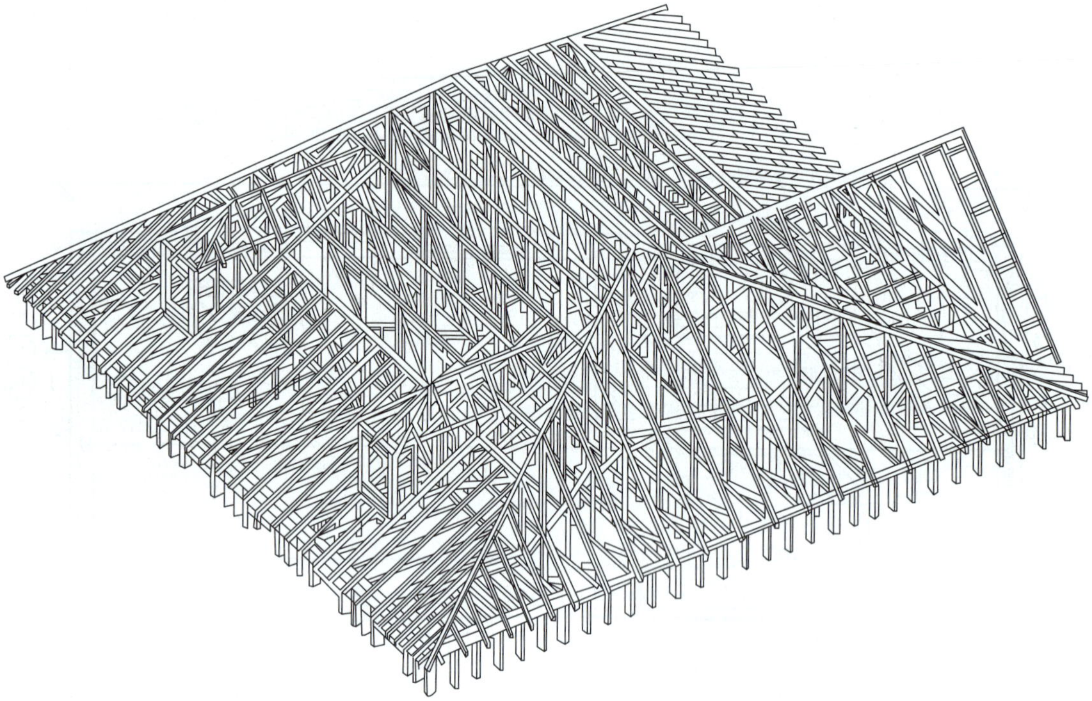

Source: Courtesy of Clint Kissoon, Chair, George Brown College, School of Architectural Studies.

FIGURE 9.30

Roof plan of Doncaster house indicating where truss, dormer, and gable roof framing is located. Also note location of second floor rooms by reviewing top plates as indicated on plan drawing. The interior second floor walls are all non-load-bearing in this design.

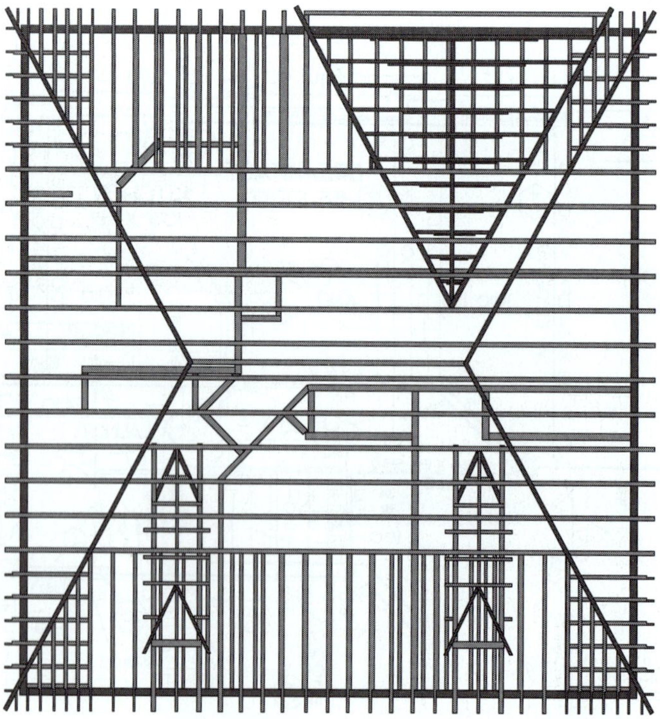

Source: Courtesy of Clint Kissoon, Chair, George Brown College, School of Architectural Studies.

FIGURE 9.31

Roof trusses being manoeuvred into place; notice scissor trusses over lower extension, which will provide a cathedral ceiling in room below.

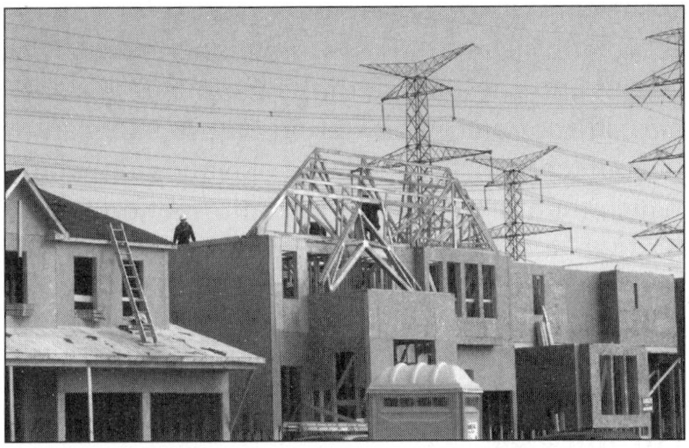

Courtesy Tom Stephenson

FIGURE 9.32

Intersecting roof; notice the main roof rests on hangers and LVL across opening. The minor roof rests on the major roof with a lay on valley.

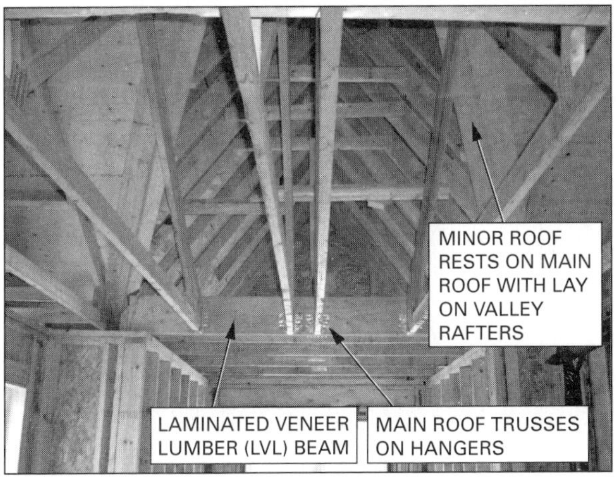

Courtesy Tom Stephenson

FIGURE 9.33

Type X fire-resistant drywall is installed on the gable end when the limiting distance between properties is less than 1.2 m. In case of fire, the fire-resistant drywall will help to slow/prevent flames from spreading to a neighbouring property.

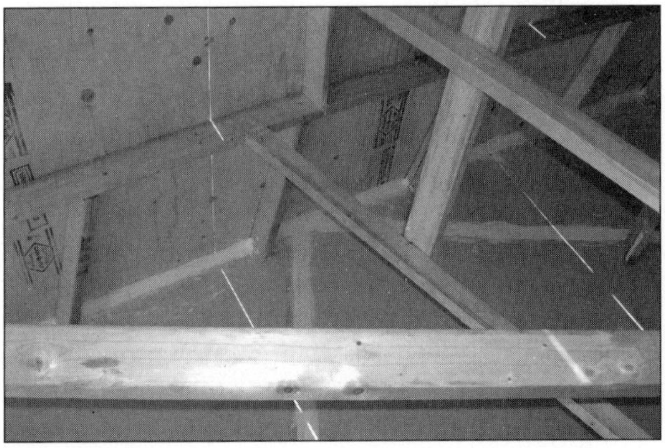

Courtesy Tom Stephenson

FIGURE 9.34

Girder truss assembled with jack trusses on the ground, ready to be lifted into place. Notice scissor portion of truss to the right of the assembly and peaked gable truss in the foreground will provide for a cathedral ceiling below.

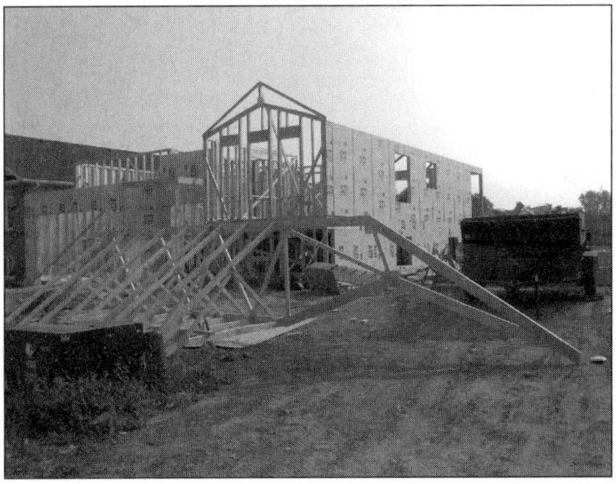

Courtesy Tom Stephenson

FIGURE 9.35

H-clip installed in roof sheathing to maintain alignment of sheathing and prevent sag between sheets of sheathing. The term H clip is used as the clip looks like the letter H when viewed in cross section.

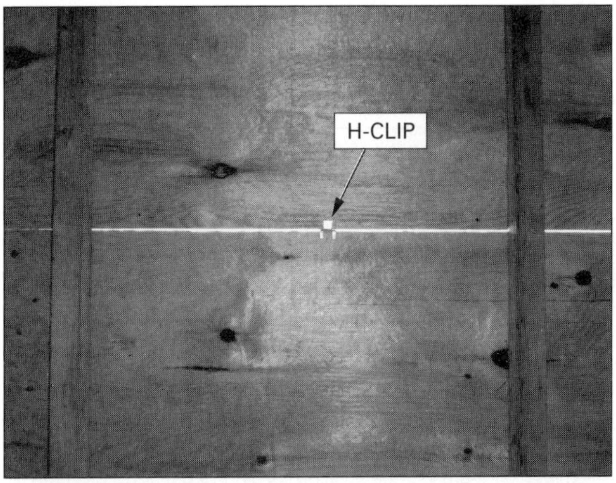

Courtesy Tom Stephenson

FIGURE 9.36
Lower roof conventionally framed; notice space left for brick veneer between wall and roof. This is a mirror image design of the Whittington house drawings supplied with the text. The floor plans are the same except they are flipped (mirrored).

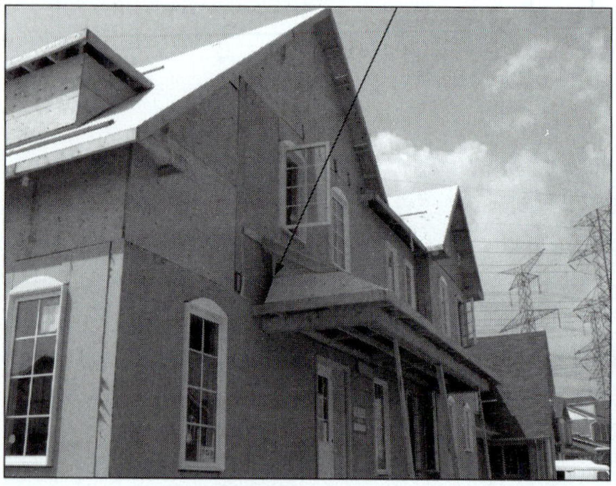

Courtesy Tom Stephenson

FIGURE 9.37
Whittington house both front and left elevations; notice that this house uses both truss roof components and conventional roof components. Review the front and left elevations of the Whittington drawings that came with the text and note that everything being constructed on the photos below is flipped (mirrored). Also note how the ledger, which is 3/4", is placed against the wall for the roof over the porch. The brick will later fill in and come up around the roof rafters. The roof is supported by the connection between the ledger and wood frame wall, not the brick.

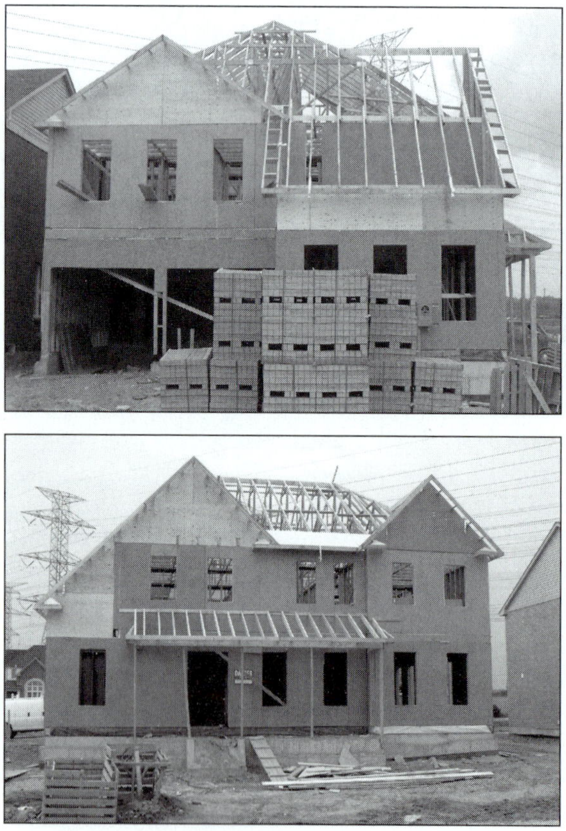

Courtesy Tom Stephenson

Roof Plans

A simple roof plan is often provided to indicate the sloped surfaces of the roof complementing the engineered truss drawings. Structural roof plans for cut roofs may be required for complicated roof layouts. These drawings indicate the layout and positioning of the various rafters. The simple roof plan indicated in Figure 9.38 can be used in conjunction with the elevation drawings, which then provides enough information for the builder and framer to visualize the slopes and intersections of the roof so they can avoid potential construction errors.

FIGURE 9.38
Example of the roof plan indicated on the Doncaster drawings. Note that smaller slopes, such as for the box window at the left side of the house and the roof directly over and to the right of the garage doors, do not necessarily indicate the roof slope with an arrow. The direction of slope can easily be determined by reviewing the elevation drawings.

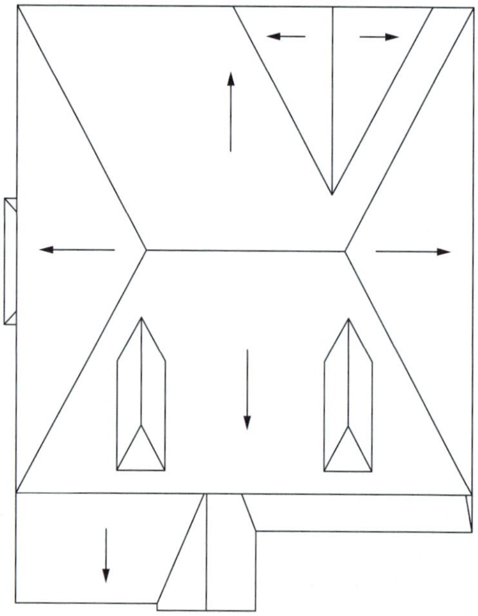

ROOF PLAN
SCALE 1–150

Source: Cassidy & Co. Architectural Technologists

FRAMING KEY POINTS

As we wrap up our sections on framing it is important to point out that the framing forms the structure and form of the home; errors in the framing or errors in how the drawings or changes to the drawings are interpreted will require rework. It is always preferable to avoid errors from occurring and if there is a problem or omission, to catch it as soon as possible in order to limit the amount of rework and avoid delays in the production schedule along with the associated extra costs. The site superintendent and trades play a key role in the overall success of the project by ensuring that the framing is installed correctly and to the highest standards of quality.

An important function of the management of the framing process is to ensure that the houses are checked on a daily basis. By checking the house, the site superintendent will have a thorough knowledge of the drawings and project requirements and see any potential problem areas as they occur. The framing trades and site superintendent must always have their eyes open for construction and safety practices that are not acceptable and act on them quickly. Setting expectations early on in the project will make certain that the project runs smoothly and that the relationship between the framers and builder is one that ensures a quality project that is on time and on budget.

The builder must ensure that the house is being built to plan and to code, and that the most up to date changes are being adhered to. It is essential to visualize all the other trades and the work that follows the framing stage. It is also important to check all the measurements of rooms, windows, and doors. Poor quality framing will transmit through to the finishes and will be difficult if not impossible to hide. High quality framing provides the layout and structural support for the other trades in a way that will make their installations easier, providing a highly desirable finished product.

Some examples of quality framing include appropriate blocking installed to support drywall at all intersections, including at stair stringers, walls and floors that are plumb, level, square, and true, and necessary support for fixtures and railings. Also, it is important to check and straighten any studs that may have twisted or bowed due to a drop in moisture content (drying of the wood). More details on these requirements will be discussed in the coming chapters.

The development of checklists can be useful to reduce errors by making sure that everything necessary has been checked and thoroughly inspected. For example, checking off that all the point loads and solid bearing points have been correctly framed and that all necessary blocking is in place can save many dollars in finish installations and warranty issues later on. It is highly recommended that a walk-through with the purchaser be done at the completion of the framing stage. That way, if there are any issues or the purchaser wants to make a minor change to a non-load-bearing wall location, it can be a relatively easy thing to do. Later on in the construction process, when the building systems are roughed in, a minor change can be a really expensive proposition, which often will make the purchaser unnecessarily unhappy.

KEY TERMS

Collar tie (p. 200)
Cut roof (p. 195)
Girder truss (p. 207)
Gusset plate (p. 203)
Hip or cottage roof (p. 200)
Line length (point length) (p. 196)
Roof joist (p. 197)
Roof rafter (p. 197)
Shop drawings (p. 195)
Total run (p. 196)
Total span (p. 196)
Truss roof (p. 195)
Unit line length (unit point length) (p. 197)
Unit rise (p. 196)
Unit run (p. 196)
Projection (p. 197)
Overhang (p. 198)

Chapter 9 Assignment 1: Doncaster Truss Drawings and Doncaster Drawings

Answer questions 1 through 10 from Figure 9.20 DoncasterTruss Drawings, and answer questions 11 to 20 using the Doncaster drawings that came attached to the text. Identify the letter that best answers the question.

____ 1. What do the truss drawings indicate that the trusses are spaced at?

a. 12"
b. 16"
c. 19.2"
d. 24"

____ 2. How many different types of truss codes are listed on the truss drawing roof plan?

a. 1
b. 5
c. 7
d. 8

____ 3. What code standard do the specifications for the truss drawings state that they are designed to meet?

a. National Building Code of Canada
b. Ontario Building Code
c. British Columbia Building Code
d. Alberta Building Code

____ 4. What truss has only one single brace in total and is indicated at the halfway point between points 5 and 15? (Hint: review the **Bracing** requirements on the individual truss drawings and also refer to elevation drawings of individual trusses on the same pages to see the actual braces in end/section view.)

a. H-03051
b. H-03050
c. H-03049
d. H-03048

____ 5. What type of roof does the Doncaster truss plan indicate that the main roof is?

a. gable roof
b. shed roof
c. hip roof
d. gambrel roof

____ 6. What size material does the truss detail drawing (third page of truss drawings: B12624A) for the hip jack (more correctly the hip rafter) require it to be manufactured out of?

a. 2" × 4"
b. 2" × 6"
c. 2" × 8"
d. 2" × 10"

____ 7. What number and size of nails are to be used to nail the roof rafters (drawings reference to jacks and trusses) to the top plate?

a. 3 – 3 ½" nails
b. 4 – 2 ½" nails
c. 4 – 3 ½" nails
d. 2 – 3 ½" nails

____ 8. What number and size of nails are used to nail the common jacks to the girder truss?

a. 3 – 3 ½" nails
b. 4 – 2 ½" nails
c. 4 – 3 ½" nails
d. 2 – 3 ½" nails

____ 9. Besides the sheathing and bottom chord bracing requirements, which one of the following truss elevations indicate a lateral bracing requirement? (Hint: read bracing requirements carefully on individual truss drawings.)

a. H-03046
b. H-03047
c. H-03048
d. H-03050

____ 10. What is the roof live load specified on the Doncaster truss drawings? (Hint: review individual truss drawing information carefully; live load is specified in pounds per square foot.)

a. 54 P.S.F
b. 31 P.S.F.
c. 27 P.S.F.
d. 8.4 P.S.F.

_____ 11. What size rafters and ceiling joists and what spacing does the roof extension over the garage use?

a. 38 mm × 184 mm SP. # 2 @ 600 OC
b. 38 mm × 140 mm SP. # 2 @ 400 OC
c. 38 mm × 89 mm SP. # 2 @ 400 OC
d. 38 mm × 235 mm SP. # 2 @ 400 OC

_____ 12. What is the size and species of the ridge board over the porch?

a. 38 mm × 89 mm SP # 2
b. 38 mm × 140 mm SP # 2
c. 38 mm × 140 mm SP # 2
d. 19 mm × 140 mm SP # 2

_____ 13. What size are the hip rafters for the roof over the living room?

a. 38 mm × 89 mm
b. 38 mm × 140 mm
c. 38 mm × 184 mm
d. 38 mm × 235 mm

_____ 14. What type of roof is over the box window in the dining room?

a. gable
b. hip
c. gambrel
d. flat

_____ 15. How many girder trusses are indicated for the main roof of the house?

a. 1
b. 2
c. 3
d. 4

_____ 16. How many hip roof type dormers are indicated on the drawings?

a. 0
b. 1
c. 2
d. 3

_____ 17. What is the width of the walls for the dormers at the front of the house (taken from outside of wall to outside of wall)?

a. 455
b. 910
c. 1060
d. 1210

_____ 18. How are the girder trusses supported for the main house roof?

a. by solid bearing on the second floor only
b. by solid bearing on the first and second floors
c. by a single top plate with no solid bearing
d. by a double top plate with no solid bearing

_____ 19. Where on the drawings is a 20 to 12 roof slope indicated?

a. the main roof
b. the gable roof at the rear
c. the front dormers
d. the box window in the dining room

_____ 20. What is the width of the roof for the dormers at the front of the house, including projection?

a. 455
b. 910
c. 1060
d. 1210

Chapter 9 Assignment 2: Roofs and the Brook Drawings

Answer all the following questions using the Brook drawings that came attached with the text. Identify the letter that best answers the question.

_____ 1. What is the minimum slope required for the flat residential roof shown over the front porch of Elevation B?

a. 0%
b. 2%
c. 3%
d. 15%

_____ 2. How many gables are shown on the front elevation of Elevation A?

a. 1
b. 2
c. 3
d. 4

_____ 3. How many gables are shown on the front elevation of Elevation B?

a. 1
b. 2
c. 3
d. 4

_____ 4. What on centre spacing are the roof trusses on the second floor to be placed at?

a. 12"
b. 16"
c. 19.2"
d. 24"

_____ 5. What is the size, species and spacing of the roof joists over the front porch of Elevation B?

a. 2" × 4" SP. #2 @ 12" OC
b. 2" × 6" SP. #2 @ 16" OC
c. 2" × 8" SP. #2 @ 16" OC
d. 2" × 10" SP. #2 @ 24" OC

_____ 6. Is there is a skylight shown on one of the upper roof slopes (2nd floor), and if so, where?

a. There is no skylight indicated on the drawings.
b. There is a skylight indicated over the master bedroom.
c. There is a skylight indicated over bedroom 4.
d. There is a skylight indicated over the walk-in closet.

_____ 7. What is the size of the solid bearing support for the Elevation B flat roof over the porch, as indicated in plan view?

a. 6" × 6"
b. 8" × 8"
c. 10" × 10"
d. 12" × 12"

_____ 8. What is the projection of the roof over the mud room entry?

a. 3"
b. 4"
c. 6"
d. 1'-0"

_____ 9. How many gable roofs are indicated on the drawings for Elevation A in total all sides, also using the upgraded rear elevation drawing?

a. 1
b. 2
c. 3
d. 4

_____ 10. How many sloping surfaces (sides) of roof in total are indicated on the second floor (only) for roof plan Elevation A (not using the upgraded rear elevation A drawing)?

a. 7
b. 10
c. 11
d. 12

_____ 11. What is the slope for the gable roof on front Elevation B?

a. 6 to 12
b. 9 to 12
c. 11 to 12
d. 12 to 12

_____ 12. How many columns is the roof over the front porch supported on?

a. 2
b. 3
c. 4
d. 6

_____ 13. What size and type of sheathing is specified for the flat roof over the porch?

a. 3/8" OSB

b. 5/8" plywood

c. 3/8" plywood

d. 1/2" plywood

_____ 14. When and where would a 2" × 4" cut roof using conventional framing be required?

a. for the roof that overhangs the garage for elevation A

b. for the roof that overhangs the garage for elevation B

c. for the gable roof at the front of elevation A

d. for the area over the gas fireplace in the Great Room for Elevation A

_____ 15. How many sloping faces (sides) of roof in total are indicated on the second floor (only) for the Elevation B roof plan?

a. 7

b. 10

c. 11

d. 12

_____ 16. How many roof vents are shown on the roof plan for Elevation A?

a. 6

b. 7

c. 9

d. 10

_____ 17. How many roof vents are shown on the roof plan for Elevation B?

a. 6

b. 7

c. 8

d. 9

_____ 18. What type of roof is required over the fireplace on the first floor?

a. truss roof, hip type

b. truss roof, shed type

c. conventional roof, shed type

d. conventional roof, hip type

_____ 19. What is the projection for the sides of the roof over the fireplace on the first floor?

a. 3"

b. 6"

c. 10"

d. 1'-0"

_____ 20. What type and thickness of sheathing is used on the roof over the second floor?

a. 3/8" plywood with H-clips

b. 1/2" plywood with H-clips

c. 5/8" tongue-and-groove plywood

d. 3/8" oriented strand board with H-clips

Chapter 9 Quiz Review: Roof Theory & Reading Roof Plans Using the Whittington Drawings

Answer questions 1 to 5 from the information provided in the chapter, and answer questions 6 to 20 using the Whittington drawings found at the rear of the textbook. Identify the letter that best answers the question.

_____ 1. Which of the following describes a girder truss?

a. it carries and supports the weight of other trusses

b. it is designed and installed exactly the same as any regular truss

c. a small truss that connects to a larger truss

d. a truss that is always cut and built on site

_____ 2. Calculate what the unit line length is for a slope of 4:12?

a. 13.49"

b. 4"

c. 12"

d. 12.65"

_____ 3. What is the minimum bearing required by the NBC for a rafter?

a. 38 mm

b. 89 mm

c. 140 mm

d. 184 mm

_____ 4. How is total run defined?

a. equal to half the roof span

b. equal to double the roof span

c. equal to the length of the ridge

d. equal to the length of the hip rafter

_____ 5. What is the minimum recommended slope of a flat residential roof listed as a percentage, as described in the chapter?

a. 0%

b. 2%

c. 3%

d. 15%

_____ 6. What structural materials is the gable end roof over bedrooms 3 and 4 to be constructed with?

a. 38 mm × 184 mm SP. #2 @ 600 mm OC rafters and ceiling joists and 38 × 235 mm ridge board

b. 38 mm × 89 mm SP. #2 @ 400 mm OC Rafters and Ceiling Joists and 38 mm × 140 mm Ridge Board

c. 38 mm × 140 mm SP. #2 @ 400 mm OC Rafters with 38 mm × 89 mm collar ties @ mid span and 38 mm × 184 mm Ridge Board

d. 38 mm × 235 mm SP. #2 @ 400 mm OC Rafters with 38 mm × 89 mm collar ties @ mid span and 38 mm × 184 mm Ridge Board

_____ 7. What is the size, species, and spacing of the rafters and ceiling joists over the porch?

a. 38 mm × 89 mm SP #2 roof rafters and ceiling joists @ 400 mm OC

b. 19 mm × 89 mm SP #1 Roof rafters and ceiling joists @ 400 mm OC

c. 38 mm × 140 mm SP #2 Roof rafters and ceiling joists @ 300 mm OC

d. 38 mm × 140 mm SP #2 Roof rafters and ceiling joists @ 400 mm OC

_____ 8. Is there is a skylight shown on one of the upper roof slopes (second floor), and if so, where is it?

a. There is no skylight indicated on the drawings.

b. There is a skylight indicated in the master bedroom.

c. There is a skylight indicated in bedroom 4.

d. There is a skylight indicated in the walk-in closet.

_____ 9. Where on the drawings is a truss roof indicated?

a. The roof over the dormer is a truss roof.

b. The roof over the porch is a truss roof.

c. The area over most of bedroom 2 is a truss roof.

d. The main section of the second floor is a truss roof.

_____ 10. How many girder trusses are shown for the main house roof?

a. 1

b. 2

c. 3

d. 4

_____ 11. How many gable roofs are indicated on the drawings in total?

a. 1

b. 2

c. 3

d. 4

_____ 12. Where are raised heel roof trusses indicated?

a. over the porch

b. over the dormer

c. over the master bedroom

d. over the family room

_____ 13. How many solid bearing points are the girder trusses over the second floor supported on?

a. 2 girder trusses supported by 4 solid bearing points

b. 2 girder trusses supported by 3 solid bearing points

c. 3 girder trusses supported by 5 solid bearing points

d. 3 girder trusses supported by 6 solid bearing points

_____ 14. What is the overall width of the dormer from outside wall to outside wall?

a. 2240 mm

b. 2440 mm

c. 2740 mm

d. 3000 mm

_____ 15. How many sloping faces (sides) of roof in total are indicated directly on the roof plan from page 6 of the drawings (include all slopes)?

a. 2

b. 8

c. 11

d. 12

_____ 16. What is one of the slopes for the roof of the dormer?

a. 11.9

b. 9

c. 6.9

d. 12.9

_____ 17. How many columns is the roof over the front porch supported on?

a. 3

b. 4

c. 5

d. 6

_____ 18. What size and type of sheathing is specified for the roof?

a. 3/8" OSB

b. 5/8" plywood

c. 3/8" plywood with H clips

d. 1/2" plywood with H clips

_____ 19. What is the overall width of the dormer from inside wall to inside wall?

a. 2240 mm

b. 2440 mm

c. 2740 mm

d. 3000 mm

_____ 20. Review the drawings carefully and determine: which sloping faces (sides) of roof are missing (but technically should have been indicated) from the roof plan on page 6?

a. the roof over the garage

b. the dormer roof

c. the roof over the bay window in the dining room

d. the roof over the bay window in the living room

10 Exterior Finishes: Walls, Roofs, and Concrete Slabs

LEARNING OBJECTIVES

After completing this chapter, you will be able to:

- Read and interpret drawings for information related to exterior finishes and their application and installation.
- Identify common exterior finish materials used in residential construction.
- Describe some of the basic methods of installation for exterior finishes.
- Describe a number of key National Building Code requirements as they apply to exterior finishes.
- Describe the basic properties of concrete and the method of pouring and finishing concrete slabs.

INTRODUCTION

This chapter discusses exterior finishes, which are the materials visible on the outside of the building. These materials protect the structure from the elements and define the look of the building. The selection of material needs to be made early in the design process because the materials may affect the structural design of the home due to size and load requirements. For example, brick or stone requires a bearing surface to rest on, and a concrete tile roof may require the roof trusses to be specially designed to carry the additional weight. The selection of an **exterior insulated finish system (EIFS)** will require special detailing requirements. The drawings will indicate on the elevations what materials are being used where and any special detailing required. Section details will describe the type of material that is to be used and the order of placement when constructing the walls, roofs, and slabs, including exterior and interior finishes. The construction notes can be referenced for a more complete listing. Specifications will provide detailed information regarding installation requirements, manufacturer, model numbers, etc. (refer to Chapter 14 for more information on specifications).

WALL FINISHES

There are numerous cladding finishes and combinations of cladding finishes used on exterior walls. The type of finish used depends on individual preferences, regional availability, suitability to climate, and architectural control requirements (dependent on municipal aesthetic requirements). Exterior finish materials include:

- Masonry
- Vinyl, prefinished composite materials, and various types of metal siding
- Wood siding and shingles
- Exterior insulated finish systems (EIFS).

Designers, general contractors, and tradespeople working with exterior wall cladding systems need to ensure that adequate planning and consideration is given to the sealing of intersections with walls, roofs, dissimilar materials, windows, doors, and service penetrations of the building envelope. Also, a mechanism must be provided to allow moisture to dry out if and when it penetrates the cladding system so that it does not damage the structure or cause unhealthy side effects such as mould.

Examples include:

- Masonry veneer with a 1" air space with properly located weep holes (refer to Figure 10.12)
- Exterior insulated finish system designed to drain moisture out by use of a drainage plane (refer to Figure 10.27)
- Exterior wood siding placed on strapping allows air to flow behind the siding (refer to Figure 10.24).

Masonry (Brick and Stone Cladding)

Walls can be constructed of masonry veneer or solid masonry, as briefly discussed in Chapter 7. Veneer is currently the most common masonry method used to clad exterior walls. This type of construction is popular for a number of reasons, including:

- Providing a high fire-resistance rating
- Providing an air space and weep holes to allow moisture that gets into the wall to dry out or exit
- Providing a long-lasting durable surface
- Requiring little maintenance
- The wide selection of colours, textures, and materials available.

Installation and Code Requirements

The possible selection of veneer material includes brick (including clay and concrete), natural stone, and manufactured stone. Brick is set in cement-based mortar and laid in a masonry bond that laps the units together to provide strength to the wall. The bond can also be used to form different decorative patterns. One of the few limitations to laying brick or stone in cement-based mortar is low temperatures and rainy or snowy weather.

The NBC states that (refer to Division B Part 9.20.14 of the NBC):

- Laying temperature of mortar and masonry:
  - Mortar and masonry shall be maintained at a temperature not below 5 degrees celsius during installation and for not less than 48 hours after installation.
  - No frozen material shall be used in the mortar.
- Protection from the weather:
  - The top surface of uncompleted masonry exposed to the weather shall be completely covered with a waterproofing material when construction is not in progress.

The above points create challenges in the cold Canadian climate. In winter, appropriate temporary enclosures and heat will be required to protect the cladding during installation and curing. This cost must be considered when pricing structures that will be constructed during the winter months.

FIGURE 10.1

Temporary enclosure and heat (note propane tanks required for temporary heating) required for masonry work done when temperature drops below 5 degrees celsius.

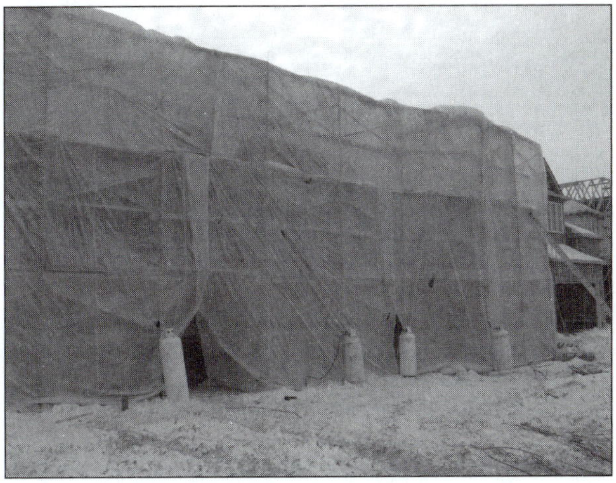

Courtesy Tom Stephenson

Veneer Cladding

The veneer cladding is tied to the frame wall with masonry ties, which provide lateral support to the wall. The wall itself carries its own dead-load weight and is not considered structurally load-bearing.

The veneer is supported over openings on a steel lintel, steel beam, or masonry arch. Arches can be used to provide traditional design detail that may not normally be achieved with a steel lintel. There are many different types of arches that can be used, each with its own unique appearance (refer to Figure 10.9 for more detail). A moisture-resistant building paper or **building wrap** such as a spunbonded polyolefin (e.g., Tyvek™) will form a barrier between the sheathing, air space, and veneer cladding. **Weep holes** are provided over lintels and beams, at the base of the wall, and underneath sills and thresholds. The weep holes act as a means of drying out the air space if moisture migrates there. In some designs, the weep holes combined with vent holes also form part of a system that acts as a rain screen. The rain screen is vented to the exterior, while the inner layer or backup wall (in most residential buildings that means frame wall) is sealed to prevent air leakage using an air barrier that can be either on the inside or outside of the frame wall, depending on the design. The design of the air and vapour barriers must be considered carefully, as it can have a huge impact on the quality of installation, the durability of materials, and the health of the occupants. Chapter 12 provides

FIGURE 10.2

Traditional pattern and bond patterns. The most commonly used patterns are simple ½ and approximately ⅓ lap bonds. Historically other patterns were used more commonly in solid masonry construction where header bricks were required to tie the wythes (rows) together. Special bonding requirements are shown on elevation drawings. The stacked bond wall requires metal reinforcement in the courses to ensure the wall is structurally sound.

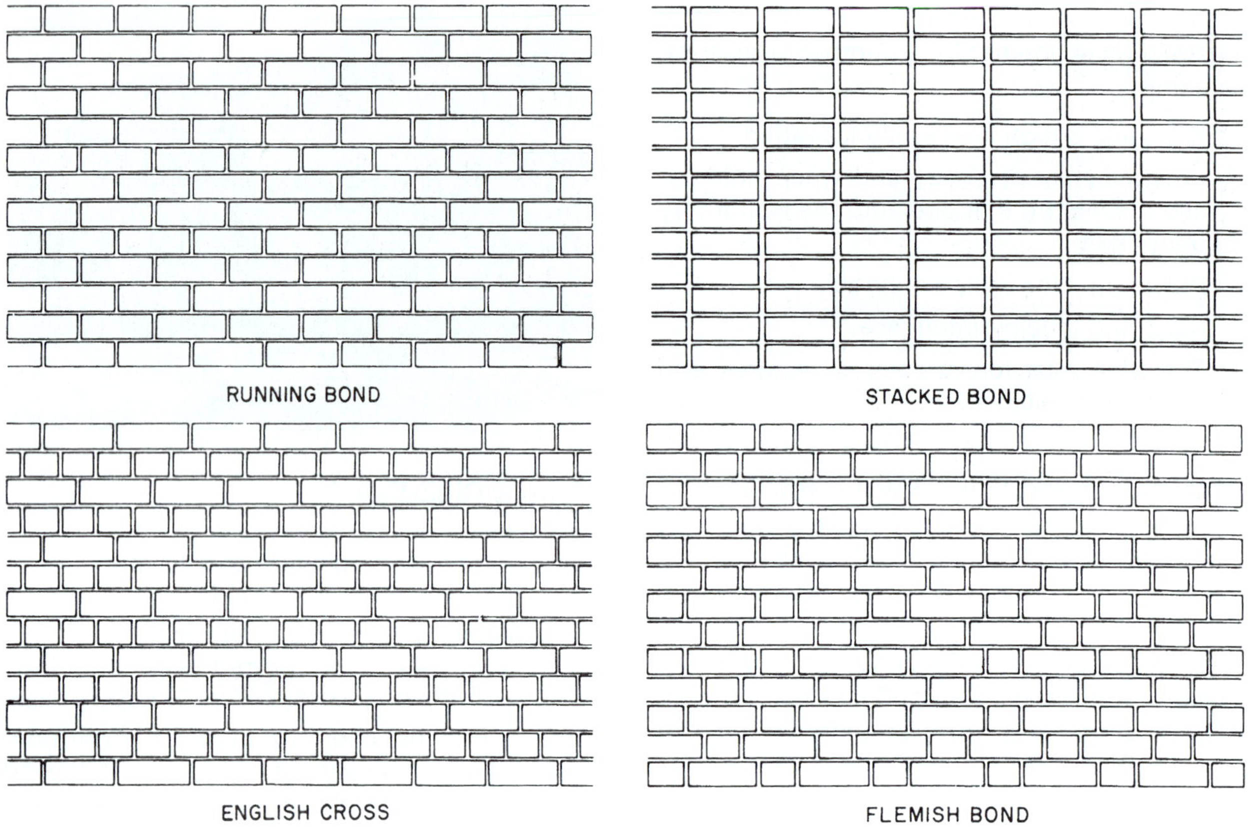

FIGURE 10.3

A Flemish bond garden wall; notice the alternating headers and stretchers. This wall is two wythes thick.

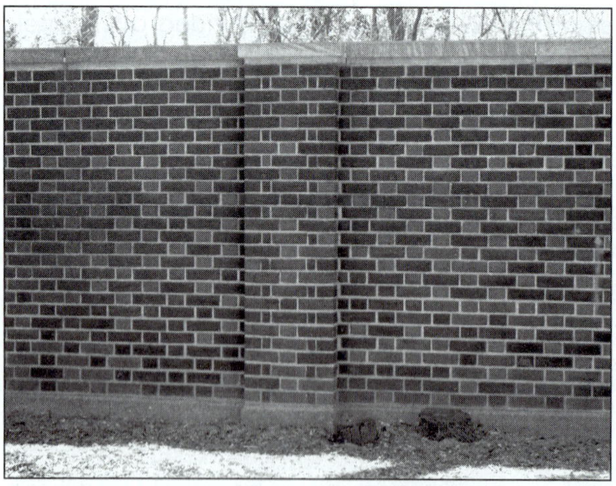

Courtesy Tom Stephenson

FIGURE 10.4

Names of brick in different positions. Often drawings will either illustrate or describe the placement/position of brick for decorative purposes using this terminology. For example, headers placed at the side of windows and soldiers over the top will be described this way. Technically any brick placed in an arch is called a voussoir, but may still be referred by its position in the drawings.

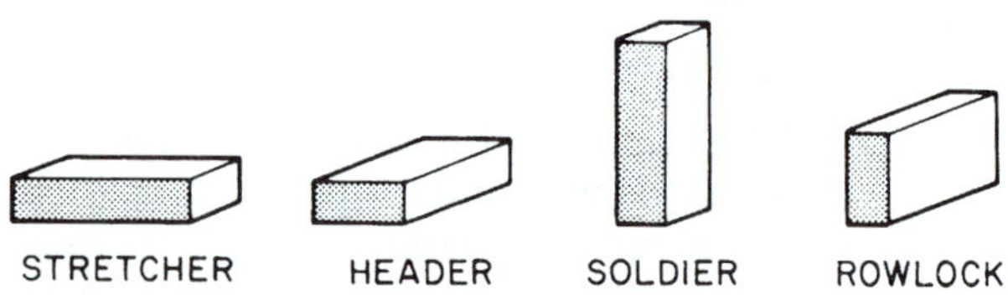

more information regarding the building envelope and building science principles.

Part 9.20 of the NBC provides the minimum acceptable requirements for installing above-grade masonry. Some key NBC requirements are as follows:

- Masonry is required to be supported on masonry, concrete, or steel, except that masonry veneer walls can be supported on preserved wood foundations (refer to Division B Part 9.20.5.1 of the NBC).

FIGURE 10.5

Steel lintel supported on wall to carry brick over opening; notice that the steel lintel must bear a minimum of 90 mm (refer to Division B Part 9.20.5.2 of the NBC) when in final position (some provincial codes, for example, Ontario, require 150 mm bearing).

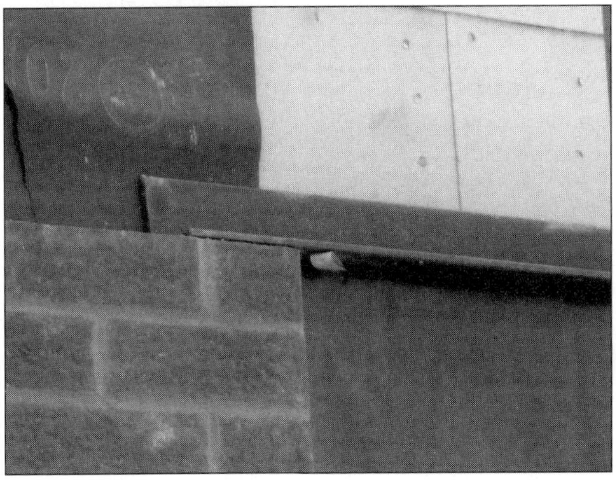

Courtesy Tom Stephenson

FIGURE 10.6

Polyethelene flashing placed over window on steel lintel to provide moisture a means of exit through the weep holes. A building paper will then be layered over the flashing to ensure water does not get behind the flashing. Ideally, the flashing should extend right out through the face of the brick.

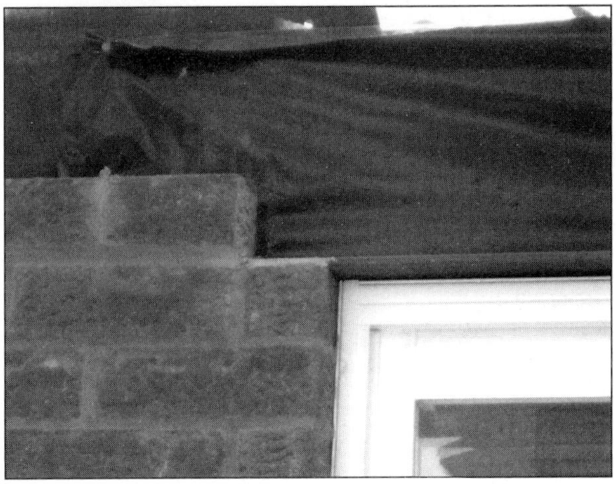

Courtesy Tom Stephenson

FIGURE 10.7

Steel lintel over window. NBC requires minimum bearing support of 90 mm for the lintel (refer to Division B Part 9.20.5.2 of the NBC).

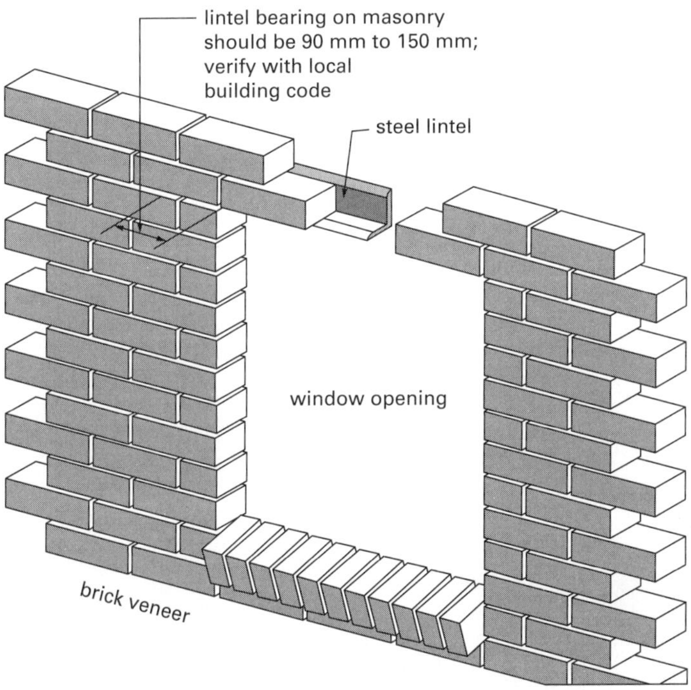

Source: Image excerpted from The Illustrated Home by Carson Dunlop www.carsondunlop.com

FIGURE 10.8

Types of mortar joints listed from most durable to least durable.

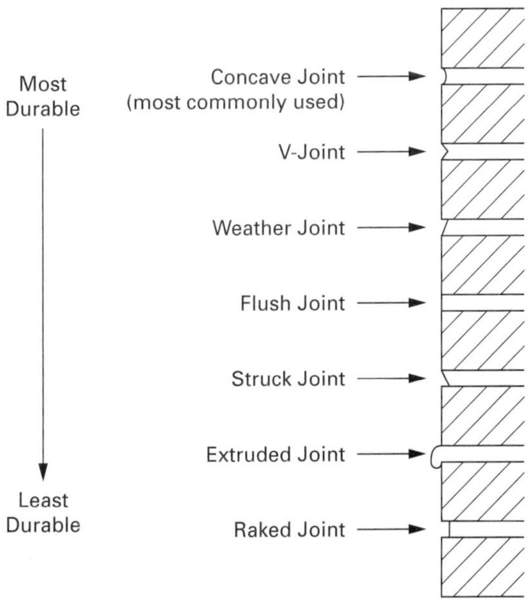

Source: Image excerpted from The Illustrated Home by Carson Dunlop www.carsondunlop.com

FIGURE 10.9

Illustrated below are some of the masonry arch types and related terminology. The bricks used in arches are called voussoirs and the centre piece of the arch is called a key brick or key stone, found only in round-top arches such as the roman, elliptical, parabolic, and segmental arches. A key joint is used for gothic and tudor arches.

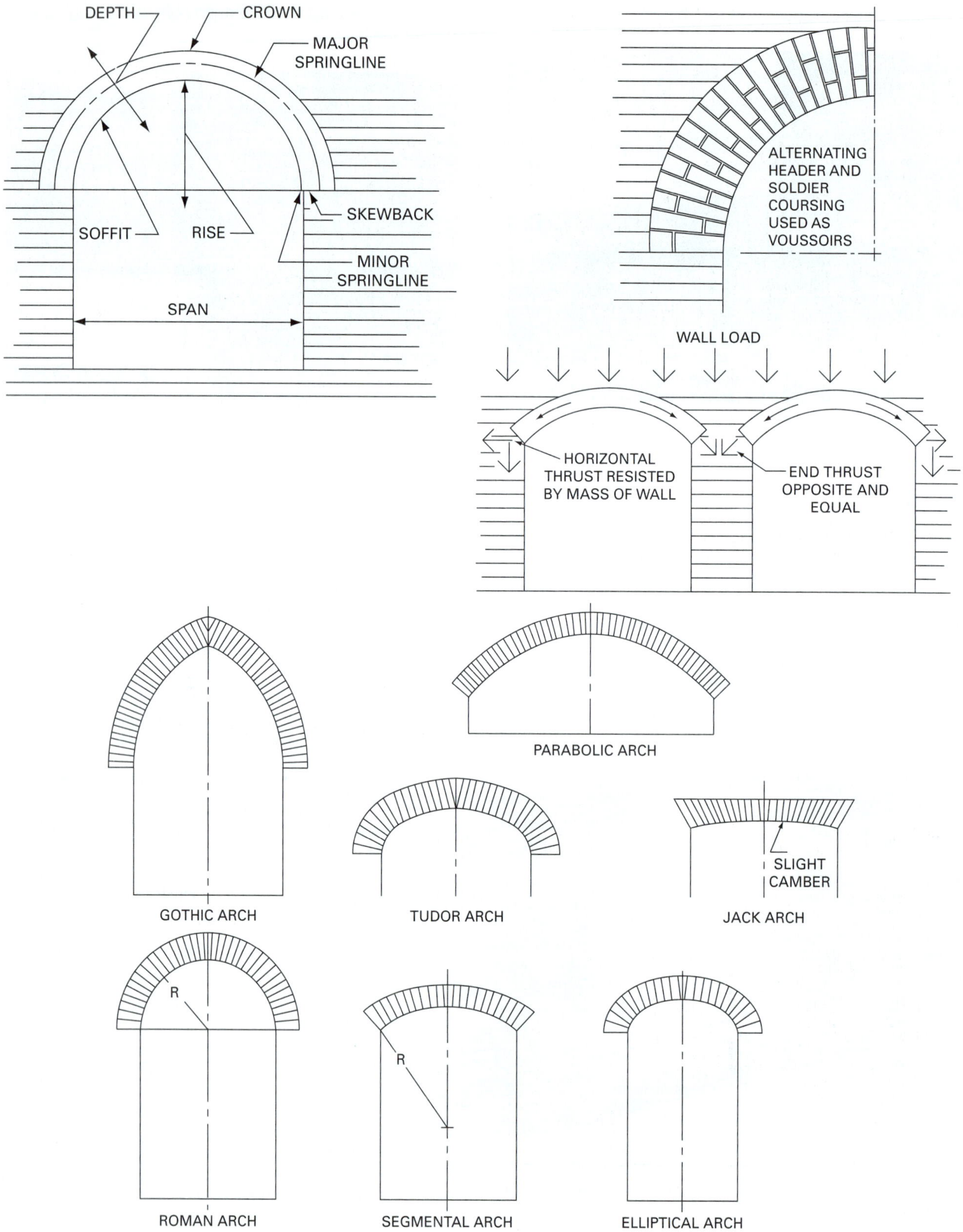

Source: From SPENCE, Construction Materials, Methods, and Techniques, 2E. © 2007 Cengage Learning.

FIGURE 10.10

An elliptical arch with a precast concrete surround. The stone in the centre is called the key stone. Note the decorative soldier course under the precast concrete sill.

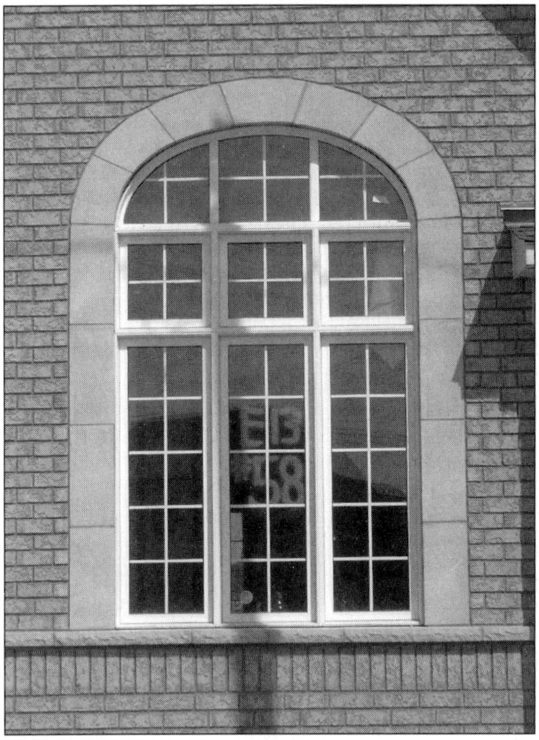

Courtesy Tom Stephenson

FIGURE 10.11

Flashing being installed in preparation for brick veneer at side of stepped foundation; detail of fold at end should create an **end dam,** a flashing process that prevents moisture migration into the structure.

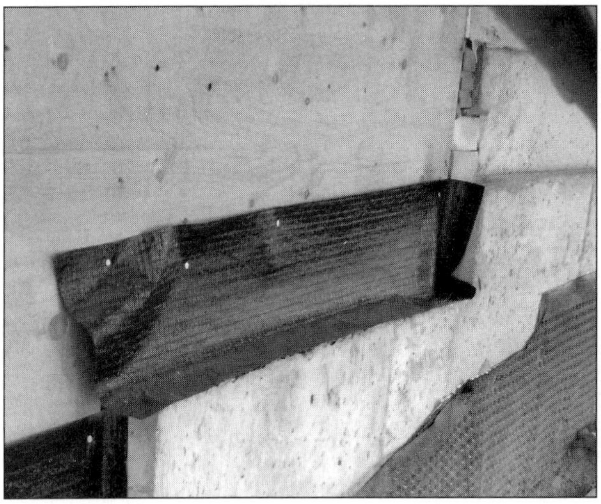

Courtesy Tom Stephenson

FIGURE 10.12

Weep holes placed at foundation wall. Notice the flashing extending up the wall; building paper will be slid to overlap the flashing. Weep holes in this case are installed every third brick. A 1" air space is left between the brick and the building paper.

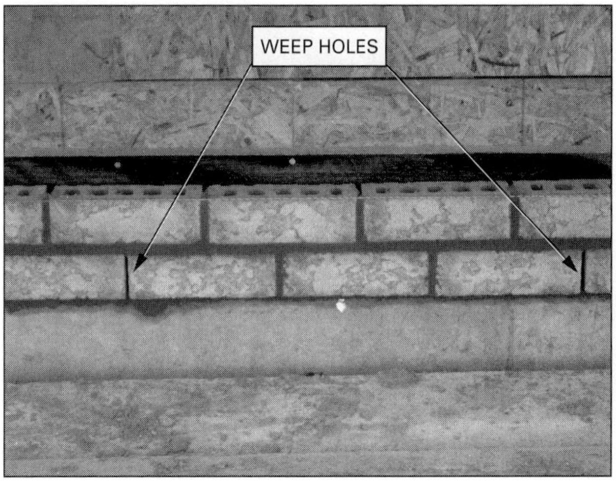

Courtesy Tom Stephenson

FIGURE 10.13

Weep hole placed below windowsill.

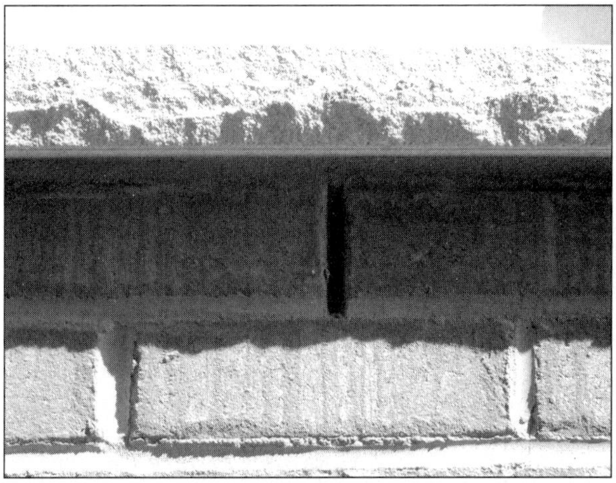

Courtesy Tom Stephenson

FIGURE 10.14

Brick toothed out at front to facilitate placement of scaffolding. Once the first wall is completed, scaffolding is moved to start brick on other side, as indicated. Care must be taken to ensure that mortar is properly compressed over top joint of toothed area.

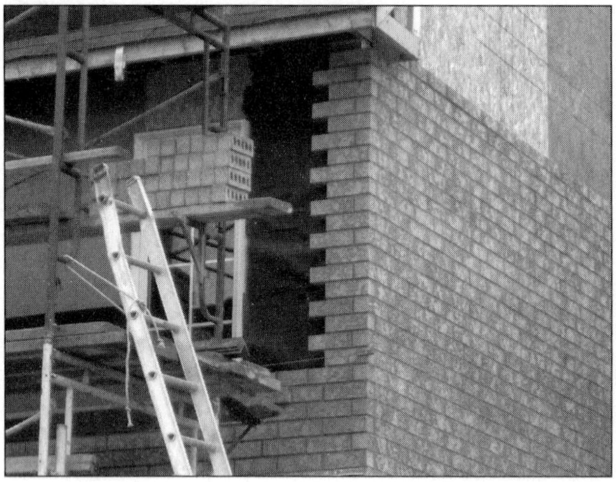

Courtesy Tom Stephenson

FIGURE 10.15

Brick ties in place ready to be folded down and embedded in mortar as brick course comes up. Ties are placed as per NBC requirements described in Figure 10.16 for metal ties 16" horizontal (to match stud spacing; if stud spacing were different the horizontal spacing would change to match and the vertical spacing would be larger or smaller to compensate) and 24" vertical. Notice that ties must be nailed or screwed through Styrofoam sheathing and into studs. Notice that the head joints in the brick are not full, which will lead to early deterioration of the wall and facilitate moisture behind the surface of the brick. This contravenes the NBC, which requires full head and bed joints.

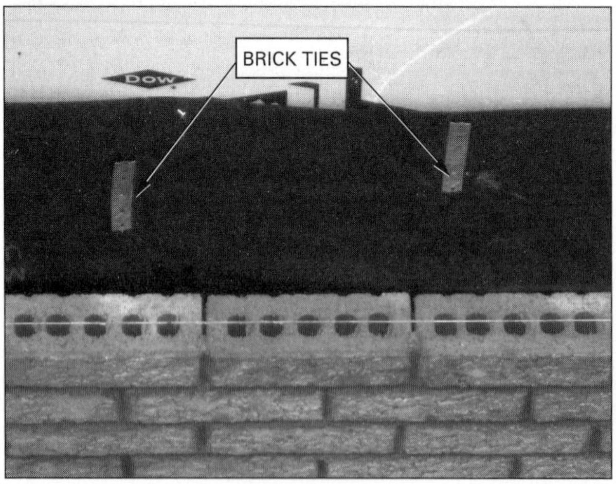

Courtesy Tom Stephenson

FIGURE 10.16

Sample specification for brick veneer from Brook construction notes; the specification outlines details about the brick, air space, placement of metal ties, weep hole openings, and base flashing.

(3) **BRICK VENEER CONSTRUCTION**

90 mm (4") OR 75 mm (3") FACE BRICK, 25 mm (1") AIR SPACE. 22 × 180 × 0.76 mm (7/8" × 7" × 22 ga.) METAL TIES @ 400 mm (16") O.C. HORIZ. 600 mm (24") O.C. VERTICAL, 0.7 kg/s.m. (No. 15) BUILDING PAPER, 28 mm (1.0") R5 RIGID INSULATION 38 × 140 (2 × 6) STUDS @ 400 mm (16") O.C., RSI 3.52 (R20) BATT INSULATION AND 6 mil. AIR / VAPOUR BARRIER, 12.7 mm (1/2") INT. DRYWALL FINISH.
PROVIDE WEEP HOLES @ 800 mm (30") O.C. HORIZ. @ BOTTOM COURSE ONLY & OVER OPENINGS.
PROVIDE BASE FLASHING UP MIN. 150 mm (6") BEHIND SHEATHING PAPER.

Source: Cassidy and Co. Architectural Technologists

FIGURE 10.17

A house façade using a mixture of masonry materials and details. This house uses a manufactured stone façade with decorative precast concrete surround on front window. Upper windows are constructed with segmental brick arch detail. The opening over the garage is constructed with a soldier arch resting on a steel plate welded to a steel beam.

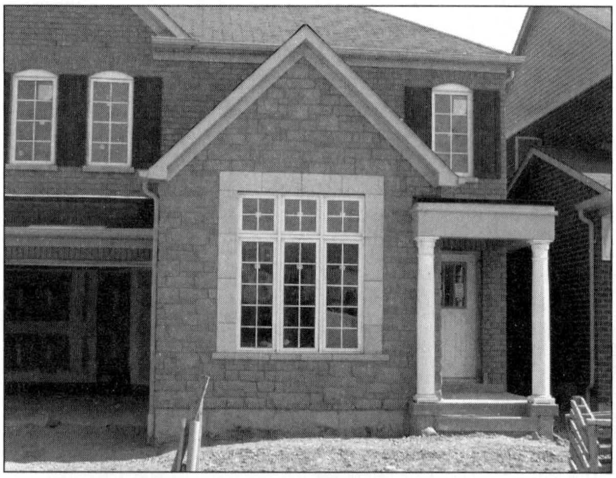

Courtesy Tom Stephenson

- Steel lintels must have a minimum bearing of 90 mm. Note that some provinces (e.g., Ontario) require a minimum bearing of 150 mm (refer to Division B Part 9.20.5.2 of the NBC).
- Masonry over openings such as windows and doors shall be supported by steel, reinforced concrete or masonry lintels, or arches that are appropriately designed (refer to Division B Part 9.20.5.2.3 of the NBC).
- Ties for masonry veneer 75 mm or more in thickness and resting on a bearing support shall be tied to masonry backup or to wood framing members. Usually the ties are placed at every stud, depending on the spacing. For example, if the studs are spaced at 400 mm, then a tie is placed horizontally on the wall at every stud and vertically not more than 600 mm, or approximately every sixth course, depending on the size of the brick (refer to Division B Table 9.20.9.5 of the NBC for a listing of spacing options).
- Weep holes must be spaced not more than 800 mm apart and shall be provided at the bottom of the cavity in a cavity wall and in masonry veneer wall construction, including the cavities above lintels, and over window and door openings required to be flashed (refer to Division B Part 9.20.13.8 of the NBC).

Siding

Siding can consist of a number of materials, with wood, prefinished composite materials (e.g. fibre cement based siding), and vinyl being the most common. Vinyl has become increasingly popular because it requires little maintenance, is durable, and is relatively inexpensive. Wood is also popular because of its versatility and its natural beauty. A wide variety of manufactured composite and prefinished wood products are readily available for use as an exterior finish.

Vinyl Siding

Vinyl siding is applied directly over a building paper or building wrap such as spun bonded polyolefin (Tyvek™). Any exterior-type code-approved sheathing is acceptable as long as the siding can be fastened to the studs underneath

FIGURE 10.18
Vinyl siding and soffit.

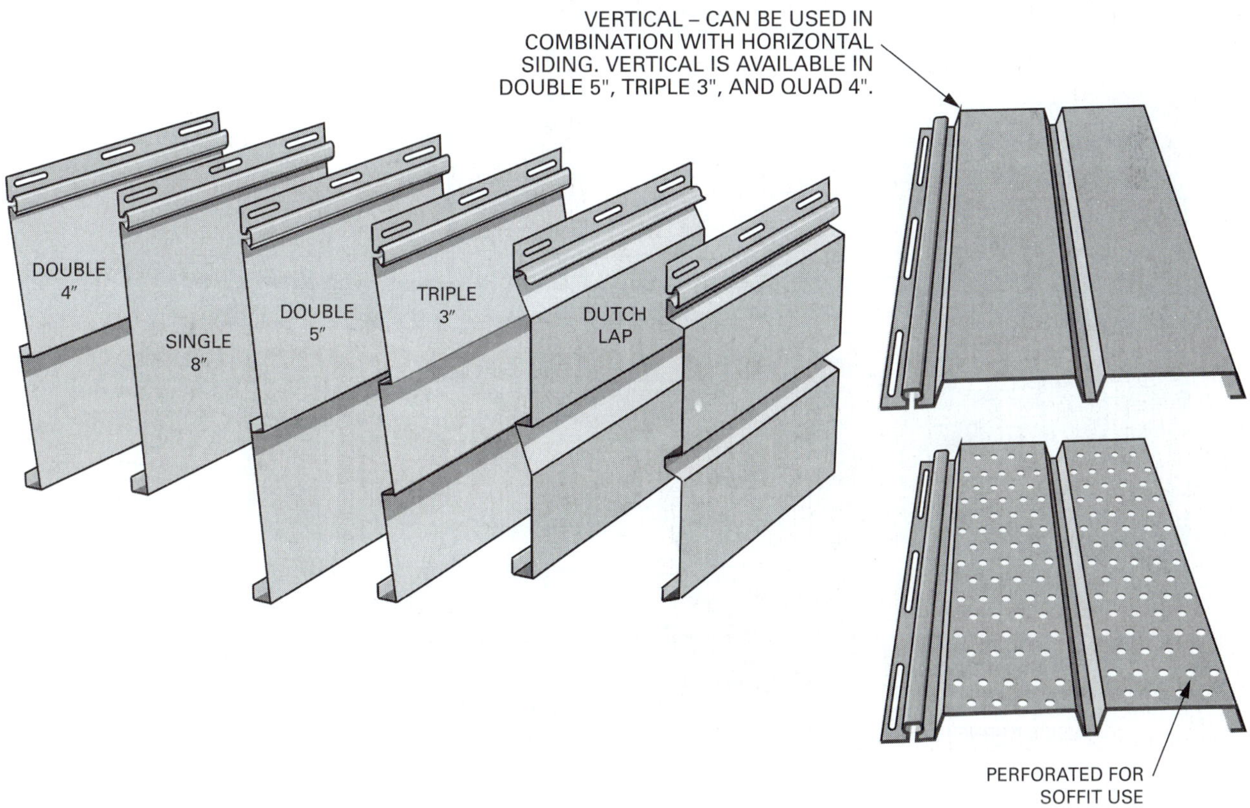

Source: From VOGT, Carpentry, 4E. © 2006 Delmar Learning, a part of Cengage Learning, Inc. Reproduced by permission. www.cengage.com/permissions

the sheathing. The siding comes with a number of trim materials that are used at starting points, intersections (e.g., around windows, and doors, and vents), and end points. Most vinyl siding attempts to replicate the look of wood and, as such, resembles bevelled, board and batten, or shingled wood sidings. Care has to be taken that the siding is not nailed too tightly, so that it can expand and contract without buckling as changes in temperature occur.

FIGURE 10.19
Vinyl siding accessories and where they are used; note perforated soffit is used to ventilate attic area in conjunction with roof vents.

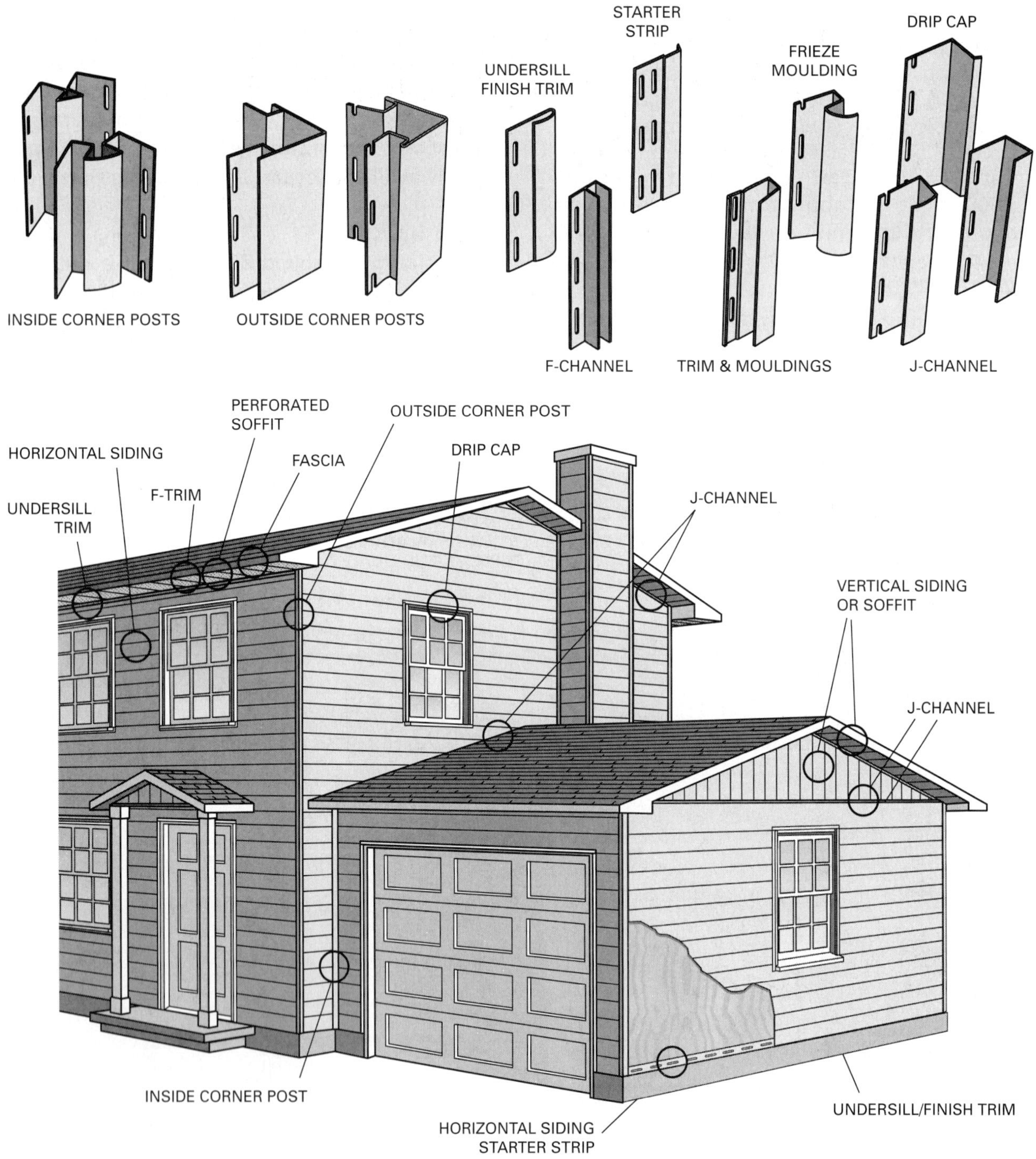

Source: From VOGT, Carpentry, 4E. © 2006 Delmar Learning, a part of Cengage Learning, Inc. Reproduced by permission. www.cengage.com/permissions

FIGURE 10.20

Vinyl siding being installed; building paper and trim pieces are applied to windows and corners in preparation for siding. Flashing for roof runs up behind the building paper.

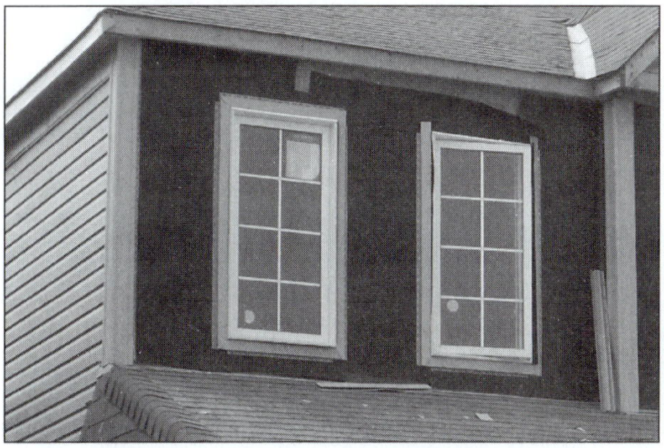

Courtesy Tom Stephenson

FIGURE 10.21

Aluminum fascia, soffit, and porch ceiling, installed with vinyl siding on walls, and polyurethane fluted columns at sides of door.

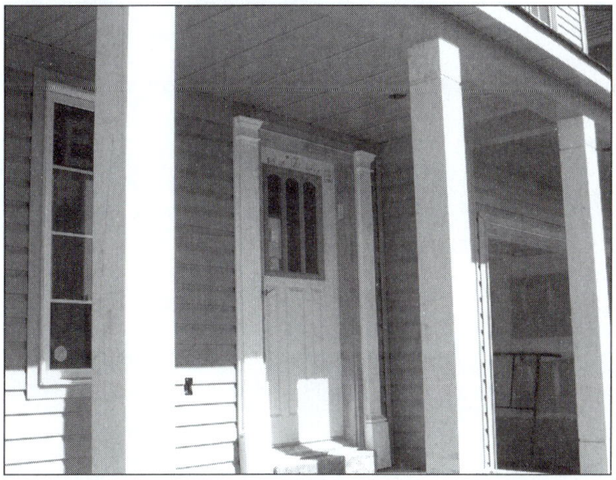

Courtesy Tom Stephenson

FIGURE 10.22

Finished vinyl siding installed around window; wide trim detail used to emulate wooden clapboard siding.

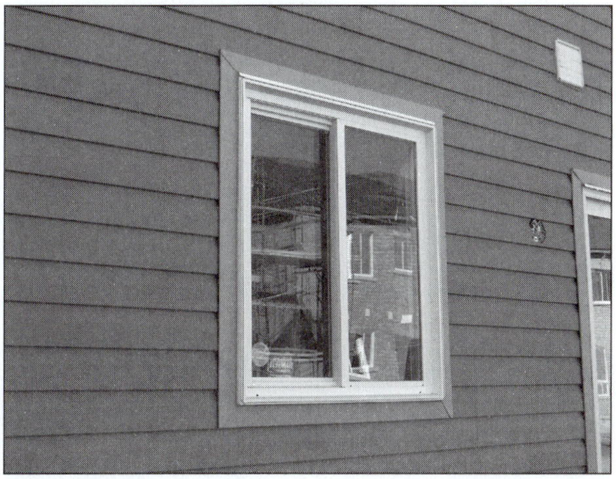

Courtesy Tom Stephenson

FIGURE 10.23

Vinyl siding being installed above masonry veneer. The gable end of the house is cantilevered so that the face of siding meets the face of the masonry veneer. Wherever dissimilar materials meet, care needs to be taken to ensure that any changes in material thickness are accounted for in the structure. For example, the gable end must be cantilevered or extended out to compensate for the air space and thickness of brick.

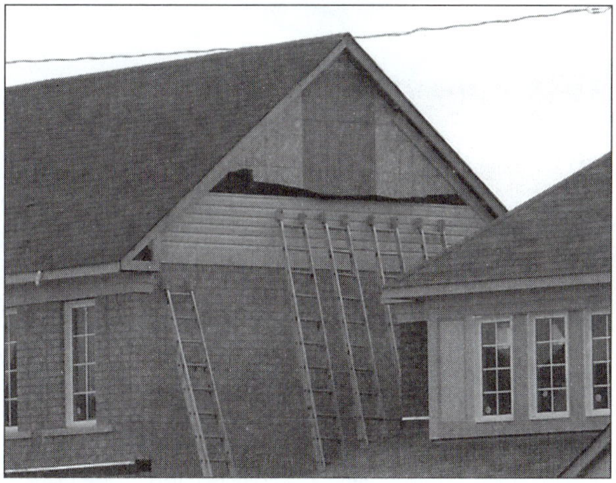

Courtesy Tom Stephenson

Wood Siding

Wood siding comes in a variety of shapes and sizes and can be placed either vertically or horizontally. It is good practice to ensure that the back side of the siding is sealed, particularly if the front side is to be painted. This will prevent the wood from absorbing moisture from the back of the siding and peeling paint on the front face of the siding. Drawing details may require the wall to be strapped out to provide an air space and ventilation behind the siding (refer to Figure 10.24).

FIGURE 10.24

An example of siding that is furred out to provide an air space for ventilation. The term furring refers to strapping of a wall for support and straightening purposes. In this case the wall is furred out to leave an air space so moisture can dry out if it penetrates through the wood siding.

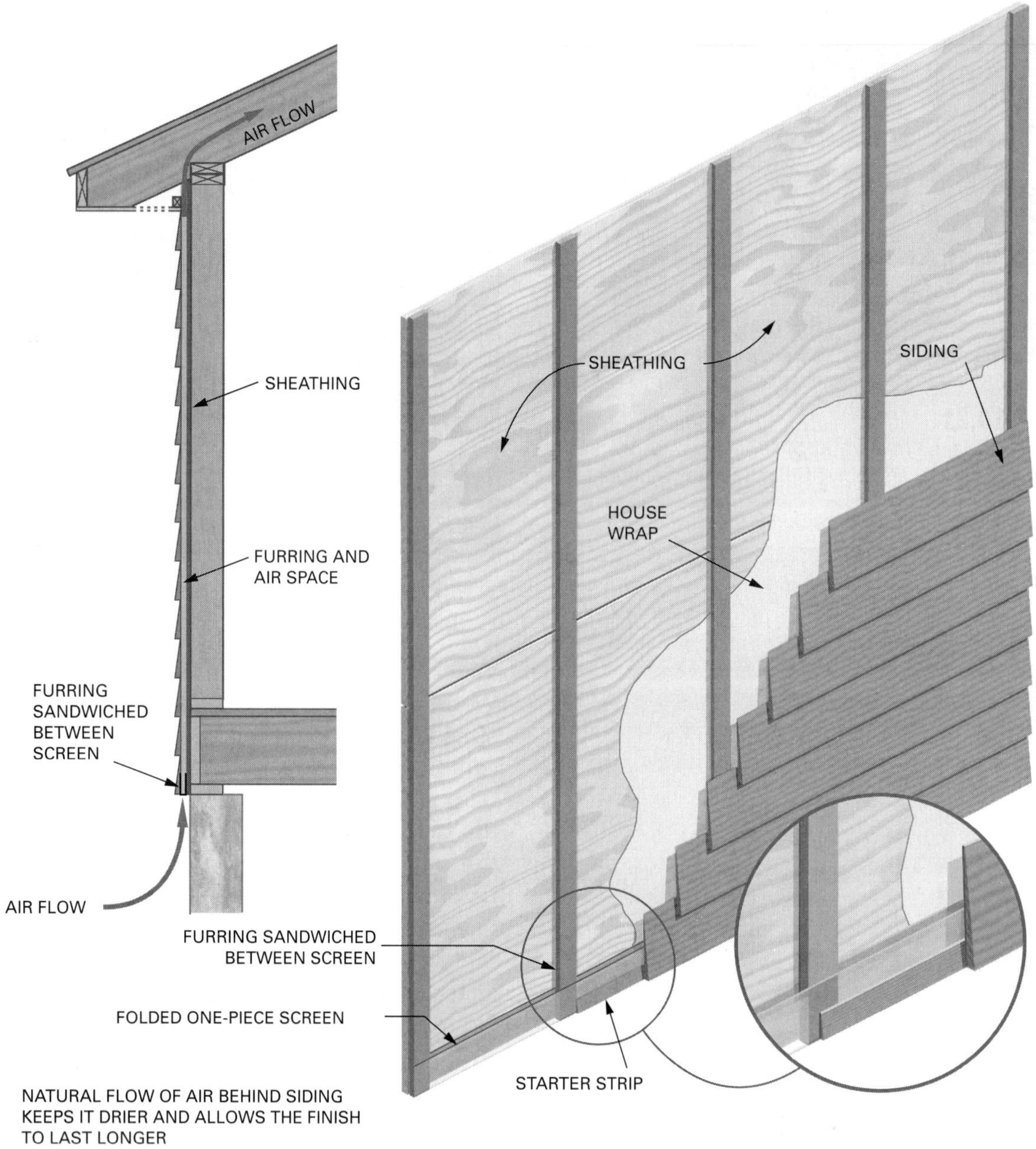

FIGURE 10.25

Vertical and horizontal siding sections and their nailing requirements.

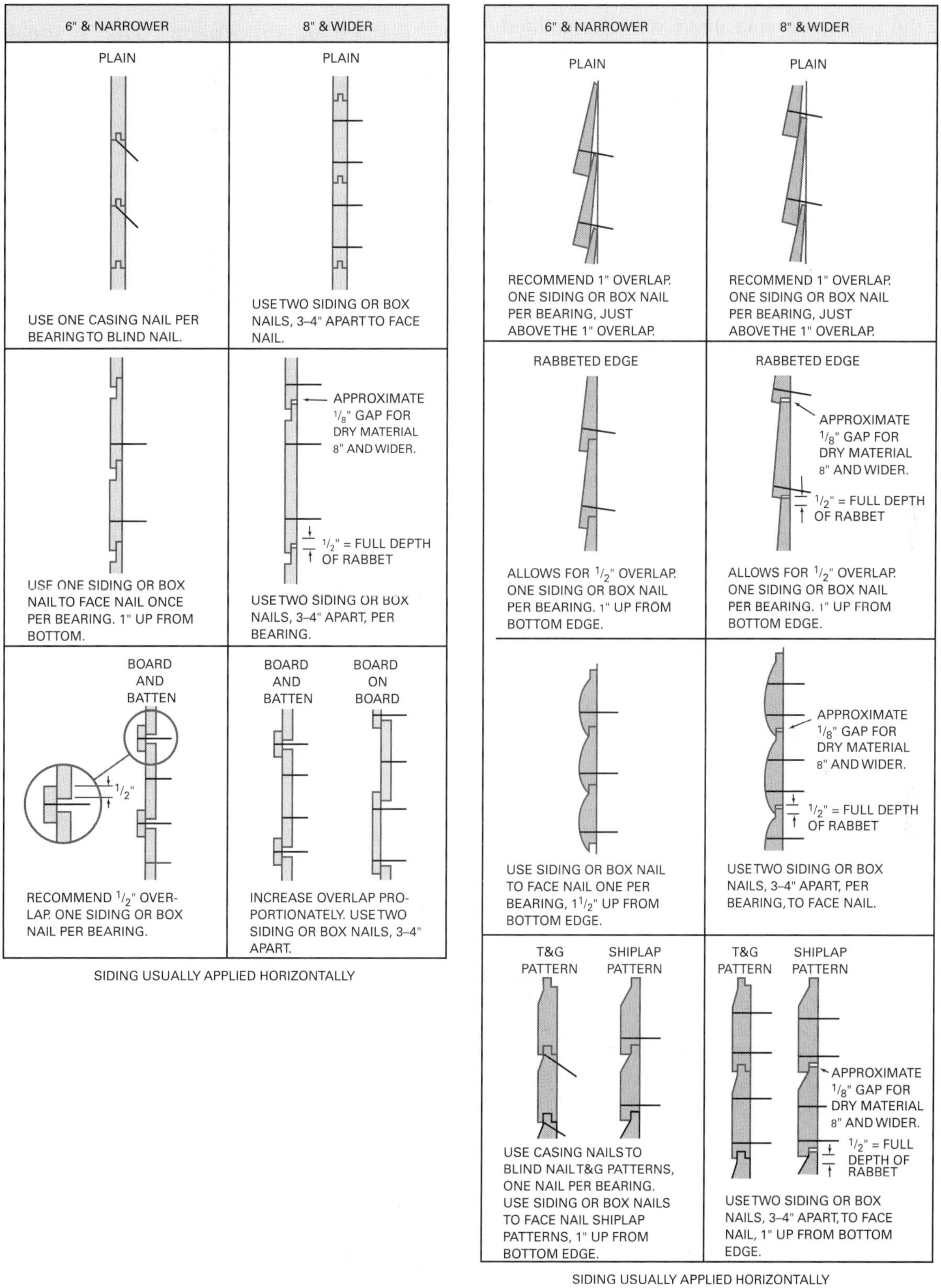

Source: From VOGT, Carpentry, 4E. © 2006 Delmar Learning, a part of Cengage Learning, Inc. Reproduced by permission. www.cengage.com/permissions

Stucco and Exterior Insulated Finish Systems

Two methods of installing stucco are commonly used, the traditional one and Exterior Insulated Finish Systems (EIFS) system, which is technically not stucco but can resemble stucco in appearance. Both methods require a sub-base that forms an effective base for the finish stucco coating to adhere to. The traditional method uses a Portland cement-based coating, while the more modern EIFS system relies on a polymer coating that is placed over a fibreglass mesh. A key characteristic of EIFS is that it uses expanded (sometimes extruded) polystyrene as a substrate. The Portland cement-based coating is much more brittle and tends to crack more over time, whereas the EIFS system is much more resistant to cracking. Both systems are comparable in cost, depending on the finishes and decorative mouldings that are applied; these costs are also comparable to brick veneer but are higher than most manufactured sidings. The appearance of stucco can be economically achieved in smaller areas by using prefinished panels that look like stucco.

Traditional Stucco Systems

Installation using the traditional stucco method involves the following steps:

- Placing building paper over a sheathing material that may consist of plywood, oriented strand board, or exterior drywall.
- Installation of a galvanized wire or fibreglass mesh, which acts as a means of attaching the concrete plaster to the wall and helps prevent cracking.
- Parging a scratch coat of mixed water, Portland cement and sand over the wire mesh to form the base for the brown and finish coats. It is called a scratch coat because it is created with a scratch tool that intentionally roughens the surface to make the next coat adhere better.
- The brown coat evens the surface and removes any imperfections.
- The finish coat involves spraying or trowelling on the stucco, depending on the desired surface texture. Sometimes the brown coat is painted and then the stucco coating is sprayed on after.

FIGURE 10.26
Traditional stucco method.

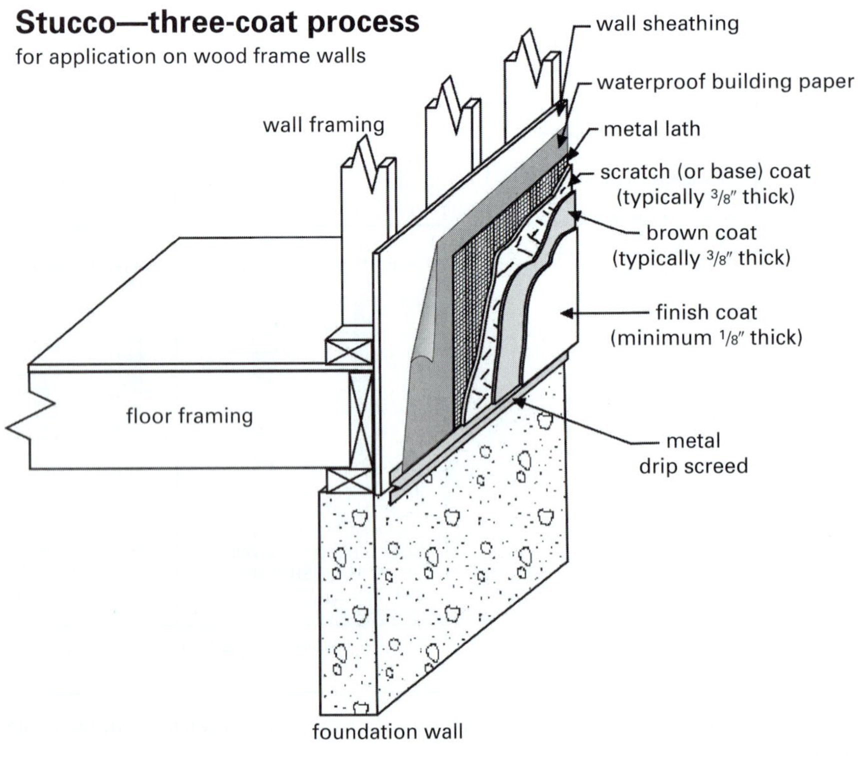

Source: Image excerpted from The Illustrated Home by Carson Dunlop www.carsondunlop.com

EIFS Systems

This system has received a lot of attention in terms of its ability to prevent water migration into the building envelope and structure, particularly in areas of high precipitation such as British Columbia. As with all cladding systems, care must be taken in choosing an installation method that meets local climatic conditions, and the relative exposure of the façade (e.g., size of overhangs and location of adjacent buildings, and the type of air barrier and vapour barriers used on the building). Particular attention should be paid to any drawing details provided by the designer and manufacturer with regard to EIFS installations. Site inspections must be done carefully to make sure that the system has been correctly installed.

FIGURE 10.27

Below is an example of an exterior insulated finish system (EIFS) manufactured by Dryvit. Note the direction and spacing of the notched trowel adhesive ribbons and also note the grooves indicated at the back and edge of the expanded polystyrene (EPS) insulation used in conjunction with the drainage strip at the base (different manufacturers' designs and systems may vary and their installation requirements must be closely adhered to).

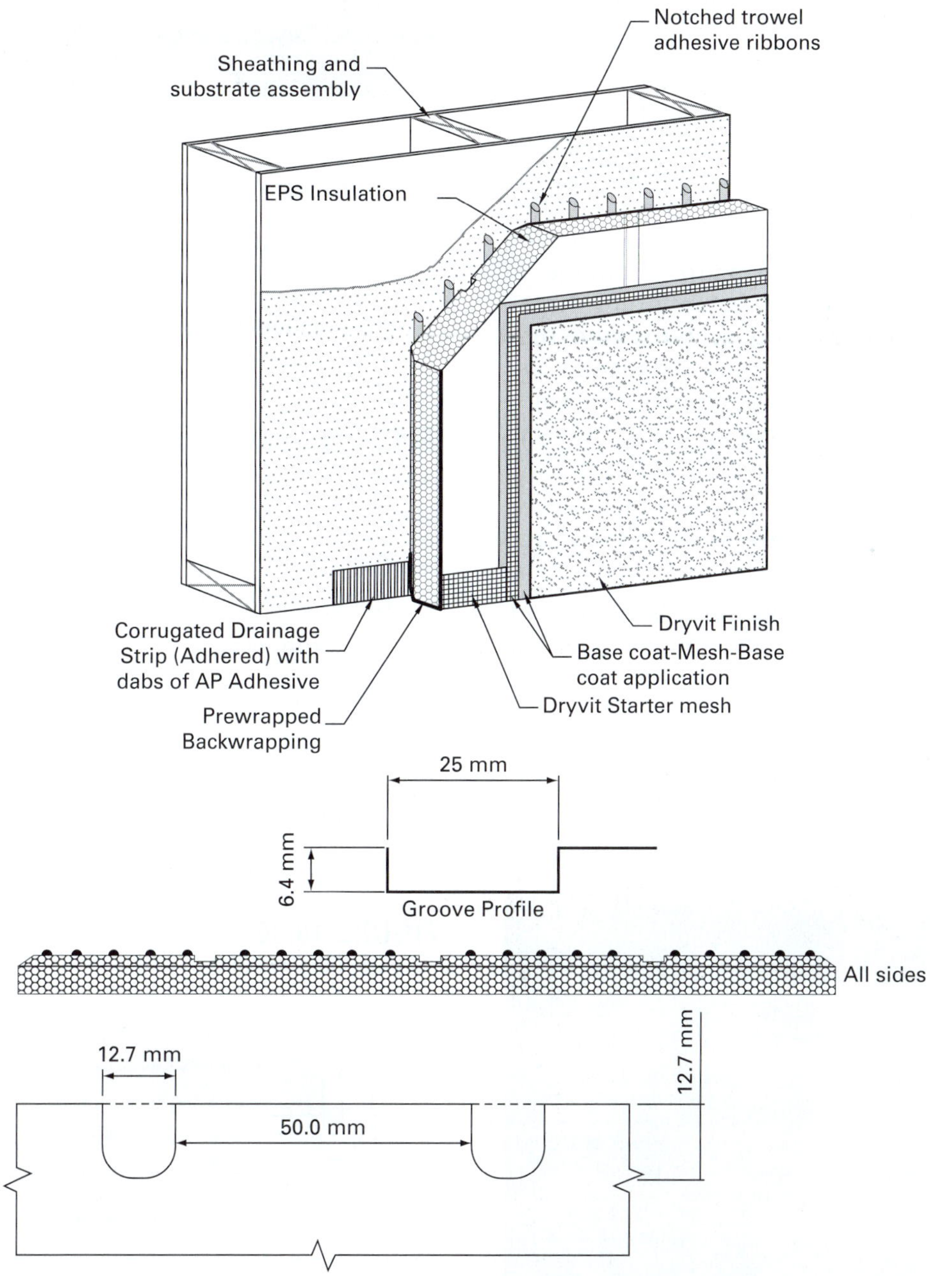

Source: Dryvit Systems Inc.

There are several installation methods, and each manufacturer's instructions should be followed. Two common methods are:

- Dual barrier source drained, which provides a secondary barrier to moisture penetration that allows excess moisture to drain at joints and intersections.
- Drainage and/or ventilated-screen EIFS, which provides small drainage outlets between the Styrofoam and sub-base to allow excess moisture to escape. This method is becoming the most often specified.

Although local codes and manufacturer's requirements may vary, installation of the face-sealed EIFS system typically requires:

- Installation of a Styrofoam base that is fastened to the sheathing with specially designed washers. The Styrofoam is also bonded to the sheathing with a specially designed adhesive applied with a notched trowel. The type of notch and its spacing may be designed to ensure drainage of moisture out of the base. Different manufacturers have varying systems and may use expanded polystyrene (EPS) that has drainage grooves cut into the backside at defined depths and spacing (refer to Figure 10.27).
- The surface of the Styrofoam is then rasped straight and smooth with specialized rasping tools.
- A fibreglass reinforcing mesh is then sandwiched into a base coat applied over the Styrofoam. The fibreglass reinforcing must be properly lapped at all joints and back wrapped at all termination points.
- A polymer finish coat is typically applied to the surface with a trowel, floating the finish to provide an even surface texture and colour.

FIGURE 10.28
Window jamb showing the back wrapping of fibreglass mesh behind the Styrofoam insulation.

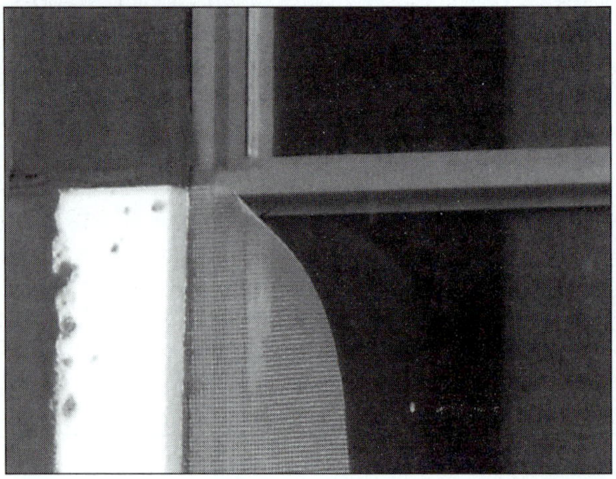

Courtesy Tom Stephenson

FIGURE 10.29
Manufactured stucco board placed in panels to emulate traditional Tudor design.

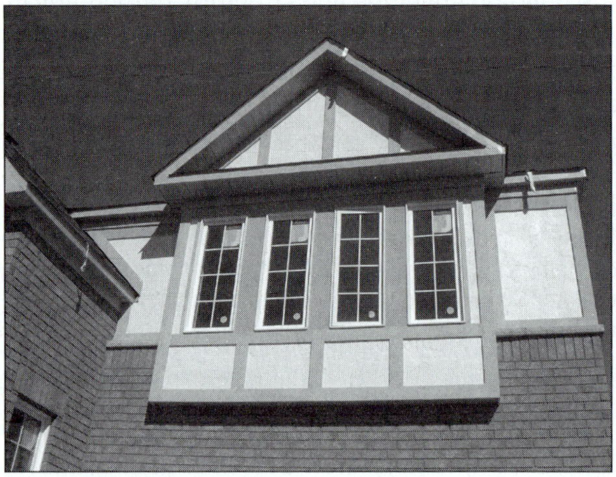

Courtesy Tom Stephenson

ROOFING MATERIALS

A number of roofing materials are used today in construction, including:

- Asphalt
- Cedar shingle and shakes
- Concrete, clay, and fibreglass
- Slate
- Steel
- Tar and gravel built-up roofs (for flat roofs)
- Membrane (for flat roofs).

FIGURE 10.30
Asphalt shingle exposure and lap.

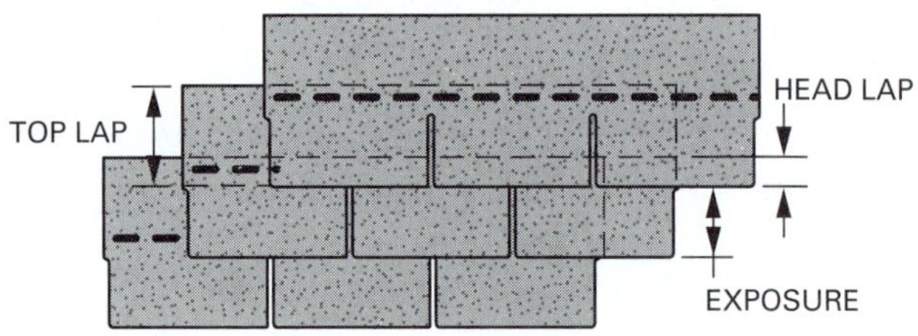

Source: From VOGT, Carpentry, 4E. © 2006 Delmar Learning, a part of Cengage Learning, Inc. Reproduced by permission. www.cengage.com/permissions

FIGURE 10.31

Step flashing installed at intersection of roof and brick wall. Step flashing is placed between each layer of shingles where a roof intersects a wall or chimney.

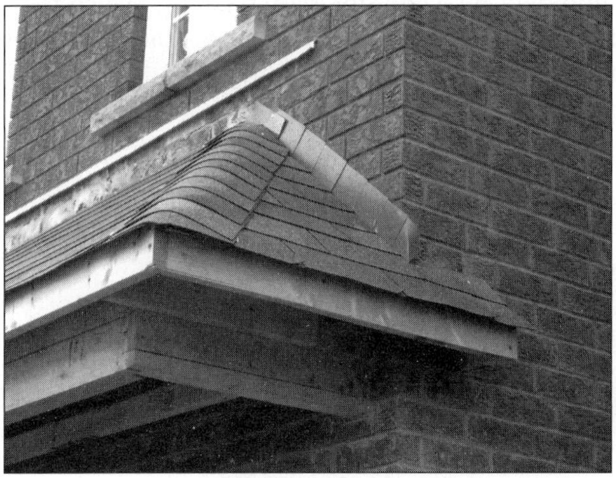

Courtesy Tom Stephenson

FIGURE 10.32

Finished counter (cap) flashing placed over step flashing and under horizontal flashing that runs horizontally across the top of roof. The counter flashing is bent at the top and tucked into a mortar joint that has either not been filled or has been ground out before sealant is applied. Also note the through-wall flashing that runs horizontally along the top of the roof. The through-wall flashing goes back behind the brick to the frame wall, where it has building paper lapped on top of it so that any moisture that accumulates can run out through the weep holes that are left open on top of the through-wall flashing.

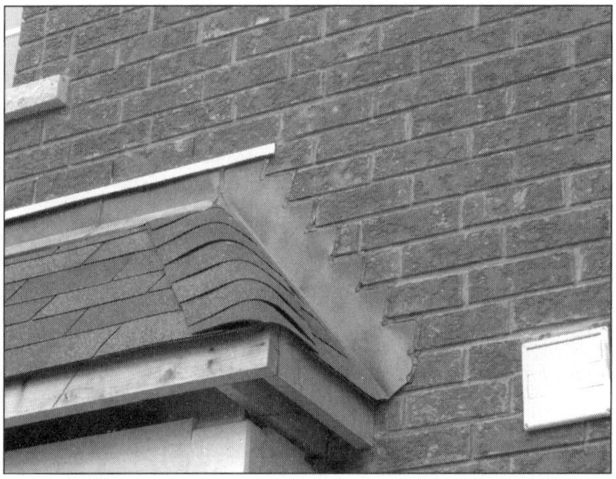

Courtesy Tom Stephenson

FIGURE 10.33

Step flashing running up roof and second-floor intersection. The wall will later have vinyl siding applied over the step flashing. In this case the siding acts as a counter or cap flashing.

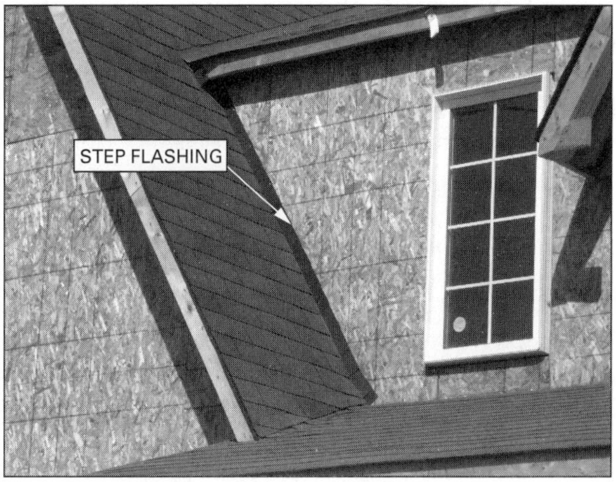

Courtesy Tom Stephenson

FIGURE 10.34

Cedar shingle roof being installed with ice and water shield installed along eaves. Discussions in Building Science may question the use of ice and water shield under cedar shingles as opposed to detailing that allows the shingles to dry from underneath. Notice that the walls have Tyvek™ house wrap with strapping. This detail involves having wood siding fastened to the strapping, allowing for an air space behind the siding for the siding to dry out and better resist warpage.

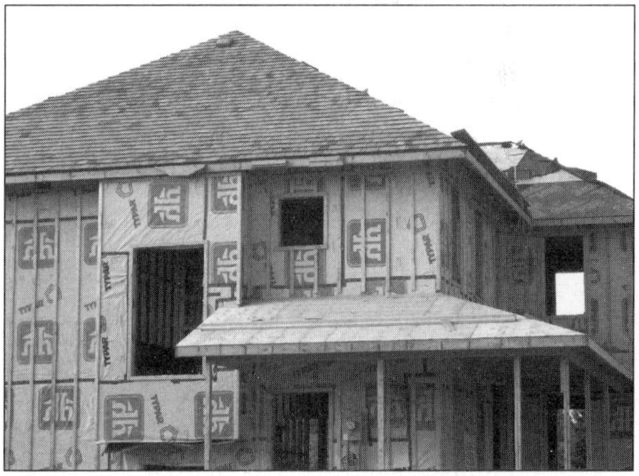

Courtesy Tom Stephenson

Asphalt is the most commonly used material due to its relative low cost and ease of installation. The life span of any roof will depend on the type and quality of material, installation method, and climatic conditions. A typical lifespan might range from 15 years for a flat roof using tar and gravel to over 50 years for a properly installed sloping cedar roof. As with wall systems, most leaks occur at intersections with roof systems. Because of this problem extra care needs to be taken at all intersections, including valleys, chimneys, dormers, and **eaves (cornice)**. Specifications may call for special products such as ice and water shield to be installed under the shingles to prevent ice damming at eaves and intersections such as valleys causing potential water leakage. (Ice and water shield is a peel-and-stick membrane that self-seals around any nails that penetrate through the shingles, providing an extra layer of protection in vulnerable areas of the roof.) **Flashing** will be required at other intersections, and detail drawings may be provided by the designer or manufacturer for more complex or unique intersections.

Flashing types may include **step flashing,** which is a series of L-shaped pieces of metal placed between the shingles at the intersection between a wall and a roof. The step flashing is then covered over by a **counter (cap) flashing,** which may be a separate piece of metal or a siding material that laps over the step flashing.

FIGURE 10.35

A wood eaves and a vinyl/aluminum eaves.

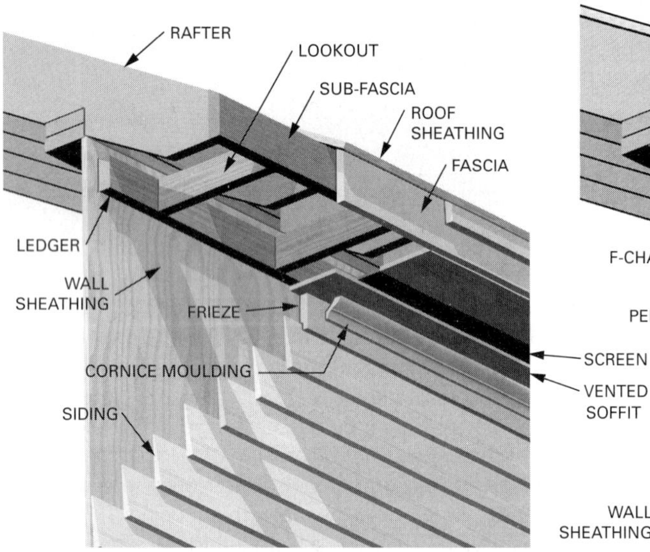

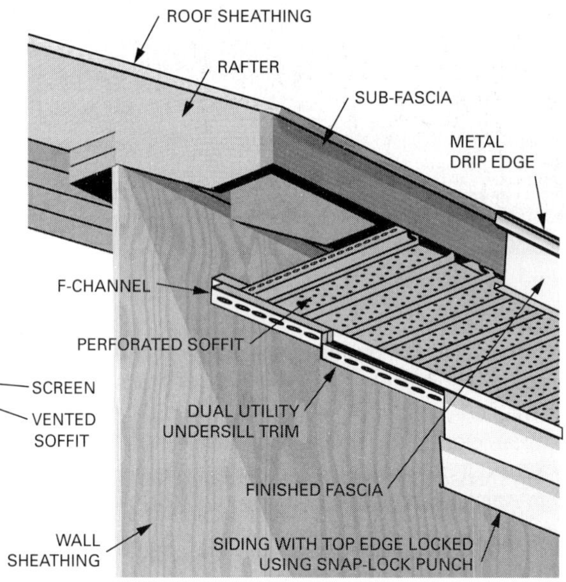

FIGURE 10.36

Section view of a wood eaves and a vinyl/aluminum eaves.

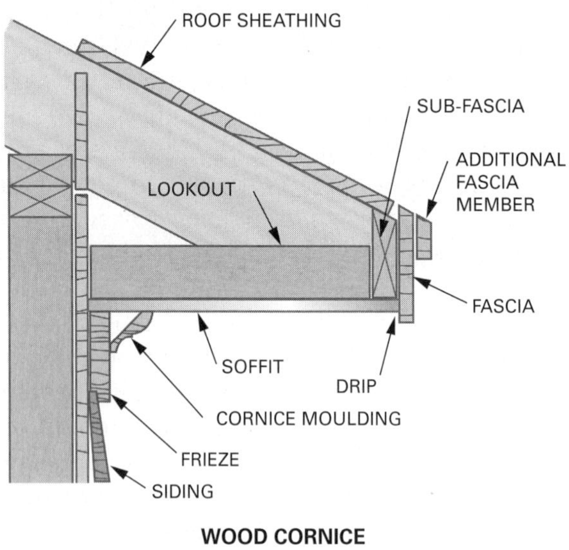

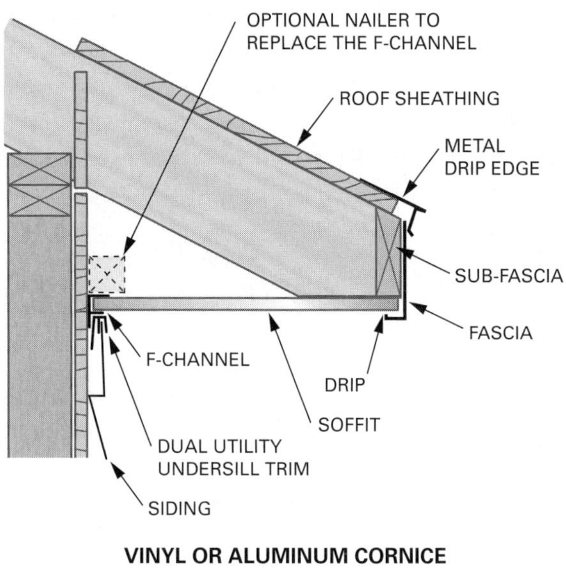

FIGURE 10.37

Aluminum fascia and soffit; notice serrations in soffit to provide ventilation in roof above.

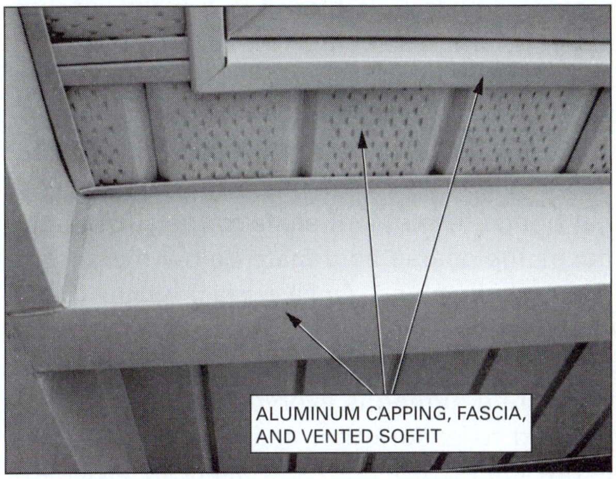

Courtesy Tom Stephenson

Eaves (Cornice)

The eaves is the portion of the roof overhang located at the intersection of the roof and exterior wall, and may include a number of different designs and materials. The fascia and soffit are part of the eaves and can be finished with wood that is later painted, or capped with aluminum sheet material and vinyl or aluminum soffit. Generally, the soffit will be vented in order to provide ventilation to the area above the insulation in the ceiling area.

FINISH CONCRETE

Besides being used to form foundation walls, concrete is also used to finish porch slabs, garage floors, sidewalks, and basement floor slabs. Concrete is one of the strongest and most durable building materials when properly installed. Concrete can be specified to have a high compressive strength, which is measured in **megapascals (MPa)** or pounds per square inch (PSI). While concrete provides high compressive strength, it provides relatively low tensile strength. Adding **reinforcing steel** (commonly referred to as rebar) to concrete greatly increases its tensile strength. Therefore, reinforcing rods are specified for conditions where higher tensile strength is required. The reinforcing rods and their installation will be indicated on the drawings. Reinforcing rod sizes are provided based on the diameter of the bar. For example, a 10M bar refers to a bar with a diameter of approximately 10 mm. Standard bar diameters go up in 5 mm intervals, starting at 10M and going up to 35M; larger sizes are also available (refer to Chapter 16 for more detail).

Successful placement of concrete relies heavily on following the designer's guidelines for strength and being mindful of weather conditions. Concrete is sensitive to both cold and hot weather conditions, and the surface will not cure or finish properly if it becomes wet due to rain or snow in the early stages of placement.

Concrete is comprised of a combination of the following materials:

- Cement
- Potable water
- Fine aggregate (concrete sand)
- Coarse aggregate (gravel).

The right proportions of the above ingredients will provide concrete with the appropriate strength, finish, and durability. In order to extend the lifespan of the concrete, air is added to concrete that will be exposed to freeze–thaw cycles and chlorides (de-icing salts). The percentage of air, known as **air entrainment,** required will be listed in the construction notes or specifications found on or with the drawings. Air is added to concrete for conditions such as porch, garage, and sidewalk slabs. A typical basement slab would not require air, as it is not exposed to chlorides or freeze–thaw cycles.

Concrete Strength

As previously described, a Mega Pascal (MPa) is a measure of the compressive strength of concrete (metric units of measurement). One MPa is equal to one million Pascals (Pa). This is equal to one million Newtons of force per square metre. The higher the MPa of concrete, the stronger the compressive strength will be. Concrete can also have its compressive strength given in pounds per square inch (imperial units of measurement). Pounds per square inch (psi) is usually easier to visualize, as it is the force in pounds applied to a square inch of concrete. ASTM provides standards that are followed to determine the compressive strength and performance requirements for concrete is met. Engineers design concrete specifications based on the individual project requirements and the loads that the concrete will face in practice. It is therefore important that contractors and trades follow the construction notes, specifications, and correct placement requirements for concrete very carefully to

TABLE 10.1

Concrete Strength in pounds per square inch (PSI)	Concrete Strength in MPa Hard Conversion	Concrete Strength in Mega Pascals (MPa) Soft Conversion (relationship factor MPa = psi × 0.006895
2,000 psi	15 MPa	13.79 MPa
3,000 psi	20 MPa	20.685 MPa
3,500 psi	25 MPa	24.1325 MPa
4,000 psi	30 MPa	27.58 MPa
5,000 psi	35 MPa	34.475 MPa
6,000 psi	40 MPa	41.37 MPa
7,000 psi	50 MPa	48.265 MPa

ensure that costly errors are avoided. Remember that when concrete strength is compromised, structural and aesthetic issues will follow.

Table 10.1 provides a comparison between some of the more common concrete strengths used for building construction. Both hard and soft metric conversions are indicated, as is the conversion factor that can be used to convert to Mega Pascals by multiplying by the PSI amount. For example, 2,000 PSI × .006895 conversion factor = 13.79 MPa soft conversion; the drawings would likely state the hard conversion amount of 15 MPa if metric was the unit of measurement used on the drawings (metric strength is most commonly used). Conversely, if 32 Mpa was given on the drawings, then divide that by the conversion factor 0.006895 which equals 4,641 psi.

Soft conversion is an exact conversion to a determined amount of precision, while in hard conversion the amount has been converted to modular metric increments (the same process used for on centre spacing, discussed in Chapter 2).

Finishing Concrete

Forms must be suitably constructed to retain and properly shape the concrete as it is poured in place. Concrete is placed in a plastic/mouldable state and is screeded to level off the surface. Once screeding has taken place, the surface is bull floated. A bull float is a large trowel-type tool that is used to prepare the surface for later finishing through a manual vibrating process that brings the cement paste to the surface and moves the coarse aggregate below the surface for easier finishing. Once surface water (bleed water) leaves the surface, then final floating and trowelling can take place. When it comes to concrete, timing is everything. Experience is required to ensure that concrete does not set before final trowelling is applied. To provide a slip-resistant surface, the concrete should be floated and, if specified, given a broom finish. If a smooth surface is required, the concrete

FIGURE 10.38
Concrete ingredients.

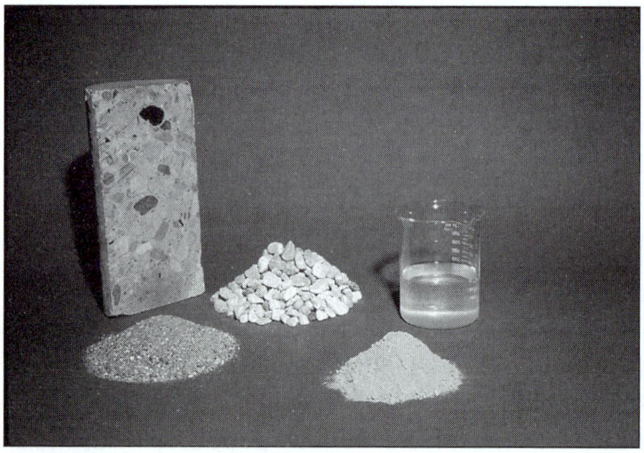

Courtesy Portland Cement Association

FIGURE 10.39
Concrete forming being constructed for an unexcavated porch and step.

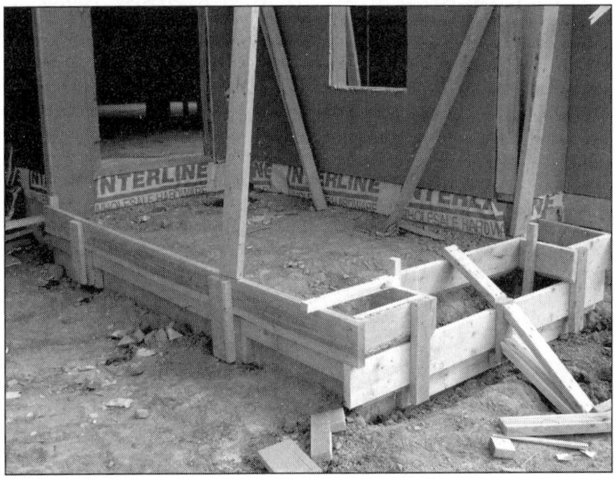
Courtesy Tom Stephenson

FIGURE 10.40
Bull-floating a concrete surface.

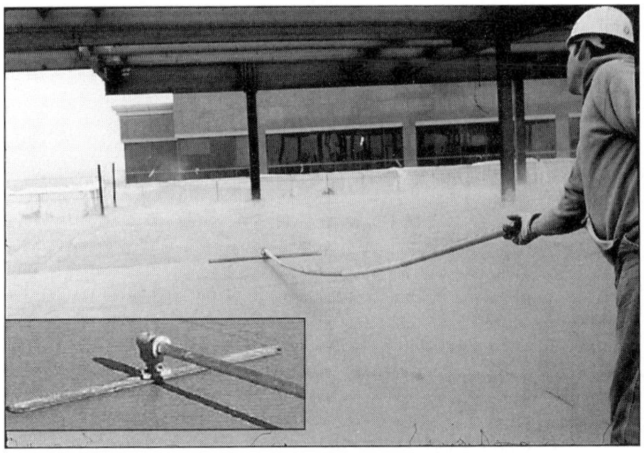
Courtesy Portland Cement Association

FIGURE 10.41
Power-floating a concrete surface.

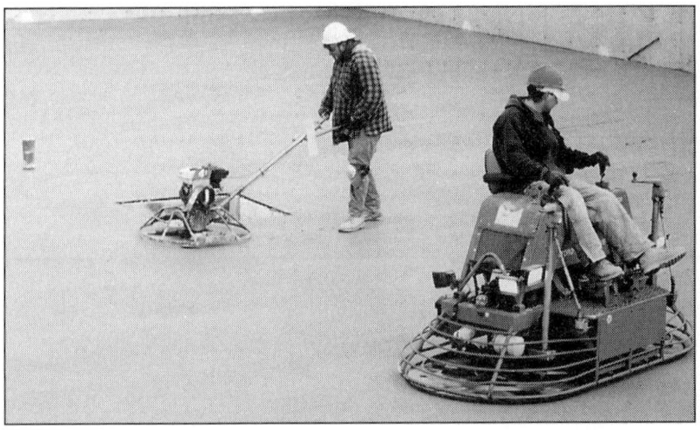
Courtesy Portland Cement Association

FIGURE 10.42
Hand-floating a concrete surface with a wooden float.

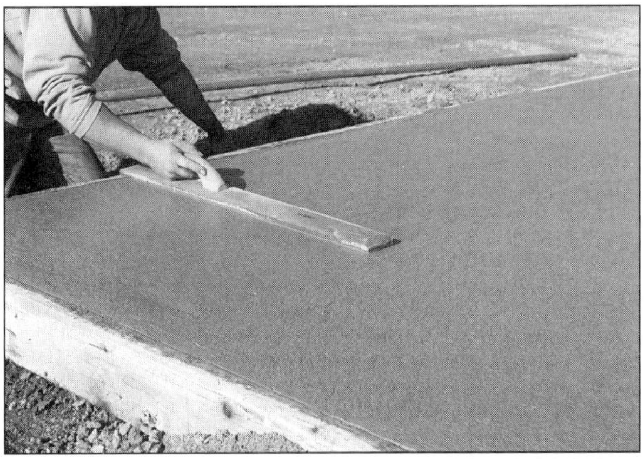
Courtesy Portland Cement Association

FIGURE 10.43
Hand-trowelling a concrete surface with a steel trowel.

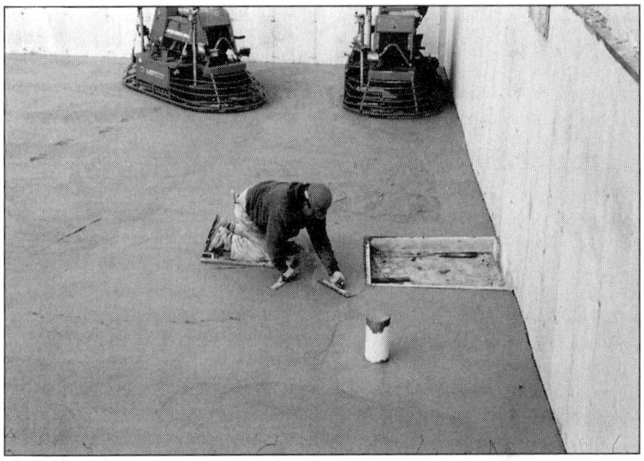
Courtesy Portland Cement Association

should be trowelled after floating. The difference is that the float provides a coarse surface, such as that found in sidewalks, while a trowel provides a smooth surface, such as that found in a basement floor slab.

EXTERIOR FINISHES AND THE DONCASTER HOUSE

The Doncaster house uses brick veneer cladding with aluminum fascia, soffit, and eavestroughs. A number of decorative details are shown on the elevations, including the following shown on the right rear elevation in Figure 10.44 below:

- Asphalt shingles
- Aluminum-clad frieze board
- Painted crezone panel (plywood with a water-resistant surface)
- Brick soldier course over window
- Decorative stack bonded header course at sides of windows. The header brick has a 10 mm projection from the other brick in the wall.
- Right casement windows open and are hinged on side of dashed-line intersection, while left windows are fixed: window sizes provided in window. FG refers to fixed glass.

FIGURE 10.44

Right rear elevation close-up of Doncaster drawings.

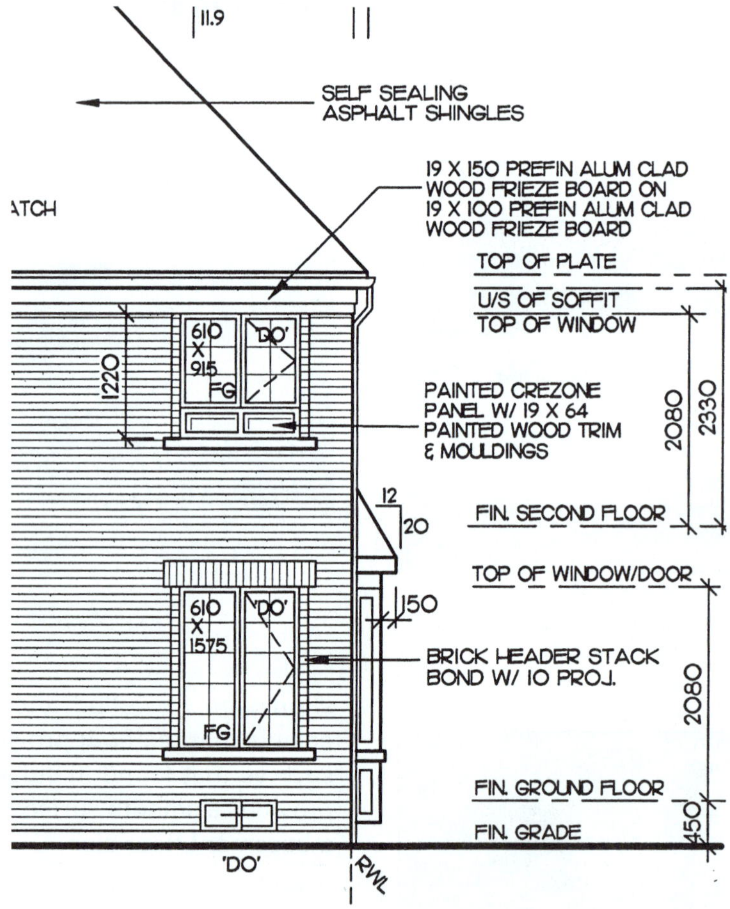

Source: Cassidy & Co. Architectural Technologists

FIGURE 10.45

Right rear elevation close-up of Doncaster house; eavestrough not yet installed (compare with drawing in Figure 10.44).

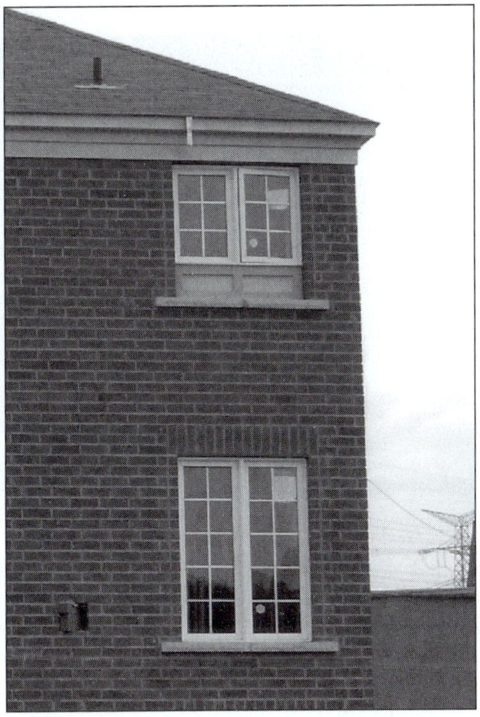

Courtesy Tom Stephenson

FIGURE 10.47

Front gable roof over porch. Compare the primed crezone panel, aluminum soffit, fascia, capped wood support beams, and wood columns with the Doncaster front elevation close-up used in Figure 10.46.

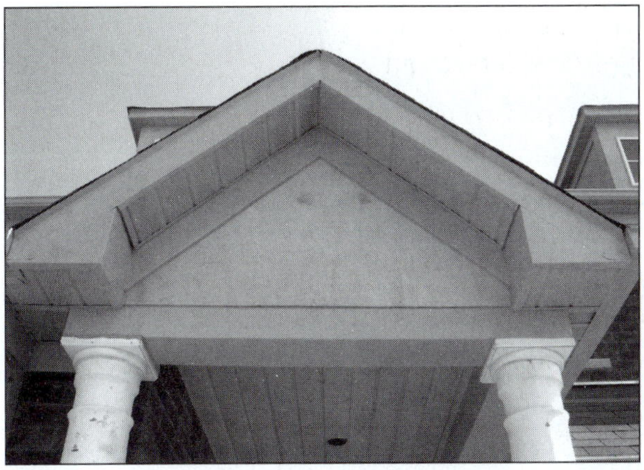

Courtesy Tom Stephenson

FIGURE 10.46

Close-up of Doncaster front elevation drawing.

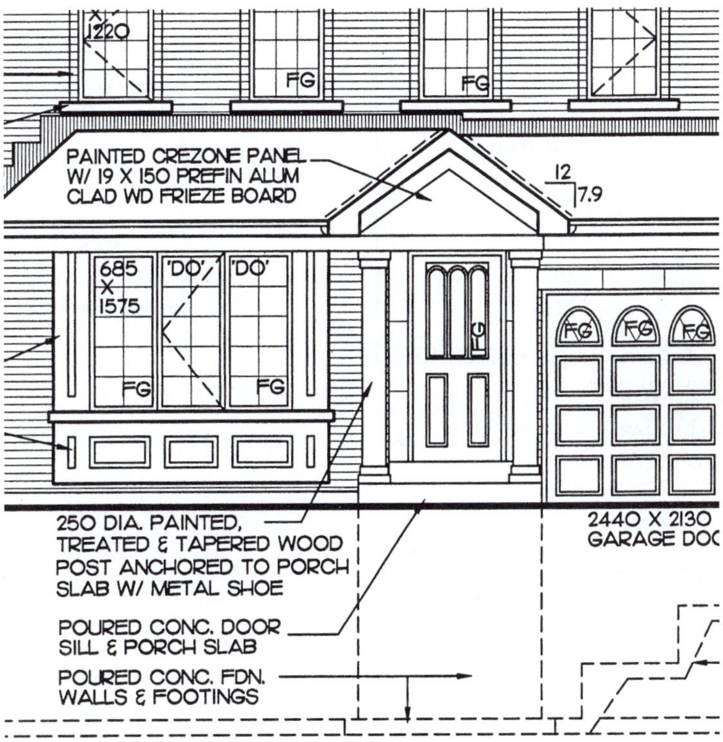

Source: Cassidy & Co. Architectural Technologists

FIGURE 10.48

Doncaster column post anchored to porch slab with metal shoe as per close-up of front elevation drawing. Metal shoe prevents post from absorbing moisture and rotting prematurely (refer to figure 10.46).

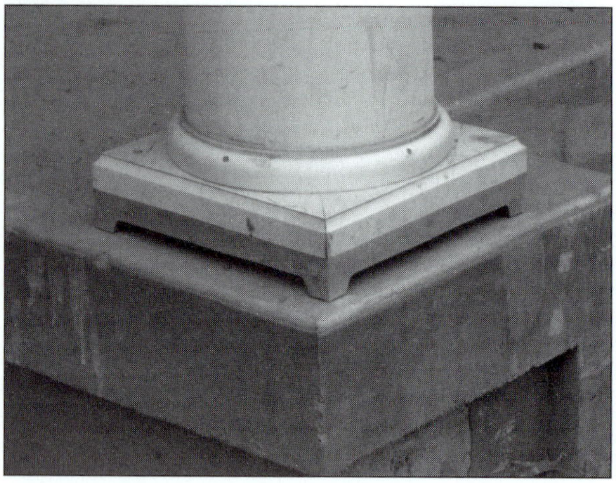

Courtesy Tom Stephenson

FIGURE 10.50

Reinforcing and 6 × 6 WWM (welded wire mesh) on gravel base; notice the pocket in the wall for the reinforcing bars and grade beam to rest on (refer to Figure 10.49).

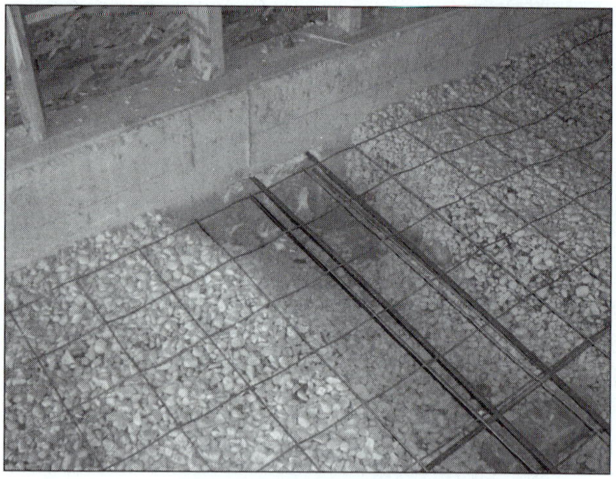

Courtesy Tom Stephenson

FIGURE 10.49

Close-up of Doncaster garage floor plan, and grade beam detail; the grade beam detail indicates 4-10M rebar (refer to figure for actual placement of rebar).

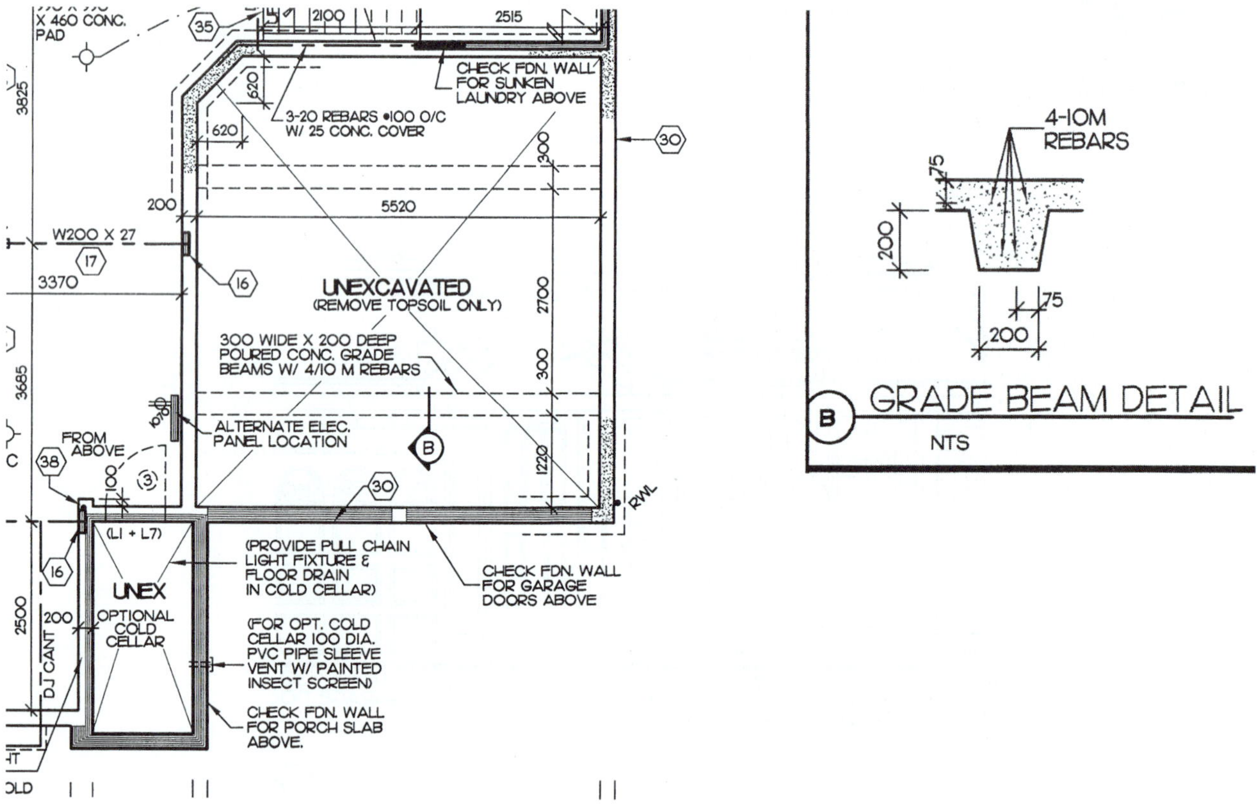

Source: Cassidy & Co. Architechural Technologists

FIGURE 10.51

Base of garage floor prepared for concrete (refer to Figure 10.49).

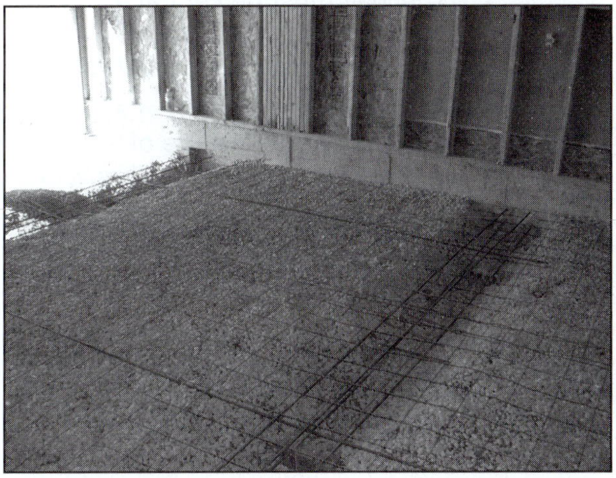

Courtesy Tom Stephenson

FIGURE 10.52

Pouring concrete in garage. Note worker lifting wire mesh so that it is fully embedded in concrete. Ideally, wire mesh should be placed on chairs or spacers that hold it up off the gravel. Grade beam has already been poured (refer to Figure 10.49).

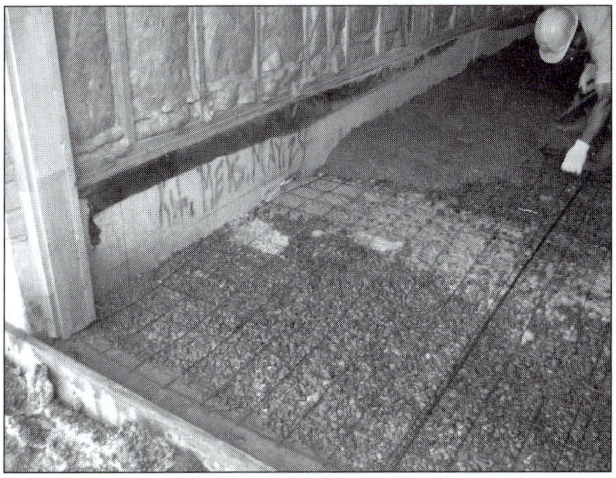

Courtesy Tom Stephenson

FIGURE 10.53

Construction note from Doncaster drawings specifying the strength, thickness, and percentage air entrainment additive for the garage slab. The slab is sloped to the front to allow the water to drain off if the car is wet or if snow or ice melts onto the floor.

(18) 100mm (4") CONC. SLAB SLOPE TO FRONT. CONC. STRG 32 Mpa (4650 PSI) WITH 5–8% AIR ENTRAINMENT. FILL BENEATH SLAB TO BE COMPACTED TO PROVIDE UNIFORM SUPPORT.

Source: Cassidy & Co. Architectural Technologists

FIGURE 10.54

Construction note from Doncaster drawings specifying strength, thickness of basement slab, and thickness of gravel (crushed stone). Note that the strength varies depending on dampproofing. Dampproofing is referring to the placement of 6 mil polyethylene prior to the pouring of the concrete. The increased strength reduces the permeability of the concrete, thus reducing the need for the dampproofing layer. Often builders will elect for the increased strength so that they don't have to place the polyethylene under the concrete, as it slows the initial set of the concrete, requiring finishers to take more time, often staying late. The gravel acts as a capillary break, preventing moisture from being absorbed and transferred through the slab by way of absorption.

(7) 80mm (3") 15Mpa (2200 PSI) CONC. SLAB ON 100mm (4") CRUSHED STONE WITH DAMPPROOFING, 25Mpa (3600 psi) CONC. WITHOUT DAMPPROOFING

Source: Cassidy & Co. Architectural Technologists

KEY TERMS

Air entrainment (p. 247)
Building wrap (p. 231)
Counter (cap) flashing (p. 246)
Eaves (cornice) (p. 245)
End dam (p. 235)
Exterior insulated finish system (EIFS) (p. 230)
Flashing (p. 246)
Reinforcing steel (p. 247)
Step flashing (p. 246)
Weep holes (p. 231)
Mega Pascal (MPa) (p. 247)

Chapter 10 Assignment 1: Exterior Finishes and Doncaster Drawings

Answer questions 1 through 14 using the material in this chapter, and answer questions 15 to 20 using the Doncaster drawings. Identify the letter that best answers the question.

_____ 1. What does EIFS refer to?

a. exterior insulated finish system
b. exterior insulated finished siding
c. an eavestrough identification flow system
d. a duct system for heating

_____ 2. How is EIFS installed?

a. with screws and a grooved trowel that places adhesive which makes grooves to drain water cut out of the back of the expanded polystyrene.
b. with masonry ties placed with screws at 400 mm OC horizontally and 600 mm vertically; all fasteners must connect to studs
c. with nails fastened through the EIFS vinyl siding
d. with nails fastened through the fibre cement siding

_____ 3. What is 20 MPa in psi when using the conversion factor 0.006895?

a. .1379 psi
b. 2900.65 psi
c. 3500.62 psi
d. 5800 psi

_____ 4. What is 4000 psi in MPa using the conversion factor of 0.006895 for converting psi to MPa?

a. 27.58 MPa
b. 35 MPa
c. 40 MPa
d. 580,131 MPa

_____ 5. What temperature does the NBC require that the wall which the brick and mortar is laid in be kept at, and for how long?

a. The wall must be kept at a minimum temperature of 0 degrees Celsius for 48 hrs.
b. The wall must be kept at a minimum temperature of 5 degrees Celsius for 48 hrs.
c. The wall must be kept at a minimum temperature of 10 degrees Celsius for 48 hrs.
d. The wall must be kept at a minimum temperature of 5 degrees Celsius for 24 hrs.

_____ 6. Which of the following is an advantage of using brick veneer?

a. It is very durable.
b. It can be installed in all weather conditions with ease and no extra preparation.
c. It is inexpensive.
d. It is relatively easy to change colours once installed.

_____ 7. What dead and live loads does the brick in a brick veneer wall support?

a. the weight of the brick and mortar above
b. the weight of the brick, mortar, and subfloor
c. the weight of the brick, mortar, framing, and roof
d. the weight of the brick, mortar, framing, roof, and snow loads

_____ 8. Which one of the following, according to the NBC, is **NOT** allowed to support a masonry opening?

a. wood lintel
b. steel lintel
c. steel beam
d. masonry arch

_____ 9. Where must weep holes be provided?

a. around the top of chimneys
b. around the top of a concrete porch slab
c. at the base of a masonry veneer wall
d. in a poured concrete wall

_____10. What is the most durable masonry joint?

a. concave joint
b. raked joint
c. extruded joint
d. struck joint

_____11. When and how is a step flashing used and placed?

a. Step flashing is always placed over the shingles where a sloping roof meets a wall.
b. Step flashing is placed between the shingles where a sloping roof meets a wall.
c. EIFS acts as a type of step flashing where a wall meets a roof.
d. Siding acts as a type of step flashing where a wall meets a roof.

_____12. Where are the eaves (cornice) vented?

a. Fascia is usually vented.
b. Soffit is usually vented.
c. Frieze board is usually vented.
d. Subfascia is usually vented.

_____13. Where is air entrained concrete used, and why?

a. Air entrained concrete has air added to it when used indoors to help it cure.
b. Air entrained concrete has air added to it to help the concrete bond to the reinforcing steel for reinforced concrete slabs.
c. Air is added to concrete to increase durability in areas where it is exposed to freeze-thaw cycles and de-icing salts.
d. Air is added to concrete to increase its flexibility so that the concrete is more resistant to cracking in areas that come under pressure, such as garage slabs.

_____14. Which masonry joint is considered the least durable?

a. concave joint
b. raked joint
c. V-joint
d. flush joint

_____15. What is the width of the patio door (sliding door)?

a. 1830 mm
b. 2000 mm
c. 2050 mm
d. 3050 mm

_____16. What is the height of the patio door (sliding door)?

a. 1830 mm
b. 2080 mm
c. 2670 mm
d. 3050 mm

_____17. What is the overall size of the boxed window in the dining room?

a. 685 mm × 1575 mm
b. 685 mm × 2080 mm
c. 2055 mm × 2080 mm
d. 2055 mm × 1575 mm

_____18. Where is the frieze board located?

a. at the base where the brick meets the foundation wall
b. where the soffit meets the brick at the front of the house
c. at the junction between the roof over the garage and the brick
d. over the front door

_____19. How many rain water leader drops are there shown entering the drain system at grade level?

a. 2
b. 3
c. 5
d. 6

_____20. How many windows open on the front elevation?

a. 2
b. 3
c. 4
d. 5

Chapter 10 Assignment 2: Exterior Finishes and the Brook Drawings

Answer questions 1 to 5 using the material in this chapter, and answer questions 6 through 20 using the Brook drawings. Identify the letter that best answers the question.

_____ 1. What is the specific technical name given to bricks that are placed in an arch?

a. stretcher
b. header
c. soldier
d. voussoir

_____ 2. Why is a dampproofing layer often specified under a basement floor slab?

a. to add support for the slab
b. to prevent cracking of the slab due to expansion and contraction of the slab
c. to prevent spalling of the slab due to the change in temperature from inside of basement to underside of slab
d. to prevent moisture transfer through the slab

_____ 3. Which of the following concrete slabs would most likely have a broom finish specified?

a. basement slab
b. garage slab
c. sidewalk slab
d. cold room slab

_____ 4. Why is it advisable to furr out an exterior wall that is using a wood siding?

a. to be able to fasten the siding to the wall
b. to provide an air space behind the siding so it can dry out uniformly
c. to provide a means for making the siding straight on a new wall
d. to provide a structural method of transferring both live and dead load back to the structure

_____ 5. What is the name given to the brick or stone that is found in the middle of a segmental arch?

a. voussoir
b. key stone or brick
c. soldier
d. rowlock

_____ 6. Where is a stone face used on the drawings?

a. rear elevation A
b. rear elevation B
c. front elevation A
d. front elevation B

_____ 7. What is the name given to the positioning of how the brick is laid over the Bedroom 3 window in Elevation B?

a. header brick
b. brick rowlock header on top of brick soldier course
c. stretcher course with a header course on top
d. rowlock course

_____ 8. What size is the garage door in Elevation B?

a. 8'-0" × 7'-0"
b. 7'-6" × 7'-0"
c. 8'-6" × 7'-0"
d. 14'-0" × 7'-0"

_____ 9. How many windows open on this house, using the Elevation A model, in total (all four elevations)?

a. 12 windows
b. 13 windows
c. 15 windows
d. 17 windows

_____10. What is the strength of concrete required for the porch slab for Elevation A?

a. 15 MPa
b. 20 MPa
c. 27 MPa
d. 32 MPa

_____11. What is the strength requirement for the garage slab floor?

a. 15 MPa

b. 20 MPa

c. 27 MPa

d. 32 MPa

_____12. How thick and what strength is the slab in the basement of the house required to be?

a. 3" thick and 20 MPa strength concrete when floor is resting on 4" of crushed stone and is dampproofed

b. 4" thick and 15 MPa strength concrete when floor is resting on 4" of crushed stone and is dampproofed

c. 2" thick and 30 MPa strength concrete when floor is resting on 6" of crushed stone

d. 3" thick and 30 MPa strength concrete when floor is resting on 4" of crushed stone

_____13. What is the outside measurement of the porch foundation walls for Elevation B?

a. 5'-0" × 5'-11"

b. 5'-0" × 3'-9"

c. 4'-11" × 5'-0"

d. 4'-2" × 4'-3"

_____14. What is the measurement from the finished first floor to the top of the transom window in the dining room?

a. 6'-8"

b. 7'-8"

c. 8'-0"

d. 8'-4"

_____15. How many basement windows are shown on the Elevation A model?

a. 2

b. 3

c. 4

d. 5

_____16. Where are cement fibre panels used as an exterior finish material?

a. at the rear elevation where the fireplace protrudes

b. at the front elevation of model Elevation B over the front porch

c. at the front elevation of model Elevation A at the face of the gable roof

d. at the right elevation of model Elevation A over the side entrance door

_____17. What type of curved arch is used over the first floor window on Elevation A?

a. roman

b. segmental

c. gothic

d. elliptical

_____18. What is the name for the direction of placement of the brick that runs directly under the sill for the bedroom 3 window in Elevation A?

a. soldier brick

b. header brick

c. rowlock brick

d. stretcher brick

_____19. What is the height of the metal railing that surrounds the flat roof on the second floor for Front Elevation B?

a. 1'-6"

b. 1'-8"

c. 3'-0"

d. 3'-6"

_____20. What is the measurement from the finished second floor to the underside of the second floor soffit?

a. 6'-10"

b. 7'-4"

c. 8'-1"

d. 8'-8"

Chapter 10 Quiz Review: Exterior Finishes Using Whittington Drawings and SWO Condition Drawing

Answer questions 1 through 16 using the Whittington drawings that came attached to the text. Answer questions 17 to 20 using the Whittington partial plans and elevations for SWO (side walkout) condition upgraded rear deck drawings based on the sloping grade condition found on the last page of the Whittington drawings. Identify the letter that best answers the question.

_____ 1. What type of shingles does the roof use?

a. self-sealing asphalt shingles
b. cedar shingles
c. slate shingles
d. clay tile shingles

_____ 2. What is the name for the direction of placement of the brick that runs up the exterior sides of the laundry room windows?

a. stack bonded soldier brick
b. stack bonded header brick
c. stack bonded rowlock brick
d. stack bonded stretcher brick

_____ 3. On what elevation of the house does the frieze board wrap/return 1000 mm along the wall and then stop?

a. the front elevation
b. the right elevation
c. the left elevation
d. the rear elevation

_____ 4. What room does the dormer window open up into (lead into)?

a. master bedroom near the walk in closet
b. bedroom 2
c. bedroom 3
d. only into closed attic space over living room

_____ 5. What is the measurement from the finished second floor to the underside of the second floor soffit?

a. 2080 mm
b. 2330 mm
c. 2460 mm
d. 2230 mm

_____ 6. What type of arch is used over the windows on the second floor that are directly over the garage door?

a. roman arch
b. segmental arch
c. gothic arch
d. elliptical arch

_____ 7. What type of panels are used on the exterior below the windows for the second floor bath?

a. decorative brick panels
b. vinyl siding panels
c. crezone panels
d. fibre cement siding panels

_____ 8. What is the name given to the positioning of how the brick is laid over the garage door?

a. header brick
b. brick rowlock on top of brick soldier course
c. stretcher course with a header course on top
d. rowlock course

_____ 9. How high is the porch railing from the top of porch to top of railing at the front entrance?

a. 900 mm
b. 1000 mm
c. 1200 mm
d. 1250 mm

_____10. How many windows on this house open, in total?

a. 10 windows
b. 11 windows
c. 12 windows
d. 14 windows

_____11. What is the strength of concrete required for the porch slab?

a. 15 MPa
b. 20 MPa
c. 27 MPa
d. 32 MPa

_____12. How thick is the porch slab over the optional cold room, if chosen?

a. 3" thick

b. 4" thick

c. 5" thick

d. 6" thick

_____13. What is the strength requirement for the garage slab floor?

a. 15 MPa

b. 20 MPa

c. 27 MPa

d. 32 MPa

_____14. How thick and what strength is the slab in the basement of the house required to be?

a. 3" thick and 15 MPa strength concrete when floor is resting on 4" of crushed stone and is dampproofed

b. 4" thick and 15 MPa strength concrete when floor is resting on 4" of crushed stone and is dampproofed

c. 2" thick and 30 MPa strength concrete when floor is resting on 6" of crushed stone

d. 3" thick and 30 MPa strength concrete when floor is resting on 4" of crushed stone

_____15. What is the outside measurement of the porch foundation walls?

a. 1100 mm x 6960 mm

b. 1100 mm x 6560 mm

c. 1300 mm x 6960 mm

d. 1300 mm x 6560 mm

_____16. What is the measurement from the finished first floor to the top of the first floor window?

a. 2080 mm

b. 2330 mm

c. 2460 mm

d. 2670 mm

_____17. How many basement windows are shown on the Whittington Partial Plans & Elevations for S.W.O. Condition (refer to page with this title)?

a. 2

b. 3

c. 4

d. 5

_____18. What size and type of wood decking is used for the deck at the rear of the house described in the Whittington Partial Plans & Elevations for S.W.O. Condition?

a. The wood decking is constructed from 38 mm × 150 mm Trex composite material.

b. The wood decking is constructed from 38 mm × 150 mm cedar.

c. The wood decking is constructed from 38 mm × 89 mm cedar.

d. The wood decking is constructed from 38 mm × 150 mm pressure-treated wood.

_____19. What do the exterior stairs from the deck rest on?

a. 2 - 600 mm × 600 mm concrete patio slabs

b. 2 - 800 mm × 800 mm concrete patio slabs

c. the stairs rest on a 6" poured concrete slab

d. the stairs rest on a 4" poured concrete slab

_____20. How is the deck header joist supported at the veneer wall?

a. The deck is directly supported by the brick veneer.

b. The deck is supported by a 38 × 184 P.T. ledger plate anchored to header joist W/12mm through bolts @ 400 OC.

c. The deck is directly supported by the concrete foundation wall.

d. The deck is supported at the veneer wall by Sono tubes filled with concrete placed close to the foundation wall.

11 Mechanical and Electrical Sub-Trades

LEARNING OBJECTIVES

After completing this chapter, you will be able to:

- Read and interpret drawings for information related to plumbing systems.
- Read and interpret drawings for information related to heating, ventilation, and air conditioning (HVAC) systems.
- Read and interpret drawings for information related to electrical systems.
- Describe the basic systems, materials, and methods of installation used for plumbing, HVAC, and electrical systems with an introduction to green sustainable systems.

INTRODUCTION

This chapter addresses building systems and how to visualize the information found on drawings and apply it to the construction process. **Remember to refer to the abbreviations and symbols provided in Chapter 4 when reviewing this chapter and answering the assignment and quiz questions.**

MECHANICAL AND ELECTRICAL SYSTEMS

The mechanical (includes plumbing, heating, ventilation, and air conditioning systems) and electrical systems are for the most part hidden from view once the building is finished. It is important therefore to make sure that all systems are installed correctly and in the right location before the drywall and finishes are installed. Once the walls and ceilings are covered over, it is expensive to relocate outlets, lights, sinks, etc., that are placed or roughed in at the wrong location. Most errors occur because the plans are not interpreted properly or because of a lack of proper consultation between the subtrades, building contractor, and/or client. Proper planning also involves understanding how the systems work and how their installation may affect the structure and other systems. Homes and buildings operate much like the human body: when one part is not functioning properly, it has a negative impact on the rest of the system. Therefore, we must carefully review and understand how the parts of our buildings work together so that we can construct safe, durable, and healthy buildings.

The duct and plumbing runs should be discussed thoroughly with the plumbing and **HVAC** subcontractors. You also want to make sure there is complete understanding about maintaining the integrity of the structure. The completed framing can be compromised when the subtrades rough in their building systems. The site superintendent needs to make sure that all building code requirements are followed.

Part 9.23.5.1 Notching and Drilling of the NBC restricts notching and drilling of roof, floor, and ceiling framing members. The NBC states that holes in framing members shall be not larger than one-quarter the depth of the member and shall be located not less than 50 mm from the edges, unless the depth of the member is increased by the size of the hole.

As an example, if a 2" × 6" member were used as a rafter, then its actual size is 1½" × 5½" and it would only be able to have maximum a 1⅜" hole drilled through it (5½" / 4 = 1⅜") and the hole could not be any closer than 2" (50 mm) to the edge.

Part 9.23.5.2 of the NBC states that floor, roof, and ceiling framing members are permitted to be notched, provided the notch is located on the top of the member within half the joist depth from the edge of bearing, and is not deeper than one-third of the joist depth, unless the depth of the member is increased by the size of the notch.

Part 9.23.5.3 of the NBC states that wall studs shall not be notched, drilled, or otherwise

FIGURE 11.1

Notching and Drilling of Floor Joists: NBC requirements Section 9.23.5.1 and Section 9.23.5.2. Closely monitor sub-trades to ensure compliance.

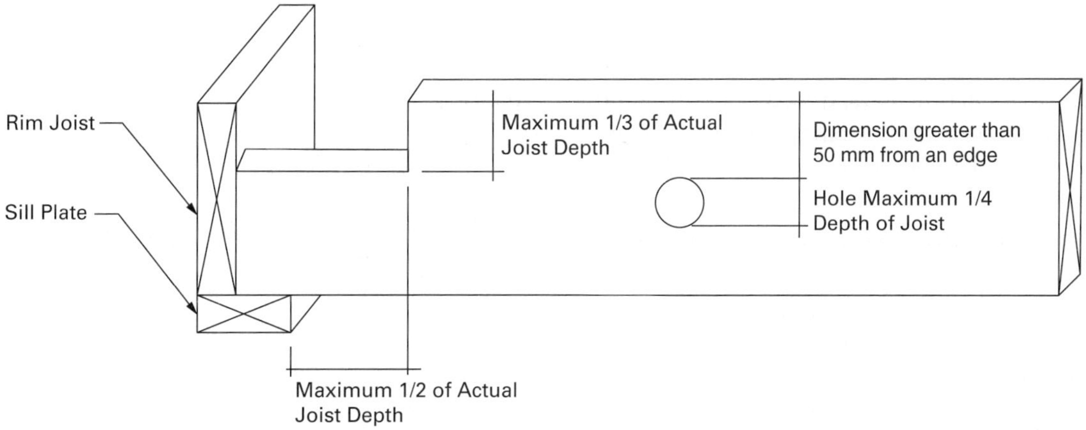

Source: Tom Stephenson

damaged in such a way that the undamaged portion of the stud is less than two-thirds the depth of the stud if the stud is load-bearing, or 40 mm if the stud is non-load-bearing, unless the weakened studs are suitably reinforced.

Part 9.23.5.4 of the NBC states that top plates in walls shall not be notched, drilled, or otherwise weakened so as to reduce the undamaged width to less than 50 mm, unless the weakened plates are suitably reinforced.

Part 9.23.5.5 of the NBC states that roof truss members shall not be notched, drilled, or otherwise weakened unless such notching or drilling is allowed for in the design of the truss.

When working with manufactured/engineered floor systems, always follow the approved individual manufacturer's guidelines regarding notching and drilling of their members, as they can differ from manufacturer to manufacturer. Consult with your subtrade to ensure that the best method is used to run the ducts and waste pipes to avoid unwanted or ugly bulkheads in inappropriate locations.

It is good practice to schedule a pre-construction meeting to go over duct and drain runs with the

FIGURE 11.2

Notching and Drilling of Wall Studs: NBC requirements Section 9.23.5.3. Closely monitor sub-trades to ensure compliance.

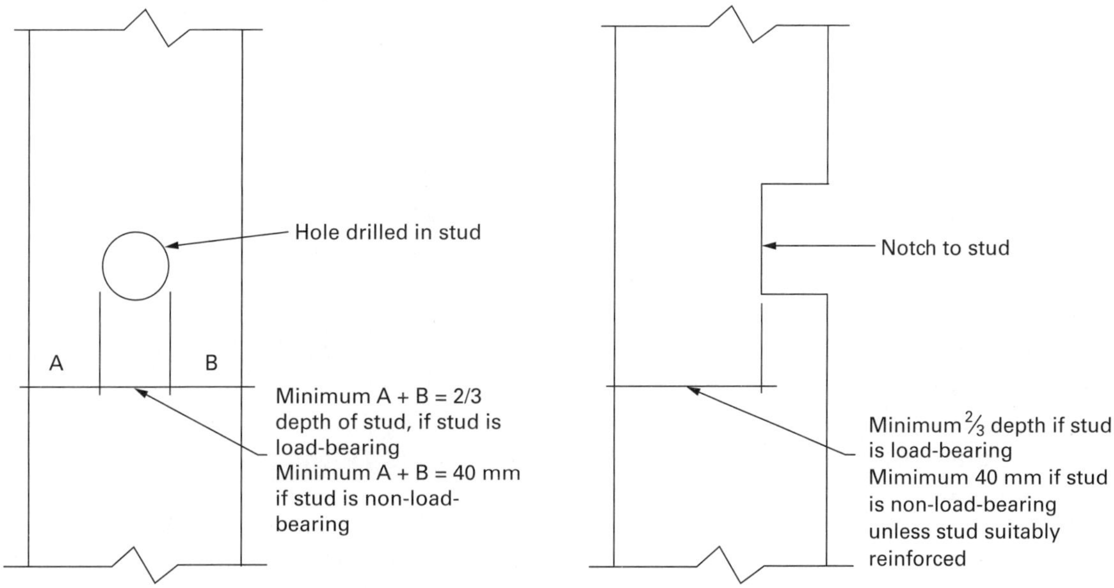

Source: Tom Stephenson

plumbing, HVAC, and electrical sub-trades to make sure potential conflicts are identified and resolved. Construction drawings do not always fully account for all the variables involved with system rough-ins and possible building out of walls or bulkheads, especially in custom homes and renovation work. For most residential drawings, the basic information regarding mechanical and electrical systems will be shown on the construction drawings; however, for more complex buildings, separate floor plan drawings will be provided specifically for the electrical and mechanical systems.

As discussed in Chapter 1, BIM provides an advantage for more complex building, structures, as a 3D representation of the building can be modelled and reviewed to determine if there are any clashes occurring between the systems themselves, as well as the structure. The review provides an opportunity to correct the clash/interference before construction begins. It also allows builders to order prefabricated materials with more confidence, and do less re-work, which can translate to lower costs and improved scheduling.

The sequence of work is equally important; generally, HVAC systems are roughed in first, followed by plumbing and then electrical. The logic for this order is that it is easier to run wires around ducts and drains than the other way around. Take care when reviewing drawings to identify walls that are drawn as being thicker; for example, a 2" × 6" wall is often used instead of a 2" × 4" wall. This may be done for several reasons, including:

1. Providing space for soil and vent stacks to run within the wall cavity
2. Providing space for heating supply and return ducts to supply the upper floors by running within the wall cavity
3. For structural purposes.

The mechanical, electrical, and building envelope needs to be considered as one integrated system, in which success or failure depends on several factors:

- Design
- Installation
- Climate conditions
- Occupant loads and lifestyles
- All of the above criteria interacting effectively as a building system. In residential building, this interaction is referred to as "the house as a system."

FIGURE 11.3
Vent terminology.

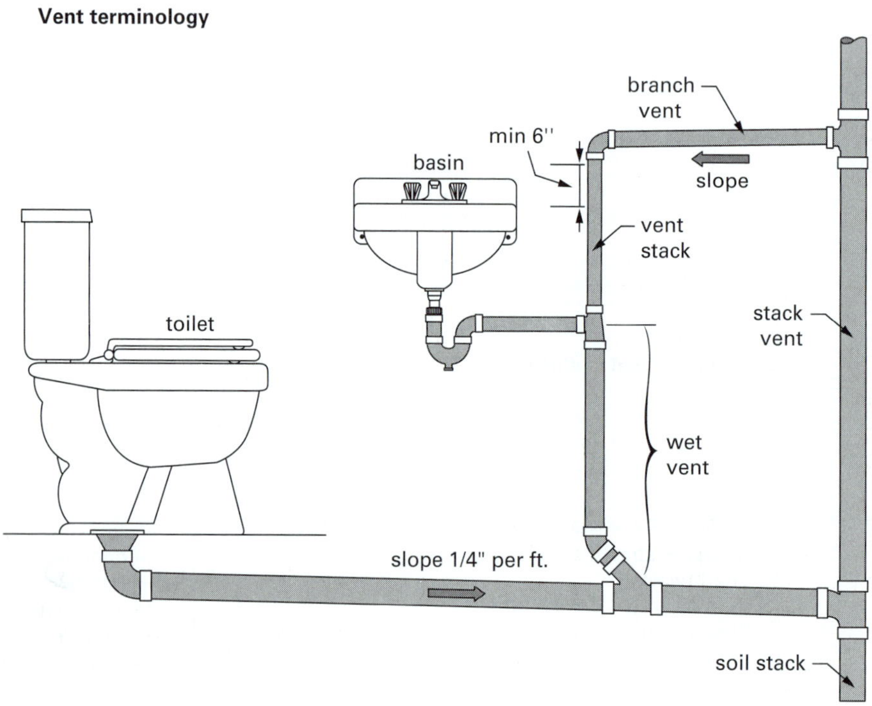

Source: Image excerpted from The Illustrated Home by Carson Dunlop www.carsondunlop.com

PLUMBING SYSTEM

The plumbing system is installed by licensed plumbers who must determine the most efficient and effective means of installation. The plumbing system consists of two main components: potable water; and the drain, waste, vent system.

Potable Water

Potable water (drinking water) is supplied to all fixtures in the structure. Water is supplied to the building in urban areas through the local municipal water supply; in rural locations, water is supplied through a well located on the property. Diameters of pipes typically used to supply residential structures are ½", ⅝", and ¾". The size of supply pipe is a key indicator of the water pressure that will be available within the building. The water supply includes both cold and hot water as required by the appropriate fixtures and appliances. Sinks, basins, showers, and tubs will require both hot and cold water supplies, while toilets, hose bibs, and refrigerators that supply drinking water will require only a cold water supply. A dishwasher requires only a hot water supply.

Drain, Waste, Vent System

The **drain, waste, vent system (DWV)** facilitates the safe and efficient removal of waste and soiled water. Depending on the location, the building may have two drain systems: one to remove the waste water and sewage, known as the **sanitary sewer,** and the other, used to remove storm water and water from the weeping or drainage tile, known as the **storm sewer.**

Most municipalities prefer to keep the storm and sanitary drains separate to eliminate treating enormous amounts of relatively clean water from the storm drains, thus ensuring that water treatment is focused on waste water and sewage

FIGURE 11.5

Stack vent extending through roof. The stack vent exhausts sewer gases and maintains atmospheric pressure in the drain system.

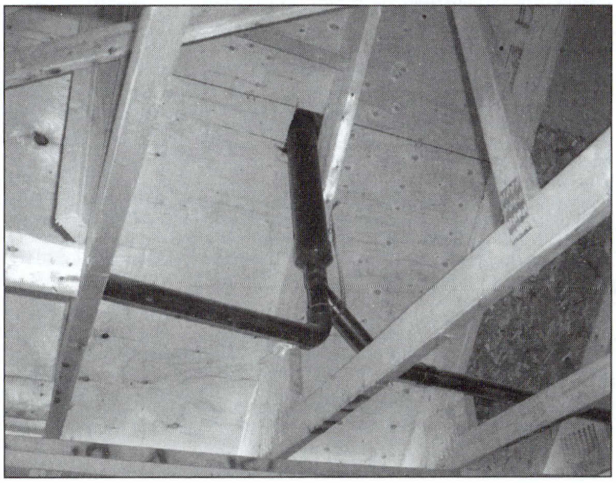

Courtesy Tom Stephenson

FIGURE 11.4

Soil stack and waste stack.

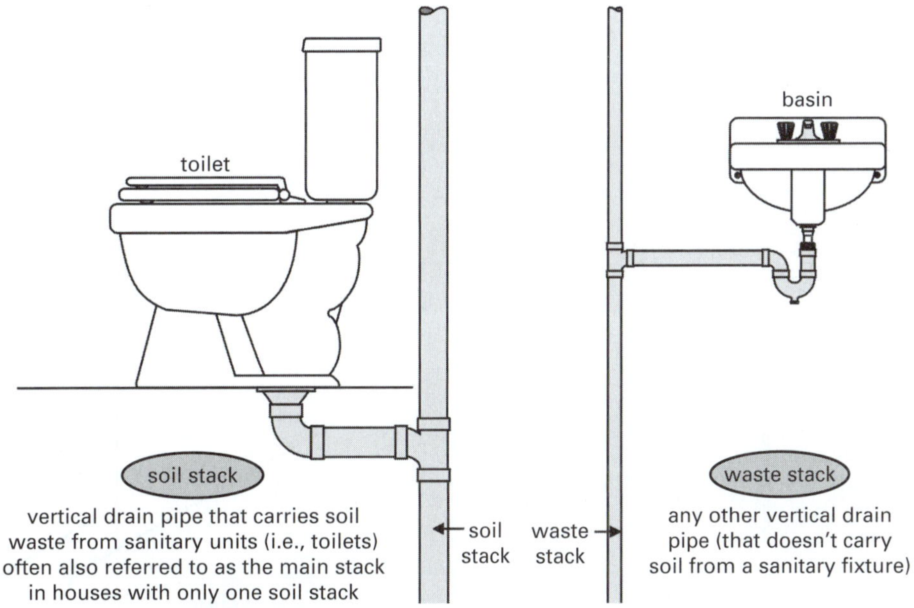

Source: Image excerpted from The Illustrated Home by Carson Dunlop www.carsondunlop.com

before it reaches rivers, lakes, or oceans. Older municipalities may have single sewer systems where there is only one combined sewer that ends at the sewage treatment plant.

In rural locations where municipal sewers are unavailable, waste water and sewage is addressed through a **septic system**, which breaks down the fecal matter through bacterial action. The liquid overflow is filtered out into the ground through absorption fields. The septic tank requires periodic cleaning to remove decomposed solids. A rural location may also require a sump pump to pump away water from the foundation to another location above ground or to a sump pit. A **sump pump** may also be required to pump away water where there is a high water table or to meet municipal requirements.

Vents are used to ventilate the drainage system and ensure that water drains quickly and efficiently. The vents also prevent water from siphoning out of the traps (traps hold water, preventing the entry of sewer gas into the building). Sewer gas contains mainly methane gas and can also contain a complex mixture of other toxic and non-toxic gases that in high concentrations can cause sickness and even death.

Drains that run vertically are called **stacks** (e.g., a soil stack carries discharge from a toilet while a waste stack carries discharge from a sink). Vertical drains that act as a vent and extend from the waste or soil stacks are called stack vents. A vent that acts as both a vent and a drain is called a wet vent.

FIGURE 11.6

An isometric drawing showing locations and sizes of the DWV system; notice that the vents are drawn using a dashed line, and the abbreviation CO stands for cleanout. Diameter of drains is noted in inches, and fixtures and quantity of fixtures dictate drain sizing.

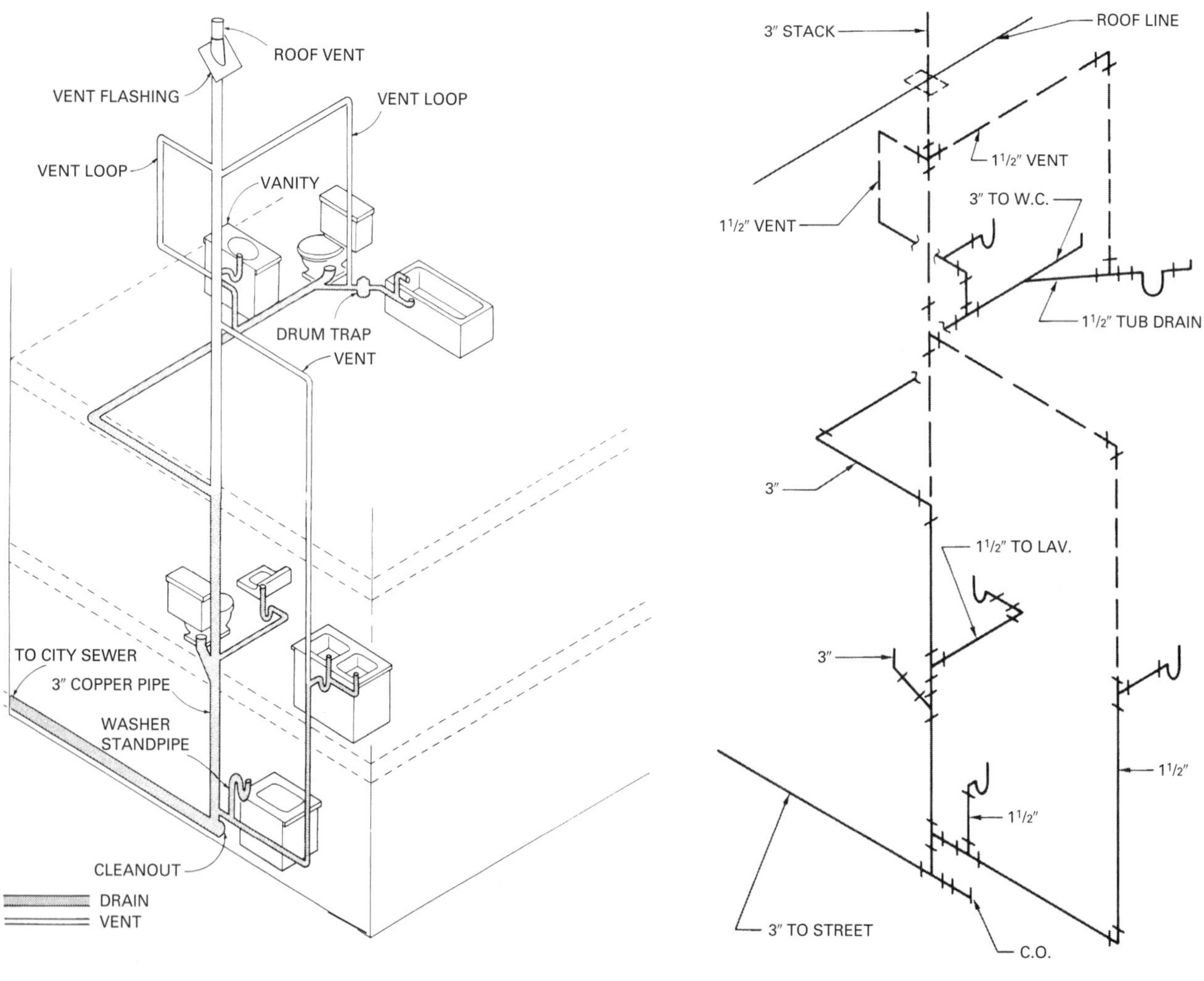

Plumbing Permit Requirements

A plumbing permit is required when installing new plumbing systems or when altering existing systems. The plumbing subcontractor is usually required to apply for this permit. As part of the application process, the plumbing subcontractor may be required to submit drawings/sketches that display how he/she intends to lay out the drain waste vent system. Depending on the extent of the project, the plumbing subcontractor may be asked to submit a floor plan detailing the layout of the drains under the basement slab. This layout will show locations and sizes of cleanouts, floor drains, traps, and stacks. It may also display the connection points for weeping tile. Drawings may also be required to display how the drain, waste, and vent pipes will be installed for the various plumbing fixtures. For example, an elevation or isometric drawing

FIGURE 11.7

The water supply system: hot and cold water piping.

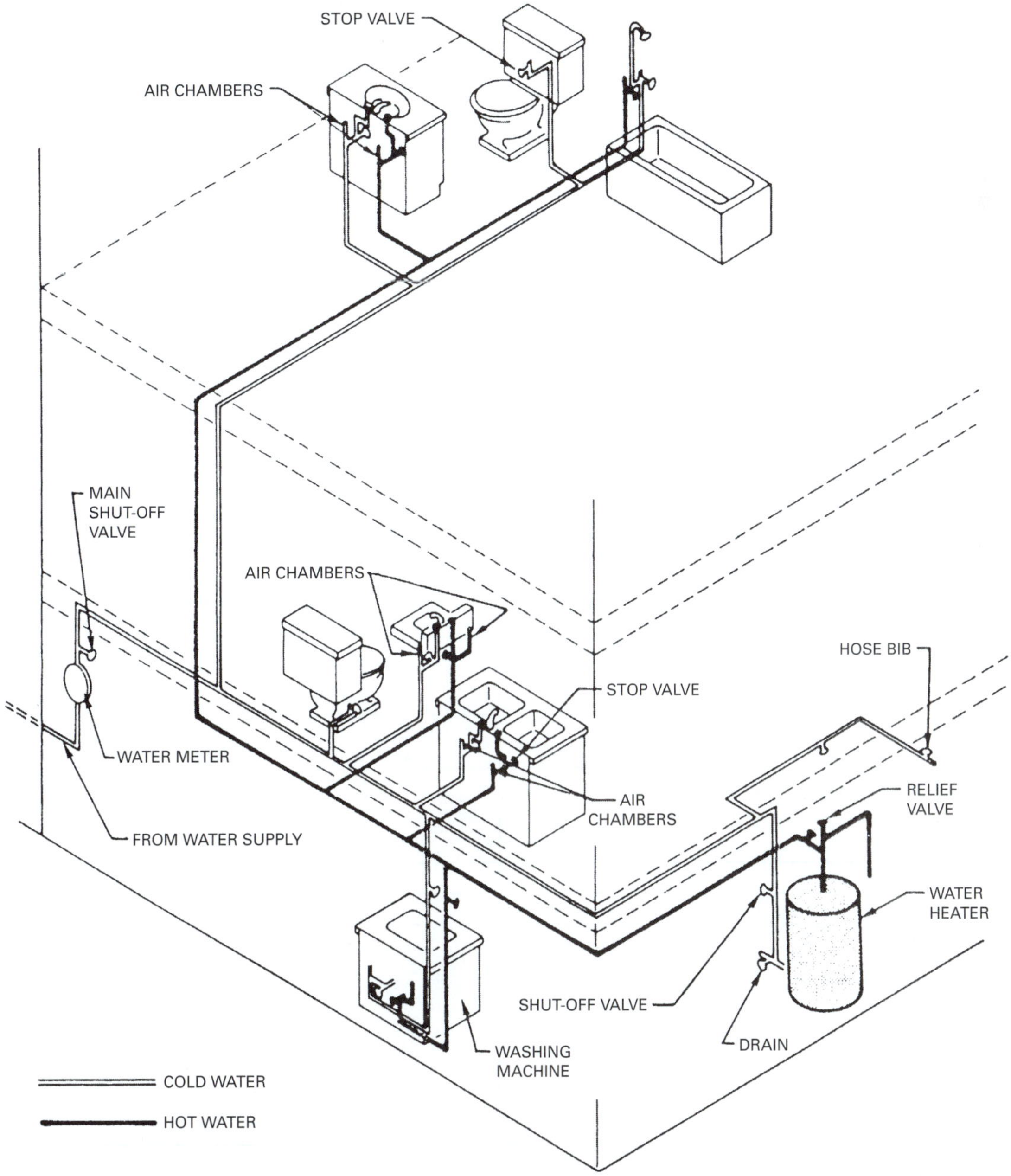

FIGURE 11.8

Floor-plan layout of drains located beneath foundation slab: combined, separate, and sub-soil connection. Note that combined is usually found in existing older homes and areas where the street has only one combined sewer system. Municipalities and provinces often require now that the downspouts be redirected out on to the ground so as to reduce the amount of storm water the sewers take in. The abbreviation FAI refers to fresh air intake. F.D. Conn. refers to field drain (weeping tile connection), S.S. refers to soil stack, C/O clean out, F.D. floor drain, and LT is laundry tub.

DRAIN DESIGN—COMBINED

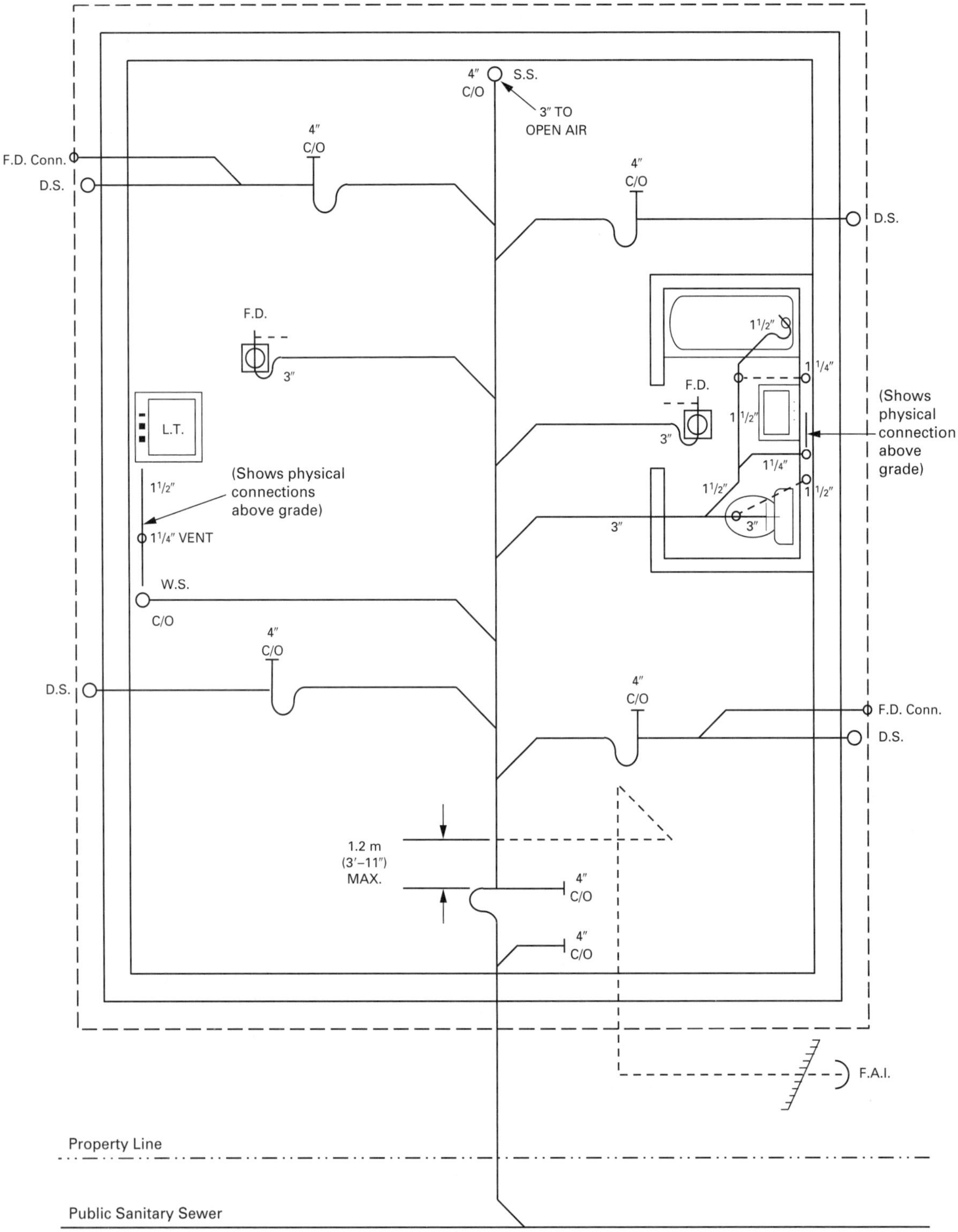

Source: Rainier Blundel, George Brown College

FIGURE 11.8 *(continued)*

New construction municipalities have separate sewer systems for sanitary and storm water. Downspouts may be directed above grade (absorbed into the ground), or if the storm sewer system is designed to carry this capacity, they may be directed below grade.

DRAIN DESIGN—SEPARATE

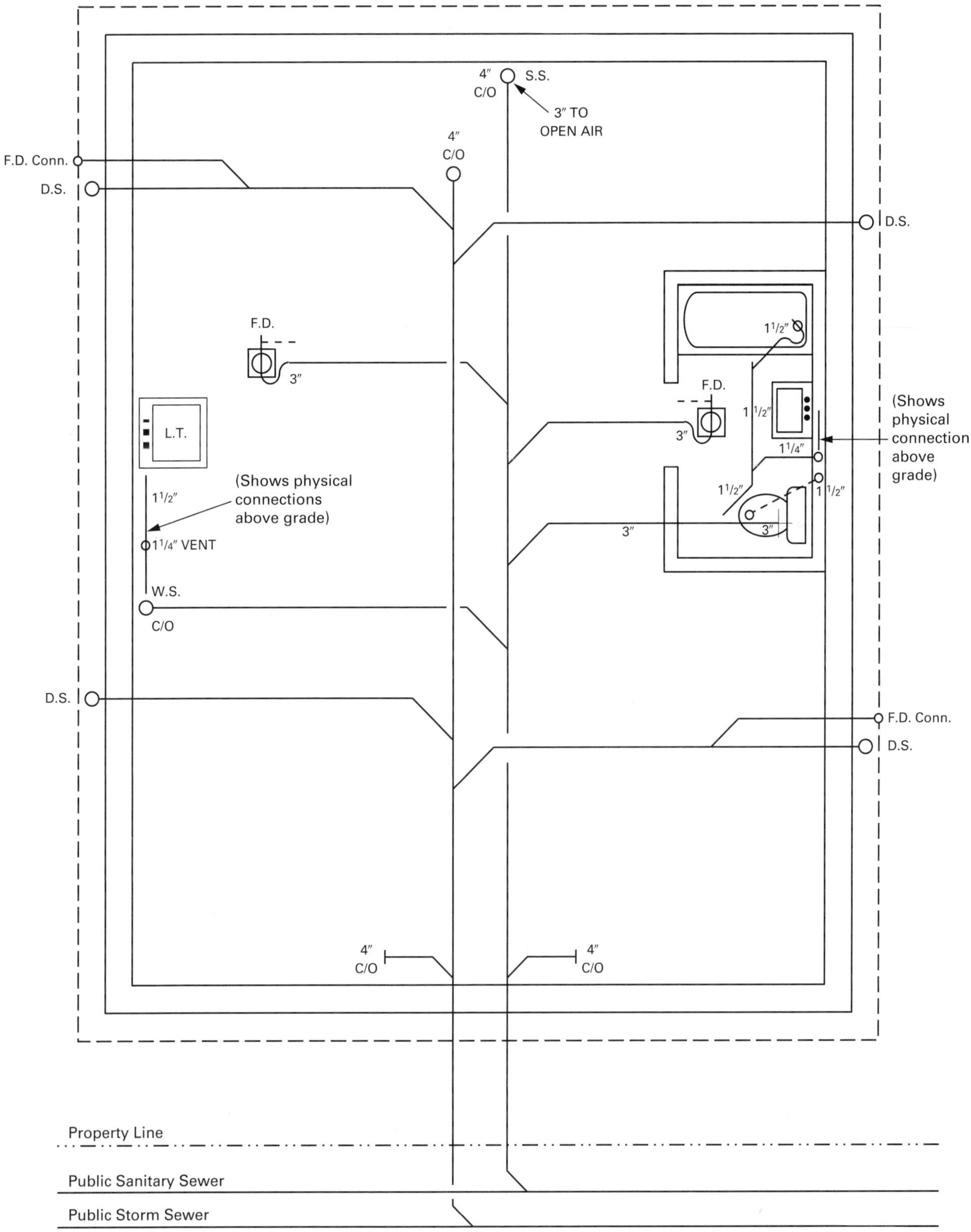

FIGURE 11.8 *(continued)*

Sump pumps are used where there is no access to storm sewers and where they are required due to town, municipal, or site requirements.

SUB-SOIL DRAINAGE CONNECTION

Discharging Sub-soil Drainage above Grade—Separate System

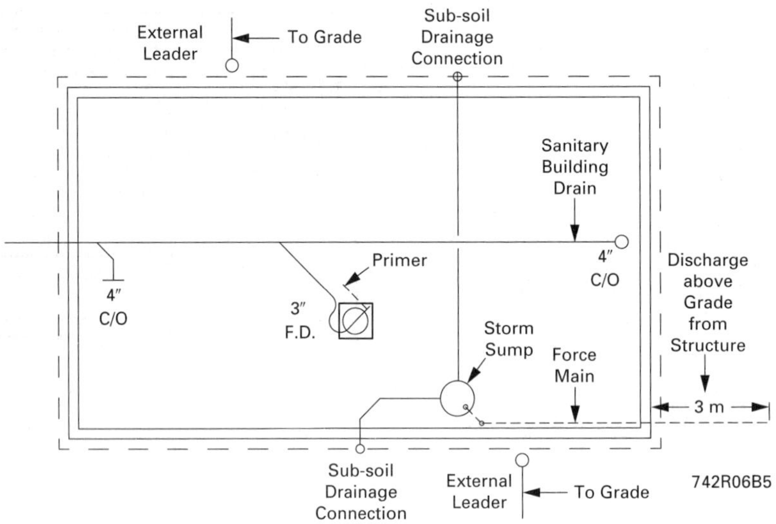

Discharging Sub-soil Drainage above Grade—Separate System

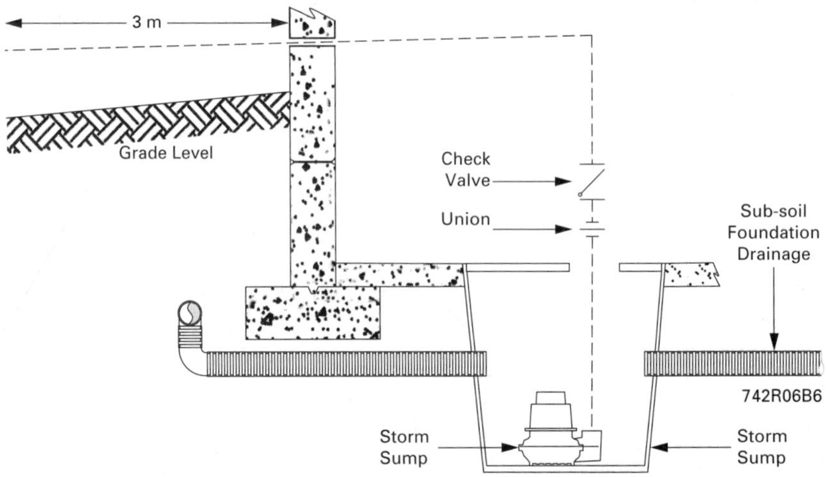

When no storm drain is available or is not allowed, the foundation drainage must discharge above grade at least 3 m from the building and must not cause a hazard.

may be used to display the layout and sizing of the drain system. Isometric drawings are particularly useful for displaying a three-dimensional perspective of how the drain system is installed. Dashed lines are used to indicate vent pipes.

Plumbing Components and Materials

Water Supply Lines

Water supply lines have traditionally used copper tubing soldered together at the joints. Copper tubing is a long-lasting and durable product. However, caution must be taken during construction because the soldered connections require the use of an open flame, which can cause a fire during construction due to the proximity of flammable products such as wood and sawdust. Care must also be taken to ensure that the copper tubing does not come into contact with dissimilar metals, which can cause deterioration of the system through electrolysis. Also, with the high cost of copper, on-site theft is an increasing problem.

As a result, new products are growing in popularity, and adoption has been occurring rapidly. One such product, **PEX,** uses cross-linked polyethylene piping (flexible plastic piping) that is suitable for both cold and hot water supplies. It is flexible and fast to install. One advantage of this system is that it requires fewer joints than copper, and there is no requirement for a torch to make the connections, thus reducing chance of fire. The system also eliminates the water hammer effect, caused by rigid pipes making a loud noise when the water is shut off.

FIGURE 11.10
Rough-in for bathroom basin to be located inside future vanity. The white tubing is a PEX water supply, and the black pipes are part of the ABS DWV system. Notice the use of metal protective plates over stud and wall plate intersections. The metal plates provide protection from nails and screws that may be used to fasten drywall and other finishes.

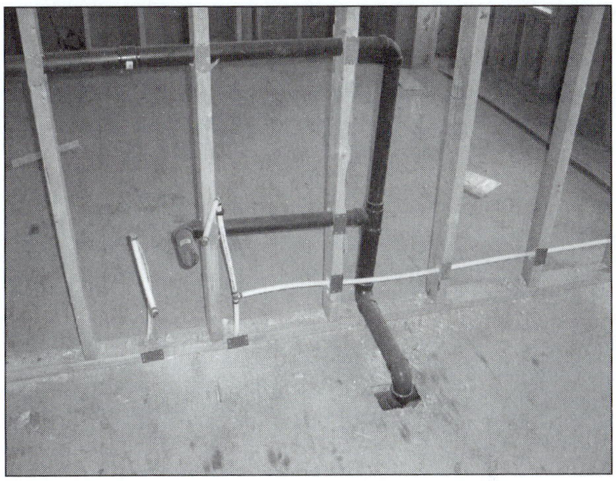

Courtesy Tom Stephenson

FIGURE 11.9
Soldered copper tubing supplying a hose bib on the outside with a shut-off valve located in the basement ceiling. If the ceiling is to be finished, an access door must be installed. This may not be indicated on the drawings.

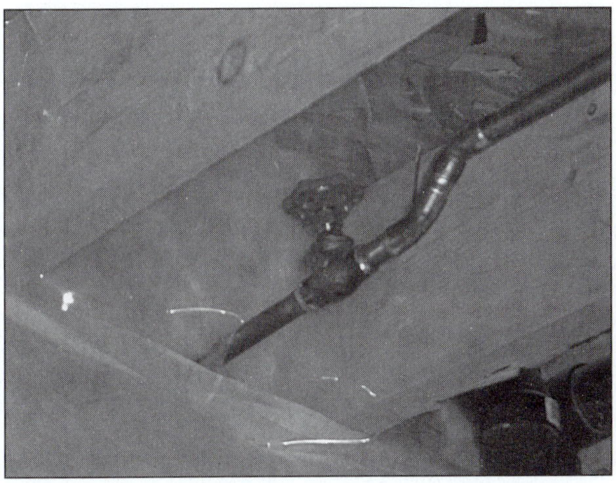

Courtesy Tom Stephenson

Drain Pipes

Drain pipes are typically constructed of one of the following materials, depending on the project and local code requirements:

- ABS (acrylonitrile butadiene styrene)
- PVC (polyvinyl chloride)
- Cast iron
- Copper
- Clay (in older applications).

FIGURE 11.11
A cleanout placed above a basement floor slab.

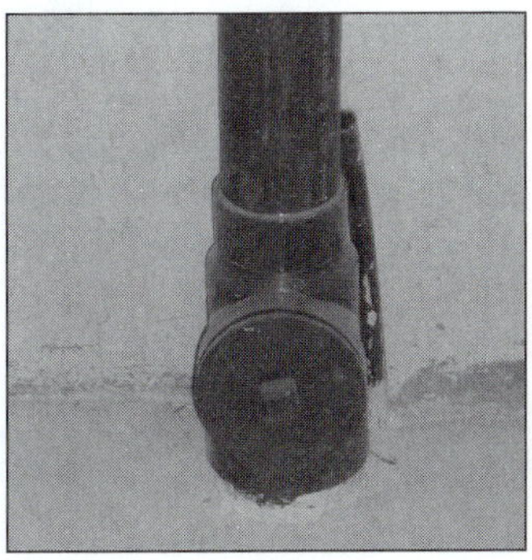

Courtesy Tom Stephenson

Cleanout

A cleanout is provided at strategic locations to ensure that a clogged drain can be cleaned using a snake without opening walls or breaking concrete floors. A cleanout on a drawing will be labelled CO or C/O.

FIGURE 11.12
Water trap.

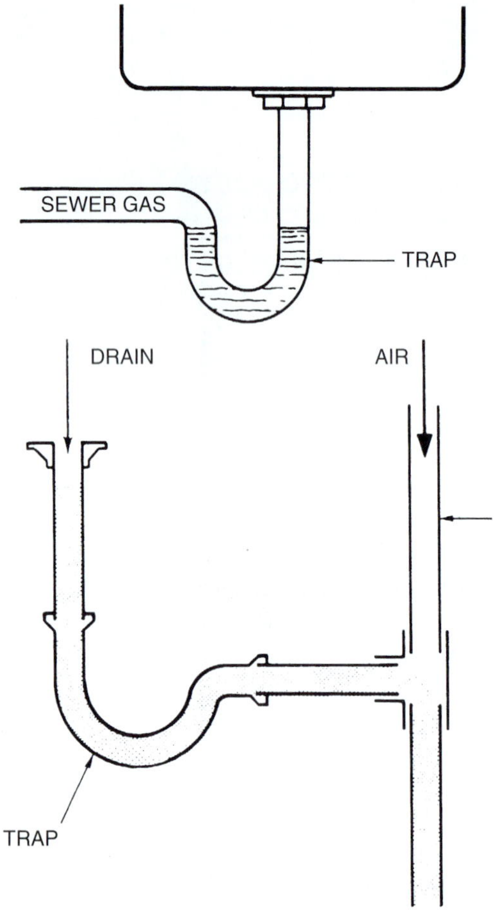

FIGURE 11.13
A water trap roughed in below the floor for a bathtub located above the floor.

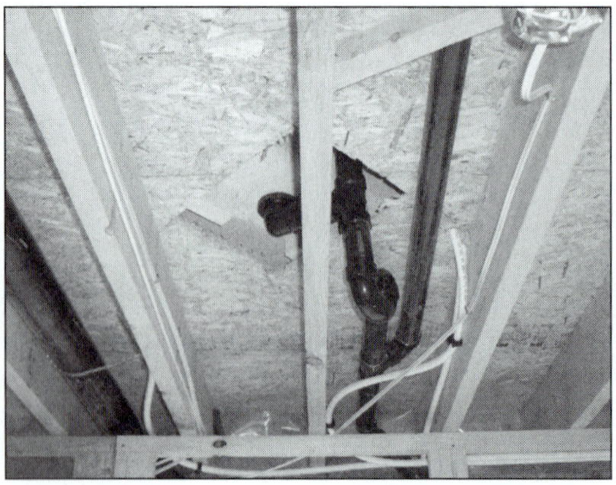

Courtesy Tom Stephenson

Water Trap

A **water trap** is used to ensure that sewer gas does not enter the building. Water traps are found under sinks, tubs, and floor drains, and are formed into the base of a toilet. This is why you see water inside the base of a toilet. The water forms a trap or seal, preventing sewer gas from entering the building. Floor drains require a trap seal primer, which is a supply connection from a faucet that sends a trickle of water to the

FIGURE 11.14
A water trap located under double sink; notice the drain connection at left ready for dishwasher, and PEX water supply connected to shut-off valves.

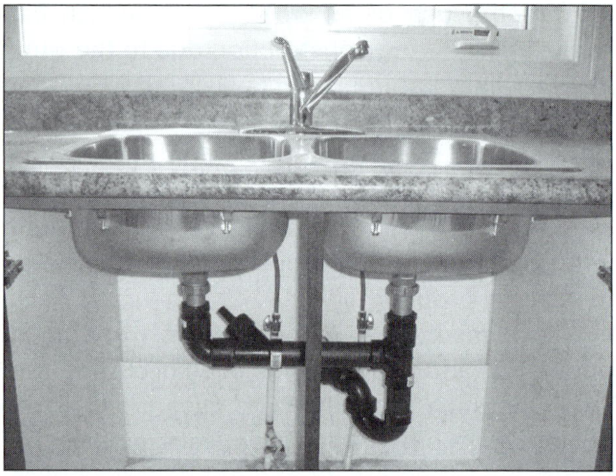

Courtesy Tom Stephenson

FIGURE 11.15
Water trap in place for a floor drain. This trap will be covered by the basement floor slab. The drain in this case is located close to the furnace to collect condensation from the furnace and air conditioning unit. The other drains shown are to collect waste and soiled water from the house. These drains will connect with the sanitary sewer or septic system. Also note the plastic water supply line located next to the trap. This pipe provides water from the street to the house.

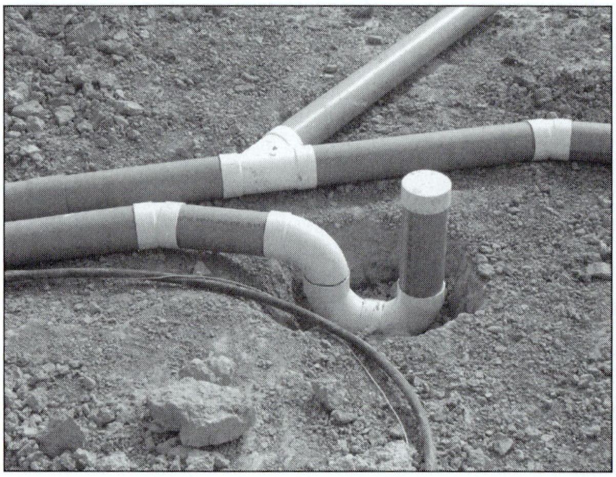

Courtesy Tom Stephenson

trap whenever the faucet is turned on. This is to make up for water that evaporates out of the floor drain trap and is not replenished like a sink trap because the floor drain only takes in water when there is a flood.

Rough-in of Fixtures

There is a wide variety of available fixtures, and the location of drains is not always standard. It is therefore wise to consult the manufacturer's specifications regarding location of the drain and water supply. Most manufacturers now provide their product specifications and drawings online, so it is relatively easy to check and print this information as required or view it directly on site on your phone or tablet. Some fixtures, such as whirlpool tubs, may require access to be provided for the motor. This information is not generally provided on the working drawings; however,

FIGURE 11.16
Example of a manufacturer's specification for a whirlpool tub and toilet.

American Standard

SCALA OVAL 5-1/2'™
WHIRLPOOL AND BATHING POOL
HIGH GLOSS ACRYLIC

SCALA OVAL 5-1/2'
WHIRLPOOL AND BATHING POOL

❑ **2645 019 Whirlpool featuring System I**

- Acrylic with fiberglass reinforcement
- Factory installed and tested whirlpool system
- 1.0 HP, .746 kw single speed self-draining pump/motor with dry run seal
- Six multi-directional and individually flow-rate adjustable, colour-coordinated body massage jets
- Two colour-coordinated air volume controls
- Deck-mounted colour-coordinated air switch On/Off control
- Anti-vortex suction fitting
- UL approved entire whirlpool system
- Safety textured tub bottom pattern
- Integral sump support feet
- Integral lumber support

❑ **2645 002 Bathing Pool only**

Actual Dimensions:
1676 x 914 x 502 mm
(66" x 36" x 19-3/4")

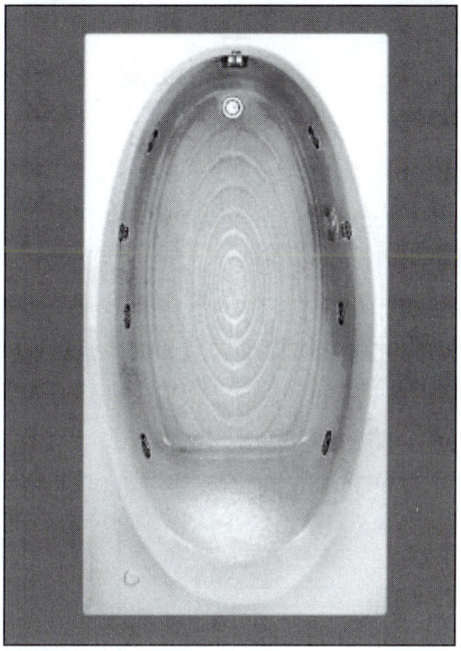

To Be Specified - Required:

❑ Colour: (specify colour)
❑ Bath Faucet* (specify finish)
❑ Rotary Drain: 1553 270 (specify finish)

To Be Specified - Optional:

❑ Metal Air Control Kit: 9850 200 (2 per kit) (specify finish)
❑ Metal Jet Trim Ring Kit: 9860 200 (2 per kit) (specify finish)
❑ Apron: 9266 018 (White, Bone or Silver only)
❑ Acrylic Apron: 9366 018 (specify colour)

* See faucet section for models available

Compliance Certifications -
Meets or Exceeds the Following Specifications:

- Underwriters Laboratories - Hydromassage Bathtubs UL 1795
- ASME A112.19.7M for Whirlpool Appliances
- ANSI Z124.1 for Plastic Bathtubs
- ASTM E162 for Flammability
- NFPA 258 for Smoke D
- CAN/CSA B45 Series
- CAN/C22.2 Number 218.2-93 Hydro massage bathtub appliances

NOTE: Roughing-in dimensions and product specifications shown on reverse side of page.

WPB-023

03/03 © 2002 American Standard Inc.

Source: © 2002 American Standard Inc. www.americanstandard.ca

FIGURE 11.16 *(continued)*

Note the pump location in plan and end elevation. Also note the centre line of drain location, and location of sub-floor cutout (9" × 12").

American Standard

SCALA OVAL 5-1/2'™ WHIRLPOOL AND BATHING POOL

HIGH GLOSS ACRYLIC

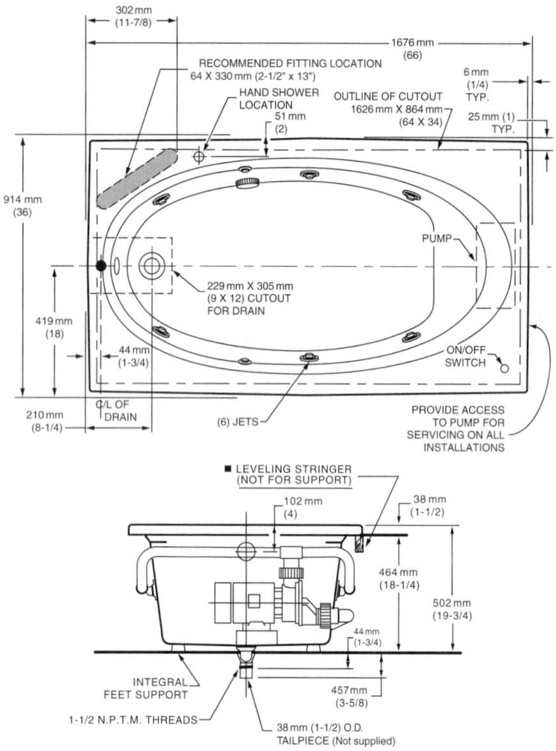

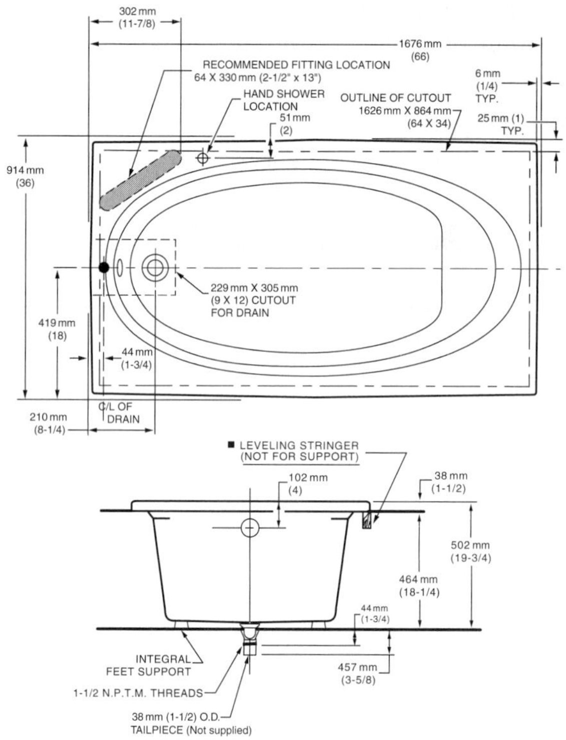

GENERAL SPECIFICATIONS FOR 2645 WHIRLPOOL

INSTALLED SIZE	66 x 36 x 19-3/4 In. (1676 x 914 x 502 mm)
WEIGHT	95 Lbs. (43 Kg.)
WEIGHT w/WATER	595 Lbs. (268 Kg.)
GAL. TO OVERFLOW	60 Gal. (227 L)
WHIRLPOOL MIN. OPERATING VOL.	55 Gal. (208 L)
BATHING WELL AT SUMP	46 x 24 In. (1168 x 610 mm)
BATHING WELL AT RIM	62 X 31-1/2 In. (1575 x 800 mm)
WATER DEPTH TO OVERFLOW	13-1/4 In. (337 mm)
FLOOR LOADING (PROJECTED AREA)	36 Lbs./Sq.Ft. (175 Kgs/Sq.m)
PTS.	25.6
CUBE (FT³)	45.1

WHIRLPOOL ELECTRICAL SPECIFICATIONS

SYSTEM I PUMP	1.0 HP, 9.9 AMPS, 120V.

GENERAL SPECIFICATIONS FOR 2645 BATHING POOL

INSTALLED SIZE	66 x 36 x 19-3/4 In. (1676 x 914 x 502 mm)
WEIGHT	75 Lbs. (34 Kg.)
WEIGHT w/WATER	575 Lbs. (259 Kg.)
GAL. TO OVERFLOW	60 Gal. (227 L)
BATHING WELL AT SUMP	46 x 24 In. (1168 x 610 mm)
BATHING WELL AT RIM	62 X 31-1/2 In. (1575 x 800 mm)
WATER DEPTH TO OVERFLOW	13-1/4 In. (337 mm)
FLOOR LOADING (PROJECTED AREA)	35 Lbs./Sq.Ft. (169 Kgs./Sq.m)
PTS.	25.6
CUBE (FT³)	45.1

NOTES:

BATH CAN BE INSTALLED EITHER ABOVE OR BELOW FLOOR LINE AS PIER, ISLAND OR PENINSULA TYPE INSTALLATION. SUMP OF BATH IS SUPPORTED BY INTEGRAL FEET. NO ADDITIONAL SUPPORT REQUIRED. ENCLOSING MATERIALS BY OTHERS.

REFER TO INSTALLATION INSTRUCTIONS SUPPLIED WITH WHIRLPOOL FOR ADDITIONAL INFORMATION.

OPTIONAL APRON 9266 018 OR 9366 018 AVAILABLE AND MUST BE ORDERED SEPARATELY.

FAUCETS NOT INCLUDED AND MUST BE ORDERED SEPARATELY.

IF INSTALLING A DECK-MOUNTED FITTING, BEFORE DRILLING ANY HOLES: CHECK FOR INTERFERENCE WITH CIRCULATING SYSTEM PIPING, DRAIN OVERFLOW, AND APRON INSTALLATION. MOUNTING HOLES DRILLED BY INSTALLER.

PROVIDE SUITABLE REINFORCEMENT FOR ALL SUPPORTS.

IMPORTANT: Dimensions of fixtures are nominal and may vary within the range of tolerances established by ANSI Standard Z124.1 / CSA Certified. These measurements are subject to change or cancellation. No responsibility is assumed for use of superseded or voided leaflet.

WPB-024

 08/03

FIGURE 11.16 *(continued)*

Note the centre line location of the toilet flange outlet and the water supply line. In cases where the toilet is located on an outside wall the supply would come up out of the floor.

American Standard

CADET III™ ROUND FRONT 13L TOILET

VITREOUS CHINA

CADET III ROUND FRONT 13L TOILET

☐ **2650 000**

- Vitreous china
- Watersaver (13.0 Lpf/2.9 gpf)
- Round front concealed trapway bowl
- Internal trapway area fully glazed 50 mm (2") ballpass
- Large 267 x 286 mm (10-1/4" x 11-1/4") water surface area
- Close-coupled tank
- Colour-matched trip lever
- Sanitary dam on bowl
- 2 bolt caps
- 100% factory flush tested

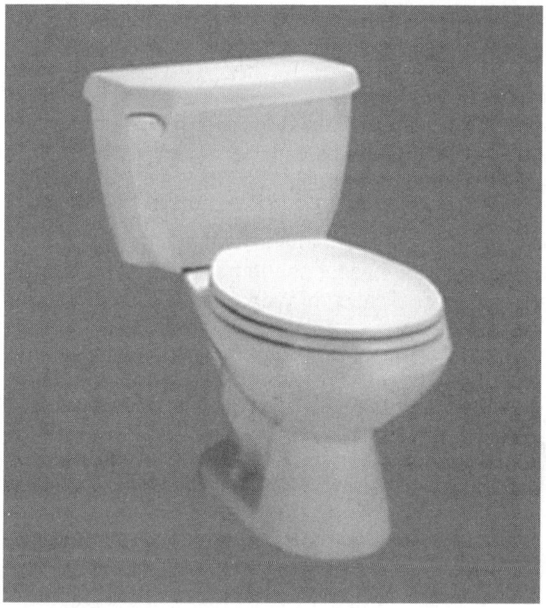

Separate Component Parts:

☐ **3650 000** Bowl
☐ **4650 000** Unlined Tank

☐ **2650 L00** with Lined Tank

Separate Component Parts:

☐ **3650 000** Bowl
☐ **4650 L00** Lined Tank

Nominal Dimensions:
759 x 518 x 692 mm (28-1/4" x 20-3/8" x 27-1/4")

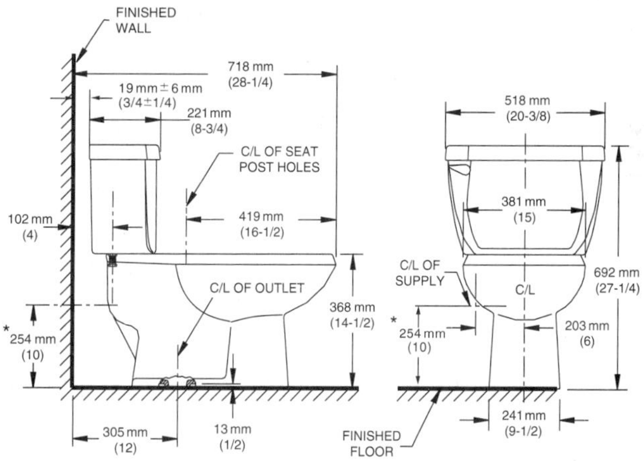

☐ **4650 B00** Unlined Tank with Bolted Cover
☐ **4650 BL0** Lined Tank with Bolted Cover

Compliance Certifications -
Meets or Exceeds the Following Specifications:

- ASME A112.19.2M (and 19.6M)
- CAN/CSA B45 series

To Be Specified

☐ Colour:
☐ Seat: American Standard #5322 011 "Rise and Shine" (with easy to clean lift-off hinge system) solid plastic closed, round front seat with cover. See price page.
☐ Alternate Seat:
☐ Supply with stop:

NOTES:
THIS TOILET IS DESIGNED TO ROUGH-IN AT A MINIMUM DIMENSION OF 305MM (12") FROM FINISHED WALL TO C/L OF OUTLET.
SUPPLY NOT INCLUDED WITH FIXTURE AND MUST BE ORDERED SEPARATELY.
* DIMENSION SHOWN FOR LOCATION OF SUPPLY IS SUGGESTED.
FOR ADDITIONAL INFORMATION REFER TO INSTALLATION INSTRUCTIONS SUPPLIED.
IMPORTANT: Dimensions of fixtures are nominal and may vary within the range of tolerance established by ANSI Standard A112.19.2
These measurements are subject to change or cancellation. No responsibility is assumed for use of superseded or voided pages.

TOI-232

Revised 03/03

the manufacturer's specifications and drawings should indicate motor location and options.

Working out those details in advance is important because access must be provided to maintain the motor of a whirlpool tub and the location of the access panel could be moved to an inconspicuous location such as a closet on the other side of the wall. Otherwise you would have the access at the front of the tile tub surround (not aesthetically pleasing). For fixture installations such as toilets, the drain location marked on the manufacturer's drawings indicates the location of the drain measured from the finished wall, because the thickness of the finishes can vary depending on variations in drywall and tile thickness.

FIGURE 11.17

The powder room floor of the Doncaster house shows a toilet rough-in that interferes with floor framing. A smaller header joist was installed to facilitate the location of the drain. If the location of the drain is communicated ahead of time by the drawings and manufacturer's details, the framer can place the header as the framing proceeds, thus eliminating duplication of work.

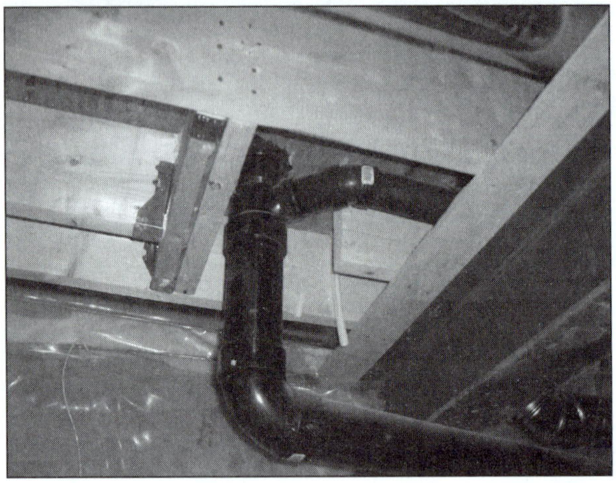

Courtesy Tom Stephenson

FIGURE 11.18

Rough-in of shower and bathtub. Notice the rubber shower tray and the positioning of the polyethylene vapour barrier (clear plastic) over the shower tray to ensure that moisture runs into the tray. The polyethylene air/vapour barrier is placed on outside walls, as discussed in more detail in Chapter 12 on the building envelope.

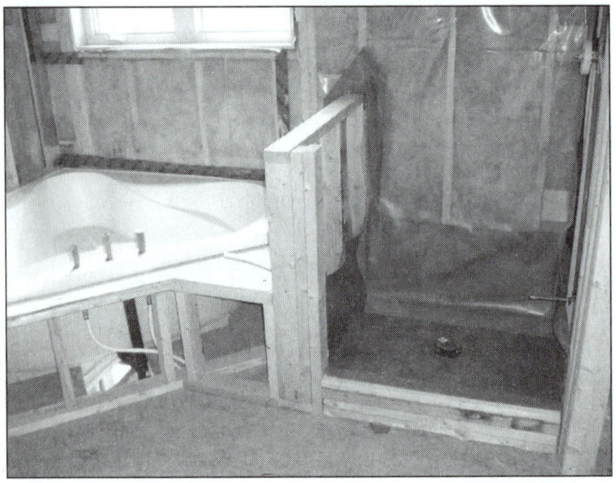

Courtesy Tom Stephenson

HEATING VENTILATION AND AIR CONDITIONING SYSTEM

The Heating, Ventilation, and Air Conditioning (HVAC) system is installed by a subcontractor who specializes in these systems. There are numerous licensing requirements for an HVAC system that may include a licensed gas installer, sheet metal worker, steamfitter, plumber, and/or electrician, depending on the type of system being installed.

Installation of a new heating system will require a permit. The heating subcontractor will be required to provide a set of heat loss calculations for the building. The heat loss calculations take into account the amount of heat lost through the building as a result of building design (e.g., number, size, and type of windows and doors, wall and ceiling area, room size and location. In a forced-air system, factors such as loss due to ventilation load are also considered). From this data, the heating subcontractor can design the appropriate size furnace and ductwork for the system. Computer programs are now generally used to make the necessary calculations. For a forced-air system, floor plan drawings or sketches are required. The drawings show the locations and sizes of the supply and return air ducts and registers. More advanced energy modelling software can be used during the design stage to predict energy costs and savings by running a variety of building envelope and insulation construction details to determine the most energy- and cost-efficient design that will often require a smaller, less expensive, heating system.

Selection of a heat distribution system is usually based on installation costs, comfort levels, availability and cost of fuel, and local climate. High-efficiency furnaces and boilers have become popular due to increasing fuel costs. These types of furnaces and boilers do not require a chimney to exhaust combustion gases; instead, a direct vent can be placed through the exterior wall (codes may have restrictions on the location of the direct vent).

FIGURE 11.19

Sample floor plans and heat loss calculations for Doncaster house. The slash in the rectangles indicates that the supply ducts are opening on a different floor. If the rectangle is clear, it is supplying or returning on this floor plan. Return ducts are denoted with an R for return, with further detail described in the legend.

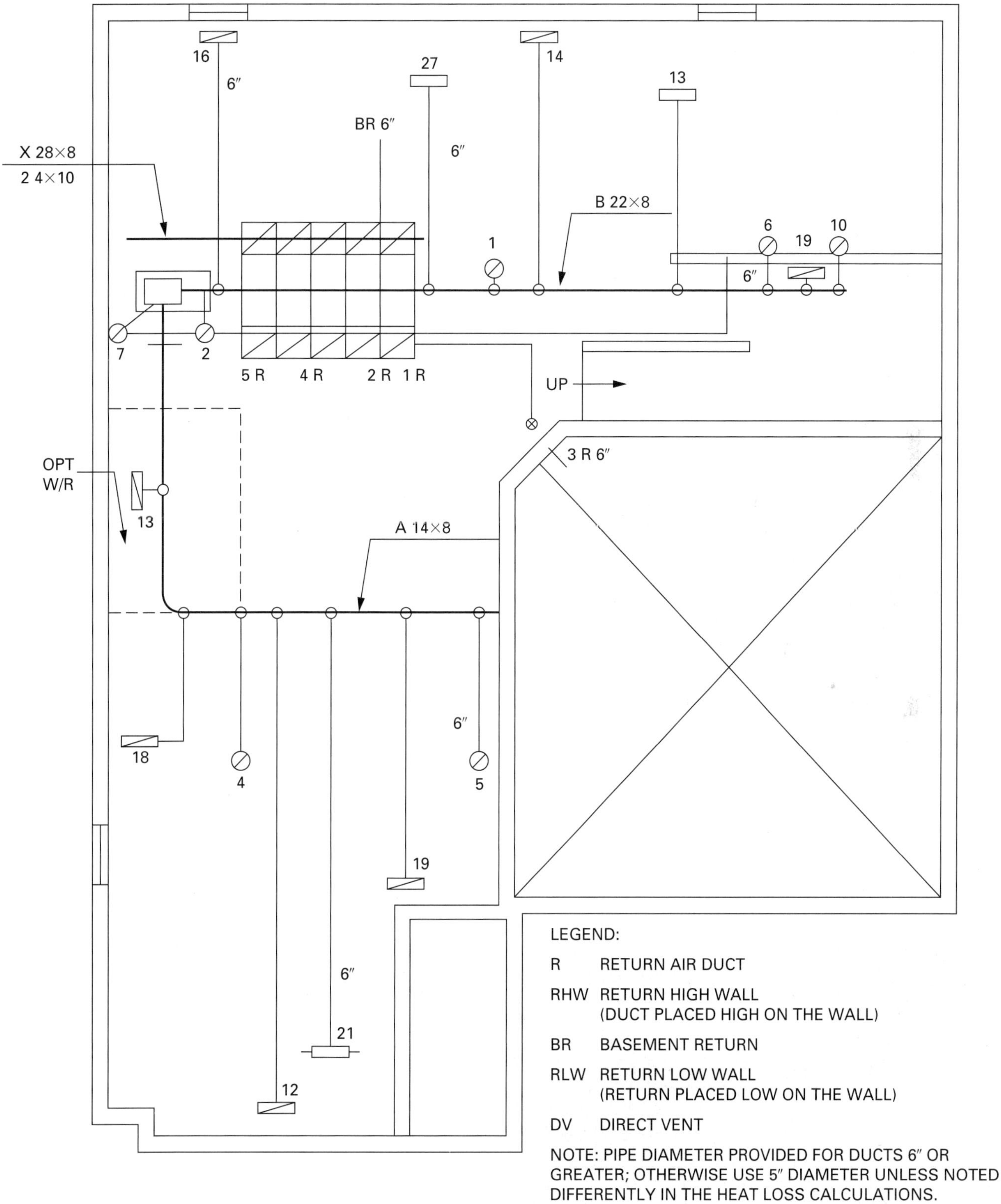

Source: HVAC Designs Ltd.

FIGURE 11.19 *(continued)*

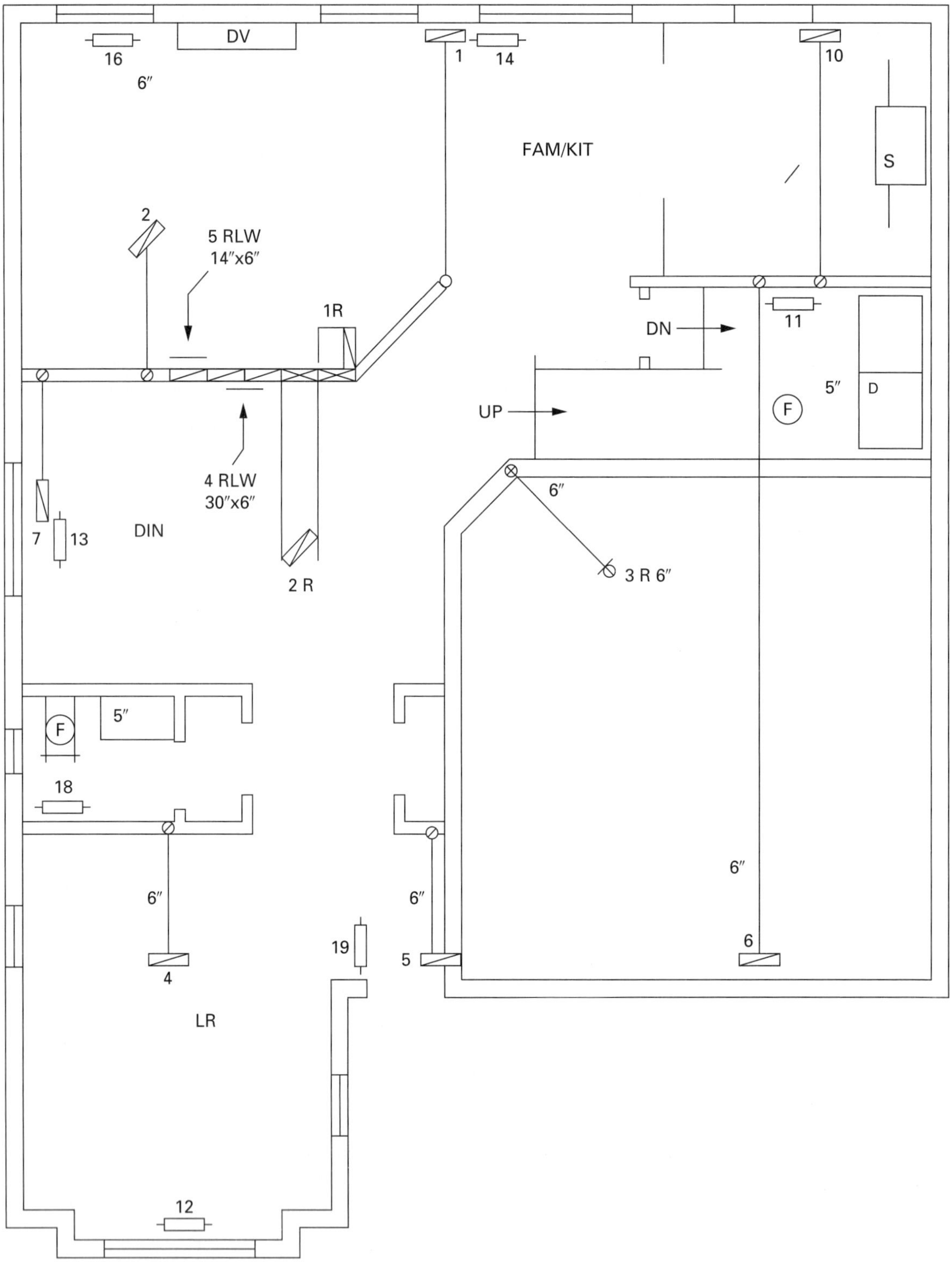

FIGURE 11.19 *(continued)*

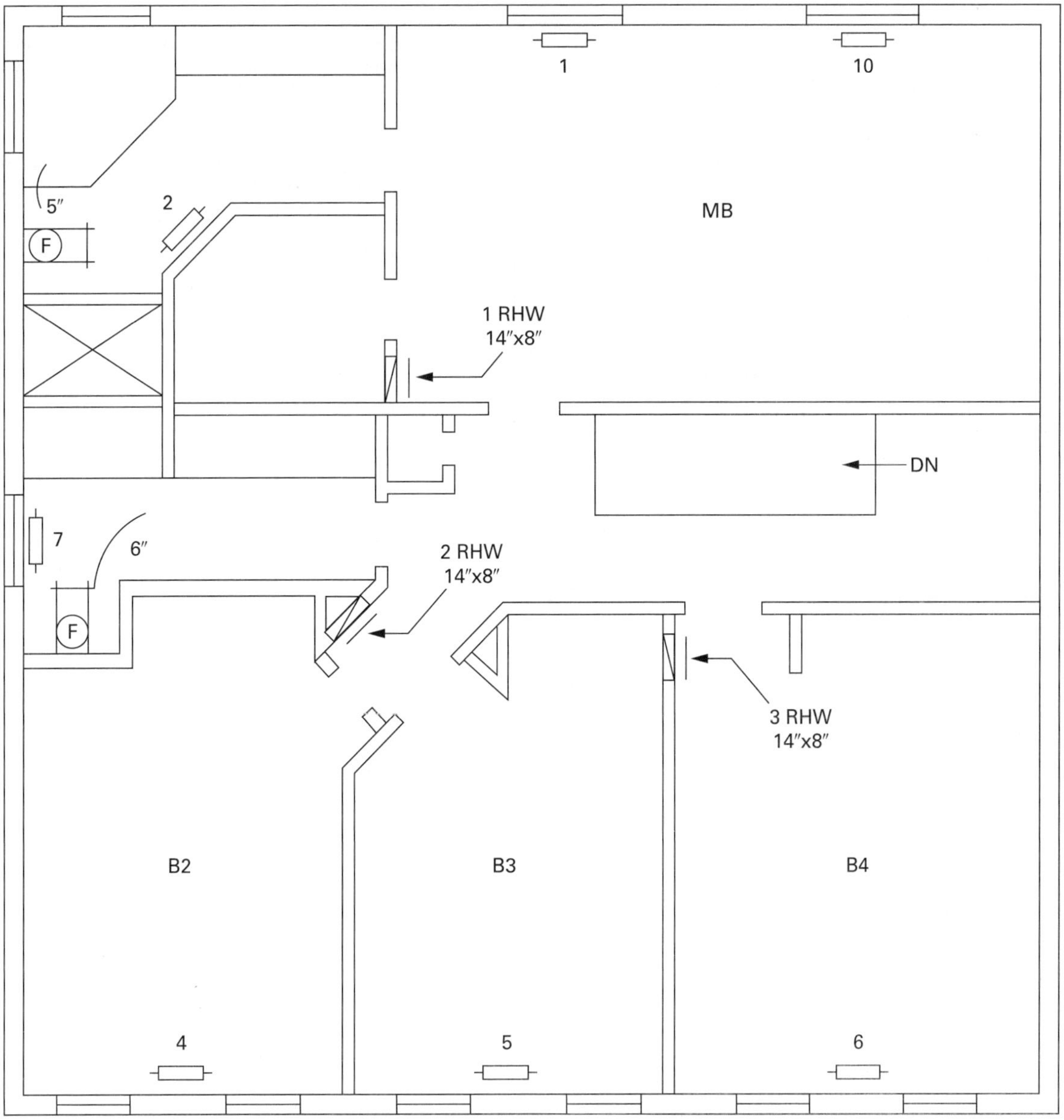

FIGURE 11.19 *(continued)*

CALCULATIONS IN ACCORDANCE WITH 1998 O.B.C.

TEMP DIFF DEG. F 76
UNFINSHED BAS'T F 76
SUMMER DEG F 15

DESIGNER:
HVAC DESIGNS LTD.
1962 wildflower dr.
pickering, ont. L1V 7B1
905-420-5300 tel.
905-420-7214 fax

PAGE 1 OF 2

BUILDER— TRIBUTE GFA— 2578 LO# 2608

TYPE— DONCASTER DATE— MARCH/2000 CALCULATION per ASHRAE FUNDAMENTALS

	EXP. WALL	COLD CLG.	ROOM AREA	WINDOW	GLASS DOOR	STEEL DOOR	COLD FLOOR	SKY-LIGHT	GRS WALL AREA	NET EXP. WALL	BASMT. WALL	BASMT. FLOOR	INFITR. 0.3 A.C.	INFITR. 0.3 A.C.	INFITR. 0.3 A.C.	SLAB ON GRADE	RM LOSS MBH.
ROOM "U"		0.03		0.5	0.5	0.4	0.04	0.77		0.05	0.11	0.03	0.03	0.05	0.05	0.53	
USE FACTOR		2.28		38	38	30.4	3.04	58.52		3.8	8.36	2.28	2.28	3.8	3.8	40.28	
MBR. CLG. HT	39	374		28	0	0	0	0	312	284				374			
8		853		1064	0	0	0	0		1079				1421			4.42
ENS. CLG. HT	26	168		22	0	0	0	0	208	186					168		
8		383		836	0	0	0	0		707					638		2.56
WIC CLG. HT	0	0		0	0	0	0	0	0	0					0		
8		0		0	0	0	0	0		0					0		0
BED-2 CLG. HT	27	176		18	0	0	0	0	216	198				176			
8		401		684	0	0	0	0		752				669			2.51
BED-3 CLG. HT	11	187		18	0	0	85	0	88	70				187			
8		426		684	0	0	258	0		266				711			2.35
BED-4 CLG. HT	30	216		18	0	0	216	0	240	222				216			
8		492		684	0	0	657	0		844				821			3.5
BED-5 CLG. HT	0	0		0	0	0	0	0	0	0				0			
8		0		0	0	0	0	0		0				0			0
BATH CLG. HT	10	120		6	0	0	0	0	80	74					120		
8		274		228	0	0	0	0		281					456		1.24
ENS-2 CLG. HT	0	0		0	0	0	0	0	0	0					0		
8		0		0	0	0	0	0		0					0		0
UPHL. CLG. HT	0	0		0	0	0	0	0	0	0				0			
8		0		0	0	0	0	0		0				0			0
LIV. CLG. HT	51	120	240	50	0	20	0	0	408	338				240			
8		274		1900	0	608	0	0		1284				912			4.98
DIN. CLG. HT	13	0	130	16	0	0	0	0	104	88				130			
8		0		608	0	0	0	0		334				494			1.44
KIT. CLG. HT	29	0	140	9	42	0	0	0	232	181				140			
8		0		342	1596	0	0	0		688				532			3.16
FAM. CLG. HT	31	0	240	40	0	0	0	0	248	208				240			
8		0		1520	0	0	0	0		790				912			3.22
LAUN. CLG. HT	14	0	48	0	0	20	0	0	112	92					48		
8		0		0	0	608	0	0		350					182		1.14
W/R CLG. HT	5	0	25	5	0	0	0	0	40	35					25		
8		0		190	0	0	0	0		133					95		0.42
FOYER CLG. HT	0	0	0	0	0	0	0	0	0	0					0		
8		0		0	0	0	0	0		0					0		0
S.ENT CLG. HT	0	0	0	0	0	0	0	0	0	0				0			
8		0		0	0	0	0	0		0				0			0
LIBR. CLG. HT	0	0	0	0	0	0	0	0	0	0				0			
8		0		0	0	0	0	0		0				0			0
DEN. CLG. HT	0	0	0	0	0	0	0	0	0	0				0			
8		0		0	0	0	0	0		0				0			0
WOB CLG. HT	0			0	0	0			0	0				0			
8				0	0	0				0				0			0
BAS'T CLG. HT	166	0	1162	9	0	0	0	0	1328		1319	1162	1162			0	
8		0		342	0	0	0	0			11027	2649	2649			0	16.67

LOSS DUE TO VENTILATION LOAD BTU/H------ 13051 TOTAL COMBINED HEAT LOSS BTU/H--- 60661 TOTAL HEAT LOSS BTU/H--- 47610

FIGURE 11.19 *(continued)*

HVAC Designs Ltd.
1962 wildflower dr.
pickering, ont. L1V 7B1
905-420-5300 tel.
905-420-7214 fax

BUILDER:	TRIBUTE
TYPE:	DONCASTER
G.F.A.	2578
DATE:	MARCH/2000
LO#	2608

FURNACE CFM ——— 1080
TOTAL HEAT LOSS —— 47610
AIR FLOW RATE (cfm) — 22.68

HEAT GAIN

	FACTO	AREA	BTU/H
GROSS WALL AREA		2288	
NET WALL AREA	0.75	1976	1482
GLASS	39.97	272	10872
SKYLIGHT	80	0	0
WOOD/METAL DOOR	3.6	40	144
EXPOSED CEILING	0.45	1361	612
EXPOSED FLOOR	0.6	301	181
INFILTRATION .5 a/c	0.75	2578	1934
PEOPLE LOAD	275	8	2200
APPLIANCES	1604	2	3208
LATEN GAIN PEOPLE	275	8	2200
INFILTRATION	1.65	2578	4254

TOTAL HEAT GAIN ——— 27087 BTUH
TOTAL HEAT GAIN ——— 2.26 TONS

VENTILATION DATA: TYPE I dwelling

Q - VENTILATION SUPPLY INLET NOT REQUIRED
E - PRINCIPAL EXHAUST FAN CAPACITY cfm
T - TIME FOR EXHAUST FAN OPERATION (hours/day)
D - MINIMUM PRINCIPAL EXHAUST DUCT SIZE (inches dia.)

PRINCIPAL EXHAUST FAN SWITCH TO BE CENTRALLY LOCATED AND IDENTIFIED AS "VENTILATION FAN" FURNACE CIRCULATION FAN LOCATED ADJACENT TO EXHAUST FAN SWITCH AND IDENTIFIED AS "CIRCULATION FAN"

EXHAUST ONLY METHOD

E 79.5 cfm
T 24 hours
D 6 inch dia.

PAGE 2of2

FURNACE DATA

BRYANT HI-BOY HI-EFFIECENCY
36080
OUTPUT 74400 BTUH
CFM @ .5" E.S.P.
FAN SPEED
LO ——— 1080
MEDLO ——— 1190
MED ——— 0
MEDHI ——— 1355
HI ——— 1545

DESIGN CFM = 1080

TEMPERATURE RISE = 64 DEG/F.

RUN #	1	2	3	4	5	6	7	8	9	10	11	12	13	14	15	16	17
ROOM NAME	MBR.	ENS.	WIC	BED-2	BED-3	BED-4	BATH	BED-5	UPHL.	MBR.	ENS-2	LIV.	DIN.	KIT.		FAM.	LAUN.
RM LOSS MBH.	2.21	2.56	0	2.51	2.35	3.5	1.24	0	0	2.21	0	2.49	1.44	3.16	0	3.22	1.14
CFM PER RUN	50	58	0	57	53	79	28	0	0	50	0	56	33	72	0	73	26
PLENUM PRESSURE	0.13	0.13	0.13	0.13	0.13	0.13	0.13	0.13	0.13	0.13	0.13	0.13	0.13	0.13	0.13	0.13	0.13
S/A DIF PRESS LOSS	0.01	0.01	0.01	0.01	0.01	0.01	0.01	0.01	0.01	0.01	0.01	0.01	0.01	0.01	0.01	0.01	0.01
ACTUAL DUCT LGH.	29	16	1	36	46	55	12	1	1	41	1	40	9	25	1	14	26
EQULVALENT LENGTH	170	190	0	120	120	130	90	0	0	110	0	130	170	130	0	170	90
TOTAL EFFECTIVE LH	199	206	1	156	166	185	102	1	1	151	1	170	179	155	1	184	116
ADJUSTED PRESSURE	0.07	0.06	13	0.08	0.08	0.07	0.13	13	13	0.09	13	0.08	0.07	0.08	13	0.07	0.11
ROUND DUCT SIZE	5	5	0	5	5	6	3	0	0	5	0	5	4	5	0	6	3
OUTLET GRILL SIZE	3	3	3	3	3	4	3	3	3	3	3	3	3	3	3	4	3
	X	X	X	X	X	X	X	X	X	X	X	X	X	X	X	X	X
OUTLET GRILL SIZE	10	10	10	10	10	10	10	10	10	10	10	10	10	10	10	10	10

RUN #	18	19	20	21	22	23	24	25	26	27	28	29	30	31	32	33	34
ROOM NAME	W/R	FOYER	S.ENT	BAS'T	BAS'T	BAS'T	BAS'T	BAS'T	BAS'T	BAS'T	BAS'T	BAS'T	BAS'T	BAS'T	BAS'T	BAS'T	BAS'T
RM LOSS MBH.	0.42	2.49	0	5.56	5.56	5.56	0	0	0	0	0	0	0	0	0	0	0
CFM PER RUN	10	56	0	126	126	126	0	0	0	0	0	0	0	0	0	0	0
PLENUM PRESSURE	0.13	0.13	0.13	0.13	0.13	0.13	0.13	0.13	0.13	0.13	0.13	0.13	0.13	0.13	0.13	0.13	0.13
S/A DIF PRESS LOSS	0.01	0.01	0.01	0.01	0.01	0.01	0.01	0.01	0.01	0.01	0.01	0.01	0.01	0.01	0.01	0.01	0.01
ACTUAL DUCT LGH.	20	38	1	39	20	30	1	1	1	1	1	1	1	1	1	1	1
EQULVALENT LENGTH	170	100	0	110	140	120	0	0	0	0	0	0	0	0	0	0	0
TOTAL EFFECTIVE LH	190	138	1	149	160	150	1	1	1	1	1	1	1	1	1	1	1
ADJUSTED PRESSURE	0.07	0.09	13	0.09	0.08	0.09	13	13	13	13	13	13	13	13	13	13	13
ROUND DUCT SIZE	3	5	0	6	6	6	0	0	0	0	0	0	0	0	0	0	0
OUTLET GRILL SIZE	3	3	3	4	4	4	3	3	3	3	3	3	3	3	3	3	3
	X	X	X	X	X	X	X	X	X	X	X	X	X	X	X	X	X
OUTLET GRILL SIZE	10	10	10	10	10	10	10	10	10	10	10	10	10	10	10	10	10

RETURN AIR #	1	2	3	4	5	6	7	8	9	BR
AIR VOLUIME	180	180	85	360	180	0	0	0	0	95
PLENUM PRESSURE	0.12	0.12	0.12	0.12	0.12	0.12	0.12	0.12	0.12	0.12
ACTUAL DUCT LGH.	35	42	47	17	15	1	1	1	1	13
EQULVALENT LENGTH	110	110	130	110	110	0	0	0	0	100
TOTAL EFFECTIVE LH	145	152	177	127	125	1	1	1	1	113
ADJUSTED PRESSURE	0.08	0.08	0.07	0.09	0.1	12	12	12	12	0.11
ROUND DUCT SIZE	7.4	7.4	5.8	9.3	7	0	0	0	0	5.4
INLET GRILL SIZE	8	8	4	8	8	0	0	0	0	4
	X	X	X	X	X	X	X	X	X	X
INLET GRILL SIZE	14	14	10	30	14	0	0	0	0	10

S/A TRUNK SIZE

	TRUNK CFM	STATIC PRESS	ROUND DUCT	RECT DUCT	
TRUNK A	391	0.06	11	14	x 8
TRUNK B	689	0.06	13.6	21	x 8
TRUNK C	0	0.06	0	0	x 8
TRUNK D	0	0.06	0	0	x 8
TRUNK E	0	0.06	0	0	x 8
TRUNK F	0	0.06	0	0	x 8
TRUNK G	0	0.06	0	0	x 8
TRUNK H	0	0.06	0	0	x 8
TRUNK I	0	0.06	0	0	x 8

RETURN AIR TRUNK SIZE

	TRUNK CFM	STATIC PRESS	ROUND DUCT	RECT DUCT	
TRUNK T	0	0.07	0	0	x 8
TRUNK U	0	0.07	0	0	x 8
TRUNK V	0	0.07	0	0	x 8
TRUNK W	0	0.07	0	0	x 8
TRUNK X	1080	0.07	15.5	27	x 8
TRUNK Y	0	0.07	0	0	x 8
TRUNK Z	0	0.07	0	0	x 8
DROP	1080	0.07	15.5	24	x 10

FIGURE 11.19 *(continued)*

TYPE: DONCASTER LO # 2608

RESIDENTIAL MECHANICAL VENTILATION DESIGN SUMMARY

For systems serving one dwelling unit & conforming to the Ontario Building Code, O.Reg 158/93

COMBUSTION APPLIANCES 9.32.3.1(1)

a) ☑ Direct vent (sealed combustion) only

b) ☐ Positive venting induced draft (except fireplaces)

c) ☐ Natural draft, B-vent or induced draft gas fireplace

d) ☐ Solid Fuel (including fireplaces)

e) ☐ No Combustion Appliances

SUPPLEMENTAL VENTILATION CAPACITY 9.32.3.5.

Total Ventilation Capacity	159	cfm
Less Principal Ventil. Capacity	90	cfm
Required Supplemental Capacity	69	cfm

PRINCIPAL EXHAUST FAN CAPACITY

Model: BROAN 684N Location: ENS

90 cfm ☑ HVI Approved

2.5 sones

HEATING SYSTEM

☑ Forced Air ☐ Non Forced Air

☐ Electric Space Heat

SUPPLEMENTAL FANS

Location	Model	cfm	HVI	Sones
ENS	650	50	✓	2.0
W/R	650	50	✓	2.0

HOUSE TYPE 9.32.1(2)

☑ I Type a) or b) appliance only, no solid fuel

☐ II Type I except with solid fuel (including fireplaces)

☐ III Any Type c) appliance

☐ IV Type I, or II with electric space heat

☐ Other: Type I, II or IV no forced air

HEAT RECOVERY VENTILATOR 9.32.3.11.

Model:

_____ cfm high _____ cfm low

_____ % Sensible Efficiency ☐ HVI Approved

@ -25 deg F

LOCATION OF INSTALLATION

Lot: Concession

Township Plan:

Address

Roll # Building Permit #

SYSTEM DESIGN OPTIONS O.N.H.W.P.

☑ 1 Exhaust only/Forced Air System

☐ 2 HRV with Ducting/Forced Air System

☐ 3 HRV Simplified/connected to forced air system

☐ 4 HRV with Ducting/non forced air system

☐ Part 6 Design

BUILDER

Name: TRIBUTE

Address:

City:

Telephone #: Fax #:

TOTAL VENTILATION CAPACITY 9.32.3.3(1)

Basement & Master Bedroom	2	@ 21.2 cfm	42.4	cfm
Other Bedrooms	3	@ 10.6 cfm	31.8	cfm
Kitchen & Bathrooms	4	@ 10.6 cfm	42.4	cfm
Other Rooms	4	@ 10. 6cfm	42.4	cfm
Table 9.32.3.A.		TOTAL	159	cfm

INSTALLING CONTRACTOR

Name: PRIMA

Address:

City:

Telephone #: Fax #:

PRINCIPAL VENTILATION CAPACITY REQUIRED 9.32.3.4.(1)

Master Bedroom	31.8 cfm
Two Bedrooms	47.7 cfm
Three Bedrooms	63.6 cfm
Four Bedrooms	79.5 cfm
Table 9.32.3.B.	TOTAL 79.5 cfm
More than 4 - Part 6	

DESIGNER CERTIFICATION

I hereby certify that this ventilation system has been designed in accordance with the Ontario Building Code.

Name: HVAC Designs Ltd.

Signature:

HRAI # DI96-0068

Date: MARCH/2000

Types of Heating Systems

The most commonly used systems are:

- **Forced air**
- Hot water
- Radiant
- Electric baseboard.

Forced Air Heating

This system distributes warm air throughout the building through **supply air ducts.** Once the air cools, it is returned to the furnace through the **return air ducts**, where it is reheated by the heat exchanger in the furnace and redistributed by a fan that forces the air through the distribution system. One of the biggest advantages of this system is that, with the addition of a central air conditioner, the same duct system can be used to distribute cold air during the warm summer months. Supply air ducts are generally located on outside walls near a window, where the greatest heat loss occurs; return air ducts are located on interior walls.

The HVAC designer must consider a number of things when designing a forced-air system. For example, an elbow in a duct can, under certain conditions, cause the same amount of resistance to air flow as 10' of straight duct, and multiple elbows will correspondingly reduce air flow. Therefore, it is important to minimize corners and elbows in installation. Larger ducts increase air flow, and round, smooth ducts provide less resistance than square ducts. Flexible ducts that have ridges reduce air flow more than smooth ducts. There is also less resistance to air flow in short duct runs.

FIGURE 11.20

A forced-air gas furnace in the Doncaster house, which corresponds with the HVAC drawings provided in Figure 11.19 (basement floor plan). Notice the following: The hot water tank is at left with gas furnace in the middle lower portion of photo. Directly above the furnace is where the warm air comes up into the supply air ducting. The large ducting at right is for the cold air return. Cold air returns to the lower portion of the furnace where it is reheated and redistributed by the furnace's heat exchanger.

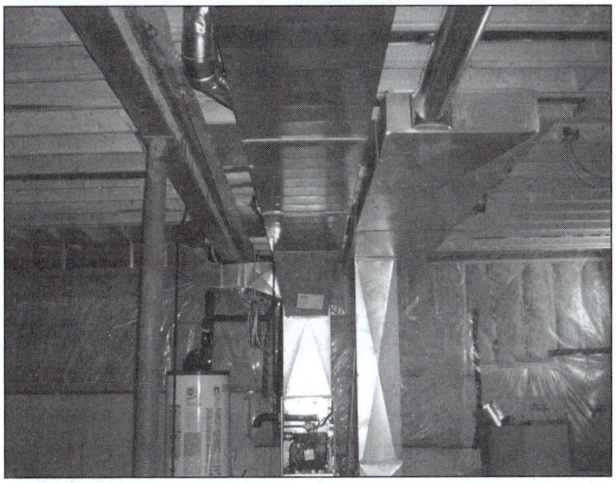

Courtesy Tom Stephenson

FIGURE 11.21

A forced-air gas furnace in the Doncaster house, which corresponds with the HVAC drawings provided in Figure 11.19. Notice that the HVAC drawings call for six returns: one in the basement, two on the first floor, and three on the second floor (see legend on page 1 of drawings). Returns on the HVAC drawings are numbered and have an R beside them (except the basement return, which is denoted by BR abbreviation). Compare the location of all of the returns and main supply air ducts in this photo to the basement floor plan on the HVAC drawings.

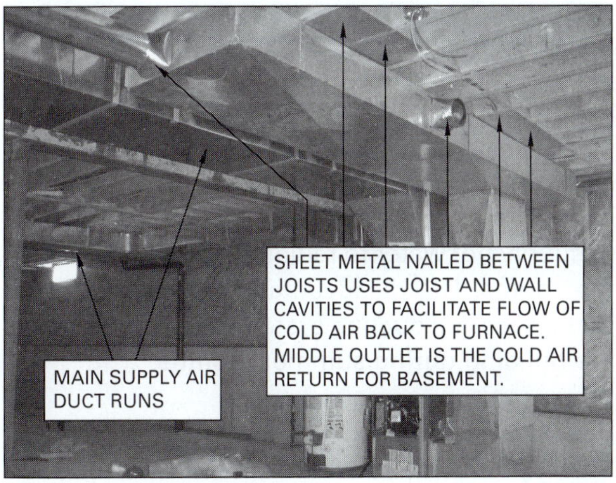

Courtesy Tom Stephenson

FIGURE 11.22
High-efficiency furnaces produce a lot of condensation and require an outlet tube to the floor drain as shown; notice the gas hookup to the hot water tank. Both furnace and tank are high efficiency and therefore do not require a chimney (no chimney is indicated on the Doncaster drawings).

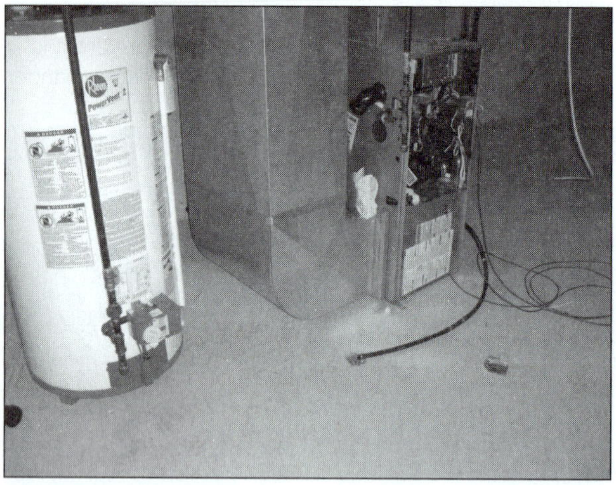

Courtesy Tom Stephenson

FIGURE 11.23
Example of a cold air return being roughed in; notice that the return flows through the stud cavity, unlike the supply air, which flows through a duct (right side of photo).

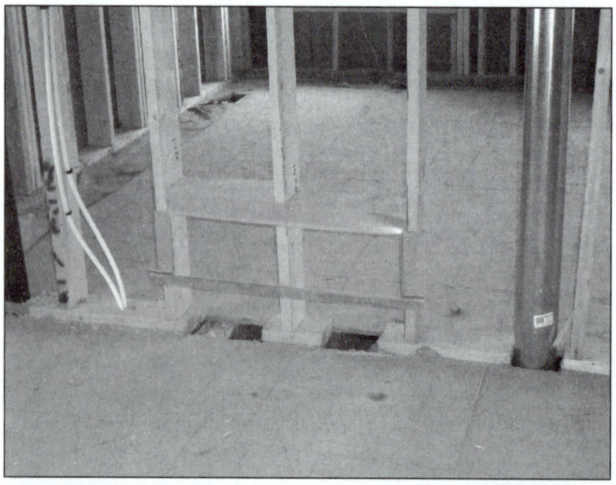

Courtesy Tom Stephenson

Hot Water Heating

This system commonly uses iron or copper tubing to supply hot water to radiators which then warm the rooms. The disadvantage to this system is that a separate ducting system is required to cool the building during summer months.

Radiant Heating

This system may use hot water tubing (also referred to as hydronic radiant heating) or a matted electrical system that is usually laid out over the floor area. The tubing is laid as a substrate to the finished floor and provides a very comfortable and evenly distributed heat. The system is generally more expensive than other systems, but is growing in popularity because of the comfort that radiant heat provides. It can be used in specific applications such as bathrooms and rooms over garages where the floor can get cold during the winter, or it can be used for heating the whole building. The heating subcontractor should consult the radiant system manufacturer for floor plan drawings and layout requirements (shop drawings).

Electric Baseboard Heating

Electric baseboards can be used to heat individual rooms or whole buildings. Thermostats can be used on individual units or from a central location. Energy costs and availability of alternate energy sources, e.g., natural gas, will dictate the popularity of this system. While electric heating is often not the most economical choice, in some regions it is used because there simply is not much choice. For example, natural gas may not be readily available, so electric heat is chosen. These areas of the country tend to lead the way in code requirements, requiring high R-values for insulation and tight envelopes. Even without the code requirements, the public would demand it in these areas due to the high cost of energy.

Central Air Conditioning

Air conditioning can be installed centrally in a forced-air system by providing a cooling coil above the plenum in the furnace, with a compressor located outside the house. If central air conditioning is to be installed, it will be noted in the specifications.

FIGURE 11.24
Air conditioner compressor located outside house.

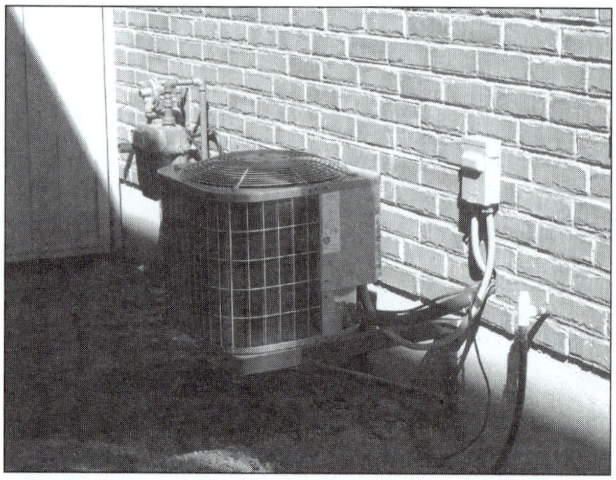

Courtesy Tom Stephenson

FIGURE 11.25

3D section view of ducts and furnace for the Doncaster house. This view provides some understanding of the logistical issues of running supply and return air ducts to different rooms of the house. Note the way the ducts have to manoeuvre around the beams and joists as indicated. You can see that the path from the furnace to bedroom 3 requires a number of adjustments so that the ductwork can find its way to the bedroom. In this case, the ductwork drops below the ceiling in the basement to find a vertical wall to move up in. This is preferable to having a bulkhead on the first floor.

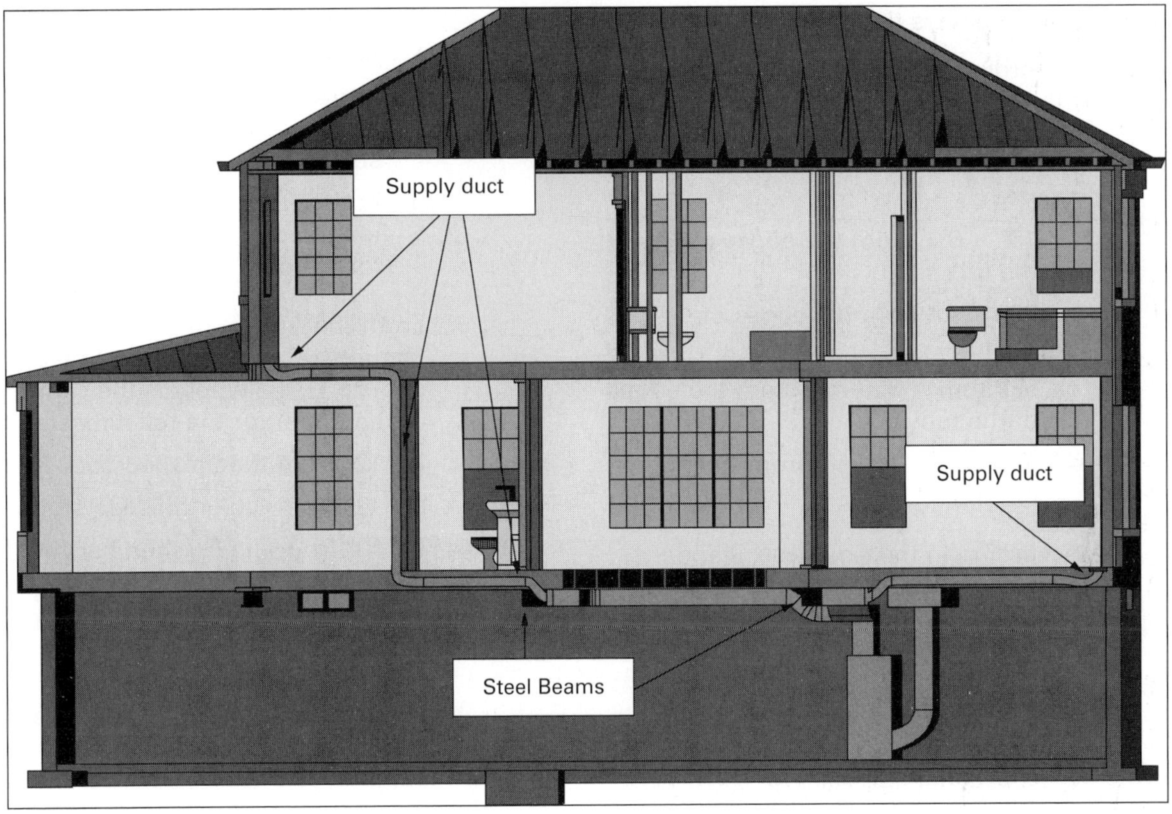

Source: Courtesy of Clint Kissoon, Chair, George Brown College, School of Architectural Studies.

Chapter 11 Assignment 1: Plumbing and HVAC Theory

Answer questions 1 through 7 using the American Standard manufacturer's drawings used in Figure 11.16. Answer questions 8 to 20 using the material in this chapter. Identify the letter that best answers the question.

_____ 1. According to the American Standard manufacturer's drawings for the **whirlpool tub** (Figure 11.16), where will the centre line of the drain be located?

a. 18" × 8¼" from the edges of the tub

b. 20" × 10" from the edges of the tub

c. 500 mm × 300 mm from the edges of the tub

d. 425 mm × 210mm from the edges of the tub

_____ 2. According to the American Standard manufacturer's drawings for the **whirlpool tub** (Figure 11.16), what is the overall size of the whirlpool tub?

a. 5'-0" long × 3'-0" wide

b. 5'-6" long × 3'-0" wide

c. 6'-0" long × 3'-0" wide

d. 5'-6" long × 2'-6" wide

_____ 3. From the American Standard manufacturer's drawing for the **whirlpool tub** (Figure 11.16), what is the height from the floor to the top of tub?

a. 464 mm

b. 502 mm

c. 546 mm

d. 572 mm

_____ 4. From the American Standard manufacturer's drawing for the **whirlpool tub** (Figure 11.16), where is the pump motor located?

a. next to the tub drain location

b. outside the tub area to be installed inside a bathroom vanity

c. centred on the right side at the opposite end of the drain location

d. the pump is located below the ceiling of where the tub is located

_____ 5. From the American Standard manufacturer's drawing for the **whirlpool tub** (Figure 11.16), where is the levelling stringer placed?

a. vertically at both ends of the tub for support

b. vertically at both ends of the tub, not for support

c. horizontally under and along the long end of the tub at the wall for support

d. horizontally under and along the long end of the tub at the wall, not for support

_____ 6. From the American Standard manufacturer's drawing for the **toilet** (Figure 11.16), where is the centre line of the outlet for the toilet located?

a. 12" from the finished back wall

b. 12" from the unfinished stud wall

c. 10" from the finished back wall

d. 10" from the unfinished stud wall

_____ 7. From the American Standard manufacturer's drawing for the **toilet** (Figure 11.16), what is the suggested height from the floor for the water supply?

a. 8"

b. 10"

c. 12"

d. 14¼"

_____ 8. What does the abbreviation HVAC refer to?

a. humidity, ventilation, air quality, and conditioning

b. heating, ventilation, and air conditioning

c. heating, ventilation, air quality, and climate control

d. humidity, ventilation, air quality, and control

_____ 9. What does the term "the house as a system" refer to?

a. the specific performance of the HVAC and mechanical system for the building

b. the specific performance of the electrical system for the building

c. the interaction that takes place between changes in temperature and the performance of the building envelope

d. the integration of all the components of the building and its systems and how well they perform together

_____10. What is the cause of **most** mechanical and electrical systems errors and disagreements during construction?

a. The building inspectors are not willing to approve the work because it does not meet the applicable code requirements.

b. Proper consultation was not made with the sub-contractors about conditions that may not be readily visible and/or the information on the drawings was not carefully reviewed by the participants.

c. The client chose the wrong system for the application and therefore it could not work.

d. The manufacturer did not make their product properly and did not provide sufficient support during the installation period for the product.

_____11. Calculate what would be the maximum hole size that could be drilled in a 1½" × 7¼" joist, according to the NBC?

a. 1 13/16"

b. 1¼"

c. 2"

d. 2 5/16"

_____12. Which of the following can be identified as having to have potable water?

a. water draining from a toilet

b. water draining from a sink

c. water from a faucet

d. water from a cistern

_____13. Which one of the following is the definition for the abbreviation DWV in plumbing terminology?

a. drain, water, valley system

b. drain, waste, vent system

c. drain, water, vent system

d. drain, waste, vex system

_____14. What is the purpose of a vent in a plumbing system?

a. to help water to siphon out of the trap

b. to allow sewer gas to ventilate to the inside of the building so the water flows more freely

c. to slow the flow of water so it does not drain to quickly

d. to prevent sewer gas from ventilating to the inside of the building

_____15. Where is Cross-linked Polyethylene piping, also referred to as PEX, used?

a. supply air ductwork

b. return air ductwork

c. water supply

d. drains

_____16. What is the abbreviation used on drawings to indicate the plumbing fitting that provides access to a clogged drain that might be located below a concrete floor?

a. ABS

b. PVC

c. CO

d. DO

_____17. Which of the following has a trap built into it and therefore does not require a separate trap?

a. bathtub

b. toilet

c. sink

d. basin

_____18. How much resistance could two elbows in a supply air duct potentially cause?

a. equal to the resistance of 5' length of straight duct

b. equal to the resistance of 10' length of straight duct

c. equal to the resistance of 20' length of straight duct

d. equal to the resistance of 30' length of straight duct

_____19. What is the maximum size hole that the NBC allows to be drilled in a 1½" × 5½" stud for a load-bearing wall if you do not want to reinforce the stud?

a. 1 13/16"

b. 2"

c. 3⅝"

d. 4"

_____20. Where are supply air ducts usually located for typical forced-air systems used in residential buildings?

a. at the floor, located under or near doors and windows on exterior walls

b. at the floor, located near interior walls

c. at the ceiling, located near interior walls

d. at the ceiling, located over doors and windows on exterior walls

ELECTRICAL SYSTEM

The wiring for a house is installed by a licensed electrician who must have applied for and received an electrical permit for the project. The **Canadian Electrical Code (CE code)** details code requirements for electrical wiring. The code is enforced through the permit and inspection process by the local hydro authority.

The wiring is generally installed after the plumbing and HVAC systems are roughed in. The process of installation is as follows:

1. The electrical boxes are located and installed based on information found on the working drawings. Some information is not always provided, such as switch and receptacle heights, unless they are noted at a specific required height on the drawings. Generally switch boxes are placed 1200 mm (4'-0") above the floor, while outlet/receptacle boxes are placed at 300 mm (12") above the floor, unless otherwise noted on the drawings. Houses designed for barrier-free access may require boxes to be located at heights different than those provided.
2. The cabling is then run from the boxes to the electrical panel. The wires are hooked up at a later time to the electrical panel.
3. The local hydro authority will provide requirements regarding the service entrance and location of the panel; this information, as well as information regarding the locations for the various lights, outlets, and switches, can be found on the working drawings (refer to Chapter 4 for a list of abbreviations and symbols).
4. The construction drawings typically do not indicate the number of lights or outlets on a particular circuit, which plays a part in calculating the circuit load. Therefore, the electrician will often draw a sketch and create a circuit schedule that indicates the number of lights and outlets on a circuit, as well as how he/she plans to run the wiring back to the panel. The sketch helps the electrician determine the most efficient way of running the wiring while also meeting code requirements.

FIGURE 11.26
Distribution of electricity throughout the house.

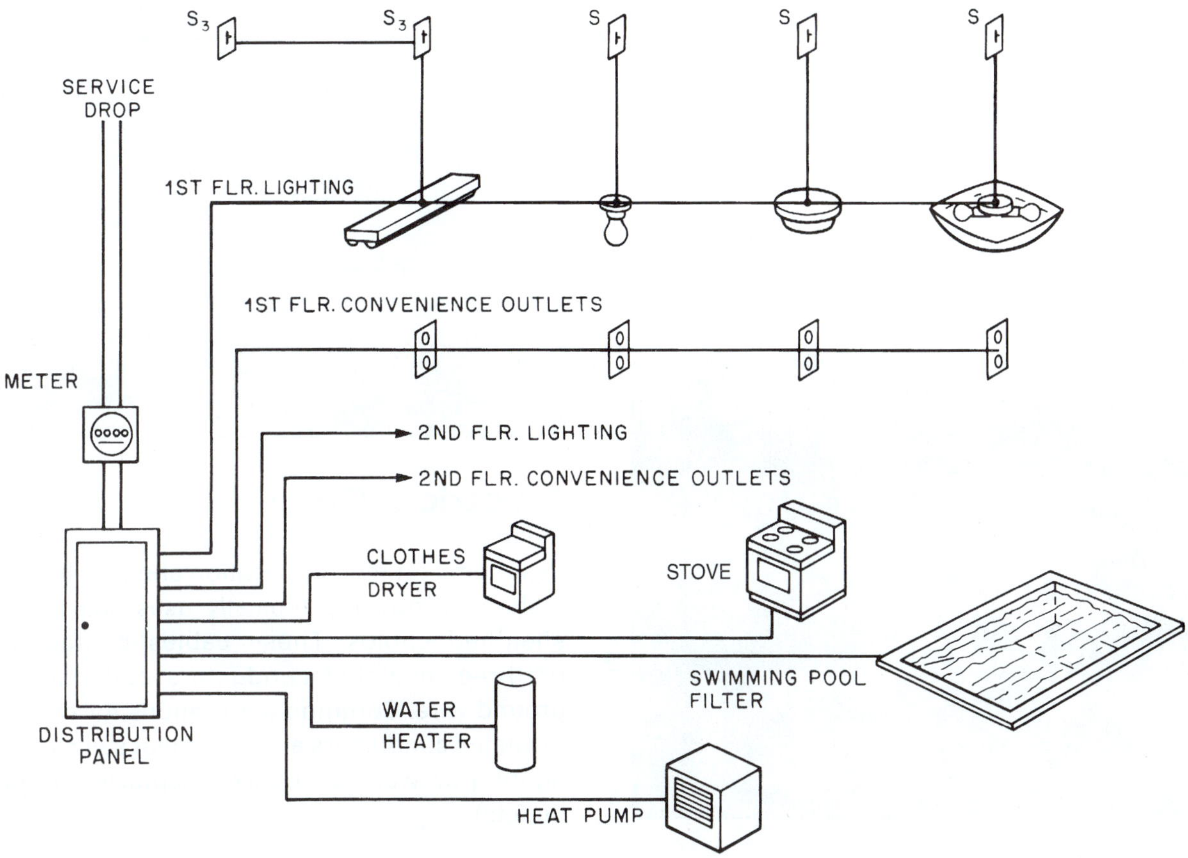

The electrical subcontractor and designer have to consider a number of factors regarding the electrical system:

- Building lighting requirements
- HVAC system requirements
- Electrical appliances to be used
- Communications (TV, network, security, telephone cabling) and alarm systems, usually roughed in by the electrical subcontractor.

The electrical subcontractor must also consider:

- The service entrance; is it above grade (from hydro poles) or below grade (conduit run underground)?
- The type of building construction; are there considerations or restrictions that need to be planned for ahead of time (e.g., is the building constructed from insulated concrete forms and, if so, how will the wiring be run)?
- The size of panel and the number of circuits required (e.g., one circuit may have six outlets and four lights on it or any combination of these, as long as it meets CE code requirements for circuit loads). The size of service the house will require: 100 amp, 200 amp, or 400 amp.
- Type of heating and cooling system; if it is electric, heat loss calculations must be performed to ensure the proper sizing of units, as discussed in the HVAC section of this chapter.

FIGURE 11.27
Wiring being roughed in throughout house and back to panel.

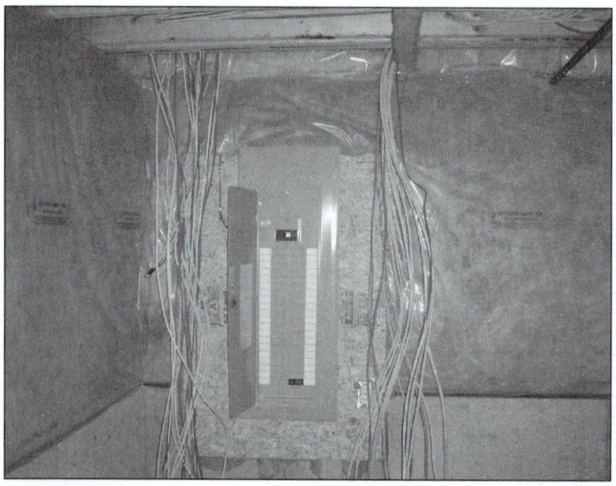

Courtesy Tom Stephenson

FIGURE 11.28
Completed panel with circuit breakers installed. Once all the wiring is run to the panel, the electrician can instal and hook up the breakers. This practice ensures that the wiring is not prematurely hooked up, causing a safety hazard during the rough-in stage.

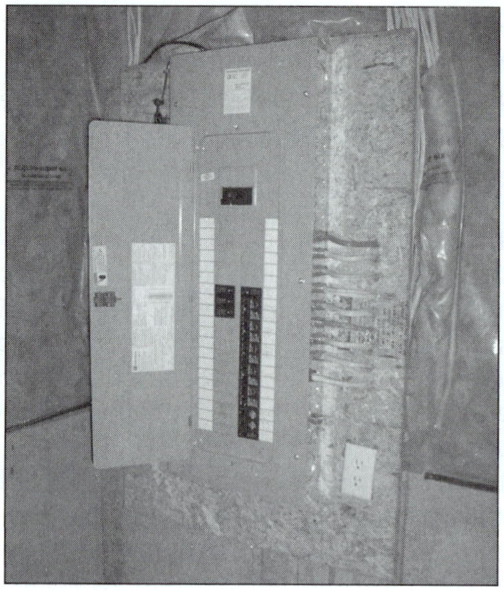

Courtesy Tom Stephenson

FIGURE 11.29
A sophisticated wiring system that includes security alarm, network cabling, and electrical panel.

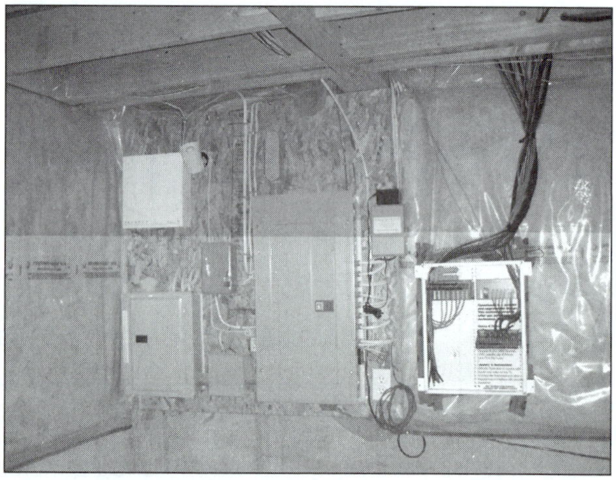

Courtesy Tom Stephenson

Electrical Cables

Electrical cables come in a variety of forms depending on their application and use. Low-rise residential housing typically uses non-metallic sheathed cables. These cables contain two or three insulated conductors and one bare ground wire. Common applications for two-wire insulated conductors are for regular switches and lighting as well as duplex receptacles (regular outlets).

FIGURE 11.30
An electrical rough-in using non-metallic sheathed cabling for one circuit providing power to several receptacles. Notice how the cable feeds the box at left and then runs to the box at right and then goes up the side of the stud to feed more lights and outlets on the circuit.

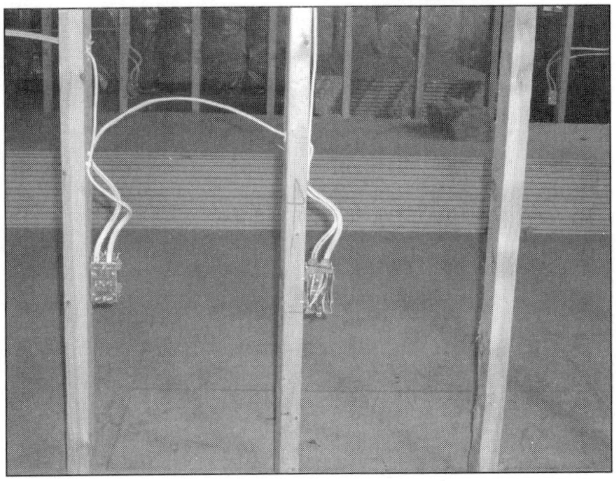

Courtesy Tom Stephenson

Three-wire insulated cables contain three insulated conductor wires and one uninsulated ground wire. Three-wire insulated cables are used in the following:

1. ***Split-switched receptacles:***
   - These receptacles are used in a room, such as a bedroom, with a lamp that is to be plugged into an outlet and operated by a switch. One of the outlets in the receptacle is left to operate normally; hence the term "split." This type of receptacle allows for a TV or alarm clock to continue to work while a lamp is turned off.
   - A three-wire is used in this case to provide two sources of power to the outlet; one to switch the outlet, the other for the regular outlet.
   - These receptacles are usually indicated by a split-switched receptacle symbol and by a wire running from the switch to the receptacle. Sometimes just the wire from the switch to the receptacle is shown, as in the Doncaster and Whittington drawings.

FIGURE 11.31
Split-switched receptacle.

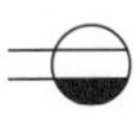

FIGURE 11.32
Split-switched receptacle.

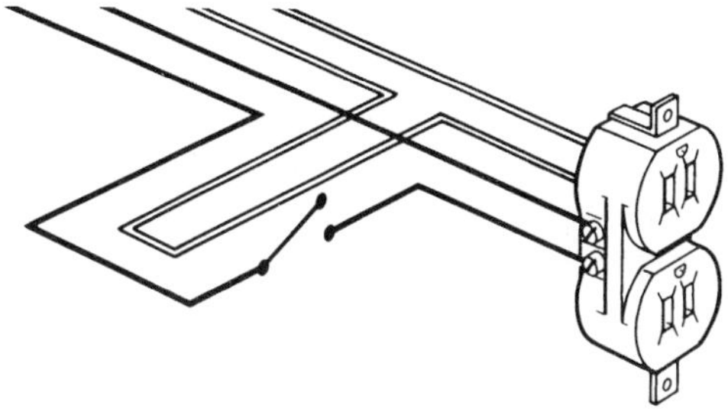

2. ***Split-circuit receptacles:***
   - A three-wire insulated conductor is used to provide two sources of power to a duplex receptacle. For example, the top outlet is on one circuit, while the bottom outlet is on another circuit. The CE code requires this for circuits in a kitchen where a user may want to plug in a toaster and kettle to the same duplex outlet. Using a split circuit ensures that there will not be any problems with overloading and tripping the circuit breaker.
   - Split-circuit receptacles are drawn similar to a regular duplex receptacle except that there are three lines instead of two in the circle. Sometimes the split-circuit receptacle is indicated like a regular duplex receptacle on the drawings; but when the electrical code requires it, the electrician will wire it as a split-circuit receptacle. This is another example of why designers will state on the drawings that construction practices must comply with applicable codes: because in many cases they do not fully know the electrical code requirements to the level of detail that the electrician must adhere to.
3. ***Three- and four-way switches:***
   - A **three-way switch** is manufactured differently from regular switches, and is used whenever a light must be operated from more than one location (e.g., a light can be turned on and off at the top and bottom of a staircase).

FIGURE 11.33
Split-circuit receptacle.

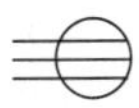

FIGURE 11.34

A three-way switch; notice how the light (device) can be turned on or off from either switch. The illustration shows the light in the off position. Changing the switch at either location will turn it on.

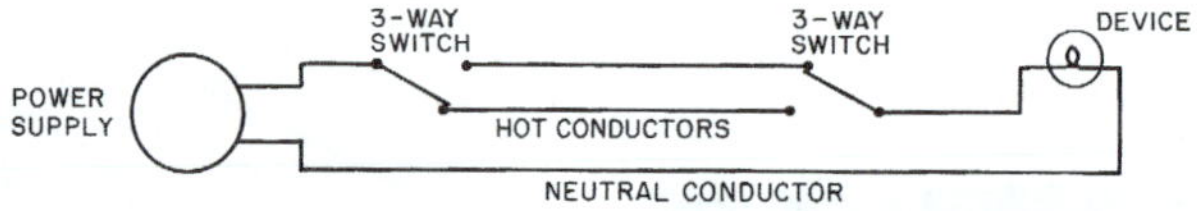

FIGURE 11.35

Electricians like to mark what each electrical box is for so that they run the proper wires to the correct locations. In this case the box will hold a three-way switch that will operate the light at the top of the stairs; there will also be a single switch that will control a pot light on the first-floor hallway.

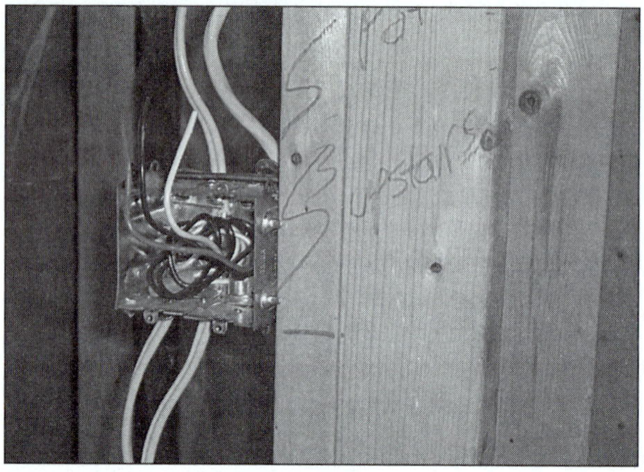

Courtesy Tom Stephenson

- A **four-way switch** is also manufactured differently from regular switches; it is used in conjunction with three-way switches when a light is to be operated from three or more locations, e.g., a living room with three entrances. Operating a light from three locations would require a three-way switch at the first location, a four-way switch at the middle location, and another three-way switch for the final location. Often drawings will indicate just three-way switches, meaning the switch operates from multiple locations. This leaves it up to the electrician to purchase the correct amount of three- and four-way switches and to wire them in the most efficient way possible.
- Three-conductor cabling is used for three- and four-way switches to act as messenger lines travelling between the switches, as indicated in Figure 11.34, either completing the circuit to turn the lights on, or opening the circuit to turn the lights off.

4 ***Stove and dryer wiring:***

- While standard house voltages use 110 or 120 volts, appliances and equipment

TABLE 11.1

Some Typical Wiring Gauges and Uses for Residential Wiring

American Wire Gauge (AWG) size of wire	Common Uses	Circuit Breaker Size
8	Stove and oven wiring	40 amps
10	Dryer and air conditioning wiring	30 amps
12	Some outlets, air conditioners	20 amps
14	Typical outlets and lights	15 amps

such as stoves and dryers require 220 or 240 volts. A three-wire insulated conductor cable is used to supply these devices. The wiring for this type of equipment is of a heavier gauge such as no. 8 AWG (American Wire Gauge) or no. 10 AWG, depending on the appliance. This wire has a much larger diameter than typical house wire, which is normally no. 14 AWG.

Typical wiring types and material include:

- Non-metallic sheathed cabling, also called Romex cabling, is used extensively in residential building where wiring is concealed.

FIGURE 11.36

Example of three-wire insulated conductor cable used to rough-in a stove.

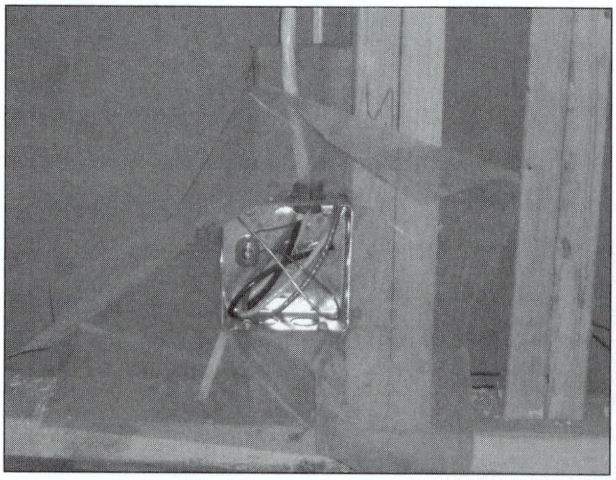

Courtesy Tom Stephenson

- Armoured cable, also known as BX cabling, is often used in residential and some commercial construction where wiring is exposed. It provides a safe, low-cost means of protecting the exposed wiring.
- Rigid conduit is generally used in institutional, commercial, and industrial applications and plastic may also be used in areas where wiring is exposed to potential damage.

FIGURE 11.37
Pot lights roughed in for ceiling. Note that because of potential for fire, care should be taken to ensure that only pot lights that are approved for insulated ceilings are installed in insulated ceilings, and that special protective boxes used for pot lights in insulated ceilings meet regulatory requirements. Pot lights are marked with the typical light symbol with the addition of a P in the centre as shown. -Ⓟ- Photo below is an example of pot lights used in an uninsulated ceiling.

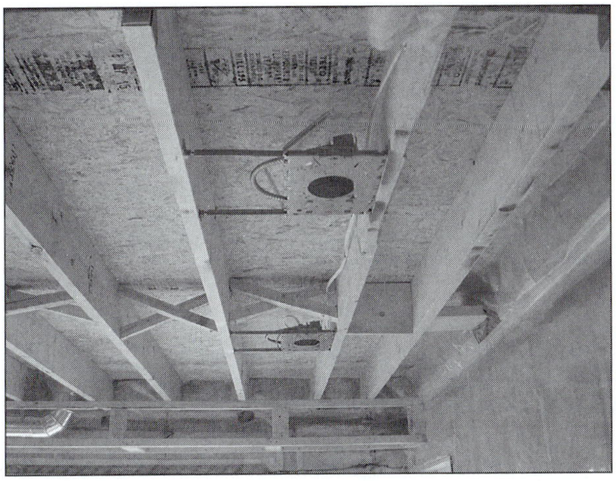

Courtesy Tom Stephenson

FIGURE 11.38
A pull-chain lamp holder typically found in basements and marked on drawings with the abbreviation PC.

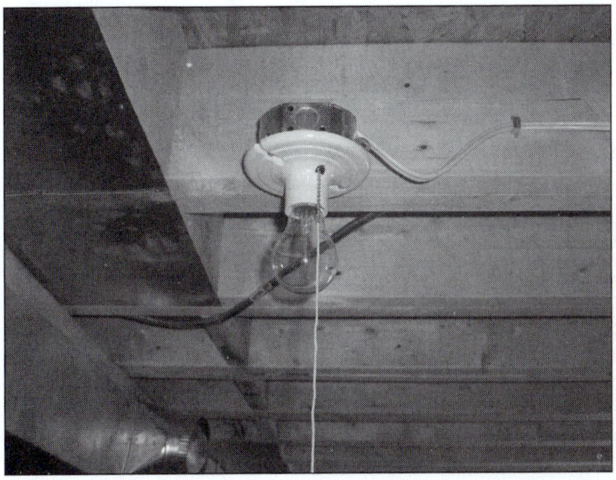

Courtesy Tom Stephenson

FIGURE 11.39
An underground hydro electric service being roughed in. Hydro services can be either above ground or below ground, depending on local requirements.

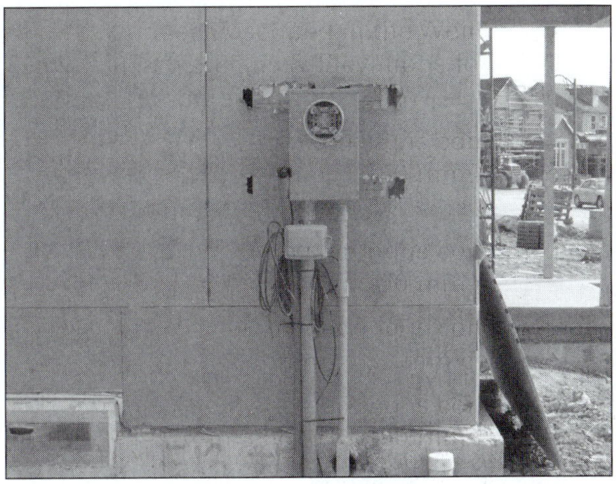

Courtesy Tom Stephenson

GREEN BUILDING SYSTEMS

As defined in Chapter 1, Green Building means building to a higher, more energy efficient and environmentally sustainable level than is currently standard practice with the intention of eventually becoming truly sustainable. New and innovative designs and technologies are becoming more commonplace and change is constant; as a result, there are many new and innovative systems besides the systems that are discussed in this chapter, including but not limited to renewable energy systems, new more efficient innovations of existing systems, and systems that use heat exchangers to heat and cool our buildings.

These systems include:

- **Photovoltaic cells** that capture energy from the sun and use it to provide electricity to the building and to help provide energy to the electrical grid
- Wind energy that utilizes **wind turbines** to produce electricity
- **Ground source heat pumps** (geothermal systems) that harness stable ground temperatures to pre-heat and pre-cool the air in the building
- **Heat recovery ventilator (HRV)** that is used to exchange stale indoor air with fresh outdoor air in an efficient manner (refer to Chapter 12 for more information on HRVs)
- Tankless hot water heaters
- **Solar hot water systems**

FIGURE 11.40

The drawing displays a typical roof-mounted solar water heating system. Photovoltaic panels are also installed on roofs, ideally designed to maximize area facing south.

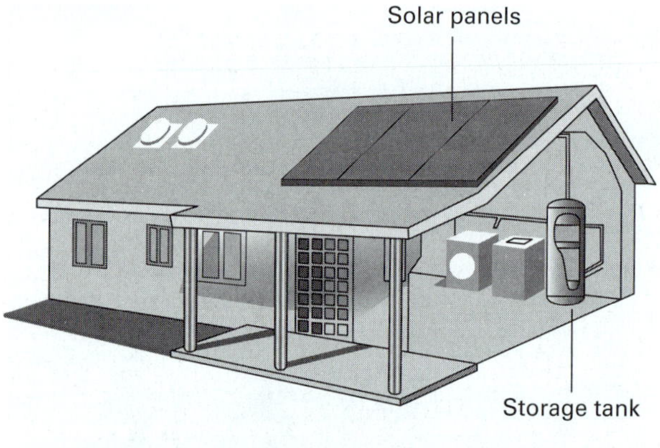

Source: From CAUFIELD, Going Green with the International Residential Code, 1E © 2011 Delmar Learning, a part of Cengage Learning, Inc. Reproduced by permission. www.cengage.com/permissions

- More efficient furnaces, such as high efficiency two-stage furnaces designed to maximize efficiencies through various temperature ranges
- **ECM blower motors** that use direct current to decrease energy use
- **Drain water heat exchangers** that extract heat from grey water and transfer it to the potable hot water tank by a copper coil attached to a copper drain pipe
- **Grey water** includes the recycling of waste water that is generated from domestic activities such as laundry, dishwashing, and bathing, which can be recycled onsite for uses such as toilet water or landscape irrigation. This is currently not allowed in some jurisdictions. Codes and regulations are changing quickly in this area, as concern over the efficient use of our water supply increases.

FIGURE 11.41

Section views of two types of ground source heat pumps (geo-thermal). Vertical ground loop and horizontal ground loop heat pumps.

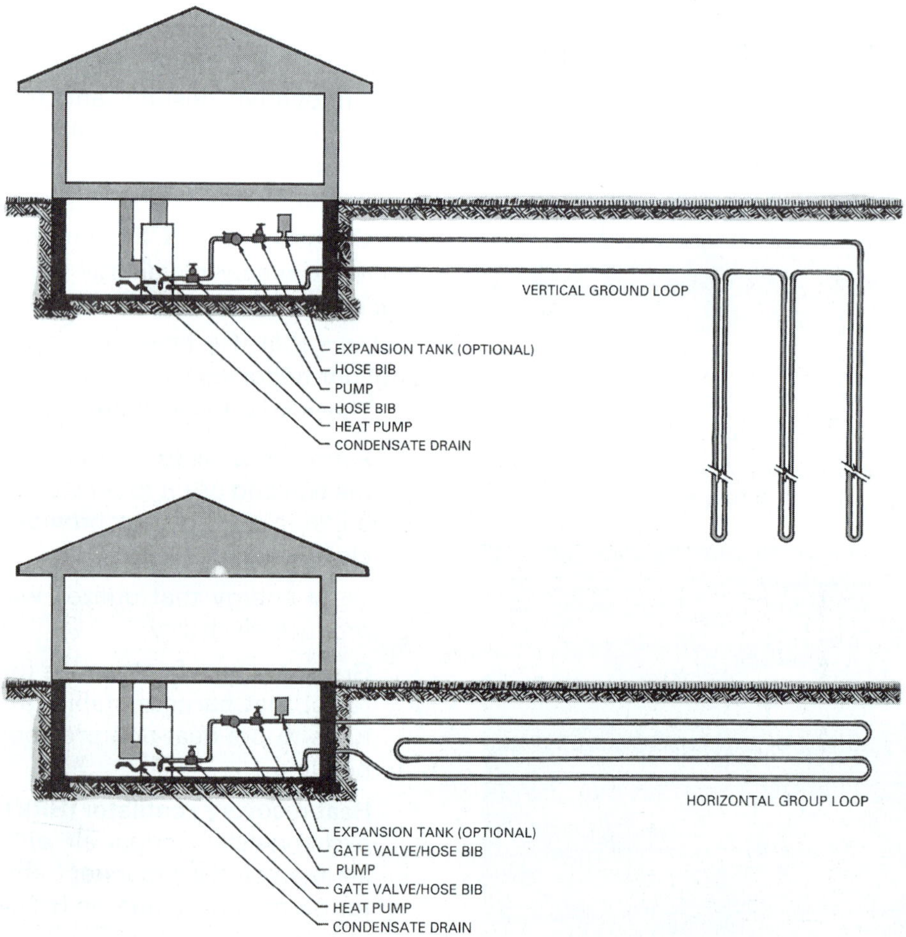

Source: Courtesy Kansas Electric Utilities Research Program, research by Mark Hannifan and Joseph King, AIA

FIGURE 11.42
An example of a grey water recycling system that substantially reduces annual water consumption by filtering grey water from a shower or other source for reuse in a toilet.

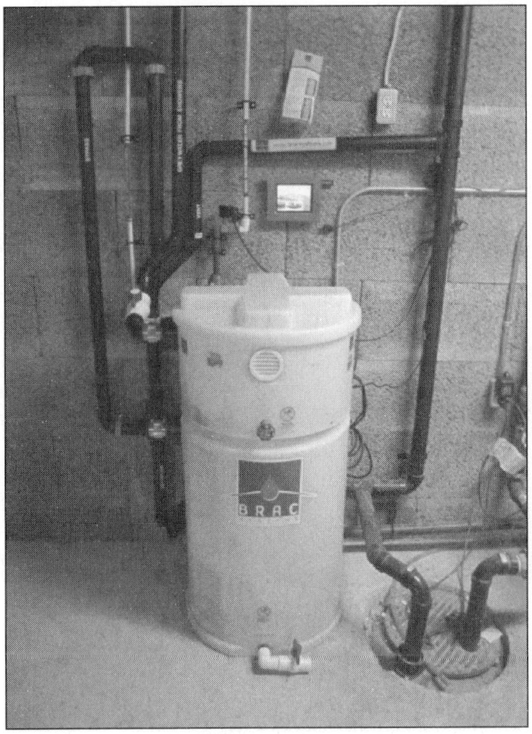

Courtesy Tom Stephenson

FIGURE 11.43
Kortright Centre Archetype Sustainable House. Note the wind turbine in the background, the photovoltaic cells on the upper roof, and the solar panels for domestic hot water and radiant floor heating on the lower roofs.

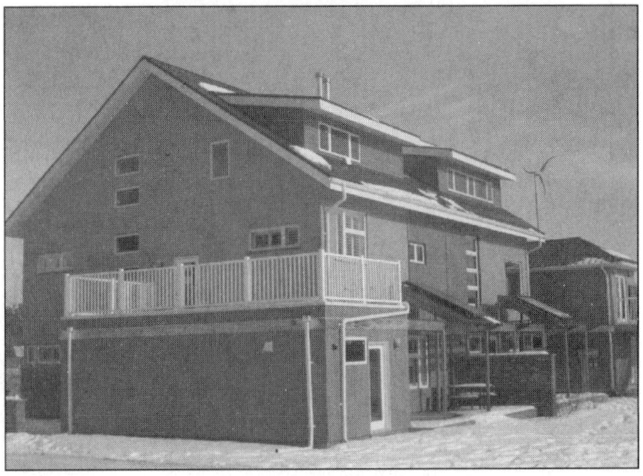

Courtesy Tom Stephenson

FIGURE 11.44
A Heat Recovery Ventilator (HRV) being tested and monitored with sensors for efficiency in the Archetype Sustainable House at the Kortright Centre.

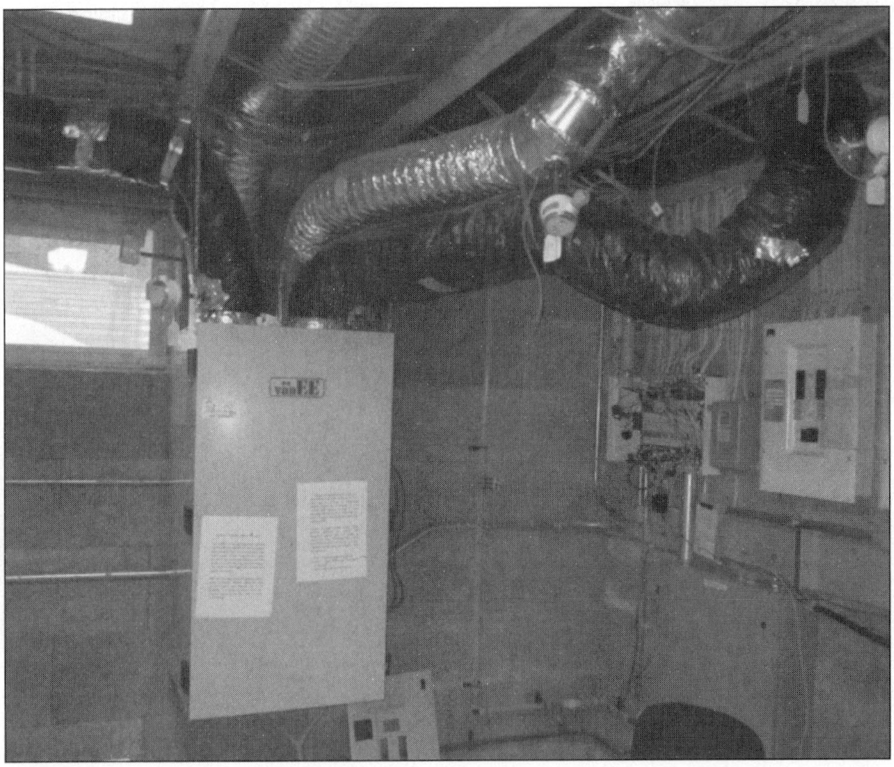

Courtesy Tom Stephenson

FIGURE 11.45

An example of a drain water heat exchanger.

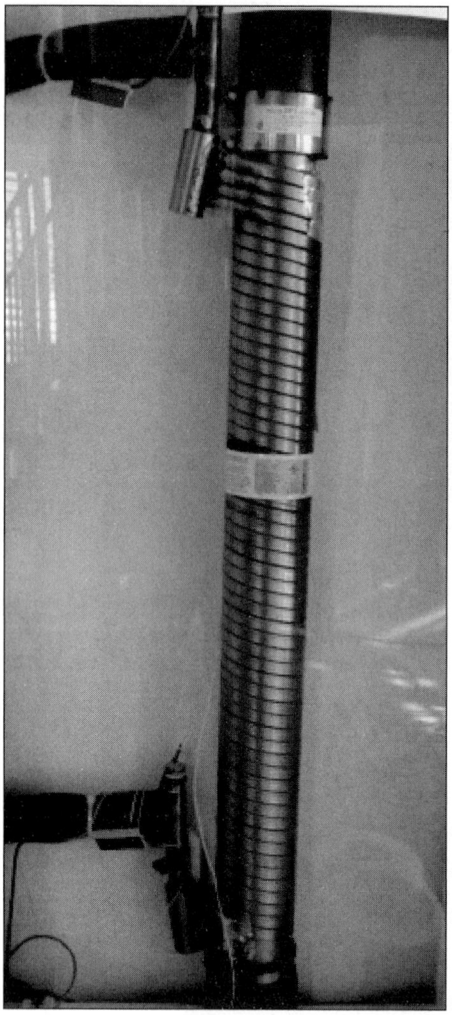

Courtesy Tom Stephenson

FIGURE 11.46

An example of a CMHC Equilibrium™ House (The NOW House retrofit). Photovoltaic cells are used to generate DC electricity that flows through an inverter that converts it to AC, which in this case is then fed to the grid (roof right). Evacuated tube solar collectors are used to heat domestic hot water and for hydronic floor heating (roof left). It is a challenge in cold climates to keep the snow off the panels so that they can work to maximum efficiency.

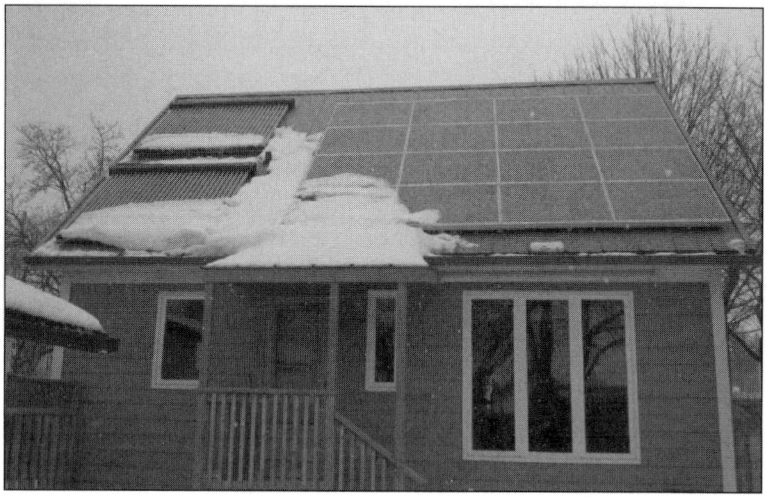

Courtesy Tom Stephenson

FIGURE 11.47

An example of 1/2 pound polyurethane foam sprayed onto a garage ceiling to form a tight air barrier with the bedroom above it. A tightly sealed and well-insulated building allows designers to downsize mechanical system requirements, thus saving on energy costs and reducing CO_2 emissions. This is an important consideration when designing energy efficient and sustainable buildings that will require the use of renewable energy systems (refer to Chapter 12 for more detail on the building envelope).

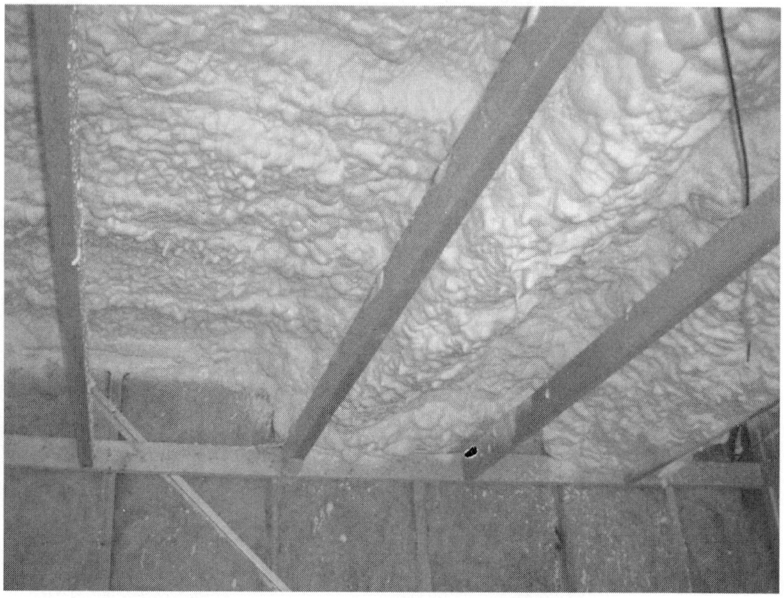

Courtesy Tom Stephenson

To be effective, the adoption and integration of these systems and innovations will require a greater focus on planning and consultation during the design process (e.g., may include careful orientation of the house and design of roof). Careful review of the specifications and drawings will ensure that the systems are installed and operating properly and that they meet their intended purpose.

KEY TERMS

Canadian Electrical Code (CE code) (p. 287)
Drain, waste, vent system (DWV) (p. 263)
Drain water heat exchanger (p. 292)
ECM blower motor (p. 292)
Forced air (distribution system) (p. 281)
Four-way switch (p. 290)
Grey water (p. 292)
Ground source heat pump (p. 291)
Heat recovery ventilator (p. 291)
HVAC (p. 260)
PEX (p. 269)
Photovoltaic cells (p. 291)
Potable water (p. 263)
Return air duct (p. 281)
Sanitary sewer (p. 263)
Septic system (p. 264)
Solar hot water system (p. 291)
Split-circuit receptacle (p. 289)
Split-switched receptacle (p. 289)
Stack (p. 264)
Storm sewer (p. 263)
Sump pump (p. 264)
Supply air duct (p. 281)
Three-way switch (p. 289)
Water trap (p. 270)
Wind turbine (p. 291)

Chapter 11 Assignment 2: Electrical, Green Systems, and Doncaster Drawings

Answer questions 1 through 10 using the information found in the chapter, and answer questions 11 to 20 using the information found in the Doncaster drawings that came attached to the text. Identify the letter that best answers the question.

_____ 1. Where are photovoltaic cells used?

a. inside forced-air furnaces
b. outdoors to capture the sun's energy and transform it into electricity
c. in security cameras to take security images and forward them to your cell phone remotely
d. below ground to capture the natural heat from the earth and transform it into space heating for the house

_____ 2. What function does a heat recovery ventilator perform?

a. purifies incoming potable water
b. acts as a humidifier for the building
c. exchanges stale indoor air for fresh outdoor air efficiently
d. acts as a mechanism to bring fresh air in through the basement floor slab

_____ 3. What is the definition of grey water?

a. fit for human consumption
b. not fit for human consumption but can be recycled for use in a toilet
c. not fit for human consumption and is water that drains from a toilet
d. water that is used in geothermal heating systems

_____ 4. Where is an electrical permit taken out from?

a. local building department
b. provincial building department
c. local hydro authority
d. local provincial board office

_____ 5. At what height from the floor are electrical outlet boxes usually installed?

a. 6"
b. 8"
c. 10"
d. 12"

_____ 6. Which of the following types of information do the floor plans usually display regarding the electrical requirements for the project?

a. how many outlets and lights will be controlled by a single circuit
b. the location of outlets, appliances, switches, and lights
c. the type of light fixtures used
d. the size and type of electrical panel

_____ 7. Which of the following types of information do the floor plans usually display regarding the electrical requirements for the project?

a. whether the switches are operated from more than one location
b. the type of wiring used
c. the loading on individual circuits
d. the type and model number of appliances used

_____ 8. What type of switch or switches will be required in order to operate a light in a stairwell from the top of the stairs and the bottom of the stairs?

a. two 2-way switches
b. two 3-way switches
c. one 3-way and one 4-way switch
d. one 2-way switch

_____ 9. What is the most common wire size for regular circuits that are used for lighting and receptacles?

a. #8 AWG cabling
b. #10 AWG cabling
c. #16 AWG cabling
d. #14 AWG cabling

_____10. What type of wire is typically used in residential applications when wiring is located inside a wall?

a. non-metallic sheathed cabling
b. rigid conduit

c. armoured cabling
d. Electrical Metallic Tubing (EMT)

_____11. Which of the following matches the Doncaster floor plan drawings in the number of receptacles indicated?
a. the dining room has four receptacles
b. the dining room has three receptacles
c. the family room has seven receptacles
d. the family room has six receptacles

_____12. How many receptacles does the garage have?
a. one
b. two
c. three
d. four

_____13. How many receptacles are there in the kitchen?
a. four
b. five
c. six
d. seven

_____14. Where is a receptacle operated from a switch indicated (split-switched receptacle)?
a. master bedroom
b. bedroom 3
c. kitchen
d. living room

_____15. What height from the floor is the duplex receptacle located over the vanity in the ensuite washroom?
a. 300 mm
b. 1070 mm
c. 1080 mm
d. 1100 mm

_____16. How many three-way switches are indicated in the hallway on the second floor?
a. one
b. two
c. three
d. four

_____17. How many three-way switches (including those in the garage if the optional door is chosen) do the Doncaster house drawings indicate are required?
a. four
b. six
c. nine
d. ten

_____18. How many pull-chain light fixtures are indicated on the Doncaster drawings?
a. two
b. three
c. four
d. five

_____19. Not including over the stove, how many exhaust fans are indicated on the drawings?
a. two
b. three
c. four
d. five

_____20. What height from the floor would the receptacle for the microwave oven need to be placed at?
a. 1070 mm
b. 1200 mm
c. 1300 mm
d. 1500 mm

Chapter 11 Quiz Review: Electrical, HVAC, Plumbing, and the Brook Drawings

Answer questions 1 to 6 from the chapter material using Figure 11.19 (review floor plans provided for this figure), answer questions 7 to 10 using Figure 11.8 as specifically identified in the question, and answer questions 10 to 24 using the Brook drawings that came attached with the text (note that the Brook drawings, page 16, have separate electrical floor plans that some of the questions will require you to review). Identify the letter that best answers the question.

_____ 1. From the HVAC floor plans in Figure 11.19, how many return air ducts are indicated on the second floor?

a. two
b. three
c. four
d. five

_____ 2. From the HVAC floor plans in Figure 11.19, how many supply air ducts are indicated on the second floor plan?

a. seven
b. eight
c. nine
d. ten

_____ 3. From the HVAC floor plans in Figure 11.19, how many supply air ducts will exit on the first floor (include laundry area)?

a. six
b. seven
c. eight
d. fourteen

_____ 4. From the HVAC floor plans in Figure 11.19, what is the diameter indicated for the supply duct to bedroom 4?

a. 3"
b. 4"
c. 5"
d. 6"

_____ 5. From the HVAC floor plans in Figure 11.19, what is the diameter indicated for the supply duct to the living room?

a. 3"
b. 4"
c. 5"
d. 6"

_____ 6. From the HVAC floor plans in Figure 11.19, what is the diameter indicated for the supply duct to the family room at the left of the fireplace?

a. 3"
b. 4"
c. 5"
d. 6"

_____ 7. From the drain design separate drawing in Figure 11.8 (continued) Drain Design Separate, how many cleanouts are indicated?

a. three
b. four
c. five
d. six

_____ 8. From the drain design separate drawing in Figure 11.8 (continued) Drain Design Separate, how many floor drains are indicated?

a. one
b. two
c. three
d. four

_____ 9. From the drain design separate drawing in Figure 11.8 (continued) Drain Design Separate, how many field drain connections are there and which sewer are they connected to?

a. one connection to the storm sewer
b. two connections to the storm sewer
c. one connection to the sanitary sewer
d. two connections to the sanitary sewer

_____10. From the drain design separate drawing in Figure 11.8 (continued) Sub-soil Drainage Connection, how far away from the structure must the discharge from the sump pump be?

a. 300 mm

b. 1 metre

c. 3 metres

d. 3.5 metres

_____11. How many floor drains are shown on the Brook drawings for the Elevation A optional basement plan?

a. one

b. two

c. three

d. four

_____12. How many smoke alarms are indicated for the whole house using the Elevation A drawings with the optional basement plan chosen?

a. one

b. two

c. three

d. four

_____13. How many switch-operated outlets (split switched receptacles) are indicated for the whole house using the Elevation A drawings with the optional basement plan chosen?

a. two

b. three

c. four

d. five

_____14. How many carbon monoxide detectors are indicated for the whole house using the Elevation A drawings with the optional basement plan chosen?

a. one

b. two

c. three

d. four

_____15. How many air changes per hour minimum does the second floor bathroom fan have to provide?

a. one air change per hour

b. two air changes per hour

c. three air changes per hour

d. four air changes per hour

_____16. How many duplex outlets are indicated in the kitchen and breakfast areas?

a. three

b. four

c. six

d. eight

_____17. How many outlets are shown in the garage?

a. two

b. three

c. four

d. five

_____18. Excluding the stove exhaust, how many exhaust fans are shown in the whole house using Elevation A with the optional lower level plan chosen?

a. three

b. four

c. five

d. six

_____19. How many hose bibs are shown on the Elevation A drawings?

a. one

b. two

c. three

d. four

_____20. On which elevation does the dryer vent exit the house?

a. front elevation

b. rear elevation

c. left elevation

d. right elevation

_____21. How many weatherproof duplex outlets are indicated on the exterior of Elevation A?

a. one

b. two

c. three

d. four

_____22. How many outdoor lights are operated by one switch at the front of the house for Elevation A?

a. one

b. two

c. three

d. four

_____23. How many outlets does the walk-in closet have?

a. none

b. one

c. two

d. three

_____24. Where is there a choice given for an optional direct vent gas fireplace?

a. rec. room one

b. rec. room two

c. master bedroom

d. great room

12 The Building Envelope

LEARNING OBJECTIVES

After completing this chapter, you will be able to:

- Read and interpret drawings for information related to the building envelope.
- Explain the qualities and function of an air barrier.
- Explain the qualities and function of a vapour barrier/retarder.
- Interpret detail drawings regarding the placement of the air barrier and vapour barrier/retarder.
- Describe the placement of the air/vapour barrier around the structure.

INTRODUCTION

This chapter introduces you to the **building envelope,** which consists of all sides of the building that directly separate the inside environment from the outside environment. This includes all exterior-facing walls including foundation walls, ceilings, and floors (e.g., the basement floor slab). This chapter also introduces you to some of the basic principles of building science.

THE BUILDING ENVELOPE

Historically, buildings were not built with much consideration given to insulation or air tightness, as heating costs were not a big concern. However, buildings today have much higher levels of insulation placed in their walls, floors, and attics, and are increasingly airtight due to environmental concerns and higher energy costs. As a result, much greater care has to be taken regarding how the building is sealed. Besides lowering energy costs and providing a very comfortable dwelling, a tighter building envelope can also bring with it an array of new challenges, such as control of indoor moisture. For example, a tight envelope means reduced air leakage and so the humidity generated by people living in the home remains inside. Other concerns arise regarding indoor air quality and pollutants that require a mechanism for exchanging stale air with fresh air.

Certification programs and ratings initiatives such as the R2000 program, Energuide for New Houses, Built Green™, LEED® for Homes, Passive House, and Energy Star® for New Homes have helped place Canada on the leading edge of energy efficiency. These programs and certifications place very stringent requirements (higher requirements than the current NBC) on home building design and construction. While most homes are not yet constructed to these high standards, many of the improvements to the NBC over the last 20 years have been and continue to be a result of adopting a number of best practices from these and similar-type programs (refer to Division B Part 9.25 of the NBC). Several provincial codes are adopting standards such as Energy Star for New Homes and will be phasing them in over the coming years. There is an ongoing debate as to how high the NBC standards should be and how higher standards may affect affordability.

From a cost perspective, it is generally accepted that the greatest return from an energy savings point of view is achieved by first increasing the airtightness of the building envelope (air barrier) and by increasing the insulation levels in the walls and ceiling during the original design and construction process. More advanced technological solutions such as PV cells can be roughed in and added later if desired, but to make the building more airtight and to increase insulation levels is generally not very easy or cost effective once the house is built. Therefore, a focus on the airtightness and design of the building envelope, including management of indoor air quality (IAQ) through the use of a heat recovery ventilator (HRV) is an important determinant of the house's energy efficiency.

Heat Flow

Heat moves from warm to cold, so in winter, warm inside air moves through the envelope toward the cold outside air. In summer when the building is air conditioned, the reverse is true; hot outside air moves toward the cooler inside air.

The three methods of heat transfer follow.

Conduction

The transfer of heat through a solid from a higher temperature to a lower temperature (heat moves from warm to cold e.g. inside to outside in winter). Heat will flow through materials at different rates. A material's capacity to resist conductive heat flow is measured by **R value** in imperial measurement and by **RSI value** for metric measurement. Insulating materials such as fibreglass will have an R and RSI value labelled on the package when it arrives on site. The NBC designates minimum acceptable standards of RSI value for the building envelope, and provincial codes may adopt further standards based on regional climatic conditions. The construction drawings will specify the type of insulation and the R or RSI value required for installation. This information is usually found in the construction notes or specifications.

The term **thermal bridging** describes the movement of heat across a solid. The effects of thermal bridging can be offset by applying extruded polystyrene (XPS, a type of Styrofoam) to the outside of the frame wall as discussed in Chapter 8. For example, studs and plates constructed out of wood form a better thermal bridge than insulation placed in a wall; therefore, if Styrofoam is placed over the entire surface area of a wall it will greatly reduce the thermal bridge through the wood, making a more energy-efficient wall.

Refer to Table 12.1 for a listing of R and RSI values for different materials. Notice that the R value of wood is much lower (1.3 per inch) than the R value of fibreglass insulation (3.2 per inch). Interpreting this information with an understanding of heat transfer helps you to understand

TABLE 12.1

R Value conversions to RSI Value. Divide RSI Value by conversion factor to determine R Value	Conversion factor 0.176 equals RSI Value. Multiply R Value by conversion factor to determine RSI Value
R5	= RSI .88
R8	= RSI 1.4
R12	= RSI 2.1
R20	= RSI 3.5
R26	= RSI 4.6
R40	= RSI 7.0
R50	= RSI 8.8

FIGURE 12.1
Methods of heat transfer.

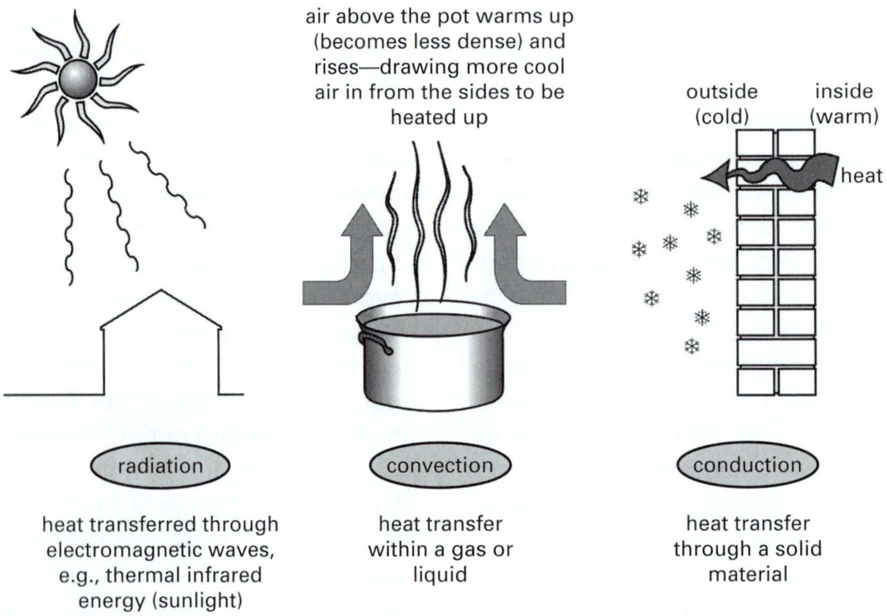

Source: Image excerpted from The Illustrated Home by Carson Dunlop www.carsondunlop.com

why the building envelope needs to be designed with consideration of the components and their placement in the envelope.

Convection

The transfer of heat in a liquid or gas by upward movement of the heated medium. Similar to a forced-air furnace system, warm air will rise and then fall as it cools. Convection currents within a wall cavity will increase the rate of heat transfer. Therefore it is important for insulation to fill all cavities and that there are no spaces between the insulation and vapour retarder where convection loops can form. The insulation must not be overly compressed, or it will lose R value.

Radiation

The transfer of heat by rays, such as are given off from the sun or from a fireplace. This type of heat transfer heats objects instead of heating air, providing a very comfortable heat that does not seem drafty to the occupants.

TABLE 12.2

Thermal Resistance Table

Material	RSI/mm	R/in.
Sheet steel	NS	NS
Concrete	0.01	0.1
Lumber	0.009	1.3
Sheet Material		
Gypsum board	0.006	0.9
Plywood	0.009	1.3
OSB	0.009	1.3
Insulating fibreboard	0.017	2.5
Cavity Insulation		
Fibreglass	0.022	3.2
Mineral fibre	0.024	3.5
Cellulose (blown in)	0.025	3.6
Urethane foam (sprayed)	0.041	6.0
Board Insulation		
Rigid fibreglass	0.029	4.2
Extruded polystyrene (Types II & III)	0.034	5.0
Polyisocyanurate	0.034	5.0
Polyurethane	0.041	6.0

NS = no significant resistance

Source: Canada Mortgage and Housing Corporation (CMHC). Best Practice Guide: Wood Frame Envelopes, 2006; Renovator's Business Guide, 2005.

Moisture Movement

Moisture moves in and out of a building in three main ways.

1. *Air leakage:* Caused by moisture-laden air moving through a building and possibly condensing in a wall cavity. When moisture-laden air reaches the **dew point** it will condense. (The dew point occurs when the relative humidity level [RH] reaches 100% and is the temperature at which the air can no longer hold all of its water vapour and the water vapour turns to liquid.)
2. ***Vapour diffusion:*** The transfer of moisture through materials. The rate of moisture movement depends on the permeability of the material.
3. *Rain penetration:* Can occur through the exterior cladding or any of the junctions and intersections of materials in the building envelope. Moisture may also enter through capillary action (the effect of being sucked in or sucked up) from contact at grade level.

Air Barriers

An **air barrier** (AB) reduces air leakage (which is difficult to totally eliminate) from the structure. The air barrier system must be continuous and as airtight as possible. Air barriers can be used on the inside or outside of a structure. An air/vapour barrier used on the inside (generally the most popular method) uses a 6 mil polyethylene sheet material that is approved for use as a vapour retarder to provide a continuous seal around all walls and the ceiling. (Note: 1 mil is equal to one thousandth of an inch, therefore the polyethylene is 6/1000 of an inch thick.) With this method it is important to seal all penetrations including electrical boxes, and around holes made through plates for wiring and plumbing vent stacks. A spun-bonded polyolefin header wrap (usually Tyvek® or Typar® building wrap) is used at floor level to ensure the integrity of the air barrier around the joists.

Another method of creating an air barrier uses a spun-bonded polyolefin wrap applied continuously to the outside walls and connected to the vapour retarder (VR) on the ceiling. It is important for air barriers applied to the outside of the wall to be constructed from a material that breathes, so as to allow any moisture that collects inside the wall to dry out. A building wrap acting as an air barrier is a much more energy efficient alternative to regular building paper (also referred to as tar paper, or 15 pound felt). Building paper placed on the outside wall is lapped, but unless it is an approved air barrier (e.g. Tyvek®) and installed as being continuous (for example, connecting through the top plates to the interior air barrier on the ceiling separating the attic), then it is only acting as a moisture barrier (MB), which is sometimes also referred to as a weather barrier (refer to Figures 12.6 and 12.7 for examples of AB, VR, and MB placement). (Refer to Division B Part 9.25.3 of the NBC.)

Vapour Barrier/Retarder

The NBC (refer to Division B Part 9.25.4) requires that a **vapour barrier** be sufficient to prevent condensation in the wall spaces, floor spaces, or attic or roof spaces. Another term that is sometimes used interchangeably with the word barrier is retarder. Currently, building scientists are working on a clearer distinction in definition between vapour barrier and **vapour retarder;** this distinction revolves around the differing permeability levels of various materials. The building envelope is a complex entity in how it behaves and acts when faced with different climatic conditions and as such requires consideration regarding the type and location of the vapour barrier/retarder. Because the terms are currently being used interchangeably, the text will also use them interchangeably with a note that a clearer definition is evolving. The vapour barrier prevents moisture from diffusing through materials into the wall. The vapour retarder is placed on the warm side of the wall; in Canada the warm side of the wall is on the interior of the house. The type of vapour barrier depends on the materials used in the wall assembly. The most common method of installing a vapour barrier is to apply a 6 mil polyethylene to the interior wall on top of the studs.

While discussion on many job sites may stress the importance of a tight vapour barrier to the envelope, it is actually the air barrier that needs to form the tightest seal. If large amounts of air flow through any openings in the envelope, large amounts of moisture-laden air can condense at the dew point within the wall, causing localized problems (e.g., mould and rot). However, this can be prevented when the air/vapour barriers are part of a combined system. A combined system is one in which the polyethelene acts as both an air and vapour barrier on the walls, forming a continuous seal, while at the floor a header wrap acts as an air barrier on the outside, and poly cut and placed between the joists covering the fibreglass insulation acts as the vapour barrier on the inside (refer to Figure 12.4 for more detail). The addition of a building wrap instead of just traditional building paper to the outside of the entire wall surface will provide a higher level of protection (best practice, not an NBC requirement yet).

Caulks and Sealants

Caulks and sealants are used to seal any points where moisture or air leakages might occur (e.g., around windows, doors, and flashings, and at joints in the air/vapour barrier). Caulks and sealants include acrylic, butyl rubber, latex, polyurethane, and silicone. Caulks and sealants have different uses (e.g., acoustic sealant is used extensively in sealing joints in polyethylene because it remains highly flexible). Outside, silicone and latex caulks are very popular because of their resistance to freeze–thaw cycles and their expansion and contraction qualities. Where large beads of caulking are required, the joint should

FIGURE 12.2

Examples of two- and three-surface bonds, and the advantage of using backer rod.

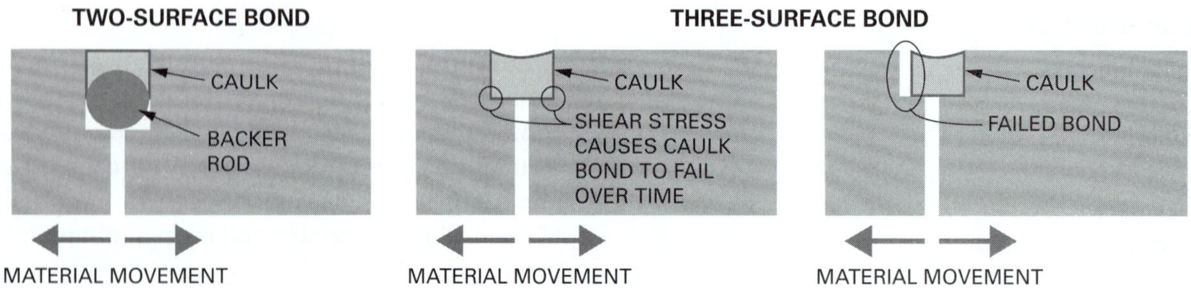

Source: From VOGT, Carpentry, 4E. © 2006 Delmar Learning, a part of Cengage Learning, Inc. Reproduced by permission. www.cengage.com/permissions

be filled with a backer rod to prevent the caulking from adhering to more than two surfaces at once, which can lead to premature failure.

Ventilation

Ventilation is required in attic space to ensure that any moisture-laden air that condenses in the attic can dry out. Section 9.19.1 of the NBC, Venting, requires that ventilation be provided in attic space at a ratio of at least 1:300. This means that for every 300 square feet of ceiling area, at least 1 square foot of ventilation must be provided. This is most effectively accomplished by providing a combination of soffit and roof vents. Specifically, the NBC requires that they be placed uniformly on opposite sides of the building, with not less than 25% of the required openings located at the top of the space, and with not less than 25% of the required openings located at the bottom of the space.

FIGURE 12.3

Ventilation provided in the soffit and by a ridge vent installed along the ridge; alternatively, individual roof vents could have been placed at a few points in the upper part of the roof instead of a ridge vent. The soffit vent would still be required.

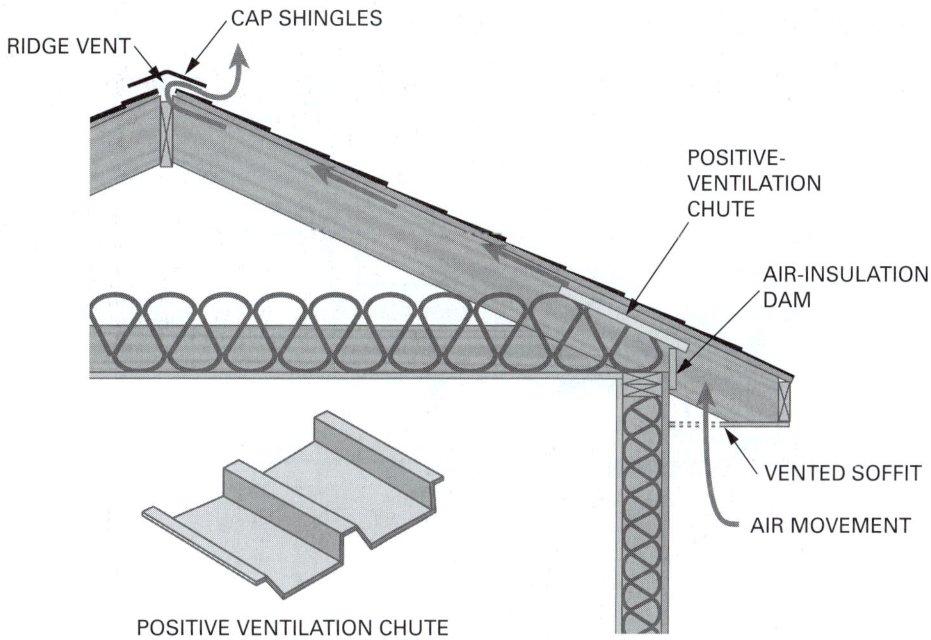

Older buildings have envelopes that leak excessively compared to new buildings, so fresh indoor air is not typically an issue in these buildings. However, in new tightly sealed buildings, fresh air must be brought in from outside; this can be accomplished by exhausting inside air out with the use of exhaust fans or by simply opening a window. Unfortunately, both defeat the purpose of energy efficient building. A better solution is to incorporate a **heat recovery ventilator (HRV)** into the HVAC system. An HRV will remove stale indoor air and exhaust it out after transferring much of the heat to thin metal plates that then transfer the heat to new fresh air being brought in at the same time on the other side of the plate (recall warm moves to cold). This greatly improves the quality of air in the building while maintaining the energy efficiencies gained by a tight envelope.

FIGURE 12.4

Section detail showing how a heat recovery ventilator (HRV) works at exchanging heat from stale indoor air with cool outdoor air.

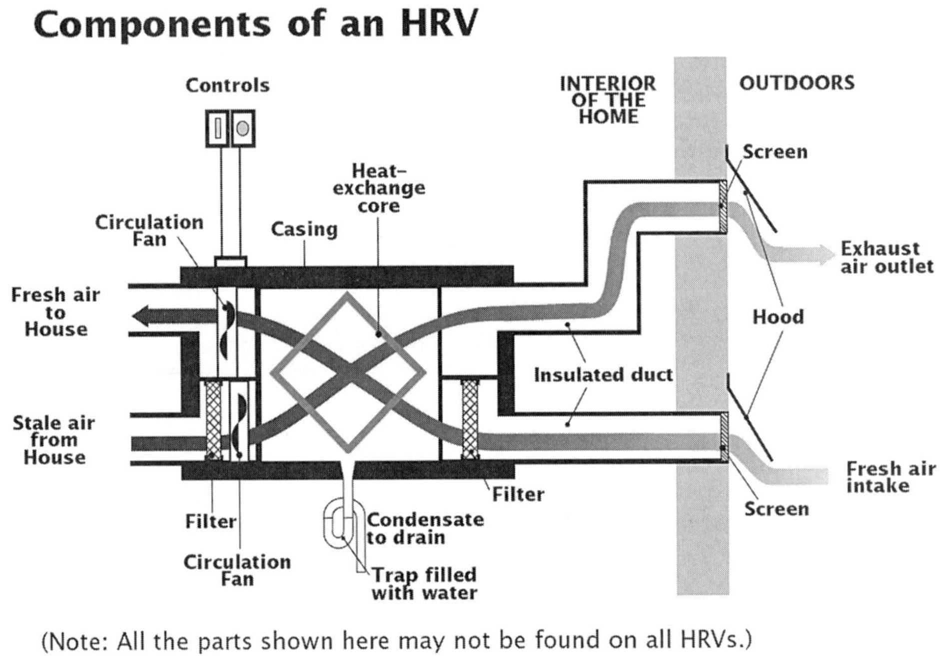

Source: © Reproduced with permission of the Minister of Natural Resources, Canada 2014.

FIGURE 12.5

Building envelope detail for foundation intersection from *CMHC Best Practice Guide: Wood Frame Envelopes*; AB refers to air barrier while VR refers to vapour retarder.

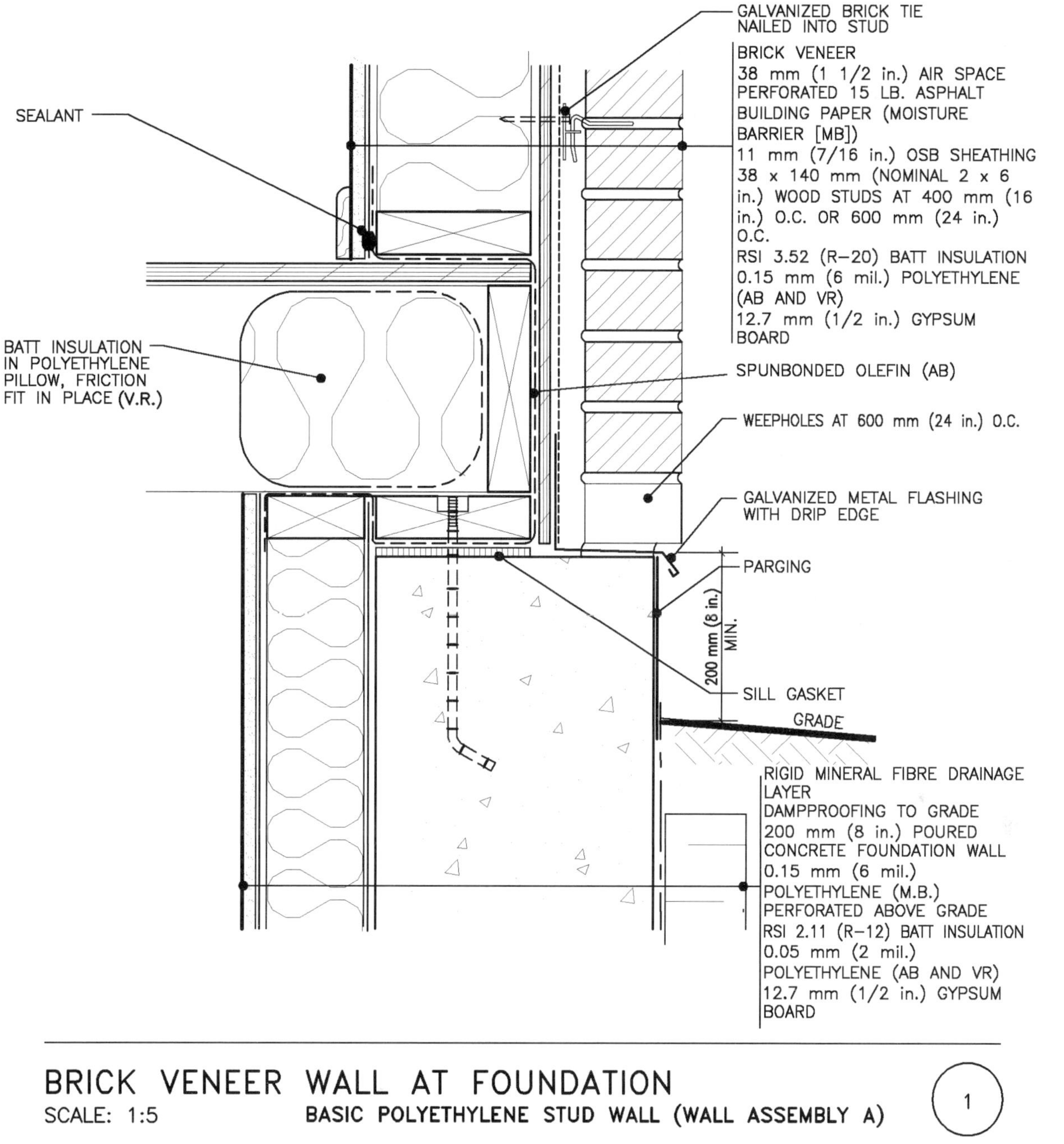

FIGURE 12.6

Building envelope detail for intersection of first-floor walls, second floor, and second-floor walls at bathtub from *CMHC Best Practice Guide: Wood Frame Envelopes.* AB refers to Air Barrier, VR refers to Vapour Retarder, and MB refers to Moisture Barrier which is also often called a weather barrier.

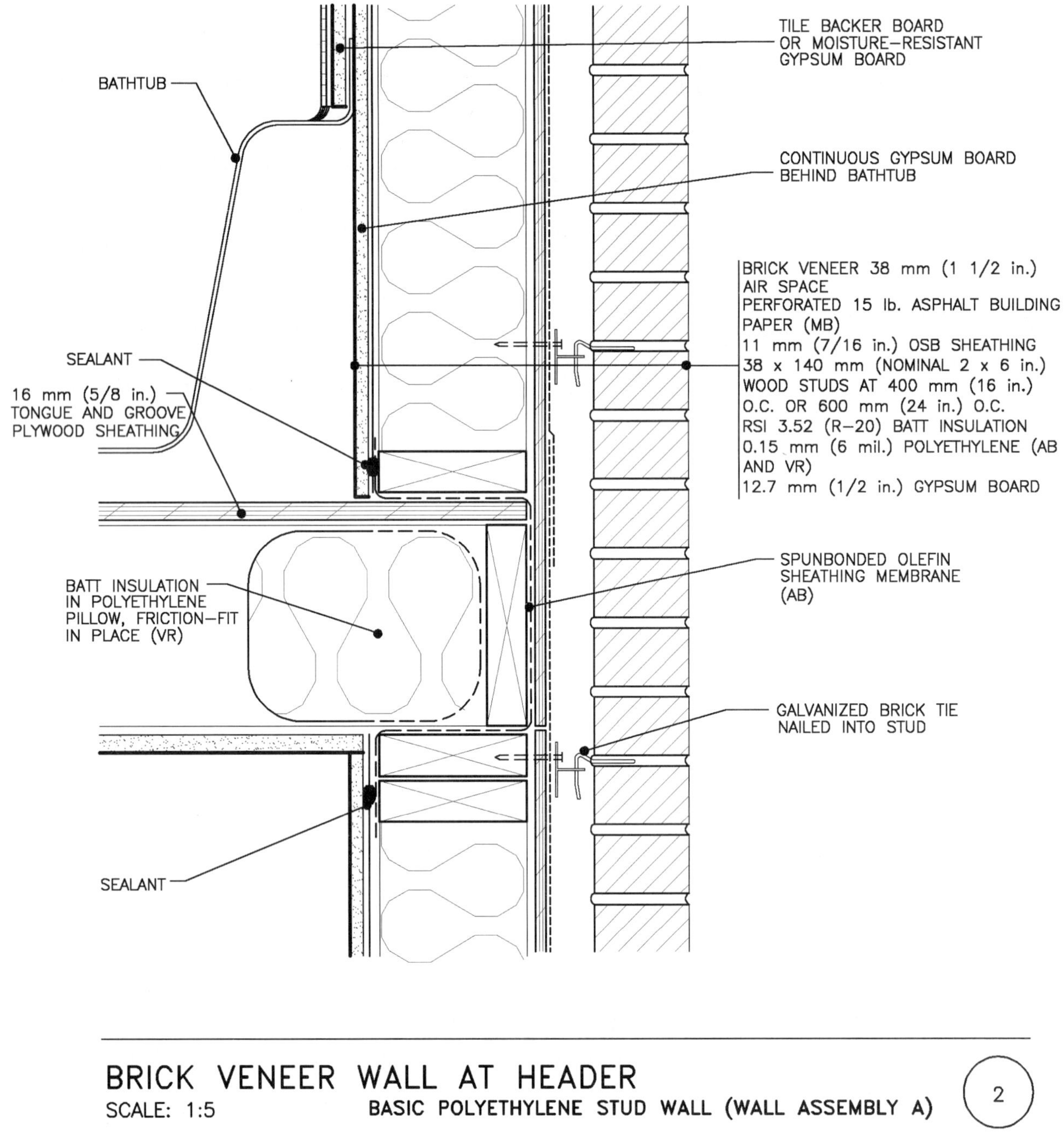

Source: Canada Mortgage and Housing Corporation (CMHC). Best Practice Guide: Wood Frame Envelopes, 2006; Renovator's Business Guide, 2005.

FIGURE 12.7

Building envelope detail for roof, exterior wall, and ceiling intersection from *CMHC Best Practice Guide: Wood Frame Envelopes.* Note the ventilation spacer that is used to ensure that the attic is ventilated to the NBC 1:300 ratios. Roof vents would also be provided to ensure even air flow. AB refers to Air Barrier, VR refers to Vapour Retarder, and MB refers to Moisture Barrier which is also often called a weather barrier.

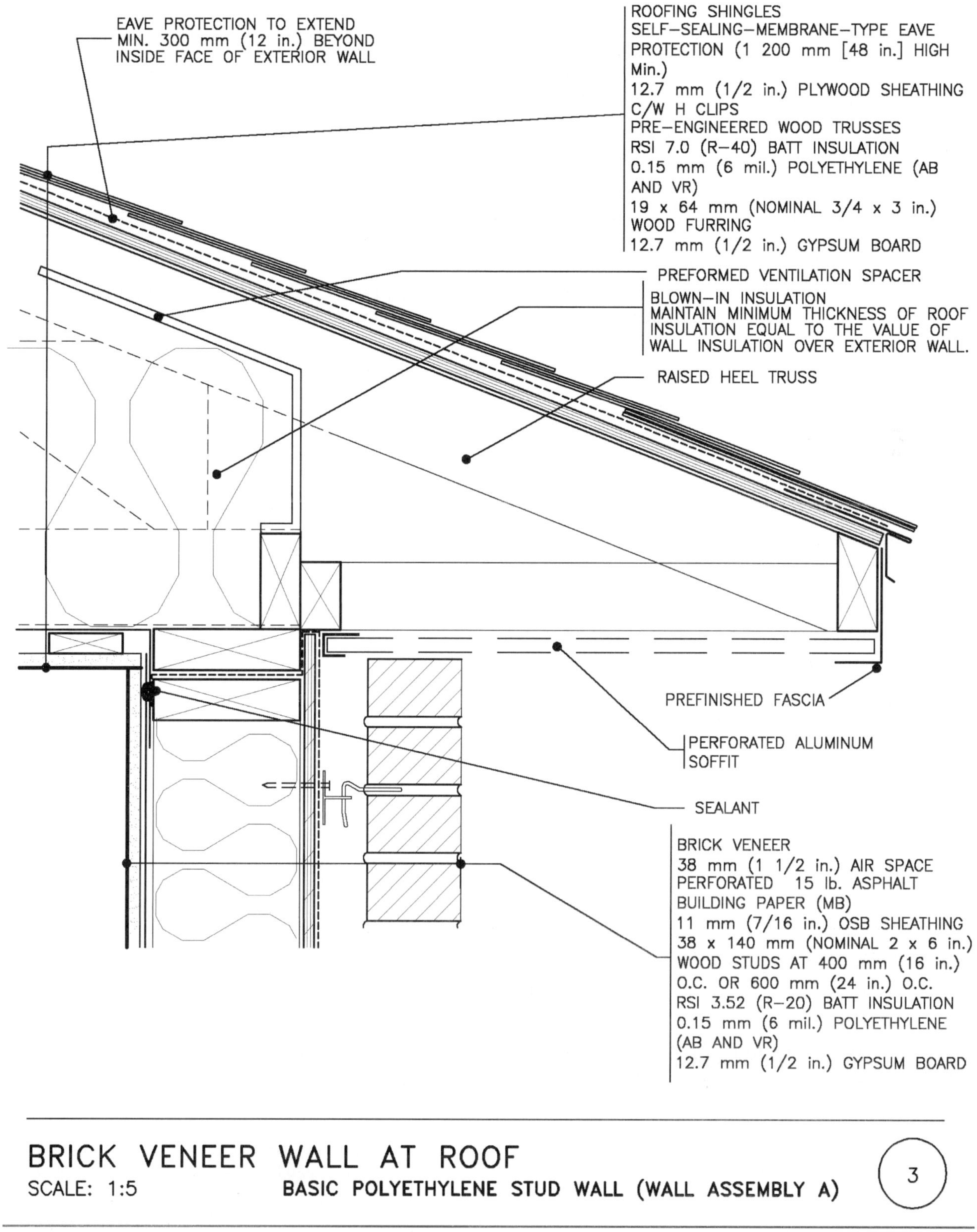

Source: Canada Mortgage and Housing Corporation (CMHC). Best Practice Guide: Wood Frame Envelopes, 2006; Renovator's Business Guide, 2005.

FIGURE 12.8

Building envelope detail for interior and exterior wall intersection from *CMHC Best Practice Guide: Wood Frame Envelopes*. Note the location of polyethylene at the wall intersection and how it is sealed, also note how the electrical penetration is sealed.

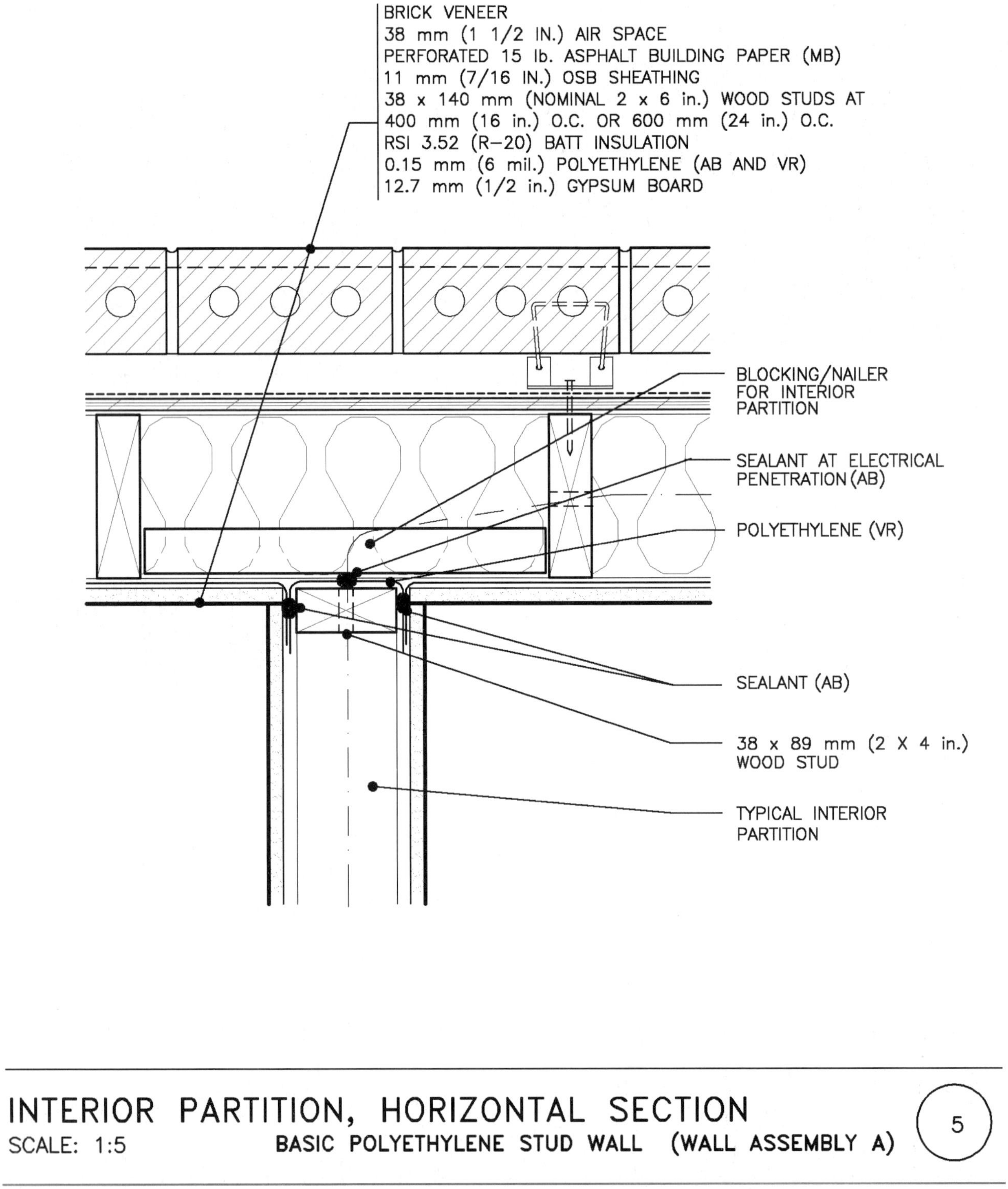

FIGURE 12.9
Electrical box on an exterior wall wrapped with 6 mil polyethylene as part of a continuous air/vapour barrier.

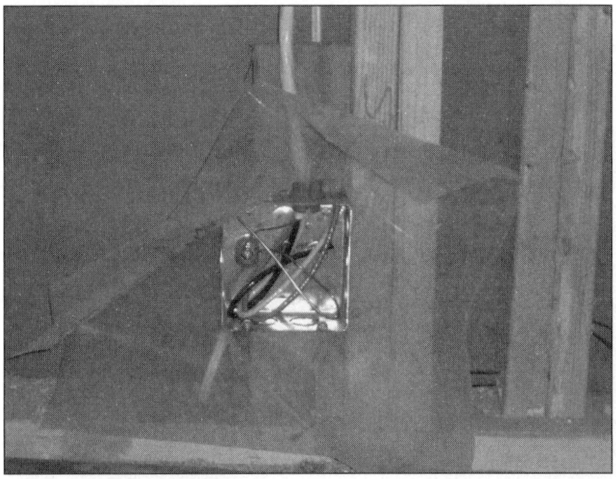

Courtesy Tom Stephenson

FIGURE 12.10
Electrical box sealed to air/vapour barrier; notice how the air/vapour barrier wraps around the box.

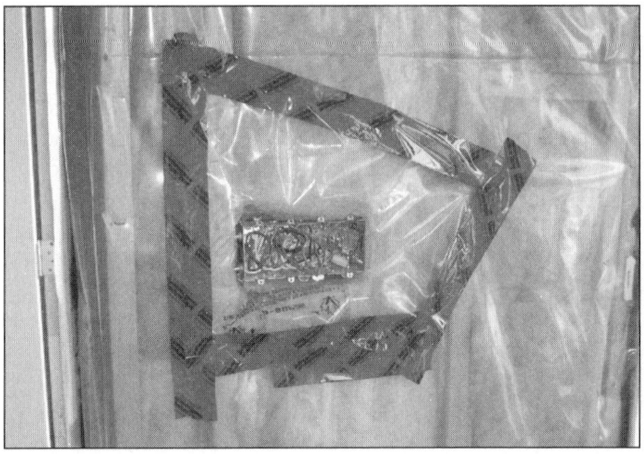

Courtesy Tom Stephenson

FIGURE 12.11
A specially designed electrical box to be used on outside walls where the air/vapour barrier can be sealed to the extended lip around the box.

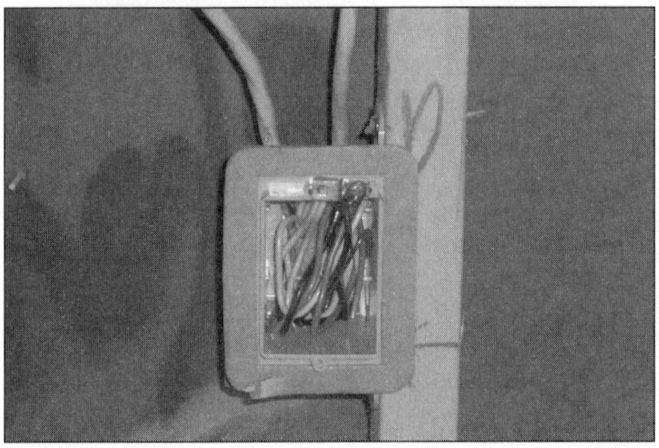

Courtesy Tom Stephenson

FIGURE 12.12
Bathroom exhaust fan covered with 6 mil polyethylene. The poly will be sealed to the polyethylene sheeting that covers the whole ceiling. Notice the black insulation covering the exhaust duct. This will be pulled down to prevent moisture-laden air from both condensing on the duct as warm inside air flows through it in the colder months of the year, and causing moisture problems in the attic.

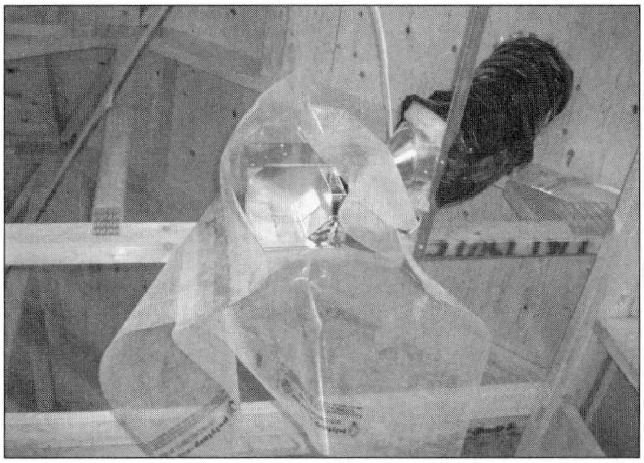

Courtesy Tom Stephenson

FIGURE 12.13
Example of a tightly sealed exhaust fan; blown-in insulation will be added after the drywall is on to ensure that the polyethylene is supported.

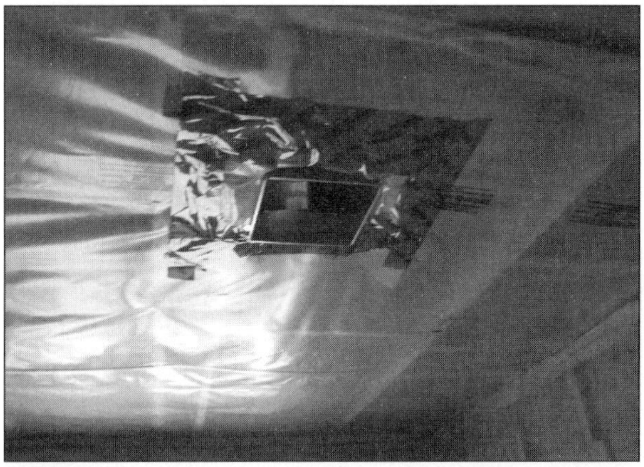

Courtesy Tom Stephenson

FIGURE 12.14
Acoustic caulking applied to holes made by wiring through top plate, which will prevent air leakage through the air barrier. All protrusions of the air barrier must be sealed to prevent air leakage.

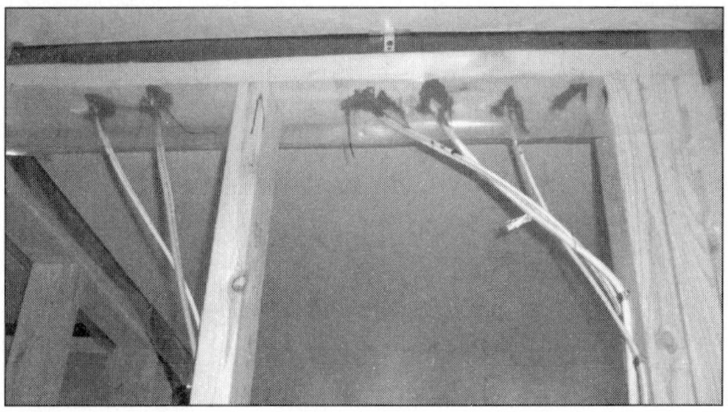

Courtesy Tom Stephenson

FIGURE 12.15

This photo taken from the Doncaster house shows fibreglass insulation installed in the wall and between joists with 6 mil air/vapour barrier installed on top of insulation. Notice the header wrap (the white material just under the floor joists), which is providing the tight seal required for the air barrier, while the polyethylene is providing the vapour retarder between the joists. It would be very difficult to provide the tight seal required for the air barrier if the header wrap were not used.

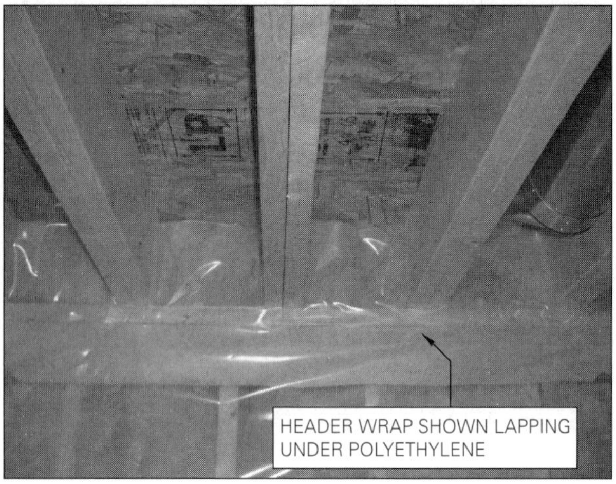

Courtesy Tom Stephenson

FIGURE 12.16

Basement foundation wall insulated to 2′ below grade with vapour retarder on warm side of wall. A number of provinces have adopted full-height basement insulation as a minimum building code requirement.

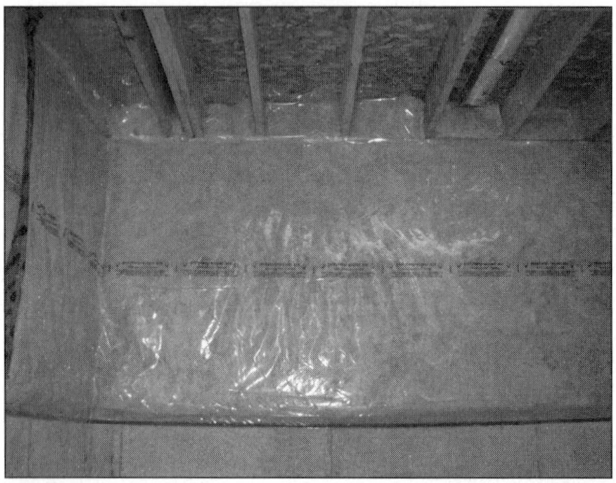

Courtesy Tom Stephenson

FIGURE 12.17

This photo taken from the Doncaster house ensuite bathroom shows that all joints in air/vapour barrier not occurring over studs or plates are sealed with appropriate tape or acoustic caulking; notice that the 6 mil polyethelene extends behind the tub and meets with the header wrap to form a continuous seal.

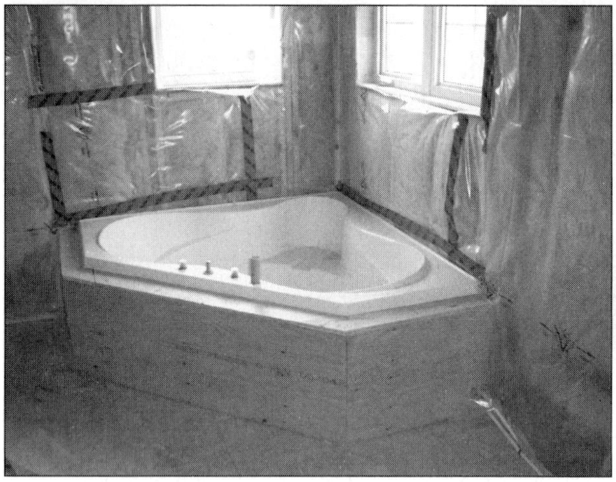

Courtesy Tom Stephenson

FIGURE 12.18

This photo is taken from the living room of the Doncaster house living room; notice that the ceiling insulation and air/vapour barrier stop halfway because the roof extends out over the front portion of the house, and that section must be insulated.

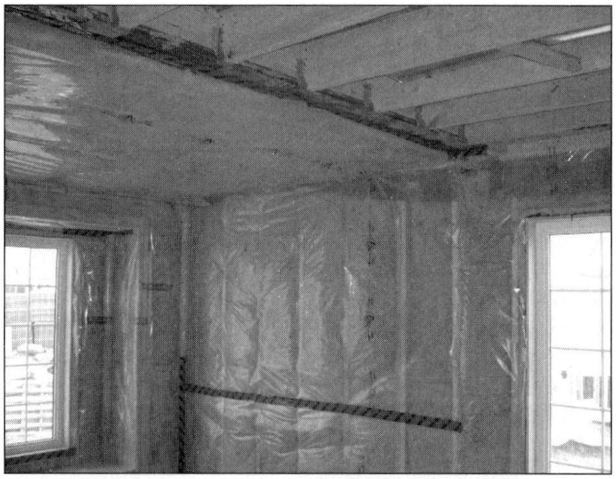

Courtesy Tom Stephenson

FIGURE 12.19

Laundry room air/vapour barrier being prepared; notice all the protrusions through the barrier, including the header wrap, intersecting wall, dryer vent, dryer outlet, washer outlet, and window, and how they have to be sealed. Note that it is best practice to seal all joints with acoustic caulking or sheathing tape, including joint at header wrap and polyethylene (red tape, e.g., Tuck Tape).

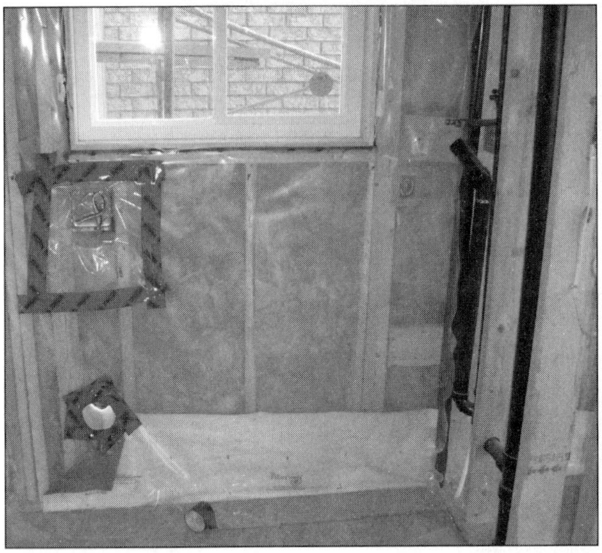

Courtesy Tom Stephenson

FIGURE 12.20

Indication for inspector of approved vapour retarder being used.

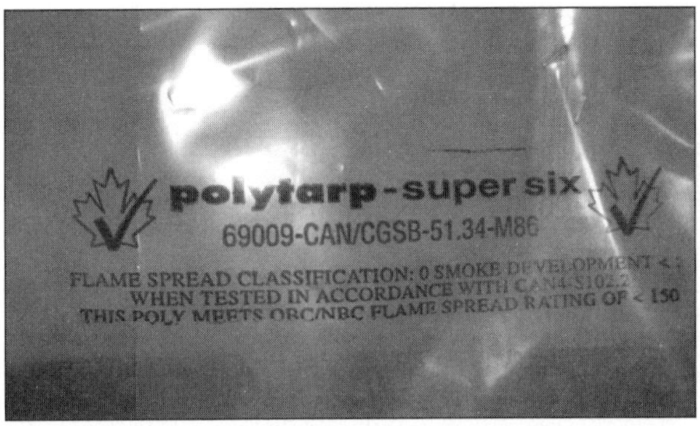

Courtesy Tom Stephenson

FIGURE 12.21

Direct vent gas fireplace installation requires careful planning to ensure that the envelope is tightly sealed. This requires timely installation of insulation, air/vapour barrier, acoustic sealant, and drywall or cement board backing. The detailing of this is rarely shown on the drawings but requires careful planning due to the many penetrations and intersections that are difficult to seal once the framing is complete.

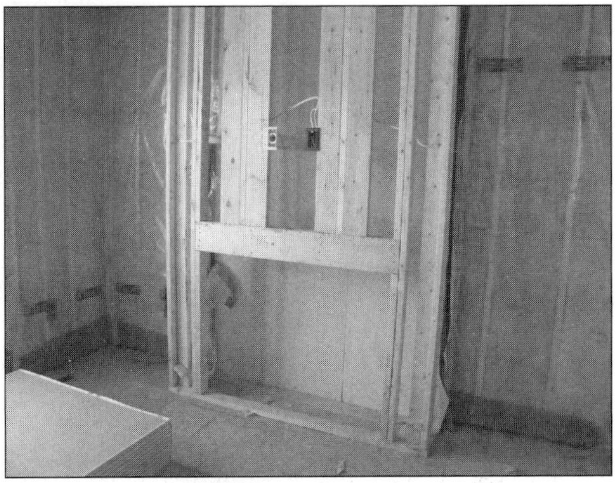

Courtesy Tom Stephenson

FIGURE 12.22

This figure indicates the detailing of sealing around wall and floor intersections. Note the black poly that provides for a continuous air/vapour barrier at the wall intersection. Also note that acoustic caulk has been used to seal all joints in the air barrier except at the ceiling where the polyethylene is acting as a vapour barrier only and where a spun bonded polyolefin header wrap (Tyvek[R], not clearly visible except from the exterior of the building) was used for the air barrier.

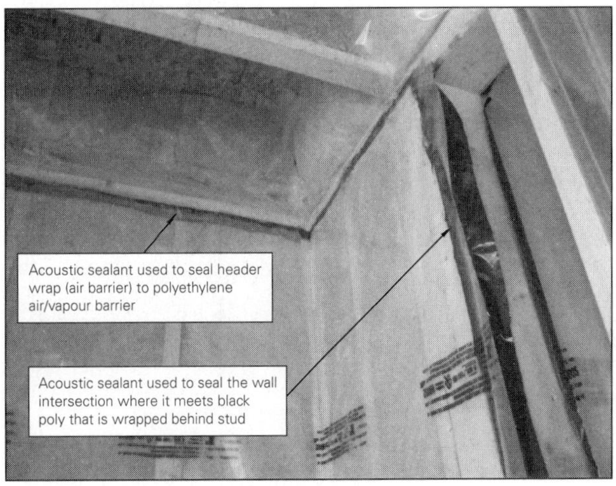

Courtesy Tom Stephenson

FIGURE 12.23

Perimetre insulation is placed to prevent blown insulation from entering into and blocking the vented soffit, which will be installed later. The ventilation baffle is shown and it will be placed around the perimeter of the house to ensure that the overall ventilation of 1:300 is met and balanced between the upper roof and soffit. Note the wood blocking strips provide fastening for the soffit and frieze board applied later.

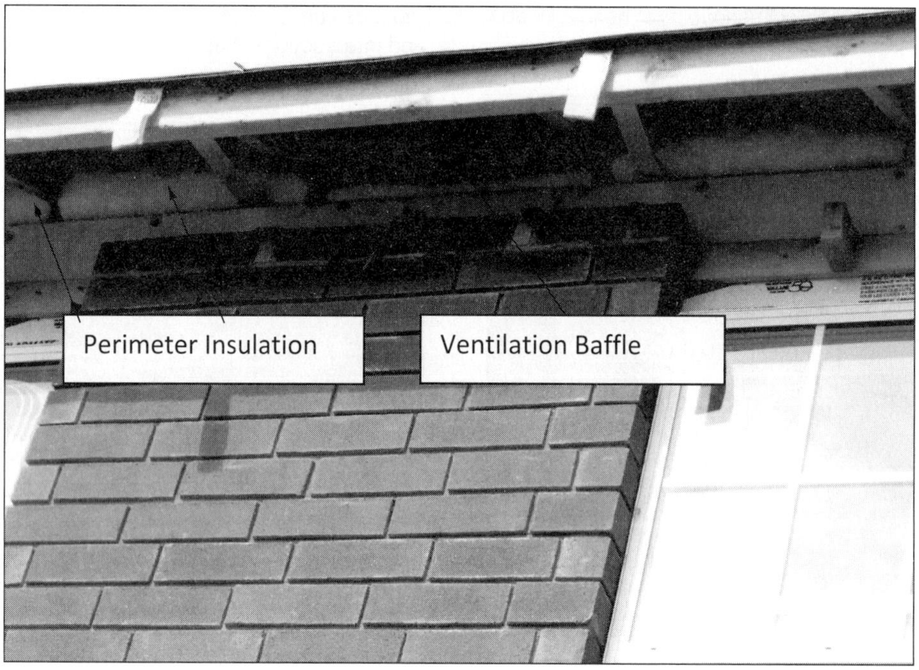

Courtesy Tom Stephenson

FIGURE 12.24

Check drawings and specifications for special requirements such as foam insulation. Increasingly, foam insulation is used for exterior walls and ceilings over garages or cantilevers, as indicated in the Brook drawings, or for other applications such as stacked townhouses. Also verify that the correct type and proper density of foam is being installed.

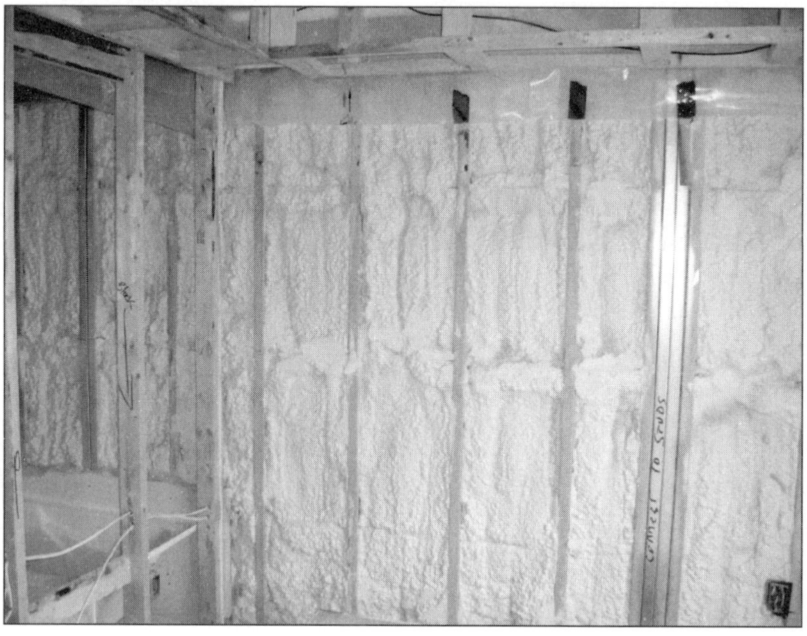

Courtesy Tom Stephenson

FIGURE 12.25

Now that you are much more familiar with all the structural and envelope materials, review the following two commonly used details closely for information regarding the air and vapour barriers and the insulation types and RSI values. Note the use of extruded polystyrene on the exterior of the wall to increase RSI values and to reduce thermal bridging through the studs. Follow the placement of the header wrap and the placement of the air/vapour barrier. Also pay particular attention to the ventilation of the attic. Review the various figures and photos in the chapter to help you visualize how the envelope is constructed.

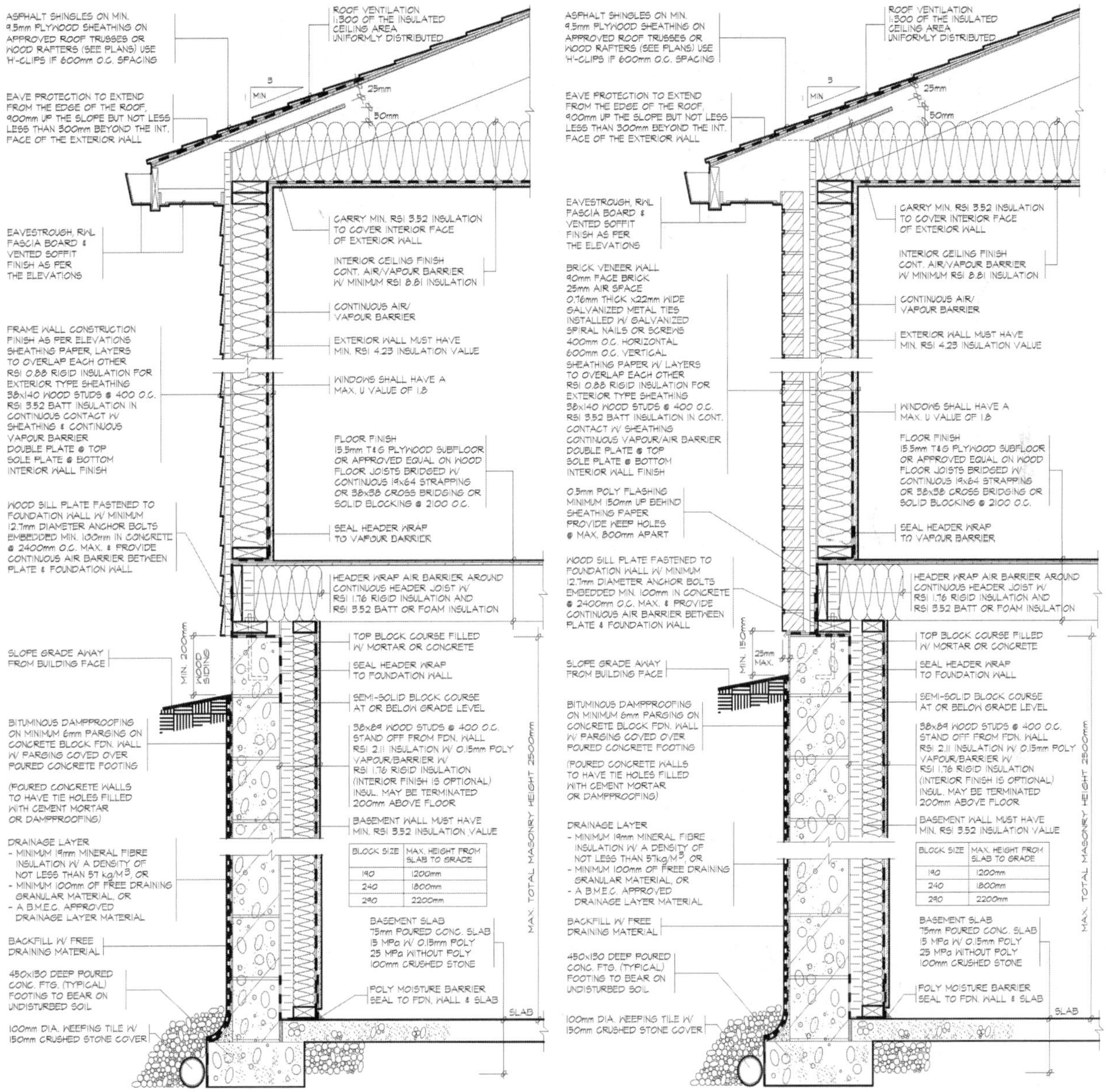

Source: Toronto Area Chief Building Officials Committee (TACBOC)

Insulation Types

There are a growing number of types of insulation materials used in construction. The type shown on the drawings will usually be as a result of meeting required NBC requirements for R value, ease of installation, cost, energy efficiency based on size and thickness of material, and the material's ability to possibly function as a air barrier or in some cases as an air and vapour barrier or whether it requires a separate air and vapour barrier.

Insulation types can generally be categorized by their method of installation and how they are formed or produced (Refer to Table 12.2 for a comparative listing of R values). The envelope will typically incorporate a number of different insulating materials.

Insulation types are as follows:

- Batt insulation, which can be made from a number of materials, including fiberglass and stone wool (e.g., Roxul®). Batts are very popular for installation in cavities for walls, or for placement between joists. Special sound attenuation batts may also be specified to reduce sound transmission in certain applications.
- Blown insulation, which can also be made from a number of materials, including fire retardant treated cellulose, fiberglass, and other materials, including recycled denim. Blown insulation is often used in ventilated attics due to the large amount required and its ease of insulation. The required R-value is usually listed in the construction notes.
- Rigid insulation, which comes in sheets and can be made from:
  - Extruded polystyrene (XPS)
  - Expanded polystyrene (EPS)
  - Polyisocyanurate (polyiso)

  Rigid insulation is often used on the outside of walls to act as an effective thermal break that provides an excellent R-value to the whole surface area. Rigid insulation is also often used to insulate foundations.
- Spray foam insulation, usually made from polyurethane, is sprayed into wall and ceiling cavities to effectively fill and seal the cavity. The two main categories of foam are open cell (1/2 lb) approx R-3.4 per inch and closed cell (2 lb) approx R-6 per inch. Open and closed cell insulation provide different R-values and different ratings regarding vapour permeability; it is best to consult with the manufacturer for specifications on their specific product.

KEY TERMS

Air barrier (p. 303)
Building envelope (p. 301)
Conduction (p. 302)
Convection (p. 303)
Dew point (p. 303)
Heat recovery ventilator (HRV) (p. 306)
Radiation (p. 303)
RSI value (p. 302)
R value (p. 302)
Thermal bridging (p. 302)
Vapour barrier (p. 304)
Vapour diffusion (p. 303)
Vapour retarder (p. 304)

Chapter 12 Assignment 1: The Building Envelope and the Doncaster Drawings

Answer questions 1 to 10 using the information from the chapter and answer questions 11 to 20 using the Doncaster drawings. Identify the letter choice that best answers the question.

_____ 1. Why is 6 mil polyethylene placed between the top plates on interior walls of the upper floor of the house?

a. to provide a continuous air/vapour barrier

b. to prevent the plates from sticking together

c. to prevent mould from forming between the plates

d. to add R value to the attic

_____ 2. What is most efficiently used in order to ensure indoor air quality levels are kept at a healthy level while maintaining energy efficiency in a tightly sealed home?

a. heat recovery ventilator

b. exhaust fan

c. open window

d. air conditioner

_____ 3. What is the best strategy to take regarding the improvement of energy efficiency from a payback perspective?

a. ensuring that the air barrier is tightly sealed and upgrading RSI (R) values in the insulation in walls and ceiling

b. installing photovoltaic cells on the roof to produce electricity for the home

c. installing a green roof

d. installing a wind-powered generator to produce electricity for the home

_____ 4. Which one of the following will most improve the airtightness of the building?

a. vapour barrier

b. air barrier

c. building paper

d. increased insulation

_____ 5. Which of the following best explains the transference of heat through a material?

a. convection

b. radiation

c. conduction

d. thermal break

_____ 6. From the reference material table (Figure 12.2) on thermal resistance located in the chapter, what is the R value per inch for lumber and plywood?

a. 2

b. .5

c. 1.3

d. 2.3

_____ 7. From a technical perspective, what is the header wrap that is placed outside of the rim joists under the sill plate and over the subfloor called?

a. air barrier

b. vapour barrier

c. weather barrier

d. moisture barrier

_____ 8. What practice is best applied when using polyethylene to act as an air/vapour barrier on the inside face of insulated frame walls?

a. Joints that don't occur over a solid surface should be lapped but not sealed.

b. It is best practice to seal all joints with acoustic caulk or approved tape.

c. Use 4 mil polyethylene material.

d. Use the polyethylene as a header wrap.

_____ 9. Which one of the following heat transfers best explains the movement of heat through the air?

a. convection

b. radiation

c. conduction

d. thermal break

_____10. When a material is said to be acting as a thermal bridge, which of the following could best be used to describe how the material is transferring heat?

a. convection

b. radiation

c. conduction

d. thermal break

_____11. What kind of insulation is used in the main floor walls?

a. Styrofoam insulation

b. batt insulation

c. cellulose insulation

d. Icynene foam insulation

_____12. What R value is required for the insulation in the exterior walls of the first and second floors?

a. R 12

b. R 19

c. R 32

d. R 20

_____13. What is the RSI value required for the insulation in the exterior walls of the first and second floors?

a. RSI 3.25

b. RSI 19

c. RSI 5

d. RSI 6

_____14. What is the R value required for the insulation in the ceiling of the second floor?

a. R40

b. R32

c. R31

d. R19

_____15. What is the RSI value of the ceiling insulation in the ceiling of the second floor?

a. RSI 31

b. RSI 5.4

c. RSI 6

d. RSI 19

_____16. What is the R value for the foundation insulation?

a. R 2.11

b. R 12

c. R 19

d. R 31

_____17. How far do the drawings indicate that the foundation insulation be placed below grade?

a. to 2' below grade

b. to 3' below grade

c. to the basement floor slab

d. none of the above are correct answers

_____18. What do the drawings indicate that the R value must be for the insulation placed between the ceiling of the garage and the bedroom floor above?

a. R 40

b. R 31

c. R 25

d. R 19

_____19. What do the drawings indicate that the ratio for ventilation of the attic is required to be?

a. 1:400

b. 1:300

c. 1:200

d. 1:500

_____20. What do the drawings indicate that the R value must be for the insulation placed between the interior garage walls and the house?

a. R 12

b. R 15

c. R 19

d. R 25

Chapter 12 Assignment 2: The Building Envelope and the Brook Drawings

Answer questions 1 through 10 using Figure 12.25 Siding and Brick Veneer Detail drawings. Answer questions 11 to 20 using the Brook drawings. Identify the letter that best answers the question.

_____ 1. From Figure 12.25, what is the RSI value of the rigid insulation that is placed on the outside of the stud wall that has siding as an exterior finish?

a. RSI .65
b. RSI .88
c. RSI 3.52
d. RSI 5

_____ 2. From Figure 12.25, what is the minimum RSI value required for the entire exterior wall that has siding as the exterior finish material?

a. RSI .88
b. RSI 3.52
c. RSI 4.23
d. RSI 5

_____ 3. From Figure 12.25, for the exterior wall that has siding used as an exterior finish, what happens where the header wrap meets the vapour barrier on the warm side of the wall (inside)?

a. It butts against it.
b. They lap over each other.
c. It is cut flush with the floor.
d. They are sealed to each other.

_____ 4. From Figure 12.25, for the exterior wall that has siding used as an exterior finish, what happens where the header wrap meets the inside of the foundation wall?

a. It seals to the foundation wall.
b. It laps the vapour barrier.
c. It laps just behind the drywall.
d. It seals to the top plate.

_____ 5. From Figure 12.25, for the exterior wall that has siding used as an exterior finish, what is the minimum RSI value that the foundation/basement wall must reach?

a. RSI 1.76
b. RSI 2.11
c. RSI 3.52
d. RSI 3.87

_____ 6. From Figure 12.25, for the exterior wall that has siding used as an exterior finish, how much space is left for ventilation under the roof sheathing?

a. 10 mm
b. 25 mm
c. 38 mm
d. 50 mm

_____ 7. From Figure 12.25, for the exterior wall that has siding used as an exterior finish, what is the minimum amount that the eave protection can extend beyond the inside face of the exterior wall?

a. 300 mm
b. 400 mm
c. 600 mm
d. 900 mm

_____ 8. From Figure 12.25, for the exterior wall that has brick veneer used as an exterior finish, what is the minimum RSI value used in the ceiling?

a. RSI 3.52
b. RSI 5.24
c. RSI 6.52
d. RSI 8.81

_____ 9. From Figure 12.25, for the exterior wall that has brick veneer used as an exterior finish, what is placed between the joists and directly behind the header joist?

a. batt insulation
b. rigid insulation
c. brick
d. air space

_____ 10. From Figure 12.25, for the exterior wall that has brick veneer used as an exterior finish, where exactly is the air vapour barrier placed?

a. on top of the insulation in the attic
b. directly underneath the roof sheathing
c. between the drywall and ceiling joists
d. the drywall in this case acts as the air and vapour barrier

_____11. Referring to the Brook drawings, what must be placed between the insulation and the foundation (basement) wall?

a. 6 mil poly

b. No.15 building paper

c. OSB

d. Tyvek air barrier

_____12. What do the Brook drawings indicate is being used as a vapour barrier for the exterior walls?

a. 6 mil polyethylene placed between the stud face and drywall

b. 1/2" drywall that is placed on the inside walls over the studs

c. building wrap Tyvek placed on the inside

d. the brick on the outside of the walls

_____13. What is the R value of the rigid insulation that is placed on the exterior walls?

a. R 3

b. R 5

c. R 12

d. R 20

_____14. What do the drawings call for as a minimum ventilation requirement for the attic?

a. 1:100

b. 1:200

c. 1:300

d. 1:500

_____15. What R value is indicated for the attic/roof insulation?

a. R 20

b. R 30

c. R 40

d. R 50

_____16. What level of insulation is required in the ceiling of the garage?

a. R 12

b. R 20

c. R 25

d. R 31

_____17. Use the data for the second floor area provided for elevation B on the drawings from page 1 to calculate the minimum amount of ventilation required for the Brook house attic area, based on the NBC requirement and listed in the drawings. What is the minimum amount needed?

a. 4.87 sq. ft.

b. 48.7 sq. ft.

c. .487 sq. ft.

d. 1461 sq. ft.

_____18. What type of insulation is used over the garage?

a. rigid insulation

b. batt insulation

c. spray foam insulation

d. blown-in fiberglass insulation

_____19. What types of insulation and what R values are typically used for the exterior walls of the house?

a. 2×4 with R12 batt insulation and 1" R5 rigid insulation on the outside

b. 2×6 with R20 batt insulation and 1" R5 rigid insulation on the outside

c. 2×4 with R19 batt insulation

d. 2×6 with R19 batt insulation

_____20. If the side door option is chosen, what type of insulation and R value is required for the roof over the side door exit at the mud room?

a. batt insulation R 50

b. foam insulation R 50

c. foam insulation R 31

d. batt insulation R 31

Chapter 12 Quiz: The Building Envelope and the Whittington Drawings

Answer questions 1 to 7 using Figures 12.5 to 12.7. Answer questions 8 to 20 using the Whittington drawings. Identify the choice that best answers the question.

_____ 1. From Figure 12.5, where is the sill gasket placed?

   a. It is placed directly between the foundation wall and the sill plate.
   b. It is placed directly between the foundation wall with the air barrier directly on top of the sill gasket, followed by the sill plate placed on the air barrier.
   c. It is placed directly between the foundation wall with the vapour retarder on top and then the sill plate placed on top of the vapour retarder.
   d. It is placed directly between the sill plate and the rim joist.

_____ 2. From Figure 12.5, where is the sealant indicated to be placed?

   a. between the vapour retarders
   b. between the air barrier from the header wrap and the air barrier / vapour retarder from the wall
   c. between the drywall that is placed over the studs on the wall and the wood bottom plate of the wall
   d. between the sill gasket and the sill

_____ 3. From Figure 12.6, what purpose does the spunbonded olefin serve at the rim joist?

   a. as a vapour barrier
   b. as an air barrier
   c. as an attachment point for the sheathing
   d. as a type of sheathing

_____ 4. From Figure 12.6, what purpose is the polyethylene behind the bathtub serving?

   a. provides additional R-value for the wall
   b. acts as an air barrier and vapour retarder
   c. acts as a building paper to prevent moisture from leaking into the building
   d. acts as a moisture barrier to prevent moisture from leaking into the building

_____ 5. From Figure 12.6, what is the RSI value of the insulation placed in the wall?

   a. RSI 3.52
   b. RSI 4.4
   c. RSI 12
   d. RSI 20

_____ 6. From Figure 12.7, what is the minimum amount that the eave protection extends up the roof beyond the inside face of the exterior wall?

   a. 300 mm
   b. 400 mm
   c. 600 mm
   d. 800 mm

_____ 7. From Figure 12.7, what purpose does the preformed spacer serve?

   a. It leaves a space between the insulation and roof sheathing so that the attic is ventilated.
   b. It insulates the attic by providing room for extra insulation to be filled into the void so that the R value of the roof is increased.
   c. It provides a sound barrier that resists sound transmission through the roof.
   d. It provides for a 1" air space between the brick and the wall sheathing.

_____ 8. Referring to the Whittington drawings, what must be placed between the insulation and the foundation (basement) wall?

   a. 6 mil poly
   b. No.15 building paper
   c. OSB
   d. Tyvek air barrier

_____ 9. What do the drawings indicate is being used as a vapour barrier for the exterior walls?

   a. 6 mil polyethylene placed between the studs and drywall is acting as the vapour barrier.
   b. 1/2" drywall placed on the inside walls over the studs is acting as the vapour barrier.

c. Building wrap, Tyvek, placed on the inside is acting as the vapour barrier.

d. The brick on the outside of the walls is acting as the vapour barrier.

_____10. What do the drawings indicate is being used as an air barrier for the exterior walls?

a. 6 mil polyethylene placed between the studs and drywall is acting as the air barrier.

b. 1/2" drywall placed on the inside walls over the studs is acting as the air barrier.

c. Building wrap, Tyvek, placed on the inside is acting as the air barrier.

d. The brick on the outside of the walls is acting as the air barrier.

_____11. What material placed on the outside of the frame walls is acting as a moisture barrier (though not referred as a moisture barrier in the drawings)?

a. exterior drywall sheathing

b. No. 15 building paper

c. OSB sheathing

d. 1/2" plywood sheathing

_____12. What do the drawings call for as a minimum ventilation requirement for the attic?

a. 1:100

b. 1:200

c. 1:300

d. 1:500

_____13. What proportion of ventilation do the drawings require to be divided between the eaves and roof?

a. 20% ventilation at the eaves therefore 80% ventilation on the roof

b. 40% ventilation at the eaves therefore 60% ventilation on the roof

c. 50% ventilation at the eaves therefore 50% ventilation on the roof

d. 60% ventilation at the eaves therefore 40% ventilation on the roof

_____14. Using the data for the second floor area provided on the drawings (construction notes), and based on the NBC requirement (and listed in the drawings), calculate the area of ventilation required for the Whittington house attic area. What is the area needed?

a. 4.6 sq. ft.

b. 460 sq. ft.

c. 13.8 sq. ft.

d. 138 sq. ft.

_____15. Where does the dryer vent exit through the envelope?

a. ceiling to the roof

b. rear elevation through the second floor joists

c. front elevation through the second floor joists

d. left side elevation through the wall

_____16. Review the living room ceiling insulation requirements and determine which of the following statements is correct.

a. The living room ceiling is required to be insulated completely.

b. The living room ceiling requires insulation from the beam towards the windows because there is no second floor above that portion of ceiling.

c. The living room ceiling is inside the house; therefore it does not require insulation because the second floor covers the entire ceiling.

d. The living room ceiling is sprayed with foam insulation.

_____17. Where is the attic access hatch located?

a. walk-in closet

b. ensuite bath

c. closet of bedroom 4

d. closet of bedroom 3

_____18. Regarding the ventilation of the roof, which of the following statements is correct?

a. There is a ridge vent specified throughout.

b. There is no mention of the method used to vent the attic.

c. There is a soffit vent specified.

d. There are numerous gable vents that fulfil all ventilation requirements.

_____19. What level of insulation is required in the ceiling of the garage?

a. R 12

b. R 20

c. R 25

d. R 31

_____20. What does the exterior frame wall construction consist of?

a. 2×4 with 1" Styrofoam on the outside

b. 2×6 with R19 batt insulation

c. 2×4 with R19 batt insulation

d. 2×6 with R24 foam insulation

13 Interior Finishes, Kitchens, and Stairs

LEARNING OBJECTIVES

After completing this chapter, you will be able to:

- Identify the different interior finish materials used in residential construction.
- Describe the basic methods and sequences of installation for interior finishes.
- Read and interpret drawings for information related to interior finishes and their application and installation.
- Describe basic concepts behind kitchen design and layout.
- Calculate the rise, run, headroom, and stairwell opening sizes for stairs.

INTRODUCTION

This chapter discusses the interior finishes of the house. Interior finishes are the coverings and finish details that you see in the completed home. They include all visible wall, floor, and ceiling finishes inside the home, including cabinetry and stairs. When reading construction drawings it is important to know what finishes will be used so that you can ensure that the location and positioning of walls, floors, and supporting materials used behind or underneath the finishes provide adequate room and support. Also from a scheduling perspective, it is important to time the installation of the various finishes so that they are not damaged by other work that will follow their installation. For example, if the finished hardwood flooring is installed too soon, it may be damaged by the installation of the ceramic tile, or there may be extensive requirements for protection of the hardwood flooring for the duration of the project.

INTERIOR FINISHES

The selection and installation of these materials will help define the space and add to the comfort of the home. The construction drawings will indicate in section details the type of materials to be used to construct the walls, including the interior finishes. The section details will often refer to the construction notes for a more complete listing. Specifications will provide more detailed information regarding manufacturer and model numbers for materials such as cabinets, windows, and trim. (Refer to Chapter 14 for more details on specifications.)

FIGURE 13.1

Portion of Section A-A of Doncaster drawings indicating construction note reference (3) for the exterior wall.

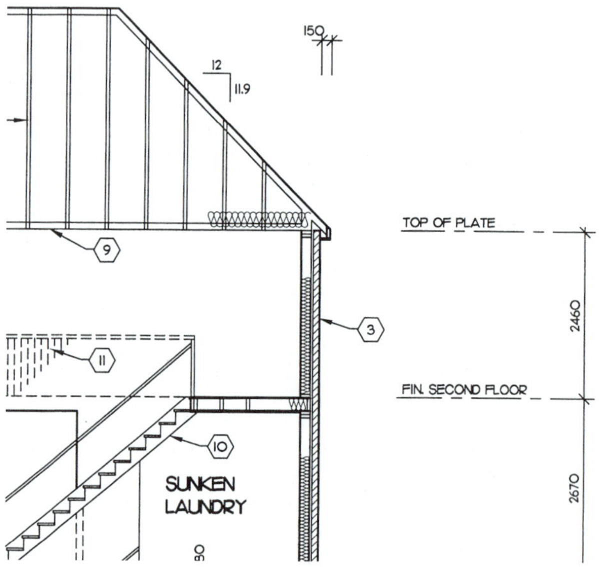

Source: Cassidy & Co. Architectural Technologists

FIGURE 13.2

Doncaster drawings construction note reference (3), as indicated in Figure 13.1; in this case the interior finish is ½" drywall and the exterior finish is brick veneer. Also check the rest of the construction notes for circumstances that may require different finishes, e.g., fire separations (refer to Chapter 15 for more details).

(3) **BRICK VENEER CONSTRUCTION** (2 X 6 STUDS)
90 mm (4") OR 75 mm (3") FACE BRICK, 25 mm
(1") AIR SPACE, 22 x 180 x 0.76 mm (7/8" x 7" x 22 ga)
METAL TIES @ 400 mm (16") O.C. HORIZ.
600 mm (24") O.C. VERTICAL, 0.7 kg/m2 (No. 15)
BUILDING PAPER, 12.7 mm (1/2") EXTERIOR TYPE SHEATHING
(UNLESS OTHERWISE SPECIFIED), 38 x 140 (2 x 6) STUDS
@ 400 mm (16") O.C., RSI 3.25 (R19) BATT INSULATION
AND 6 Mil AIR/VAPOUR BARRIER, 12.7 mm (1/2")
INT. DRYWALL FINISH. PROVIDE WEEP HOLES @ 800 (30") o.c.
HORIZ. @ BOTTOM COURSE ONLY & OVER OPENINGS, PROVIDE
BASE FLASHING UP MIN. 150 mm (6") BEHIND SHEATHING PAPER.

Source: Cassidy & Co. Architectural Technologists

Gypsum Board Panels (Drywall)

The most commonly used interior wall finish is gypsum board panels, more commonly referred to as drywall. Other finishes include plaster placed on top of gypsum panels and wood panelling, including solid wood and manufactured panels. The type and thickness of drywall will depend on a number of factors, including the following:

- ½" regular drywall is used for most common applications where stud and joist spacing are 16" or less.
- ½" moisture-resistant drywall is used in areas such as bathrooms and basements where moisture is a concern.
- In areas facing direct and constant moisture contact, such as shower stalls and bathtubs, cement board or other waterproof products are used.
- Mould and indoor air quality have become bigger health concerns in the last few years. As we build our envelopes tighter, greater care needs to be taken in the selection of materials in areas where moisture may be present. Mould resistant drywall can/should be used for areas such as bathrooms and basements. This drywall uses a non-organic fibreglass matt on the outside surfaces of the gypsum. The mould-resistant fibreglass matt replaces the typical mould-attracting paper face of the drywall, making it an excellent choice where mould growing conditions may occur.

FIGURE 13.3

Shower area constructed with cement board in lower area inside shower, and water-resistant drywall on upper area (this shower stall is shown in the master ensuite on the Doncaster drawings).

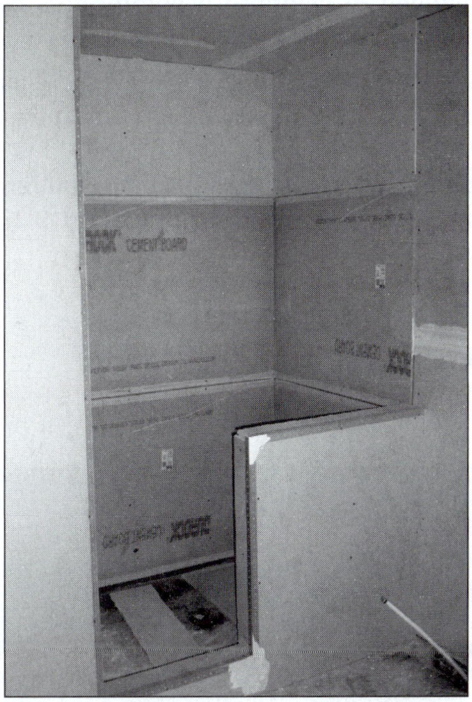

Courtesy Tom Stephenson

FIGURE 13.4

Example of a roof truss bottom chord spaced at 24" OC; compare to wall studs below to see the difference in spacing. ⅝" drywall will be used for the ceiling and ½" drywall for wall below.

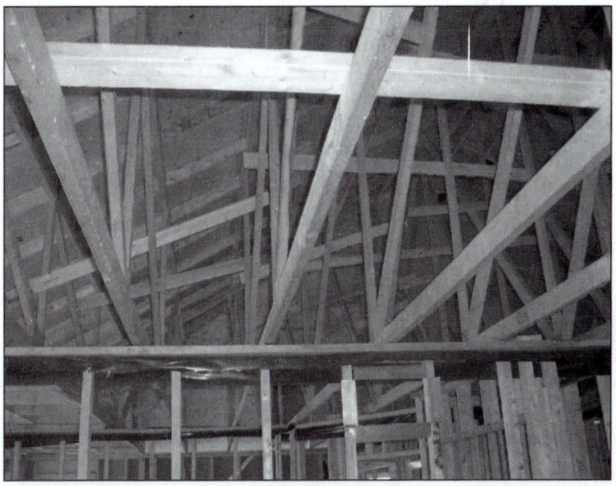

Courtesy Tom Stephenson

- ⅝" drywall may be used on ceilings and wide truss or joist spacings, for example, 19.2" or 24" OC. The extra thickness provides extra rigidity and prevents sagging between the trusses.
- Type X drywall (a type of fire-resistant drywall) may be required between walls, ceilings, and exposed roof gables where limiting

distance and fire code regulations require special fire-resistant detailing. Drywall may also be required to be placed in double layers to meet building and fire code regulations.

- ¼" flexible drywall may be required to be installed in layers to accommodate curved and irregular surfaces.

Take care that the type of drywall used is appropriate for its intended purpose. Carefully check drawings and specifications to verify this information and to ensure that all code requirements are satisfied (refer to Division B Part 9.29.5 of the NBC).

FIGURE 13.5
Exterior wall less than 1.2 m to property line requires type X drywall to provide 45 min. fire resistance rating.

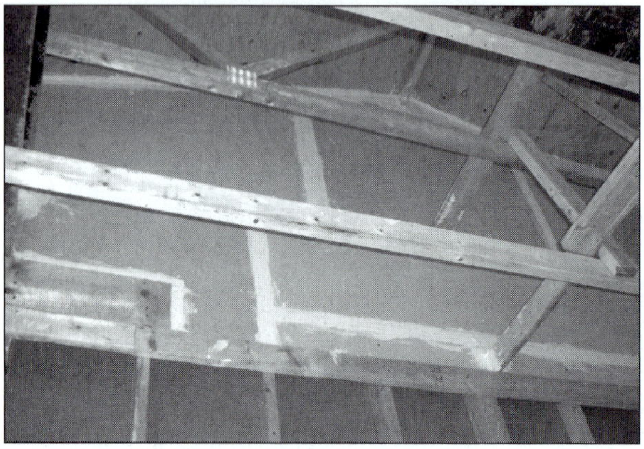

Courtesy Tom Stephenson

FIGURE 13.6
An exterior wall less than 1.2 m to property line requires extra precautions due to the proximity of the neighbouring home. Type X drywall is placed on the exposed framing inside garage gable end and above future garage ceiling; the rest of the wall will be drywalled after the garage ceiling is insulated and drywalled.

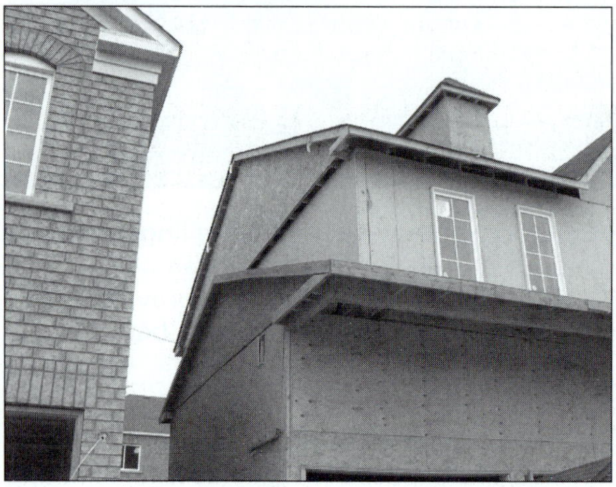

Courtesy Tom Stephenson

FIGURE 13.7
Garage with finished drywall on walls and dropped ceiling.

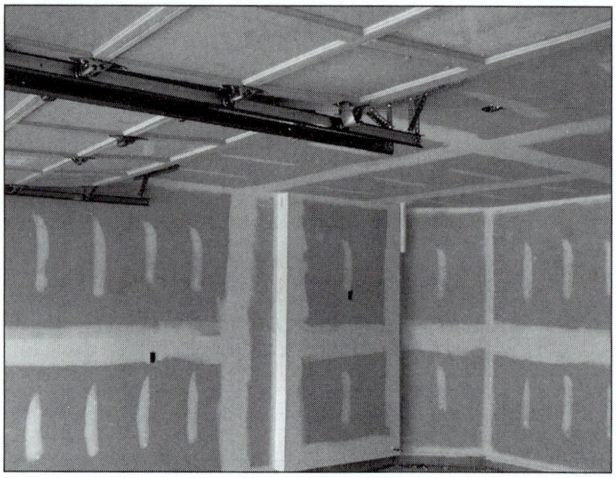

Courtesy Tom Stephenson

FIGURE 13.8
Construction notes from Doncaster drawings, explaining condition shown in Figure 13.5. The 45-min. flame resistance rating is a requirement of the NBC (refer to DivisionB Part 9.2.3 of the NBC).

(44) **EXTERIOR WALL LESS THAN 1.2 M TO PROPERTY LINE (45 MIN. F.R.R.)**
BRICK VENEER WALL OR FRAME WALL CONSTRUCTION
CONSTRUCTION OF WALLS AS PER NOTES (2)(3)(40) (42)
EXCEPT AS PER THE FOLLOWING NOTES.
INSULATION WITH A MASS OF 0.032 kg/m2 PER 1 mm OF THICKNESS AND 12.7mm (1/2") TYPE 'X' INT. DRYWALL FINISH.

(45) **EXTERIOR COMBUSTIBLE CLAD WALLS LESS THAN 0.6 M TO PROPERTY LINE (45 MIN. F.R.R.)**
CONSTRUCTION OF WALLS AS PER NOTES (2) (40) (42)
EXCEPT AS PER THE FOLLOWING NOTES.
PROVIDE 12.7mm (1/2") GYPSUM EXT. SHEATHING,
INSULATION WITH A MASS OF 0.032 kg/m2 PER 1 mm OF THICKNESS AND 12.7mm (1/2") TYPE 'X' INT. DRYWALL FINISH.

Source: Cassidy & Co. Architectural Technologists

Drywall Installation and Finishing

The installation and finishing of drywall (drywall is often also referred to as gypsum board or gypsum wall board) is a five-part process. On larger projects this process may require two different crews or subtrades, one to install the drywall while another crew finishes the drywall. The five-part process is as follows.

- *Step 1—Installation of drywall panels:* Panels are cut with a drywall knife and installed with drywall screws (nails are sometimes used to hold panels in place for later screwing down). Panels come in widths of 4' (sometimes available at 4'–6" for 9' ceilings) and lengths of 8', 10', and 12'. Panel sizes and lengths are chosen to stagger and minimize the number of end joints, thus reducing possible visual imperfections. Panels are fastened to joists and studs using drywall screws at approximately 12" intervals. (Refer to NBC Section 9.29.5.9 for minimum exact spacing requirements under various conditions.) Metal or plastic corners are installed to protect seams. The design layout and specifications may also require a number of other specialty trim pieces.

The finishing of drywall comprises the remaining four steps in the five-part process:

- *Step 2—Coat 1, taping and filling:* Narrow strips of paper tape are embedded into a thin coat of taping compound that adheres to the drywall. The tape is installed along all joints and acts as a reinforcement to minimize the possibility of cracking. The drywall has an indentation along its length to help ensure that there are no visible ridges or humps left after taping. Screw holes are also filled at this time with taping compound.

FIGURE 13.9

⅝" thick drywall panels installed on ceiling with staggered joints.

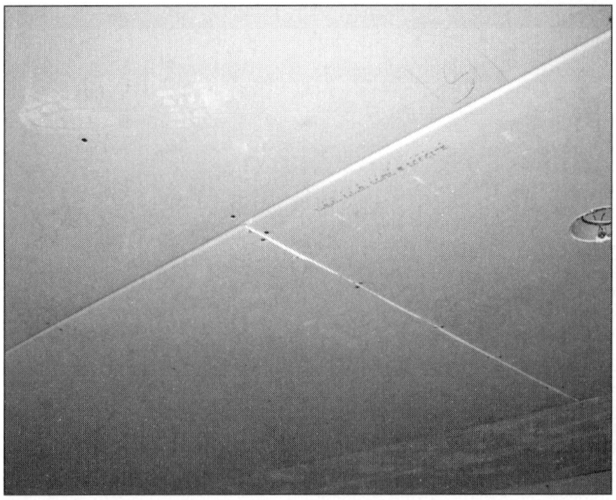

Courtesy Tom Stephenson

FIGURE 13.10

Metal corner bead placed on corners to protect drywall. It is best practice to screw drywall; use NBC minimums for length of screws and quantity of screws. Fewer and shorter screws means fewer screws to pop due to lumber shrinkage (refer to Division B Part 9.29.5 of the NBC).

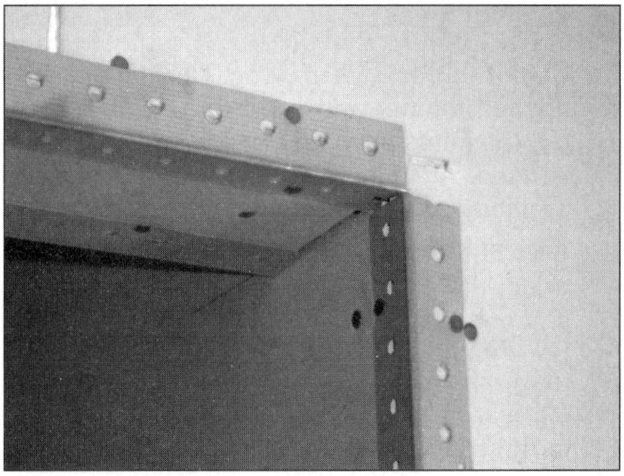

Courtesy Tom Stephenson

FIGURE 13.11

Flexible plastic corner bead placed on radius of corner to protect drywall and provide uniform curve.

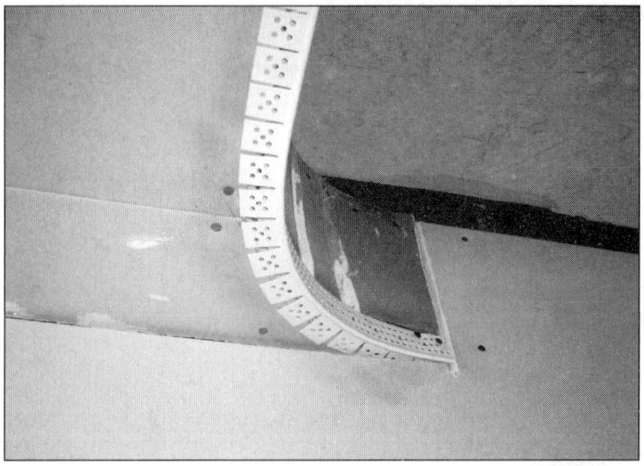

Courtesy Tom Stephenson

FIGURE 13.12

Drywall is installed on ceiling first, then on walls to hold up drywall; drywall joints over windows should not line up with edge of window jamb because cracks will occur more frequently due to loading and shrinkage of lintels.

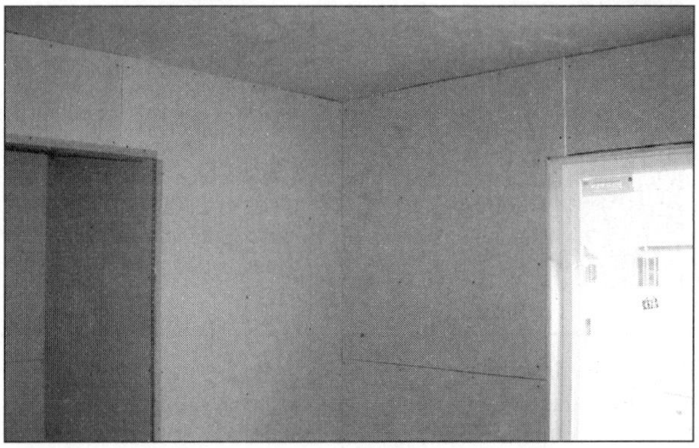

Courtesy Tom Stephenson

- *Step 3—Coat 2, feathering:* Once the first coat is dry, a second coat of compound is applied. The compound is applied in a wider, thinner strip and feathered out so that when the drywall is painted it will not be visible. This second coat is also applied to all screw holes to level the shrinkage that has occurred from the first coat.
- *Step 4—Coat 3, feathering:* The third coat is applied thinner and wider to further feather out the joints. A wider drywall knife or trowel is used. At this stage, any imperfections in previous coats are covered with a final coat.
- *Step 5—Final sanding:* After the third coat has dried, the drywall is given a final sanding to smooth the surface. If the finishing has been done well, there will not be a lot of sanding required. Best practice is to then prime the walls as soon as possible so that any defects, such as shadow lines and bumps, are visible and easily fixed.

Stipple finishes may be added to ceilings once the drywall is finished and primed. A stipple finish is applied by spraying the stippling compound onto the ceiling surface.

FIGURE 13.13
Taping and finishing process.

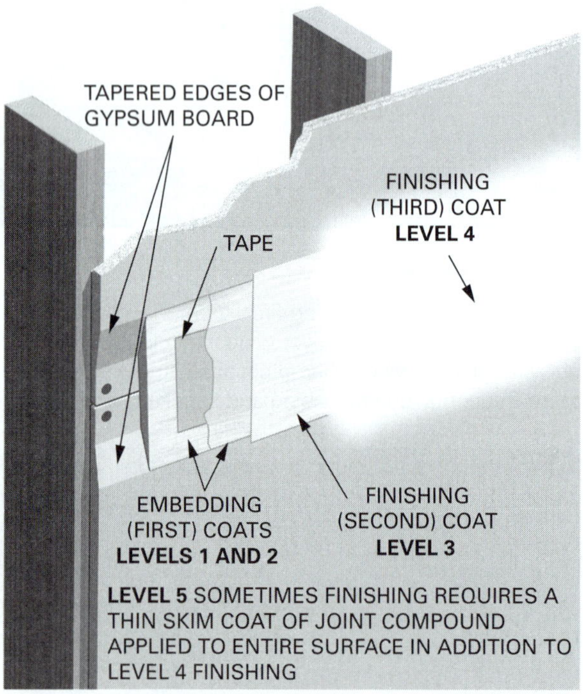

Source: From VOGT, Carpentry, 4E. © 2006 Delmar Learning, a part of Cengage Learning, Inc. Reproduced by permission. www.cengage.com/permissions

FIGURE 13.14
Drywall installed and finished in Doncaster kitchen.

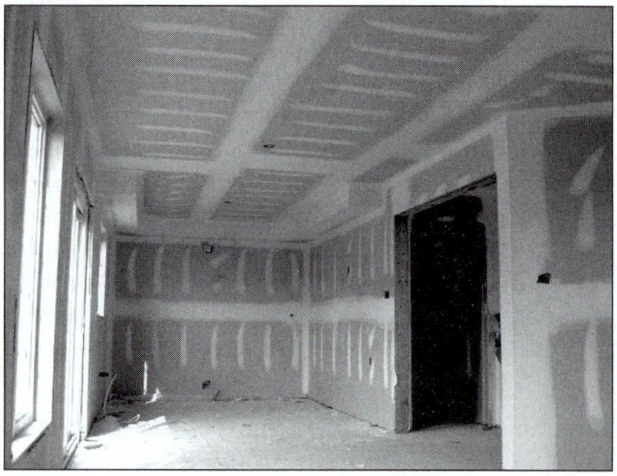

Courtesy Tom Stephenson

FIGURE 13.15
Ceiling with a stipple finish.

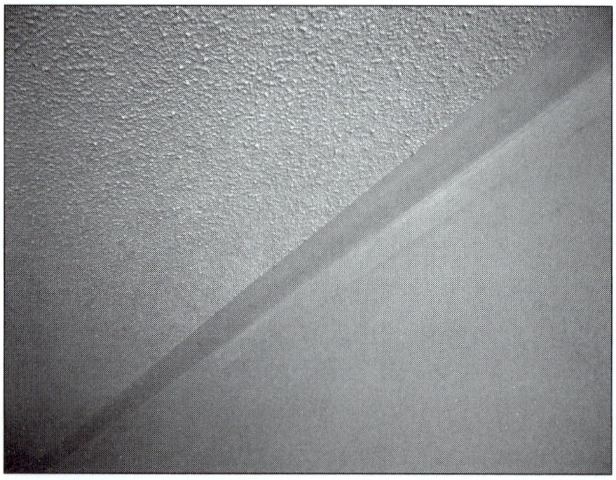

Courtesy Tom Stephenson

Millwork and Trim

Millwork and trim includes the following:

- Doors
- Windows
- **Casing,** the moulding that finishes the edge between the door, window frames, and wall surface
- **Baseboard,** the moulding that finishes the floor and wall intersection
- **Crown moulding,** decorative moulding placed at the intersection between the walls and ceiling
- Built-in cabinetry including kitchen, bath, home office, closet shelving
- Wall panelling to accentuate a room.

Selecting trim materials can be arduous for clients because there are literally hundreds of different options and materials. The type and location of mouldings are shown in the room schedule if one is provided; otherwise, details are provided in the specifications (refer to Chapter 14 for more details on specifications).

FIGURE 13.16
Section details of common moulding patterns.

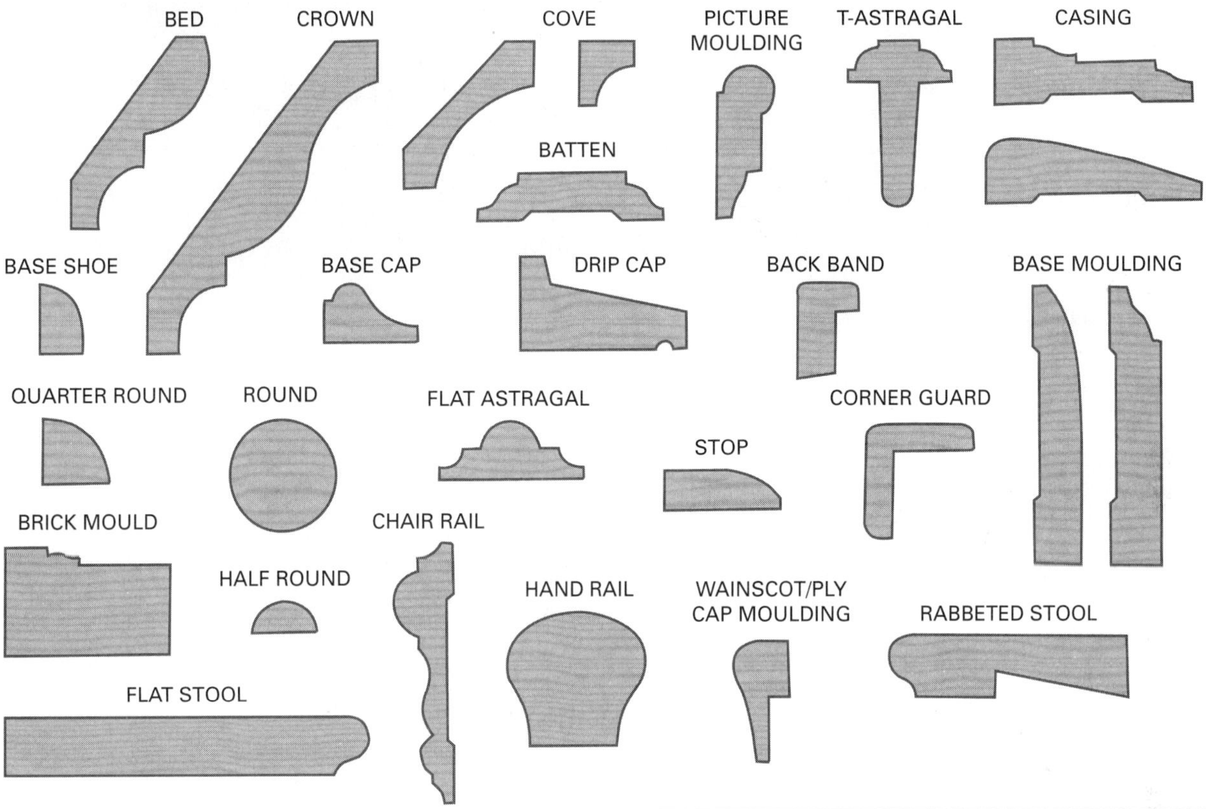

Source: From VOGT, Carpentry, 4E. © 2006 Delmar Learning, a part of Cengage Learning, Inc. Reproduced by permission. www.cengage.com/permissions

FIGURE 13.17
Oblique drawing of typical crown mouldings.

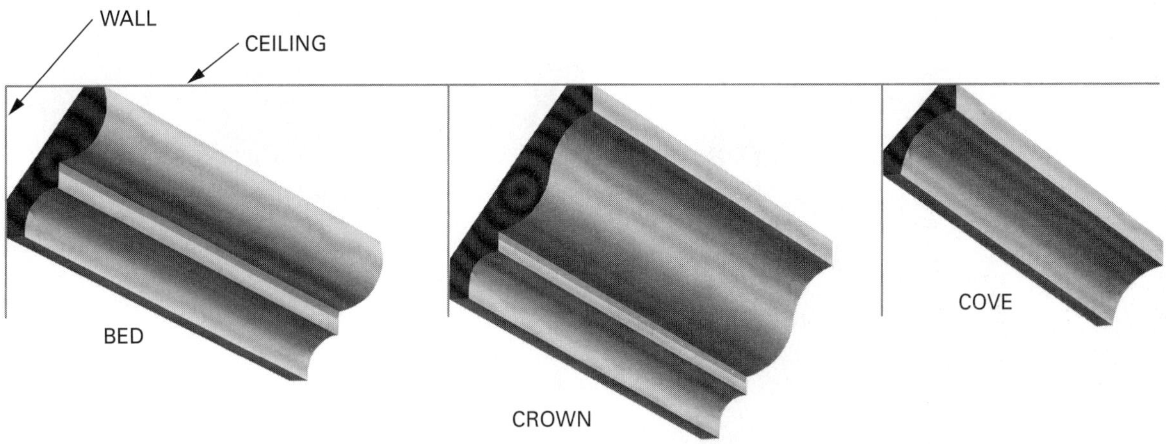

Source: From VOGT, Carpentry, 4E. © 2006 Delmar Learning, a part of Cengage Learning, Inc. Reproduced by permission. www.cengage.com/permissions

FIGURE 13.18

Window and door trim components.

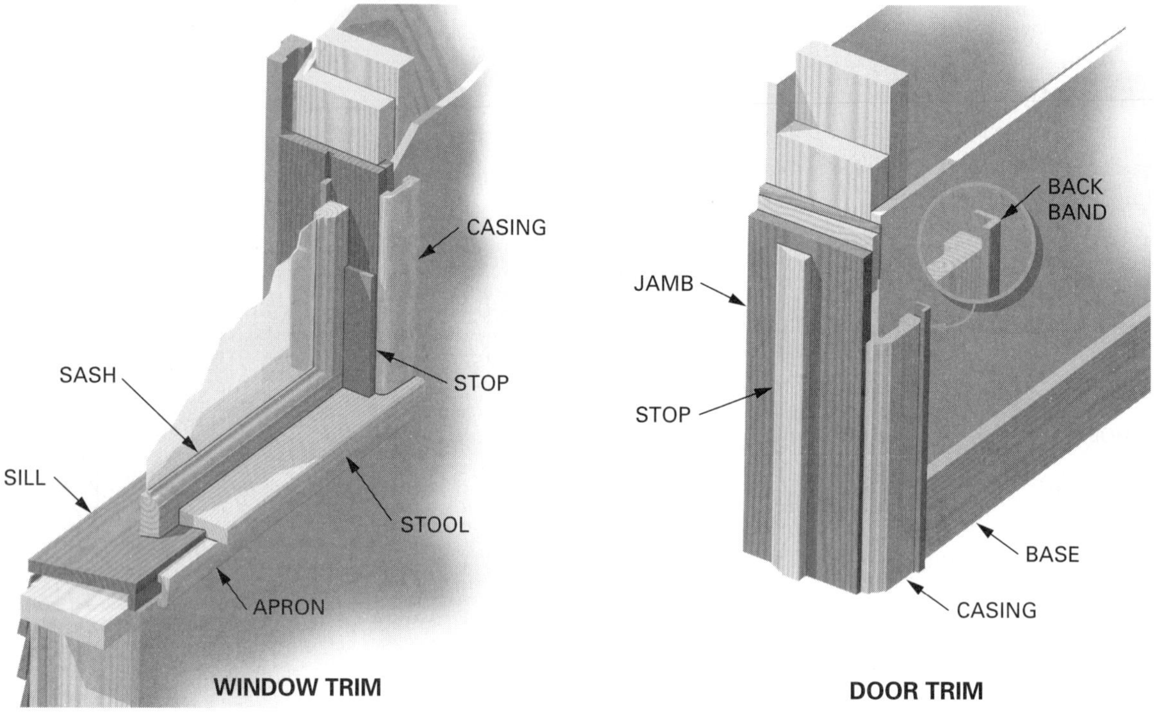

Source: From VOGT, Carpentry, 4E. © 2006 Delmar Learning, a part of Cengage Learning, Inc. Reproduced by permission. www.cengage.com/permissions

FIGURE 13.19

Crown moulding applied at ceiling, casing around door, baseboard at floor, and built-in cabinet including glass shelves and decorative panels in powder room of Doncaster model home.

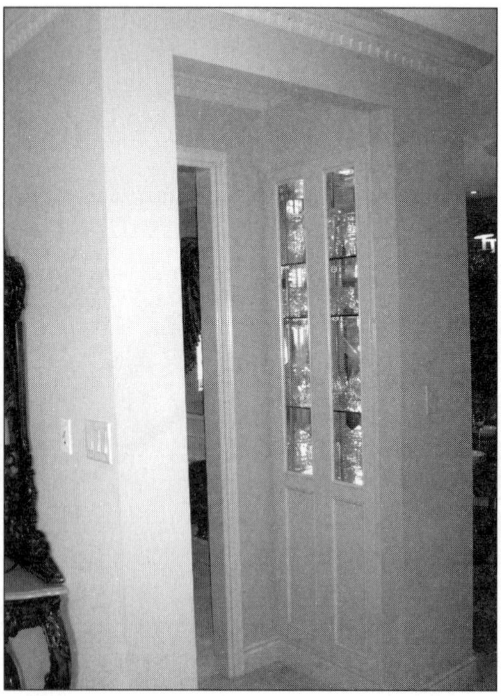

Courtesy Tom Stephenson

FIGURE 13.20

Bathroom vanity in Doncaster master bathroom; compare with floor plan close-up in Figure 13.21.

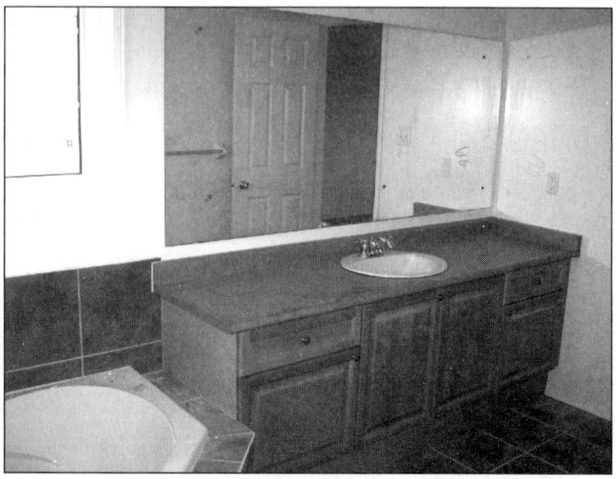

Courtesy Tom Stephenson

FIGURE 13.21
Floor plan of ensuite to master bedroom.

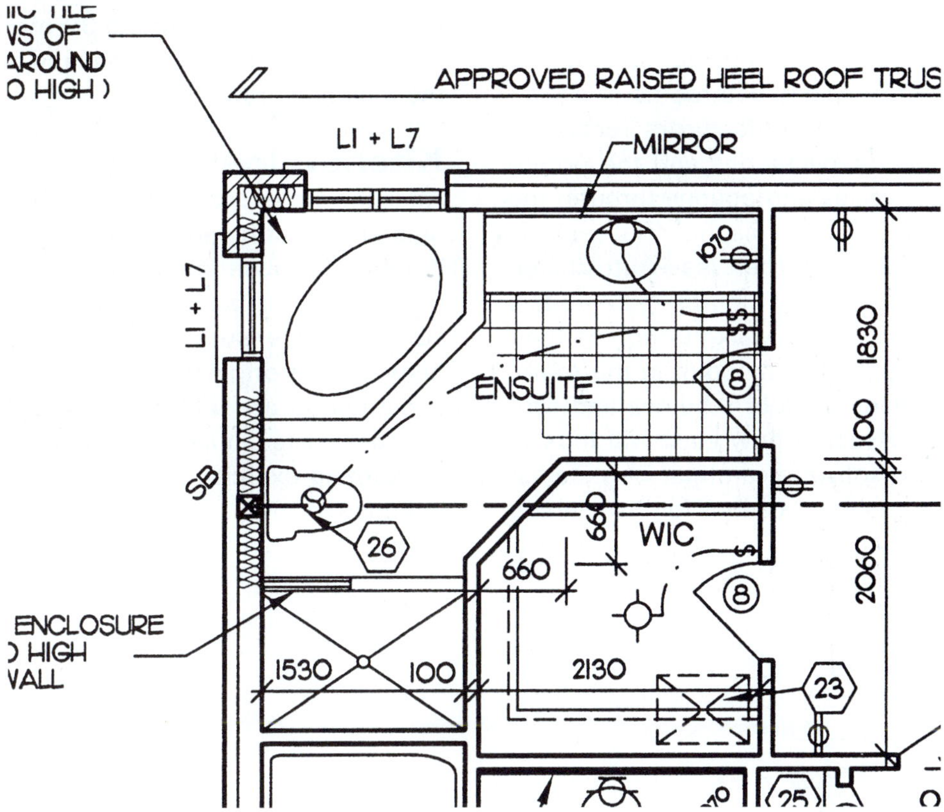

Source: Cassidy & Co. Architectural Technologists

KITCHENS

Clients spend a lot of time in the kitchen; as a result, they have high expectations of this room. They also expect to spend more money on this room than others. Construction drawings provide the basic floor plan of the kitchen. Clients select their cabinets from a kitchen cabinet manufacturer or designer referred by the builder. A new set of kitchen plans are drawn by the kitchen cabinet manufacturer or designer; these drawings will more accurately reflect the options and extras of the client, as well as the standard size of cabinets that the manufacturer uses.

Kitchen designers apply many design principles to the layout of the kitchen, including some concepts based on National Kitchen and Bath Association guidelines:

- The work triangle provides a layout of the major appliances and sink that makes it easier for the cook to work in the kitchen. The triangle is formed between the sink, stove, and refrigerator.
- Avoid placing a stove under a window to prevent drapes catching fire; use exhaust fans.
- Try to locate the sink under a window, so the client can look out the window while doing dishes.
- Place the dishwasher as close as possible to the sink.
- Consider how the refrigerator will be used:
  - How will groceries be loaded?
  - Is there counter space beside the refrigerator?
  - Will the door open out and away from the triangle, so that it does not block or interfere with the work area?
- To avoid breaking up the counter space, locate pantries and refrigerators at the end of a cabinet run.

Many Canadian and U.S. kitchen and bath designers and manufacturers are members of the National Kitchen and Bath Association. This organization certifies kitchen designers and sets industry standards for the design, layout, and installation of kitchens and bathrooms.

Many of today's manufacturers recognize clients' increased expectations and the potential of increasing sales and options through the use of advanced CAD systems that offer realistic 3-D renderings of the kitchen (see Autokitchen drawings and renderings in Figures 13.22–13.26). Clients can have a virtual walk-through of their kitchen and view it from any point in the room. They can try different colour schemes and cabinet types, and visualize how each will look in their kitchen. The software can then create a list of cabinet numbers and locations for installers to use. In its most advanced form, the kitchen software integration works as a building information model (BIM), providing estimates and cut lists for cabinetry, providing the 3D model to ensure everything will fit, and scheduling production for the manufacturer.

Kitchen Installation

After the walls are framed but before the cabinets are constructed, a true final measurement is taken because some manufacturers require the drywall to be installed before final measuring. This ensures that any inconsistencies between the original drawings and the as-built construction are addressed when the cabinets are constructed. It can be a very costly and time-consuming exercise if cabinets don't fit in their intended location.

FIGURE 13.22

Perspective drawing of cabinets using Autokitchen software.

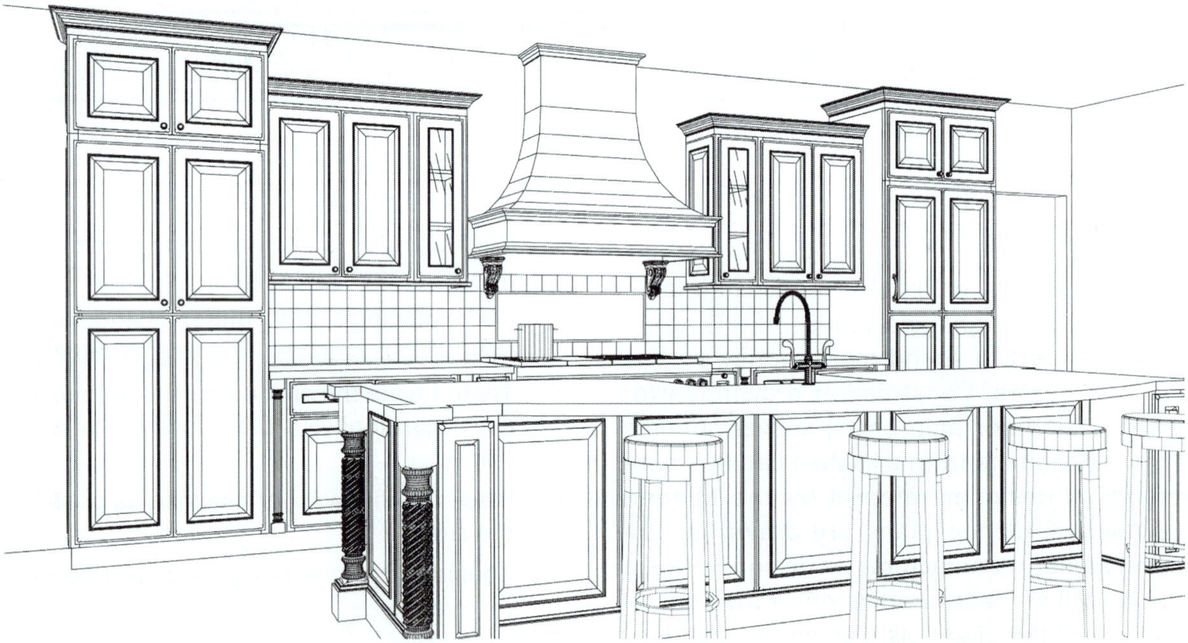

Source: Courtesy of Autokitchen

FIGURE 13.23

Bird's-eye perspective plan view of kitchen drawn using Autokitchen software; compare with Figure 13.22.

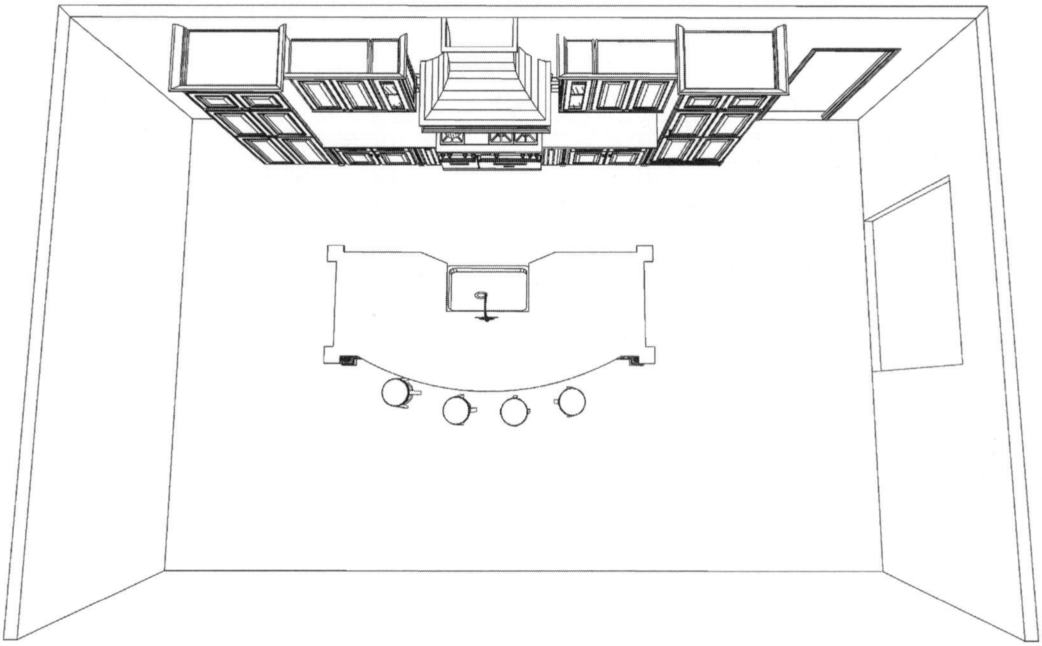

Source: Courtesy of Autokitchen

FIGURE 13.24

Detailed floor plan of kitchen drawn from Figures 13.22 and 13.23 using Autokitchen software. Notice the outline of the individual cabinets that are grouped together to form the overall cabinet layout.

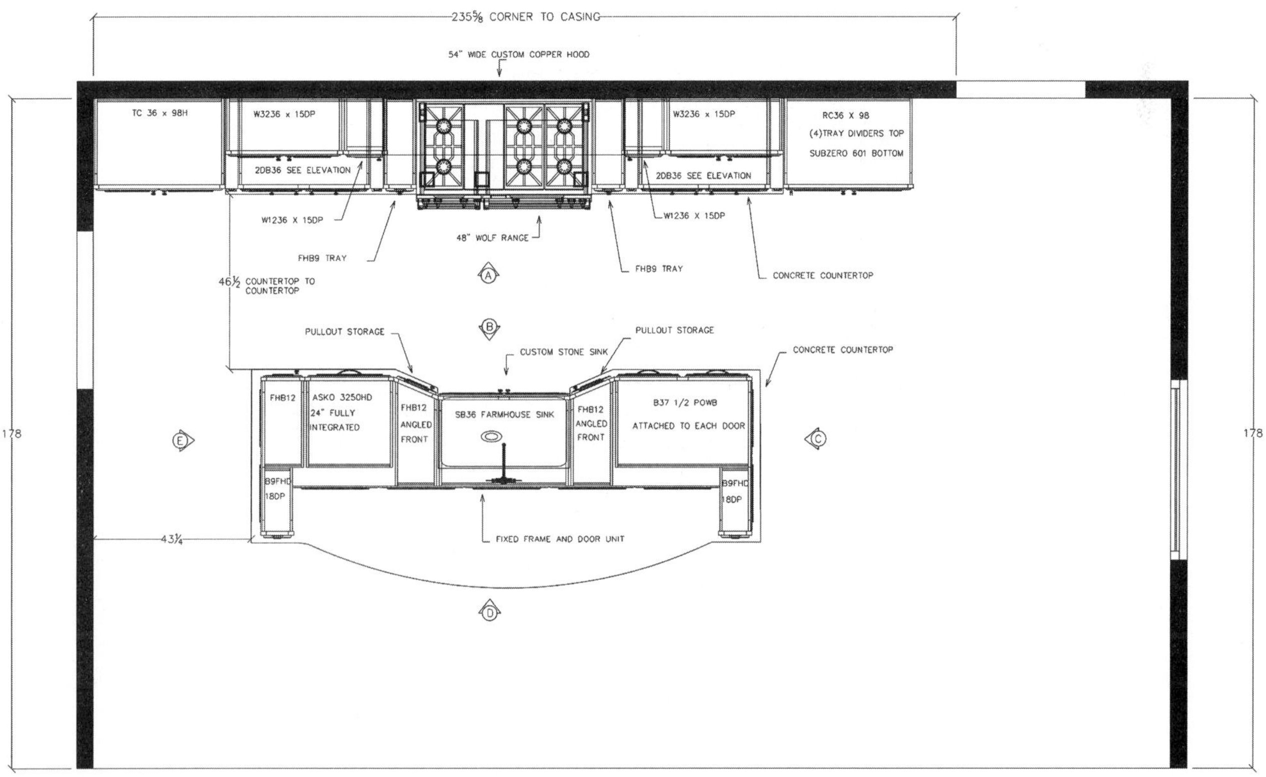

Source: Courtesy of Autokitchen

FIGURE 13.25

Orthographic elevation drawings of island showing front, rear, right, and plan views, drawn using Autokitchen software.

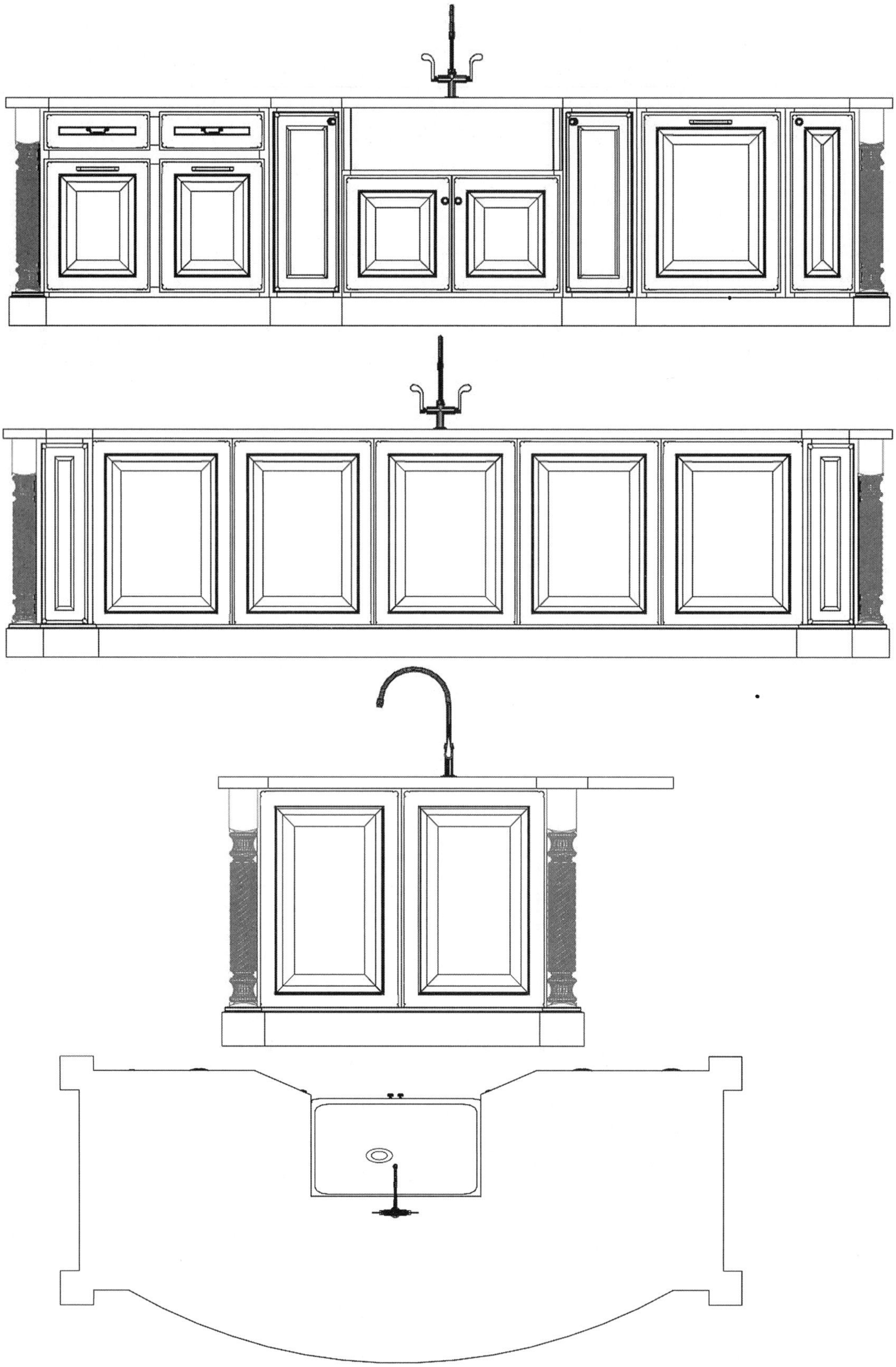

Source: Courtesy of Autokitchen

FIGURE 13.26

Elevation view A of cabinets drawn based on the plan view in Figure 13.24 using Autokitchen software.

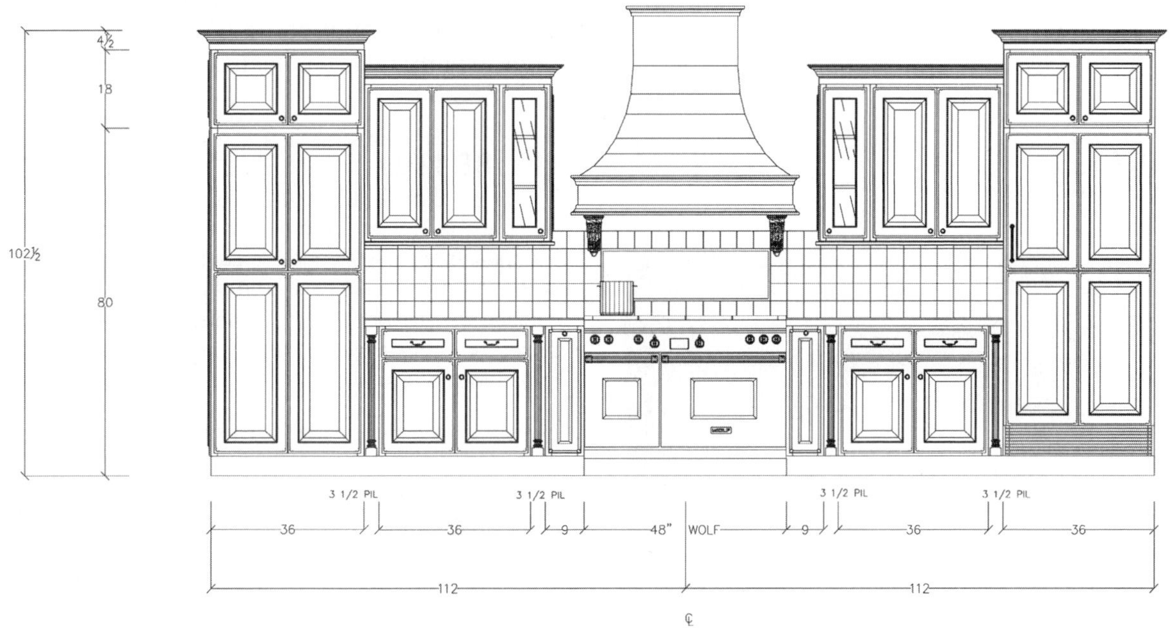

Source: Courtesy of Autokitchen

FIGURE 13.27

Doncaster kitchen floor plan; notice the location of the major appliances and how they form a work triangle. Also note the location of the dishwasher next to the sink for easy loading. The numbers beside the outlets reflect the height of the boxes, which is important to ensure that outlets are placed at the proper height. For example, the microwave outlet is located 1500 mm above the floor so that it will be above the micro-shelf. The upper cabinets are displayed with a dashed line. A ceramic tile floor is shown with the grid lines.

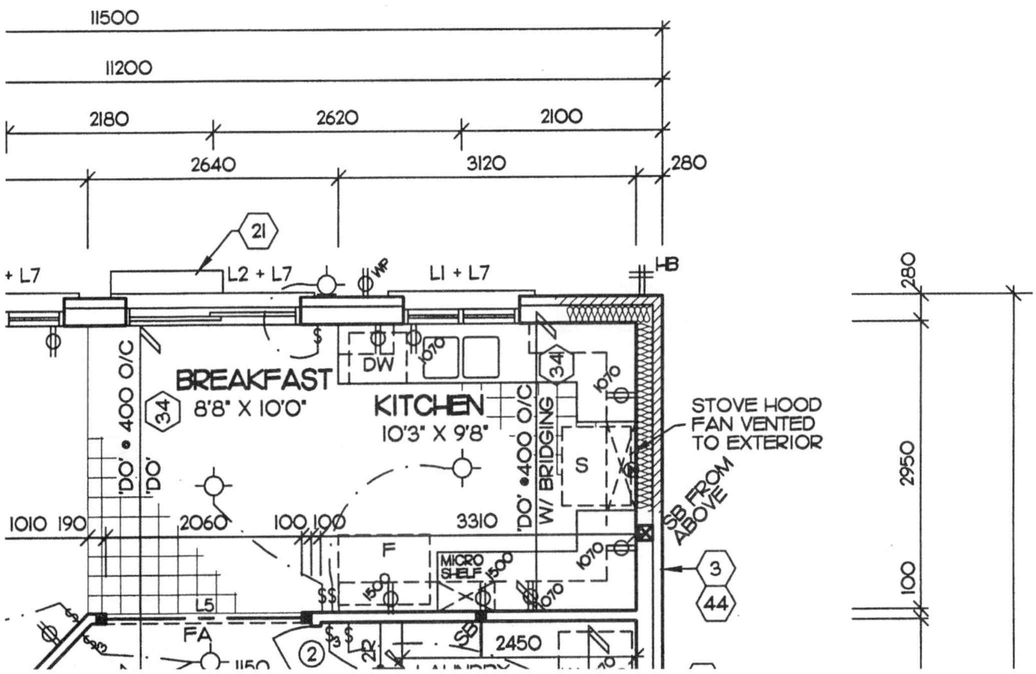

Source: Cassidy & Co. Architectural Technologists

The Doncaster Kitchen

The Doncaster kitchen floor plan is shown in Figure 13.27 and the following photographs (Figures 13.28 to 13.30) provide a view of the finished kitchen for the Doncaster drawings. They differ from those shown in the virtual walk-through in Chapter 3; those photos were of the model home, which had numerous upgrades. See if you can spot some of the differences and upgrades that were selected for this kitchen.

FIGURE 13.28

Doncaster kitchen with tile floor.

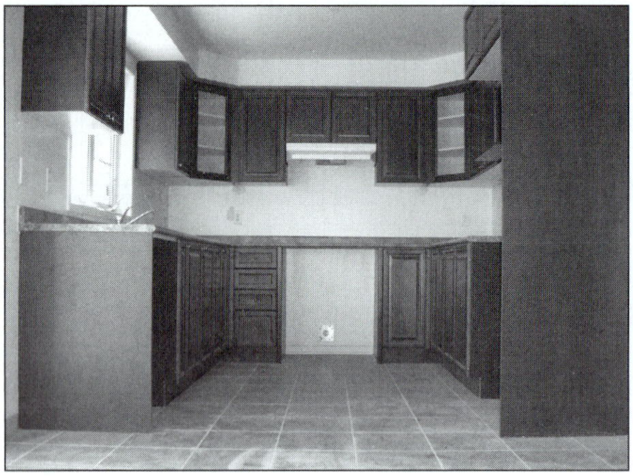

Courtesy Tom Stephenson

FIGURE 13.29

Micro-shelf and refrigerator opening; notice height of receptacle outlet and compare to drawing.

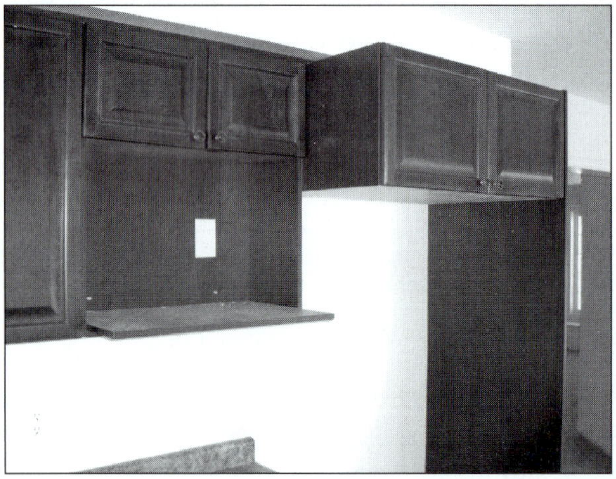

Courtesy Tom Stephenson

FIGURE 13.30

These are **European-style cabinets** that are designed without a face frame. The hinges are referred to as European-style hinges and are adjustable and concealed when the doors are closed. The cabinet doors of the cabinet are real wood-raised panels, while the **cabinet carcass** (cabinet box) is constructed out of pre-finished melamine panels.

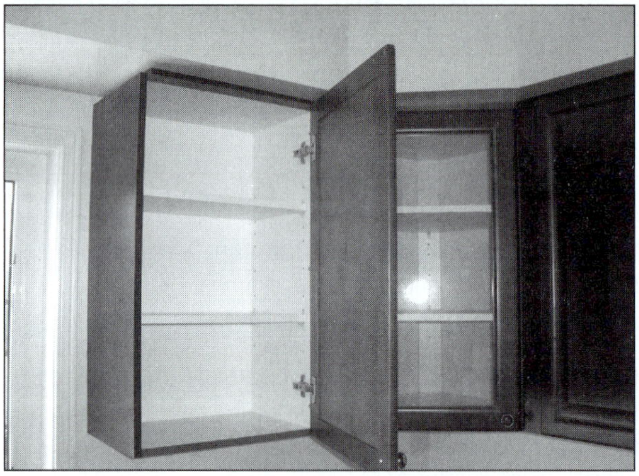

Courtesy Tom Stephenson

Chapter 13 Assignment 1: Interior Finishes, Kitchens, Baths, and Doncaster Drawings

Answer questions 1 through 10 using the material in this chapter, and answer questions 11 to 20 using the Doncaster drawings. Identify the letter that best answers the question.

_____ 1. Where is information generally found regarding the type of moulding to be used?

a. in the floor plans
b. in the section detail drawings
c. in the specifications
d. in the construction notes

_____ 2. What is the name of the association that certifies kitchen designers?

a. The Canadian Home Builders Association
b. The Canadian Construction Association
c. The National Kitchen and Bath Association
d. Canadian Association of Architects

_____ 3. Which kind of drywall provides the highest fire resistance rating?

a. 1/2" regular drywall
b. 1/2" water resistant drywall
c. 1/2" Type X drywall
d. 3/8" Type Y drywall

_____ 4. What does casing intersect with (meet or join up with)?

a. Casing is found at the intersection where the ceiling meets the wall.
b. Casing is found around doors and windows between the frames and the walls.
c. Casing is found at the intersection where the wall meets the floor.
d. Casing is found as a filler piece or scribe that is placed around the kitchen cabinets.

_____ 5. What type of drywall is the most common finish material for walls and ceilings with stud and joist spacing of 16" or less?

a. 3/8"
b. 1/2"
c. 5/8"
d. 3/4"

_____ 6. Where is fiberglass matte finish drywall (non-organic) most suitable?

a. for use where a higher fire rating is required
b. for general use on walls and ceilings
c. for use where sound transmission is a concern (not recommended for basements)
d. for use in indoor areas potentially exposed to moisture (not including showers)

_____ 7. What is the most suitable regular type of drywall to use on ceilings using trusses spaced at 24" OC?

a. 5/8"
b. 1/2"
c. 3/8"
d. 1/4"

_____ 8. How many coats of taping compound does the finishing of drywall require (excluding sanding)?

a. 1
b. 2
c. 3
d. 5

_____ 9. What does the work triangle for kitchen design consist of?

a. sink, stove, and dishwasher
b. sink, stove, and microwave
c. sink, dishwasher, and stove
d. sink, stove, and refrigerator

_____10. According to NKBA design guidelines, which of the following should be placed under a window if possible?

a. sink
b. counter
c. stove
d. dishwasher

_____11. Which of the following is true regarding the kitchen layout on the floor plan drawings?

a. There is a shelf for the microwave oven.
b. The kitchen does not have a dishwasher.
c. There is a stove fan vented to the interior.
d. There is a double sink and a bar sink.

_____12. Which of the following is true regarding the kitchen design and layout on the floor plan drawings?

a. Upper cabinets are drawn with a solid line.

b. Upper cabinets are drawn with a dashed line.

c. The stove is placed under the window.

d. The dishwasher is placed under the microwave shelf.

_____13. Which room has a vanity cabinet indicated?

a. the master bedroom

b. the ensuite bathroom

c. the kitchen

d. the family room

_____14. How many shelves does the linen closet have?

a. 2 shelves

b. 3 shelves

c. 4 shelves

d. 5 shelves

_____15. Unless otherwise noted, what does the typical interior wall partition use?

a. 1/2" drywall on both sides

b. 5/8" drywall on both sides

c. type X drywall on both sides

d. type Y drywall on both sides

_____16. What size is the opening for the attic access hatch?

a. 500 mm × 600 mm

b. 500 mm × 700 mm

c. 600 mm × 800 mm

d. 600 mm × 900 mm

_____17. Review the kitchen plan and determine the length of the base cabinets from back wall to end of cabinets that run under the window. What is the length?

a. 3050 mm

b. 3100 mm

c. 3120 mm

d. 3310 mm

_____18. Review the kitchen plan and determine how far the ceramic tiles come out from the back wall, where the stove is, to where they end at the family room. How far do they extend?

a. 2640 mm

b. 2060 mm

c. 5760 mm

d. 6240 mm

_____19. How many mirrors are indicated on the drawings?

a. 2

b. 3

c. 4

d. 5

_____20. What size door is used for the linen closet, and, if it uses a 3/4" door frame, what should the rough opening width be?

a. 1'-6" × 6'-8" rough opening width is 1' 8 1/4"

b. 1'-6" × 6'-8" rough opening width is 1' 9 1/4"

c. 2'-6" × 6'-8" rough opening width is 1' 8 1/4"

d. 2'-6" × 6'-8" rough opening width is 1' 9 1/4"

STAIRS

Stairs must provide a safe, convenient way of moving from one level to another within a home or building. The National Building Code provides minimum standards for the safe design of a staircase (refer to Division B Part 9.8 of the NBC). You should note that there are differences in code requirements for private dwelling stairs and public stairs. There are also commonly used conventions that help the designer calculate the most comfortable layout for a person ascending or descending a staircase. Stair types include straight run, circular, curved, and winder stairs. This section will look at some of the basic fundamentals of straight-stair layout and design, including calculating stairwell size and **headroom** (the size of opening for a stairs, and headroom clearance). The Doncaster plans will serve as an example of how to read and interpret information for stairs from the drawings. All three residential drawings will be used for the assignments and quizzes in this chapter.

Stair Terminology

As in roof construction, specific terms are used to define various components of stairs. To clarify, "for stairs" has been added to some previously defined roof construction terms.

Stair definitions include the following:

- ***Total rise for stairs:*** The distance from one finished floor level to another. This measurement must include finished flooring materials. For example, if ¾" hardwood is used on the first floor and ceramic tile with a 1" sub-base is used on the second floor, both must be included in the total rise. The maximum vertical height for any single flight of stairs must not exceed 3.7 m (refer to Division B Part 9.8.3.3 of the NBC). If the height exceeds 3.7 metres, the stairs would then require a landing.
- ***Total run for stairs:*** The distance from the face of the bottom riser of the stairs to the face of the top riser of the stairs is the total run. This measurement is important in determining whether the stairs will fit and meet code requirements in situations where the run is limited. For example, if there is a wall at the end of the stairs, there must be enough room at the base of the stairs to provide access.
- ***Unit rise for stairs:*** The distance from top of tread to top of tread is known at the unit rise. The unit rise must be consistent for all steps in a given run of stairs. Studies have shown that if one unit rise is out by even a small amount people will consistently stumble on the inconsistent step. The NBC requires a

FIGURE 13.31

Common layout configurations of different straight run stairs. Note how a winder stair is used when space is limited—more steps in a shorter overall run (also more difficult to climb and descend).

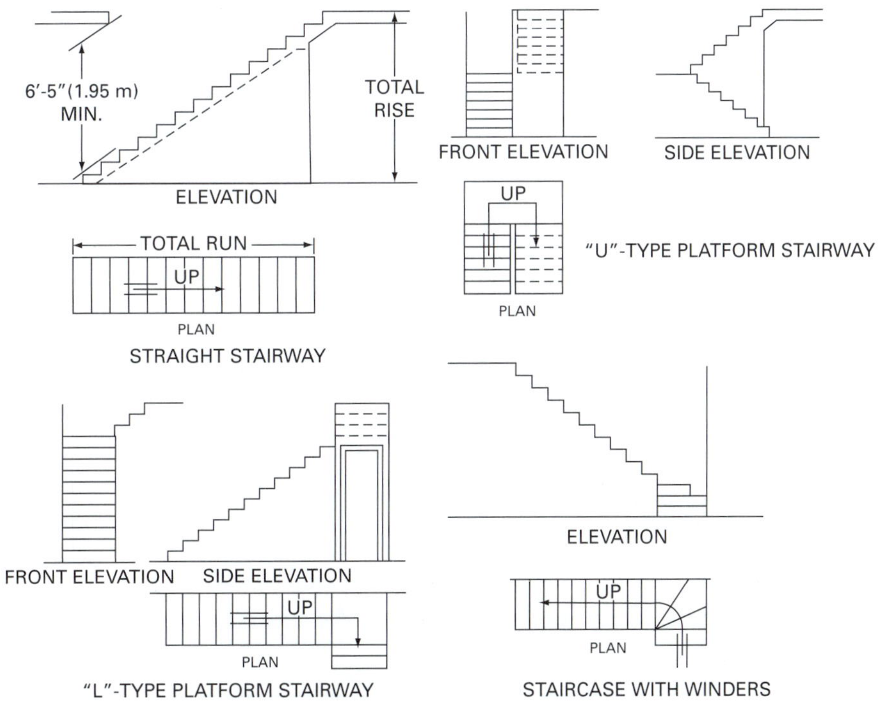

maximum tolerance of 10 mm between the tallest and shortest risers in a flight and maximum of 5mm difference between adjacent treads or landings in a flight (refer to Division B Part 9.8.4.4 of the NBC).

- ***Unit run for stairs:*** The distance from face of riser to face of riser. Unit run is calculated based on total run available, code requirements, and standard rules for a comfortable stair. Similar to the unit rise, the unit run must also be consistent for a given stair or people will stumble (refer to Division B Part 9.8.4.4 of the NBC).
- ***Nosing:*** The portion of stairs that protrudes from the face of the riser. An equivalent of a nosing can be made by back sloping the face of the riser on an angle to allow the tread to stick out past the base of the riser. For calculation purposes, the nosing is not considered part of the unit run.
- ***Tread:*** The distance from face of riser to edge of nosing is called the tread. It includes the length of step available to rest your foot on as you climb the stairs.

  Unit Run + Nosing = Tread.

  There is always one fewer tread than there are risers; for example, a staircase with 11 risers will have 10 treads.

- ***Riser:*** The vertical board that fits between the treads.
- ***Headroom:*** This measurement ensures that people will not bump their heads as they ascend or descend stairs. Headroom is the vertical distance from a line drawn through the nosings of the stairs. The NBC minimum headroom requirement for a stair is 1950 mm for a dwelling unit, 1850 mm for stairs that are located under beams and ducting in secondary suites, and 2050 mm for all other stairs (refer to Divison B Part 9.8.2.2 and 9.8.6.4 of the NBC).
- ***Stairwell opening size:*** This measurement determines the size of opening that must be framed based on meeting code requirements and the layout constraints for the floor plan. The length of stairwell opening is critical because an incorrect length will result in headroom not meeting building code requirements.
- ***Stair stringer:*** Stringers, also referred to as carriages, are the main support for the stairs. There are two types of stringers:
  - *Open stringer:* An open stringer is cut out for the treads and risers, and is usually used where the staircase is not placed against a wall.

FIGURE 13.32

Terms used in stair construction.

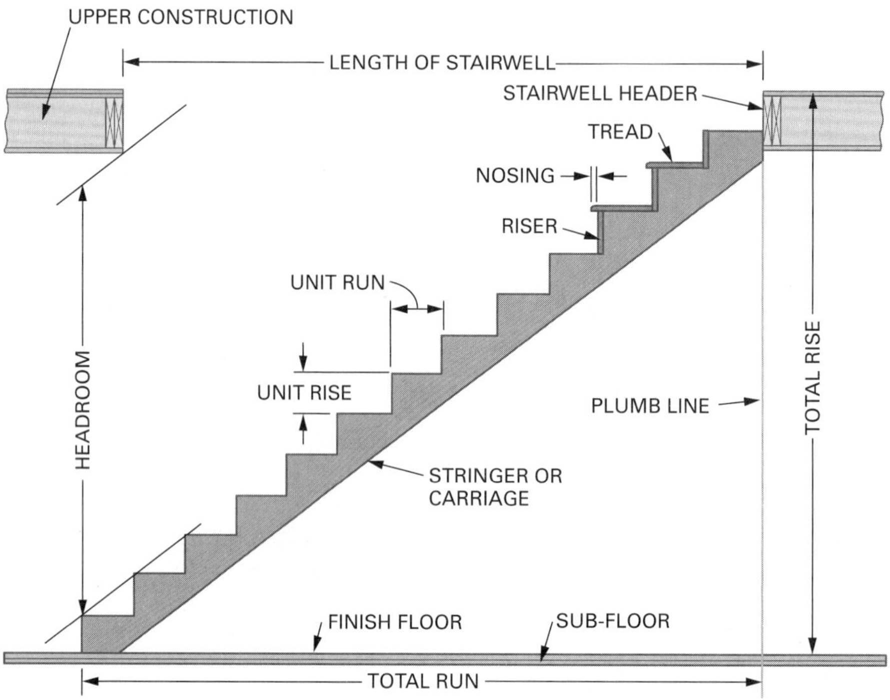

Source: From VOGT, Carpentry, 4E. © 2006 Delmar Learning, a part of Cengage Learning, Inc. Reproduced by permission. www.cengage.com/permissions

FIGURE 13.33

Unit run, unit rise, tread, riser, and nosing locations on a stair.

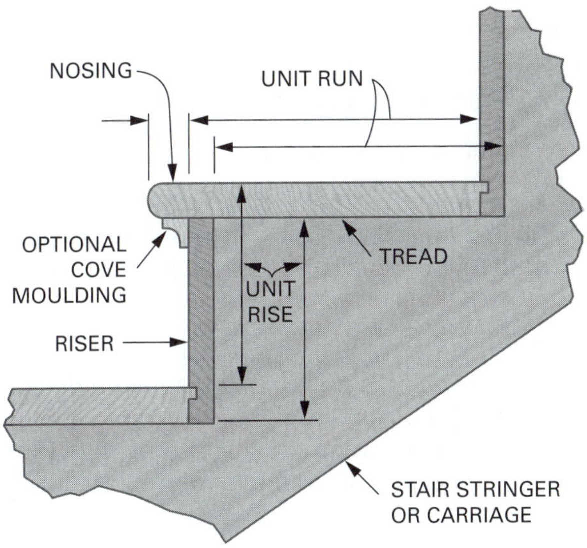

Source: From VOGT, Carpentry, 4E. © 2006 Delmar Learning, a part of Cengage Learning, Inc. Reproduced by permission. www.cengage.com/permissions

FIGURE 13.34

Headroom over stairs NBC minimum requirement for X is 1950 mm for stairs within dwelling units, and 2050 mm for stairs not within dwelling units (refer to Division B Part 9.8.2.2 of the NBC).

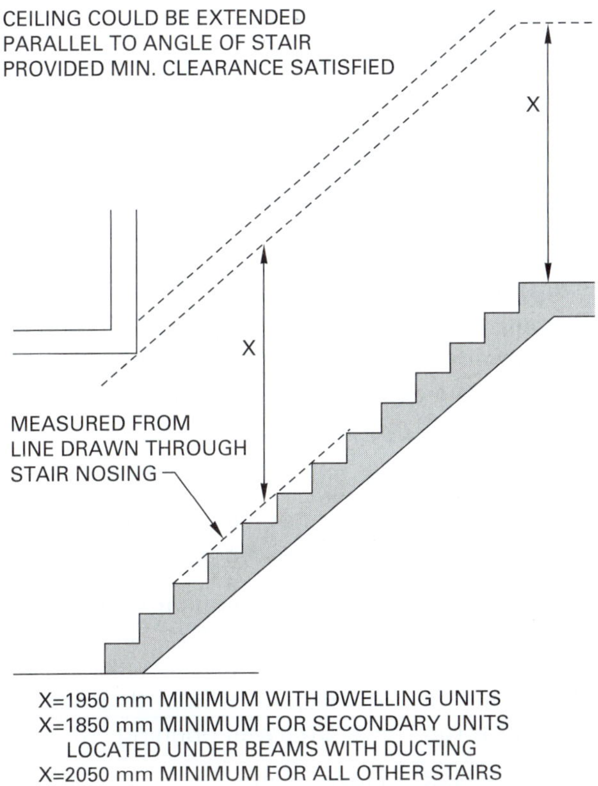

FIGURE 13.35

A first tread with a bullnose. These stairs have an open stringer on the left side and a closed stringer on the right side. The stairs will be carpeted, as indicated by the quality of materials used.

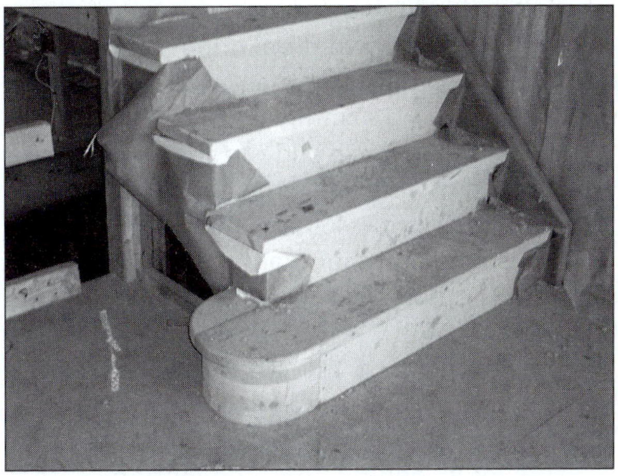

Courtesy Tom Stephenson

- *Closed stringer:* A closed stringer has the profile of the treads and risers dadoed out so they fit neatly inside it. The treads and risers are then glued and wedged into place. This type of stringer is generally used where the staircase fits up against a wall.

- ***Handrail:*** These form the protective barrier to ensure public safety. There are specific code requirements regarding the safe height and placement of handrails and railings. The NBC requires that the height of handrails on stairs and ramps shall be not less than 865 mm, and not more than 965 mm. Check your own provincial code, as this is an area where numerous modifications have been made that vary from province to province. (Refer to Division B Part 9.8.7 of the NBC.)
- ***Guard:*** This term is used by the NBC to describe protection around openings in floors, stairs, landings, and balconies; for example, balusters used in combination with railings. A guard must be provided if there is a difference in elevation of more than 600 mm between the walking surface and the adjacent surface, for example, the top of a deck and the ground. (Refer to Division B Part 9.8.8 of the NBC.)

Stair Design and Calculations

The designer calculates the unit rise and unit run for a staircase to determine if the stairs will work with the floor plan layout. Measurements should be checked during the framing stage to ensure that the stairwell opening size shown on the drawings is large enough to provide the proper headroom and that the stairs will fit properly and meet all code requirements.

A standard convention often used in the design of a comfortable stair is to have a unit rise close to 7" or 180 mm, and a unit run close to 10" or 250 mm. The rise and run should add up to 17" or 430 mm: imperial 7" + 10" = 17" or metric 180 mm + 250 mm = 430 mm.

For example, if the total rise requires a unit rise of 6½", then the ideal run would be 10½" (total 17").

Or in metric, if the total rise requires a 190 mm unit rise, then the ideal run would be 240 mm (total 430 mm).

The logic behind this convention is that the lower the unit rise, the closer the step resembles walking and the easier it is to have a longer unit run. The higher the unit rise the more difficult it is to step forward; therefore, a shorter unit run makes for a more comfortable step. It should be noted that ideal rise and run don't always work and that often the unit rise and unit run that are used are adjusted within the tolerances of building code minimum and maximum requirements to allow the designer to fit the stairs within the floor plan when space is at a premium. You can review examples of minimum and maximum requirements in the construction notes of the drawings used with this text.

Calculating Unit Run and Unit Rise

Example 1—Metric For a stair with a total rise of 2600 mm (from finished floor to finished floor) and an unlimited run, what would be the ideal unit run and unit rise?

1. Divide 2600 by the standard unit of rise of 180 mm; the result rounded will provide the correct number of risers.
   a. 2600/180 = 14.444 or 14 risers
2. Divide 2600 by 14 risers to provide the unit rise: 185.7 or 186 mm.
3. To determine the ideal unit run use the standard convention total of 430 mm and subtract 186 mm from the calculated unit rise.
   a. 430 – 186 = 244 mm is the unit run.

Therefore this staircase will have:

14 risers with a unit rise of 186 mm

13 treads with a unit run of 244 mm.

Example 2—Imperial For a stair with a total rise of 8'-10" and an unlimited run, what would be the ideal unit run and unit rise?

1. Convert 8'–10" to inches.
   a. 8 × 12 = 96" + 10" = 106"
2. Divide the total rise in inches by the standard unit of rise of 7"; the result rounded will provide the correct number of risers.
   a. 106/7 = 15.14 or 15 risers
   b. Divide 106 by 15 risers to provide the unit rise 7.06"; multiply .06 by 16 to determine the number to the nearest 1⁄16". (The result will tell you how many 16ths there are. For example, .5" × 16 = 8; therefore there are 8⁄16" in .5".)

      .06 × 16 = .96 rounded up equals 1 for 1⁄16".

      Therefore each unit rise is 7 1⁄16".
3. To determine the ideal unit run, use the standard convention total 17" (7" unit rise + 10" unit run) and subtract 7 1⁄16" from the calculated unit rise.

   17 − 7 1⁄16" = 9 15⁄16" is the unit run for the stairs.

Therefore, this stairs will have:

15 risers with a unit rise of 7 1⁄16"

14 treads (always one less tread than risers) with a unit run of 9 15⁄16".

Calculating Stairwell Opening Size

Example 3—Metric Calculate the stairwell opening size based on a headroom of 1950 mm and provided with the following information regarding the design of the rise and run:

14 risers with a unit rise of 186 mm
13 treads with a unit run of 244 mm

Thickness of floor 300 mm (measurement taken from finished ceiling to finished floor).

The solution is based on similar triangles as indicated in Figures 13.36 and 13.37.

$$\frac{\text{Unit rise}}{\text{Unit run}} = \frac{\text{Headroom} + \text{Floor thickness}}{\text{Stairwell opening}}$$

$$\frac{186}{244} = \frac{(1950 + 300)}{X}$$

FIGURE 13.36
Calculating stairwell opening size.

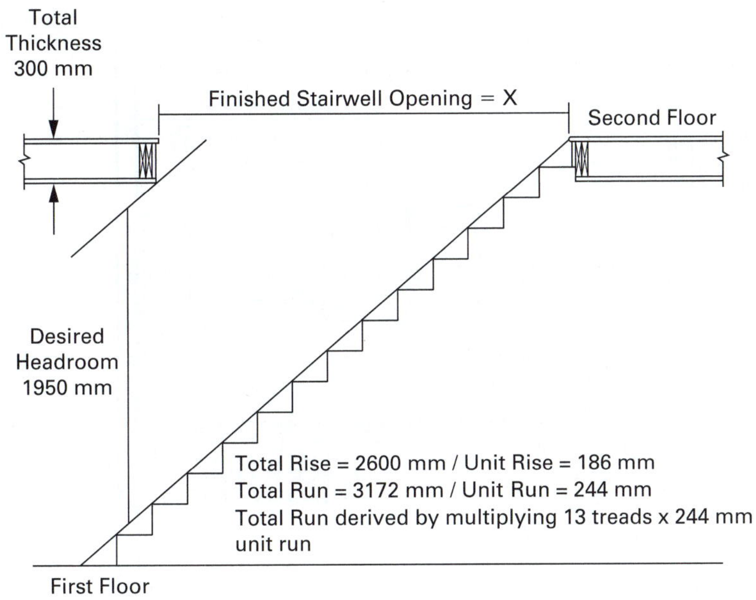

FIGURE 13.37
Similar triangles are used in calculating stairwell size; notice how the rise and run triangle fits in the headroom + floor thickness to stairwell opening triangle. (Both triangles have the same angles; the rise and run triangle is proportionately smaller. The opening is calculated using these ratios as indicated in the sample calculation in the text.)

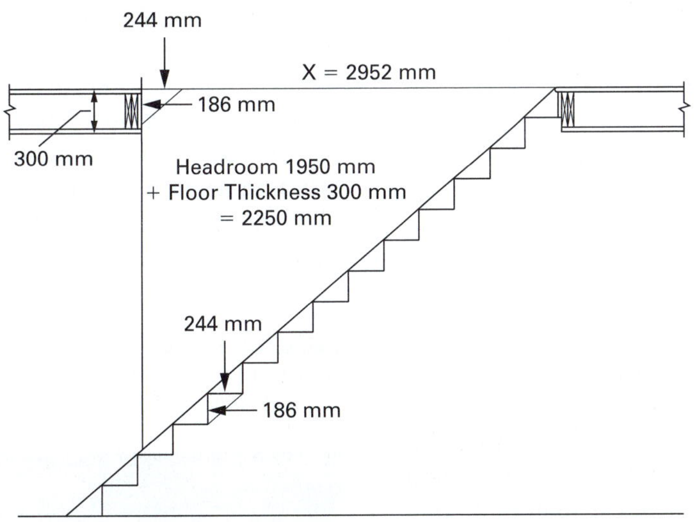

Cross multiply to solve for X which is the minimum stairwell opening size.

$$\frac{186}{244} = \frac{2250}{X}$$

$$X = \frac{549{,}000}{186}$$

$$X = 2951.61$$

Therefore, the stairwell opening length must be at least 2952 mm, including finishes such as drywall and top riser facing material to maintain a headroom of 1950 mm. Also note that the floor thickness used in the calculation includes floor and ceiling finish thicknesses (e.g., drywall and flooring). An alternate plug-in formula that will provide the same result is as follows:

FIGURE 13.38

Close-up of foundation floor plan indicating 13 risers up to first floor, 11 to laundry landing, and two to first floor, Doncaster drawings.

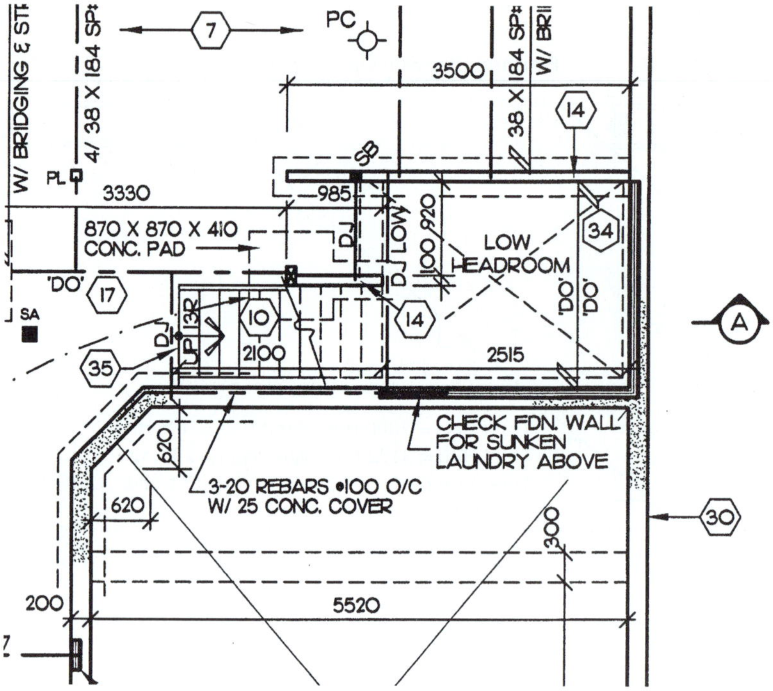

Source: Cassidy & Co. Architectural Technologists

$$\textit{Stairwell opening} = \frac{\textit{Unit run} \times (\textit{Headroom} + \textit{Floor thickness})}{\textit{Unit rise}}$$

Calculating Headroom

Sometimes the stairwell opening may already be framed and you need to determine if the headroom will be sufficient for a given unit rise and unit run. Don't always assume the original drawings outlining the size of the stairwell opening are correct. The ability to verify the layout provided on the drawings is important. The sooner you verify that the layout is correct, the less costly it will be to make corrections. The method used to calculate headroom is similar to that of calculating a stairwell opening size except the stairwell opening size is known and the desired headroom is not. Refer to Appendix D for an example of this process.

Stairs and Doncaster Drawings: How to Reference Information from Drawings

You will need to reference a number of places on the drawings to determine the required information for the layout and construction of the stairs and openings. The stairwell location and size can be determined by reading the floor plans; the limits of overall total run can also be determined from the floor plans. The total rise for the stairs can be found on the elevation drawings and/or building section detail; remember, however, that final measurements for total rise should always be taken from the built structure to ensure that

FIGURE 13.39

Doncaster basement stairs installed; notice the staircase has a closed stringer on each side.

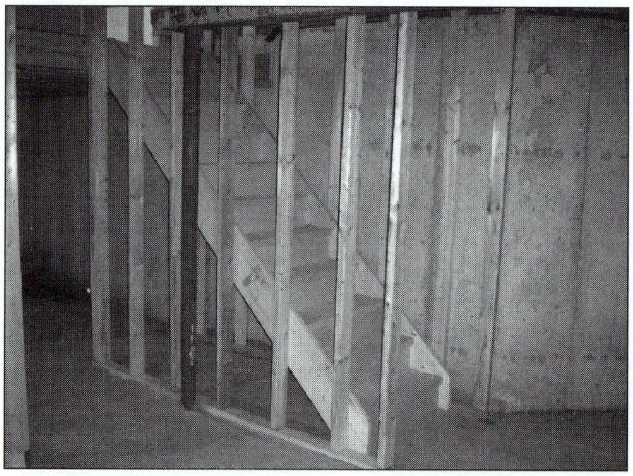

Courtesy Tom Stephenson

all risers are equal in height and consideration has been taken for the thickness of finish floor materials. Details referring to unit rise and unit run requirements can be found by referencing the construction notes and NBC requirements.

Refer to the Doncaster drawings and you will notice that this design includes one staircase from the basement to the laundry room area with 11 risers, and one smaller staircase with one tread, two risers from the laundry area to the first floor. The total number of risers from basement to first floor is 13, which can be misleading when reading the foundation plan. That plan indicates 13 risers, which could mistakenly be interpreted as 13 risers to the laundry room instead of first floor. There are 14 risers on the stairs that run from the first floor to the second floor.

FIGURE 13.40

Close-up of first-floor plan stairs from Doncaster drawings; notice there are 11 risers down to the basement, two risers up from the laundry room, and 14 risers up to the second floor.

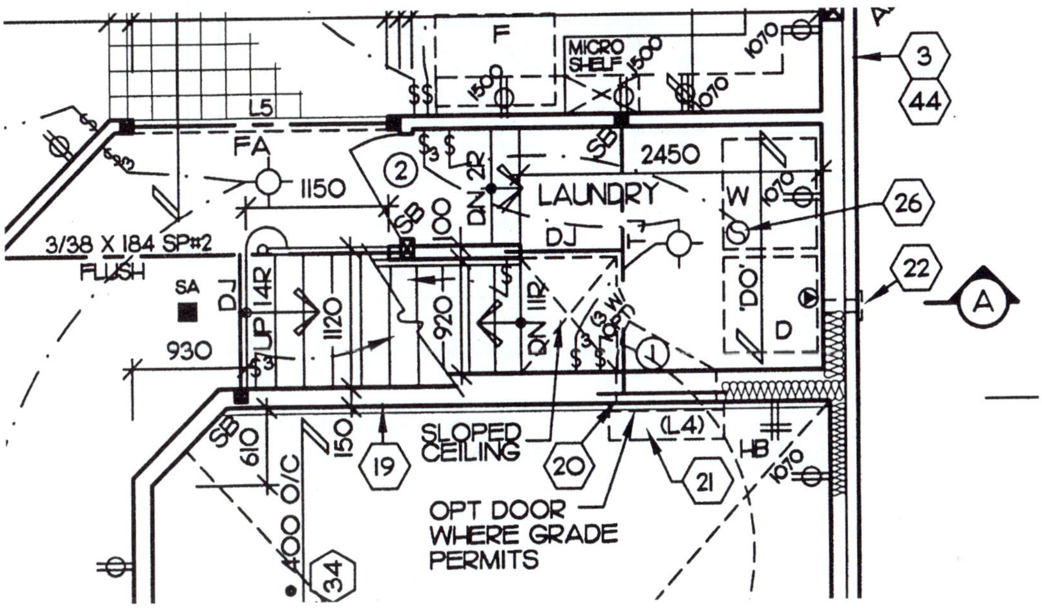

Source: Cassidy & Co. Architectural Technologists

FIGURE 13.41

Doncaster stairs at first floor; notice the open stringer on the left side with bullnose tread at the base and closed stringer on the right side. The left stringer becomes closed at the sixth riser.

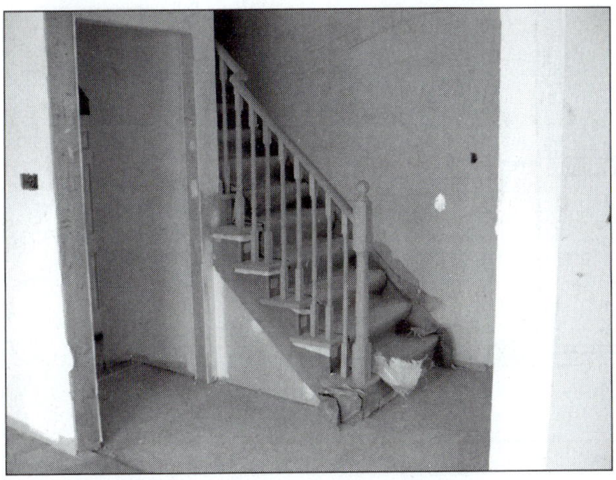

Courtesy Tom Stephenson

FIGURE 13.42

Close-up of second-floor Doncaster floor plan; notice the width and length of the stairwell opening can be clearly seen.

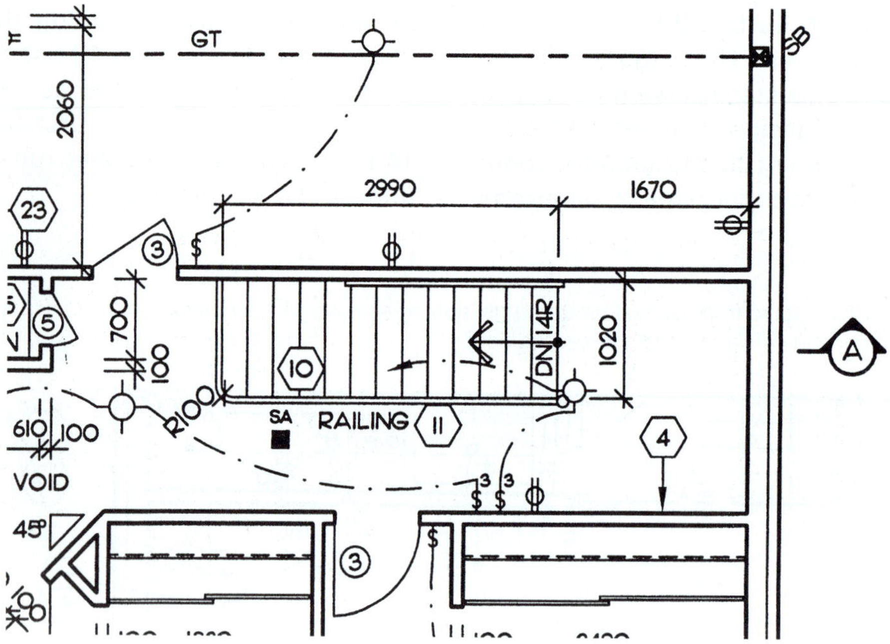

Source: Cassidy & Co. Architectural Technologists

FIGURE 13.43

Close-up of Section A-A Doncaster drawing; the total rise for each stair can be taken from this drawing. However in practice it is best to measure the actual total rise to ensure accuracy in case there is a slight variance between the drawings and what is actually built.

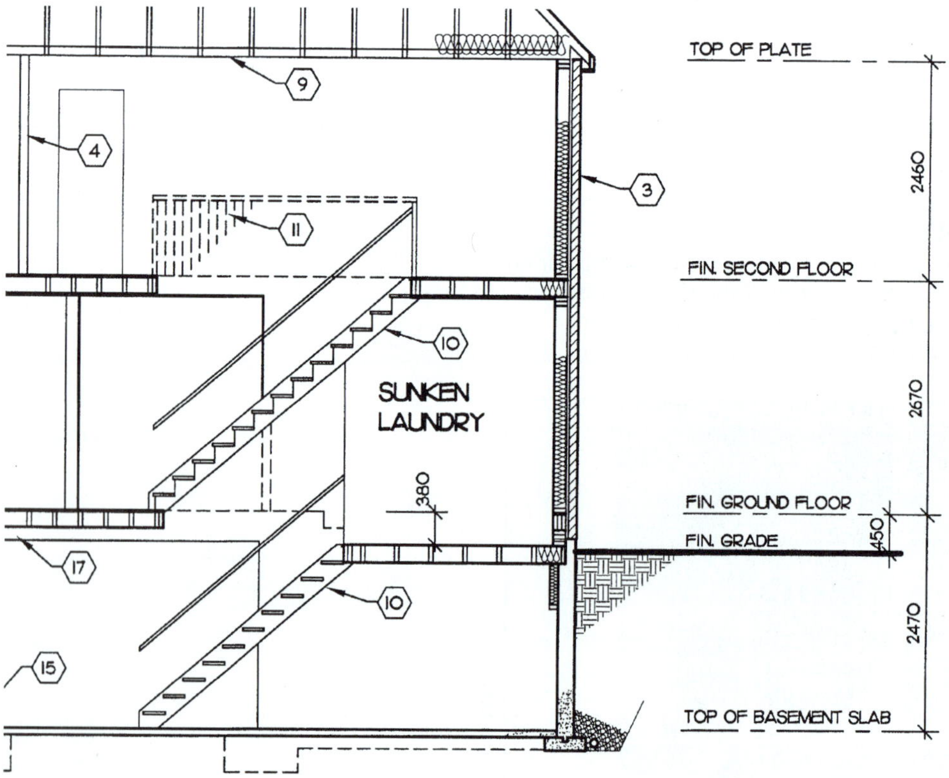

Source: Cassidy & Co. Architectural Technologists

FIGURE 13.44

Doncaster second-floor rough stairwell opening with finished stairs installed.

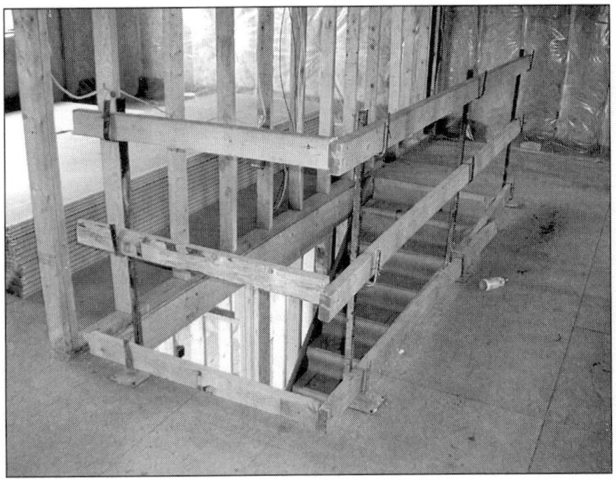

Courtesy Tom Stephenson

FIGURE 13.45

Doncaster drawings construction notes concerning stair construction. The notes reference code requirements for rise, run and guardrail heights. These are current Ontario code requirements but don't meet all the recently changed 2010 NBC requirements, for example, minimum width of stairs is 900 mm and maximum unit rise is dependent on whether the stairs is a private or public stairs. The requirements indicated below are for a private stairs. National and Provincial codes change, so always confirm that what you are building meets current code requirements for your province.

(10) ALL STAIRS (EXTERIOR & INTERIOR)

MIN. RISE = 125 (4 7/8")		MAX. RISE = 200 (7 7/8")
MIN. RUN = 210 (8 1/4")		MAX. RUN = 355 (14")
MIN. TREAD = 235 (9 1/4")		MAX. TREAD = 355 (14")

FOR CURVED STAIRS..........

MIN. RUN	= 150	(5 7/8")
MIN. AVERAGE RUN	= 200	(7 7/8")
MIN. HEADROOM	= 1950	(6'–5")
MIN. WIDTH	= 860	(2' 10")
NOSING (Max. curved or bevelled edge)	= 25	(1")

(11) GUARDRAIL/HANDRAIL

RAIL @ INT. LANDING	= 900	(2' 11")
RAIL @ INT. STAIR	= 800	(2' 7")
RAIL @ EXT. LANDING (Greater than 1800 above finish grade)	= 1070	(3' 6")
RAIL @ EXT. LANDING	= 900	(2' 11")
RAIL @ EXT. STAIR	= 900	(2' 11")

WOOD PICKETS MAX 100 mm (4") BETWEEN

Source: Cassidy & Co. Architectural Technologists

FIGURE 13.46

An oak winder staircase provides a means of providing extra space when there is a limited run. Notice the protective covering so that the stair does not become damaged. The sequencing of finish materials is important; if something is installed too soon it is more likely to be damaged by the rest of the construction process.

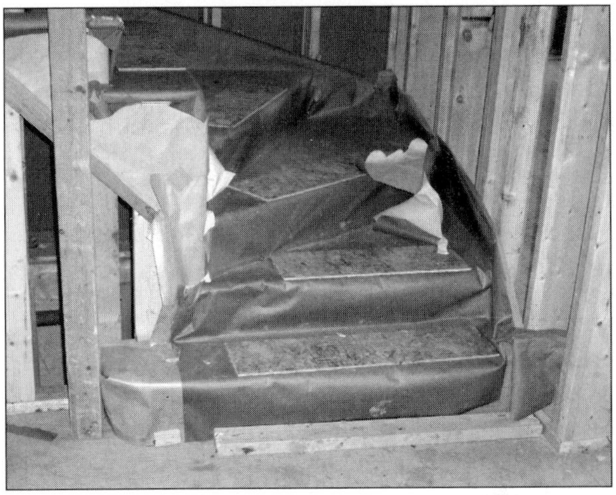

Courtesy Tom Stephenson

OTHER INTERIOR FINISHES

There are numerous other interior finishes that may be listed or shown in construction drawings, such as a variety of floor and wall finishes, which are referred to in the specifications and may also be listed in a room schedule (refer to Figure 15.9 for an example of a room schedule). There are also a variety of different types of fireplaces available today and their location is shown on the floor plan and referenced in detail drawings, specifications, and/or manufacturer specifications and shop drawings. Consideration must be given to the timing of the installation of these finishes to ensure that other work does not damage them.

FIGURE 13.47

Ceramic tile floor installed with wire mesh and thin-set cement placed on subfloor; grout in process of being cleaned.

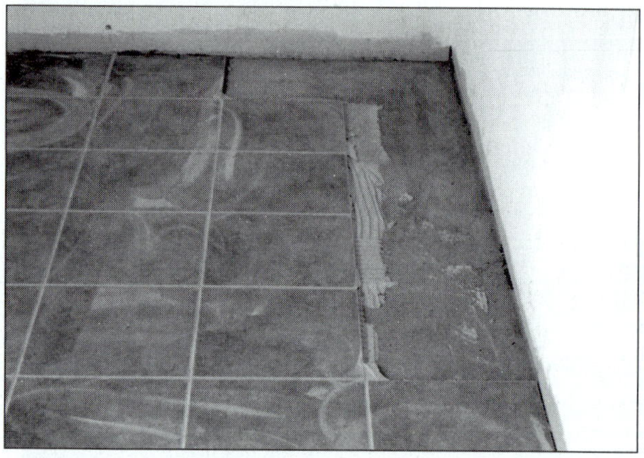

Courtesy Tom Stephenson

Floor Finishes

There are a wide variety of floor finishes used in residential construction; some will require a special sub-base or **substrate** to be installed in order for them to perform properly, while others require no special treatment other than the subfloor. Final floor finish selections should be made early in the construction phase so that any special structural or sub-base decisions can be made before it is too late or too costly to change. Detailed information about the floor finishes can be found in the specifications or in the options selections provided by the builder.

Types of Floor Finishes

- *Carpet:* requires no special substrate
- *Hardwood flooring:* requires no special substrate
- *Ceramic tile, slate, marble, granite:* require a special substrate that may include a number of options such as wire mesh and thin-set cement, double layers of plywood, dry pack concrete, and a myriad of new synthetic sub-base products. Joists may also be placed at narrower spacing such as 12" OC to reduce deflection and potential cracking.
- *Laminate flooring:* requires no special substrate.

Fireplaces

The fireplace has long been an attractive feature of many homes. It is also a feature that adds value to the home. Types of fireplaces that are available include:

- *Traditional solid masonry fireplaces:* Although generally not as efficient as other types of fireplaces the masonry fireplace is popular because it burns wood and provides a traditional look to the home. Constructed from solid masonry, it requires a chimney that extends through the roof. Due to its massive weight, a masonry fireplace also requires a full footing and foundation.
- *Manufactured fireplaces:* There is a variety of manufactured wood- and pellet-burning fireplaces available. They have metal fireboxes and some models can be constructed using wood framing methods (often called zero clearance). The installation of these fireplaces requires close attention to both the manufacturer's directions and shop drawings regarding placement and clearances to combustible surfaces.
- *Gas fireplaces:* This type of fireplace is especially popular in urban centres and areas where natural gas is available. Propane is another option in areas where natural gas is not available. The fireplace is connected to the gas line and can be turned on and off with a remote control or switch. The fireplace is usually designed as high efficiency and does not typically require a chimney; instead, it can be direct vented through the wall, saving the cost of installing a full masonry or metal chimney.

FIGURE 13.48

Solid masonry fireplace section and plan views (refer to Division B Part 9.22 of the NBC).

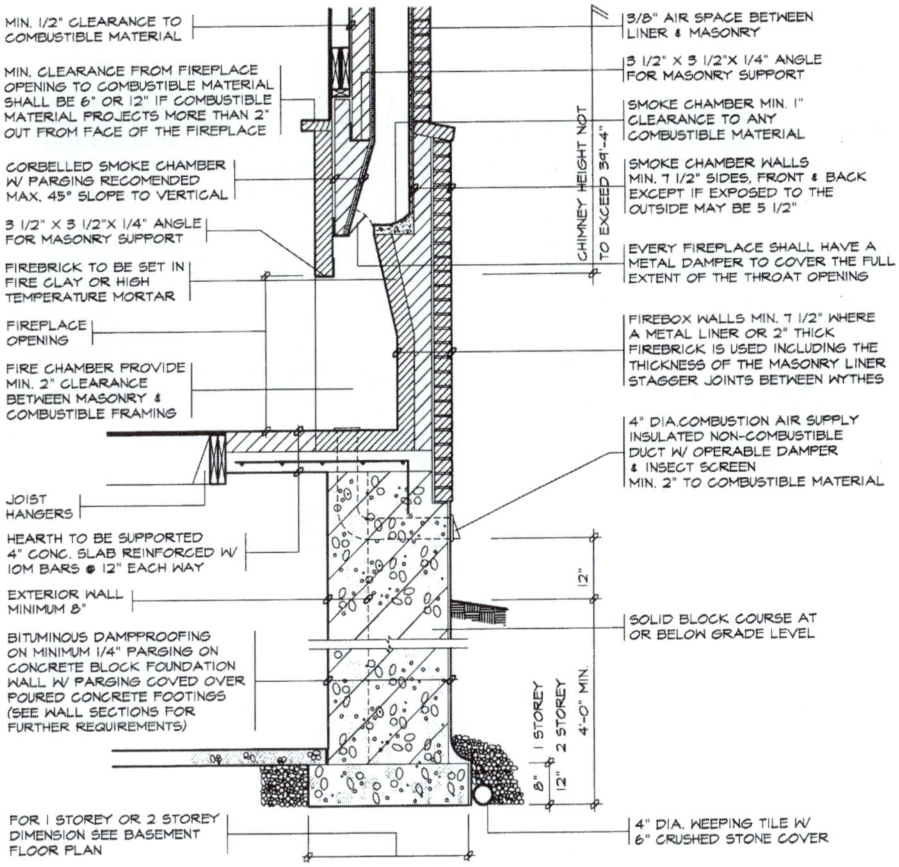

SECTION

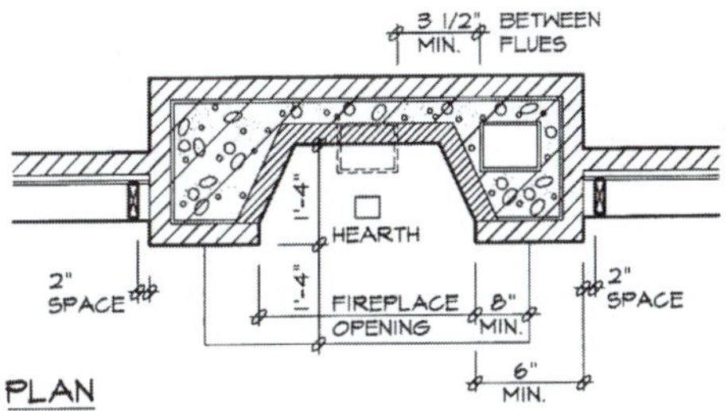

PLAN

FIREPLACE OPENING IN SQ. FT.	CHIMNEY HEIGHT IN FEET							
	9.8 - 14.8		14.8 - 19.4		19.4 - 29.2		29.2 - 39.4	
	FLUE SIZES							
	MIN.	MAX.	MIN.	MAX.	MIN.	MAX.	MIN.	MAX.
0 - 1.6	8"X8"	8"X8"	4"X8"	4"X8"	4"X8"	4"X8"	4"X8"	4"X8"
1.6 - 2.7	8"X8"	8"X8"	8"X8"	8"X8"	8"X8"	8"X8"	8"X8"	8"X8"
2.7 - 3.8	8"X12"	8"X12"	8"X8"	8"X12"	8"X8"	8"X8"	8"X8"	8"X8"
3.8 - 5.4	12"X12"	12"X12"	8"X12"	8"X12"	8"X12"	8"X12"	8"X8"	8"X12"
5.4 - 7.0	12"X12"	12"X16"	12"X12"	12"X12"	12"X12"	12"X12"	8"X12"	8"X12"
7.0 - 8.6	12"X16"	12"X16"	12"X12"	12"X16"	12"X12"	12"X12"	12"X12"	12"X12"
8.6 - 10.8	16"X16"	16"X16"	12"X16"	12"X16"	12"X16"	12"X16"	12"X12"	12"X12"
10.8 - 12.9	16"X16"	16"X16"	16"X16"	16"X16"	12"X16"	12"X16"	12"X16"	12"X16"
12.9 - 15.1			16"X16"	16"X16"	16"X16"	16"X16"	12"X16"	12"X16"
15.1 - 17.2					16"X16"	16"X16"	16"X16"	16"X16"
17.2 - 19.4							16"X16"	16"X16"
19.4 - 21.5							16"X16"	16"X16"
COLUMN 1.	2.	3.	4.	5.	6.	7.	8.	9.

GENERAL NOTES

1. JOISTS OR BEAMS SUPPORTED ON CHIMNEY FLUE SHALL BE SEPARATED BY 12" OF SOLID MASONRY
2. MAXIMUM ANGLE OF SLOPE FOR SMOKE CHAMBER IS 45° FROM VERTICAL
3. COMBUSTIBLE FLOORING, SUB FLOORING & CEILING FINISHES SHALL HAVE A MINIMUM 1/2" CLEARANCE TO MASONRY CHIMNEY.
4. SADDLE NOT REQUIRED IF FLASHING USED THAT EXTENDS UP THE CHIMNEY TO HEIGHT EQUAL TO NOT LESS THAN 1/6 THE WIDTH OF THE CHIMNEY BUT NOT LESS THAN 6" UP THE ROOF SLOPE TO A POINT EQUAL IN HEIGHT TO THE FLASHING ON THE CHIMNEY, BUT NOT LESS THAN 1 1/2 TIMES THE SHINGLE EXPOSURE. PROVIDE COUNTERFLASHING AT THE CHIMNEY.
5. EXCEPT AS REQUIRED IN SENTENCE (2) FIREPLACES SHALL HAVE A NONCOMBUSTIBLE HEARTH EXTENDING NOT LESS THAN 16" IN FRONT OF THE FIREPLACE OPENING MEASURED FROM THE FACING & NOT LESS THAN 8" BEYOND EACH SIDE OF THE FIREPLACE OPENING.
6. WHERE THE HEARTH IS ELEVATED MORE THAN 6" ABOVE THE HEARTH EXTENSION. THE WIDTH OF THE HEARTH EXTENSION SHALL BE INCREASED BY:
   (A) 2" FOR AN ELEVATION ABOVE 6" & NOT MORE THAN 12" &
   (B) AN ADDITIONAL 1" FOR EVERY 2" IN ELEVATION ABOVE 12"
7. INSTALL A CARBON MONOXIDE DETECTOR CONFORMING TO CAN/CSA-6.19 OR UL 2034

Source: Toronto Area Chief Building Officials Committee

FIGURE 13.49

Solid masonry fireplace chimney requirements (refer to Division B Part 9.21 of the NBC).

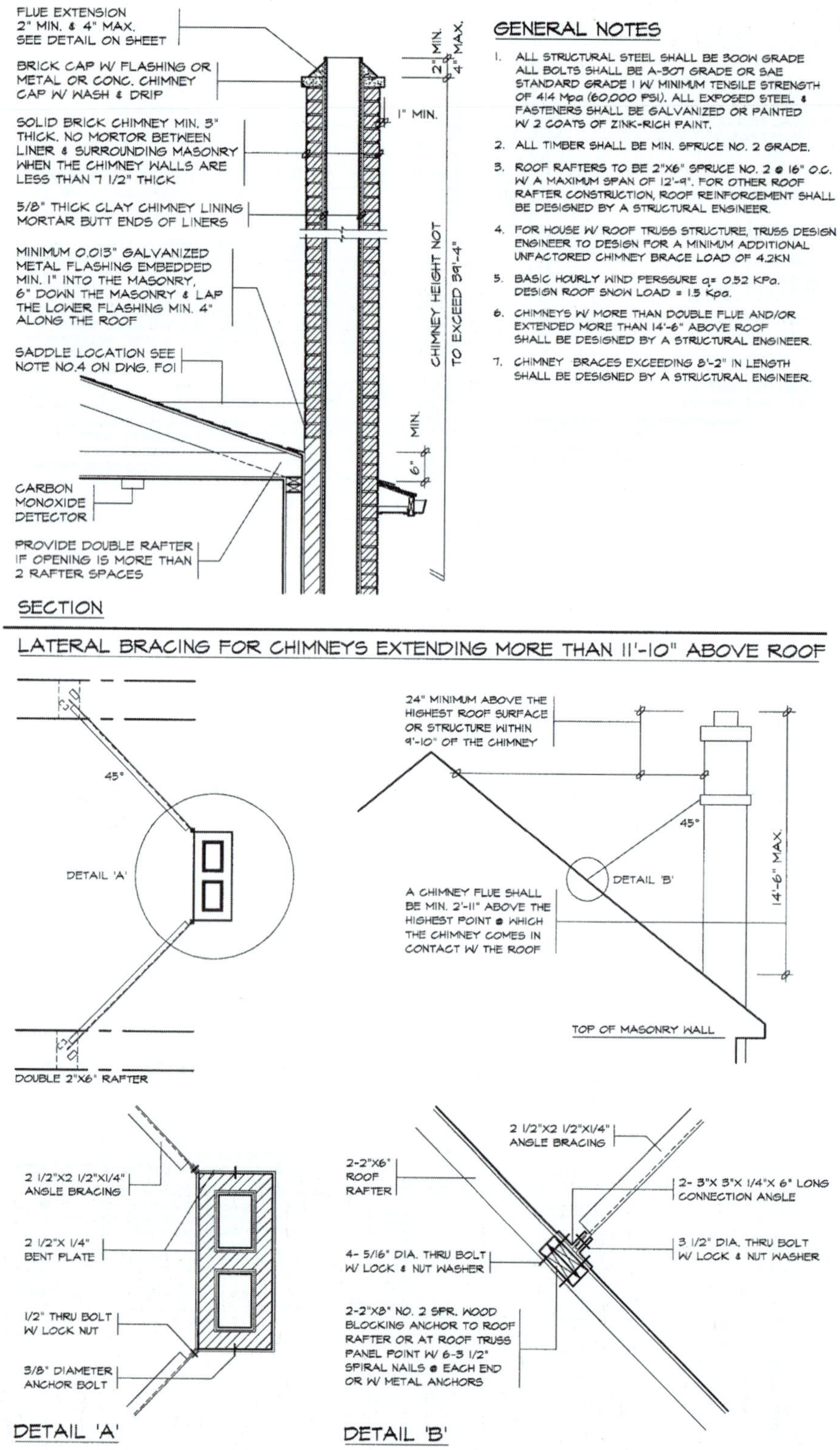

Source: Toronto Area Chief Building Officials Committee

Doncaster Gas Fireplace

The Doncaster drawings indicate a direct-vent gas fireplace in the family room. The fireplace is a separate unit installed by a licensed HVAC contractor. After installation, a decorative mantel or cabinet is added, based on the client's preference and budget.

FIGURE 13.50
Close-up of Doncaster drawings fireplace from first-floor plan.

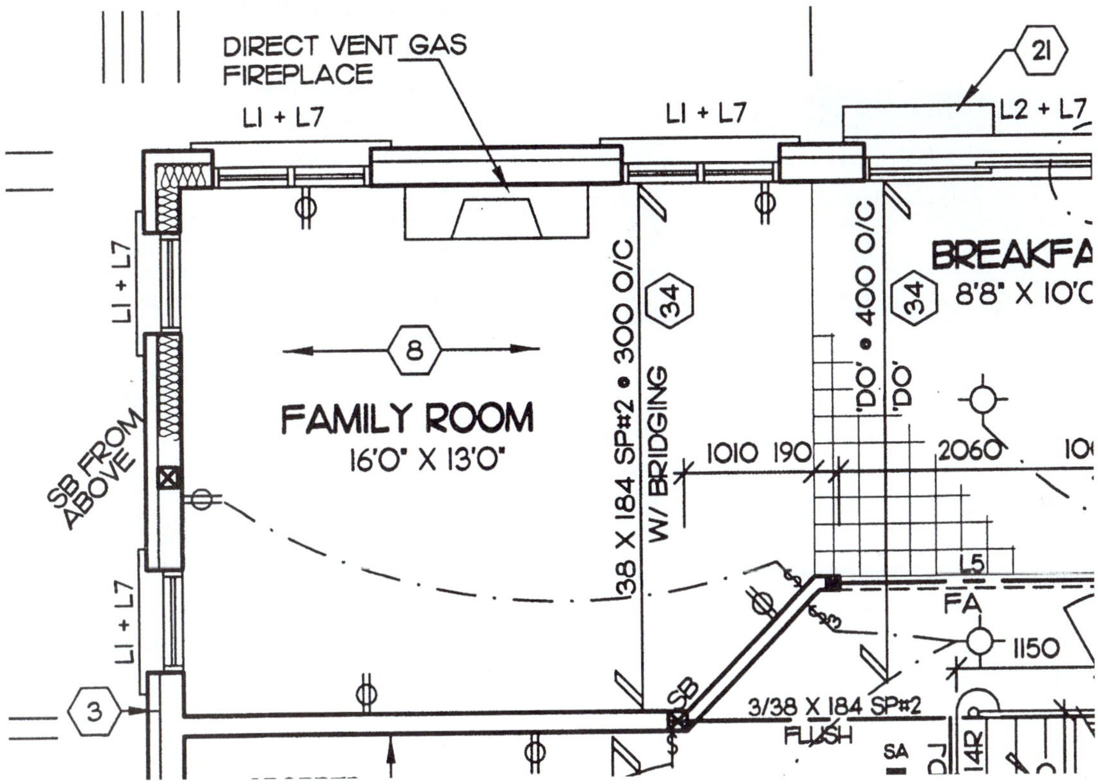

Source: Cassidy & Co. Architectural Technologists

FIGURE 13.51
Doncaster drawings direct-vent gas fireplace, which vents directly out the wall, and requires an electrical (for the fireplace fan) and gas hookup. The switch is located in upper right of screen. Manufacturer's instructions must be followed for installation purposes.

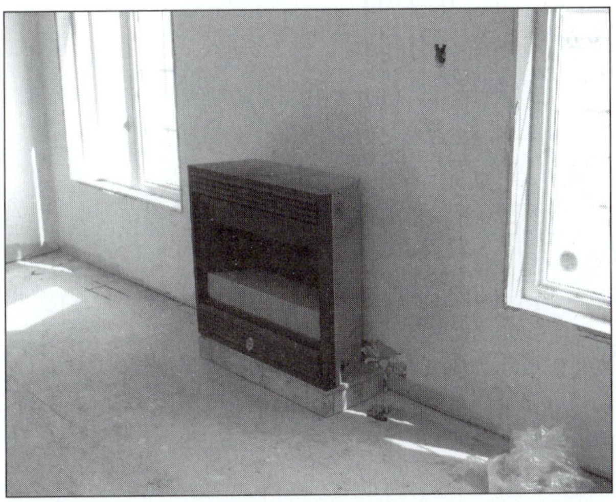

Courtesy Tom Stephenson

FIGURE 13.52
Direct vent with cap not installed yet. Notice that there are two pipes; the inner pipe is used to exhaust the combusted air, while the outer pipe brings in fresh air for combustion.

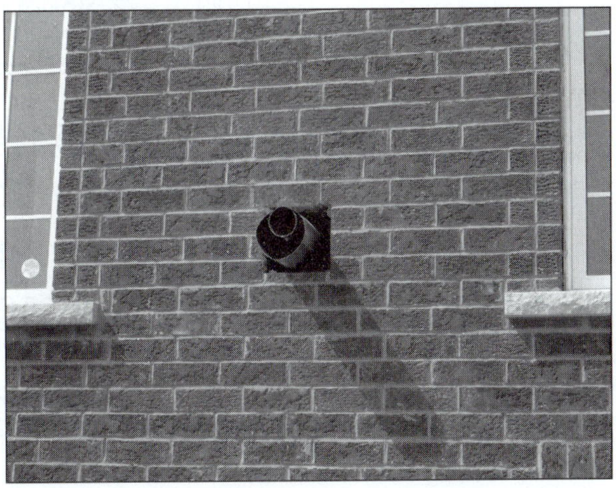

Courtesy Tom Stephenson

FIGURE 13.53
Doncaster finished fireplace with client-selected wood mantel.

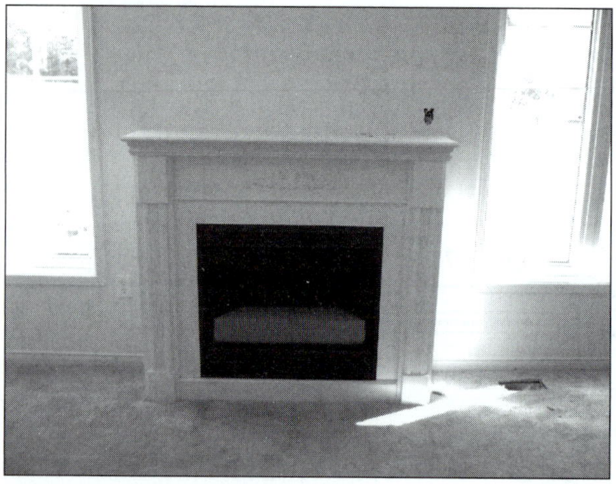

Courtesy Tom Stephenson

SCHEDULES

As discussed in Chapter 8, schedules are used to provide information in table form about a variety of topics, from structural items such as lintels to finishes such as room, window, and door schedules. The Doncaster drawings have a door schedule; the doors are referenced by numbers that are provided in the doorways on the various floor plans. The hall closet door indicated with number 2 would have a door size of 815 × 2030 × 35 (2'–8" × 6'–8" × ¾"). Size is indicated as width × height × thickness. Minimum doorway opening sizes must conform to Division B Part 9.5.5 of the NBC.

FIGURE 13.54
Doncaster door schedule.

DOOR SCHEDULE

MAIN ENTRANCE DOOR TO BE OPENABLE FROM INSIDE WITHOUT KEY.

1. EXTERIOR DOOR 815 x 2030 x 45 (2'8" x 6'8" x 1 3/4") INSULATED MIN RSI 0.7, (R4)
2. DOOR 815 x 2030 x 35 (2'8" x6'8" x 1 3/8")
3. DOOR 760 x 2030 x 35 (2'6" x 6'8" x 1 3/8")
4. DOOR 610 x 2030 x 35 (2'0" x 6'8" x 1 3/8")
5. DOOR 460 x 2030 x 35 (1'6" x 6'8" x 1 3/8")
6. EXTERIOR FRENCH OR GARDEN DOOR 815 x 2030 x 45 (2'8" x 6'8" x 1 3/4")
7. DOOR 815 x 2030 x 45 (2'8" x 6'8" x 1 3/4") SOLID CORE WOOD DOOR W/20 MIN. FIRE PROTECTION RATING OR EQUAL AND 38 (1 1/2") THICK WOOD FRAME. PROVIDE SELF CLOSING DEVICE.
8. DOOR 710 x 2030 x 35 (2'4" x 6'8" x 1 3/8")
9. EXTERIOR DOOR 910 x 2030 x 45 (3'0" x 6'8" x 1 3/4") INSULATED MIN RSI 0.7, (R4)

Source: Cassidy & Co. Architectural Technologists

FIGURE 13.55
Doncaster floor plan displaying door numbering for door schedule.

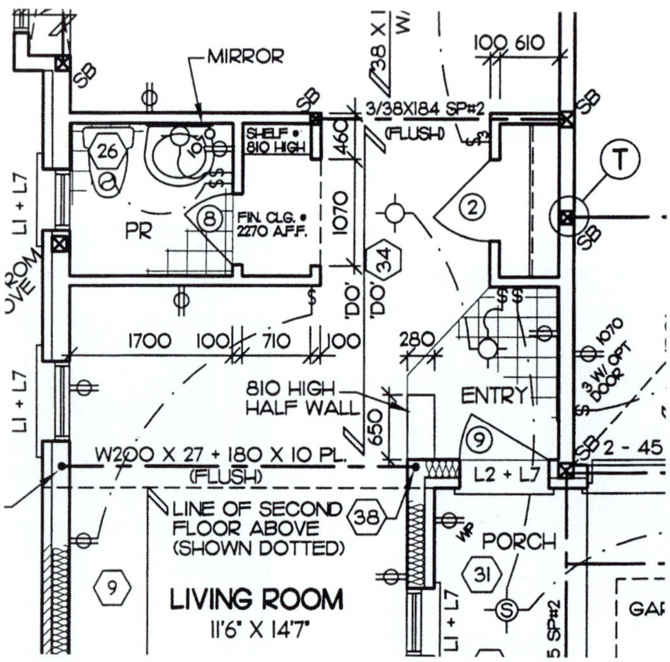

Source: Cassidy & Co. Architectural Technologists

KEY TERMS

Baseboard (p. 328)
Cabinet carcass (p. 336)
Casing (p. 328)
Crown moulding (p. 328)
European-style cabinet (p. 336)
Guard (p. 341)
Handrail (p. 341)
Headroom (p. 339)
Nosing (p. 340)
Riser (p. 340)
Stair stringer (p. 340)
Stairwell opening size (p. 340)
Substrate (p. 348)
Total rise for stairs (p. 339)
Total run for stairs (p. 339)
Tread (p. 340)
Unit rise for stairs (p. 339)
Unit run for stairs (p. 340)

Chapter 13 Assignment 2: Stairs Using Doncaster Drawings

Except for Question 1, answer the following questions using the Doncaster drawings. Identify the letter that best answers the question.

_____ 1. What type of stringers does the stairs shown in the photo below have?

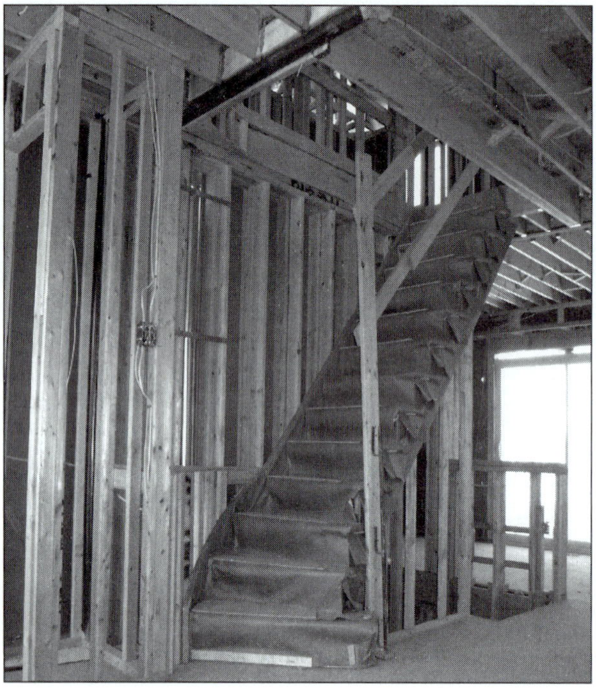

Courtesy Tom Stephenson

a. a closed stringer on both sides
b. an open stringer on both sides
c. a closed stringer on the left side and an open stringer on the right side
d. a closed stringer on the right side and an open stringer on the left side

_____ 2. What is the minimum acceptable unit rise for all stairs, according to the drawings?

a. 125 mm
b. 180 mm
c. 200 mm
d. 230 mm

_____ 3. What is the total run of the basement stairs?

a. 1150 mm
b. 2100 mm
c. 2515 mm
d. 2990 mm

_____ 4. From the drawings, what do the stairs that go from the first floor to the second floor have?

a. a closed stringer on both sides all the way up the stairs
b. an open stringer on both sides all the way up the stairs
c. a quadrant (1/4 rounded on one end of the tread) shape on the first tread
d. a bullnose (rounded on one end of the tread) shape on the first tread

_____ 5. What do the construction notes show as the minimum width of the stairs?

a. 2'-8"
b. 2'-10"
c. 3'-0"
d. 3'-6"

_____ 6. What is the minimum headroom allowed, according to the drawings and the NBC?

a. 1950 mm
b. 2050 mm
c. 2100 mm
d. 2200 mm

_____ 7. From the drawings, what is the unit run of each tread on the basement stairs? (Review the drawings carefully and calculate your answer based on the information from your review.)

a. 175 mm
b. 200 mm
c. 210 mm
d. 250 mm

_____ 8. From the drawings, how many risers are there from the basement to the laundry room?

a. 2
b. 11
c. 13
d. 14

_____ 9. Review the drawings and determine how many **treads** there are from the first floor to the second floor. How many are there?

a. 11
b. 12
c. 13
d. 14

_____10. From the drawings, what is the maximum unit rise allowed for a stairs?

a. 100 mm

b. 150 mm

c. 200 mm

d. 300 mm

_____11. According to the drawings, what is the minimum height requirement for the railing at an interior landing?

a. 800 mm

b. 900 mm

c. 1070 mm

d. 1100 mm

_____12. According to the drawings, what is the maximum space permitted between the wood pickets?

a. 75 mm

b. 100 mm

c. 120 mm

d. 150 mm

_____13. Calculate the unit rise for the stairs leading from the basement to the laundry room (assume that the measurement provided includes floor finishes). What is it?

a. 186.6 mm

b. 190 mm

c. 200 mm

d. 210.4 mm

_____14. Calculate the unit rise for the stairs leading from the first floor to the second floor (assume that the measurements provided on the plans include floor finishes, which in reality they don't). What is it?

a. 186.6 mm

b. 190.7 mm

c. 200 mm

d. 210 mm

_____15. Calculate the unit rise for the stairs leading from the laundry room to the first floor (assume that the measurements provided on the drawings include floor finishes). What is it?

a. 186.6 mm

b. 190 mm

c. 200 mm

d. 210.4 mm

_____16. From the drawings, what is the width of the stairs from outside stringer to outside stringer at the fourth tread for the stairs that go from the first floor to the second floor?

a. 920 mm

b. 1020 mm

c. 1120 mm

d. 1200 mm

_____17. From the drawings, what is the width of the basement stairs?

a. 850 mm

b. 920 mm

c. 1020 mm

d. 1120 mm

_____18. From the drawings, what is the width of the stairs from outside stringer to outside stringer at the top of the second floor stairs?

a. 920 mm

b. 1000 mm

c. 1020 mm

d. 1120 mm

_____19. From the drawings, what is the total run of the stairs that lead from the first to the second floor?

a. 1950 mm

b. 2990 mm

c. 3040 mm

d. 3123 mm

_____20. From the drawings, calculate the unit run for the stairs that go from the first to second floor. What is it?

a. 213.6 mm

b. 217 mm

c. 230 mm

d. 233.8 mm

Chapter 13 Quiz Review: Interior Finishes, Kitchens, And Stairs Using The Brook Drawings

Answer questions 1 to 14 below using the Brook drawings that came attached to the text. Calculate the answers for questions 15 to 20. Identify the letter that best answers the question.

_____ 1. From the drawings, how many risers are there from the basement slab to the sunken landing at the top of the stairs?

a. 13
b. 12
c. 14
d. 15

_____ 2. From the drawings, how much headroom is indicated for the stairs to the basement?

a. 6'-4"
b. 6'-6"
c. 6'-8"
d. 6'-10"

_____ 3. From the drawings, how much headroom is indicated for the stairs from the first floor to the technology centre?

a. 6'-4"
b. 6'-6"
c. 6'-8"
d. 6'-10"

_____ 4. From the drawings, how many treads are there from the Technology Centre down to the 1st floor?

a. 11
b. 12
c. 13
d. 14

_____ 5. From the drawings, what is the unit rise for the stairs from the basement slab to the top of the landing (review drawings carefully and calculate)?

a. 13 risers @ 7⁵⁄₁₆"
b. 13 risers @ 7⅞"
c. 14 risers @ 6⅛"
d. 14 risers @ 6¹³⁄₁₆"

_____ 6. From the drawings, what is the total rise from the finished main floor to the finished floor of the technology centre?

a. 7'-4"
b. 7'-6"
c. 7'-8"
d. 8'-4"

_____ 7. From the drawings, calculate what the unit rise is from the finished main floor to the finished floor of the technology centre, based on the total rise and number of risers required. What is the unit rise?

a. 7"
b. 7¹⁄₁₆"
c. 7¹¹⁄₁₆"
d. 8"

_____ 8. From the drawings, what is the unit run for the stairs from the technology centre to the master bedroom area hallway?

a. 8"
b. 9"
c. 10"
d. 2'-3"

_____ 9. From the drawings, what size is the door to the laundry room, and, if you used 3/4" frames, what would the rough opening width need to be?

a. 36" and the rough opening width is 38¼"
b. 36" and the rough opening width is 40"
c. 34" and the rough opening width is 36"
d. 34" and the rough opening width is 38¼"

_____ 10. What is the minimum distance up the wall that the ceramic tiles need to be placed around the top of the oval tub in the ensuite bathroom?

a. 12"
b. 16"
c. 24"
d. 30"

_____ 11. From the drawings, which one of the following abbreviations is correctly used and applied?

a. "W" means window
b. "D" means dryer

c. "T" means toilet

d. "S" means sink

_____12. Which of the following abbreviations is true based on how they are used on the drawings?

a. "B" means basin type sink

b. "F" means refrigerator

c. "S" means sink

d. "T" means trimmer

_____13. From the drawings, how far is the top of the bookshelf niche above the finished floor of the technology centre?

a. 2'-3"

b. 3'-6"

c. 5'-9"

d. 6'-4"

_____14. How many shelves do the drawings require the linen closet to have?

a. 3

b. 4

c. 5

d. 6

Stair Calculation Questions: Answer the following questions based on the following criteria. Stairs that have a total rise from finished floor to finished floor of 8'-10", including a floor thickness (joist, subfloor, flooring material and drywall all included) of 10". Assume that there is an unlimited run when calculating the ideal rise and run.
Calculate the following, and then proceed to answer the next six questions based on your answers:

Closest number to the ideal run (10") and number of treads

Closest number to the ideal rise (7") and number of risers

Total run based on the closest number to the ideal run

Minimum stairwell opening that will provide headroom of 6'-6"

_____15. What is the **closest** ideal rise for a stairs that has unlimited run, based on the information provided in the question narrative above?

a. 6⅝"

b. 7$\frac{1}{16}$"

c. 7½"

d. 7¾"

_____16. What is the **closest** number to the ideal run for a stairs that has unlimited run, based on the information provided in the question narrative above?

a. 7$\frac{1}{16}$"

b. 8$\frac{9}{16}$"

c. 9$\frac{15}{16}$"

d. 10½"

_____17. How many risers does the stair require, if using the closest number to the ideal rise and run using the information provided in the question narrative above?

a. 12

b. 13

c. 14

d. 15

_____18. What is the **total** run for the stairs, based on the information provided in the question narrative above, using the ideal tread calculation information?

a. 139⅛"

b. 148⅞"

c. 155¼"

d. 160"

_____19. How many treads does the stairs require if using the closest number to the ideal rise and run, using the information provided in the question narrative above?

a. 13

b. 14

c. 15

d. 16

_____20. Calculate the minimum stairwell opening, based on your calculations using the information provided in the question narrative above, and assuming that you want to achieve minimum headroom of 6'-6" and that the floor thickness is 10" (answer should be within 3/8" depending where you rounded your numbers).

a. 110"

b. 123$\frac{13}{16}$"

c. 126¼"

d. 128$\frac{9}{16}$"

14 Specifications and Contracts

LEARNING OBJECTIVES

After completing this chapter, you will be able to:

- Understand the purpose of specifications.
- Identify the appropriate locations for the information provided on construction drawings and specifications.
- Understand the organization and layout of the CSI MasterFormat specification.
- Identify the different types and components of a typical construction contract.

INTRODUCTION

This chapter discusses construction specifications and construction contracts. Both specifications and contracts are essential to the successful completion of any construction project.

SPECIFICATIONS

Construction projects require the use of thousands of different products and materials. Because projects must be built to a budget, construction companies bid on or negotiate based on the scope of work. The **scope of work** defines what is to be included in the work done by a general contractor or subcontractor. Therefore, in order to avoid future conflicts the scope of work must be clearly defined. You have already learned about reading and interpreting construction drawings, which include:

- Site plans
- Floor plans
- Mechanical and electrical plans (if not shown on floor plans)
- Section drawings (building and wall sections)
- Detail drawings
- Abbreviations and construction notes.

The information on the drawings helps to define the shape and geometry of the project and includes the necessary details and dimensions to construct it. However, the drawings do not provide enough information to allow contractors to accurately price and bid on these projects. A set of specifications is required to complete the information required to build.

Specifications are designed to complement drawings. Specifications provide the information that is not provided on the drawings and add clarification. Because of this, information found in the specifications will generally not conflict with the drawings. For example, a set of drawings will show where the toilet and bathtub are to be placed in a bathroom, but will not indicate the make and model number of these fixtures. The specifications, however, will provide the contractor with information regarding the make and model number of the toilet (water closet) and bathtub, and will also provide special details or restrictions regarding their installation.

Specifications help ensure both that the client's expectations are met and that contractors bidding on a project are all bidding on the same thing. Continuing the example of a toilet, without specifications, one contractor might review the drawings and price a $100 toilet while another contractor prices a $200 toilet; meanwhile, the client might have a $300 toilet in mind. The client may accept the lowest price without understanding that the product visualized is not the product being quoted.

Specifications should be written in an enforceable way. For example, a specification using wording such as "Polyurethane foam insulation must be of the best standard or of top quality" is difficult to enforce. One person's notion of quality may be completely different from another's. It is best to provide an industry-accepted standard such as "Polyurethane foam insulation must meet CAN/ULC-705.1." Poorly written specifications inevitably cause misunderstandings that can end up in disputes that lead to litigation; therefore, have a clearly defined set of specifications when jobs are being competitively bid on.

Specification Types

Specifications can take many forms, depending on the format that the designer uses and the level of detail used, which also depends on the contractual arrangement between the client and contractor. In residential construction, designers may use their own unique type of specification. Contractors have to be careful when reviewing specifications because the organization and layout can differ between designers. It is important to review a set of specifications carefully because if they are incorrectly read or misinterpreted it can cost you thousands of dollars. If you make a wrong assumption and incorrectly install the wrong product you will have to correct it, or if you quote based on a wrong assumption you will either lose money because you quoted too low or you will not get the project because you quoted too high.

As mentioned above, not all specifications are standardized; however, by far the most widely used standard format in North America is the MasterFormat, developed by the Construction Specification Institute (CSI) along with its sister organization Construction Specifications Canada (CSC). These organizations have standardized specifications, and currently CSI has two formats in use. One is known as the **16 Division MasterFormat;** this format was standard for over 40 years and is still in use but is slowly being phased out and replaced by the **MasterFormat 2004 Edition,** which has 50 divisions numbered from 0 to 49. The 16-division format is being replaced by the revised and expanded 50-division version to better reflect the new products and technologies that are being used in construction; however, many of the divisions have

FIGURE 14.1

Wood trim and subtitles section taken from the MasterFormat 2014 Edition.

06 46 00	**Wood Trim**
06 46 13	Wood Door and Window Casings
06 46 16	Wood Aprons
06 46 19	Wood Base and Shoe Moldings
06 46 23	Wood Chair Rails
06 46 26	Wood Cornices
06 46 29	Wood Fasciae and Soffits
06 46 33	Wood Stops, Stools, and Sills
06 46 91	Splicing of Wooden Components

Source: The MasterFormat (R) in this book are from MasterFormat(R), published by CSI and Construction Specifications Canada (CSC), and are used with permission from CSI. For those interested in a more in-depth explanation of MasterFormat(R) and its use in the construction industry visit www.csinet.org/masterformat or contact:
CSI 110 South Union Street, Suite 100
Alexandria, VA 22314
800-689-2900: 703-684-0300
www.csinet.org

TABLE 14.1

16 Division MasterFormat Division Numbers and Titles

Division 01	General Requirements
Division 02	Site Work and Utilities
Division 03	Concrete
Division 04	Masonry
Division 05	Metals
Division 06	Wood and Plastics
Division 07	Thermal and Moisture Protection
Division 08	Doors and Windows
Division 09	Finishes
Division 10	Specialties
Division 11	Equipment
Division 12	Furnishings
Division 13	Special Construction
Division 14	Conveying Systems
Division 15	Mechanical
Division 16	Electrical

Source: The MasterFormat (R) in this book are from MasterFormat(R), published by CSI and Construction Specifications Canada (CSC), and are used with permission from CSI. For those interested in a more in-depth explanation of MasterFormat(R) and its use in the construction industry visit www.csinet.org/masterformat or contact:
CSI 110 South Union Street, Suite 100
Alexandria, VA 22314
800-689-2900: 703-684-0300
www.csinet.org

TABLE 14.2
MasterFormat Groups, Subgroups, and Divisions

Procurement And Contracting Requirements Group
Division 00 – Procurement and Contracting Requirements
Introductory Information
Procurement Requirements
Contracting Requirements
Specifications Group
General Requirements Subgroup
Division 01 – General Requirements
Facility Construction Subgroup
Division 02 – Existing Conditions
Division 03 – Concrete
Division 04 – Masonry
Division 05 – Metals
Division 06 – Wood, Plastics, and Composites
Division 07 – Thermal and Moisture Protection
Division 08 – Openings
Division 09 – Finishes
Division 10 – Specialties
Division 11 – Equipment
Division 12 – Furnishings
Division 13 – Special Construction
Division 14 – Conveying Equipment
Division 15 – Reserved for Future Expansion
Division 16 – Reserved for Future Expansion
Division 17 – Reserved for Future Expansion
Division 18 – Reserved for Future Expansion
Division 19 – Reserved for Future Expansion
Facility Services Subgroup
Division 20 – Reserved for Future Expansion
Division 21 – Fire Suppression
Division 22 – Plumbing
Division 23 – Heating, Ventilating, and Air-Conditioning (HVAC)
Division 24 – Reserved for Future Expansion
Division 25 – Integrated Automation
Division 26 – Electrical
Division 27 – Communications
Division 28 – Electronic Safety and Security
Division 29 – Reserved for Future Expansion
Site and Infrastructure Subgroup
Division 30 – Reserved for Future Expansion
Division 31 – Earthwork
Division 32 – Exterior Improvements
Division 33 – Utilities
Division 34 – Transportation
Division 35 – Waterway and Marine Construction
Division 36 – Reserved for Future Expansion
Division 37 – Reserved for Future Expansion
Division 38 – Reserved for Future Expansion
Division 39 – Reserved for Future Expansion
Process Equipment Subgroup
Division 40 – Process Interconnections
Division 41 – Material Processing and Handling Equipment
Division 42 – Process Heating, Cooling, and Drying Equipment
Division 43 – Process Gas and Liquid Handling, Purification, and Storage Equipment
Division 44 – Pollution and Waste Control Equipment
Division 45 – Industry-Specific Manufacturing Equipment
Division 46 – Water and Wastewater Equipment
Division 47 – Reserved for Future Expansion
Division 48 – Electrical Power Generation
Division 49 – Reserved for Future Expansion

Source: The MasterFormat (R) in this book are from MasterFormat(R), published by CSI and Construction Specifications Canada (CSC), and are used with permission from CSI. For those interested in a more in-depth explanation of MasterFormat(R) and its use in the construction industry visit www.csinet.org/masterformat or contact:
CSI 110 South Union Street, Suite 100
Alexandria, VA 22314
800-689-2900: 703-684-0300
www.csinet.org

kept similar numbering in the new edition. Tables 14.1 and 14.2 list the division numbers and titles for the 16 Division MasterFormat and the MasterFormat 2004 Edition; compare the numbering between both versions.

Division 01 provides the general information that applies to and complements the rest of the divisions. This is also the section that provides information on payment procedures that a subtrade would refer to regarding progress billing information. It is also the main division that would describe the project's environmental goals, for example, LEED certification requirements, environmental procedures including resource efficiency, energy efficiency, indoor air quality, building component reuse, and recycling. Specific related requirements would then be cross-referenced and found in the appropriate remaining divisions.

MasterFormat organizes the specifications so that information can be found more readily. Once

TABLE 14.3
Items Typically Found in Division 01 General Requirements

• 01 10 00 Summary
• 01 20 00 Price and Payment Procedures
• 01 30 00 Administrative Procedures
• 01 40 00 Quality Requirements
• 01 50 00 Temporary Facilities and Controls
• 01 60 00 Product Requirements
• 01 70 00 Execution and Closeout Requirements
• 01 80 00 Performance Requirements
• 01 90 00 Life Cycle Activities

Source: The MasterFormat (R) in this book are from MasterFormat(R), published by CSI and Construction Specifications Canada (CSC), and are used with permission from CSI. For those interested in a more in-depth explanation of MasterFormat(R) and its use in the construction industry visit www.csinet.org/masterformat or contact:
CSI 110 South Union Street, Suite 100
Alexandria, VA 22314
800-689-2900: 703-684-0300
www.csinet.org

you are familiar with the format and the specific categories, you know in which section to find information regarding specific topics. For example, information on wood trim is in division 06 under section 06 04 00 (numbering is always consistent). Trim suppliers go to this section to find detailed information about the casings and baseboards that are specified for a particular project. Alternatively, plumbers use division 22 for the 50-division format and division 15 for the 16-division format, for plumbing information and masonry subcontractors use division 04 for masonry information.

Construction specifications are generally organized using a three-part format:

- *Part 1—General:* Includes general procedural and administrative information about the section being specified
- *Part 2—Products:* Includes product-specific information such as acceptable testing standards that must be met, and make/manufacturer and model number if applicable
- *Part 3—Execution:* Provides requirements regarding installation and may include schedules.

Sample Specification

Figure 14.2 is a sample specification taken from the *CMHC Best Practice Guide: Wood Frame Envelopes*. The sample specification uses the 16 Division MasterFormat and specifies information regarding the insulation, vapour retarders, and air barriers. Pay attention to the numbering system and how the specification is organized; compare the division numbers located at the top right-hand corner of the specification sample pages to the division numbers provided for the 16 Division MasterFormat (Table 14.1). For example, insulation is located at section 07 200; the *07* refers to Division 7 Thermal and Moisture Protection.

Specifications for Production Home Builders

For projects that are not being competitively bid on directly with clients, such as production homes, the specifications will typically take a different form. Production builders will generally sell the home with a base level of options; the client then will select the final finishes within a range of products and extras offered by the builder. The builder will then produce a final specification itemizing the client's selections; this document is then used on site to ensure that the client's selections are installed.

The builder will also produce sub-contracts for the sub-trades that reflect the overall design and client's selections. Each subcontract will include a scope of work, which is a specification similar to the sample specification shown previously. The scope of work will include only the work required to be performed by that individual subtrade as part of their contract.

FIGURE 14.2

Sample specification from *CMHC Best Practice Guide: Wood Frame Envelopes.*

CMHC Best Practice Guide Wood Frame Envelopes	INSULATION	Section 07200 Page 1 of 4

PART 1–GENERAL

1.1	Related Work Specified Elsewhere	.1	Rough Carpentry–Section 06100
		.2	Subgrade Insulation–Section 07212
		.3	Sealants, Gaskets and Air-Barrier Membranes–Section 07900
1.2	Handling and Storage of Materials	.1	Store packaged materials in original undamaged containers with manufacturer's labels and seals intact. Deliver to site in sealed packages.
		.2	Prevent damage to materials during handling, storing and erection. Damaged materials will be rejected.

PART 2–PRODUCTS

2.1	Installation Material	.1	Batt insulation shall be unfaced friction-fit fibrous glass wool insulation and shall conform to CAN/ULC-S702-97 Mineral Fibre Thermal Insulation for Buildings, Type 1A standards.
		.2	Batt insulation shall be mineral wool insulation and shall conform to the standards set out in CAN/ULC-S702-97 Mineral Fibre Thermal Insulation for Buildings, Type 1.
		.3	Sprayed-in-place cellulose fibre wall insulation shall conform to CAN/ULC-S703 Cellulose Fibre Loose Fill Thermal Insulation C (previously CAN/CGSB-51.60-M90), complete with jute mesh backing as supplied by the manufacturer.
		.4	Attic insulation shall be blown-in fibreglass insulation conforming to the standards set out in CAN/ULC-S702-97 Mineral Fibre Thermal Insulation for Buildings and installed in accordance with CGSB 51-GP-45MP Manual for Installers of Mineral Fibre Blown Loose Fill Thermal Insulation.
		.5	Attic insulation shall be blown-in cellulose fibre insulation conforming to the standards set out in CAN/CGSB 51-GP-44MP Manual for Installers of Cellulose Fiber Thermal Insulation.

Source: Canada Mortgage and Housing Corporation (CMHC). Best Practice Guide: Wood Frame Envelopes, 2006; Renovator's Business Guide, 2005.

FIGURE 14.2 *(continued)*

CMHC Best Practice Guide Wood Frame Envelopes	INSULATION	Section 07200 Page 2 of 4

.6 Polyurethane foam insulation conforming to the standards set out in CAN/ULC-705.1 Thermal Insulation-Spray Applied Rigid Polyurethane Foam, Medium Density, Material–Specification Class 1. Installed to the standards of CAN/ULC-S705.2 Installer's Responsibilities–Specification.

.7 Vapour retarder, when not employed also as an air-barrier material, shall be a .05-mm (2-mil.)-thick polyethylene sheet conforming to CAN/CGSB-51.34-M86, Vapour Barrier, Polyethylene Sheet for Use in Building Construction.

PART 3–EXECUTION

3.1 Polyethylene Vapour Retarder

.1 Unless otherwise indicated, apply polyethylene to the warm side of exterior walls and ceilings using wide pieces to result in the fewest joints possible.

.2 Make joints over solid bearing and lap 50 mm (2 in.). Do not seal. Staple in position at not more than 200 mm (8 in.) o.c. on all edges and as required to assist installation.

.3 Repair damaged or torn membrane with pressure-sensitive tape recommended for the purpose by the tape manufacturer, unless in the opinion of the architect, damage is too great to be repaired, in which case, a new undamaged membrane must be provided.

.4 At all openings in the exterior walls, such as windows and doors, the polyethylene shall be cut neatly around the opening and fastened against the frame of the opening or as shown.

.5 Where pipe duct shafts are located against exterior walls, install polyethylene and drywall continuously prior to enclosing the shafts.

.6 Polyethylene at ceilings shall be installed with all joints lapped by 100 mm (4 in.) and all joints occurring at framing members.

.7 Extend the ceiling vapour barrier over the top of all interior partitions.

FIGURE 14.2 *(continued)*

CMHC Best Practice Guide Wood Frame Envelopes	INSULATION	Section 07200 Page 3 of 4

3.2 Batt Wall Insulation

.1 Install insulation to maintain continuity of the thermal protection to the building elements and spaces.

.2 Fit the insulation tight to electrical boxes, plumbing and heating pipes and ducts, and around external doors and windows and protrusions.

.3 Cut and trim insulation neatly to fit spaces. Use insulation free of ripped edges. Fill all voids.

3.3 Foam Sealant Installation

.1 Install polyurethane foam sealant at the perimeter of windows, doors and other exterior wall penetrations between the unit and the rough frame in accordance with the manufacturer's instructions.

.2 Ensure continuity of installation against air leakage.

.3 Ensure that window frames, door frames or other items are not damaged or distorted by foam.

.4 Trim neatly, flush with frame.

3.4 Sprayed-in-Place Cellulose Wall Insulation

.1 Mechanically fasten mesh to wall studs with staples.

.2 Cut holes in the mesh where required for insertion of a pneumatic insulation delivery tube.

.3 Pneumatically place insulation in the walls successively between studding to provide minimum specified thermal resistance value.

.4 Keep insulation a minimum of 75 mm (3 in.) from heat-emitting devices such as recessed light fixtures, and a minimum 50 mm (2 in.) from sidewalls of Type A chimneys, according to CAN/ULC-S604-M91 Standard for Factory Built Type A Chimneys, and for Types B and L vents according to CAN/ULC-B149.1 Natural Gas Installation Code and CAN/ULC-149.2 Propane Installation Code.

FIGURE 14.2 *(continued)*

CMHC Best Practice Guide Wood Frame Envelopes	INSULATION	Section 07200 Page 4 of 4

	.5	Remove insulation material spilled during installation, and leave work ready for the application of wallboard.
3.5 Attic Insulation	.1	Install blown-in attic insulation to applicable standard.
	.2	Install preformed vent channels at all truss spaces above the exterior wall.
	.3	Ensure that insulation is not placed on ventilated soffits.
	.4	Use batt insulation for all sloped ceilings.
	.5	Insulate and weatherstrip the attic hatch.

FIGURE 14.2 *(continued)*

CMHC Best Practice Guide Wood Frame Envelopes	SUBGRADE INSULATION	Section 07212 Page 1 of 1

PART 1–GENERAL

1.1	<u>Related Work Specified Elsewhere</u>	.1	Insulation–Section 07200
1.2	<u>Handling and Storage of Materials</u>	.1	Store packaged materials in original undamaged containers with the manufacturer's labels and seals intact. Deliver to the site in sealed packages.
		.2	Store packaged materials and manufactured items off the ground, and cover with weatherproof covering.
		.3	Do not leave polystyrene insulation exposed to sunlight. Cover polystyrene insulation with light-coloured tarpaulins at all times.
1.3	<u>Preparations</u>	.1	Surfaces to receive insulation shall be level.

PART 2–PRODUCTS

2.1	<u>Materials</u>	.1	Extruded polystyrene conforming to the standards set out in CAN/ULC-S701-97 Type 3, Thermal Insulation, Polystyrene, Boards and Pipe Covering.
		.2	Rigid mineral fibre conforming to the standards set out in CAN/ULC-S702-97 Mineral Fibre Thermal Insulation for Buildings.

PART 3–EXECUTION

3.1	<u>Installation</u>	.1	Install insulation as shown on the drawings.
		.2	Butt the insulation boards tightly together.
		.3	Ensure that the insulation is not displaced during backfilling and slab-pouring operations.

FIGURE 14.2 *(continued)*

CMHC Best Practice Guide Wood Frame Envelopes	SEALANTS, GASKETS AND AIR BARRIER MEMBRANES	Section 07900 Page 1 of 4

PART 1–GENERAL

1.1	Related Work Specified Elsewhere	.1	Insulation–Section 07200
		.2	Gypsum Wallboard–Section 09250
1.2	Delivery, Storage and Handling	.1	Deliver and store materials in original wrappings and containers with the manufacturer's seals and labels intact. Protect sealants from freezing, moisture and water.
1.3	Environmental and Safety Requirements	.1	Comply with the requirements of the Workplace Hazardous Materials and Safety Information System (WHMIS) regarding the use, handling, storage and disposal of hazardous materials and regarding labelling and the provision of material safety data sheets acceptable to Labour Canada.
		.2	Conform to the manufacturer's recommended temperatures, relative humidity and substrate moisture content for the application and curing of sealants, including special conditions governing use.
1.4	Guarantee	.1	Provide a written guarantee, signed and issued in the name of the owner, stating that caulking work of this section is guaranteed against leakage, cracking, crumbling, melting, shrinkage, running, loss of adhesion, staining adjacent surfaces or other failure, for a period of three years from the date of Certificate of Substantial Performance.

PART 2–PRODUCTS

2.1	Sealants	.1	Cladding Joint Sealant: One-component acrylic base, solvent curing compound that conforms to the standards set out in CAN/CGSB 19GP-5M.
		.2	Cladding Movement Joint Sealant: One-component elastomeric (urethane) chemical curing compound that conforms to the standards set out in CAN/CGSB 19.13-M87.

FIGURE 14.2 *(continued)*

CMHC Best Practice Guide Wood Frame Envelopes	SEALANTS, GASKETS AND AIR BARRIER MEMBRANES	Section 07900 Page 2 of 4

.3 Interior Air-Barrier Element Sealant: One-component elastomeric (urethane) chemical curing compound that conforms to the standards set out in CAN/CGSB 19.13-M87 (not for use on polyethylene).

.4 Glazing Tape: 100 per cent solid, cross-linked butyl preformed sealant (non-shimmed).

2.2 Sealant Accessories

.1 Primer, where required, shall be as recommended by the sealant manufacturer.

.2 Back-up Materials: polyethylene, urethane, neoprene or vinyl foam

- .1 Extruded closed-cell foam backer rod.
- .2 Size: oversize 30 to 50 per cent
- .3 Bond breakers, where required, shall be polyethylene tape which will not bond to sealant.
- .4 Cleaning solvent shall be coal tar naphtha equal to "Xylol" or as recommended by the sealant manufacturer.

2.3 Gaskets

.1 Sill Gasket: Closed-cell polyethylene strip gasket sized to suit plate.

.2 Air Barrier Component Gasket: Closed-cell polyvinylchloride (PVC) self-adhesive tape 12.7 x 6-mm (1/2 x 1/4-in.) thick or as shown (can only be used in clean, dry conditions).

.3 Air Barrier Component Gasket: Closed-cell Ethylene Propylene Diene Monomer (EPDM) 8 mm (5/16 in.) diameter hollow tubular gasket with stapling flange. (For use when lumber is wet or in freezing weather or with the EASE system.)

2.4 Membranes

.1 Self-adhesive Rubberized Asphalt Membrane: Composite sheet of rubberized asphalt integrally bonded to a film of cross-laminated polyethylene that is a minimum of 1 mm (1/32 in.) thick, for peel and stick application.

FIGURE 14.2 *(continued)*

CMHC Best Practice Guide Wood Frame Envelopes	SEALANTS, GASKETS AND AIR BARRIER MEMBRANES	Section 07900 Page 3 of 4

.2 Primer and mastics as recommended for application by the manufacturer.

.3 Rubber Membrane: EPDM rubber membrane 1.6 mm (1/16 in.) thick.

PART 3–EXECUTION

3.1 Preparation of Joint Surfaces

.1 Examine joint sizes and conditions to establish correct depth-to-width relationship for installation of back-up materials and sealants.

.2 Remove harmful substances, such as dust, rust, oil grease, and other matter which may impair work, from bonding joint surfaces.

.3 Do not apply sealants to joint surfaces treated with sealer, curing compound, water repellent or other coatings unless tests have been performed to ensure compatibility of materials. Remove coatings as required.

.4 Ensure joint surfaces are dry and frost-free.

.5 Prepare surfaces in accordance with the manufacturer's directions.

3.2 Priming

.1 Where necessary to prevent staining, mask adjacent surfaces before priming and caulking.

.2 Prime sides of joints in accordance with the sealant manufacturer's instructions immediately prior to caulking.

3.3 Back-up Material

.1 Apply bond breaker tape where required according to the manufacturer's instructions.

.2 Install joint filler to achieve correct joint depth and shape.

3.4 Sealant Application

.1 Sealant

.1 Apply sealant in accordance with manufacturer's instructions.

.2 Apply sealant in continuous beads.

.3 Apply sealant using a gun with the proper size nozzle.

FIGURE 14.2 *(continued)*

CMHC Best Practice Guide Wood Frame Envelopes	SEALANTS, GASKETS AND AIR BARRIER MEMBRANES	Section 07900 Page 4 of 4

.4 Use sufficient pressure to fill voids and joints solid.

.5 Form the surface of the sealant with a full bead that is smooth, free from ridges, wrinkles, sags, air pockets and embedded impurities.

.6 Tool exposed surfaces to give them a slightly concave shape on exposed joints.

.7 Remove excess compound promptly as work progresses and upon completion.

.2 Curing

.1 Cure sealants in accordance with the sealant manufacturer's instructions.

.2 Do not cover up sealants until proper curing has taken place.

.3 Cleanup

.1 Clean adjacent surfaces immediately, and leave work neat and clean.

.2 Remove excess and droppings as work progresses, using recommended cleaners.

.3 Remove masking tape after initial set of sealant.

3.7 <u>Gasket and Glazing Tape Application</u>

.1 Install gaskets and tape in accordance with the manufacturer's instructions.

.2 Staple the gasket or tape at 300 mm (12 in.) o.c. to maintain it in place while abutting materials are installed.

.3 Ensure that gasket or tape is compressed by the installation of abutting materials.

3.8 <u>Membrane Application</u>

.1 Install membranes where shown and in accordance with the manufacturer's recommendations.

CONTRACTS

A construction contract produces a binding agreement between two parties. It should be written in a way that is fair to both parties, clarifying risks and defining responsibilities. A well-written contract will help avoid disputes and will help resolve any disputes.

Typical contract arrangements can be between:

- Owner and architect
- Owner and contractor
- Owner and project manager
- Contractor and subcontractor
- Contractor and supplier.

There are numerous types and variations of contracts used in the construction industry; some of the more common types used in residential construction are:

- ***Fixed-price contract*** *(also called lump sum or stipulated sum):*
    - This is the most commonly used contract because the parties, with the exception of changes, know exactly what the final cost will be before starting work on the project.
    - Contractors bid on the project based on the information found in the **tender documents,** which includes the instructions to bidders, tender form, agreement, drawings and specifications, and any amendments to the contract documents that took place during the tender period. It is crucial for fixed-price contracts to have a complete and thorough set of drawings and specifications so bids will properly reflect the scope of work and the client expectations.
- ***Cost-plus contracts*** *(also called time and materials):*
    - There are numerous variations of the cost-plus contract; however, it is based on the direct cost of labour and materials used on the project plus a fixed percentage for management, overhead, and profit.
    - Cost-plus contracts are not as widely used as fixed price because the final cost of the project is not known by the client or contractor before the project commences. This can make it difficult for the client to budget costs ahead of time.
    - Cost-plus contracts are generally used when time is tight and a complete set of plans and specifications have not yet been developed. The project is fast-tracked by starting the work without totally completed drawings and specifications.
    - Cost-plus contracts are also used when the complexity of the project makes accurate pricing difficult. Contractors forced to bid a fixed price in these instances may unnecessarily increase their prices to protect themselves from underbidding.
- ***Unit-price contracts:***
    - Unit-price contracts are used when the work can be easily measured or quantified. Usually this type of contract is made between a contractor and subcontractor when suitable. Unit-price contracts are based on measured amounts such as a fixed fee per square foot of work. For example, a drywall contractor may receive a fixed price per square foot of drywall installed and taped, as the walls can be easily measured to verify the number of square feet and total cost. The unit-price amount is based on a minimum amount of work or area; otherwise the project may not be profitable for the subcontractor. In projects that include a higher-than-normal number of arches, bulkheads, and/or corners, additional costs may be applied.

Contract Components (Fixed Price)

There are four main parts of a construction contract (refer to sample contract Figure 14.3 as you read this section):

1. The parties involved in the contract
    - No contract exists without knowledge of all parties (the individuals and/or companies that form the contract).
    - This section includes the names and contact information of the clients.
2. A description of the work to be performed (the description defines the scope of work for the project)
    - This section includes an overview of the work performed, and attaches the drawings and specifications to the contract so that all the work described becomes part of the contract. It is important that the drawings and specifications are dated and initialed as revisions done at a later date may introduce confusion as to which drawings and specifications were part of the original contract.

- From a legal perspective, where there is a conflict in the description between any of the contract documents, the contract is taken as being most accurate, followed by the specifications, then the drawings. For example, if the contract stated that painting was not included and the specifications said it was, a court would most likely find in favour of the contract over the specifications. If the drawings called for Pella windows and the specifications called for Anderson windows, the court would most likely find in favour of the specifications. **If you see a discrepancy in the documents, notify the appropriate person immediately. When cases reach court the outcome is never certain; therefore it should be avoided if at all possible.** Changes and errors do happen, and it is by paying attention to detail that conflicts are avoided.

3. The schedule for the project
   - Clients need to be assured of when their project will be completed. The contractor has to be mindful that purchasers may have sold their existing home and may not have anywhere to go if their new home is not completed on time. Provincial warranty programs may have specific criteria regarding closing dates and penalties (e.g., hotel costs) for not making those dates or for not advising the client in a timely manner that the home will not finish as scheduled.
   - A client's business may have closed for renovation work; every day it is closed it is losing revenue. In this case the contract may include a clause known as **liquidated damages** to limit potential losses should the project not be completed on time. Usually the liquidated damages clause is more prevalent in the Institutional, Commercial, and Industrial sectors. The clause is meant to represent the losses the owner will incur, not to act as a penalty to the contractor that would exceed the owner's losses.
4. Payment schedule
   - The payment schedule lists when payments are to be made for completed work.
   - After usually requiring a deposit upon signing of the contract, the contractor creates a payment schedule based on either dates or milestones. A **milestone** is a specific point in the project that is easily recognized. Inspection approvals are often used as milestones (e.g., framing inspection approval). The contractor would have stated an amount due when the project reaches this stage.
   - The contractor must consider cash flow when developing the payment schedule. The contractor's goal is to ensure that he/she is not paying out more money than he/she is taking in. The contractor also must ensure that no payments are unnecessarily large, to avoid placing the contractor at risk should the client be unable or unwilling to pay for the work completed. For example, it is better to have a number of smaller payments made at shorter intervals than to wait until a large portion of the project is completed before getting paid.
   - A 10% holdback for provincial lien legislation is required, and cannot be written out of a construction contract. Holdback money is placed in trust to ensure that the contractor, subcontractors, suppliers, and employees are paid for the work performed. Lien legislation protects people working in construction from non-payment for work performed; it can be used as a tool to help protect their work until the conflict is resolved (e.g., the owner will not be able to sell the property until the lien is resolved). There is a time limit for when a lien can be placed, usually 45 days after work is substantially completed, ended, or abandoned. Legal advice should be sought regarding lien legislation and the process of applying a lien because it is complex and varies between provinces.
   - **Substantial completion**, also referred to as **substantial performance**, is a legal term used to describe when the building is mostly complete. What "mostly" means can be a bit ambiguous from a legal perspective, but it generally means that the building can be used for its intended purpose. This is used in regards to establishing lien placement timing, and also the time frame for a liquidated damages clause. Substantial completion is typically achieved once a certificate of occupancy has been provided by the municipal authority having jurisdiction, and once the architect has issued a certificate of completion. The date of substantial completion is important because it usually determines the date at which liquidated damages end if the project ran long.

Other Items

There are several other items typically covered in a contract.

Proof of Insurance

Proof of insurance, including liability and workers' compensation, must be provided.

Contingencies and allowances

Contingencies are used for components of the work that may be difficult to determine until work commences.

Allowances are used for products or finishes that may not yet be determined; however, the client wants their approximate cost reflected in the price. For example, if the client wants hardwood flooring but is unsure of the type, the contractor may include an allowance of $8 per square foot. If the client later selects hardwood that is more expensive, he/she will have to pay the difference; if the chosen hardwood is less than $8 per square foot, he/she will receive a credit.

Standards of work

The contract defines who is responsible for protection of the property.

The contract may refer to standard performance guidelines that list tolerances of acceptable work. Many residential provincial warranty programs use this document to set the minimum standards for quality (e.g., the Tarion Warranty Program in Ontario has developed Construction Performance Guidelines that are available on line).

The contract clarifies who is responsible for garbage and cleanup, which is particularly important in subcontract agreements.

Warranties

The contract defines the warranty coverage and time period.

Many provinces have warranty programs for new home construction. For example:

1. Ontario has the mandatory Tarion Warranty Program
2. British Columbia has a mandatory warranty program that uses the Homeowner Protection Office to license builders and to monitor the performance of third-party home warranty insurance systems
3. Atlantic Home Warranty Program
4. Saskatchewan New Home Warranty Program
5. Local home builder member programs such as the Renomark™ Renovator Program offered through numerous Canadian local home building associations in connection with the Canadian Home Builders, Association requires members to offer a minimum two-year warranty.

Permits

The contract explains who is responsible for applying for and receiving various permits.

Dispute mechanism

Many construction contracts now have an alternate dispute resolution (ADR) clause built in, which states that in the event of a dispute the parties will turn to an arbitrator to resolve the dispute. Generally this process is much faster and less costly than the court system.

Change Orders

Construction projects are complex, and rarely does a project remain unchanged from start to finish. If a change is necessary, it may require the drawings to be updated and will require a signed change-order form. The change-order form will list the changes, deletions or additions to the original contract, in detail. The change-order form will also include the extra cost or credit and will provide the new revised contract amount based on the extra or credit. Change-order forms should be completed with any updated sketches or drawings and signed by the client before proceeding with the change; otherwise the change and/or charge may be disputed later on. Change order forms should be numbered so they can be tracked and to ensure that the revised total amount of the contract reflects that sum of all previous changes.

Sample Contract Documents

A series of sample contract documents is shown for your review in Figure 14.3. It includes:

- A sample fixed-cost agreement for a contractor and client
- A sample change-order form
- A sample subtrade contract.

These documents show a simple construction contract. Always consult legal counsel before developing contracts. Many home builder organizations have standardized contracts that they provide to their members; as well, the **Canadian Construction Documents Committee (CCDC)** has developed numerous contracts and forms that can be purchased for use.

CCDC documents include the standard types of contract forms, standard forms, guides, and

FIGURE 14.3
Sample renovation contract

Sample Renovation Contract

Reliable Renovators

123 Anywhere Street, Boom Town, Canada, I4Y 2R1 (789) 765-4321

This **AGREEMENT** made this ____________ day of ____________________ 19 __ ,

BETWEEN :

____

Owner

hereafter known as the "Owner";

AND

RELIABLE RENOVATORS

123 Anywhere Street, Boom Town, Canada, I4Y 2RI

hereafter known as the "Contractor";

WHEREAS the Owner is registered as owner of

____________________________________ ("the land"), and is desirous of having certain renovations done to the residence upon such land;

AND WHEREAS the Contractor has agreed to perform certain services and/or to supply certain materials in connection with such renovations as more particularly described herein;

NOW THEREFORE this agreement witnesseth as follows:

I. The Contractor agrees to perform such services and work, and to supply such materials as are described in the "Contractors responsibilities" in **Schedule A** annexed hereto, the plans/drawings and the specification. The Owner shall be responsible for "Owners responsibilities" described in **Schedule B** annexed hereto. Any alteration or deviation from the specifications described therein will only be executed by written change orders reflecting agreed cost adjustments.

II. The Owner agrees to pay to the Contractor the sum of $__________ for such services and materials which payment shall be as follows:

a) upon execution of this agreement	$________.00
b) upon completion of framing	$________.00
c) upon completion of drywall	$________.00
d) upon completion of air conditioner	$________.00
e) upon completion of deck	$________.00
f) upon completion of contractors final duties	$________.00

II. a) LIEN ACT

The Contractor acknowledges the right and responsibility of the Owner to withhold ten (10) % of each payment as per the provincial Lien Act.

II. b) INTEREST

The Owner further agrees to pay to the Contractor interest on all sums due by the Owner to the Contractor at the rate of ________ % per annum from the date each sum is due to the date of the sum is paid.

III. QUALITY STANDARD

The Contractor agrees to use only good quality materials and that all work shall be completed in a good and professional fashion using only skilled, experienced and competent personnel and sub-trades to perform the work."

Source: Canada Mortgage and Housing Corporation (CMHC). Best Practice Guide: Wood Frame Envelopes, 2006; Renovator's Business Guide, 2005.

FIGURE 14.3 *(continued)*

IV. INSURANCE

The Owner agrees to be responsible for all fire insurance for the home including the new work or any part thereof.

V. PERMITS

The Contractor shall obtain and pay for all necessary building permits.

VI. START AND COMPLETION

The Contractor agrees that it shall commence work and shall proceed expeditiously to complete the work and provide the materials required hereunder. The Contractor shall use its best efforts to complete its responsibilities, subject to delays beyond the contractor's control. The Contractor agrees that it will commence work by ______________ and be fully completed by ____________.

VII. WORKERS COMPENSATION AND LIABILITY INSURANCE

The Contractor shall maintain such insurance and pay required assessments to protect the Contractor and the Owner from claims under the Workers Compensation Act and from any other claims for damages or personal injury, from its operation under this contract.

VIII. INSPECTION OF THE WORK

The Owner or anyone acting on their behalf with their authority shall have the right of access to the work, but in so doing neither the Owner or their agent shall hinder, alter, or interfere with the work being performed by the Contractor.

IX. SIGNS

The Contractor shall be entitled to display a sign on the property until the date of actual completion of all work.

X. WARRANTY

The Contractor shall be responsible for the proper performance of the work only to the extent that the plans and specifications permit such performance, and agrees to correct promptly, at its own expense, defects or deficiencies in the work which appear prior to and during the period of one year from the date of substantial completion. Notice must be in writing to the Contractor, within this time period. The Contractor shall not be responsible for any damage or loss whatsoever as a result of cracking, shifting, breaking or other damage to concrete foundations, walls, floors, caused by frost or soil shifting, expansion or heaving.

XI. ARBITRATION

Disputes between the parties as to the interpretation, application or administration of this agreement or any failure to agree when agreement is called for shall be determined by arbitration and the provisions of The Arbitration Act for the Province of __________________ shall apply. If the parties cannot agree upon the selection of an arbitrator, then either of the parties hereto may apply to the Court of Queen's Bench, in the Province of ________________, for the appointment of a single arbitrator. The award and determination of the arbitrator shall be binding on the parties hereto, and their respective heirs, executors, administrators, successors and assignees. Refusal by one part to submit to arbitration shall entitle the other part to terminate this agreement upon five days written notice to that party.

XII. WCB CLEARANCE

Prior to receiving final payment the Contractor will provide to the Owner a Worker's Compensation Board Clearance certifying that all payments in relation to the contract, as are lawfully required, have been made.

______________________________ ______________________________

Owner Reliable Renovators

FIGURE 14.3 ***(continued)***

Schedule A

Contractor's Responsibilities

Owner

Reliable Renovators

FIGURE 14.3 *(continued)*

Schedule B

Owner's Responsibilities

Owner	Reliable Renovators

FIGURE 14.3 ***(continued)***

Change Order Agreement

Reliable Renovators

123 Anywhere Street
Boom Town, Canada, I4Y 2R1
(789) 765-4321

Date:____________________ Change Order #____________________

Customer____

The following changes are to be made to the job located at____________________

____

____

Description	**Extra**	**Credit**
Total		

Previous Contract Amount	**$**	________
Extra/Credit	**$**	________
Revised Contract Amount	**$**	________

The Customer hereby approves the above changes and authorizes Reliable Renovators to process and to adjust the Contract Price as shown above.

AGREED

Customer ____

Reliable Renovators____________________________________

Copies: 1– Customer 2– Job file 3– Accounting

FIGURE 14.3 ***(continued)***

Sample Subtrade Contract

This agreement made this ____________ day of ____________ 19 __.

Reliable Renovators
123 Anywhere Street
Boom Town, Canada

(hereinafter called the "Contractor")

and

ABC Heating
642 Main Ave.
Boom Town, Canada

(hereinafter called the "Subcontractor")

agree to the terms and conditions listed below.

1. The Subcontractor shall provide the materials and labour to do the following work:

____
____
____
____
____
____
____
____
____

2. The above noted work shall take place at a residence located at:

____
____
____

3. The Contractor agrees to pay the Subcontractor the amount of $__________ in the following manner:

   - 90% upon completion of the job; and
   - 10% forty five days after completion of the job and upon receiving a Declaration from the Subcontractor that all suppliers and workers have been paid.

FIGURE 14.3 ***(continued)***

4. The above price includes all taxes and permits, with the amount of GST/HST specifically being identified.

5. Any changes to the work noted above shall be documented by a written change order signed by both parties.

6. The Subcontractor agrees to provide all manuals for equipment installed and shall provide a one year warranty on all work.

7. The work shall be done in a professional manner and all work shall meet all applicable codes and standards. If the work does not meet current standards, it shall be modified to meet those standards. The Subcontractor agrees to protect the property from damage and agrees to keep the job site clean at all times.

8. The workers agree to maintain good relations with the customer/occupant.

9. If the Subcontractor should fail to complete the work or become insolvent, the Contractor shall have the right to complete the work and deduct the cost of such work from any moneys owing the Subcontractor. If the Subcontractor is unable to do the work, the Contractor may assign the work to another Subcontractor and have no further obligation under this contract. In the case of a dispute, both parties shall agree on a third party who will make a binding decision.

10. The Subcontractor agrees to providing evidence of the following:
    - Comprehensive General Liability Insurance
    - Automobile Liability Insurance
    - WCB coverage both before and after the job

11. Cooperation and flexibility is expected in the execution of the work described in this contract.

Both parties agree to the above and witness such with their signatures below.

______________________	______________________
Reliable Renovators	ABC Heating

bulletins. These documents are intended for use by all construction industry participants, including owners, design professionals, construction managers, legal professionals, contractors, and subcontractors; and are designed to be fair to all parties involved in the agreement. The documents are the most widely used in the industrial, commercial, and institutional (ICI) sectors and may also be used in new custom residential construction, and renovation.

The following are standard CCDC type contracts:

- CCDC 2 is a standard lump-sum (stipulated price) contract
- CCDC 3 is a cost-plus contract
- CCDC 4 is a unit-price type contract
- CCDC 14 is a design build stipulated price contract (between owner and contractor also providing design services).

KEY TERMS

16 Division MasterFormat (p. 359)
Allowance (p. 373)
Canadian Construction Documents Committee (CCDC) (p. 373)
Contingency (p. 373)
Cost-plus contract (p. 371)
Fixed-price contract (p. 371)
Liquidated damages (p. 372)
MasterFormat 2004 Edition (p. 359)
Milestone (p. 372)
Scope of work (p. 358)
Specifications (p. 358)
Substantial Completion/Substantial Performance (p. 372)
Tender documents (p. 371)
Unit-price contract (p. 371)

Chapter 14 Assignment 1: Specifications and Contracts

Answer the questions below using the material found in this chapter. Identify the letter that best answers the question.

_____ 1. What is the main division that describes the project's environmental goals?

a. Division 01
b. Division 02
c. Division 09
d. Division 50

_____ 2. What is the Division that provides the general information that applies to and complements the rest of the divisions in the Masterformat Specification?

a. Division 00
b. Division 01
c. Division 02
d. Division 50

_____ 3. Which of the following statements is correct in the event of a conflict in the contractual documentation for the project?

a. Drawings govern over specifications.
b. Specifications govern over the contract.
c. Contract governs over detail drawing.
d. Detail drawing governs over specifications.

_____ 4. Why are allowances often used in a contract?

a. to protect the client from errors in the contract documents
b. to include costs for products or finishes that may not yet be determined so the project represents a realistic overall cost that can be adjusted based on future selections
c. to provide additional required costs to a contract that will ensure that the contractor is able to meet their targeted profit margins
d. to protect the client from omissions in the contract documents by placing an allowance that covers an estimated amount for the omissions that typically take place during the estimating process

_____ 5. What does the abbreviation CCDC stand for?

a. a specific type of change order form
b. a specific type of specification
c. the Canadian Construction Documents Committee
d. the Construction Committee for Documents and Contracts

_____ 6. How many divisions does the Masterformat 2004 specification have?

a. 15
b. 16
c. 48
d. 50

_____ 7. In terms of specification writing, what does the term "skilfully fitted" represent?

a. a well-written specification that is enforceable because it clearly represents the intention of the specification
b. a poorly written specification that is not enforceable: one person's idea of skilfully fitted may be different than the next person's
c. a specification that is easily understood from a quality perspective
d. a specification that was written to skilfully fit the requirements of the project

_____ 8. What does Division 9 of the Masterformat 2004 edition refer to?

a. finishes
b. openings
c. wood and plastics
d. masonry

_____ 9. What are the components that CSI Construction specifications use for their three-part headings?

a. General, Products, Manufacturers
b. General, Products, Installation
c. Overview, Products, and Execution
d. General, Products, and Execution

_____10. Regarding specifications, what type of format does a production home builder typically follow?

a. They will typically use the 50 division Masterformat set of specifications because they bid competitively for the houses.

b. They will produce a separate options document that outlines the client's selections based on a predefined offering set up by the builder.

c. They will typically use the 16 division Masterformat set of specifications because they have always used that traditionally.

d. They will use the construction notes on the drawings and therefore do not require any further information.

_____11. What type of contract requires a total price for the entire project that is quoted through the tender process?

a. construction management contract

b. lump sum contract

c. cost plus contract

d. unit cost contract

_____12. What type of contract is priced based on the cost of job as it is being constructed?

a. lump sum contract

b. fixed cost contract

c. cost plus contract

d. unit cost contract

_____13. What type of contract is based on measurements such as "$1.50 per square foot of drywall installed"?

a. lump sum contract

b. total price contract

c. cost plus contract

d. unit cost contract

_____14. What do we call it when something new is added to an existing contract between the client and contractor?

a. subcontract agreement

b. lump sum price

c. change order

d. addendum

_____15. What is the most common type of contract used in construction?

a. lump sum contract

b. time and materials contract

c. cost plus contract

d. unit price contract

_____16. Why should a holdback be included in the payment schedule?

a. to legally protect the owners against faulty workmanship

b. to legally protect the contractor from the owner

c. to legally fulfil the requirements of lien legislation

d. to legally protect the suppliers from the contractor

_____17. What would the holdback requirement be for a milestone payment of $10,000 for framing?

a. $500

b. $1,000

c. $50

d. $100

_____18. What is the time limit for placing a lien on a property?

a. 20 days after substantial completion or when work has ended

b. 30 days after substantial completion or when work has ended

c. 45 days after substantial completion or when work has ended

d. 60 days after substantial completion or when work has ended

_____19. In what type of project is the liquidated damages clause typically used?

a. in new residential homes to encourage the contractor to complete the work on time
b. in residential renovations to encourage the contractor to complete the work on time
c. in residential restoration work to encourage the contractor to complete the work on time
d. in commercial work to encourage the contractor to complete the work on time

_____20. What does the term "substantial completion" refer to?

a. work that is completed
b. work that is complete to where the building can be used for its intended purpose
c. work that has all the structural and building systems completed but where the finishes have not yet begun to be installed
d. work that is mostly complete but has not had occupancy granted by the town or municipality

Chapter 14 Quiz Review: Specifications and Contracts

Answer the questions below using the material in this chapter. Follow the directions in some of the questions regarding which documents to use from the chapter in answering questions. Identify the letter that best answers the question.

_____ 1. What type of contract is a CCDC 2?

a. cost plus contract
b. construction management
c. unit cost
d. lump sum

_____ 2. What does Division 26 of the Masterformat 2004 edition refer to?

a. masonry
b. finishes
c. electrical
d. plumbing

_____ 3. What does Division 4 of the Masterformat 2004 edition refer to?

a. masonry
b. finishes
c. electrical
d. plumbing

_____ 4. What are specifications designed to do for construction drawings?

a. replace drawings
b. complement drawings
c. circumvent drawings
d. provide missing dimensions

_____ 5. What is a payment schedule usually based on?

a. completion of milestones or key components of work
b. always four equal payments based on the total contract amount
c. percentage of the project, always in payments made in 10% increments
d. seasons of the year

_____ 6. What standard must the blown-in fibreglass attic insulation conform to, based on the sample specifications provided in the chapter?

a. CAN/ULC-S703
b. CAN/ULC-S702-97
c. CSGB 51-GP-44MP
d. CSGB 51-GP-45MP

_____ 7. According to the sample specifications provided in the chapter, where and how much must the polyethylene vapour retarder be lapped where it meets at the ceiling?

a. 100 mm and all joints occurring at a framing member
b. 150 mm and all joints occurring at a framing member
c. 200 mm and all joints occurring at a framing member
d. 400 mm and all joints occurring at a framing member

_____ 8. According to the sample specifications provided in the chapter, what type of insulation is to be installed around the window frames?

a. polyurethane foam insulation
b. fibreglass insulation
c. cellulose insulation
d. rigid insulation

_____ 9. According to the sample specifications provided in the chapter, what type of material is used for the sill gasket?

a. fibreglass
b. polyurethane foam
c. closed-cell polyethylene strip
d. rigid sprayed rock wool strip

_____10. According to the sample specifications provided in the chapter, how long is the caulking work required to carry a written guarantee for?

a. All caulking work carries a written guarantee of 1 year.
b. All caulking work carries a written guarantee of 3 years.
c. All caulking work carries a written guarantee of 5 years.
d. All caulking work carries a written guarantee of 10 years.

_____11. According to the sample specifications provided in the chapter, where should the pre-formed vent channels be placed?

a. between the studs below the wall plates
b. between the trusses above the wall plates
c. between the joists above the wall plates
d. between the sill plate and foundation wall

_____12. What is the amount of holdback required by the *Lien Act*, based on the sample renovation contract and change order agreement provided in the chapter?

a. 10 percent
b. 15 percent
c. 20 percent
d. 50 percent

_____13. What type of contract is the sample renovation contract included in the chapter?

a. cost plus contract
b. lump sum
c. unit cost
d. construction management

_____14. What is the contractor entitled to display on the property until completion of the work, based on the sample renovation contract and change order agreement provided in the chapter?

a. building permit
b. sign
c. no parking or blocking of driveway posting
d. fire exit sign

_____15. Who is responsible for obtaining the building permits, based on the sample renovation contract provided in the chapter?

a. the architect
b. the engineer
c. the contractor
d. the owner

_____16. From the sample renovation contract, what is the title for the clause in the contract that is used when there is a disagreement between the owner and contractor?

a. mediation clause
b. dispute clause
c. disagreement clause
d. arbitration clause

_____17. How long will the warranty for the work be, based on the sample subtrade contract provided in the chapter?

a. one year
b. two years
c. three years
d. four years

_____18. What is the contractor's responsibility regarding payment to the subcontractor based on the sample subtrade contract provided in the chapter?

a. The contractor pays 90% of the subcontractor's work upon completion of the job and the contractor pays the remaining 10% forty-five days after completion of the work and upon receiving a declaration that all suppliers and workers have been paid.
b. The contractor pays 80% of the subcontractor's work upon completion of the job and the contractor pays the remaining 20% forty-five days after completion of the work and upon receiving a declaration that all suppliers and workers have been paid.
c. The contractor pays 70% of the subcontractor's work upon completion of the job and the contractor pays the remaining 30% forty-five days after completion of the work and upon receiving a declaration that all suppliers and workers have been paid.
d. The contractor pays 50% of the subcontractor's work upon completion of the job and the contractor pays the remaining 50% forty-five days after completion of the work and upon receiving a declaration that all suppliers and workers have been paid.

_____19. What does the change order agreement include regarding costs based on the sample change order agreement provided in the chapter?

a. The change order agreement includes the cost of the change order only.

b. The change order agreement includes the original contract amount only.

c. The change order agreement includes the most recent previous contract amount (includes running total of previous changes) plus the extra/credit amount, thus providing the owner with an up-to-date revised contract amount.

d. The change order agreement includes the most recent previous contract amount plus a detailed listing of all previous changes plus the new change listing, thus providing the owner with a revised contract amount.

_____20. Who is required to sign the change order agreement, based on the sample change order agreement provided in the chapter?

a. The change order agreement must be signed by the client.

b. The change order agreement must be signed by the owner.

c. The change order agreement must be signed by both the owner and client.

d. The change order agreement does not need to be signed to be valid.

15 An Introduction to Drawings and Details for Renovation and Multiple-Unit Buildings

LEARNING OBJECTIVES

After completing this chapter, you will be able to:

- Read and interpret drawings for renovation projects.
- Analyze and distinguish some of the key elements of a renovation project with regards to tying into an existing structure.
- Read and interpret drawings for multiple-unit residential projects.
- Analyze and distinguish some of the key elements of a multiple-unit building with regards to fire separations, noise reduction, and constructability.

INTRODUCTION

This chapter introduces you to renovation drawings and multiple-unit buildings. The skills you have learned for understanding new, single-unit construction drawings can be applied here, but renovation and multiple-unit drawings each have their own unique features and challenges.

RENOVATION DRAWINGS

Success in renovations requires experience and a working understanding of buildings that have been built during different time periods. You must be able to identify the type of construction that was used and analyze how you can tie more current methods of construction to an existing structure. One of the challenges of renovations is the fact that no two projects are alike and, no matter how much experience you have, surprises lurk behind existing walls and ceilings.

Renovations are more labour-intensive than new construction, due mainly to the extra work required to demolish and remove parts of the existing structure. Tying in and matching new components with old can also be challenging given that many products may no longer be produced. Renovating also provides an opportunity to upgrade the building envelope and make improvements to energy efficiency.

When reading and interpreting renovation drawings use the following steps:

1. Review the drawings, looking for unusual details and intersections. Earlier in the text you were asked to try to visualize the finished project in three dimensions when reviewing drawings. For renovation drawings, try to also visualize what parts of the building will remain untouched and how the new components will intersect and tie to the existing components.
2. Where renovation drawings require the removal of walls or the widening of openings, review the drawings for the types of beams and support columns that are shown or listed, and visualize how they will be installed while maintaining the integrity of the structure. Remember, with a renovation project you must work around the existing structure, which is significantly different from new construction where you are building the entire structure from scratch. You must make allowances for temporarily supporting the existing structure while work proceeds. For example, if you are opening up an exterior load-bearing wall to make a connection between the existing house and a new addition, you may simply construct a temporary support wall. However, in other cases you may need to provide temporary support by **needling beams,** which is the placing of temporary beams supported on columns in

cutouts above a floor to temporarily support masonry walls until a permanent beam is placed below the masonry.

3. Ask yourself if any of the existing materials are to be saved and used elsewhere in the project. It is sometimes difficult, if not impossible, to match materials in renovations. This problem can sometimes be overcome by reusing materials where a match is not possible. For example, if the rear of the house is being opened for an addition and an addition is being added over the garage that connects to the front of the house, it may make sense to save and clean the brick from the rear for use at the front of the house. This will ensure that there is a close match in texture and colour where the addition over the garage meets the front of the house; it is generally not as important to have a perfect match at the side and rear. Don't assume that the designer has necessarily thought all these details through. The reuse of the existing brick may be listed on the elevation drawings and/or in the specifications, or the specifications may simply state "match existing brick," which is easier said than done.
4. As you review the drawings, make a list of questions and concerns that are not clear from the drawings. This is known as a request for information (RFI).
5. Visit and inspect the site to try to answer your questions. Inspect the house from the basement through the roof. Don't assume anything; check as much of the structure as possible. If you will be involved in structural work, check which walls are load-bearing and non–load-bearing. Determine the type of construction used (e.g., wood frame, solid masonry) and the materials and methods of installation used. For example, are the walls drywall or plaster? If they are plaster, are they wood lath and plaster or gyproc lath and plaster? Check to see whether the site is accessible and whether there is space for storage and disposal boxes. All of these factors will affect your costs and methods of installation, and most of these items will not be readily shown on your drawings. Also, you may discover that the drawings indicate things that are not correct or are dimensioned incorrectly. Note any errors and inform the designer. (Recall in Chapter 14 we discussed the importance of accurate specifications so that contractors all bid on the same thing.)
6. Check the electrical and mechanical systems. Remember renovation specifications may call for using an existing system. Make sure that the existing system will be able to handle the additional loads and that there are no code issues with expanding the system. For example, is the electrical service large enough to handle its new requirements? Will the electrical panel be large enough? Is the wiring up to date? If not, upgrading must be done while the walls are open. You may also want to ask questions such as the following: Is the furnace appropriately sized for the new addition? How will the new ductwork tie into the existing? Is the existing water supply sufficient? Are the existing supply lines up to date? These are just some examples of potential challenges to consider when working on renovation projects; each project will have its own unique circumstances. Be sure to include prices for any extra options that might be required in your quote to the client.

Sample Renovation Drawings

Figures 15.1 to 15.9 show a sample set of drawings for a one-storey addition. The renovation work also includes a number of interior alterations, including the installation of a new bathroom and kitchen. Most of the new construction elements of an addition are the same as those discussed throughout the text. However, wherever there are intersections between new and old, a number of considerations will be necessary. Usually you can interpret this information directly from the drawings but, as stated above, confirm your interpretations at the actual site.

In the drawings, note these key features:

- These drawings are in imperial measurement.
- The site plan includes information for zoning about coverage, gross floor area, height, and setbacks for both the existing and proposed structures (Figure 15.1).
- The site plan clearly indicates what is existing and what is proposed (Figure 15.1).
- The drawings use a dotted/dashed line on the floor plans to indicate the location where existing walls are to be removed (Figures 15.2 and 15.3).
- Detailed dimensions are provided for addition on foundation plan; dimensions of existing building are not shown (Figure 15.2).
- Location and size of existing furnace are shown, as well as connections of new supply air ducts (Figure 15.2).

FIGURE 15.1

Khalmur renovation drawings, site plan and zoning information.

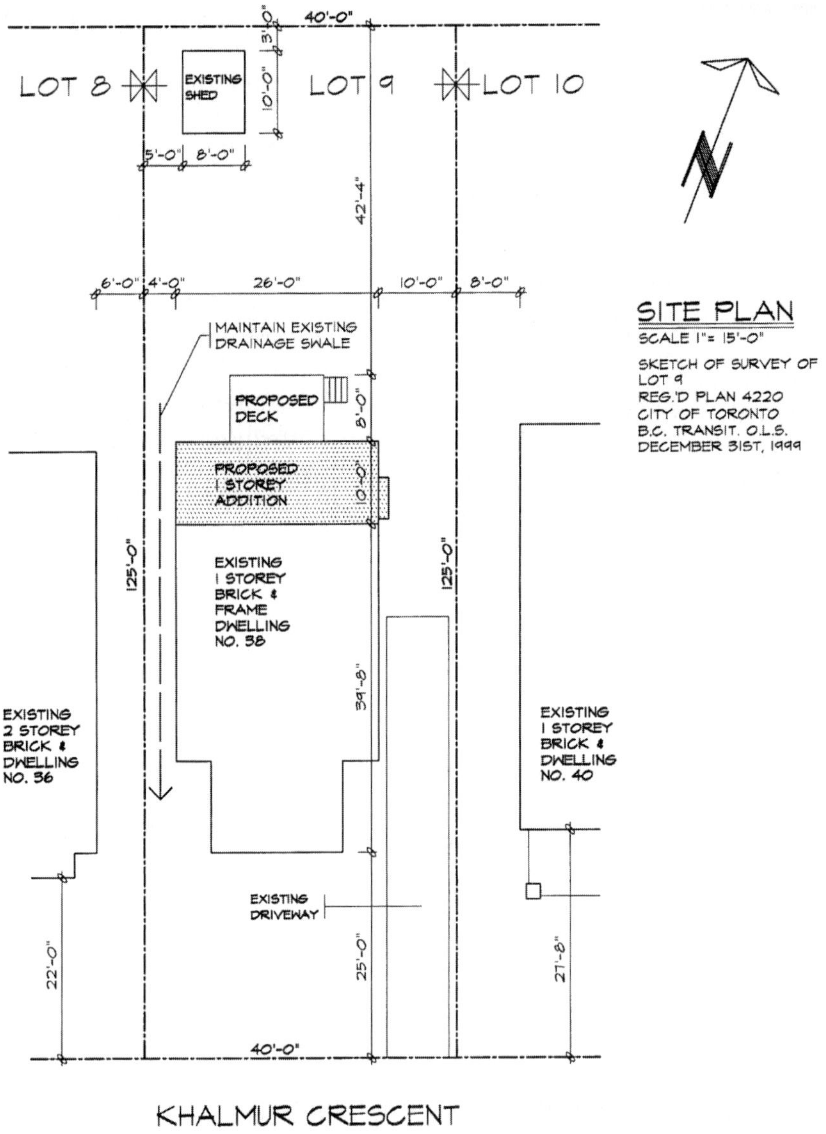

ZONING	LOT NO:	PLAN NO:	LOT AREA	LOT FRONTAGE	LOT DEPTH
R2 Z0.6	LOT 9	4220	6250.00 S.F.(580.64)	50.00' (15.24)	125.00' (38.10)

DESCRIPTION	EXISTING	ADDITION	TOTAL	%	ALLOWED	%
LOT COVERAGE	931.26 S.F. (86.52)	260.00 S.F. (24.15)	1191.00 S.F. (110.65)	19.0	---------	
GROSS FLOOR AREA	931.26 S.F. (86.52)	260.00 S.F. (24.15)	1191.00 S.F. (110.65)	19.0	3750.00 S.F. (348.39)	60.0
LANDSCAPED AREA	------------	------------	------------		---------	
NO. OF STORIES HEIGHT	1 STOREY 14'-11" (4.55)	1 STOREY 14'-11" (4.55)	1 STOREY 14'-11" (4.55)		32'-10" (10.00)	
WIDTH	26'-0" (7.93)	26'-0" (7.93)	26'-0" (7.93)		------------	
DEPTH	39'-8" (12.09)	10'-0" (3.50)	49'-8" (15.14)		55'-9" (17.00)	
PARKING	------------	------------	------------		------------	

SETBACKS	EXISTING	PROPOSED
FRONT YARD	25'-0" (7.62)	25'-0" (7.62)
REAR YARD	60'-4" (18.39)	42'-4" (12.90)
INTERIOR SIDE (east)	10'-0" (3.50)	10'-0" (3.50)
INTERIOR SIDE (west)	4'-0" (1.20)	4'-0" (1.20)
EXTERIOR	------------	------------

NOTE: ZONING RESTRICTIONS VARY IN EVERY MUNICIPALITY. CONTACT YOUR LOCAL MUNICIPAL OFFICE FOR SPECIFIC SETBACKS AND OTHER LIMITATIONS IN YOUR AREA.

Source: Toronto Area Chief Building Officials Committee

FIGURE 15.2

Khalmur renovation drawings, basement plan.

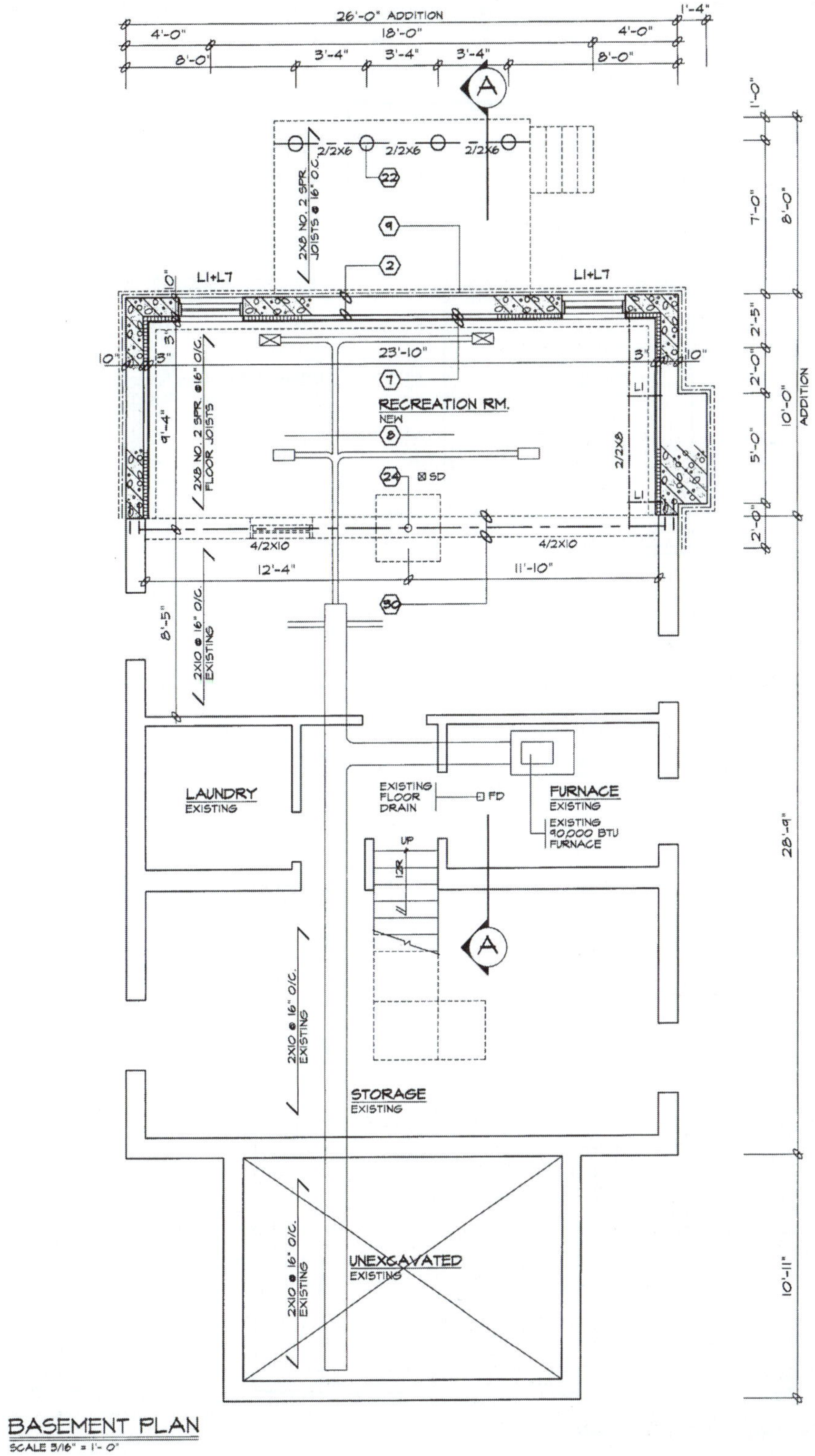

Source: Toronto Area Chief Building Officials Committee

FIGURE 15.3

Khalmur renovation drawings, ground floor plan.

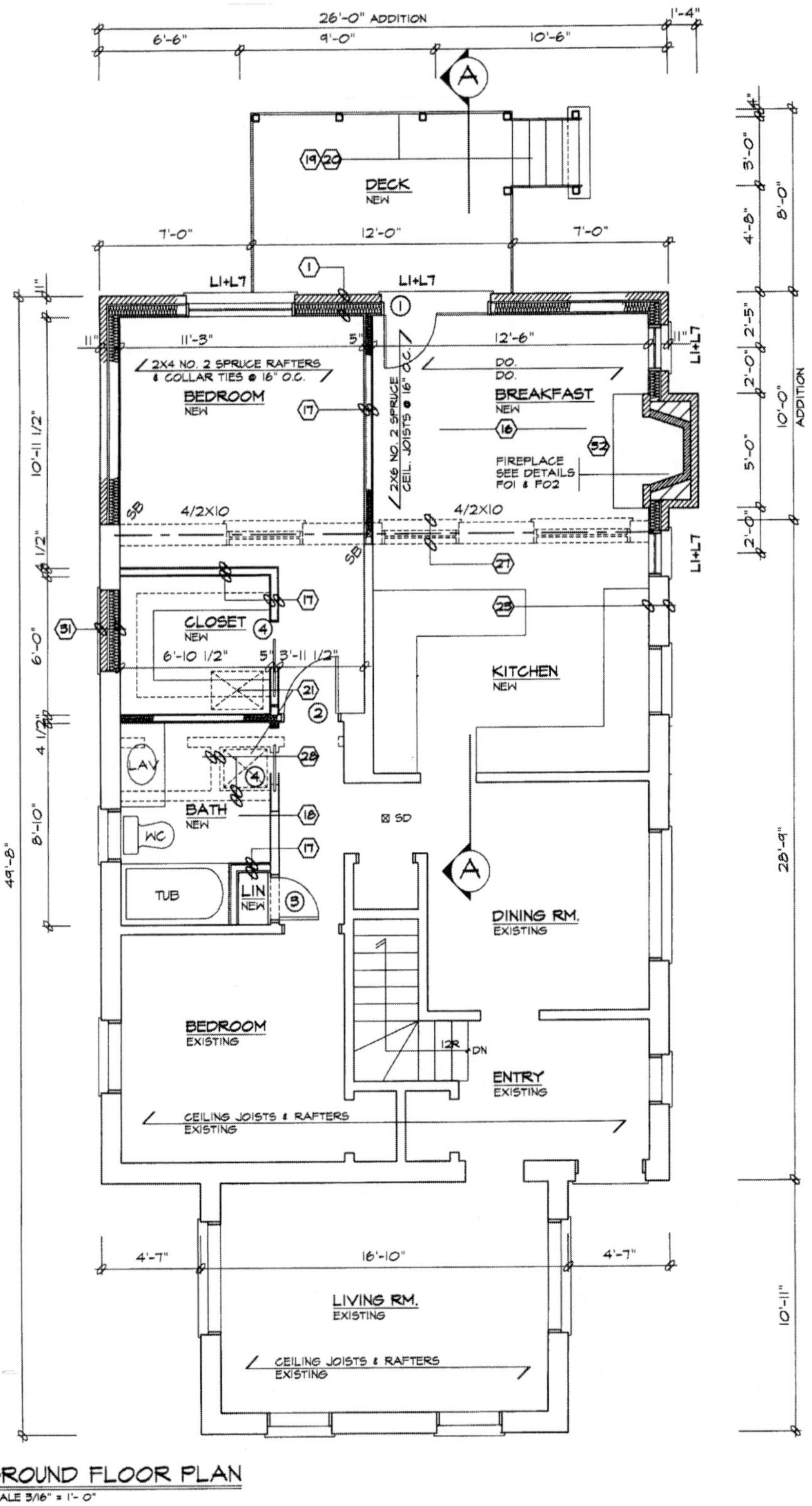

Source: Toronto Area Chief Building Officials Committee

- Review the position of the existing floor joists and visualize how you would need to support the existing floor when you remove the foundation wall at the rear of the house (Figure 15.2).
- Do the drawings show how the roof is framed? Can you tell whether the existing roof is a cut or truss roof? What actions might you have to take when opening the back wall of the first floor of the house? (Refer to Figures 15.3, 15.4, 15.5, and 15.6.)
- Review the elevations to determine if there are any potential areas where brick removed from the rear wall could be reused (Figure 15.6).

FIGURE 15.4

Khalmur renovation drawings, north elevation.

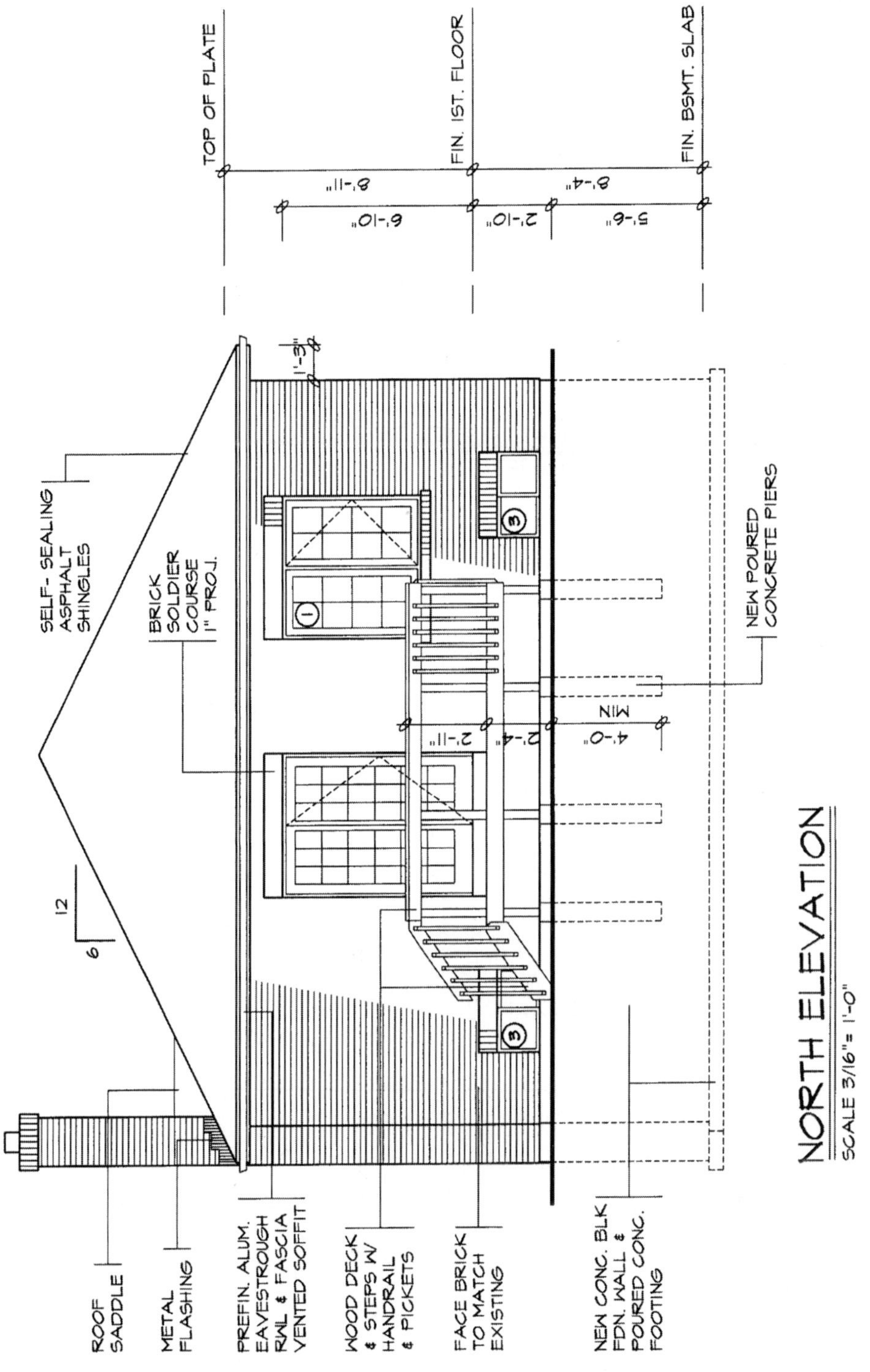

Source: Toronto Area Chief Building Officials Committee

FIGURE 15.5
Khalmur renovation drawings, east elevation.

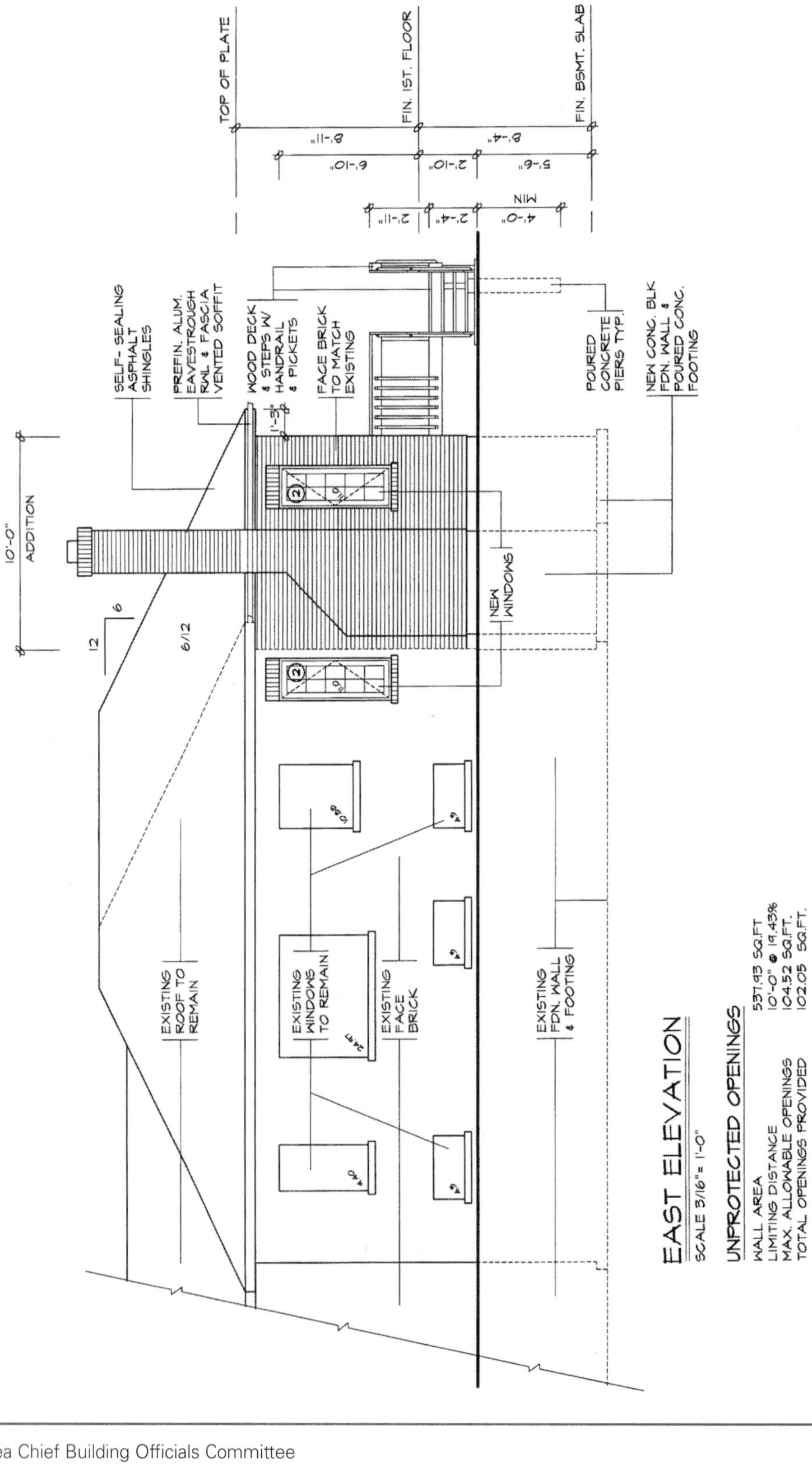

Source: Toronto Area Chief Building Officials Committee

FIGURE 15.6
Khalmur renovation drawings, west elevation.

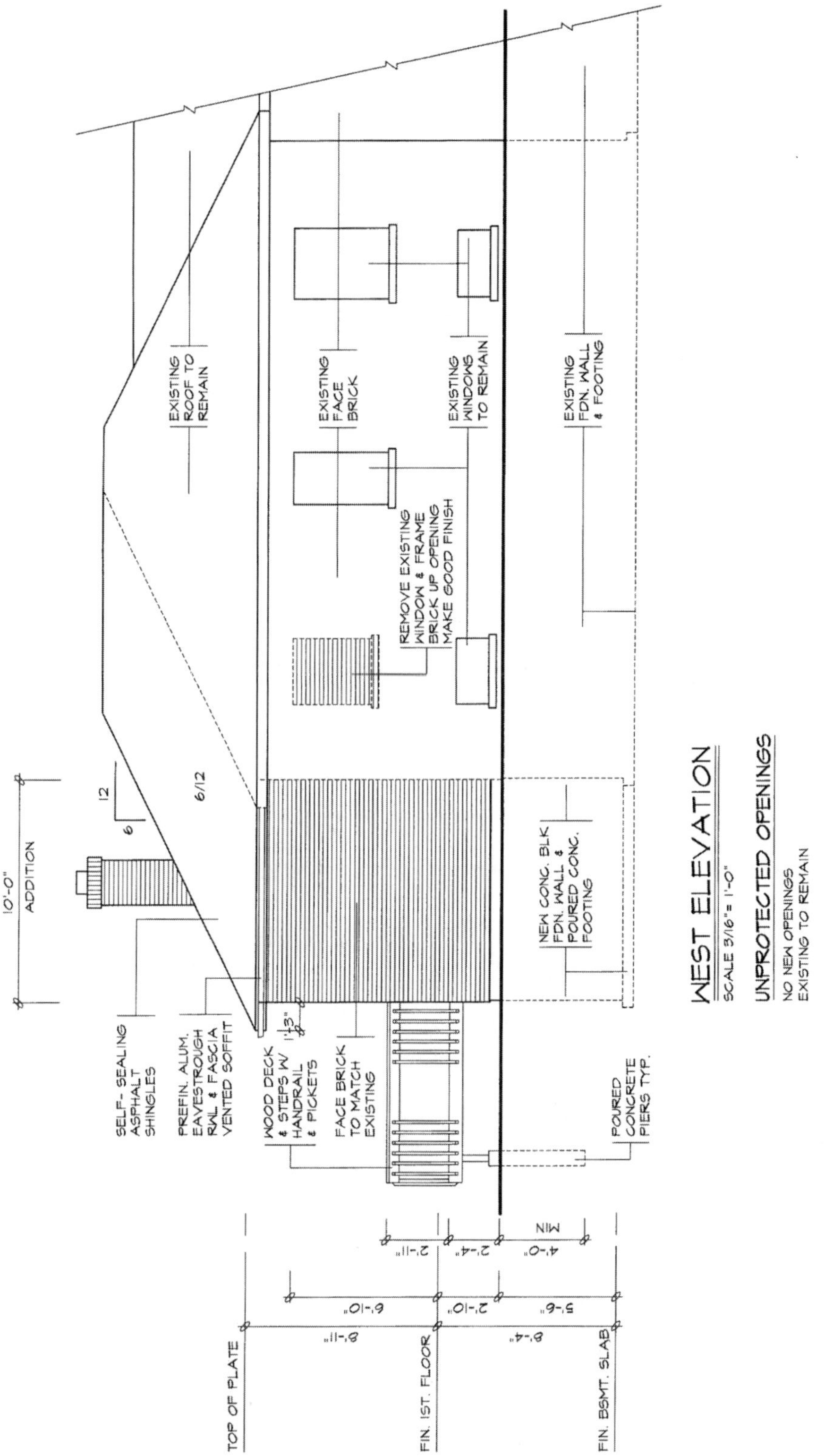

Source: Toronto Area Chief Building Officials Committee

- Review the drawings and consider the connection and tie-in points between the old and new structural connections and finishes.
- Notice the hexagon symbols referencing the construction notes/specifications. Check item 30 in the specification for information about the demolition of part of the foundation wall (Figure 15.8).
- Review the room schedule for the different room finishes used (Figure 15.9).

FIGURE 15.7

Khalmur renovation drawings, cross-section.

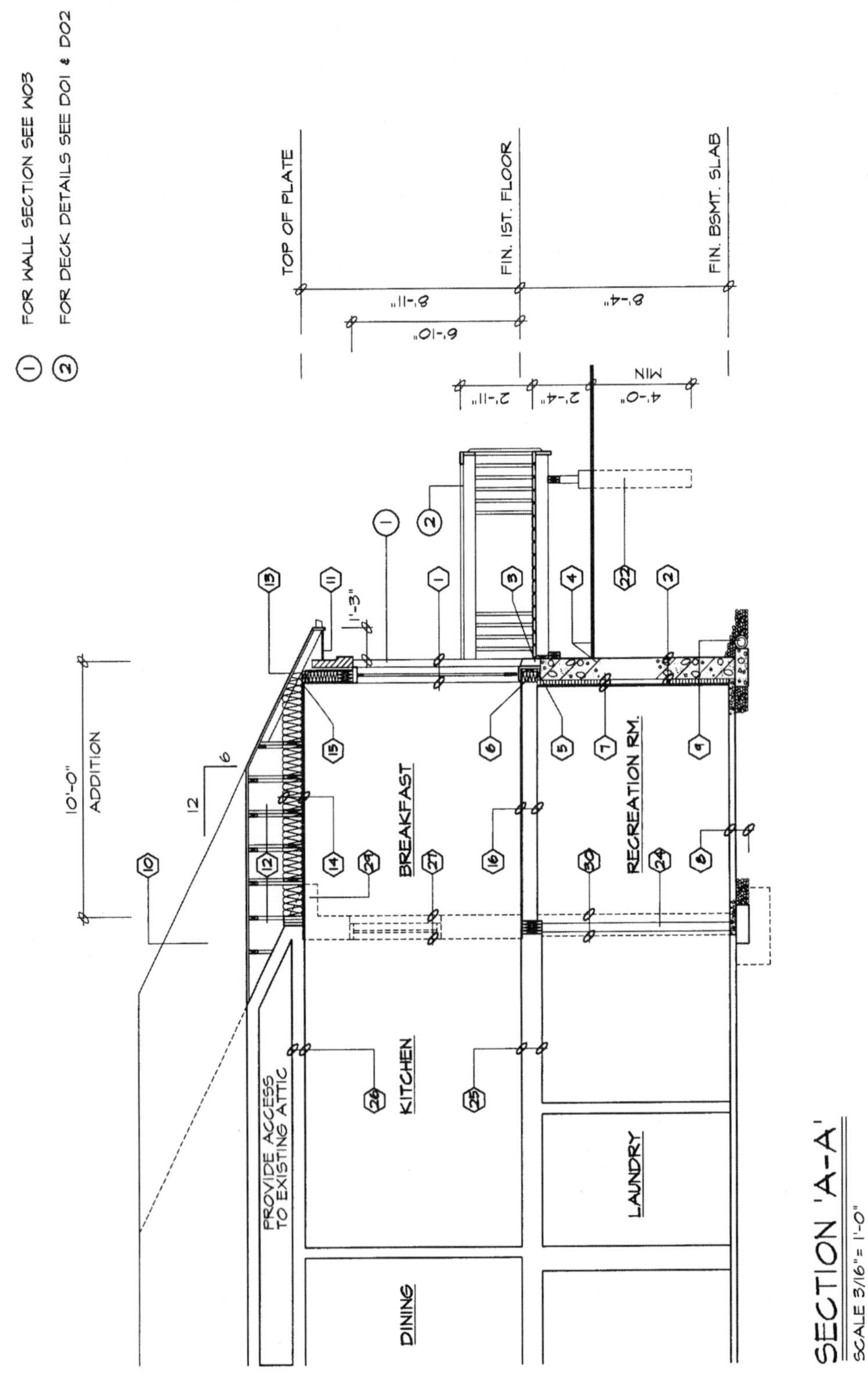

Source: Toronto Area Chief Building Officials Committee

FIGURE 15.8

Khalmur renovation drawings, construction specifications.

CONSTRUCTION SPECIFICATIONS

(1) BRICK VENEER WALL

4" FACE BRICK, 1" AIR SPACE
0.03 THICK X 7/8" WIDE
GALVANIZED METAL TIES
INSTALLED W/ GALVANIZED
SPIRAL NAILS OR SCREWS
32" O.C. HORIZ., 16" O.C. VERT.
SHEATHING PAPER, LAYERS
TO OVERLAP EACH OTHER
EXTERIOR TYPE SHEATHING
2"x6" WOOD STUDS @ 16" O.C.
R 17 BATT INSUL. IN CONTINUOUS
CONTACT W/ EXTERIOR SHEATHING
CONTINUOUS AIR / VAPOUR BARRIER
1/2" INTERIOR DRYWALL FINISH
DOUBLE PLATE @ TOP
SOLE PLATE @ BOTTOM

(2) FOUNDATION WALL

BITUMINOUS DAMPPROOFING ON
MINIMUM 1/4" PARGING ON
CONCRETE BLOCK FDN. WALL
TOP BLOCK COURSE FILLED
W/ MORTAR OR CONCRETE
PROVIDE PARGING COVED OVER
18"X 6" POURED CONC. FOOTING
TO BEAR ON UNDISTURBED SOIL
PROVIDE DRAINAGE LAYER
- MIN. 3/4" MINERAL FIBRE
INSULATION W/ A DENSITY OF
NOT LESS THAN 3.6 LB./FT. OR
- MIN. 4" OF FREE DRAINING
GRANULAR MATERIAL OR
- A B.M.E.C. APPROVED
DRAINAGE LAYER MATERIAL

(3) BRICK VENEER @ FDN. WALL

20 MIL POLY FLASHING MINIMUM
6" UP BEHIND SHEATHING PAPER
WEEP HOLES @ MIN. 2'-7" APART

(4) GRADE

SLOPE GRADE AWAY FROM
BUILDING FACE & PROVIDE
SEMI-SOLID BLOCK COURSE
AT OR BELOW GRADE LEVEL

(5) SILL PLATE

2"X6" SILL PLATE FASTENED
TO FOUNDATION WALL WITH
MIN. 1/2" DIA. ANCHOR BOLTS
EMBEDDED MIN. 4" IN CONCRETE
@ 7'-10" O/C. MAX. & PROVIDE
CAULKING OR GASKET BETWEEN
PLATE & FOUNDATION WALL

(6) FLOOR INSULATION

CONTINUOUS HEADER JOIST WITH
R 17 BATT INSULATION, EXTEND
VAPOUR / AIR BARRIER & SEAL
TO JOIST AND SUBFLOOR

(7) FOUNDATION INSULATION

1/2" INTERIOR DRYWALL FINISH
2"X3" WOOD STRAPPING @ 16" O/C.
MIN. R8 INSULATION W/ 6 MIL POLY
AIR / VAPOUR BARRIER FULL HEIGHT.
MOISTURE BARRIER TO HEIGHT OF
EXTERIOR GRADE BETWEEN
FOUNDATION WALL & WOOD FRAMING

(8) BASEMENT SLAB

3" POURED CONCRETE SLAB
(3600 PSI CONC. STRENGTH)
4" CRUSHED STONE BELOW

(9) DRAINAGE

4" DIA. WEEPING TILE W/
6" CRUSHED STONE COVER

(10) ROOF CONSTRUCTION

20 YEAR ASPHALT SHINGLES ON MIN.
3/8" EXTERIOR PLYWOOD SHEATHING
ON APPROVED ROOF TRUSSES OR
CONVENTIONAL FRAMING (SEE PLANS)
USE 'H' CLIPS IF 24" O.C. SPACING

(11) OVERHANG CONSTRUCTION

PREFINISHED ALUMINUM FASCIA,
EAVESTROUGH & RAIN WATER LEADERS
TO MATCH EXISTING FINISHES. PROVIDE
DRIP EDGE AT FASCIA & VENTED SOFFIT
EXTEND DOWNSPOUTS TO GRADE LEVEL

(12) ROOF VENTILATION

1:300 OF THE INSULATED CEILING
AREA UNIFORMLY DISTRIBUTED.

(13) EAVES PROTECTION

EAVES PROTECTION MEMBRANE TO
EXTEND FROM THE EDGE OF THE
ROOF, 36" UP THE SLOPE BUT NOT
LESS THAN 12" BEYOND THE INTERIOR
FACE OF THE EXTERIOR WALL

(14) CEILING CONSTRUCTION

5/8" INTERIOR DRYWALL FINISH
CONTINUOUS AIR / VAPOUR BARRIER
W/ MINIMUM R 31 BATT INSULATION

(15) WALL/CEILING INSULATION

CARRY MIN. R12 INSULATION
TO COVER THE INTERIOR FACE
OF THE EXTERIOR WALL

(16) FLOOR CONSTRUCTION

5/8" T&G PLYWOOD SUBFLOOR
2X8 FLOOR JOISTS @ 16" O/C.
FLOOR JOISTS BRIDGED W/
CONTINUOUS 1"X3" STRAPPING OR
2 ROWS OF 2"X2" CROSS BRIDGING
OR SOLID BLOCKING

(17) INTERIOR STUD PARTITION

1/2" DRYWALL FINISH BOTH SIDES OF
2"X4" WOOD STUDS @ 16" O/C
2 TOP PLATES & 1 BOTTOM PLATE
PROVIDE SOUND ATTENUATION
INSULATION IN BATHROOM WALLS
& WHERE INDICATED ON PLAN

(18) MECHANICAL VENTILATION

PROVIDE MIN. 1 AIR CHANGE
PER HOUR IN ROOMS SPECIFIED
TO BE MECHANICALLY VENTED
80 CFM FOR BATH PRIMARY VENTS

(19) STAIRS INTERIOR/EXTERIOR

MAXIMUM RISE	= 7 7/8"
MINIMUM RISE	= 4 7/8"
MINIMUM RUN	= 8 1/4"
MAXIMUM RUN	= 14"
MINIMUM TREAD	= 9 1/4"
MAXIMUM TREAD	= 14"
MAXIMUM NOSING	= 1"
MINIMUM WIDTH	= 2'-10"
MINIMUM HEADROOM	= 6'-5"

(20) GUARDS

INTERIOR LANDINGS	= 2'-11"
EXTERIOR BALCONY	= 3'-6"
INTERIOR STAIRS	= 2'-11"
EXTERIOR STAIRS	= 2'-11"
MAX. BETWEEN PICKETS	= 4"

GUARD HEIGHT IF
DECK TO GRADE IS:

GREATER THAN 5'-11"	= 3'-6"
5'-11" OR LESS	= 2'-11"

NO MEMBER OR ATTACHMENT
BETWEEN 4" & 2'-11" HIGH
SHALL FACILITATE CLIMBING

(21) ATTIC ACCESS

PROVIDE ATTIC ACCESS
MIN. 20"X 28" W/ INSULATION
& WEATHER STRIPPING

(22) PIERS

PROVIDE 8" DIA. SONO TUBE
FOR POURED CONCRETE PIERS
MINIMUM 4'-0" BELOW GRADE

(23) EXISTING SOLID MASONRY
EXTERIOR WALLTO REMAIN.

(24) 3 1/2" DIA. PIPE COLUMN W/
6X6X3/8" TOP & BOTTOM PLATE
38"X38"X16" CONCRETE FOOTING

(25) EXISTING FLOOR STRUCTURE
TO REMAIN.

(26) EXISTING CEILING STRUCTURE
TO REMAIN.

(27) REMOVE EXISTING EXTERIOR WALL
AS SHOWN DOTTED

(28) REMOVE EXISTING INTERIOR STUD
PARTITIONS AS SHOWN DOTTED

(29) REMOVE EXISTING ROOF OVERHANG
AS SHOWN DOTTED

(30) REMOVE EXISTING FOUNDATION WALL
AS SHOWN DOTTED

(31) REMOVE EXISTING WINDOW & FRAME
MAKE GOOD OPENING W/ BRICK TO
MATCH EXISTING ON THE EXTERIOR

(32) INSTALL A CARBON MONOXIDE
DETECTOR CONFORMING TO
CAN/CGA-6.19 OR UL 2034

Source: Toronto Area Chief Building Officials Committee

FIGURE 15.9

Khalmur renovation drawings, schedules.

ROOM FINISH SCHEDULE

RM. NO.	ROOM NAME	FLOOR		BASE		WALLS		CEILING			REMARKS
		MATERIAL	FINISH	MATERIAL	FINISH	MATERIAL	FINISH	MATERIAL	FINISH	HEIGHT	
	FIRST FLOOR										
1.	KITCHEN	CERAMIC TILE	-----------	WOOD	PAINT	DRYWALL	PAINT	DRYWALL	PAINT	8'-11"	
2.	BREAKFAST	WOOD	STAIN	WOOD	PAINT	DRYWALL	PAINT	DRYWALL	PAINT	8'-11"	MAPLE TO MATCH EXISTING
3.	BEDROOM	WOOD	STAIN	WOOD	PAINT	DRYWALL	PAINT	DRYWALL	PAINT	8'-11"	MAPLE TO MATCH EXISTING
4.	CLOSET	WOOD	STAIN	WOOD	PAINT	DRYWALL	PAINT	DRYWALL	PAINT	8'-11"	MAPLE TO MATCH EXISTING
5.	BATH	CERAMIC TILE	-----------	WOOD	PAINT	DRYWALL	PAINT	DRYWALL	PAINT	8'-11"	
	BASEMENT										
6.	REC. ROOM	CONC.	CERAMIC TILE	WOOD	PAINT	DRYWALL	PAINT			7'-8"	

DOOR SCHEDULE

NO.	TYPE	SIZE	QTY.	REMARKS
1.	EXTERIOR	5'-0"X 6'-8"	1.	FRENCH DOOR
2.	SLAB	2'-6"X 6'-8"	1.	800 SERIES
3.	SLAB	2'-0"X 6'-8"	1.	800 SERIES
4.	POCKET DOOR	2'-0"X 6'-8"	2.	

LINTEL SCHEDULE

NO.	DESCRIPTION
L1	2-2X8 SPRUCE
L2	3-2X8 SPRUCE
L3	2-2X10 SPRUCE
L4	3-2X10 SPRUCE
L5	2-2X12 SPRUCE
L6	3-2X12 SPRUCE
L7	3 1/2"X 3 1/2"X 1/4" L
L8	3 1/2"X 3 1/2"X 5/16" L
L9	4"X 3 1/2"X 1/4" L

LEGEND

- WP — DUPLEX OUTLET (WEATHERPROOF)
- DUPLEX OUTLET (HGT. ABOVE FLR.)
- DUPLEX OUTLET (12" ABOVE FLR.)
- EXHAUST FAN
- SWITCH
- HB — HOSE BIB
- SD — SMOKE DETECTOR
- HEAVY DUTY OUTLET
- LIGHT FIXTURE (WALL MOUNTED)
- LIGHT FIXTURE (CEILING MOUNTED)
- P — POT LIGHT FIXTURE
- R — LIGHT FIXTURE (WATER RESISTANT)
- LIGHT FIXTURE (CAPPED)
- FL — FLUORESCENT LIGHT FIXTURE
- SOLID WOOD BEARING
- FD — FLOOR DRAIN
- TV CABLE OUTLET
- TELEPHONE OUTLET
- C — COMPUTER OUTLET
- DE — DRYER EXHAUST

WINDOW SCHEDULE

ONE WINDOW PER FLOOR TO HAVE AN UNOBSTRUCTED OPEN PORTION W/ A MIN. AREA OF 0.35m2 W/ NO DIMENSION LESS THAN 380mm & MAXIMUM SILL HEIGHT OF 1M ABOVE FLOOR

NO.	TYPE	SIZE	QTY.	REMARKS
1.	CASEMENT	5'-0"X 5'-0"	1.	
2.	CASEMENT	2'-0"X 5'-0"	2.	
3.	SLIDER	3'-0"X 1'-6"	2.	
4.				
5.				
6.				

Source: Toronto Area Chief Building Officials Committee

Chapter 15 Assignment 1: Renovation Drawings Using Khalmur Drawings

Answer the questions below using the Khalmur drawings (Figures 15.1 to 15.9). Identify the letter that best answers the question.

_____ 1. Where is a drainage swale indicated?

a. A drainage swale runs at the back of the North lot line.

b. A drainage swale runs at the front of the South lot line.

c. A drainage swale runs at the side of the West lot line.

d. A drainage swale runs at the side of the East lot line.

_____ 2. What is the setback from the addition (excluding the deck) to the rear lot line?

a. 13'-0"

b. 42'-4"

c. 50'-4"

d. 60'-4"

_____ 3. What is the maximum allowed gross floor area on the lot?

a. 260 square feet

b. 931.26 square feet

c. 1191 square feet

d. 3750 square feet

_____ 4. How far is the west side of the addition set back from the lot line?

a. 4'-0"

b. 10'-0"

c. 25'-0"

d. 42'-0"

_____ 5. What size piers is the deck supported on?

a. 6"

b. 8"

c. 10"

d. 12"

_____ 6. What size wood beam is used to support the floor of the deck?

a. a triple 2" × 6" wood beam

b. a triple 2" × 8" wood beam

c. a double 2" × 6" wood beam

d. a double 2" × 8" wood beam

_____ 7. What is the minimum that the piers that support the deck must be placed below grade?

a. 3'-6"

b. 3'-10"

c. 4'-0"

d. 5'-0"

_____ 8. What is the maximum height that is allowed for the house/addition?

a. 14'-11"

b. 20'-6"

c. 26'-0"

d. 32'-10"

_____ 9. What type and size of beam is used to open up the basement recreation room between the addition and the existing house?

a. W200 × 27 steel beam

b. 2-45 mm × 200 mm PSL beam

c. 2-45 mm × 200 mm LVL beam

d. 4-2" × 10" built-up wood beam

_____ 10. From the drawings, what size floor joists and spacing were used on the existing first floor?

a. 2" × 6" @ 16" OC

b. 2" × 8" @ 16" OC

c. 2" × 10" @ 16" OC

d. 2" × 12" @ 16" OC

_____ 11. From the drawings in the chapter, what size floor joists and spacing will be used for the addition's first floor?

a. 2" × 6" @ 16" OC

b. 2" × 8" @ 16" OC

c. 2" × 10" @ 16" OC

d. 2" × 12" @ 16" OC

_____ 12. How is a wall that is to be removed indicated on the drawings?

a. wall is indicated with a solid object line

b. wall is indicated with a centre line

c. wall is indicated with extension line

d. wall is indicated with a dashed line

_____13. How does the ceiling of the existing basement line up with the ceiling of the addition?

a. The ceiling of the basement and floor of the first floor will line up perfectly at the addition.

b. The ceiling of the basement addition will have an elevation 2" higher than the existing basement ceiling.

c. The ceiling of the basement addition will have an elevation 3.5" higher than the existing basement ceiling.

d. The ceiling of the basement addition will have an elevation 3" higher than the existing basement ceiling.

_____14. Review the room schedule and determine which of the following finishes is specified for the addition. Which one is it?

a. ceramic tile used on the floor in the breakfast area

b. wood stained maple to match existing in the breakfast area

c. natural slate used on the floor in the bathroom

d. hardwood flooring used in the bathroom

_____15. On which elevation does a window get filled in?

a. on the west side of the house

b. on the south side of the house

c. on the east side of the house

d. on the north side of the house

_____16. On which elevation of the ones provided is the roof saddle indicated?

a. north elevation

b. south elevation

c. east elevation

d. west elevation

_____17. How many different sizes of window are used for the addition?

a. 2

b. 3

c. 4

d. 5

_____18. What is the size of the window for the breakfast area?

a. 2'-0" × 3-0"

b. 2'-0" × 5-0"

c. 1'-6" × 3-0"

d. 5'-0" × 5-0"

_____19. If you were a contractor pricing the addition, what question would you ask concerning the deck, based on the information not being in the drawings or specifications?

a. What size joists are used for the deck?

b. What size are the piers used for the deck?

c. What depth below grade must the piers be, for example 4'-0" or 5'-0"?

d. What type of finish material is being used for the deck surface, for example, is it pressure treated, Trex, or cedar?

_____20. What do the drawings specify as the R value for the insulation over the main part of the ceiling?

a. R12

b. R20

c. R31

d. R50

MULTIPLE-UNIT BUILDINGS: SEMI-DETACHED AND TOWNHOUSES

This section reviews some of the differences in drawings and construction of wood frame low-rise multiple-unit buildings. Multiple-unit buildings are becoming more popular due to the increased cost of land and services. This is particularly true in large urban centres such as Calgary, Edmonton, Vancouver, and Toronto. As a result, many municipalities are increasing density requirements to maximize the use of services for builders (e.g., infrastructure, buses and subways), resulting in dramatic increases in demand for this type of design and construction. This type of housing is also more affordable for low- and middle-income earners.

Increased density has the added benefit that it is more green because there are more units per acre, shared interior walls, and efficiencies in transportation and infrastructure. As a further note, in order to increase affordability, mid-rise six-storey wood frame buildings have been approved and are being adopted and built in British Columbia. Ontario is currently in the final stages for the approval of six-storey wood frame buildings that could replace some of the concrete buildings in the mid-rise range (currently the maximum is four-storey, see Figure 15.10). These buildings, which fall outside of Part 9 of the NBC, are all sprinklered and require extensive engineering and review of fire separations to ensure the safety of occupants. The framing techniques often utilize manufactured wall and floor panels that are assembled on site, thus shortening on-site structural schedules. Consideration in the design of multi-storey wood frame buildings must account for shrinkage that will occur in the structure with accumulations based on the number of floors.

FIGURE 15.10

Example of stacked wood frame condominiums being constructed using manufactured floor and wall panels. Note floor panel being installed.

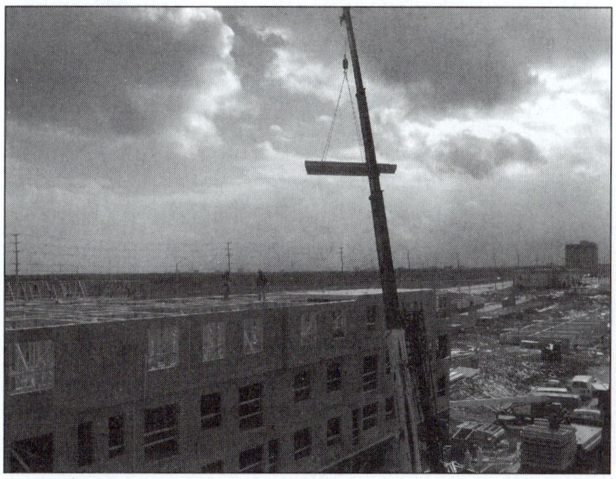

Courtesy Tom Stephenson

Semi-detached homes are one building comprising two homes or units. The building is built in the same way as a regular home, with the exception of the centre separating wall, which must provide a **fire separation** to protect the occupants of both homes should either one catch fire. Preventing the start of a fire and managing one if necessary are two prime considerations driving building and fire code requirements. The code considers how a fire might spread and the fire resistance ratings that are required to stop it from spreading, taking into account the occupancy and the difficulties that people may have exiting a building. This is why the requirements are much more complicated for a high-rise office building or a football stadium than for a townhouse or single-family unit. Although townhouses are low rise with numerous openings easily accessible by rescue workers, building and fire code requirements still must consider that residents may be sleeping or be otherwise distracted when a fire starts in a neighbouring unit. Consequently, there are numerous requirements regarding smoke detectors, window and door openings, and fire separations. As in single-family dwellings, the drawings for multiple-unit buildings indicate the location of smoke detectors/alarms, doors, and windows on the floor plan. The wall that separates the units is known as a **party wall** and operates as a continuous fire separation between the units, from the foundation to the underside of the roof sheathing, including attic spaces and between floor levels, or through the roof to form a **parapet** (a wall that extends through the roof).

The fire separation must be strong enough to withstand the collapse of members against it from the fire side to prevent its collapse. The fire separation is constructed with non-combustible materials assembled so they meet the required building code fire resistance ratings.

Sound transmission is another consideration when constructing party walls and other walls and floors through which sound transmission

FIGURE 15.11

Different types of wall construction and the STC ratings.

SOUND INSULATION OF SINGLE WALLS

DETAIL	DESCRIPTION	STC RATING
A 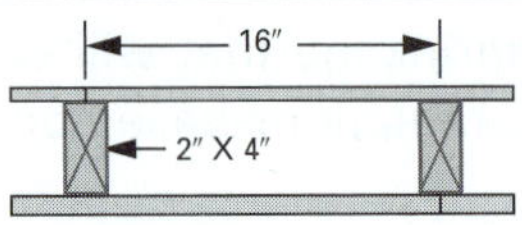	1/2" GYPSUM WALL-BOARD 5/8" GYPSUM WALL-BOARD	32 37
B 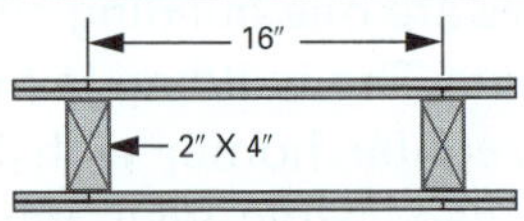	3/8" GYPSUM LATH (NAILED) PLUS 1/2" GYPSUM PLASTER WITH WHITECOAT FINISH (EACH SIDE)	39
C 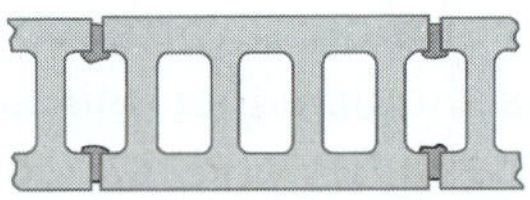	8" CONCRETE BLOCK	45
D 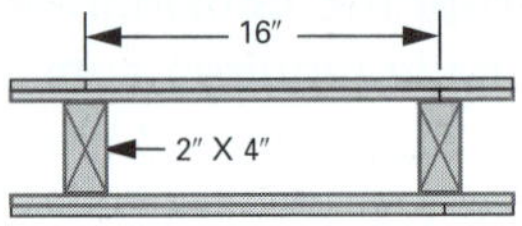	1/2" SOUND DEADENING BOARD (NAILED) 1/2" GYPSUM WALLBOARD (LAMINATED) (EACH SIDE)	46
E 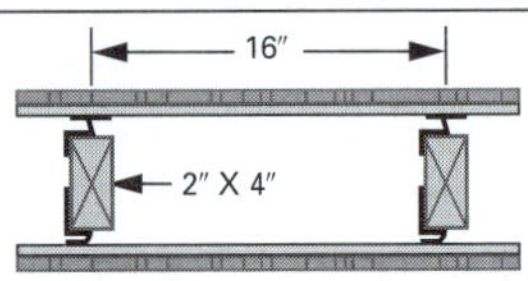	RESILIENT CLIPS TO 3/8" GYPSUM BACKER BOARD 1/2" FIBREBOARD (LAMINATED) (EACH SIDE)	52

SOUND INSULATION OF DOUBLE WALLS

DETAIL	DESCRIPTION	STC RATING
A 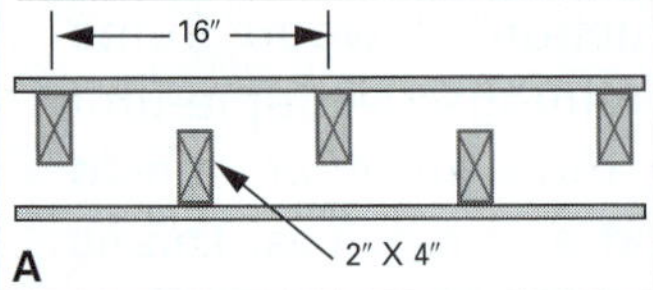	1/2" GYPSUM WALLBOARD	38
B 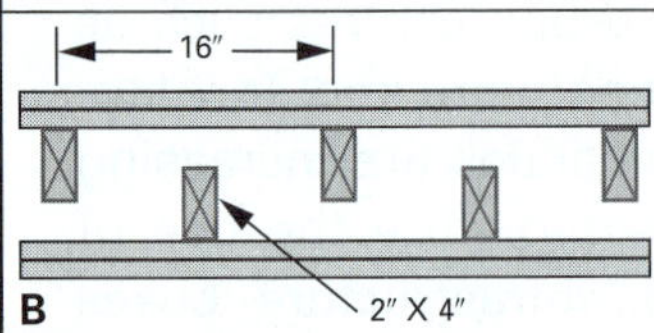	5/8" GYPSUM WALLBOARD (DOUBLE LAYER EACH SIDE)	45
C 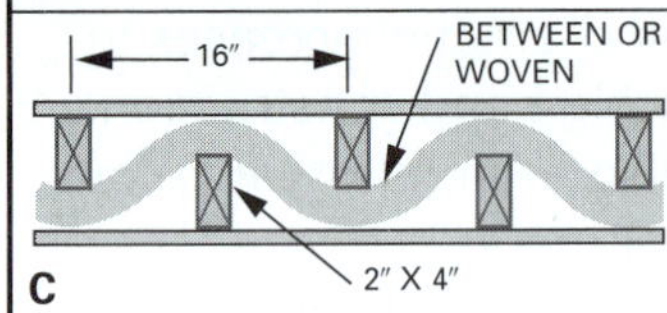	1/2" GYPSUM WALLBOARD 1 1/2" FIBROUS INSULATION	49
D 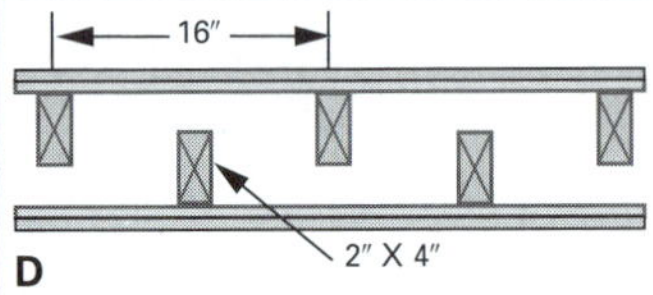	1/2" SOUND DEADENING BOARD (NAILED) 1/3" GYPSUM WALLBOARD (LAMINATED)	50

must be reduced. Sound transmission is rated by **sound transmission class (STC);** the higher the number the better the sound barrier, e.g., an STC rating of 25 will allow normal speech to be heard while a rating of 50 will make loud speech inaudible (refer to Figure 15.11).

Sound transmission can be limited by layering of drywall and by constructing independent walls with offsetting of studs. Fortunately, many of the construction details required to construct a fire separation also help prevent sound transmission. In addition, using insulation in the walls, offsetting electrical boxes and studs, and using sealant at all joints all help to reduce sound transmission. If specified, **resilient channels** (metal strips designed to isolate the drywall from the adjoining stud, and placed over the stud walls before the drywall is installed) will further limit sound transmission. Refer to the NBC Division B Appendix A Table A-9.10.3.1.A and B for a thorough listing of fire and sound resistance ratings of walls, floors, ceilings, and roofs.

FIGURE 15.12

Example of resilient channels being used to improve STC ratings, thus mitigating sound transmission between different floors of a stacked wood frame condominium building. A double layer of Type X drywall was later added to this application for fire separation purposes and to reduce sound transmission.

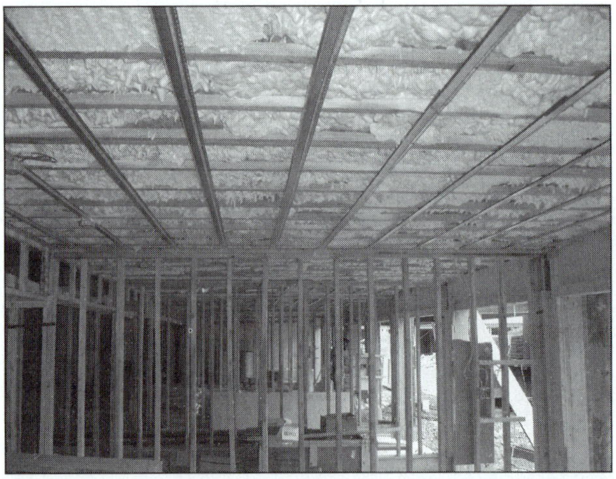

Courtesy Tom Stephenson

Party Wall (Fire Separation) Details

Figures 15.13 to 15.15 show examples of party wall construction. Figure 15.13 illustrates typical party wall detailing, from footing through the roof for a two-storey house. Detailing requirements will vary, based on the fire resistance rating required and the materials used. Follow the approved detail drawings carefully, remembering that you are creating a complete separation between the walls; if in doubt consult with the designer and building inspector for clarification. Notice the following conditions in Figure 15.13:

- The joists resting on party wall at foundation must have a fire cut (an angled cut at the joists). The dashed line on angle indicates a fire cut, which will allow joists to fall free without disrupting the fire separation. Space between joists must be filled solid; usually this means filling with concrete.
- The electrical boxes must be offset.
- The drywall (note the terms drywall and gypsum board have the same meaning and are often used interchangeably) is continuous behind bulkheads for ducts and piping.
- Strips of drywall are placed between the floor levels and at the ceiling to provide fire stopping.
- If double roof trusses are installed over the party wall, they must have three layers of ⅝" type X drywall (gypsum board), one on each outer face and one sandwiched between the trusses. If the trusses are not installed over the party wall, then a 2" × 4" wall must run up to the underside of the roof with ⅝" type X drywall (gypsum board) on each side.

The detail provides examples of three possible alternative designs. The design shows different layouts based on different conditions:

- Alternative 1 illustrates the detail for joists running parallel to the party wall and supported on a steel beam that is supported on the party wall.
- Alternative 2 illustrates the detail for joists running parallel to the party wall and supported on a wood beam. Notice the steel plate used under the beam and the fire cut made to the beam.
- Alternative 3 illustrates the detail for joists running perpendicular (at 90 degrees) to the party wall with the joists resting on the foundation wall and built out with blocking, and ⅝" type X drywall cut in between the joists and taped and sealed. A strip of drywall between the separation provides a fire stop.

Figure 15.14 illustrates a party wall intersection where the roof trusses run perpendicular to the party wall (detail states this as "normal to the party wall"). Elevation A shows both an end elevation (top left) and a front elevation (bottom left). Figure 15.14 also illustrates a number of intersecting conditions by providing a horizontal section of a party wall (a section in plan view shown to the right—review carefully using the visualization skills that you have developed throughout the text). From Figure 15.14 notice the following conditions:

- The detail states "vertical trusses normal to party wall," referring to the trusses being at a right angle or 90 degrees to the party wall (normal=at right angle).
- Three layers of type X drywall are used to fill in between the trusses, one on each outside surface of each knee wall and one in between the double constructed knee walls that are built between the trusses with the type X drywall cut around the chords of the trusses. Note the arrow on the top left detail indicating how elevation A view is taken directionally.
- The dashed lines shown on the drawings at the lower left represent knee walls built between the trusses to support the ⅝" type X drywall.
- The horizontal section illustrates how the ⅝" type X drywall must be continuous behind intersecting walls and pipe chases.
- Electrical box cutouts must be insulated and sealed around openings

Figure 15.15 is a vertical section of a drain pipe moving between floors. The drawing details how the fire separation is maintained. The drawing at top right shows the fire separation where an electrical panel will be placed on the party wall where it meets an outside wall. The drawing at bottom right illustrates the fire separation where pipe chases are located at the intersection of the party wall and outside wall. From Figure 15.15, notice the following conditions:

FIGURE 15.13

Party wall vertical section. Note that gypsum board and drywall mean the same thing.

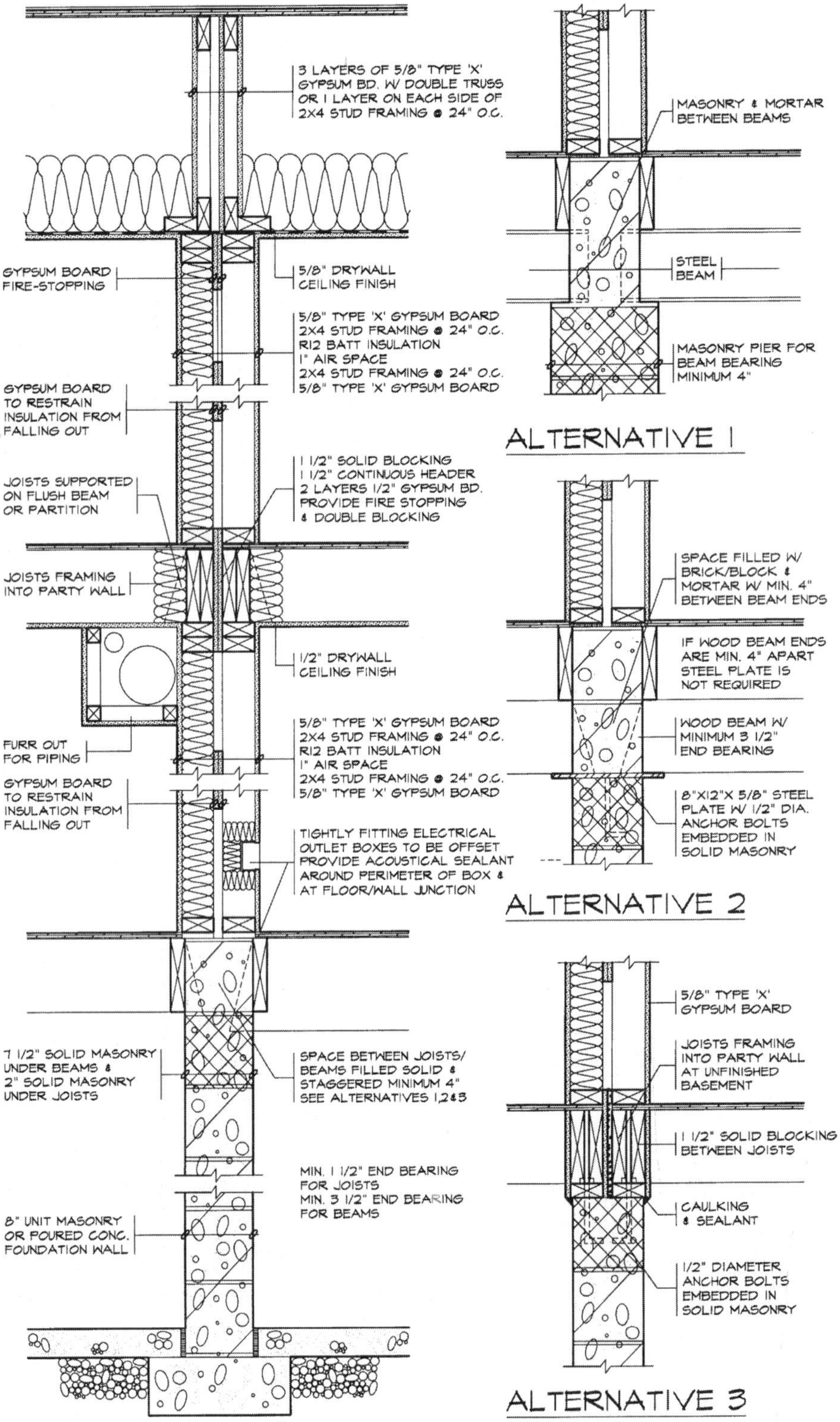

Source: Toronto Area Chief Building Officials Committee

FIGURE 15.14
Party wall details.

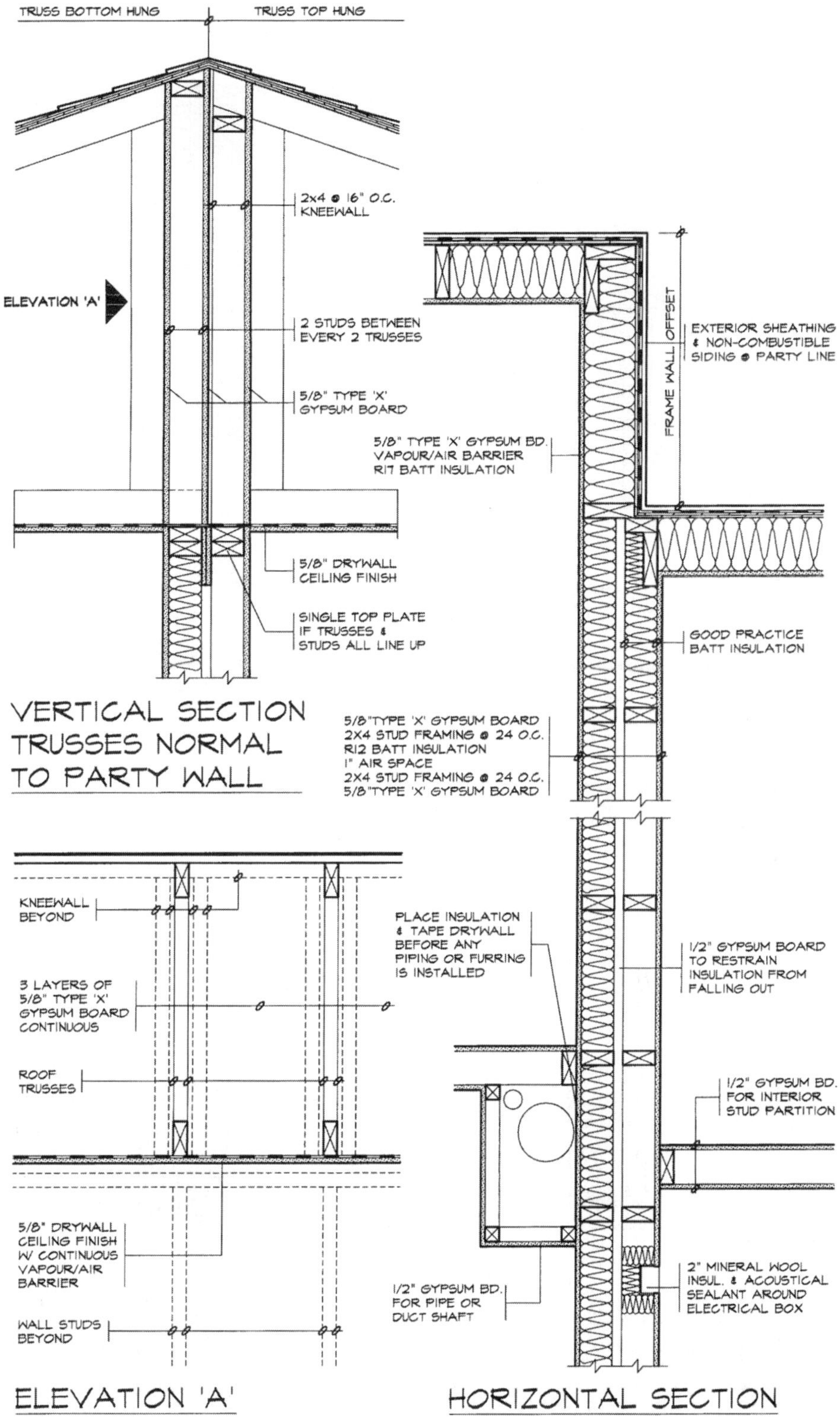

Source: Toronto Area Chief Building Officials Committee

FIGURE 15.15

Party wall service details.

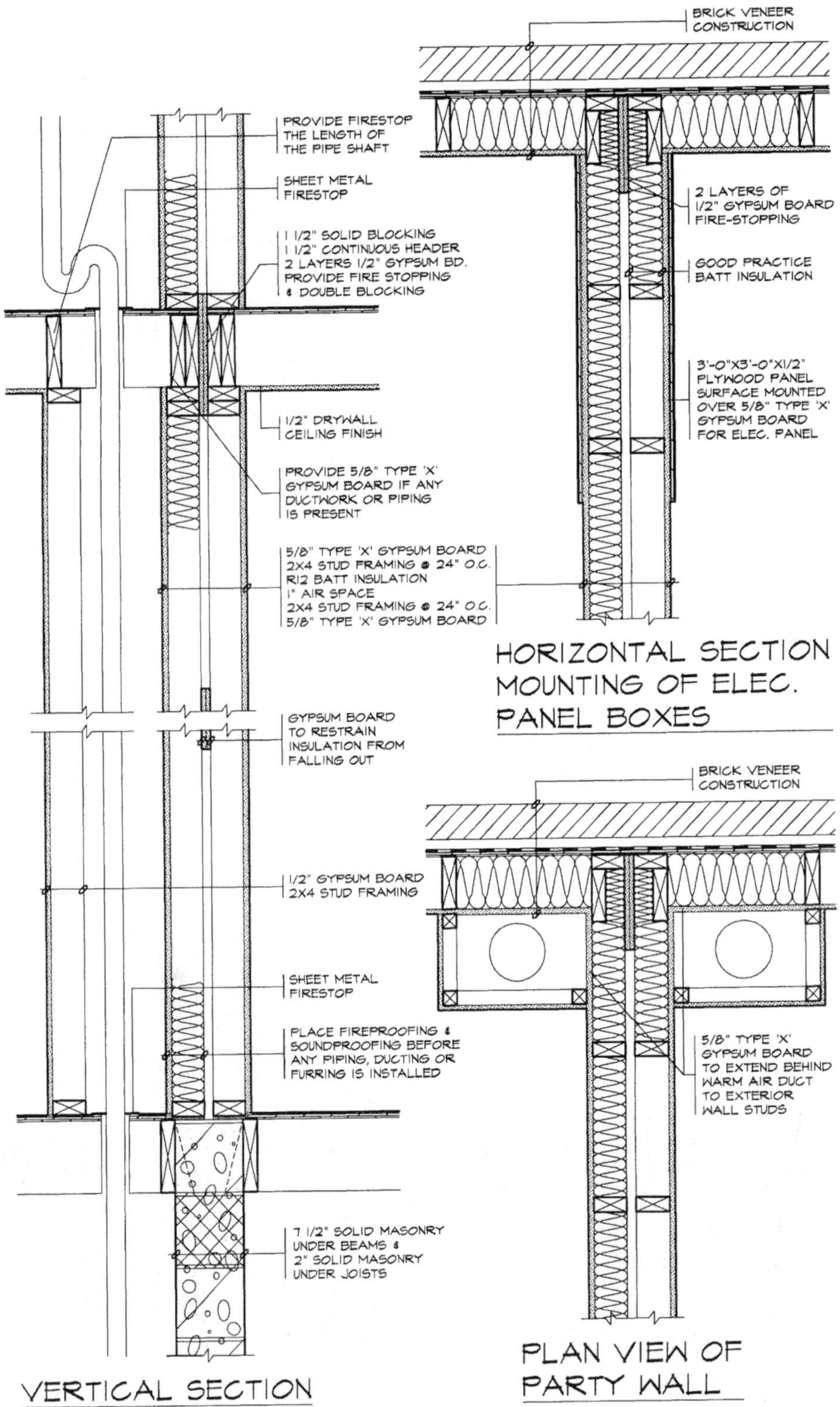

Source: Toronto Area Chief Building Officials Committee

- Sheet metal fire stops are used where the pipe runs through the floor. Always check the specifications, as there may be special fire-rated couplings listed.
- Where the electrical panel plywood is applied, it is good practice and may be a requirement to run the drywall through to the outside wall to provide a continuous break (not shown in the electrical panel or pipe chase detail). Refer to Figure 15.16 to see drywall extending to the outside.

FIGURE 15.16

Drywall extending through the party wall to the outside; this photo is taken at the party wall between two adjoining garages.

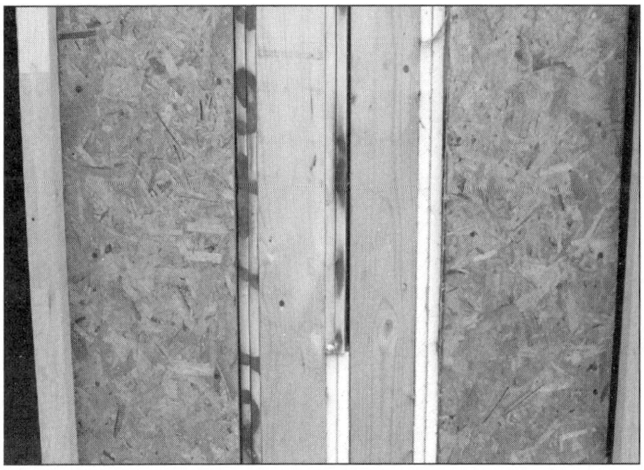

Courtesy Tom Stephenson

FIGURE 15.17

Party wall with intersecting girder truss (trusses running normal to party wall; at right angle), similar to the detail indicated in Figure 15.14.

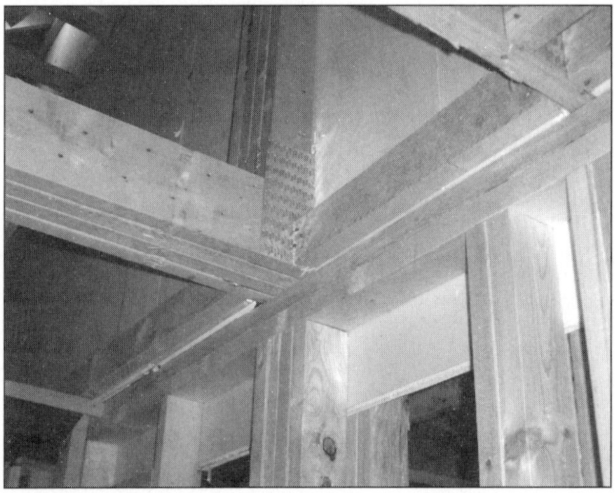

Courtesy Tom Stephenson

FIGURE 15.18

Fire separation between party wall, outside wall, and roof rafters.

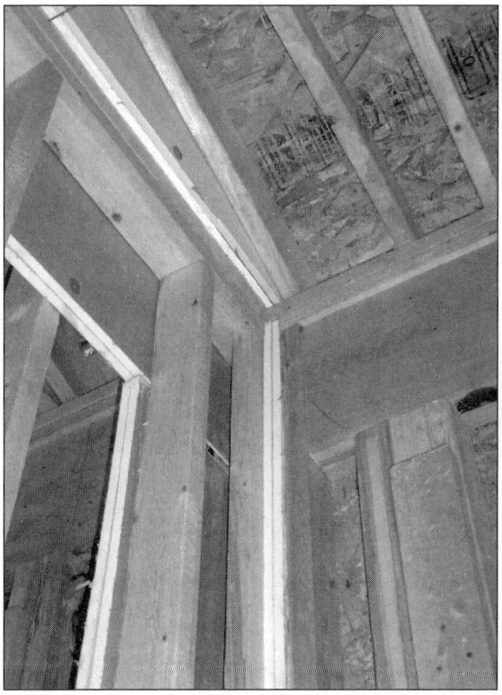

Courtesy Tom Stephenson

FIGURE 15.19

Fire separation at floor level runs through the joists; notice that the studs are offset, which helps reduce sound transmission and ensure that electrical boxes do not line up between the walls.

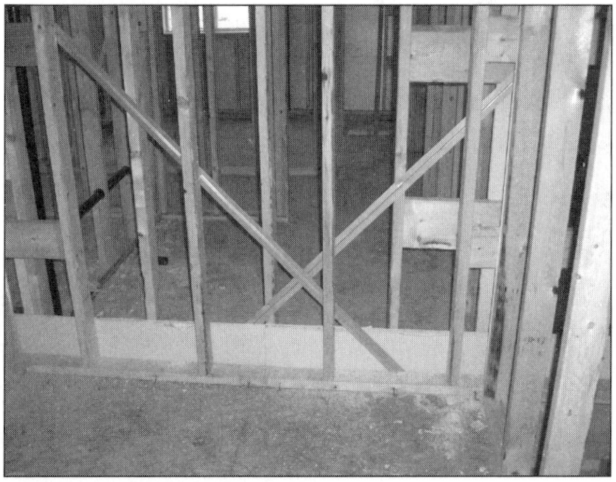

Courtesy Tom Stephenson

FIGURE 15.20
A fire separation extending down between the walls from the floor above; notice marking on drywall identifying thickness and type.

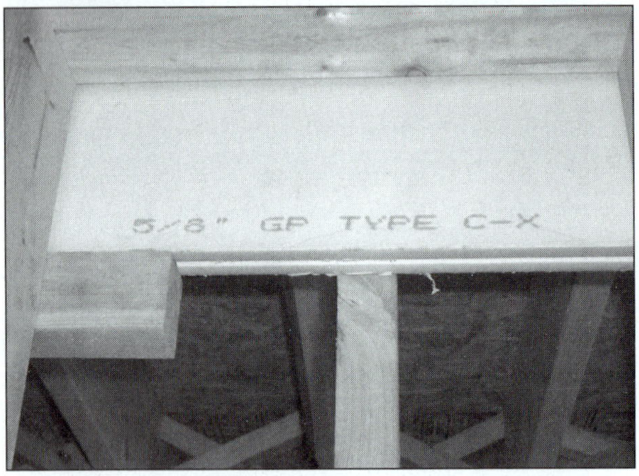

Courtesy Tom Stephenson

KEY TERMS

Fire separation (p. 401)
Needling beams (p. 388)
Parapet (p. 401)
Party wall (p. 401)
Resilient channel (p. 402)
Sound transmission class (STC) (p. 402)

Chapter 15 Quiz Review: Multiple-Unit Buildings and Renovations

Answer questions 1 to 10 using material and details as described found in the chapter, and answer questions 11 to 20 using the Khalmur Drawings, Figure 15.1. Identify the letter that best answers the question.

_____ 1. How many units are in a semi-detached building?

a. two
b. three
c. four
d. five

_____ 2. What does the abbreviation STC refer to?

a. seismic transmission code
b. speed transmission code
c. smoke transmission class
d. sound transmission class

_____ 3. What is the wall that separates two units in a townhouse or semidetached house called?

a. separation wall
b. party wall
c. sound transmission class wall
d. unit separation wall

_____ 4. Why are resilient channels sometimes used in wood frame construction?

a. to prevent smoke transmission
b. to limit sound transmission
c. to add lateral strength to the ceiling
d. to add lateral strength to the walls

_____ 5. What is the STC rating for an 8" concrete block wall?

a. 32
b. 37
c. 39
d. 45

_____ 6. What is the STC rating for a 2"× 4" uninsulated wall with 1/2" gypsum wall board (drywall) on each side?

a. 25
b. 29
c. 32
d. 37

_____ 7. How much does the STC rating for a 2"× 4" uninsulated wall with 1/2" gypsum wall board (drywall) on each side improve when 5/8" drywall is used instead of 1/2"?

a. The STC rating improves by 2.
b. The STC rating improves by 4.
c. The STC rating improves by 5.
d. The STC rating improves by 10.

_____ 8. What is a unique feature for joists that rest on a masonry party wall at the foundation?

a. They must rest on a steel beam placed in the middle of the party wall.
b. They must be pressure treated to prevent rot from condensation caused by moisture reaching the dew point.
c. They must be isolated from the party wall so as not to make direct contact.
d. They must have a fire cut to prevent disruption to the party wall in case of fire.

_____ 9. From the section detail drawing in Figure 15.13 for the bulkhead (a bulkhead is a furred out box covered in drywall used to conceal ducts and pipes), what type and what thickness of gypsum board is installed on the wall behind the bulkhead?

a. One layer of 5/8" type X gypsum board is installed before the bulkhead is constructed.
b. Two layers of 5/8" type X gypsum board are installed before the bulkhead is constructed.
c. One layer of 1/2" gypsum board is installed before the bulkhead is constructed.
d. Two layers of 1/2" gypsum board are installed before the bulkhead is constructed.

_____ 10. From the section detail drawing in Figure 15.15, what is required where the pipe moves through the floor?

a. opening
b. wood fire stop
c. sheet metal fire stop
d. asbestos fire stop

_____11. From the Khalmur drawings (Figures 15.1 to 15.9), what thickness is the foundation wall of the addition?

a. 6" thick

b. 8" thick

c. 10" thick

d. 12" thick

_____12. From the Khalmur drawings (Figures 15.1 to 15.9), what is specified regarding the furnace?

a. existing furnace to remain in its original location

b. existing furnace to be relocated

c. new furnace to be installed in the same location as the existing furnace

d. new furnace to be installed in a new location in the basement directly under the bedroom

_____13. From the Khalmur drawings (Figures 15.1 to 15.9), what size is the lintel that supports the brick over the bedroom window?

a. 3½" × 3½" × 1/4" L

b. 4" × 3½" × 1/4" L

c. 2-2" × 6" spruce

d. 2-2" × 8" spruce

_____14. From the Khalmur drawings (Figures 15.1 to 15.9), what size is the lintel that supports the brick over the basement window?

a. 3½" × 3½" × 1/4" L

b. 4" × 3½" × 1/4" L

c. 2-2" × 6" spruce

d. 2-2" × 8" spruce

_____15. From the Khalmur drawings (Figures 15.1 to 15.9), what is the elevation of the sliding door lintel from the first floor?

a. 6'-8"

b. 6'-10"

c. 7'-0"

d. 8'-4"

_____16. From the Khalmur drawings (Figures 15.1 to 15.9), what is the height of the top of the railing for the deck?

a. 2'-10"

b. 2'-11"

c. 3'-0"

d. 3'-6"

_____17. From the Khalmur drawings (Figures 15.1 to 15.9), what are the ventilation requirements for the attic as a ratio of ventilation to ceiling area?

a. 1:200

b. 1:350

c. 1:300

d. 1:400

_____18. From the Khalmur drawings (Figures 15.1 to 15.9), what size are the footings for the addition?

a. 12" × 4"

b. 18" × 6"

c. 24" × 8"

d. 35" × 12"

_____19. From the Khalmur drawings (Figures 15.1 to 15.9), how many smoke detectors are shown?

a. two

b. three

c. four

d. five

_____20. From the Khalmur drawings (Figures 15.1 to 15.9), what is the size and spacing of wood studs used for the interior stud walls?

a. 2" × 4" @ 16" OC

b. 2" × 4" @ 19.2" OC

c. 2" × 6" @ 16" OC

d. 2" × 6" @ 19.2" OC

16 Introduction to Small Buildings and Institutional Commercial Industrial (ICI) Type Drawings

LEARNING OBJECTIVES

After completing this chapter, you will be able to:

- Read basic information regarding institutional commercial and industrial drawings.
- Interpret and orient the main structural elements in relationship to the reference (structural) grid.
- Identify the cross referencing detail designations that were used and be able to effectively cross reference the plan and elevation drawings with the appropriate detail drawings.
- Identify the difference between engineering (structural) drawings and architectural drawings and apply that knowledge when seeking specific construction information.
- Identify reinforcing requirements in simple concrete reinforced walls and footings.
- Identify the basic elements and construction materials in a structural steel building.
- Perform a simple volume calculation for concrete based on information taken from the construction drawings.

INTRODUCTION TO SMALL BUILDINGS AND ICI TYPE DRAWINGS

If you have worked through the textbook to this point, you will know that we have focused on residential buildings as outlined in Part 9 of the National Building Code of Canada. Part 9 also, however, includes "**small buildings**" that are three storeys and 600 square metres or less in size and used for major occupancies classified as residential occupancies (as reviewed in the text), business and personal services occupancies, mercantile occupancies, or medium and low hazard industrial occupancies as defined in the NBC (e.g., Part 9 would not apply to a school or church but a small retail store could fit the criteria; refer to Division A Part 1.3.3.3 of the NBC). This chapter will introduce some of the differences in drawing standards that are typically applied to the design of small buildings and will act as an introduction to and basis for the understanding of construction drawings for larger building projects such as condominium buildings, box stores, and schools, to name just a few, which are designed based on the requirements of Part 3 of the NBC.

Small buildings may be made out of wood, but given that our discussion also includes an introduction to **institutional, commercial, and industrial (ICI)** buildings that go beyond the scope of Part 9, they will include buildings constructed out of other materials such as steel and reinforced concrete. Steel and reinforced concrete buildings provide much greater latitude to architects and engineers when designing large building structures, allowing for greater spans and much higher loading. The material covered in Chapters 1 through 5 acts as a primer for material found in this chapter, as many of the concepts and practices are the same or similar.

With the introduction to other types of building projects, you will see that there are a lot of other construction methods and materials used in the field of construction and that there are many opportunities available to people with a wide and diverse range of skills.

SIMILARITIES BETWEEN RESIDENTIAL AND INSTITUTIONAL, COMMERCIAL, AND INDUSTRIAL DRAWINGS

Similarities between construction drawings include:

- Drawing types, for example, elevations, floor plans, section details, site plans are drawn orthographically
- Line and dimension types and their use are similar or the same
- The process of visualizing from orthographic drawings to three-dimensional buildings must become second nature. You will find the process is the same except that it will take more time as the complexity of the structure increases. Spend extra time reviewing the drawings and make a list of questions for clarification or possible correction. The submittal of formal questions to the designer is often termed a request for information (RFI).
- Scales: Many of the symbols are the same although for larger buildings the size of paper will increase, as will the reduction size of scale
- Schedules, as described in earlier chapters, are used to reference many components in ICI drawings, including rooms, doors, windows, columns, beams, piles, and caissons.
- A number of symbols and abbreviations that we have used will remain the same, although there will be a number of new and more specialized symbols that you will learn when working with concrete and steel.

You are encouraged to take the same approach as we did in the earlier chapters to become familiar with the drawings and to imagine yourself walking around the building and through the building, all the while visualizing the finished project, the construction details, and the order of assembly of the materials.

TYPICAL DIFFERENCES BETWEEN RESIDENTIAL AND INSTITUTIONAL, COMMERCIAL, AND INDUSTRIAL DRAWINGS

This section will attempt to identify some of the differences and new methods that will typically be found in the way that information is provided in ICI drawings and why those differences exist. Residential drawings, while being potentially complicated and requiring a lot of information, are nonetheless simpler to describe in terms of the information required to build, and as a result require less reference information and typically fewer levels of drawings. ICI drawings, however, depending on the building, can be quite complex, with a great deal of information required for the general contractor and the subtrades. Also, because of their size and scope, finding a particular detail or piece of information can be confusing if some sort of order and method of reference is not applied as part of the method of communication for the drawings.

The first thing that you will notice is that there is a set of **architectural drawings** and a set of **engineering drawings**. The engineering drawings define the structural fabric of the building. If the building is constructed out of reinforced concrete, information will be provided regarding the size, spacing, and type of columns and slabs, including the thickness and strength of the concrete and the reinforcing steel requirements. If the building is constructed out of structural steel, information will be provided regarding the size, spacing, and type of steel columns and beams and the type of open web steel joists and decking that is to be used. Also, information regarding mechanical (plumbing and HVAC) and electrical will have their own drawings, whereas residential will usually include the plumbing and electrical all on the same drawings while the HVAC has their own set of drawings, as discussed in Chapter 11.

The following is a list of potential reference coding standards that may be used in ICI construction drawings (note that while there is no mandatory standard, the following represents coding standards as outlined in the CAN3-B78.3-M77 National Standard of Canada):

- Coding of floors
- Coding of rooms and spaces
- Coding of components
- Reference grids
- Cross referencing of drawings.

Coding of Floors

In single dwelling residential, it is relatively easy to identify floor levels because there are so few of them, but in larger high rise buildings care must be taken to ensure that they are easily identifiable.

The floor numbering usually goes upward and downward from the ground floor, for example:

Floor 05

Floor 04

Floor 03

Floor 02

Ground Floor (Main Floor, Floor 01)

Basement 1 (or Parking P1)

Basement 2 (or Parking P2)

Coding of Rooms and Spaces

Rooms and spaces in large or complex buildings must be numbered in such a way that they are easily identifiable. For example, buildings with more than two stairwells should have them numbered. The same goes for HVAC shafts. Condominium buildings will need to reference floors and unit numbers so as not to confuse similar rooms on different floors. For example, 608 references unit 8 on the sixth floor.

Coding of Components

A method of coding of components becomes important because there may be a large number of different size components that require identification and referencing through the use of schedules. As a result, a consistent and accurate coding system must be developed. Similar to residential drawings, the coding may vary depending on what component is being referenced. For example, when a reference is made for a door on a particular floor other than the floor plan that it is on, the floor number should precede the door number (if you are on the 15th floor plan then it is obvious what floor the door is on). As an example, the door might be identified with the letter D for Door followed by the floor number, e.g., 15, and the door number on floor 15 might be number 10 and so would be seen on the door schedule as D 15 10. When reviewing new drawings, it is advisable to check how the drawing is coded so that it is easier for you to find your way around.

Reference Grids

The design and construction of ICI type buildings includes not only the structural requirements of the building but also detailed information about the envelope, building systems, and finishes. These elements or components must be coordinated and designed by a dimensional discipline that recognizes the modular nature of the components and ensures that the construction process is simplified (as far as reasonable for the design) to provide maximum repetition and standardization. From a drawing and design perspective, this means quite simply that the building can be laid out in a grid format that provides for the orderly transference of loads from beams to columns to the foundations.

As discussed in earlier chapters, a building must be able to support both live and dead loads and the structural frame of the building must be capable of transmitting these loads from their points of origin to the foundation. The design combines structural efficiency with minimum impact on economy and function. An important difference between typical residential drawings and ICI drawings is that a grid identification system is used to simplify the ease with which the drawings can be designed and interpreted. That's not to say that residential drawings don't often also use a structural grid, as it can be advantageous for larger residential buildings. The grid, called a **reference grid, structural grid, or planning grid**, looks similar to how main arterial roads are often laid out in cities, forming a grid system of roads that run north and south and east and west with regular spacing. Before GPS became prevalent, we used maps that were laid out on a grid. We were able to locate streets based on their proximity to the reference grid. The choice of structural grid defines the centre line column positions and spacing, which are a key element in design decisions.

The structural grid is principally defined by the location of the columns, which is based on the main beams and/or types of slabs they are supporting. On the drawings, there are two sets of grid lines that are normally perpendicular (at 90 degrees) to each other. The horizontal grid lines running left to right on one axis of the plan drawings are typically lettered while the vertical grid lines running up and down on the other axis of the plan drawing page are typically numerical (note: numbering and lettering may be reversed on some drawings).

The following points are considered when designing the structural grid:

- The columns are normally positioned or centred at the intersection of two grid lines
- A beam is normally positioned along every grid line

FIGURE 16.1

Illustrates a structural grid layout with numbers and letters encased in a circle at the outside extents of the drawing page.

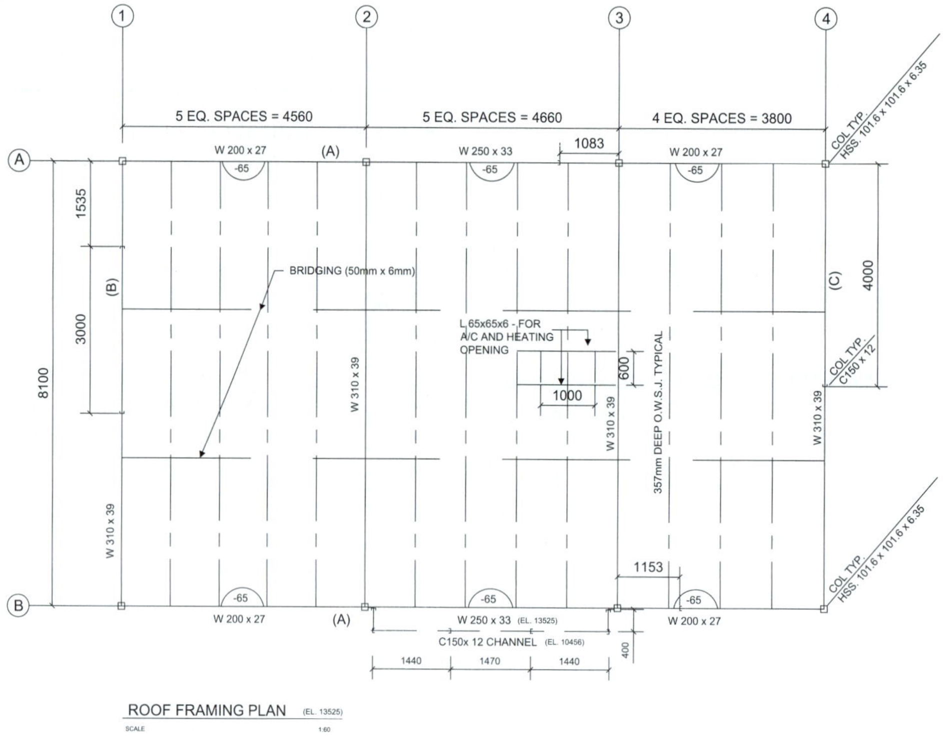

Source: Courtesy of Clint Kissoon, Chair, George Brown College, School of Architectural Studies.

- The grid lines normally are at right angles (orthogonal) to each other, making a square or rectangular grid
- The shape of the building and the site can require deviation from the above and may require some irregular spacing; however, these are usually the exception, while the main part of the building can be laid out in accordance with the structural grid
- A regular grid improves construction efficiency of load transfer while having the added benefit of allowing forming components to be quickly moved from one floor to the next (known as fly forms); when using reinforced concrete or when using structural steel the steel can be more readily prefabricated, maximizing repetition and speeding construction. This system works particularly well in multi-storey buildings such as office and condominium buildings.

The following points are helpful when reviewing drawings that have a structural grid:

- Note the orientation and identification of the grid intersections shown on the floor plans
- Reference the grid intersections when moving from the architectural drawings to the engineering drawings
- Check the elevation drawings and when the grid is indicated on the elevations you can use the grid to orient yourself when moving from plans to elevations and also from floor to floor. This can be helpful when reviewing larger drawings.
- Note how the grid is referenced in column schedules. For example, the details about the column centred at grid axis D5 would be found on the column schedule reference to D5.

Cross Referencing of Drawings

You will recall from Chapter 2 that we discussed **cutting-plane lines** and the use of a **detail designation**, sometimes referred to as a bubble, to identify the number or letter of the reference drawing. In more complex drawings, where there are many details, the detail designation will not only indicate the number of the detail but also the page number of where the detail can be found.

FIGURE 16.2

There are two cutting-plane lines indicated below; the lower one displays the number of the detail as 1 and that it can be found on page A4, where the A refers to the architectural drawings page 4, and that there are 6 pages in total. The upper detail designation indicates that detail 2 can be found on page A3 (closeup taken from attached Mini Mart drawings).

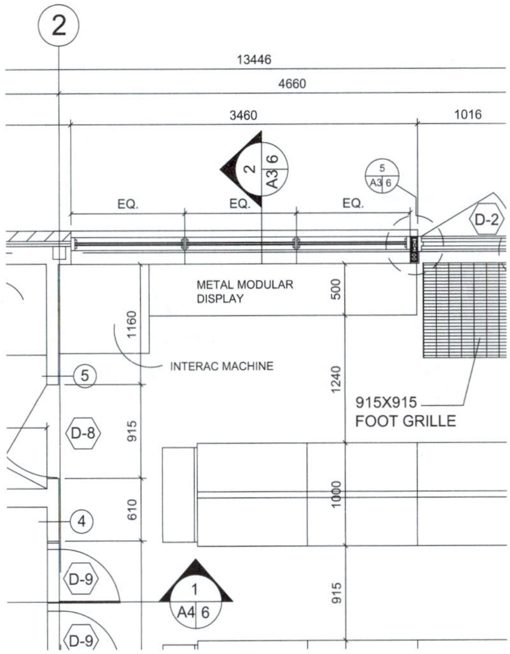

Source: Courtesy of Clint Kissoon, Chair, George Brown College, School of Architectural Studies.

Engineering Drawings

The design of larger ICI type buildings requires that they be engineered such that they are designed to span greater distances and support much higher loads. These loads include both the **live loads** (furnishings, people, wind, snow, etc.) and **dead loads** (the weight of the structure itself) that are typically placed on them.

The engineering drawings indicate the structural elements of the building and provide detail regarding their placement, the types of components, and the necessary connections and reinforcing. Schedules may be provided to identify specific requirements of certain components such as columns, beams, or girders. Shop drawings used for construction are then prepared by the structural steel/reinforcing steel manufacturer/supplier or subcontractor of the general contractor for the project (who prepares the drawings depends on the type of building, e.g., is it reinforced concrete or structural steel, or a combination of both). These drawings are based on the information provided in the engineering drawings and specifications and ensure that the components will be able to be correctly assembled on site and that they meet the engineering

FIGURE 16.3

The drawing below indicates the section detail as indicated in Figure 16.2.

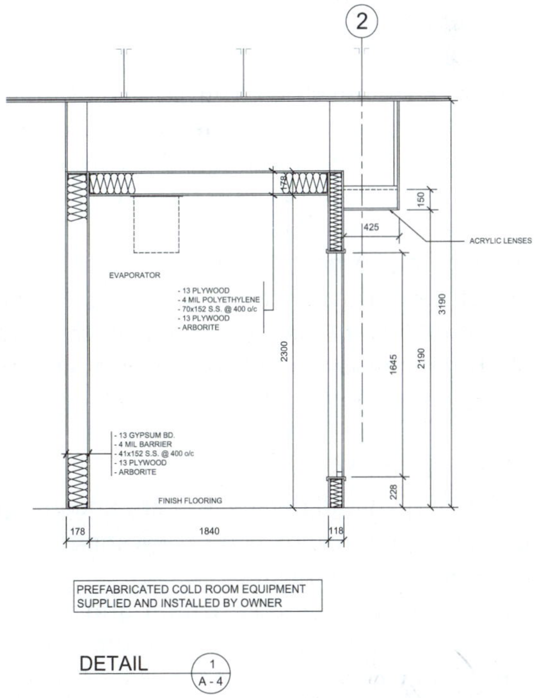

Source: Courtesy of Clint Kissoon, Chair, George Brown College, School of Architectural Studies.

FIGURE 16.4

Common structural steel shapes and their designations as they are indicated on the drawings.

Shape	Letter Designation	Size Designation
Wide Flange	W	depth of web x kilograms per metre
Standard Beam	S	depth of web x kilograms per metre
Channel or Miscellaneous Channel	C or MC	depth of web x kilograms per metre
Hollow structural section	HSS	width x depth
Tee	T	depth of web x kilograms per metre
Angle	L	vertical leg x horizontal leg x thickness

Source: Courtesy of Clint Kissoon, Chair, George Brown College, School of Architectural Studies.

requirements for the project. The shop drawings will indicate specifics regarding connection requirements, e.g., are the connections bolted, welded, or both.

Mini Mart Drawings

The Mini Mart drawings included with the textbook (10 pages titled Mini Mart & Gas Bar), which were produced for educational purposes, will be used to introduce you to a small building as defined by the NBC that incorporates some of the principles of reinforced concrete and structural steel construction. The drawings will also introduce you to how information is shown on architectural and engineering drawings and how they correlate together. As noted on the cover page to the drawings, the six pages of architectural drawings are denoted with an A, e.g., A1 is the floor plan, and the three pages of engineering drawings are denoted with an S, e.g., S1 is the foundation plan and details. You are encouraged to pull out the drawings and review them as you read through the chapter.

The drawings use a structural grid that has numbers on the horizontal axis of the plan and letters on the vertical axis of the plan. The drawings include numerous examples of reference keying for various details and cutting plane lines. While it is beyond the scope of this textbook to provide a complete understanding of reinforced concrete and structural steel construction, this chapter will introduce you to some of the terminology and methods used for this type of construction, with a focus on how to interpret information from these types of drawings.

From the architectural drawings (A1–A6), you can see that the building is an architectural block veneer building that has open web steel joists with metal decking that supports the roof loads, insulation, built up flat roof, and pea gravel that is placed on top of the roof. The roof system is supported on a structural steel frame with lightweight steel studs filling in the openings but not supporting the structural load of the building. You can also see from the wall sections the steel beam and open web steel joists (O.W.S.J.) in section view, but you don't see the actual steel frame details, for example, the size and type of steel beams and columns, as these details are indicated on the engineering drawings (S1–S3). Details are provided explaining the different steel stud interior partition wall types, and are indicated by a number reference system shown on the floor plan. Walls around the fridge are insulated.

The foundation of the Mini Mart is constructed using reinforced concrete footings and walls. You can find detailed information regarding the reinforcing in Section A-A on page 3 of the structural drawings, denoted S3 at bottom right of page. Also note how the top of the foundation slab changes at the various door openings and that there are three reference section details (A, B, C) indicated at the right side of drawing S1. Section A is where there is no door opening, section B is where there is a regular door opening to the men's and women's washrooms and the storage area. Note how you have to refer to the architectural drawings, to see what is going on and why the concrete is different in those locations. Get used to referring back and forth between the architectural and structural drawings just as you have gotten used to referring back and forth between the plans, elevations and details. Section C is where there is a door opening that leads to a recessed floor that has a foot grill (the grill catches snow and slush from people entering the store in winter).

Recall from Chapter 10 that reinforcing rods are indicated on drawings based on diameter; for example, a 10M bar refers to a reinforcing bar with a diameter of approximately 10 mm. Refer to Table 16.1 for a detailed description of

FIGURE 16.5

Typical types of metal decking. Note that the metal roof deck with insulation is similar to the Mini Mart roof section (found on A3 Detail 1 of the Mini Mart drawings). **Steel decking** sizes and gauge will vary depending on loading and span requirements.

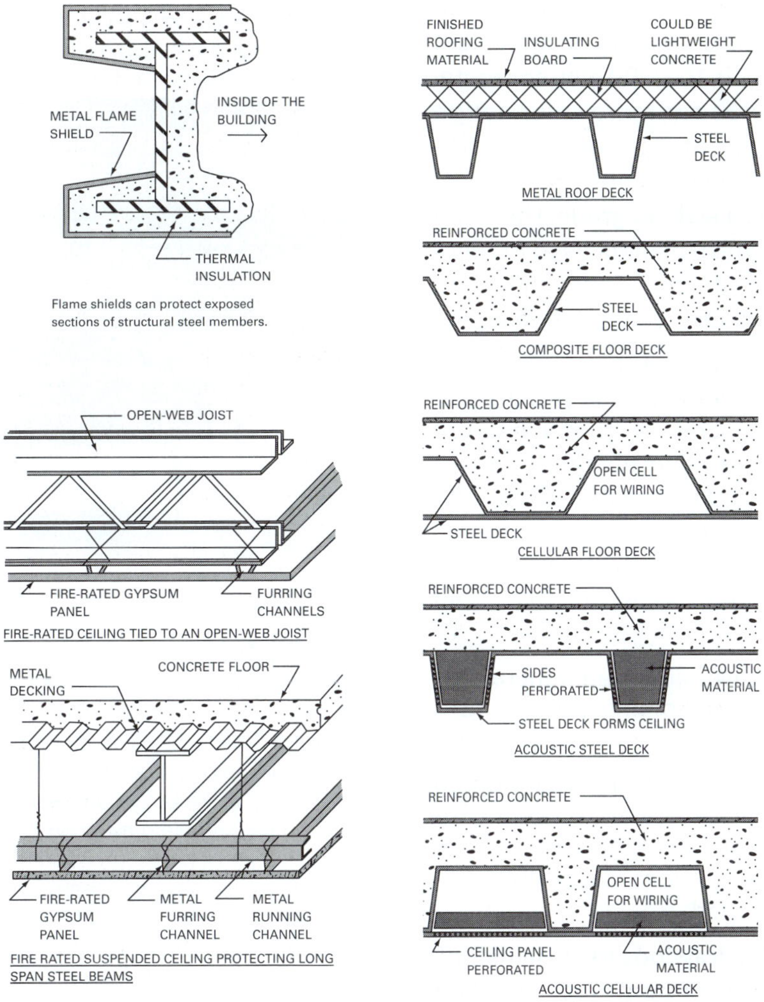

Source: From Spence, Construction Materials, Methods, and Techniques, 2E.

TABLE 16.1

Reinforcing Bar Sizing

Reinforcing Steel Bars				
		Nominal Dimension		
Bar Designation Number (Bar Size)	Mass kg/m	Diameter mm	Cross Sectional Area mm²	Circumference mm
10M	0.785	11.3	100	35.5
15M	1.570	16.0	200	50.1
20M	2.355	19.5	300	61.3
25M	3.925	25.2	500	79.2
30M	5.495	29.9	700	93.9
35M	7.850	35.7	1000	112.2
45M	11.75	43.7	1500	137.3
55M	19.625	56.4	2500	177.2

standard reinforcing bars. The strength of concrete is measured in **megapascals (MPa)**, a pressure unit that indicates the compressive strength of the concrete. A comparative strength in imperial measurement is **pounds per square inch (psi)**; for example, 20 MPa concrete is equivalent in compressive strength to 2900 psi concrete. Refer back to Chapter 10 for more information on concrete and concrete strength requirements.

The Mini Mart building does not have a basement and therefore has a slab on grade. There are a number of different types of joints used in preparation and pouring of concrete floors slabs and walls. The types of slabs include the following:

- **Control joints**, produced by a saw cut placed in the concrete, which provides a weak spot in the concrete so that instead of cracking unevenly leaving unsightly cracks the concrete will crack evenly along the predetermined cut lines
- **Expansion joints**, which allow for the expansion of concrete so that it does not heave and crack
- **Isolation joints**, which allow the concrete to be separate from its adjoining surface so it can move up or down
- **Construction joints**, which are planned to allow work to finish at a planned point for that day so that it can later continue on from the stop point, that is, from where the concrete pour was stopped.

Being able to review the drawings for dimensions is important for a number of purposes, including calculating concrete volumes. For example, concrete is ordered in **cubic metres** and must be calculated by figuring out the length × width × thickness, which can found on the drawings.

Concrete Volume Calculation Example

The required amount of concrete for a concrete floor slab that measures 10500 mm × 6400 mm × 150 mm thick can be calculated by changing the units of measurement to metres from millimetres. To do this we move the decimal place three positions to the left (one space to change from mm to cm, another space to change from centimetres (cm) to decimetres (dm), and the third space to change from decimetres (dm) to metres (m)). 10500 mm becomes 10.5 m, and 6400 mm becomes 6.4 m, and 150 mm becomes .150 m.

FIGURE 16.6

The photograph shows a structural steel braced frame building under construction. Note the O.W.S.J. resting on the beams and the metal decking placed over the joists; there is also some cross bridging installed in the top right corner. Also observe the grid that the columns make, and the bracing at the end and left side.

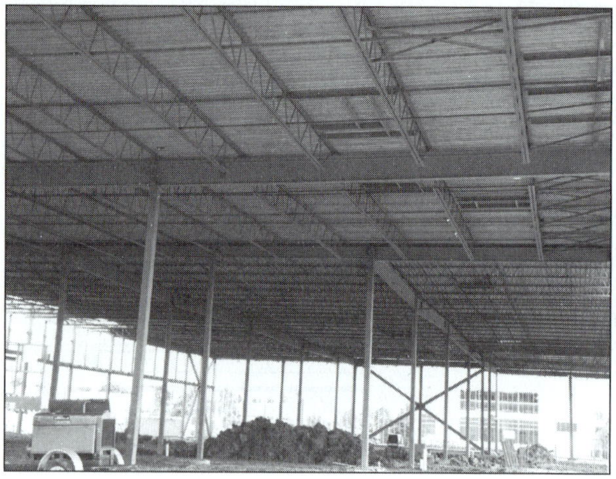

Courtesy Tom Stephenson

FIGURE 16.7

The photograph indicates how a typical O.W.S.J. rests on the beam. Note how the top chord of the joist rises above the beam. This is why the drawings on steel structures will usually indicate a negative number, which displays the level for the beam below the required elevation, as indicated on the sample drawing in Figure 16.8 with a −65 meaning 65 mm below the required elevation indicated in the semi-circle above the beam size. The beams that the joists are not resting on will be at the same elevation as the top of the joists, so that the decking is at one level. This is more evident in Figure 16.8.

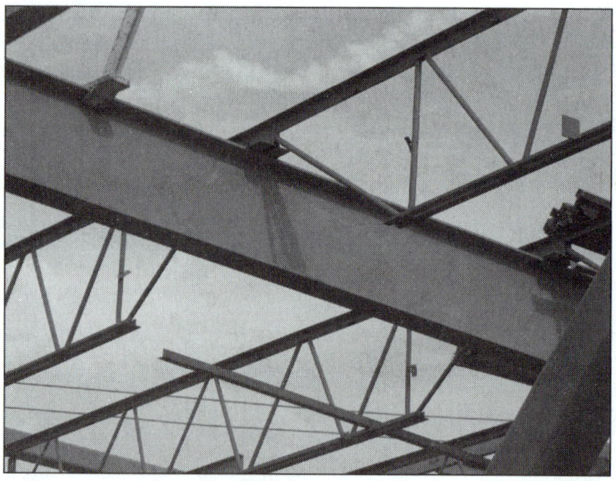

Courtesy Tom Stephenson

10.5 m × 6.4 m × .150 m = 10.08 cubic metres of concrete required (usually rounded up to the nearest .5 cubic metre e.g. 10.5 cubic metres, to ensure that there is enough concrete during placement).

FIGURE 16.8

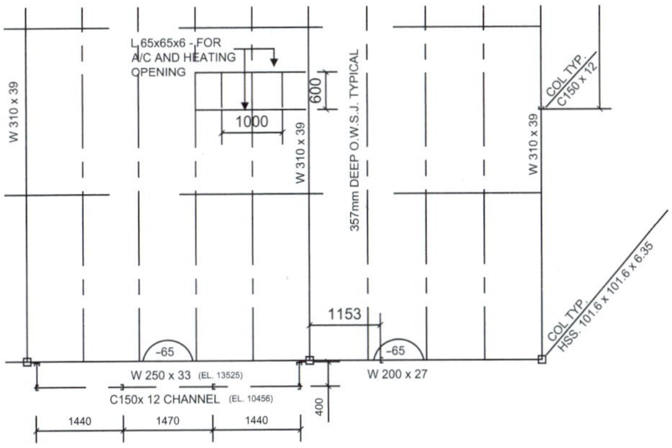

Source: Courtesy of Clint Kissoon, Chair, George Brown College, School of Architectural Studies.

FIGURE 16.9

A small steel structure building under construction. Note how the top of the O.W.S.J. are in alignment with the beam to the left of corner column. In our Mini Mart drawings, the beam at right carrying the joists would be placed at an elevation of 65 mm lower than the beam at left (see Figure 16.8).

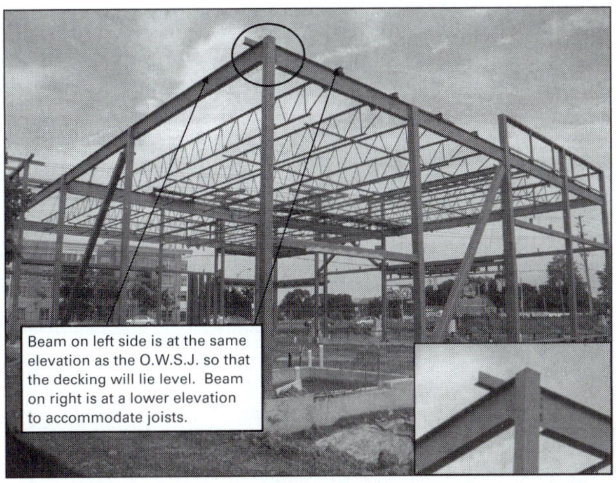

Courtesy Tom Stephenson

FIGURE 16.10

Photo shows the bolt connections taking place between the columns, beams, and braces. Note the temporary connecting bar aligning the connection point at the brace.

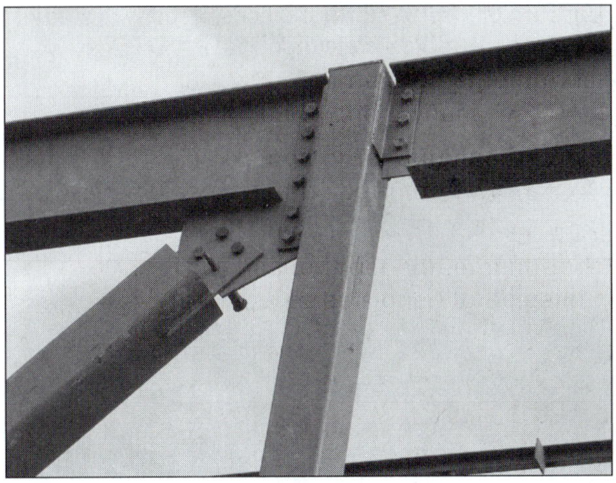

Courtesy Tom Stephenson

FIGURE 16.11

Lightweight structural steel studs are filled in around supporting structural steel. These studs are non-load-bearing in this use.

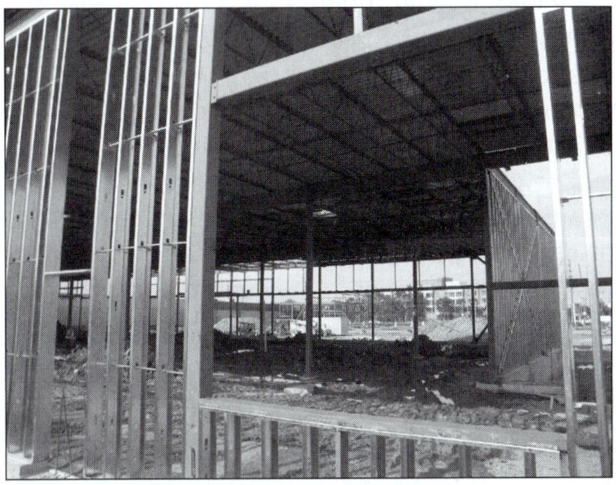

Courtesy Tom Stephenson

Mini Mart Drawings: Structural Steel

The Mini Mart is constructed out of structural steel with the main column supports laid out on the structural grid (refer to S2 Mini Mart & Gas Bar drawings). You can see the grid laid out from 1 to 4 horizontally and A to B vertically. A series of columns around the perimeter support the steel beam, which supports the steel joists, called **open web steel joists (O.W.S.J.)**. Recall that steel beams are identified by a letter, e.g., W wide flange I beam, then the height, and then the weight per metre listed in kilograms (refer back to Chapter 7, Figure 7.27). W250 × 33 would refer to a wide flange I beam that is 250 mm in height and weighs 33 kilograms per metre. Structural steel members are often joined by welding or may also be joined by bolting the members together. Structural steel buildings can be categorized into two basic types of construction: a braced frame and an unbraced frame. A braced frame uses diagonal bracing in required locations to stabilize the structure, while the unbraced frame must use the stiffness of the load-carrying sections in the assembled frame to resist the horizontal movement that is predicted.

FIGURE 16.12
Lightweight structural steel studs top track has vertical slots to ensure that studs do not take weight from deflection in roof. Another method is to use a second track that allows the first track to slide up and down based on deflection of steel. A similar method is used with stud walls in reinforced concrete buildings.

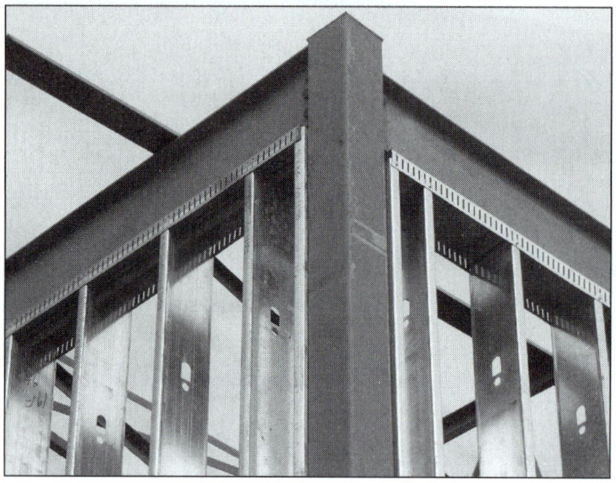

Courtesy Tom Stephenson

The Mini Mart is a braced building, and the locations of the braces are indicated in the plan view with a letter in brackets (A), (B), (C) on the roof framing plan S2; and the reference elevation is shown on S3 elevation (A), (B), (C). You can also see along the A1 to A4 and B1 to B4 grid axis that there is a notation indicated by a semicircle with a –65 inside. This indicates that the beam will be kept down 65 mm from the beams at the side so that the top of the O.W.S.J. that rest on them will end up level with the beams that run along axis A1 to B1 and A4 to B4. This ensures that the decking will lay flat and be adequately supported; these notations can vary from drawing to drawing and are used to indicate relative variances between beams and girders (refer to Figures 16.7 to 16.9).

KEY TERMS

Architectural drawings (p. 412)
Construction joint (p. 418)
Control joint (p. 418)
Cubic metres (p. 418)
Cutting-plane line (p. 414)
Dead loads (p. 415)
Detail designation (p. 414)
Engineering drawings (p. 412)
Expansion joint (p. 418)
Institutional, commercial, and industrial (ICI) (p. 411)
Isolation joint (p. 418)
Live loads (p. 415)
Megapascal (MPa) (p. 418)
Open web steel joists (O.W.S.J.) (p. 419)
Pounds per square inch (psi) (p. 418)
Reference grid, structural grid, or planning grid (p. 413)
Small buildings (p. 411)
Steel decking (p. 417)

Chapter 16 Assignment: Small Buildings, ICI Theory, and Mini Mart Drawings

Answer questions 1 to 6 using the material found in this chapter, and answer questions 7 to 20 using the Mini Mart & Gas Bar drawings that came with the textbook. Identify the letter that best answers the question.

_____ 1. What purpose does a reference grid serve when used in construction drawings?

a. It is used to define and identify the locations of the structural elements such as columns.
b. It is used to reference cutting plane line details.
c. It is used to reference elevation heights for structural beams that carry the loads of floor systems.
d. It is used to reference the property lines on site plans.

_____ 2. How is a structural steel frame building categorized?

a. braced or unbraced
b. shear supported
c. anchored or free floating
d. laterally wedged

_____ 3. Regarding the structure of the building, what is the photo below an example of?

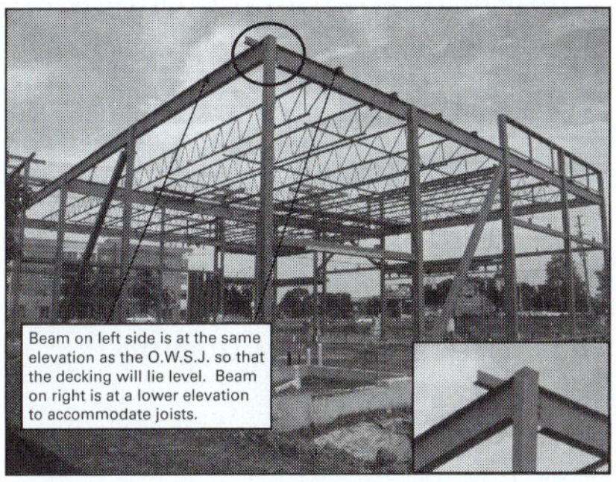

Courtesy Tom Stephenson

a. unbraced structural steel building
b. braced structural steel building
c. shear supported
d. laterally wedged

_____ 4. What is the weight of a C200 × 20 that is 3 metres long?

a. 600 lbs.
b. 600 kg
c. 60 kg
d. 60 lbs.

_____ 5. What is the figure below an example of?

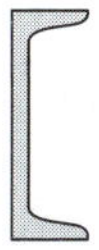

a. channel designated with a C on the drawings
b. I beam designated with a C on the drawings
c. wide flange I beam designated with a C on the drawings
d. C beam designated with a C on the drawings

_____ 6. What is the height (top of flange to top of flange) of a C150 × 12?

a. 150 inches
b. 150 mm
c. 12 inches
d. 12 mm

_____ 7. What size are the foot grilles used on the Mini Mart drawings?

a. 800 mm × 800 mm
b. 915 mm × 915 mm
c. 1000 mm × 1000 mm
d. 1020 mm × 10200 mm

_____ 8. What type and swing of door is used for the women's washroom that enters from outside the building?

a. left hand swing door
b. right hand swing door
c. pocket door
d. double action swing door

_____ 9. What is the size of the door labelled D-2?

a. 915 mm × 2190 mm × 91 mm
b. 915 mm × 2032 mm × 45 mm
c. 711 mm × 2032 mm × 45 mm
d. 915 mm × 2190 mm × 51 mm

_____10. How thick is the foundation wall for the Mini Mart?

a. 150 mm
b. 200 mm
c. 250 mm
d. 300 mm

_____11. How many recesses are there for foot grilles indicated on the drawings?

a. one

b. two

c. three

d. four

_____12. What size are the hollow structural steel columns that support the roof structure?

a. 100 mm × 100 mm × 6 mm

b. 101.6 mm × 101.6 mm × 6.35 mm

c. 150.6 mm × 150.6 mm × 10 mm

d. 200 mm × 200 mm × 15 mm

_____13. What size steel beam carries the roof structure over the right and left elevation exterior walls?

a. W200 × 20

b. W200 × 27

c. W250 × 33

d. W310 × 39

_____14. From the Mini Mart drawings, from an elevation perspective, how are all of the roof framing beams that surround the exterior of the building placed in relation to the O.W.S.J.?

a. The beams that the joists rest on are placed 65 mm higher.

b. The beams that the joists rest on are placed 65 mm lower than the beams that they don't rest on.

c. The beams are placed at the same elevation so that the joist will bear evenly on all the beams.

d. The beams that the joists don't rest on are placed 65 mm lower than the beams that they do rest on.

_____15. What material is used as an exterior finish for the Mini Mart?

a. Concrete ribbed block is used.

b. Hardy board siding is used.

c. Clay brick is used.

d. Stone veneer is used.

_____16. What is the size of the base plate that is welded to the HSS steel columns?

a. 175 mm × 150 mm × 10 mm

b. 200 mm × 200 mm × 20 mm

c. 300 mm × 300 mm × 10 mm

d. 326 mm × 210 mm × 11.6 mm

_____17. What is the diameter of the anchor bolts used to attach the steel columns to the foundation walls?

a. 10 mm

b. 20 mm

c. 25 mm

d. 30 mm

_____18. What is the spacing of the rungs for the steel ladder to the roof?

a. 200 mm

b. 300 mm

c. 400 mm

d. 600 mm

_____19. From the Mini Mart drawings, how are the galvanized metal ties placed in the ribbed concrete blocks?

a. 2nd block course @ 300 mm OC

b. 3rd block course @ 400 mm OC

c. 4th block course @ 400 mm OC

d. 5th block course @ 300 mm OC

_____20. Regarding the structure of the building, what is the Mini Mart an example of?

a. unbraced structural steel building

b. braced structural steel building

c. reinforced concrete building, including walls and roof

d. precast steel building

Chapter 16 Quiz Review: Small Buildings, ICI Theory, and Mini Mart Drawings

Answer questions 1 to 5 using the material found in this chapter, and answer questions 6 to 20 using the Mini Mart & Gas Bar drawings that came with the textbook. Identify the letter that best answers the question.

_____ 1. Why is a control joint made in poured concrete slabs and walls?

a. to allow for expansion that will prevent concrete from heaving and cracking

b. to control cracks so that they happen in planned locations

c. to prevent any cracking of the concrete

d. to control the end point of a pour so that it can be completed the next day

_____ 2. How is the diameter of reinforcing bars indicated on the drawings?

a. D indicating diameter, for example 20 mm diameter bar will state 20 D

b. CM indicating diameter, for example a 20 cm diameter bar will state 20 CM

c. DM indicating diameter, for example a 20 mm diameter bar will state 20 DM

d. M indicating diameter, for example a 20 mm diameter bar will state 20 M

_____ 3. What does the abbreviation HSS stands for?

a. hollow structural section

b. hollow structural steel

c. wide flange steel straight beam

d. height of structural steel

_____ 4. What purpose is served by the slots in the top track of the steel stud wall shown below?

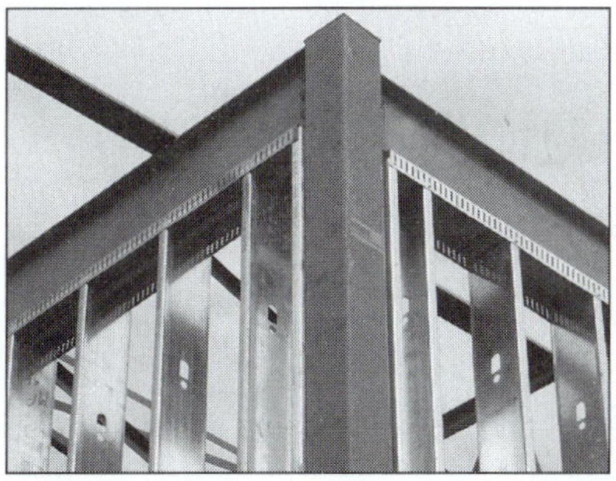

Courtesy Tom Stephenson

a. to help ventilate the wall when moisture laden air moves from warm to cold

b. to prevent lateral movement of the steel frame wall when wind forces hit the wall

c. to ensure that the wall does not absorb the weight of the steel when it deflects

d. to facilitate fastening the plates to the studs

_____ 5. Calculate the volume of concrete required for a concrete slab that is 12,000 mm × 8,500 mm × 100 mm thick?

a. 10.2 cubic metres of concrete

b. 102 cubic metres of concrete

c. 1,020 cubic metres of concrete

d. 10,200,000,000 cubic metres of concrete

_____ 6. What size of cant strip is used between the parapet and the roof of the Mini Mart?

a. 50 mm

b. 75 mm

c. 100 mm

d. 150 mm

_____ 7. What is the size of the exterior frame wall studs and what are they constructed out of?

a. 89 mm SS (steel studs) @ 400 OC

b. 89 mm SS (steel studs) @ 600 OC

c. 152 mm SS (steel studs) @ 400 OC

d. 152 mm SS (steel studs) @ 600 OC

_____ 8. What size is the air space between the block veneer and Styrofoam insulation?

a. 20 mm

b. 25 mm

c. 30 mm

d. 50 mm

_____ 9. What size reinforcing bars are placed horizontally in the concrete footings?

a. 5 - 20 M

b. 5 - 25 M

c. 4 - 20 M

d. 4 - 25 M

_____10. What strength of concrete is used for the slab on grade?

a. 20 MPa

b. 17.5 MPa

c. 25 MPa

d. 30.5 MPa

_____11. What is the typical R-value shown in the exterior walls of the Mini Mart?

a. R15

b. R12

c. R20

d. R22

_____12. What is the elevation for the top of the foundation wall? (This measurement, which is indicated on the drawings, is based on a benchmark that would be shown on the site plan, which is not included.)

a. 10000 mm

b. 10050 mm

c. 13525 mm

d. 12400 mm

_____13. For the beams that run parallel (at the edges of the building) with the O.W.S.J. where the joists don't actually bear on the steel beams, what is the elevation at the top of the steel?

a. 10000 mm

b. 10050 mm

c. 13525 mm

d. 13460 mm

_____14. What is the thickness of the slab on grade (in the locations where it is placed on grade/gravel)?

a. 100 mm thick

b. 75 mm thick

c. 125 mm thick

d. 150 mm thick

_____15. What size is the steel lintel over the exterior door to the storage area?

a. 90 mm × 90 mm × 8 mm

b. 90 mm × 150 mm × 12 mm

c. 100 mm × 100 mm × 12 mm

d. 90 mm × 100 mm × 10 mm

_____16. What size and material is used for the wall that is framed between the washrooms?

a. 152 mm steel stud

b. 152 mm wood stud

c. 89 mm steel stud

d. 89 mm wood stud

_____17. From the Mini Mart structural drawings, what size steel column is used at corner A1?

a. HSS 101.6 mm × 101.6 mm × 6.35 mm

b. HSS 150 mm × 150 mm × 8 mm

c. C 150 mm × 12

d. W200 mm × 27

_____18. What size steel support is used around the A/C and Heating opening in the roof?

a. 65 mm × 65 mm × 6 mm

b. 100 mm × 100 mm × 8 mm

c. 90 mm × 90 mm × 8 mm

d. 150 mm × 150 mm × 10 mm

_____19. What is the difference in elevation between the top of the foundation wall and the grade?

a. 300 mm

b. 350 mm

c. 400 mm

d. 600 mm

_____20. What is the height from the finished floor to the dropped ceiling?

a. 2800 mm

b. 2990 mm

c. 3190 mm

d. 4000 mm

APPENDIX A
Assignment Answers

Chapter 1 Assignment: Orthographic Views

(Answers 1–5; see your instructor for answers 6–10)

1.

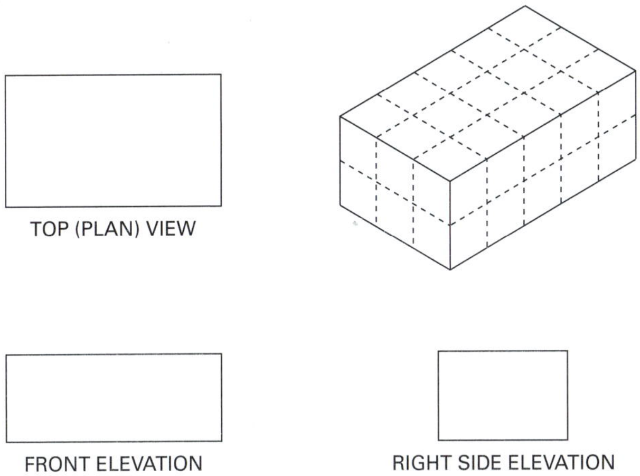

2.

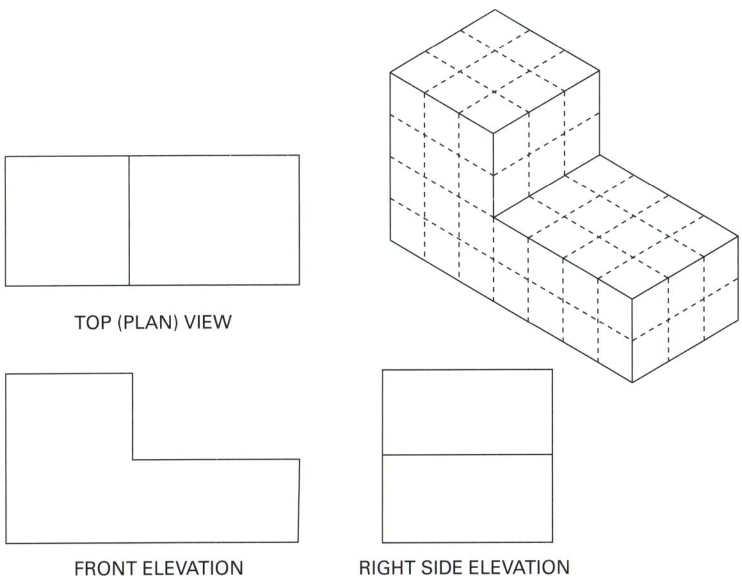

3.

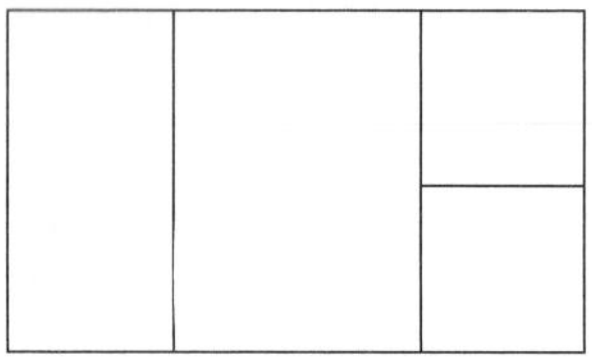

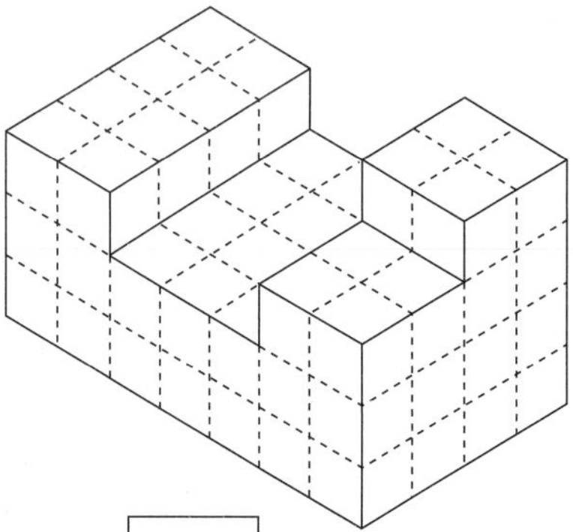

TOP (PLAN) VIEW

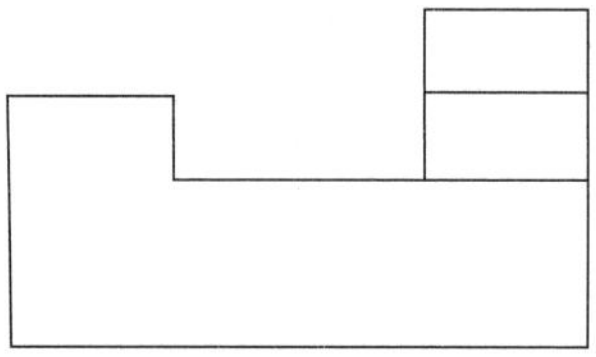

FRONT ELEVATION

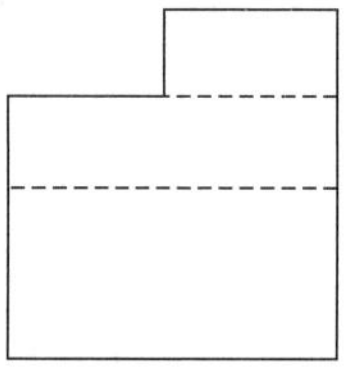

RIGHT SIDE ELEVATION

4.

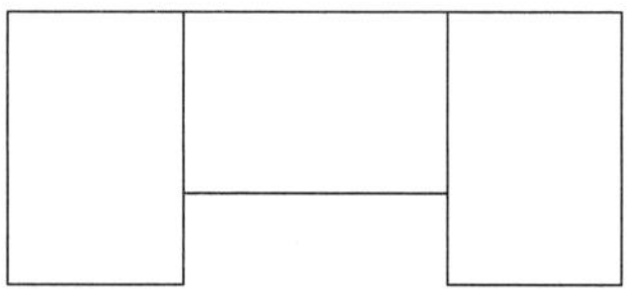

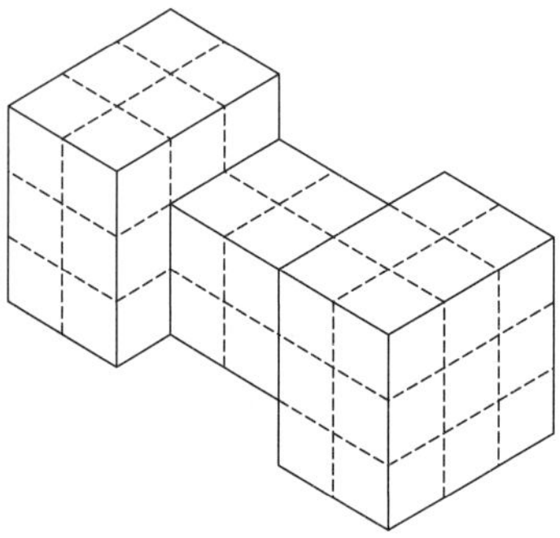

TOP (PLAN) VIEW

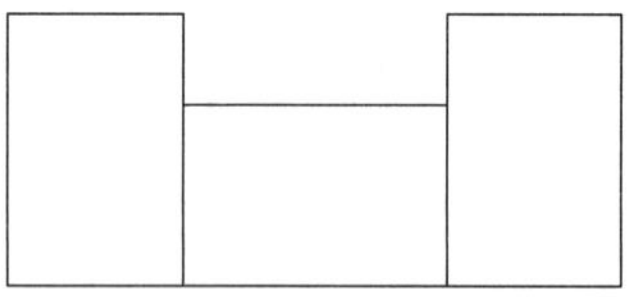

FRONT ELEVATION

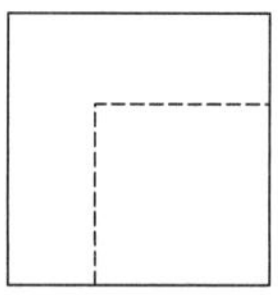

RIGHT SIDE ELEVATION

Note: Three Squares are in the middle section

5.

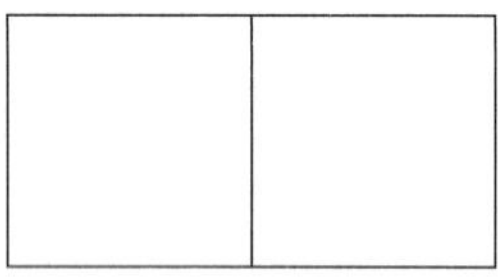

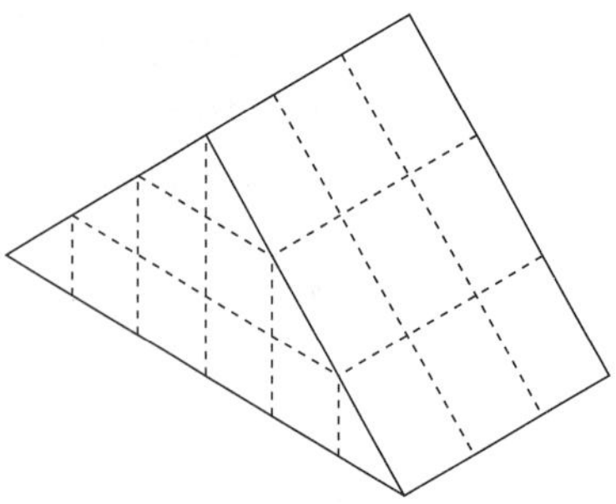

TOP (PLAN) VIEW

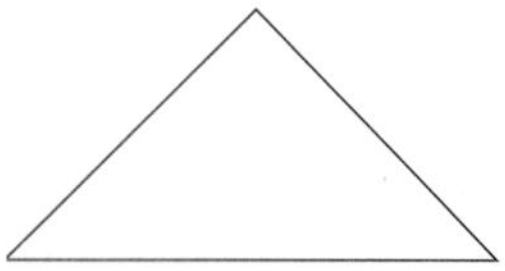

FRONT ELEVATION

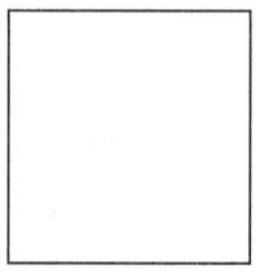

RIGHT SIDE ELEVATION

Chapter 2 Assignment: Reading Metric Scales

1. 210 mm
2. 108 mm
3. 62 mm
4. 2840 mm
5. 2140 mm
6. 1300 mm
7. 7 m
8. 5.550 m
9. 2.5 m
10. 1420 mm
11. 1050 mm
12. 660 mm

Chapter 2 Assignment: Reading Imperial Scales

1. 52'–6"
2. 40'
3. 41'–10"
4. 32'–10"
5. 22'–2"
6. 10'–9"
7. 5'–6"
8. 3'
9. 6'–10"
10. 4'–6"
11. 2'–2"

Chapter 3 Assignment: Elevation Drawings Using Doncaster Drawings

1. ANS: A
2. ANS: C
3. ANS: D
4. ANS: B
5. ANS: A
6. ANS: A
7. ANS: C
8. ANS: A
9. ANS: B
10. ANS: C
11. ANS: D
12. ANS: D
13. ANS: A
14. ANS: D
15. ANS: B
16. ANS: C
17. ANS: C
18. ANS: C
19. ANS: A
20. ANS: B

Chapter 4 Assignment: Abbreviations, Hatchings, Symbols, and Construction Notes

1. ANS: C
2. ANS: B
3. ANS: B
4. ANS: C
5. ANS: D
6. ANS: C
7. ANS: C
8. ANS: B
9. ANS: A
10. ANS: B
11. ANS: B
12. ANS: A
13. ANS: C
14. ANS: B
15. ANS: C
16. ANS: C
17. ANS: B
18. ANS: D
19. ANS: A
20. ANS: B

Chapter 5 Assignment: Reading Site Plan Drawings

1. ANS: B
2. ANS: A
3. ANS: A
4. ANS: B
5. ANS: D
6. ANS: C
7. ANS: A
8. ANS: C
9. ANS: C
10. ANS: D
11. ANS: B
12. ANS: C
13. ANS: B
14. ANS: B
15. ANS: C
16. ANS: B
17. ANS: D
18. ANS: A
19. ANS: D
20. ANS: C

Chapter 6 Assignment 1: Foundations

1. ANS: B
2. ANS: D
3. ANS: C
4. ANS: D
5. ANS: C
6. ANS: B
7. ANS: B
8. ANS: C
9. ANS: C
10. ANS: B
11. ANS: A
12. ANS: A
13. ANS: A
14. ANS: C
15. ANS: B
16. ANS: A
17. ANS: C
18. ANS: A
19. ANS: B
20. ANS: C

Chapter 6 Assignment 2: Foundations Using the Brook Drawings

1. ANS: A
2. ANS: C
3. ANS: B
4. ANS: C
5. ANS: A
6. ANS: C
7. ANS: D
8. ANS: C
9. ANS: B
10. ANS: A
11. ANS: D
12. ANS: B
13. ANS: B
14. ANS: A
15. ANS: C
16. ANS: B
17. ANS: C
18. ANS: B
19. ANS: C
20. ANS: A

Chapter 7 Assignment 1: Methods of Residential Construction

Answer Section

Multiple Choice

1. ANS: C
2. ANS: A
3. ANS: B
4. ANS: D
5. ANS: C
6. ANS: D
7. ANS: A
8. ANS: A
9. ANS: D
10. ANS: D
11. ANS: C
12. ANS: B
13. ANS: B
14. ANS: C
15. ANS: B
16. ANS: B
17. ANS: B
18. ANS: A
19. ANS: D
20. ANS: D

Chapter 7 Assignment 2: Floor Framing Using the Brook Drawings

1. ANS: C
2. ANS: C
3. ANS: B
4. ANS: A
5. ANS: D
6. ANS: C
7. ANS: D
8. ANS: D
9. ANS: B
10. ANS: C
11. ANS: B
12. ANS: C
13. ANS: B
14. ANS: B
15. ANS: C
16. ANS: C

17. ANS: B
18. ANS: C
19. ANS: A
20. ANS: C

Chapter 8 Assignment 1: Wall Framing and Doncaster Drawings

1. ANS: B
2. ANS: A
3. ANS: C
4. ANS: A
5. ANS: D
6. ANS: D
7. ANS: C
8. ANS: C
9. ANS: B
10. ANS: C
11. ANS: D
12. ANS: B
13. ANS: A
14. ANS: A
15. ANS: B
16. ANS: B
17. ANS: B
18. ANS: C
19. ANS: B
20. ANS: C

Chapter 8 Assignment 2: Floors and Walls, the Brook Drawings

1. ANS: B
2. ANS: D
3. ANS: C
4. ANS: C
5. ANS: A
6. ANS: D
7. ANS: A
8. ANS: A
9. ANS: D
10. ANS: C
11. ANS: B
12. ANS: C
13. ANS: B
14. ANS: D
15. ANS: A
16. ANS: B
17. ANS: A
18. ANS: A
19. ANS: D
20. ANS: D

Chapter 9 Assignment 1: Doncaster Truss Drawings and Doncaster Drawings

1. ANS: D
2. ANS: C
3. ANS: A
4. ANS: C
5. ANS: C
6. ANS: A
7. ANS: C
8. ANS: C
9. ANS: D
10. ANS: C
11. ANS: C
12. ANS: B
13. ANS: B
14. ANS: B
15. ANS: B
16. ANS: C
17. ANS: B
18. ANS: B
19. ANS: D
20. ANS: D

Chapter 9 Assignment 2: Roofs and the Brook Drawings

1. ANS: B
2. ANS: C
3. ANS: A
4. ANS: D
5. ANS: C
6. ANS: A
7. ANS: B
8. ANS: C
9. ANS: D
10. ANS: C
11. ANS: D
12. ANS: A
13. ANS: B
14. ANS: D
15. ANS: B
16. ANS: B
17. ANS: C
18. ANS: C
19. ANS: A
20. ANS: A

Chapter 10 Assignment 1: Exterior Finishes and Doncaster Drawings

1. ANS: A
2. ANS: A
3. ANS: B
4. ANS: A
5. ANS: B
6. ANS: A
7. ANS: A
8. ANS: A
9. ANS: C
10. ANS: A
11. ANS: B
12. ANS: B
13. ANS: C
14. ANS: B
15. ANS: A
16. ANS: B
17. ANS: D
18. ANS: B
19. ANS: B
20. ANS: C

Chapter 10 Assignment 2: Exterior Finishes and the Brook Drawings

1. ANS: D
2. ANS: D
3. ANS: C
4. ANS: B
5. ANS: B
6. ANS: C
7. ANS: B
8. ANS: A
9. ANS: C
10. ANS: D
11. ANS: D
12. ANS: A
13. ANS: A
14. ANS: C
15. ANS: B
16. ANS: C
17. ANS: B
18. ANS: A
19. ANS: B
20. ANS: B

Chapter 11 Assignment 1: Plumbing and HVAC Theory

1. ANS: A
2. ANS: B
3. ANS: B
4. ANS: C
5. ANS: D
6. ANS: A
7. ANS: B
8. ANS: B
9. ANS: D
10. ANS: B
11. ANS: A
12. ANS: C
13. ANS: B
14. ANS: D
15. ANS: C
16. ANS: C
17. ANS: B
18. ANS: C
19. ANS: A
20. ANS: A

Chapter 11 Assignment 2: Electrical, Green Systems, and Doncaster Drawings

1. ANS: B
2. ANS: C
3. ANS: B
4. ANS: C
5. ANS: D
6. ANS: B
7. ANS: A
8. ANS: B
9. ANS: D
10. ANS: A
11. ANS: A
12. ANS: B
13. ANS: D
14. ANS: D
15. ANS: B
16. ANS: C
17. ANS: D
18. ANS: B
19. ANS: C
20. ANS: D

Chapter 12 Assignment 1: The Building Envelope and the Doncaster Drawings

1. ANS: A
2. ANS: A
3. ANS: A
4. ANS: B
5. ANS: C
6. ANS: C
7. ANS: A
8. ANS: B
9. ANS: A
10. ANS: C
11. ANS: B
12. ANS: B
13. ANS: A
14. ANS: C
15. ANS: B
16. ANS: B
17. ANS: A
18. ANS: C
19. ANS: B
20. ANS: C

Chapter 12 Assignment 2: The Building Envelope and the Brook Drawings

1. ANS: B
2. ANS: C
3. ANS: D
4. ANS: A
5. ANS: C
6. ANS: B
7. ANS: A
8. ANS: D
9. ANS: B
10. ANS: C
11. ANS: B
12. ANS: A
13. ANS: B
14. ANS: C
15. ANS: D
16. ANS: C
17. ANS: A
18. ANS: C
19. ANS: B
20. ANS: B

Chapter 13 Assignment 1: Interior Finishes, Kitchens, Baths, and Doncaster Drawings

1. ANS: C
2. ANS: C
3. ANS: C
4. ANS: B
5. ANS: B
6. ANS: D
7. ANS: A
8. ANS: C
9. ANS: D
10. ANS: A
11. ANS: A
12. ANS: B
13. ANS: B
14. ANS: C
15. ANS: A
16. ANS: B
17. ANS: C
18. ANS: C
19. ANS: B
20. ANS: A

Chapter 13 Assignment 2: Stairs Using Doncaster Drawings

Answer Section

1. ANS: C
2. ANS: A
3. ANS: B
4. ANS: D
5. ANS: B
6. ANS: A
7. ANS: C
8. ANS: B
9. ANS: C
10. ANS: C
11. ANS: B
12. ANS: B
13. ANS: B
14. ANS: B
15. ANS: B
16. ANS: C
17. ANS: B
18. ANS: C

19. ANS: B
20. ANS: C

Chapter 14 Assignment 1: Specifications and Contracts

1. ANS: A
2. ANS: B
3. ANS: C
4. ANS: B
5. ANS: C
6. ANS: D
7. ANS: B
8. ANS: A
9. ANS: D
10. ANS: B
11. ANS: B
12. ANS: C
13. ANS: D
14. ANS: C
15. ANS: A
16. ANS: C
17. ANS: B
18. ANS: C
19. ANS: D
20. ANS: B

Chapter 15 Assignment 1: Renovation Drawings Using Khalmur Drawings

1. ANS: C
2. ANS: C
3. ANS: D
4. ANS: A
5. ANS: B
6. ANS: C
7. ANS: C
8. ANS: D
9. ANS: D
10. ANS: C
11. ANS: B
12. ANS: D
13. ANS: B
14. ANS: B
15. ANS: A
16. ANS: A
17. ANS: B
18. ANS: B
19. ANS: D
20. ANS: C

Chapter 16 Assignment: Small Buildings, ICI Theory, and Mini Mart Drawings

1. ANS: A
2. ANS: A
3. ANS: B
4. ANS: C
5. ANS: A
6. ANS: B
7. ANS: B
8. ANS: B
9. ANS: D
10. ANS: D
11. ANS: B
12. ANS: B
13. ANS: D
14. ANS: B
15. ANS: A
16. ANS: A
17. ANS: B
18. ANS: B
19. ANS: B
20. ANS: B

APPENDIX B
Math Review

The following formulas and examples provide a simple review of mathematics typically used in construction layout and estimating.

Review: The Metric System

1 kilometre (km) = 1000 metres (m)
1 centimetre (cm) = .01 metre (m)
1 millimetre (mm) = .001 metre (m)

Example 1

Express 5250 mm as metres.

Divide 5250 by 1000 = 5.250 m (note there are 1000 mm in a metre).

An easier way to do the above example is to simply move the decimal place to the left three spaces, the same as the number of zeros in 1000.

Example 2

Express 3.25 m as millimetres.

Multiply 3.25 by 1000 = 3250 mm.

Alternately, move the decimal place to the right three spaces, the same as the number of zeros in 1000.

Example 3

Express 4750 mm as centimetres.

Divide 4750 by 10 = 475 cm.

Alternately, move the decimal place to the left one space, the same as the number of zeros in 10.

Note when calculating measurements in metric for perimeters, areas, and volumes, you always convert all measurements to the same units (e.g., convert all to millimetres, centimetres, or metres). In some cases it may be necessary to convert to the imperial system; therefore, standard conversion factors are **http://www.wsdot.wa.gov/reference/metrics/factors.htm**.

Conversion Factors for Imperial to Metric and for Metric to Imperial

1 in. = 25.4 mm Remembering this one conversion factor can be very useful.

Example 1

Express 16 in. as millimetres.

Multiply 16 × 25.4 = 406.4 mm.

You can also use the 25.4 conversion factor to convert from metric to imperial.

Example 2

Express 400 mm as inches.

Divide 400 by 25.4 = 15.75 in.

Review: Expressing Common Fractions as Decimal Fractions

A common fraction can be expressed as a decimal fraction by dividing the numerator by the denominator.

Example 1

Express $\frac{5}{8}$ as a decimal fraction.

Write $\frac{5}{8}$ as indicated and divide 5 by 8. $8\overline{)5}$

Place a decimal point after the 5 and add zeros to the right of the decimal point. $8\overline{)5.000}$

Place the decimal point for the answer directly above the decimal point in the dividend. $\begin{array}{r} . \\ 8\overline{)5.000} \end{array}$

Divide. $\begin{array}{r} 0.625 \\ 8\overline{)5.000} \end{array}$

A common fraction that will not divide evenly is expressed as a repeating decimal.

Example 2

Express $\frac{1}{3}$ as a decimal fraction.

Write $\frac{1}{3}$ as indicated in the division example. $3\overline{)1}$

Place a decimal point after the 1 and add zeros to the right of the decimal point. $3\overline{)1.0000}$

Place the decimal point for the answer directly above the decimal point in the dividend. $\begin{array}{r} . \\ 3\overline{)1.0000} \end{array}$

Divide. $\begin{array}{r} 0.3333 \\ 3\overline{)1.0000} \end{array}$

Review: Expressing Decimal Fractions as Common Fractions

To express a decimal fraction as a common fraction, write the number after the decimal point as the numerator of a common fraction. Write the denominator as 1 followed by as many zeros as there are digits to the right of the decimal point. Express the common fraction in lowest terms.

Example 1

Express 0.9 as a common fraction.

Write 9 as the numerator.

Write the denominator as 1 followed by one zero. The denominator is 10. $\frac{9}{10}$

Example 2

Express 0.125 as a common fraction.

Write 125 as the numerator.

Write the denominator as 1 followed by 3 zeros. The denominator is 1000.

$\frac{125}{1000}$

Express the fraction in lowest terms.

$\frac{125}{1000} = \frac{1}{8}$

Review: Expressing Inches as Feet and Inches

To express inches as feet and inches, divide the given length in inches by 12 to obtain the number of whole feet. The remainder is the extra number of inches. The answer is the number of whole feet plus the remainder in inches.

Example 1

Express 176 $\frac{7}{16}$ in. as feet and inches.

Divide 176 $\frac{7}{16}$ in. by 12.

There are 14 ft. plus a remainder of 8 $\frac{7}{16}$ in.

4 (feet)
14)176 $\frac{7}{16}$
12
56
48
8 $\frac{7}{16}$ ← Remainder (feet)

14 feet 8 $\frac{7}{16}$ inches

Example 2

Express 54.2 in. as feet and inches.

Divide 54.2 in. by 12.

There are 4 ft. plus a remainder of 6.2 in.

4 (feet)
12)54.2
48
6.2 ← Remainder (feet)

14 feet 6.2 inches

Review: Expressing Feet and Inches as Inches

To express feet and inches as inches, multiply the number of feet in the given length by 12. To this product, add the number of inches in the given length.

Example 1

Express 7 ft. 9 $\frac{3}{4}$ in. as inches.

Multiple 7 ft. by 12. There are 84 in. in 7 ft.

Add 9 $\frac{3}{4}$ in. to 84 in.

$7 \times 12 = 84$
7 ft. = 84 in.
84 in .+ 9 $\frac{3}{4}$ in. = 93 $\frac{3}{4}$ in.

Review: Expressing Inches as Decimal Fractions of a Foot

An inch is $\frac{1}{12}$ of a foot. To express whole inches as a decimal part of a foot, divide the number of inches by 12.

Example 1

Express 7 in. as a decimal fraction of a foot.

Divide 7 by 12.

$7 \div 12 = 0.58$ ft.

To express a common fraction of an inch as a decimal fraction of a foot, express the common fraction as a decimal, then divide the decimal by 12.

Example 2

Express $\frac{3}{4}$ in. as a decimal fraction of a foot.

Express $\frac{3}{4}$ as a decimal.

$3 \div 4 = 0.75$

Divide the decimal by 12.

$0.75 \div 12 = 0.06$ ft.

Example 3

Express $4\frac{3}{4}$ in. as a decimal fraction of a foot.

Express $4\frac{3}{4}$ as a decimal. $4 + \frac{3}{4} = 4 + 0.75 = 4.75$

Divide the decimal inches by 12. $4.75 \div 12 = 0.39666$ ft. or rounded to .40 ft.

Review: Expressing Decimal Fractions of a Foot as Inches

To express a decimal part of a foot as decimal inches, multiply by 12.

Example 1

Express 0.62 feet as inches.

Multiply 0.62 by 12. $0.62 \times 12 = 7.44$

7.44 in.

To express a decimal fraction of an inch as a common fraction, see Review: Expressing Decimal Fractions as Common Fractions.

Review: Perimeter Measurement

Perimeter measurement is the linear measurement around the outside edge of an object. Therefore, the sum of the sides will equal the perimeter.

Rectangle

The formula for the perimeter of a rectangle is:

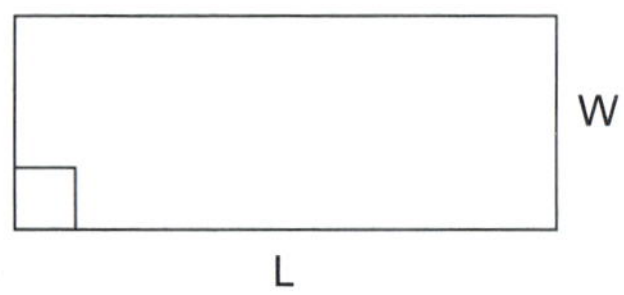

$P = 2(L + W)$

or

$P = 2L + 2W$

Example 1

Find the perimeter of a rectangle with a length of 8 ft. and a width of 6 ft.

$P = 2(L + W)$

$P = 2(8 + 6)$

$P = 28$ ft.

Parallelogram

Perimeter of a parallelogram is expressed as:

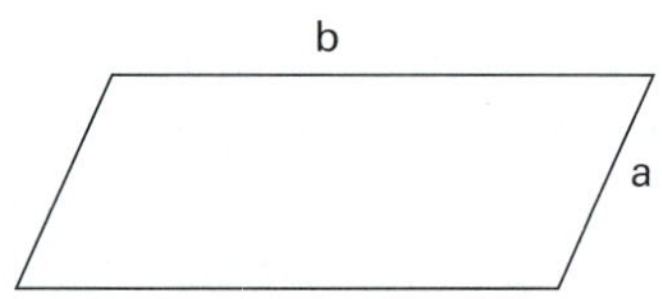

$P = 2a + 2b$

Example 2

Find the perimeter of a parallelogram where side a is 11 ft. and side b is 7 ft.

$P = 2a + 2b$

$P = 2(11) + 2(7)$

$P = 36$ ft.

Circle

The perimeter measurement of a circle is called circumference and is expressed as:

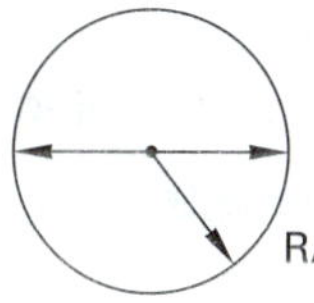

$C = \pi d$

or

$C = 2\pi r$

Note: π (pronounced "pi") is approximately equal to 3.14.

Example 3

A circle with a radius of 3 ft. will have a circumference of

$C = \pi d$

$C = 2(3.14)3$

$C = 18.84$ ft.

Review: Area Measure

Surface area is measured by determining the number of surface units contained in a surface. A surface is two-dimensional; it has a length and width, but no thickness or height. A cube for example would have six surfaces; each surface area would be added together to provide the total surface area of the cube. Both length and width must be expressed in the same unit of measure. Area is expressed in square units (e.g., 5 ft. by 8 ft. equals 40 sq. ft.).

Equivalent Units of Area Measure

1 sq. ft. = 12 in. × 12 in. = 144 sq. in.

1 sq. yd. = 3 ft. × 3 ft. = 9 sq. ft.

To express a given unit of area as a larger unit of area (e.g., to express square inches as square feet), divide the given area by the number of square units contained in one of the larger units.

Example 1

Express 648 sq. in. as square feet.

Since 144 sq. in. = 1 sq. ft., divide 648 by 144.

$\frac{648}{144} = 4.5$ sq. ft.

648 sq. in. = 4.5 sq. ft.

648 ÷ 144 = 4.5
648 sq. in. = 4.5 sq. ft.

Example 2

Express 28.8 sq. ft. as square yards.

Since 9 sq. ft. = 1 sq. yd., divide 28.8 by 9.

28.8 ÷ 9 = 3.2
28.8 sq. ft. = 3.2 sq. yd.

To express a given unit of area as a smaller unit of area (e.g., to convert square feet to square inches), multiply the given area by the number of smaller square units contained in one of the larger units (e.g., 144 sq. in. in 1 sq. ft.).

Example 3

Express 7.5 sq. ft. as square inches.

Since 144 sq. in. = 1 sq. ft., multiply 7.5 by 144.

7.5 × 144 = 1080
23 sq. yd. = 207 sq. ft.

Example 4

Express 23 sq. yd. as square feet.

Since 9 sq. ft. = 1 sq. yd., multiply 23 by 9.

$23 \times 9 = 207$
23 sq. yd. = 207 sq. ft.

Computing Areas of Common Geometric Figures

Rectangle: A rectangle is a four-sided plane figure with four right (90°) angles.

The area of a rectangle is equal to the product of its length and width.

Area = length × width ($A = l \times w$)

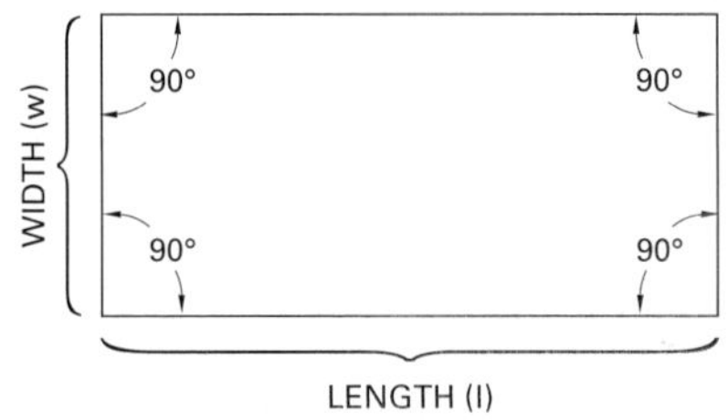

Example 1

Find the area of a rectangle 24 ft. long and 13 ft. wide.

$A = l \times w$

A = 24 ft. × 13 ft.

A = 312 sq. ft.

Triangle: A triangle is a plane figure with three sides and three angles.

The area of a triangle is equal to one-half the product of its base and height.

$A = \frac{1}{2}$ base × height

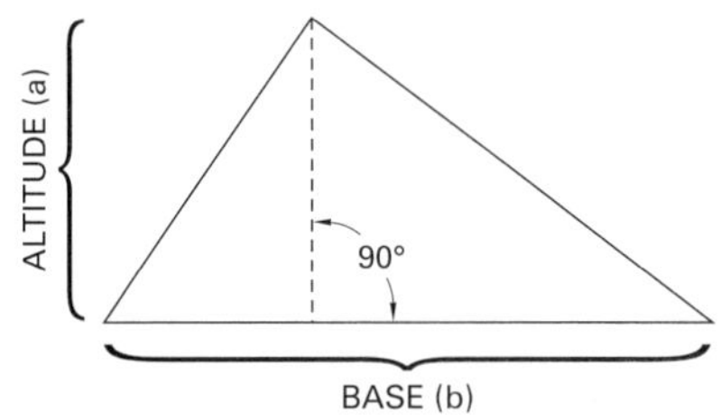

Example 2

Find the area of a triangle with a base of 16 ft. and a height of 12 ft.

$A = \frac{1}{2} b \times a$

$A = \frac{1}{2} \times$ 16 ft. × 12 ft.

A = 96 sq. ft.

Circle: The area of a circle is equal to p times the square of its radius.

Area = $\pi \times$ radius^2 ($A = \pi \times r^2$)

Note: π (pronounced "pi") is approximately equal to 3.14. Radius squared (r^2) means $r \times r$.

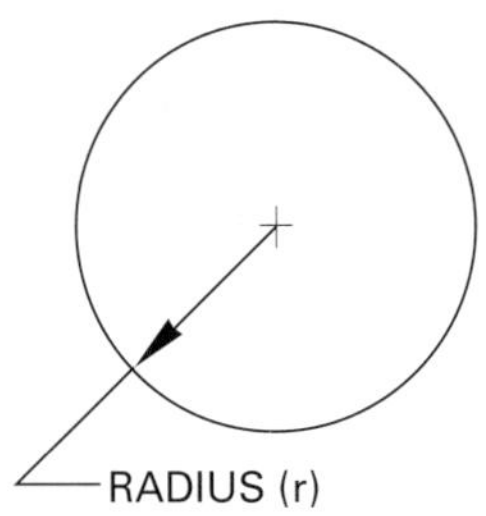

Example 3

Find the area of a circle with a 15-in. radius.

$A = \pi \times r^2$

A = 3.14 × (15 in.)^2

A = 3.14 × 225 sq. in.

A = 706.5 sq. in.

Review: Volume Measure

A solid is measured by determining the number of cubic units contained in it. A solid is three-dimensional; it has length, width, and thickness or height. Length, width, and thickness or height must be expressed in the same unit of measure. Volume is expressed in cubic units (e.g., 3 ft. × 5 ft. × 10 ft. = 150 cu. ft.).

Equivalent Units of Volume Measure

1 cu. ft. = 12 in. × 12 in. × 12 in. = 1728 cu. in.

1 cu. yd. = 3 ft. × 3 ft. × 3 ft. = 27 cu. ft.

To express a given unit of volume as a larger unit of volume (e.g., to convert cubic inches to cubic feet), divide the given volume by the number of cubic units contained in one of the larger units (e.g., 1728 cu. in. = 1 cu. ft.; therefore divide by 1728).

Example 1

Express 6048 cu. in. as cubic feet.

Since 1728 cu. in. = 1 cu. ft., divide 6048 by 1728.

6048 ÷ 1728 = 3.5
6048 cu. in. = 3.5 cu. ft.

Example 2

Express 167.4 cu. ft. as cubic yards.

Since 27 cu. ft. = 1 cu. yd., divide 167.4 by 27.

167.4 ÷ 27 = 6.2
167.4 cu. ft. = 6.2 cu. yd.

To express a given unit of volume as a smaller unit of volume (e.g., to convert cubic feet to cubic inches), multiply the given volume by the number of smaller cubic units contained in one of the larger units (e.g., 1728 cu. in. = 1 cu. ft.; therefore multiply by 1728).

Example 1

Express 1.6 cu. ft. as cubic inches.

Since 1728 cu. in. = 1 cu. ft., multiply 1.6 by 1728.

1.6 × 1728 = 2764.8
1.6 cu. ft. = 2764.8 cu. in.

Example 2

Express 8.1 cu. yd. as cubic feet.

Since 27 cu. ft. = 1 cu. yd., multiply 8.1 by 27.

8.1 × 27 = 218.7
8.1 cu. yd. = 218.7 ft.

Computing Volumes of Common Solids

A prism is a solid with two identical faces called bases and parallel lateral edges. In a right prism, the lateral edges are perpendicular (at 90°) to the base. The altitude or height (h) of a prism is the perpendicular distance between its two bases. Prisms are named according to the shapes of their bases.

The volume of any prism is equal to the product of the area of its base and altitude or height.

Volume = area of base × altitude ($V = A_B \times h$)

Right Rectangular Prism

A right rectangular prism has rectangular bases.

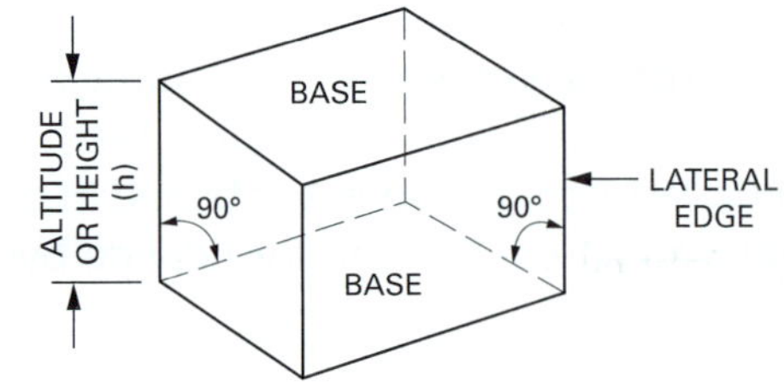

Volume = area of base × altitude

$V = A_B \times h$

Example 1

Find the volume of a rectangular prism with a base length of 20 ft., a base width of 14 ft., and a height (altitude) of 8 ft.

$V = A_B \times h$

Compute the area of the base (A_B):

Area of base = length × width

A_B = 20 ft. × 14 ft.

A_B = 280 sq. ft.

Compute the volume of the prism:

$V = A_B \times h$

V = 280 sq. ft. × 8 ft.

V = 2240 cu. ft.

Right Triangular Prism

A right triangular prism has a triangular base in which the joining edges and faces are perpendicular to the base edges and are rectangular.

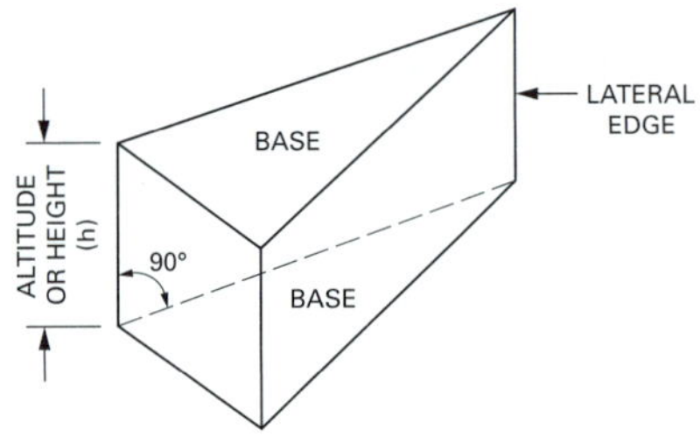

Volume = area of base × height

$V = A_B \times h$

Example 2

Find the volume of a right triangular prism in which the base of the triangle is 5 ft., the altitude of the triangle is 3 ft., and the altitude (height) of the prism is 4 ft. Refer to the accompanying figure.

Volume = area of base × height

$V = A_B \times h$

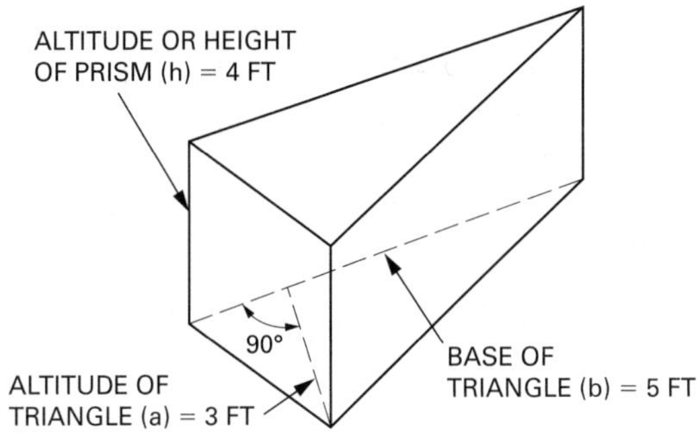

Compute the area of the base:

Area of base = $\frac{1}{2}$ base of triangle × height of triangle

$A_B = \frac{1}{2} b \times h$

$A_B = \frac{1}{2} 5 \text{ ft.} \times 3 \text{ ft.}$

$A_B = 7.5$ sq. ft.

Compute the volume of the prism:

$V = A_B \times h$

$V = 7.5$ sq. ft. × 4 ft.

$V = 30$ cu. ft.

Right Circular Cylinder

A right circular cylinder has circular bases.

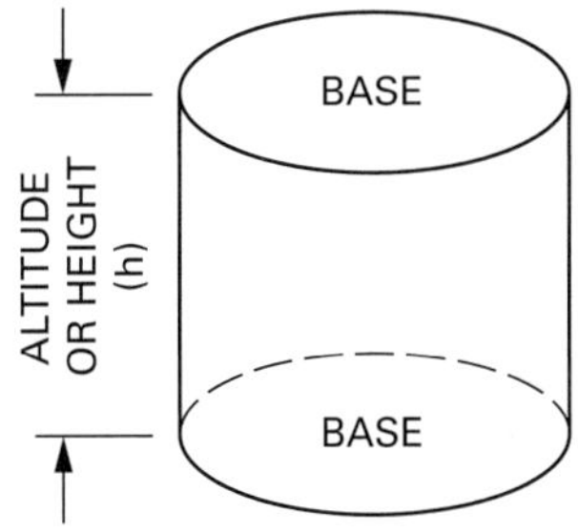

Volume = area of base × altitude

$V = A_B \times h$

Example 3

Find the volume of a circular cylinder 1 ft. in diameter and 10 ft. high.

Note: radius = ½ diameter.

$V = A_B \times h$

Compute the area of the base:

Area of the base = p × radius squared

$A_B = 3.14 \times (0.5)^2$

$A_B = 3.14 \times 0.5$ ft. × 0.5 ft.

$A_B = 3.14 \times 0.25$ sq. ft.

$A_B = 0.785$ sq. ft.

Compute the volume of the cylinder:

$V = A_B \times h$

$V = 0.785$ sq. ft. × 10 ft.

$V = 7.85$ cu. ft.

Review: Finding an Unknown Side of a Right Triangle, Given Two Sides

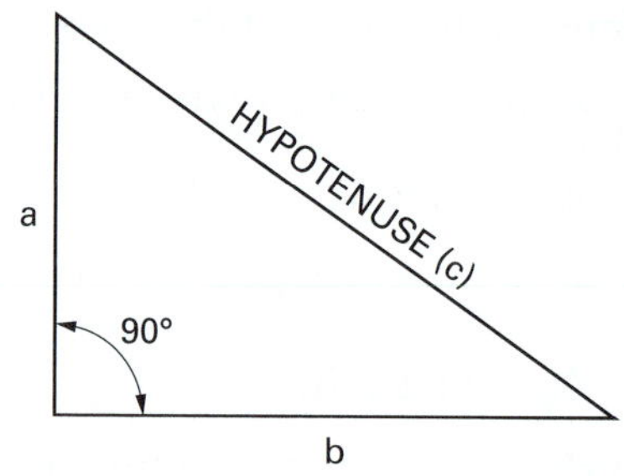

If one of the angles of a triangle is a right (90°) angle, the figure is called a right triangle. The side opposite the right angle is called the hypotenuse. In the figure shown, c is opposite the right angle; therefore c is the hypotenuse.

In a right triangle, the square of the hypotenuse is equal to the sum of the squares of the other two sides:

$$c^2 = a^2 + b^2$$

If any two sides of a right triangle are known, the length of the third side can be determined by one of the following formulas:

$$c = \sqrt{a^2 + b^2}$$

$$a = \sqrt{c^2 - b^2}$$

$$b = \sqrt{c^2 - a^2}$$

Example 1

In the right triangle shown, a = 6 ft., b = 8 ft.; find c.

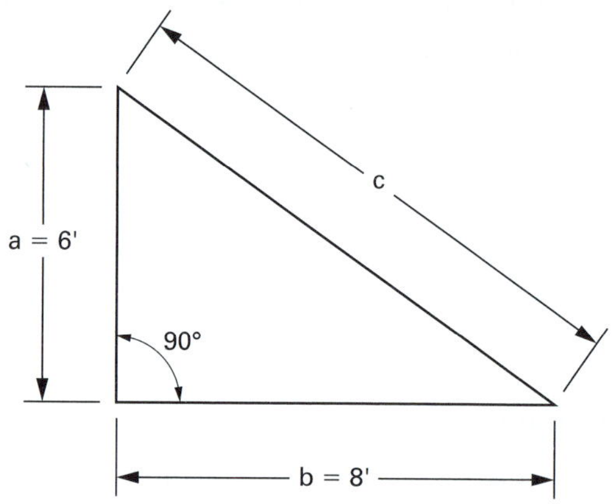

$$c = \sqrt{a^2 + b^2}$$

$$c = \sqrt{6^2 + 8^2}$$

$$c = \sqrt{36 + 64}$$

$$c = \sqrt{100}$$

$$c = 10 \text{ feet}$$

Example 2

In the right triangle shown, c = 30 ft., b = 20 ft.; find a.

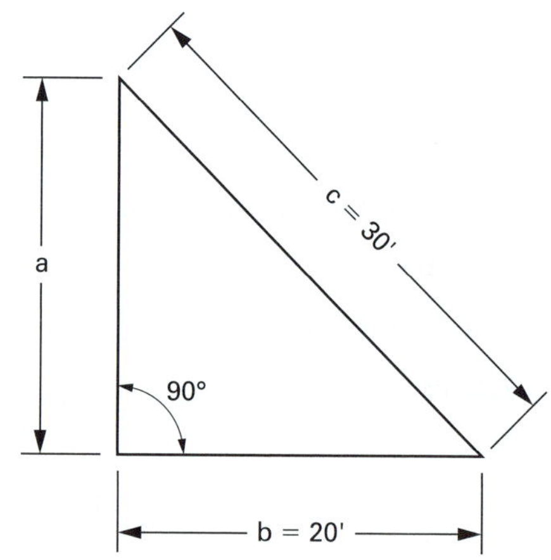

$$a = \sqrt{c^2 - b^2}$$

$$a = \sqrt{30^2 - 20^2}$$

$$a = \sqrt{900 - 400}$$

$$a = \sqrt{500}$$

$$a = 22.36 \text{ feet (to 2 decimal places)}$$

Example 3

In the right triangle shown, c = 18 ft., a = 6 ft.; find b.

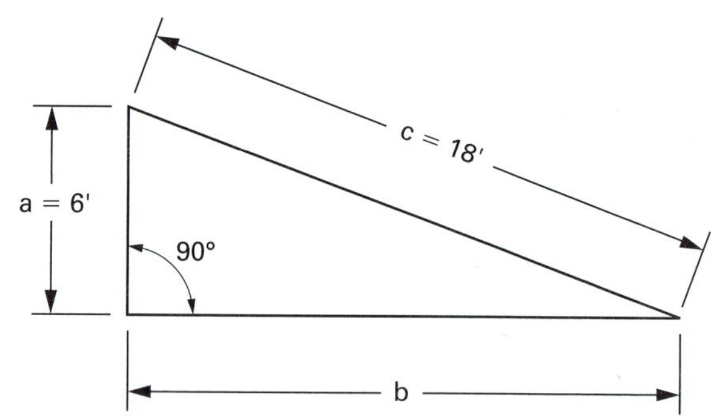

$$b = \sqrt{c^2 - a^2}$$

$$b = \sqrt{18^2 - 6^2}$$

$$b = \sqrt{324 - 36}$$

$$b = \sqrt{288}$$

$$b = 16.97 \text{ feet (to 2 decimal places)}$$

APPENDIX C
Squaring a Corner Using the Pythagorean Formula

The Pythagorean formula states that for a right-angled triangle $A^2 + B^2 = C^2$, or the square root of $\sqrt{A^2 + B^2} = C$.

A common method used for squaring corners is the 3, 4, 5 method. This method applies the Pythagorean formula using the constant 3, 4, 5. If length A is 3', length B is 4', and it is a right-angled triangle, length C—the hypotenuse—must equal 5. On larger corners a multiple of these numbers can be used (e.g., 6', 8', 10' or 9', 12', 15').

To double-check for square, measure corner to corner and then measure the opposite corner to corner. If the measurements are the same and the tape was held approximately level, then the building is laid out square.

Layout of addition using two blocks of wood and a string and the 3, 4, 5 method of squaring a corner.

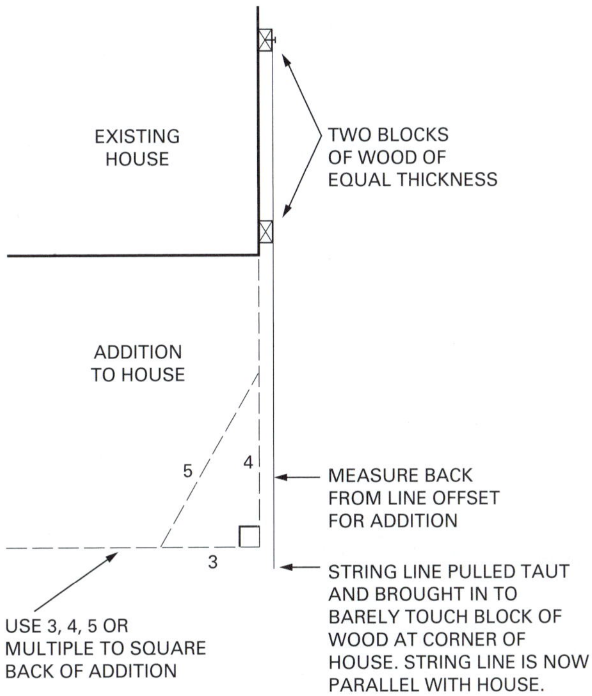

APPENDIX D
Calculating Headroom for Stairs

(Further Example of Stairwell Opening from Chapter 13)

Sometimes the opening may already be framed for the stairwell and you need to determine if the headroom will be sufficient for a given unit rise and unit run (for example, a stair builder subcontractor will check to ensure that the framed opening will actually work before they go ahead and build the stairs). The method is similar to that of calculating a stairwell opening, except the stairwell opening size is known and the resulting headroom is not. Once calculated, the headroom can be checked to make sure that it meets the minimum code requirement. If it does not meet minimum code, the length of the treads could be reduced or the number of risers could be reduced as long as the sizes stay within the NBC requirements in Section 9.8.4.1 and 9.8.4.2.

Example

Calculate the headroom size based on a stairwell opening size of 3175 mm and:

14 risers with a unit rise of 186 mm

13 treads with a unit run of 254 mm

Thickness of floor 300 mm (measurement taken from finished ceiling to finished floor).

Headroom + Floor thickness = X

$$\frac{\text{Unit run}}{\text{Unit rise}} = \frac{\text{Stairwell opening size}}{X}$$

$$\frac{254}{186} = \frac{3175}{X}$$

$$254X = 590\ 550$$

$$X = \frac{590\ 550}{254}$$

$$X = 2325$$

Next subtract the floor thickness to provide the headroom.

$$2325 - 300 = 2025 \text{ mm}$$

The headroom for this stairs is legal for a dwelling unit as it is above the minimum 1950 mm NBC requirement (refer to NBC Division B Part 9.8.2.2). A more simplified plug-in formula that will provide the same result is as follows:

$$\text{Headroom} = \frac{(\text{Unit rise} \times \text{Stairwell opening})}{\text{Unit run}} - \text{Floor thickness}$$

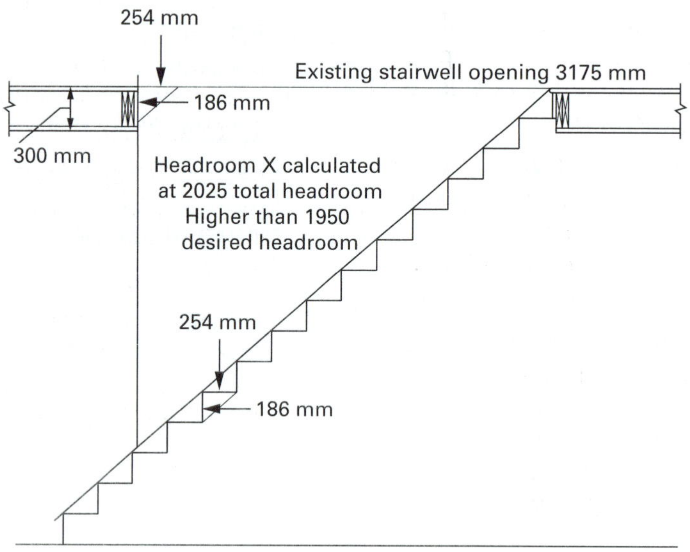

APPENDIX E
Web References

The following websites provide helpful information for researching and developing your knowledge of the construction industry:

An excellent measurement conversion website that also has a free downloadable conversion software program: **http://www.wsdot.wa.gov/reference/metrics/factors.htm**

Product manufacturer Weyerhaeuser, has a engineered joist system called Trus-Joist that provides detailed information regarding the design and installation of its manufactured wood products such as joist systems: **http://www.woodbywy.com/**

Canada Mortgage and Housing Corporation (CMHC) has a virtual library of information regarding all aspects of the Canadian housing industry, from trends to building methods: **http://www.cmhc-schl.gc.ca**

The National Research Council Canada (NRC) and its construction research arm, the Research Institute in Construction, provide both current and historical information (*Canadian Building Digest*, a discontinued series of articles published between 1960 and 1990, contains scientific, practical, and technical principles on virtually every aspect of building design, materials, and construction in Canada). The organization is also responsible for providing technical and administrative support for the development of the model building and fire codes used in Canada: **http://irc.nrc-cnrc.gc.ca**

Canadian Wood Council (CWC) provides information on the design, specification, construction, and inspection of all types of buildings constructed of wood and wood-manufactured products: **http://www.cwc.ca**

Canadian Home Builders' Association (CHBA) supplies information on the business of home building and renovating. The site also provides links to provincial and local associations, apprenticeships, training and education, and job postings: **http://www.chba.ca/**

Home Builder Magazine delivers information on the home building industry in Canada and includes articles on new products, methods of construction, and design layout for homes and condominiums. The site has access to back and current issues as well as hundreds of links to manufacturers and suppliers: **http://www.homebuildercanada.com/index.htm**

The Canadian Construction Association (CCA) provides information for the non-residential sectors of construction; the association also represents the ICI sector: **http://www.cca-acc.com**

Canadian Institute of Quantity Surveyors (CIQS) furnishes information regarding becoming a professional quantity surveyor (management of construction costs) and provides links to other sites: **http://www.ciqs.org**

Toolbase Services provides a wealth of information and links on all aspects of the home building and renovation sectors:

Advanced Buildings Technology and Practices is an excellent resource for information regarding advanced building systems and building science: **http://www.advancedbuildings.org**

Buildingscience.com is an excellent website that provides a wealth of free information regarding building science and green sustainable building practices: **http://buildingscience.com**

Canada Green Building Council (CAGBC) provides information regarding green and sustainable building requirements for LEED certification: **http://www.cagbc.org/**

Natural Resources Canada website has information on products and building practices connected to the Energy Star certification programs: **http://www.nrcan.gc.ca/home**

Glossary

50 (0 to 49 Divisions) Division MasterFormat The most recognized standard for specifications; replaced the 16 Division MasterFormat. (p. 358)

Abbreviation Short forms for descriptions that would otherwise clutter the drawings and make them more difficult to read. (p. 61)

Air barrier Placed on either the inside or outside of the wall, it forms a tight seal preventing air leakage from the building. It can be constructed in combination with a vapour retarder or as a separate entity using a house-wrap material such as Tyvek. When placed on the outside of the wall, the material must be able to breathe. (p. 302)

Air entrainment The introduction of tiny air bubbles to concrete by an air entraining agent. The purpose of air entrainment in concrete is to increase durability, particularly where concrete is exposed to freeze thaw cycles. (p. 246)

Allowance Values included in the contract for items that have not been defined or finalized (e.g., an allowance of $5 per square foot of floor tile). (p. 372)

Anchor bolt Used to fasten the sill plate to the foundation wall, thus connecting the framing to the foundation. (pp. 118, 154)

Architectural control A design requirement used by many towns and cities to ensure that development fits in with its environment and the aesthetic plan for the jurisdiction. Architectural control actually goes further than typical zoning requirements by ensuring that cookie cutter subdivisions are avoided by requiring a variety of designs, colours, and finish materials that might be otherwise omitted if the developer is left to their own devices. (p. 44)

Architectural drawings When describing ICI type drawings, the architectural drawings describe the design, size, dimension, and layout for the building without providing details regarding the structural requirements. (p. 411)

Architectural rendering A life-like drawing of an object drawn in perspective. (p. 5)

Back framing A separate step in the construction schedule that involves going back and framing things like bulkheads, bathtubs, and showers once the final details and mechanical rough-ins have been worked out. (p. 178)

Balloon construction Older method of construction that uses single-piece studs from sill plate to the second-floor roof; this method is rarely used in new construction. Joists rest on ribbon boards that are notched into studs. (p. 142)

Barrier-free access This type of design may respond to NBC requirements regarding accessibility as per Part 3.8 of the NBC, it may also go beyond code minimum standards to ensure that occupants that may have disabilities are able to function fully in their homes with minimum restriction. (p. 50)

Baseboard The moulding that finishes the floor and wall intersection. (p. 327)

Beam A horizontal structural member used to support floors and wall systems. (p. 142)

Bearing The support provided for a structural member. With reference to the NBC, bearing refers to the minimum amount that a structural member must be supported or sitting on a wall or column (e.g., the minimum bearing for a steel beam is 89 mm, and minimum bearing for a joist is 38 mm). (p. 149)

Benchmark A starting point elevation to which other elevations on the construction site are referenced from. Often the benchmark is 100 (metres or feet) to avoid negative numbers for drops in elevation. (pp. 61, 91)

Blocking Used sometimes in wall construction to strengthen the wall or act as a firestop. In some cases blocking may be an NBC requirement. (p. 170)

Break line A line used in construction drawings to cut off an object that continues but would not fit on the page or where another object may be located directly over it. (p. 32)

Brick veneer construction Construction that uses a non-load-bearing veneer wall (one wythe) on the exterior façade of the building. The brick is separated from the frame walls by a minimum 1" (25 mm) air space. (p. 141)

Building envelope All surfaces of the building that directly separate the inside environment from the outside environment. (p. 300)

Building Information Modelling (BIM) The convergence of 3D models with estimating, scheduling and facilities management software and databases used to more effectively manage construction projects. (p. 11)

Building permit Required when performing any structural work and work requiring a permit for a variety of other reasons. A permit will not be issued unless work meets all provincial code requirements. The building permit is provided and enforced by the local town or municipality. Other permits may be required based on the work being executed. (p. 105)

Building wrap A special breathable material used to wrap the outside of the house to protect from moisture penetration and air leakage. (p. 230)

Built-up wood beam A wood beam comprised of multiple pieces of dimensional or manufactured lumber. (p. 150)

Cabinet carcass The frame of the cabinet. (p. 335)

CAD Computer-aided design drafting. (p. 4)

Canadian Construction Documents Committee (CCDC) This national joint committee is responsible for the development, production, and review of standard Canadian construction contracts, forms, and guides. These documents are intended for use by all construction industry participants, including owners, design professionals, construction managers, legal professionals, contractors and subcontractors; and are designed to be fair to all parties involved in the agreement. The documents are the most widely used in the industrial, commercial, and institutional (ICI) sectors and may also be sometimes used in new custom residential construction and renovation projects. (p. 372)

Canadian Electrical Code (CE code) The CE code details code requirements for electrical wiring. The code is enforced through the permit and inspection process by the local hydro authority. (p. 286)

Cantilever The portion of a building or structure that projects out beyond its support (e.g., a bay window or deck may be cantilevered). (p. 45)

Casing The moulding that finishes the edge between the door or window frame and the wall surface. (p. 327)

Centre line A long line with a dash followed by a long line, used in construction drawings to indicate where the centre of an object is. (p. 32)

Collar tie A structural member used to tie rafters together; often placed at approximately mid-span in the roof triangle. A collar tie reduces span for design purposes. (p. 200)

Conduction A method of heat transfer through materials. The rate of conduction will vary depending on the type of material. (p. 301)

Construction joint A place where work finishes for the day at a planned point, so that it can continue on later from the stop point, that is, from where the concrete pour was stopped. (p. 417)

Construction notes Written information describing details about the construction requirements that would be too cumbersome to be placed directly on the drawings. On simpler projects the construction notes may take the place of a separate set of specifications. (p. 61)

Contingency A provision made for an unknown aspect of the project. (p. 372)

Contour lines Lines that may be shown on a site plan to provide information about changes in grade and elevations on the site. A point anywhere on a contour line will have approximately the same elevation. (p. 89)

Control joint Produced by a saw cut in the concrete, which provides a weak spot in the concrete so that instead of cracking unevenly, leaving unsightly cracks, the concrete will crack evenly along the predetermined cut lines. (p. 417)

Convection A method of heat transfer created by air movement and currents that are created by varying temperatures. (p. 302)

Cost-plus contract Also known as "time and materials," this type of contract comes in many forms but is generally used to fast-track projects or on projects that are difficult to define completely. (p. 370)

Counter (cap) flashing A finished flashing that prevents water from migrating into the building (e.g., a counter flashing will cover a step flashing). (p. 245)

Crown moulding A decorative moulding placed at the intersection between the walls and ceiling. (p. 327)

Cubic metres Used to measure volume, for example, a cubic metre of concrete would fill a container 1 metre wide by 1 metre wide by 1 metre high. (p. 417)

Cut roof The traditional method of cutting roofs. This method requires cutting the rafters on site, also referred to as a conventional roof. Still used extensively in the renovation sector and for portions of, or complete, new roofs. (p. 194)

Cutting-plane line A line used in construction drawings to cut through an object in a specific direction so that a cross-sectional view can be provided to better illustrate the components that comprise the object. (pp. 30, 413)

Dampproofing Sealing all penetrations in the foundation wall left by form ties; concrete block walls are parged with a sand and cement-based coating and then coated with a bituminous or other approved coating. (p. 129)

Dashed line A line used in construction drawings for a number of reasons, including: a hidden line indicating an object covered by another object; the location of existing walls that are to be removed; something that may be installed in the future; an x pattern showing a change in the elevation of the floor or ceiling overhead; upper cabinets in a kitchen; a bulkhead or coffered ceiling above the floor that is being viewed; or whether a window opens and how it is hinged. (p. 30)

Dead load The weight of the structure itself and all attached structural and nonstructural materials and components. (pp. 110, 414)

Descriptive line A line used in construction drawings to indicate the location and direction of structural elements. (p. 32)

Detail designation In construction drawings, provides the reference information on where the section detail that the cutting plane line cuts through can be found. (pp. 31, 413)

Detail drawings Type of drawing usually drawn at a larger scale than floor plans and elevations; focuses on specific components of construction (e.g., wall sections, beam and joist intersections, cabinet sections, etc.). (p. 73)

Dew point Occurs when the relative humidity level (RH) reaches 100% and moisture-laden air condenses. (p. 302)

Dimension line A line used in construction drawings to indicate the point-to-point length of the object being dimensioned. (p. 32)

Dimensional lumber Lumber that is milled and graded to standardized sizes and quality for use in construction. (p. 144)

Drain, waste, vent system (DWV) Facilitates the safe and efficient removal of waste and soiled water. (p. 262)

Drain Water Heat Exchanger Extracts heat from hot grey water from showers and transfers it to the potable hot water tank by a copper potable water coil attached to a copper drain pipe. (p. 291)

Drainage pipe/weeping tile Also referred to as weeping tile or drainage tile, drainage pipe is placed along the side of the footing. Gravel is then placed over the pipe at the base of the footing and next to the foundation wall. (p. 129)

Easement The right given to a third party to use the property for a specific purpose (e.g., an easement for hydro might allow hydro lines to cross over the property). (p. 90)

Eaves (cornice) The section of roof overhang that meets the exterior wall and includes the eavestrough, soffit, and fascia. (p. 244)

ECM blower motor Uses direct current and is used on furnaces to greatly reduce energy usage and improve efficiency. (p. 291)

Economies of scale Economic concept achieved in construction when processes are repeated, as is the case in production home building. Through repetition and specialization, construction becomes more efficient. Material costs also decrease due to volume discounts. (p. 44)

Elevation drawing Type of orthographic drawing that is a straight-on view of the object (e.g., a front elevation of a house). (p. 36)

End dam A flashing process that prevents moisture migration into the building. (p. 234)

Engineering drawings These drawings provide details about the structural requirements for the construction of the building and conform to the requirements of the architectural drawings. (p. 411)

European-style cabinet A cabinet that does not have a face frame and uses European- style hinges that are not visible when the cabinet doors are closed. (p. 335)

Expansion joint Allows space for the concrete to expand so it does not heave and crack. (p. 417)

Extension line A line used in construction drawings to indicate from where a dimension is taken. (p. 32)

Exterior insulated finish system (EIFS) A stucco system that consists of rigid insulation and synthetic materials. (pp. 45, 229)

Fire separation As defined by the NBC, "a construction assembly that acts as a barrier against the spread of fire." (p. 400)

Fixed-price contract Also known as lump sum, informs the client of exact costs before startup (excluding extras and changes that may occur during the course of the project). (p. 370)

Flashing Used at intersections of materials to prevent moisture from migrating into the building (includes valley flashing, step flashing, drip flashing used over doors and windows, etc.). (p. 249)

Floor joists The horizontal structural members found in a floor system. (p. 19)

Floor-plan drawing A horizontal cut through a building that allows the designer to dimension and describe detailed information about the location of walls, windows, and doors as well as the structural requirements for the building process, including information about the floors. (p. 56)

Forced air distribution system A heating and cooling system that uses supply and return air ducts to distribute warm or cool air throughout the building. (p. 280)

Four-way switch Switch used in conjunction with three-way switches to operate lights from three or more locations. (p. 289)

Frost line The depth below which frost will not penetrate. Footings placed above the frost line will heave. (p. 111)

Girder truss A truss that supports other trusses; identified by the abbreviation GT. (p. 206)

Green Building Building to a higher, more energy efficient and environmentally sustainable level than is currently standard practice with the intention of eventually becoming truly sustainable. (p. 3)

Grey water Water discharged from a sink, shower, or bathtub that can potentially be reused as non-potable water once filtered, for example, for use in a toilet. (p. 291)

Ground source heat pump Takes advantage of the thermal energy stored in the ground or ground water at a depth where the temperature is relatively stable year round. The ground source heat pump can be used for both heating and cooling and can greatly reduce energy costs. (p. 290)

Guard The term used by the NBC to describe protection around openings in floors, stairs, landings, and balconies (e.g., balusters used with railings). (p. 340)

Gusset plate The mechanical fastener that connects wood members of a truss together. (p. 202)

Handrail A protective barrier to ensure safety. Specific code requirements regulate the safe height and placement of handrails. (p. 340)

Hard conversion Not an exact conversion from one system of measurement to another, but one that is easier to remember and resembles the measurement in the original system (e.g., converting 16″ to metric produces a soft conversion of 406.4 mm; a hard conversion is 400 mm). (p. 19)

Hatching An area of an object that is filled in to provide identification for the reader (e.g., batt insulation has a particular hatching so readers can quickly identify the material). (p. 65)

Headroom The headroom clearance when ascending or descending stairs. The clearance is taken from a line drawn through the front nosing from step to step measured vertically to the lowest point of the ceiling above the stairwell opening. NBC minimum is 1950 mm for stairs within a dwelling unit. (p. 338)

Heat recovery ventilator (HRV) is a heat exchanger that exchanges the heat from stale, warm inside air and transfers it to inside-coming cool outside air. An HRV provides essential fresh air and improved climate control in a energy efficient manner for tightly sealed homes. (pp. 290, 305)

Hexagon and reference number A systematic way of referencing points on the drawings to the construction notes, using numbered hexagons on the drawings that relate to corresponding numbers on the construction notes. Other similar methods may be used depending on the designer's preference. (p. 72)

Highest and best use A real estate term indicating what developers try to obtain with each property they develop. When a property reaches its highest and best use, it offers the greatest return on investment. (p. 3)

Hip or cottage roof A roof that slopes in more than one direction (e.g., a rectangular building with a hip roof has four sloping sides). (p. 199)

HVAC Heating, ventilation, and air conditioning. (p. 259)

Hydrostatic pressure The pressure exerted by water at rest. In the case of a foundation wall where water is found to be substantial enough to place pressure on the outside foundation wall, waterproofing treatment must be taken. (p. 129)

ICI An acronym that identifies the industrial, commercial, and institutional construction sectors. (pp. 2, 410)

Imperial system of measurement The British system of measurement that is still commonly used in Canada to lay out and construct buildings. Most commonly used in residential construction; however, many materials are manufactured using the metric system. (p. 19)

Institutional, commercial, and industrial (ICI) This wording and abbreviation is commonly used to describe the sector for which construction is taking place. (pp. 2, 410)

Insulated concrete forms (ICF) Insulated panels that snap together to act as forms for the concrete walls. These forms are left in place after the concrete is poured to create an insulated wall to which interior and exterior finishes can be fastened. (pp. 115, 143)

Isolation joint Separates the concrete from its adjoining surfaces so it can move up or down. (p. 417)

Isometric drawings Drawn with horizontal lines at 30 degrees and vertical lines vertical. These drawings provide a three-dimensional view that can be drawn to scale. (p. 5)

Jack stud The vertical member of the wall that supports a lintel. (p. 171)

Joist A horizontal structural member used to span and support a floor system. (pp. 19, 141)

Laterally supported An NBC term for design purposes referring to a foundation wall that is adequately supported to resist lateral earth pressure within certain limitations. (p. 126)

Laterally unsupported An NBC term for design purposes referring to a foundation wall that is not adequately supported to resist lateral earth pressure and will require design considerations. (p. 126)

Leader line A line used in construction drawings to point to an object so that a description can be given. (p. 32)

Legend A reference table of abbreviations and symbols that the designer used in the drawings. (p. 61)

Lightweight structural steel framing Sheet steel studs and joists used to construct the structure. Currently being adopted and used by some builders. (p. 142)

Limiting distance The distance from an exposing building face to a property line measured at right angles to the exposing building face. The NBC lays out specific requirements for the fire protection of buildings based on limiting distance requirements. (p. 104)

Line length (point length) The theoretical length of a rafter before deductions are made. (p. 196)

Lintel A type of beam used over door and window openings in a wall as defined in the NBC. (p. 171)

Liquidated damages In the case of a construction contract, an amount predetermined by the parties in the contract as the compensation required to be paid to an injured party should the other party breach the contract. (p. 371)

Live load A variable load that is not part of the structure and includes people, furniture, rain, and snow loads. (pp. 110, 414)

Manufactured beam and joist systems Beams that are engineered to be structurally sound and span long distances. They can be joined to span greater distances and carry large loads. LVL (laminated veneer lumber) and PSL (parallel strand lumber) are two common types of manufactured beams. (p. 150)

MasterFormat 2004 Edition The new standard format for specifications; it has 49 divisions numbered from 0 to 48. (p. 358)

Megapascal (MPa) This is a metric pressure unit that is often used to indicate the compressive strength of concrete. (pp. 247, 417)

Metric system of measurement The official Canadian system of measurement. The NBC uses the metric system. Most ICI drawings are drawn in metric, and many residential drawings are drawn in metric depending on designer and regional preferences. (p. 18)

Milestone A point in time that is easily recognizable (e.g., completion of a building inspection for the foundation). (p. 371)

Minor variance A process that involves applying for a change to the existing zoning for a particular property. (p. 104)

National Building Code of Canada (NBC) This is the model building code of Canada; many provinces adopt their own codes from the model code or simply adopt the model code. (p. 1)

Needling beams Placing temporary beams supported on columns in cutouts above a floor to temporarily support masonry walls until a permanent beam is placed below the masonry. (p. 387)

Non-load-bearing A portion of the structure that is not carrying the live or dead loads from other portions of the structure (e.g., an interior wall that is not supporting the floor or ceiling above it). (p. 81)

Nosing The portion of a tread that extends beyond the face of the riser to form a lip. (pp. 56, 339)

Not to scale This phrase is usually found in the title block and means that the drawings are not drawn to scale. The designer is clarifying that readers should not scale directly from the drawings. NTS is also sometimes used. (p. 28)

Object line A darker or more heavily weighted line used in construction drawings to identify the object being constructed. (p. 30)

Oblique drawings An orthographic view that a side view running out at a 45 degree angle. This drawing provides a three-dimensional look that can be drawn to scale. (p. 7)

Official plan Usually set out by the municipality, looks forward over a long period of time (10 to 20 years) and establishes how the town or city is to grow in the most beneficial and prosperous way for the community as a whole. (p. 104)

Open web steel joists (O.W.S.J.) This joist used as part of a floor and roof system is constructed out of steel forming a truss with parallel chords and a triangulated web system that spans between bearing points. The O.W.S.J. provides direct support for roof or floor decks transferring the loads to the structural frame, e.g., beams to columns or in some cases directly to block. (p. 418)

Optimum value engineering (OVE) Also termed advanced framing techniques, is a method of design and construction that strives to minimize the use of lumber in the building process while reducing thermal bridge points and increasing the energy efficiency of the building. (p. 173)

Oriented strand board (OSB) A panel material used for subflooring, wall sheathing, and roof sheathing; sometimes referred to as waferboard. (p. 162)

Orthographic drawings Most commonly used method of displaying information for construction drawings. Orthographic drawings are drawn as a straight-on view (e.g., front view, or plan view). (p. 8)

Overhang The sloping length of the projection. In some areas it may be used with the same meaning as projection. (p. 197)

Parapet A wall that extends up through a roof. A parapet can be used as a fire separation but may also be used for other purposes. (p. 400)

Party wall A wall shared by two units or owners that separates properties (e.g., a wall between two townhouse units is a party wall). (p. 400)

Permafrost Ground found in cold climates that remains frozen all year round. (p. 111)

Perspective drawings Provide a realistic view of an object, with lines that move toward one or more vanishing points. (p. 5)

PEX A water supply line constructed of cross-linked polyethylene piping (a type of flexible plastic piping that is suitable for both cold- and hot-water supplies). (p. 269)

Photovoltaic cells Also called solar cells, they are used as a method to generate renewable electricity, converting solar radiation into direct current electricity. The production of electricity is accomplished by configuring solar panels that are made up of a number of cells into arrays that can be placed on the south-facing side of a roof or as a sun tracking system on the ground. (p. 290)

Piles Designed as part of the foundation system where soil conditions are considered to be inadequate to provide the proper bearing capacity under normal circumstances. Piles take many forms and can be used in numerous applications including provision for bearing support (bearing pile) and for lateral support (soldier pile). (p. 119)

Plotter A printer that can print large drawings. Drawings printed using a plotter are defined as having been "plotted." (p. 4)

Point load A load placed at a particular point that is being transferred from that point (e.g., a load being placed by a beam that is transferred by a column to the floor will create a point load at that spot). (pp. 155, 176)

Potable water Water fit for human consumption. (p. 261)

Pounds per square inch (psi) This is an imperial pressure unit that is often used to indicate the compressive strength of concrete. (p. 417)

Projection The horizontal distance that the roof projects out from the wall of the house. (p. 196)

Production home building The construction of multiple homes on lots that have been subdivided. The term also refers to homes that are manufactured in plants and transported to a site. (p. 42)

Racking The force placed on a wall that makes it move diagonally. Racking can be resisted by using diagonal bracing or structural sheathing. (p. 178)

Radiation A method of heat transfer created by the sun or other means. Radiant heat provides a comfortable environment because it warms objects instead of air. (p. 302)

Reference grid, structural grid, or planning grid The grid is principally defined by the location of the columns, which is based on the main beams and/or types of slabs they are supporting. On the drawings, there are two sets of grid lines that are normally perpendicular (at 90 degrees) to each other. The horizontal grid lines running left to right on one axis of the plan drawings are typically lettered, while the vertical grid lines running up and down on the other axis of the plan drawing page are typically numerical. (Note: numbering and lettering may be reversed on some drawings). (p. 412)

Reinforcing steel Bars of steel (commonly referred to as rebar) that are added to concrete to greatly increase its tensile strength. (p. 247)

Resilient channel A specially designed metal strap that minimizes sound transmission from drywall walls or ceilings to studs or joists. (p. 401)

Return air duct This duct returns warm air or cool air (depending on whether a furnace or air conditioner is being used) to various rooms. The return air duct is located on an interior wall. (p. 280)

Riser The vertical board that fits between the treads. (pp. 56, 339)

Roof joist A rafter with a slope of less than 1:4, as specified by the NBC for design purposes, and/or a rafter that is also supporting a finished ceiling. (p. 196)

Roof rafter An individual structural member of a conventional roof. NBC defines a rafter as having a slope of 1:4 or more for design/structural purposes. (p. 196)

Rough opening The opening that is framed for windows and doors. This opening is framed larger to accommodate for warping and twisting of the lumber and to provide room for shimming to ensure the doors and windows are placed plumb and level. (p. 175)

Rough sill The wall member that is placed below a window opening. (p. 171)

R value An imperial measure of a material's capacity to resist conductive heat flow. The higher the value, the greater the resistance. (pp. 170, 301)

RSI value The metric measure for the resistance of a material to conduct heat flow. The higher the value, the greater the resistance. (pp. 170, 301)

Sanitary sewer A drainpipe used to remove waste water and fecal matter. (p. 262)

Scale The proportional shrinking or enlarging of an object. Construction drawings are generally drawn to a smaller scale so they can fit the paper they are to be plotted on. (pp. 5, 18)

Schedule Usually in table form, used for construction drawings to provide information that can be quickly referenced (e.g., window, lintel, door, and room schedules). (pp. 45, 176)

Scope of work What is to be included in the work done by a general or a subcontractor. (p. 357)

Septic system A sewage treatment system common in rural areas with no connection to town or municipal sewage pipes. (p. 262)

Sheathing The outer panels that are applied to walls and roofs. Sheathing can be made of a variety of materials including plywood and oriented strand board. (p. 178)

Shop drawings Drawings that clarify installation and material requirements are provided by subtrades and manufacturers as requested/required by the designer, builder, and municipality. (p. 194)

Sill plate A 2″ × 4″ or 2″ × 6″ member that forms the connection point from the framing to the foundation wall. The sill plate is fastened with anchor bolts. (p. 154)

Site plan Shows the location and setbacks of the building on the property, and provides information regarding elevations and services. (p. 89)

Small buildings Refers to the definition used for Housing and Small Buildings in Part 9 of the NBC. Buildings three storeys or less in height having a building area not exceeding 600 square metres and used for major occupancies as classified in the NBC (refer to Division A Part 1.3.3.3) but not for assembly purposes. (p. 410)

Soft conversion Used when calculating an exact conversion from one system of measurement to another. (p. 19)

Solar hot water system A system for heating or preheating the hot water supply using solar power, often mounted on a roof. (p. 290)

Solid masonry construction Structural walls constructed of blocks or multiple wythes of bricks and blocks that are mechanically tied or bonded together. (p. 143)

Sound transmission class (STC) A numbered method used to rate partitions, doors, and windows for their effectiveness in blocking sound. (p. 401)

Specifications Document that provides the written detail about materials, items, and installation tolerances and methods that cannot be placed on the drawings themselves. (p. 357)

Split-circuit receptacle Where the top and bottom receptacles on a duplex receptacle are connected to different circuits to prevent overload. The CE code requires this over kitchen countertops. (p. 288)

Split-switched receptacle A duplex receptacle where one part is operated from a switch (e.g., a lamp operated from a wall switch). (p. 288)

Stack A vertical drainpipe. (p. 263)

Stair stringer The main support for the stairs; also referred to as "carriages." (p. 340)

Stairwell opening size The opening that must be framed based on code requirements and the layout constraints for the floor plan. (p. 339)

Steel beams Commonly used in wood frame construction to span long distances and carry large loads. Wide flange I-beams (W) are the most commonly used in residential construction. (p. 152)

Steel decking The corrugated material that is generally used in structural steel construction to act as the base for the floor or roof. (p. 416)

Step flashing Used on a sloping roof intersection with a wall or chimney. It is comprised of folded metal pieces that are inserted between the shingles that lap over each other and step up the roof against the wall. (p. 245)

Stepped footing A footing that transcends to different elevations; generally required due to changes in grade, maintaining depth below frost line, or meeting the requirement of resting on undisturbed soil. (p. 114)

Storm sewer A drainpipe to remove surface water from drainage tile, rain water leaders, and floor drains. (p. 262)

Strip footing Long footings that support walls in a building; also called spread footings. (p. 112)

Studs Vertical members used in wall construction, usually constructed of 2" × 4" or 2" × 6" pieces of dimensional lumber. (pp. 19, 141, 170)

Subfloor Horizontal sheets that form the floor and are attached to the floor joists. The subfloor is located beneath the finished floor and underlay, if required by the floor finish. (p. 19)

Substantial Completion/Substantial Performance A legal term used to describe when the building is mostly complete. (p. 371)

Substrate The support material under a finish (e.g., cement board is a substrate to ceramic tile). (p. 347)

Sump pump A sump pump is a pump that is used to remove water that has collected in a sump pit having drained there from around the foundation. Often used to keep basements dry in areas with high water tables and where there may not be a municipal storm sewer. (p. 262)

Supply air duct Distributes warm or cool air (depends on whether furnace or air conditioner is being used) to various rooms. The supply air duct is usually located on an exterior wall, preferably below or near window and door openings. (p. 280)

Supported joist length (SJL) The length of joist that a beam is supporting. This measurement is taken as half the span of the joist on each side of the beam added together. Also referred to as supported length (SL). (p. 149)

Sustainable development The World Commission on Environment and Development developed a definition of sustainable development that was included in its 1987 report: "Humanity has the ability to make development sustainable—to ensure that it meets the needs of the present without compromising the ability of future generations to meet their own needs." (p. 3)

Swale An indentation in the grade that is designed to facilitate the movement of excess water to flow to storm sewers and flood control collection areas. (p. 92)

Symbol Frequently used in construction drawings to identify objects (e.g., switches, outlets, vents, diffusers, etc.). (p. 65)

Tender The process of asking for bids for construction projects. (p. 3)

Tender documents Documents used in the bidding process that include the instructions to bidders, tender form, agreement, drawings and specifications, and any amendments to the contract documents that took place during the tender period. (p. 370)

Thermal break Usually created by using a material that greatly reduces or resists heat flow paths through an object such as a window frame (aluminum window frames will have a non-conductive thermal break in the middle of the frame) or a wall assembly. (p. 117)

Thermal bridge A conductive pathway that allows heat to move to cold relatively easily. (p. 117)

Thermal bridging The ability of a material to transfer heat to cold (e.g., wood will provide a better thermal bridge than insulation). Reducing thermal bridging increases energy efficiency. (p. 301)

Three-way switch This type of switch is used where a light needs to be controlled from more than one location (e.g., a three-way switch can be used at the top and bottom of stairs). (p. 288)

Timber frame construction Uses a series of posts and beams to make up the structure of the building. (p. 142)

Total rise for stairs The distance from one finished floor level to another. (p. 338)

Total run Half of the span for calculating roof rafters. Another definition for stair construction includes total length from face of first riser to face of last riser. (p. 195)

Total run for stairs The distance from the face of the bottom riser to the face of the top riser. (p. 338)

Total span Depending on the purpose of use either (1) the distance from outside of frame wall to outside of frame wall used for calculating rafter line lengths or (2) the clear span from support to support that the member is spanning without support, as is the case for a floor joist. (p. 195)

Tread The distance from the face of riser to the front edge of nosing. (pp. 56, 339)

Truss roof A system of trusses that are engineered and manufactured off-site, and installed on-site following the manufacturer's requirements. Truss roofs are the most common method of constructing roofs in new construction but still may require portions of the roof to be cut on-site. (p. 194)

Underpinning A method of lowering an existing foundation, usually for the purpose of increasing headroom. Also may be required to prevent frost from forming under a basement walkout constructed as an addition. (p. 122)

Undisturbed soil Soil that has not been moved or altered from its original condition. (p. 110)

Unit line length (unit point length) The hypotenuse of unit rise and unit run. (p. 196)

Unit-price contract Used extensively with subcontractors where work can be defined by units (e.g., a brick layer may be paid by the square foot or by the brick laid). (p. 370)

Unit rise The vertical unit used in relation to the unit run to calculate slope for rafters and stairs. (pp. 45, 195)

Unit rise for stairs The distance from top of tread to top of tread. (p. 338)

Unit run The horizontal unit used in relation to the unit rise to calculate slope for rafters and stairs. (pp. 45, 195)

Unit run for stairs The distance from face of riser to face of riser. (pp. 339)

Value The concept that combines quality and cost to drive the use of materials and methods used in construction. (p. 140)

Vapour barrier/retarder A barrier formed on the warm side of the building envelope to prevent moisture-laden air from entering the wall cavity and condensing when it reaches the dew point; usually constructed using an approved 6 mil polyethylene sheeting under the drywall. (p. 303)

Vapour diffusion The transfer of moisture through materials. The rate of moisture movement depends on the permeability of the material. (p. 303)

Visualization The ability to look at a series of orthographic drawings and visualize the three-dimensional finished product. (p. 8)

Waterproofing Ensuring that water cannot penetrate; an NBC requirement when hydrostatic pressure is present. (p. 129)

Water trap Used in toilets, sinks, and drains to ensure that sewer gas does not enter the building. (p. 269)

Weep holes Spaces that are left open between some units of brick to allow any moisture that has migrated into the air space to dry or exit. Weep holes are typically required at the base of foundation wall, over lintels, and under sills. (p. 230)

Wind turbine Also called a wind generator, it converts kinetic energy from the wind into mechanical energy that is used to produce renewable electricity. (p. 291)

Window well Used to permit basement windows to be located partially below grade, thus increasing light into the basement; usually constructed of galvanized steel, concrete, or pressure-treated wood. (pp. 45, 121)

Wood frame construction The most common method of construction used in Canada. this method uses dimensional lumber to construct the floors, walls, and roof of the building using standardized spacing. (p. 140)

Wythe A single row of masonry. Two rows placed side by side are considered two wythes, and so on. (pp. 81, 141)

Zoning A system of regulating land use that separates zones and designates uses for the land within these zones. (p. 103)

Index

D

E

F

G

H

I

J

K

L

M

N

O

P

R

S

T

U

V

W

Z